Climate Change, Air Pollution and Global Challenges

Understanding and Perspectives from Forest Research

Developments in Environmental Science

Volume 12 **Forest Monitoring**
Edited by
Marco Ferretti, Richard Fischer

Volume 11 **Alberta Oil Sands: Energy, Industry and the Environment**
Edited by
Kevin E. Percy

Volume 10 **Cross-Border Resource Management, 2nd edition**
Rongxing Guo

Volume 9 **Air Quality and Ecological Impacts**
Edited by
Allan Legge

Volume 8 **Wildland Fires and Air Pollution**
Edited by
Andrzej Bytnerowicz, Michael Arbaugh, Allen Riebau, Christian Andersen

Volume 7 **Persistent Organic Pollutants in Asia**
Edited by
An Li, Shinsuke Tanabe, Guibin Jiang, John Giesy, Paul Lam

Volume 6 **Air Pollution Modeling and its Application XVIII**
Edited by
Carlos Borrego, Eberhard Renner

Volume 5 **Concepts and Applications in Environmental Geochemistry**
Edited by
Dibyendu Sarkar, Rupali Datta, Robyn Hannigan

Volume 4 **Cross-Border Resource Management, 1st edition**
Rongxing Guo

Volume 3 **Air Pollution, Global Change and Forests in the New Millennium**
Edited by
D.F. Karnosky, K.E. Percy, A.H. Chappelka, C. Simpson, J. Pikkarainen

Volume 2 **Ozone Air Pollution in the Sierra Nevada - Distribution and Effects on Forests**
Edited By
A. Bytnerowicz, R. Alonso, M. Arbaugh

Volume 1 **Air Pollution Science for the 21st Century**
Edited By
J. Austin, P. Brimblecombe, W. Sturges

Developments in Environmental Science
Volume 13

Climate Change, Air Pollution and Global Challenges

Understanding and Perspectives from Forest Research

Edited by

R. Matyssek
Freising, Germany

N. Clarke
Ås, Norway

P. Cudlin
Ceské Budejovice, Czech Republic

T.N. Mikkelsen
Roskilde, Denmark

J.-P. Tuovinen
Helsinki, Finland

G. Wieser
Innsbruck, Austria

E. Paoletti
Florence, Italy

ELSEVIER

AMSTERDAM • BOSTON • HEIDELBERG • LONDON • NEW YORK • OXFORD
PARIS • SAN DIEGO • SAN FRANCISCO • SINGAPORE • SYDNEY • TOKYO

Front cover image sources from left to the right:

Eddy tower and platform in snow covered forest in Hyytiälä, Finland (Mikkelsen)
Eddy tower and platform in Sorø beech forest, Denmark (Lund)
Free-Air ozone canopy fumigation in Kranzberg beech forest Germany (Paoletti)
Free Air O_3 fumigation for beech, birch and oak in northern Japan (Koike)

Elsevier
The Boulevard, Langford Lane, Kidlington, Oxford, OX5 1 GB, UK
Radarweg 29, PO Box 211, 1000 AE Amsterdam, The Netherlands
225 Wyman Street, Waltham, MA 02451, USA
525 B Street, Suite 1800, San Diego, CA 92101-4495, USA

First edition:

British Library Cataloguing in Publication Data
A catalogue record for this book is available from the British Library

Library of Congress Cataloging-in-Publication Data
A catalog record for this book is available from the Library of Congress

For information on all **Elsevier** publications
visit our web site at store.elsevier.com

ISBN: 978-0-08-098349-3
ISSN: 1474-8177

Contents

Part III
Significance of Biotic Processes in Forest Ecosystem Response

Part VII
Knowledge Transfer and Socio-Economic Aspects

Contributors

Numbers in Parentheses indicate the pages on which the author's contributions begin.

Jana Albrechtová (521), Faculty of Science, Charles University in Prague, Prague, Czech Republic

Michael Aspinwall (179), Hawkesbury Institute for the Environment, University of Western Sydney, Richmond, New South Wales, Australia

Marko Bajc (151), Slovenian Forestry Institute, Ljubljana, Slovenia

Olaf Bastian (521), Leibniz Institute of Ecological Urban and Regional Development, Dresden, Germany

Manuela Baumgarten (19), Technische Universität München, Ecophysiology of Plants, Hans-Carl-von-Carlowitz-Platz 2, D-85354 Freising-Weihenstephan, Germany

Johan Paul Beukes (429), Environmental Sciences and Management, North-West University, Potchefstroom, South Africa

Patricia Brito (19), Division of Plant Biology, University of La Laguna, Tenerife, La Laguna, Spain

Mercedes M.C. Bustamante (411), Departamento de Ecologia, Instituto de Ciências Biológicas, Campus Universitário Darcy Ribeiro, Universidade de Brasília, Brasília, Brazil

Andrzej Bytnerowicz (333), USDA Forest Service, Pacific Southwest Research Station, Riverside, California, USA

Kgaugelo Chiloane (429), Sustainability and Innovation, Environmental Sciences Department, Eskom, Johannesburg, South Africa

Stan Cieslik (19), Joint Research Centre, Ispra, Italy

Nicholas Clarke (3, 475, 591), Norwegian Forest and Landscape Institute, Norway

John J. Couture (227), Department of Forest and Wildlife Ecology, University of Wisconsin–Madison, Madison, Wisconsin, USA

Pavel Cudlin (3, 521, 591), Global Change Research Centre, Academy of Sciences of the Czech Republic, Ceske Budejovice, Czech Republic

Alina Danielewska (475), Meteorology Department, Poznan University of Life Sciences, Poznan, Poland

Wim de Vries (497), Alterra, Wageningen University and Research Centre, and Environmental Systems Analysis Group, Wageningen University, P.O. Box 47, Wageningen, The Netherlands

Jochen Dieler (251), Forest Growth and Yield Science, Wissenschaftszentrum Weihenstephan, Technische Universität München, Hans-Carl-von-Carlowitz-Platz 2, D-85354 Freising-Weihenstephan, Germany

Pierre Dizengremel (267), Université de Lorraine, and INRA, UMR 1137 EEF, Champenoux, France

Charles Driscoll (333), Department of Civil and Environmental Engineering, Syracuse University, Syracuse, New York, USA

Dieter Ernst (289), Helmholtz Zentrum München, Neuherberg, Germany

Werner Eugster (37), Department of Environmental Systems Science, Institute of Agricultural Sciences, ETH Zürich, Zürich, Switzerland

Silvano Fares (57), Consiglio per la ricerca e la sperimentazione in agricoltura (CRA), Research Center for the Soil–Plant System (RPS), Rome, Italy

Mark Fenn (333), USDA Forest Service, Pacific Southwest Research Station, Riverside, California, USA

Richard Fischer (475), Thünen Institute for World Forestry, Hamburg, Germany

Frank Fleischmann (103), Pathology of Woody Plants, Technische Universität München, Freising, Germany

R. Funada (371), Institute of Agriculture, Tokyo University of Agriculture and Technology, Fuchu, Tokyo 183-8509, Japan

Tine Grebenc (151), Slovenian Forestry Institute, Ljubljana, Slovenia

Winfried Grimmeisen (251), Bayerische Landesanstalt für Wald und Forstwirtschaft, Hans-Carl-von-Carlowitz-Platz 1, D-85354 Freising-Weihenstephan, Germany

Ludger Grünhage (103, 251), Department of Plant Ecology, Justus-Liebig University, Giessen, Germany

Matthias Haeni (37), Department of Environmental Systems Science, Institute of Agricultural Sciences, ETH Zürich, Zürich, Switzerland

Andreas Hahn (569), Technische Universität München, Center of Life and Food Sciences Weihenstephan, Department of Ecology and Ecosystem Management, Institute of Forest Management, Freising, Germany

Hannele Hakola (77), Finnish Meteorological Institute, Helsinki, Finland

Karl-Heinz Häberle (251), Technische Universität München, Ecophysiology of Plants, Hans-Carl-von-Carlowitz-Platz 2, D-85354 Freising-Weihenstephan, Germany

Yasutomo Hoshika (371), Silviculture and Forest Ecological Studies, Hokkaido University, Sapporo, Hokkaido 060-8589, Japan, and JSPS Research Fellow

Andreas Ibrom (497), Centre for Ecosystems and Environmental Sustainability (ECO), Department of Chemical and Biochemical Engineering, Technical University of Denmark, Denmark

Takeshi Izuta (371), Institute of Agriculture, Tokyo University of Agriculture and Technology, Fuchu, Tokyo 183-8509, Japan

Kerneels Jaars (429), Environmental Sciences and Management, North-West University, Potchefstroom, South Africa

Yves Jolivet (267), Université de Lorraine, and INRA, Champenoux, France

Miroslav Josipovic (429), Environmental Sciences and Management, North-West University, Potchefstroom, South Africa

Per Erik Karlsson (77), IVL Swedish Environmental Research Institute, P.O. Box 53021, Gothenburg, Sweden

Markku Keinänen (309), Department of Biology, University of Eastern Finland, P.O. Box 111, Joensuu, Finland

Sarita Keski-Saari (309), Department of Biology, University of Eastern Finland, P.O. Box 111, Joensuu, Finland

John King (179), Department of Forestry and Environmental Resources, North Carolina State University, Raleigh, North Carolina, USA

Mitsutoshi Kitao (371), Department of Plant Ecology, Forestry and Forest Products Research Institute, Matsunosato 1, Tsukuba 305-8687, Japan

Thomas Knoke (569, 591), Technische Universität München, Center of Life and Food Sciences Weihenstephan, Department of Ecology and Ecosystem Management, Institute of Forest Management, Freising, Germany

Takayoshi Koike (371), Silviculture and Forest Ecological Studies, Hokkaido University, Sapporo, Hokkaido 060-8589, Japan

Sari Kontunen-Soppela (309), Department of Biology, University of Eastern Finland, P.O. Box 111, Joensuu, Finland

Alessandra R. Kozovits (411), Departamento de Biodiversidade, Evolução Meio Ambiente, Universidade Federal de Ouro Preto, Campus Morro do Cruzeiro, Ouro Preto, Brazil

Hojka Kraigher (151), Slovenian Forestry Institute, Ljubljana, Slovenia

Jörg Kruse (131), Chair of Tree Physiology, Institute of Forest Sciences, University of Freiburg, Freiburg, Germany

Lauri Laakso (429), Finnish Meteorological Institute, Helsinki, Finland, and Environmental Sciences and Management, North-West University, Potchefstroom, South Africa

Casper Labuschagne (429), Cape Point GAW Observatory, South African Weather Service, Stellenbosch, South Africa

K.S. Larsen (497), Centre for Ecosystems and Environmental Sustainability (ECO), Department of Chemical and Biochemical Engineering, Technical University of Denmark, Denmark

Didier Le Thiec (267), Université de Lorraine, and INRA, Champenoux, France

Michael Leuchner (251), Ecoclimatology, Technische Universität München, Wissenschaftszentrum Weihenstephan, Hans-Carl-von-Carlowitz-Platz 2, D-85354 Freising-Weihenstephan, Germany

Richard L. Lindroth (227), Department of Entomology, University of Wisconsin–Madison, Madison, Wisconsin, USA

Lingli Liu (179), State Key Laboratory of Vegetation and Environmental Change, Institute of Botany, Chinese Academy of Sciences, Beijing, China

Francesco Loreto (57), National Research Council (CNR), IPP, Florence, Italy

Michal Marek (521), Global Change Research Centre, Academy of Sciences of the Czech Republic, Ceske Budejovice, Czech Republic

Hideyuki Matsumura (371), Environmental Science Research Laboratory, Central Research Institute of Electric Power Industry, Abiko, Chiba 270-1194, Japan

Rainer Matyssek (3, 19, 103, 251, 497, 591), Technische Universität München, Ecophysiology of Plants, Hans-Carl-von-Carlowitz-Platz 2, D-85354 Freising-Weihenstephan, Germany

Steven McNulty (333), USDA Forest Service, Southern Station, Raleigh, North Carolina, USA

Tom Meixner (333), University of Arizona, Tucson, Arizona, USA

Annette Menzel (251), Ecoclimatology, Technische Universität München, Wissenschaftszentrum Weihenstephan, Hans-Carl-von-Carlowitz-Platz 2, D-85354 Freising-Weihenstephan, Germany

Teis Nørgaard Mikkelsen (3, 475, 497, 591), Centre for Ecosystems and Environmental Sustainability (ECO), Department of Chemical and Biochemical Engineering, Technical University of Denmark, Denmark

David Miller (547), The James Hutton Institute, Aberdeen, Scotland, United Kingdom

Maria Nijnik (547), The James Hutton Institute, Aberdeen, Scotland, United Kingdom

Elina Oksanen (309, 391), Department of Biology, University of Eastern Finland, P.O. Box 111, Joensuu, Finland

Vivek Pandey (391), Plant Ecology and Environmental Science, National Botanical Research Institute (CSIR-NBRI), Lucknow, India

Elena Paoletti (3, 497, 591), IPP-CNR, Florence, Italy

Jacobus J. Pienaar (429), Environmental Sciences and Management, North-West University, Potchefstroom, South Africa

Jan Pokorný (521), Enki Ltd., Třeboň, Czech Republic

Afshin Pourmokhtarian (333), Department of Civil and Environmental Engineering, Syracuse University, Syracuse, New York, USA

Kurt S. Pregitzer (209), College of Natural Resources, University of Idaho, Moscow, Idaho, USA

Hans Pretzsch (251), Forest Growth and Yield Science, Wissenschaftszentrum Weihenstephan, Technische Universität München, Hans-Carl-von-Carlowitz-Platz 2, D-85354 Freising-Weihenstephan, Germany

Annamaria Ranieri (267), Department of Agriculture, Food and Environment, University of Pisa, Pisa, Italy

Stephan Raspe (251), Bayerische Landesanstalt für Wald und Forstwirtschaft, Hans-Carl-von-Carlowitz-Platz 1, D-85354 Freising-Weihenstephan, Germany

Heinz Rennenberg (131), Chair of Tree Physiology, Institute of Forest Sciences, University of Freiburg, Freiburg, Germany

Matthias Schröder (251), Department of Plant Ecology, Justus-Liebig University, D-35392 Giessen, Germany

Josef Seják (521), Faculty of Environment, J.E. Purkyně University in Ústí nad Labem, Ústí nad Labem, Czech Republic

Yussuf Serengil (497), Faculty of Forestry, Istanbul University, Istanbul, Turkey

Chhemendra Sharma (391), Radio and Atmospheric Sciences Division, National Physical Laboratory, New Delhi, India

Judy Simon (131), Chair of Tree Physiology, Institute of Forest Sciences, University of Freiburg, Freiburg, Germany

David Simpson (77), EMEP MSC-W, Norwegian Meteorological Institute, P.O. Box 43, Oslo, Norway, and Chalmers University of Technology, Gothenburg, Sweden

Nandita Singh (391), Plant Ecology and Environmental Science, National Botanical Research Institute (CSIR-NBRI), Lucknow, India

Mikhail Sofiev (451), Adjunct Professor, Senior Scientist, Air Quality Department, Finnish Meteorological Institute

Alan F. Talhelm (209), College of Natural Resources, University of Idaho, Moscow, Idaho, USA

Juha-Pekka Tuovinen (3, 19, 77, 429, 497, 591), Finnish Meteorological Institute, Helsinki, Finland

Andrée Tuzet (267), UMR INRA/AgroParisTech EGC, Thiverval-Grignon, France

Ville Vakkari (429), Finnish Meteorological Institute, and Department of Physics, University of Helsinki, Helsinki, Finland

Pieter Gideon Van Zyl (429), Environmental Sciences and Management, North-West University, Potchefstroom, South Africa

Andrew Venter (429), Environmental Sciences and Management, North-West University, Potchefstroom, South Africa

Makoto Watanabe (371), Silviculture and Forest Ecological Studies, Hokkaido University, Sapporo, Hokkaido 060-8589, and Institute of Agriculture, Tokyo University of Agriculture and Technology, Fuchu, Tokyo 183-8509, Japan

Gerhard Wieser (3, 19, 103, 251, 497, 591), Department of Alpine Timberline Ecophysiology, Federal Office and Research Centre for Forests, Innsbruck, Austria

Fengming Yuan (333), Climate Change Science Institute and Environmental Science Division, Oak Ridge National Laboratory, Oak Ridge, Tennessee, USA

I. Yurtseven (497), Faculty of Forestry, Istanbul University, Istanbul, Turkey

Lothar Zimmermann (251), Bayerische Landesanstalt für Wald und Forstwirtschaft, Hans-Carl-von-Carlowitz-Platz 1, D-85354 Freising-Weihenstephan, Germany

Preface

The need for this book arises from the significant anthropogenic pressures that forests currently face from climate change and air pollution, reflecting the outcome of the scientific discussion and synthesis from the *COST Action FP0903* (*Climate Change and Forest Mitigation and Adaptation in a Polluted Environment*, MAFor), which was active during 2009 through 2013 (http://cost-fp0903.ipp.cnr.it/). COST, standing for *European Cooperation in Science and Technology*, represents a funding instrument of the European Union (EU) in support of network cooperation in science and technology within the EU and associated countries, also catalysing joint publications like the one presented here.

The editors gratefully acknowledge this kind of support by the EU, including scientific workshops in Antalya (Turkey), Rome (Italy), Prague (Czech Republic), Kaunas (Lithuania), Kahramanmaras (Turkey), Florence (Italy), Copenhagen (Denmark) and Brussels (Belgium) in fostering the scientific communication, as well as partial coverage of the publishing costs.

MAFor pursued two main objectives: (i) to increase understanding of the state and potential of forest ecosystems to mitigate and adapt to climate change in a polluted environment and (ii) to reconcile process-oriented research, long-term monitoring and applied modelling at comprehensive forest research sites (supersites) to be established. To mirror these objectives, 26 author teams were invited worldwide to contribute chapters to this book project. Manuscripts were peer reviewed by two experts each of internationally high scientific reputation in the respective research fields, meeting the high standards of refereed scientific journals. The breadth of contributions has become comprehensive, starting from interactions between trace gases, climate change factors and vegetation, highlighting the significance of biotic processes in forest ecosystem response to climate change and air pollution, demanding mechanistic and diagnostic understanding for risk assessment, elucidating the global dimension of air pollution as part of climate change, promoting the 'supersite' concept for research on forest ecosystems in a changing environment and suggesting strategies for knowledge transfer to and use within the socio-economic dimension of the book topic.

In such terms, the book seeks readerships not only within the scientific research community and academic teaching but also in the areas of environmental policy making, socio-economics, non-governmental organisations and journalism (public media) to stimulate over-arching communication and set perspectives towards the post-Kyoto debate. Hence, the benefits of this book are manifold: Apart from providing a timely update of the evidence and knowledge on the topic and featuring upcoming environmental scenarios along with means of future integrated research, guidelines are provided for the required innovative concept development in risk assessment, socio-economic implications and stakeholder-oriented decision making.

In summary, the book not only updates the process-based understanding from forest research of climate change and intrinsically associated air pollution but also offers perspectives for meeting the addressed global challenges. It is now up to the politicians and society to create supportive policies for the 'next generation' of research. These will ultimately be for the sake of mankind, given the significance of forest ecosystems in mitigating climate change and air pollution effects. The editors hope that this book will promote such beneficial capacities in research and policy making.

R. Matyssek, Freising, Germany
N. Clarke, Ås, Norway
P. Cudlin, Ceské Budejovice, Czech Republic
T.N. Mikkelsen, Roskilde, Denmark
J.-P. Tuovinen, Helsinki, Finland
G. Wieser, Innsbruck, Austria
E. Paoletti, Florence, Italy
May 2013

Part I

Introduction into the Scope and Structure of the Book

Chapter 1

Climate Change, Air Pollution and Global Challenges: Understanding and Perspectives from Forest Research

Rainer Matyssek*[,1], Nicholas Clarke[†], Pavel Cudlin[‡], Teis Nørgaard Mikkelsen[§], Juha-Pekka Tuovinen[¶], Gerhard Wieser[||] and Elena Paoletti[#]

[*]*Technische Universität München, Ecophysiology of Plants, Hans-Carl-von-Carlowitz-Platz 2, D-85354 Freising-Weihenstephan, Germany*

[†]*Norwegian Forest and Landscape Institute, Norway*

[‡]*Global Change Research Centre, Academy of Sciences of the Czech Republic, Ceske Budejovice, Czech Republic*

[§]*Centre for Ecosystems and Environmental Sustainability (ECO), Department of Chemical and Biochemical Engineering, Technical University of Denmark, Denmark*

[¶]*Finnish Meteorological Institute, Helsinki, Finland*

[||]*Department of Alpine Timberline Ecophysiology, Federal Office and Research Centre for Forests, Innsbruck, Austria*

[#]*IPP-CNR, Florence, Italy*

[1]*Corresponding author: e-mail: matyssek@wzw.tum.de*

Chapter Outline

1.1 WHY WRITE THIS BOOK?

We have become used to acknowledging that forest ecosystems worldwide are under pressure, and that this pressure will increase in the future. This idea is substantiated through assessments which demonstrate a global shrinkage in the forested area by about 50% since the last glaciation (Figure 1.1). The

Developments in Environmental Science, Vol. 13. http://dx.doi.org/10.1016/B978-0-08-098349-3.00001-3

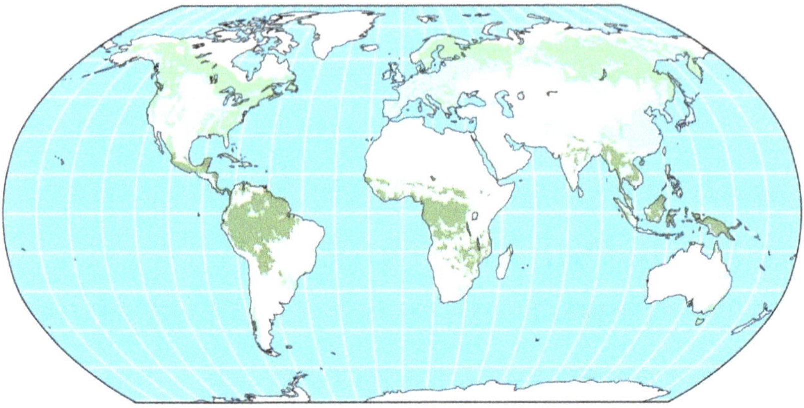

FIGURE 1.1 Global distribution of original and remaining forests: dark green, tropical current; light green, tropical original; medium green, temperate and boreal current; light blue, temperate and boreal original; 'original' refers to geohistorically modern, post-Pleistocene climatic conditions, but before the spread of human influence; 'current' denotes the result largely of human activity at the beginning of the twenty-first century. *After Food and Agriculture Organization of the United Nations, FAO.*

forest loss is especially meaningful for the global carbon (C) storage capacity of forest ecosystems, currently comprising 80% of the terrestrial aboveground C and more than 70% of the soil organic C (Dixon et al., 1994). This storage slows the anthropogenic increase in atmospheric CO_2 concentration substantially, highlighting the significance of forests in mitigating global warming. As compared with other terrestrial ecosystems and types of land use, forests until the present have been functioning at the global and long-term scales as C sinks (Luyssaert et al., 2008; Schulze et al., 2009). However, can we take it for granted that this sink will persist in the future?

Obviously, the pressure on forest ecosystems is anthropogenic and intrinsically generated by the dramatic increase in the population of mankind, for which no stabilization is in sight (Figure 1.2A; Bengtson et al., 2006; Cohen, 2003). Such a development may end catastrophically, both in terms of the finite feeding capacity of the Earth (Lüttge, 2013) and the increasing pressure on the environment, gradually threatening the ecological foundations of human existence (Lovelock, 2009). This kind of crisis appears to be imminent, as the human population has more than doubled since the 1950s to about seven billion at present, and is expected to reach nine billion people by 2050. The increase is dramatic and unprecedented in mankind's history, given that reaching the first billion people on Earth by the early nineteenth century took as much as 40,000 years (Fritsch, 1990).

Although being of paramount significance for the global C balance, forest ecosystems represent a major casualty of the development of civilization. Land area has been exploited for feeding and settling, and increasingly for

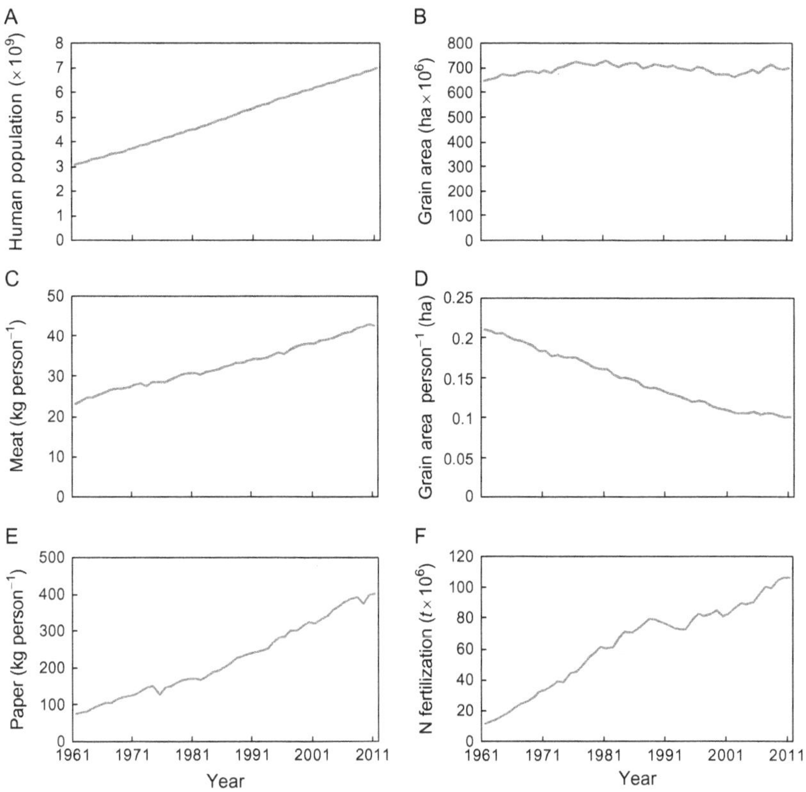

FIGURE 1.2 Time courses between 1961 through 2011 of (A) world population, (B) world grain area harvested, (C) per capita world meat production, (D) per capita world grain area harvested, (E) per capita consumption of paper and paper board and (F) nitrogen fertilizer use. *Data derived from Food and Agriculture Organization of the United Nations, FAO.*

providing biogenic energy, and all such demands are met at the expense of forest ecosystems. Given the steady population growth of mankind, the resource of arable land has become exhausted, as reflected by the plateau in the harvested grain area since the 1950s (Figure 1.2B). Nevertheless, the C release into the atmosphere through agricultural practices and land-use change, often through forest burning, has remained high (Figure 1.3). In particular, the ratio of CO_2 emission versus atmospheric CO_2 removal is large in Latin America, Africa and Southeast Asia (Figure 1.3). In parallel, the per capita consumption of natural resources worldwide, exemplified by paper and meat production, has clearly increased (Figure 1.2C and E). Hence, the demand for resources has been more than proportional relative to the global population increase. Such enhancements in resource use, in particular through agriculture, have become achievable only through tremendously increased fertilization (Figure 1.2F), given the limitation in arable land. This limitation is strikingly mirrored by the decline in harvested grain area per capita since the 1950s (Figure 1.2D).

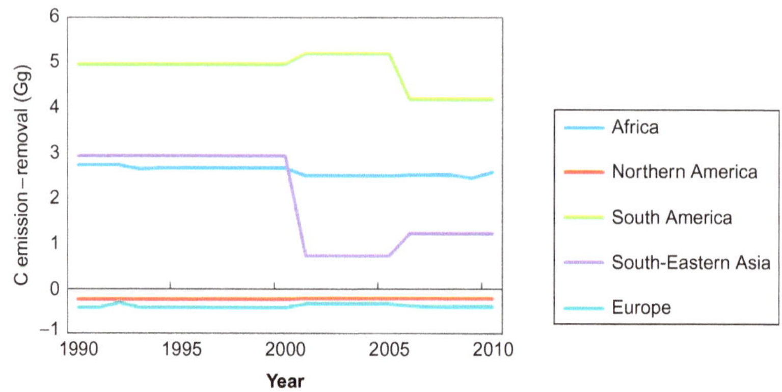

FIGURE 1.3 Time course of the net balance of continental carbon (C) emission into and removal from the atmosphere related to ecosystem conversion; negative numbers denote removal higher than emission. *Data from Food and Agriculture Organization of the United Nations, FAO.*

Intensive agriculture in combination with forest destruction has substantially enhanced the releases of methane (CH_4) and nitrous oxide (N_2O) into the atmosphere, in addition to CO_2 release from the altered ecosystems and usage of fossil energy sources (Figure 1.4). CO_2, CH_4 and N_2O are the main greenhouse gases that drive climate change. In addition, high gaseous N emissions from agriculture and fossil fuel burning cause eutrophication of forests, bearing progressive risks of damage by nutritional imbalances, acidification decrease in frost hardiness and raised susceptibility to pathogens and herbivores. Damage will also arise from atmospheric warming and the associated soil water limitation. For example, the area of tropical rain forests is predicted to be reduced by 25% through changing climate during the upcoming decades (Figure 1.5), which exacerbates the loss by land conversion and burning. However, the effects of atmospheric warming and altered precipitation patterns may develop quite variably depending on latitude and longitude, topography and zonobiome (Figure 1.6). This regional variation in climate change effects generates uncertainty about the future shifts in C sequestration and storage versus C release.

Apart from its immediate relevance for climate warming, forest burning is a major source of air pollution at the global scale. In Brazil, for example, about 75% of emissions can be related to land-use change (MCT, 2009). The fire-released emissions comprise precursors of secondary pollutant formation, which *inter alia* contribute to enhanced formation of ground-level ozone (O_3). Remarkable is the congruency, in particular in the tropics and subtropics, between regions of distinct enhancement in O_3 levels predicted towards the end of the twenty-first century (Figure 1.7; Sitch et al., 2007) and extensive forest burning. However, with the exception of some urban areas, little is known across the low latitudes worldwide about O_3 regimes

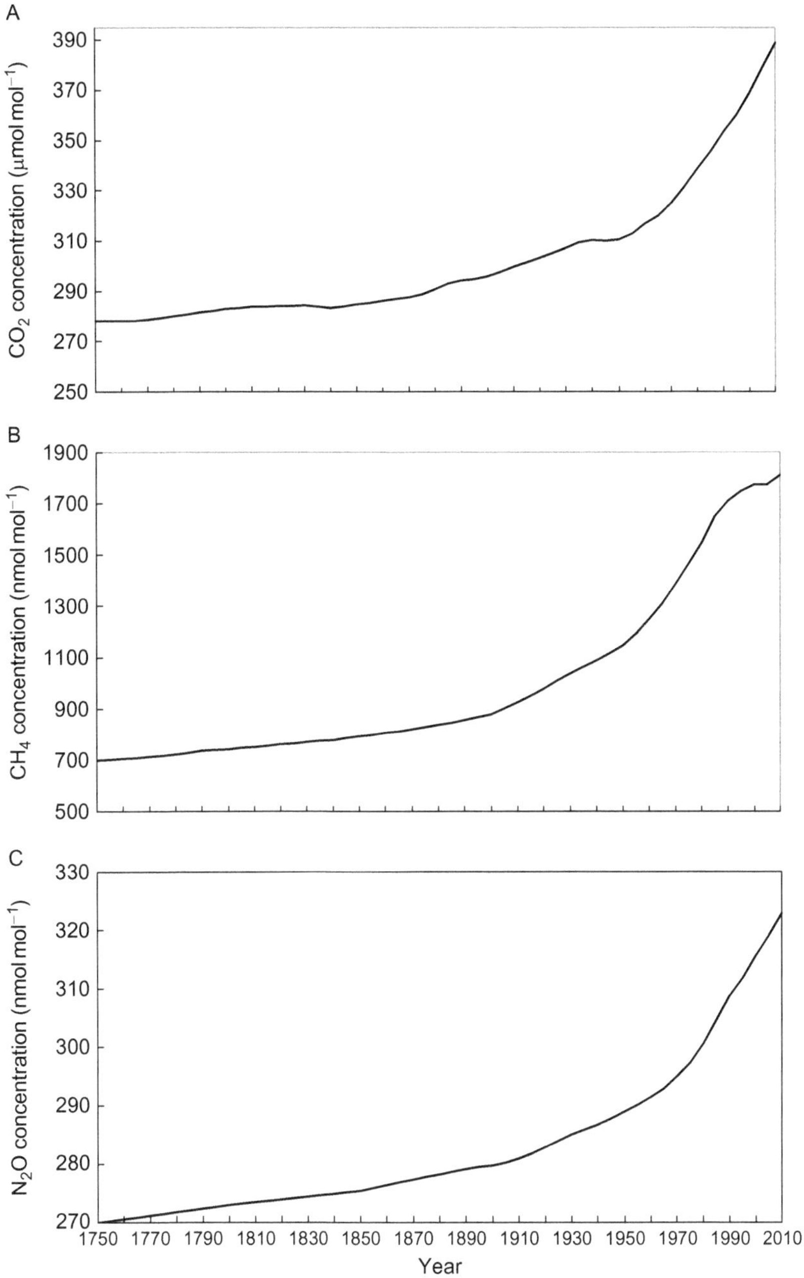

FIGURE 1.4 Global atmospheric concentration of (A) carbon dioxide, (B) methane and (C) nitrous oxide from 1750 through 2010. *Data provided by EEA (http://www.eea.europa.eu/legal/copyright).*

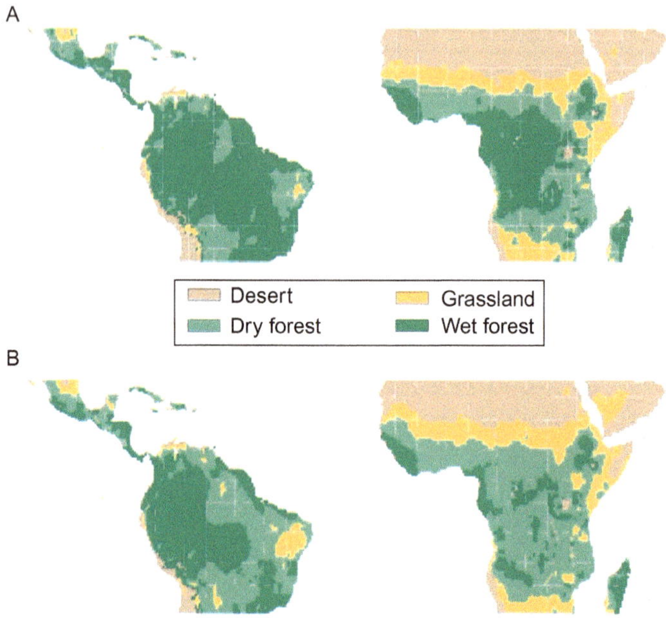

FIGURE 1.5 (A) Current and (B) predicted future expansions of wet and dry forests, grasslands and deserts in the tropics and subtropics of South America and Africa. *After Smith et al. (1992).*

and especially about their effects on vegetation. This deficit is grave in view of O_3 being the potentially most detrimental air pollutant Matyssek and Sandermann (2003). O_3 further weakens the capacity of forests for C sequestration and storage, in addition to the destruction by land-use change. Hence, increasing O_3 levels may counteract potentially stimulating effects on vegetation of the anthropogenic CO_2 enrichment in the atmosphere (Kubiske et al., 2007; Matyssek et al., 2010).

The estimated loss in gross primary production of forest ecosystems worldwide due to O_3 amounts to 50–100 Gt of C from the second half of the nineteenth century until the present (Sitch et al., 2007). Under the current rate of CO_2 enrichment in air, only somewhat more than 1 Gt of C is presumed to be additionally fixed each year by photosynthesis worldwide from the total amount of CO_2 released into the atmosphere by anthropogenic activity. That latter amount of C per year remains in the global balancing of C fluxes as the 'missing carbon' (IPCC, 2007). A recent estimate of forest C balances shows that in the tropics the large sink is more than balanced by deforestation and that the net forest sink of 1.1 Gt C year^{-1} is due to temperate and boreal forests (Pan et al., 2011). Such considerations mediate the vulnerability to air pollution, in particular that of tropospheric O_3, as an intrinsic component of climate change (Matyssek et al., 2013), and if including ecosystem conversion, of global change. Enhanced O_3 concentrations in the troposphere also drive warming, as O_3 is a greenhouse gas.

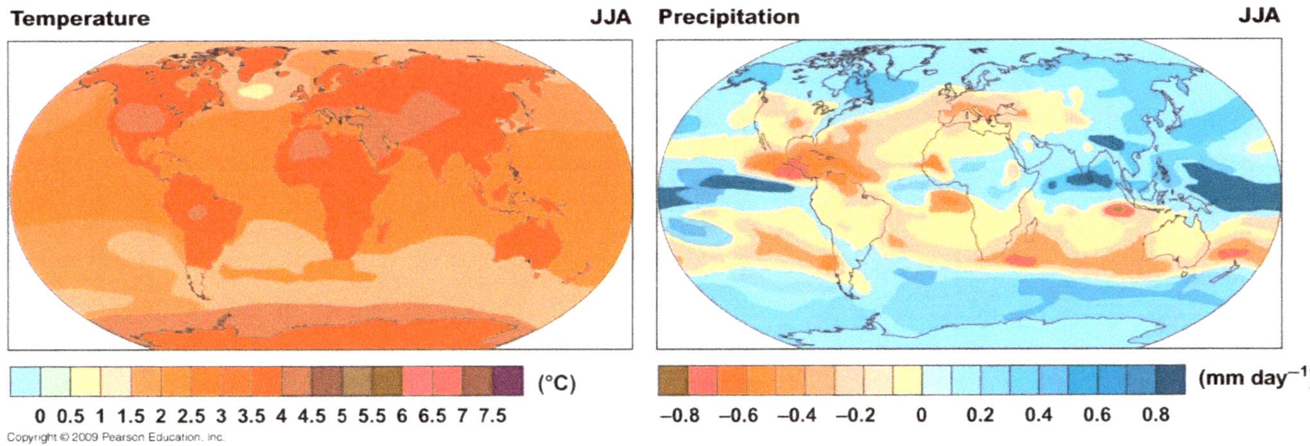

FIGURE 1.6 Global distributions of changes in air temperature (left) and precipitation (right) predicted for the summer months of the northern hemisphere; patterns denote the average of predictions provided by several global circulation models; predictions of changes were made for the period 2080 through 2099 as related to 1980 through 1999. *After IPCC (2007).*

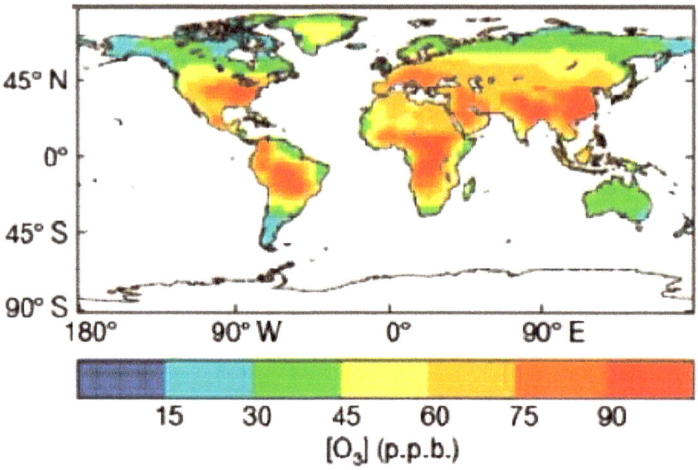

FIGURE 1.7 Global distribution of tropospheric ozone (O_3) concentration predicted towards the end of the twenty-first century. *After Sitch et al. (2007).*

As a result of the excursus given above, we can see that forest ecosystems are crucial components within climate change scenarios. Still, the effects of climate change on C sequestration and storage capacities of forest ecosystems, along with influences on forest hydrology and freshwater resources, are apparently complex and difficult to predict. Although this book must acknowledge the apparently ungovernable global population growth with its ecological consequences, capacities will be explored in forest research by which the anthropogenic pressure on forests during the upcoming decades by climate change and air pollution can be understood and tackled. We feel that the need for process-based understanding prior to effectively adopting mitigation strategies provides well-reasoned arguments for writing this book. To this end, the current state of understanding forest ecosystem functioning will be examined as a key pre-requisite towards stress and risk mitigation. Upcoming research needs and innovative methodological concepts will be explored, along with implications for socio-economy and environmental policy making. Given this kind of direction, the subsequent sections of this chapter will introduce the scope and aims, and the chapter structure of this book.

1.2 AIMS, SCOPE AND RATIONALE

The need for writing this book arises from the ecologically significant, anthropogenic pressures forests currently face. In summary, such pressures result in a loss of forested area (Figures 1.1 and 1.5) and forest capacity for carbon sequestration and storage, mediated, for example, through exacerbating water limitation and air pollution reducing forest functionality. Consequences altogether contribute to the excessive atmospheric CO_2 contamination. Other climate-effective gases apart from CO_2 are also released (Figures 1.3 and 1.4) which through enforcing temperature increase and drought (Figure 1.6) curtail the photosynthetic CO_2 resorption from the atmosphere. O_3 in rising concentrations (Figure 1.7) both as a climate-effective gas and as a noxious agent weakens photosynthesis and, by this, adds to atmospheric overheating. Land-use changes apparently are determinants of climate change and air pollution, and as drivers of ecosystem–atmosphere interactions they are part of the conflicting ecological and socio-economic demands. On such grounds, the aims pursued by this book emerge as follows:

a. clarifying mechanistic, that is, cause–effect, bases of the above relationships between anthropogenic pressures, forest ecosystem responses and atmospheric regimes for comprehending the current state and potential of forests to acclimate and adapt to climate change in a polluted environment, and to mitigate adverse impacts;
b. comprehensively reconciling process-oriented research, long-term monitoring and modelling of environmental changes and forest ecosystem responses;
c. next, introducing the novel 'forest Supersites for Research' concept for promoting and integrating ecosystem monitoring and empirical knowledge acquisition on soil–plant–atmosphere process interactions;
d. and finally, preparing grounds for mechanistically anchored and policy-oriented modelling tools with scientifically sound risk indication for atmospheric changes and ecosystem responsiveness, and eventually, for sustaining ecosystem services within prevalent socio-economic contexts.

In pursuing such aims, the scientific scope of topics covered by the book must be wide. In view of (b) and (c), one important topic is the exploration of the information potential which is presently available in the databases of large-scale programmes and projects, but overlooked or unused because of non-standardized database structures and, hence, impeded information flow between and access to databases (Clarke et al., 2011; Danielewska et al., 2013). What are in view of (a) and (d), bottlenecks for proficient, integrative information drain across databases and the means for overcoming present limitations? Progress here will facilitate the identification of current knowledge gaps and emerging research needs, which is a prominent task of this book from the perspectives of both empirical analysis and theoretical modelling approaches (see a, b and d; Matyssek and Mohren, 2012). How far have we come in reaching integrative knowledge related to the topic of this book, and where do further efforts need to be directed to for augmenting and consolidating evidence (see a and c)? Emerging directions will provide and sharpen arguments for using or developing novel conducive methodologies and integrated, ecologically meaningful research concepts. This will guide interpretation of findings and help to prioritize research needs.

For prioritization, it is necessary to be aware that the previous acquisition and accumulation of knowledge related to the book's subject has been focused mainly on forest ecosystems of the northern hemisphere, in particular those of the temperate zonobiomes, whereas a high need for clarification exists in the southern hemisphere (Figures 1.3 and 1.5; Calfapietra et al., 2009; Matyssek et al., 2012a). Which are the needs and means for optimizing the geographical focus of research activities towards achieving globally balanced acquisitions of evidence and holistic integration at the global scale, given the worldwide distribution of woody-plant systems (Figure 1.1)? This latter question indeed poses a major challenge towards understanding planetary ecology and its spatial finiteness and, hence, limitations in exploitable resources, which is crucial in face of the human population growth and its resource demands (Figure 1.2). One part of this book is dedicated, therefore, to exploring southern hemispheric needs for approaching a balanced planetary scope. The global perspective brings us to the socio-economic context (see d), with implications for raising efficacy in impacting research and environmental policies.

Having elucidated the scientific scope of the book, its innovation is borne by the overall rationale towards integrating modelling, monitoring and empirical experimentation, and in such view, stimulating scientific networking across northern and southern hemispheres on ecological analysis, socio-economic relevancy and policy making. Political farsightedness in decision making is to be encouraged, based both on what we know today and what we still need to find out through support of innovative research, for the sake of ensuring ecosystem integrity, and hence, the survival of mankind. On such grounds, the ambition of the book is to develop new perspectives which also support the post-Kyoto debate. The book's rationale has originated from the *COST Action FP0903* (*Climate Change and Forest Mitigation and Adaptation in a Polluted Environment, MAFor*, which was active from 2009 through 2013; http://cost-fp0903.ipp.cnr.it/). COST standing for *European Cooperation in Science and Technology* represents a funding instrument of the European Union (EU) in support of network cooperation in science and technology within the EU and associated countries. The book represents the concluding publication from *MAFor*, which pursued two main objectives, namely

i. to increase understanding of the state and potential of forest ecosystems to mitigate and adapt to climate change in a polluted environment, and
ii. to reconcile process-oriented research, long-term monitoring and applied modelling at comprehensive forest research sites (Supersites).

Regarding these objectives, including the newly introduced Supersite concept (see c), *MAFor* gave rise to the publications by Clarke et al. (2011), Fischer et al. (2011), Paoletti and Tuovinen (2011), Matyssek and Mohren (2012), Matyssek et al. (2012a) and Danielewska et al. (2013). These publications, in particular that by Matyssek et al. (2012a) on collocating gaps in understanding and future directions for research, provided inspiration for this book in exploring the addressed objectives in depth and towards an integrative rationale. The essence of that rationale is mirrored by the chapter structure of this book, as shown below.

1.3 OVERVIEW OF THE BOOK'S STRUCTURE

After introducing the concept and rationale of this book (Part I, this chapter), the book's first major focus will be on interactions between trace gases, climate change and vegetation (Part II), relevant for the aspects shown in Figures 1.3–1.7 (see a). Key issues are the mechanistic grounds of interactions between CO_2, O_3, different N species and VOCs as well as the atmosphere–biosphere exchange of these compounds, the arising significance for aerosol formation and dynamics related to the formation of O_3 in the troposphere from both anthropogenic and biogenic precursor emissions. What kinds of feedbacks exist with regard to climate change? Five chapters will be dedicated to such considerations.

Another six chapters will elucidate the significance of biotic processes in forest ecosystem response to climate change and air pollution (Part III), underlying the large-scale phenomena of Figures 1.3–1.7, while being drivers related to (a). Factorial biotic–abiotic associations impacting on C cycling and C sink formation in forest ecosystems in concert with water and nutrient relations will be elucidated at the tree and stand level. This implies consideration of trace gas fluxes as determined by stomatal regulation and respiratory processes along with interactions between elevated CO_2 and O_3 regimes in tree and ecosystem response. Particular attention is directed to influences through the genotype, ontogeny and phenology of trees, as well as competitive facilitative, mutualistic and parasitic above- and belowground interactions, weighting the role of edaphic factors (e.g. soil moisture and nutrients) within the analysis.

Part IV is the extension of Part III leading over to highlighting the mechanistic and diagnostic understanding of ecosystem response for conducting risk assessment and tree-ecosystem process-scaling. Five chapters are assigned here, dealing with the empirical and modelling perspectives on the subject. Empirical assessment to integrate the molecular level of metabolic control and the biochemical and physiological process levels of metabolic activity will be stressed, looking for biological markers of environmental impact (Matyssek et al., 2012b; Sandermann and Matyssek, 2004). Such integration will yield the mechanistic grounds for process up-scaling in spatio-temporal terms to the ecosystem level and beyond, by promoting tool development for differential stress diagnosis and process-based risk assessment (see a and d). Empirical research is stressed to provide databases for establishing and validating numeric models in order to enable their crucial function in research: as models representing hypotheses, guidance is provided for empirical research towards mechanistic consolidation in understanding (Priesack et al., 2012).

Part V is dedicated to the global dimension of air pollution as part of climate change, being one key concern of this book, in comprehending rationales in relation to Figures 1.1 and 1.5–1.7. Five chapters will contribute to a global perspective (see a), with particular attention to 'hot spots' in climate change and air pollution arising during the course of the twenty-first century. Scenarios will be exemplified from South America, South Africa, India and East Asia and set into perspective with the knowledge presently available from the temperate zonobiome of the northern hemisphere. Major gaps in information will be identified worldwide on the book's subject and the potential of the newly introduced concept of 'Supersites' for research on forest ecosystems (see c; Fischer et al., 2011; Matyssek et al., 2012a). The potential of research Supersites will be explored in Part VI, as represented by two chapters, and reference will be made to facilitating information flow between databases across sites and to research networking across hemispheres.

Parts V and VI will lead us to the three chapters of Part VII, examining the socio-economic context of the book's subject and the means of knowledge transfer to practice and policy making (see d). Rationales are related to Figures 1.1, 1.2, 1.5 and 1.7. Ways of translating findings from empirical research and process-based, numeric modelling into silvicultural, political and socio-economic dimensions are key issues in this part of the book. Perspectives are on post-Kyoto policies, cost benefit relationships of ecosystem services, and human health. It will be stressed that research outcomes can become effective only within and for the sake of society through appropriate communication to the public. Special attention will be directed to the expectations and needs of policy makers and various stakeholders.

In Chapter 27 of Part VIII, we will arrive at the conclusions and perspectives derivable from the book. Can unification of air pollution and climate change research at 'Supersites' in forest ecosystems, interlinked within global research networks, provide breakthroughs in understanding the resilience of the ecosystem Earth under the increasing anthropogenic pressure? Will aims and scopes as propagated by this book enable strengthening of communication between experimentalists, monitoring experts, modellers, policy makers and stakeholders—for the sake of forest ecosystems, and ultimately, mankind? These are the key questions to which this book and its conclusions intend to give compelling answers.

ACKNOWLEDGEMENTS

The constructive suggestions by Allan Legge on the manuscript are highly appreciated.

REFERENCES

Bengtson, M., Shen, Y., Oki, T., 2006. A SRES-based gridded global population data set for 1990–2100. Popul. Environ. 28, 113–131.

Calfapietra, C., Ainsworth, E.A., Beier, C., De Angelis, P., Ellsworth, D.S., Godbold, D.L., Hendrey, G.R., Hickler, T., Hoosbeek, M.R., Karnosky, D.F., King, J., Körner, C., Leakey, A.D.B., Lewin, K.F., Liberloo, M., Long, S.P., Lukac, M., Matyssek, R., Miglietta, F., Nagy, J., Norby, R.J., Oren, R., Percy, K.E., Rogers, A., Scarascia Mugnozza, G., Stitt, M., Taylor, G., Ceulemans, R., 2009. Challenges in elevated CO_2 experiments on forests. Trends Plant Sci. 15, 5–10.

Clarke, N., Fischer, R., de Vries, W., Lundin, L., Papale, D., Vesala, T., Merilä, P., Matteucci, G., Mirtl, M., Simpson, D., Paoletti, E., 2011. Availability, accessibility, quality and comparability of monitoring data for European forests for use in air pollution and climate change science. iForest 4, 162–166.

Cohen, J.E., 2003. Human population: the next half century. Science 302, 1172–1175.

Danielewska, A., Clarke, N., Olejnik, J., Hansen, K., de Vries, W., Lundin, L., Tuovinen, J., Fischer, R., Urbaniak, M., Paoletti, E., 2013. A meta-database comparison from various European Research and Monitoring Networks dedicated to forest sites. iForest 6, 1–9.

Dixon, R.K., Brown, S., Houghton, R.A., Solomon, A.M., Trexler, M.C., Wisniewski, J., 1994. Carbon pools and flux of global forest ecosystems. Science 263, 185–191.

Fischer, R., Aas, W., De Vries, W., Clarke, N., Cudlin, P., Leaver, D., Lundin, L., Matteucci, G., Matyssek, R., Mikkelsen, T.N., Mirtl, M., Öztürk, Y., Papale, D., Potocic, N., Simpson, D., Tuovinen, J.-P., Vesala, T., Wieser, G., Paoletti, E., 2011. Towards a transnational system of supersites for forest monitoring and research in Europe—an overview on present state and future recommendations. iForest 4, 167–171.

Fritsch, B., 1990. Mensch, Umwelt, Wissen. Evolutionsgeschichtliche, London.

IPCC, 2007. Climate change 2007: synthesis report. Summary for policymakers. WG1, AR4. Available from: http://ipcc-wg1.ucar.edu/wg1/wg1-report.html.

Kubiske, M.E., Quinn, V.S., Marquardt, P.E., Karnosky, D.F., 2007. Effects of elevated atmospheric CO_2 and/or O_3 on intra- and interspecific competitive ability of Aspen. Plant Biol. 9, 342–355.

Lovelock, J., 2009. A Final Warning: The Vanishing Face of Gaia. Penguin Books, London.

Lüttge, U., 2013. The planet Earth: can it feed nine billion people? Nova Acta Leopold. 391, 345–364.

Luyssaert, S., Schulze, E.-D., Börner, A., Knohl, A., Hessenmöller, D., Law, B.E., Grace, J., Ciais, P., 2008. Old-growth forests as global carbon sinks. Nature 455, 213–215.

Matyssek, R., Mohren, G.M.J., 2012. Special topic: integrating modelling and experimentation. Trees. http://dx.doi.org/10.1007/s00468-012-0778-4.

Matyssek, R., Sandermann, H., 2003. Impact of ozone on trees: an ecophysiological perspective. Progress in Botany, vol. 64. Springer Verlag, Heidelberg, pp. 349–404.

Matyssek, R., Wieser, G., Ceulemans, R., Rennenberg, H., Pretzsch, H., Haberer, K., Löw, M., Nunn, J.J., Werner, H., Wipfler, P., Oßwald, W., Nikolova, P., Hanke, D., Kraigher, H., Tausz, M., Bahnweg, G., Kitao, M., Dieler, J., Sandermann, H., Herbinger, K., Grebenc, T., Blumenröther, M., Deckmyn, G., Grams, T.E.E., Heerdt, C., Leuchner, M., Fabian, P., Häberle, K.H., 2010. Enhanced ozone strongly reduces carbon sink strength of adult beech (*Fagus sylvatica*)—resume from the free-air fumigation study at Kranzberg Forest. Environ. Pollut. 158, 2527–2532.

Matyssek, R., Wieser, G., Calfapietra, C., de Vries, W., Dizengremel, P., Ernst, D., Jolivet, Y., Mikkelsen, T.N., Mohren, G.M.J., le Thiec, D., Tuovinen, J.-P., Weatherall, A., Paoletti, E., 2012a. Forests under climate change and air pollution: gaps in understanding and future directions for research. Environ. Pollut. 160, 57–65.

Matyssek, R., Schnyder, H., Oßwald, W., Ernst, D., Munch, J.C., Pretzsch, H. (Eds.), 2012b. Growth and defence in plants—resource allocation at multiple scales. In: Ecological Studies, vol. 220. Springer, Heidelberg, p. 467.

Matyssek, R., Kozovits, A.R., Schnitzler, J., Pretzsch, H., Dieler, J., Wieser, G., 2013. Forest trees under air pollution as a factor of climate change. In: Tausz, M., Grulke, N. (Eds.), Trees in a Changing Environment: Ecophysiology, Adaptation and Future Survival. Springer, Heidelberg (in press).

MCT, 2009. Ministério de Ciência e Tecnologia, Inventário brasileiro das emissões e remoções antrópicas de gases de efeito estufa: informações gerais e valores preliminares. http://www.mct.gov.br/index.php/content/view/310922/Segundo_Inventario_Brasileiro_de_Emissoes_e_Remocoes_Antropicas_de_Gases_de_Efeito_Estufa.html.

Pan, Y., Birdsey, R.A., Fang, J., Houghton, R., Kauppi, P., Kurz, W.A., Phillips, O.L., Shvidenko, A., Lewis, S.L., Canadell, J.G., Ciais, P., Jackson, R.B., Pacala, S.W., McGuire, A.D., Piao, S., Rautiainen, A., Sitch, S., Hayes, D., 2011. A large and persistent carbon sink in the world's forests. Science 333, 988–993.

Paoletti, E., Tuovinen, J.-P., 2011. COST Action FP0903: research, monitoring and modelling in the study of climate change and air pollution impacts on forest ecosystems. iForest 4, 160–161.

Priesack, E., Gayler, S., Rötzer, T., Seifert, T., Pretzsch, H., 2012. Mechanistic modelling of soil–plant–atmosphere systems. In: Matyssek, R., Schnyder, H., Oßwald, W., Ernst, D., Munch, J.C., Pretzsch, H. (Eds.), Growth and defence in plants—resource allocation at multiple scales. Ecological Studies, vol. 220. Springer, Heidelberg, pp. 335–354.

Sandermann, H., Matyssek, R., 2004. Scaling up from molecular to ecological processes. In: Sandermann, H. (Ed.), Molecular Ecotoxicology of Plants. Ecological Studies, vol. 170. Springer Verlag, Heidelberg, pp. 207–226.

Schulze, E.D., Luyssaert, S., Ciais, P., Freibauer, A., Janssens, I.A., et al., 2009. Importance of methane and nitrous oxide for Europe's terrestrial greenhouse-gas balance. Nat. Geosci. 2, 842–850.

Sitch, S., Cox, P.M., Collins, W.J., Huntingford, C., 2007. Indirect radiative forcing of climate change through ozone effects on the land-carbon sink. Nature 448, 791–795.

Smith, T.M., Leemans, R., Shugart, W.H., 1992. Sensitivity of terrestrial carbon storage to climate change: Comparison of four scenarios based on general circulation models. Clim. Change. 21, 367–384.

Part II

Interactions Between Trace Gases, Climate Change and Vegetation

Chapter 2

Gaseous Exchange Between Forests and the Atmosphere

Stan Cieslik*, Juha-Pekka Tuovinen[†], Manuela Baumgarten[‡], Rainer Matyssek[‡], Patricia Brito[§] and Gerhard Wieser[¶,1]

*Joint Research Centre, Ispra, Italy

[†]Finnish Meteorological Institute, Helsinki, Finland

[‡]Technische Universität München, Ecophysiology of Plants, Hans-Carl-von-Carlowitz-Platz 2, D-85354 Freising-Weihenstephan, Germany

[§]Division of Plant Biology, University of La Laguna, Tenerife, La Laguna, Spain

[¶]Department of Alpine Timberline Ecophysiology, Federal Office and Research Centre for Forests, Innsbruck, Austria

[1]Corresponding author: e-mail: Gerhard.Wieser@uibk.ac.at

Chapter Outline

2.1 INTRODUCTION

Forests and plants, in general, are powerful engines that exchange various gases between themselves and the atmosphere. The term 'exchange' is used here because gases can both be released by plants into air and be taken up by plants from air. The processes involved here represent essential steps in biogeochemical cycling and constitute a fundamental link between the biosphere and the atmosphere, thereby having an impact on weather, climate and the environment (Arneth et al., 2010). While the present chemical

Developments in Environmental Science, Vol. 13. http://dx.doi.org/10.1016/B978-0-08-098349-3.00002-5

composition of the atmosphere has largely developed through the emergence of photosynthetic plants, it is noteworthy that the release and uptake of different gases affect atmospheric composition on much shorter timescales as well. For example, the emissions of biogenic volatile organic compounds (BVOCs), such as isoprene and monoterpenes, play an important role in atmospheric chemistry, especially in the formation of ozone (O_3) in the troposphere (Loreto and Fares, 2013, this vol.; Monks et al., 2009).

However, gas exchange is not only a chemical phenomenon but also entails physical aspects. In particular, evaporation and condensation of water involve energy transfer, and atmospheric turbulence exerts a strong control on surface fluxes of mass and energy (Garratt, 1992). It is also now well established that release of greenhouse gases (GHGs), most importantly carbon dioxide (CO_2) and methane (CH_4), affects climate on a global scale, since these gases modify radiative transfer and thus change the Earth's energy balance (IPCC, 2007). Ecophysiological functioning of forest ecosystems is inherently involved in the exchange of these GHGs as well as other gases. For example, the production of BVOCs is linked to plant metabolism (Loreto and Fares, 2013, this vol.). Plant growth and distribution not only influence but also are influenced by exchange of various gases, some of which are essential for plant functioning (CO_2, oxygen, water vapour), while some, for example, O_3 and various nitrogen compounds, may act as air pollutants with detrimental effects on plant physiology (Matyssek et al., 2012).

In this chapter, we will address some selected topics related to the gaseous exchange between forest ecosystems and the atmosphere. First, we will present a short overview of the emissions of key GHGs from the forest floor. The following section deals with forest fires. While this book includes a full chapter on this topic (Sofiev, 2013, this vol.), we present complementary material by focusing on forest fires as emission sources of GHGs and pollutants and by highlighting how the fires affect CO_2 efflux from soils. As O_3 is considered the most important phytotoxic air pollutant (Ashmore, 2005), we next summarise the current understanding of O_3 fluxes to forests, noting that the stomatal uptake is a key determinant of O_3-induced damage (Dizengremel et al., 2013, this vol.). The BVOC emissions from forest canopies are discussed in a separate chapter (Loreto and Fares, 2013, this vol.) and are here only briefly touched upon in a section that presents examples of the interactions between forests, air quality and climate. The final section outlines directions for the multi- and interdisciplinary research that is increasingly needed in the future for understanding these complex interactions.

2.2 TRACE GAS EMISSIONS FROM THE FOREST FLOOR

In terrestrial ecosystems, soil CO_2 efflux (E_S) is considered as the major flux of carbon (C) from the ecosystem to the atmosphere and hence is an important component of the global carbon balance (Kruse et al., 2013, this vol.; Schimel, 1995). For example, while the mean gross primary productivity of

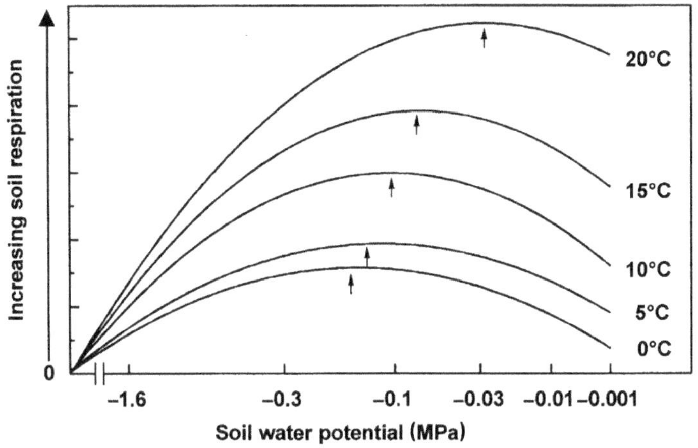

FIGURE 2.1 Influence of soil water potential on soil CO_2 efflux over a range of temperatures in a Canary Island Pine forest at treeline, Las Canadas 2070 m a.s.l. Arrows depict the position of optimum soil water availability. *Redrawn after data from Brito et al. (2013).*

European forests is about 1400 g C m^{-2} yr^{-1}, as much as 55% of this is released back to the atmosphere by soil respiration (Janssens et al., 2001). On a global scale, E_S is mainly controlled by soil temperature (Raich and Potter, 1995; Raich and Schlesinger, 1992). Soil water availability is the second abiotic factor that exerts a strong influence over E_S (Janssens et al., 2003; Lavigne et al., 2004). Soil water availability can limit E_S from the forest floor in two ways: either by limiting aeration, and thus the diffusivity of the air, when water is abundant, or by limiting soil microbial communities, root and mycorrhizal respiration under soil drought (Figure 2.1), as has been demonstrated for temperate and Mediterranean forest ecosystems (Brito et al., 2013; Schindlbacher et al., 2009; Wieser, 2004).

Forest soils also release nitrous oxide (N_2O) and nitric oxide (NO) (Smith et al., 2003). On a global scale, the natural emissions of N_2O and NO total 15 and 5 Tg N yr^{-1}, respectively (Cape, 2002). The N_2O and NO fluxes observed at the soil–atmosphere interface are a result of simultaneous production and consumption processes due to two contrasting microbial processes: nitrification of ammonium to nitrite and, subsequently, nitrate and denitrification of nitrate to N_2O and, eventually, to molecular nitrogen (N_2). N_2O and NO emission rates depend on biological and physico-chemical processes in the soil (Smith et al., 2003), which are driven by abiotic environmental factors such as temperature, moisture, pH, ammonium and nitrate levels of soils, as well as atmospheric N deposition and the soil C/N (Conrad, 1996). In addition, tree species and forest type also determine nitrogenous trace gas emissions from the soil to the atmosphere. On a global scale, the contribution of tropical forest ecosystems, which are dominated by N-fixing leguminous tree species (Erickson et al., 2001), to the global N_2O budget is approximately 3.6 Tg N yr^{-1} (Breuer et al., 2000). Temperate and boreal forest soils in contrast have been considered to be weak sources of N_2O, with a contribution of approximately 0.5 Tg N yr^{-1} (Potter et al., 1996).

Forest soils serve as both net sinks and net sources of atmospheric CH_4 (Butterbach-Bahl, 2002). Approximately 5–10% of the atmospheric CH_4 is removed from the atmosphere by diffusion into aerobic top soils (Smith et al., 2003), in which oxidation occurs by methanotrophic bacteria. Conversely, CH_4 is produced in soils by microbial breakdown of organic compounds under anaerobic conditions at low redox potentials. Formation and oxidation of CH_4 within different horizons of soil profiles is controlled by various environmental factors, with oxygen availability along with edaphic and climatic conditions being the most important ones (Butterbach-Bahl, 2002; Smith et al., 2003). Forest ecosystems which are seasonally or permanently flooded are strong sources of CH_4 irrespective of the zonobiome, and the maximal CH_4 emission from such ecosystems can reach 20 mg CH_4 m^{-2} h^{-1}. Uptake rates of atmospheric CH_4 by forest soils are generally less than 0.3 mg CH_4 m^{-2} h^{-1}, which is approximately one magnitude lower than the rates of CH_4 emission from flooded forest soils under corresponding climatic conditions (Butterbach-Bahl, 2002 and references therein).

2.3 EFFECTS OF FOREST FIRES

Forest fires emit large quantities of gases and particulate matter to the atmosphere (for a review, see Bytnerowicz et al., 2009). Gaseous emissions include GHGs (CO_2, CH_4) and numerous air pollutants (e.g. CO, VOCs, SO_x, NO_x and NH_3). These emissions produce local-scale pollution plumes that may cause severe air quality deterioration, as was the case during the extreme air pollution episode in the Moscow region in 2010 (Konovalov et al., 2011). On the other hand, the emissions can be transported over thousands of kilometres, especially in the case of very intense fires that inject emissions above the atmospheric boundary layer (Sofiev, 2013, this vol.).

During the transport, O_3 is typically produced from its precursors and significantly elevated O_3 concentrations may result from forest fires (Jaffe and Wigder, 2012). For example, while heat waves *per se* provide conditions that are conducive to net O_3 production, the forest fires triggered by heat waves in southern Europe have been observed to strongly enhance O_3 concentrations on a regional scale (Hodnebrog et al., 2012; Solberg et al., 2008). Jaffe and Wigder (2012) estimate that wildfire emissions produce globally as much as 170 Tg O_3 yr^{-1}, which is 3.5% of the total O_3 production in the troposphere.

Forest fires can cause long-term changes in species composition depending on succession or can even replace forests by other vegetation types such as steppe, chaparral or cerrado (Kozovits and Bustamante, 2013, this vol.). Fires can also have a direct effect on carbon stocks (Conrad and Solomon, 2009). Van der Werf et al. (2010) estimated that the global emissions from biomass burning were on average 2.0 Pg C yr^{-1} in 1997–2009, varying annually within 1.6–2.8 Pg C yr^{-1}. This corresponds to about 4% of the total terrestrial net primary productivity. About 15% of these emissions originate from forest fires, which are mostly confined to temperate and boreal regions (van der Werf et al., 2010).

Forest fires are known to influence soil processes, in particular, the CO_2 efflux from soils. Depending on their severity, fires may significantly affect the organic carbon content of soils. Although the net effect of wildland fires on soil respiration varies widely between ecosystems, several studies have reported large reductions in E_S (46–64%; Conrad and Solomon, 2009 and references therein). E_S has been shown to decline with increasing severity of the fire (Figure 2.2). In parallel, soil microbial biomass as a measure of microbial activity is also depressed with increasing fire intensity (Figure 2.2). Consequently, the post-fire dynamics may also influence the exchange of N_2O, NO and CH_4 between the soil and the atmosphere, and hence ultimately atmospheric chemistry.

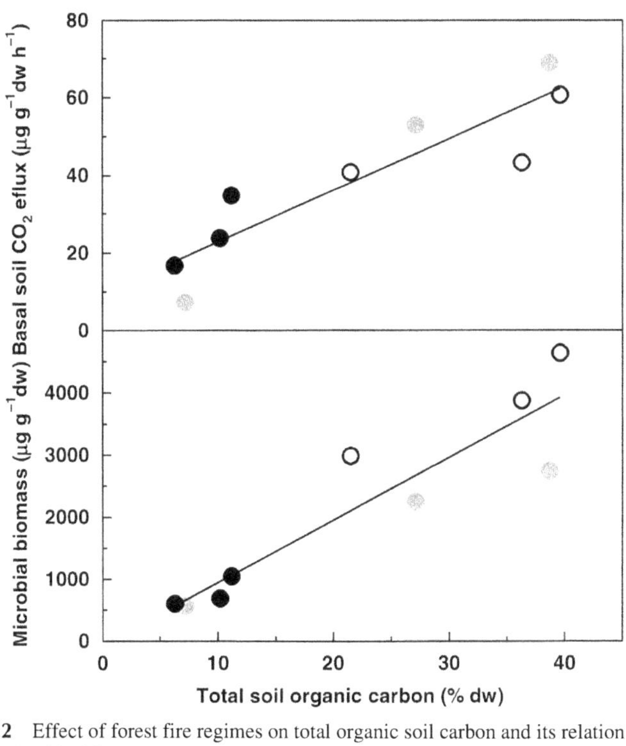

FIGURE 2.2 Effect of forest fire regimes on total organic soil carbon and its relation to soil CO_2 efflux and microbial biomass: open symbols fire-free control, grey symbols moderately sensitive to fire, black symbols highly sensitive to fire. *Wieser and Bahn unpublished data collected in a Canary Island Pine forest, Chinyero Tenerife 1400 m a.s.l., in September 2007, 1 month after a severe forest fire.*

Model projections indicate that atmospheric temperature is likely to play an increasingly important role in driving the global trends of wildfire occurrence in the future (Pechony and Shindell, 2010). Northern American observations provide evidence that the frequency and activity of fires have already increased and that this increase is associated with warming in spring and summer and an earlier spring snowmelt (Westerling et al., 2006).

2.4 OZONE DEPOSITION

The foliage acts as the dominant sink in the atmosphere–forest interaction of O_3, and thus the canopy structure has a noticeable effect on the O_3 exposure and uptake by the foliage (Wieser, 2002). The O_3 levels within dense canopies decline with canopy depth due to stomatal uptake and reactions with external plant surfaces (Karlsson et al., 2006), as well as atmospheric O_3 scavenging by reactive gases such as VOCs and NO_x, which can be emitted from the foliage and the underlying soil (Fowler et al., 2009; Loreto and Fares, 2013, this vol.). The formation of the vertical O_3 flux and concentration profiles depends on the distribution of these sinks and is strongly controlled by the intensity of turbulent mixing within and below the canopy. This means that the development of total leaf area and its distribution, possibly stimulated by elevated CO_2 but reduced by enhanced O_3 levels (King et al., 2013, this vol.), must not be overlooked when estimating the O_3 fluxes to forests.

2.4.1 Stomatal Uptake

Stomatal O_3 uptake is highly variable, reflecting the dynamics of atmospheric O_3 concentrations and stomatal regulation under internal and external controls. For evergreen conifers of temperate-boreal zonobiomes, stomatal conductance and O_3 uptake show a pronounced seasonal variation with maxima during the growing season and decrease during fall upon incidents of freezing temperatures (Altimir et al., 2006). During winter, O_3 uptake is almost completely suppressed, until photosynthesis is recovered with increasing irradiance and temperature in spring (Altimir et al., 2006; Wieser et al., 2000). In deciduous trees, by contrast, the temperature-induced limitations are less important, as stomatal conductance rather is influenced by seasonal variations in leaf area and leaf age (Iacobelli and MacCaughery, 1993; Löw et al., 2006; Squire and Black, 1981). Indications exist that even moderate soil drought can significantly reduce O_3 uptake (Löw et al., 2006; Wieser and Havranek, 1993, 1995) upon stomatal closure of the drought-sensitive mesomorphic leaves (Hinckley et al., 1978). In Mediterranean climates, long-lasting droughts can substantially decrease the seasonal O_3 uptake (Gerosa et al., 2008; Inclan et al., 1999; Wieser et al., 2006).

During warm and humid periods, that is, in the absence of soil drought, stomatal O_3 influx is determined by irradiance and the leaf–air mole fraction difference of water vapour (Δw) (Wieser, 2002), as has been shown for adult trees by Wieser and Havranek (1993, 1995). The formation of high O_3 concentrations often correlates with Δw (Grünhage and Jäger, 1994), which limits O_3 influx upon stomatal closure (Wieser and Havranek, 1993, 1995; Wieser et al., 2012).

As tree canopies are not homogeneous in terms of crown architecture, complex horizontal and vertical environmental gradients may occur within stands. Such structural features shape the micro-environment as a driver of the within-canopy variations in conductance and O_3 uptake (Löw et al., 2006; Wieser, 2002). On a total leaf area basis, conifers, except for deciduous larch with its mesomorphic needles, possess a lower stomatal conductance than deciduous broadleaved trees (Körner et al., 1979; Reich et al., 1992) and, as a consequence, are subjected to lower O_3 uptake rates (Reich, 1987). Leaf conductance has also been shown to decrease by shading within canopies (Dang et al., 1997; Leverenz et al., 1982). In addition to species and genotype-related specifics, conductance is a function of both tree age (Kolb et al., 1997; Wieser et al., 2000) and leaf age (Leverenz et al., 1982; Reich et al., 1992; Wieser et al., 2000).

2.4.2 Non-Stomatal Sinks

As mentioned above, the total O_3 flux to a forest ecosystem results from different non-stomatal sinks, in addition to stomatal uptake. While physiological responses to O_3 essentially depend on stomatal uptake, the non-stomatal fluxes must be studied as a component of O_3 mass balance. The removal of O_3 at external plant surfaces cannot be explained by a single mechanism and has been suggested to result from various chemical interactions including (i) thermal decomposition of O_3 that is mediated by waxes and other substances on the surface, (ii) surface- and VOC-mediated O_3 photolysis and (iii) aqueous reactions in water films present on leaf surfaces (Fowler et al., 2009). Thus, these processes are controlled by the same environmental factors as stomatal uptake, that is, temperature, irradiance, atmospheric humidity and precipitation. The scavenging potential of biogenic VOC and NO emissions depends on tree species, soil properties and climate and thus shows strong geographical variations (Fowler et al., 2009).

Significant (20–80% of total) non-stomatal O_3 fluxes have been observed in different forests in southern European conditions (Cieslik, 2004, 2009), which often limit stomatal gas exchange. Similar observations have been made in cooler and more humid environments in the temperate, boreal and subarctic regions as well (Coe et al., 1995; Mikkelsen et al., 2004; Tuovinen et al., 2001). Recent long-term micrometeorological datasets provide deeper insights into the stomatal/non-stomatal flux partitioning patterns and the related controls. For example, 10-year-long measurements in a boreal Scots pine forest in Finland showed that the fraction of the total non-stomatal

O_3 deposition in the daytime during the growing season varied within 26–44% (Rannik et al., 2012). The non-stomatal conductance correlated with atmospheric humidity, and additionally a chemical in-canopy sink was observed, even if not identified, while the NO emissions from soil were too small to perturb the O_3 fluxes. Another decade-long dataset collected in a mixed temperate forest in Belgium showed larger non-stomatal fractions, exceeding 60% even during the daytime in summer (Neirynck et al., 2012). Significant fluxes occurred at this site also in winter and at night-time, especially to rain-wetted canopies and in well-mixed conditions. In a warmer climate, the large non-stomatal flux component observed above a Ponderosa pine plantation in California was largely attributed to atmospheric reactions with BVOCs (Fares et al., 2010).

2.4.3 Up-Scaling and Measurements

The traditional focus of plant ecophysiological research is organism centred. Assessing O_3 influx via stomata into leaves by gas exchange measurements is the primary stage for further up-scaling to the tree, stand/ecosystem and landscape level (Wieser et al. 2008). The stomatal O_3 influx as a proportion to the total flux into stands can quantitatively be related to the leaf level and derived from cuvette or fumigation chamber studies, using a mass-balance approach (Havranek and Wieser, 1994).

Since transpiration and O_3 influx are coupled through stomatal regulation, sap-flow techniques that determine crown transpiration can be used to assess tree-level O_3 uptake (Köstner et al., 2008; Wieser et al., 2003). Whole-tree O_3 uptake rates can be combined with stand density data so that stomatal stand-level O_3 uptake can be expressed per unit of ground surface area (Wieser et al., 2006). The sap-flow-based O_3 flux assessment can be combined with the eddy covariance technique (Nunn et al., 2010). The latter methodology inherently assesses the total O_3 flux, that is, the sum of stomatal and non-stomatal O_3 fluxes at the stand level. In addition, the canopy-scale stomatal conductance can be estimated from the eddy covariance measurements of water vapour and sensible heat fluxes based on the Penman–Monteith equation (Coe et al., 1995).

Figure 2.3 provides an example of diurnal courses of total and stomatal O_3 fluxes obtained in a mixed beech/spruce forest in Kranzberg, Germany. Eddy covariance-derived total stand O_3 flux displayed its maximum already before those of Δw and O_3, which was also the case for stomatal O_3 flux derived from the sap-flow approach. Conversely, stomatal O_3 flux derived from the Penman–Monteith algorithm reached its maximum around noon. The sap-flow-based stomatal O_3 flux was 33% of the total O_3 flux, whereas the Penman–Monteith approach resulted in 47%. Such observed differences in stomatal O_3 flux may be due to calculations of evapotranspiration based on micrometeorological methods as they, in contrast to the sap-flow approach,

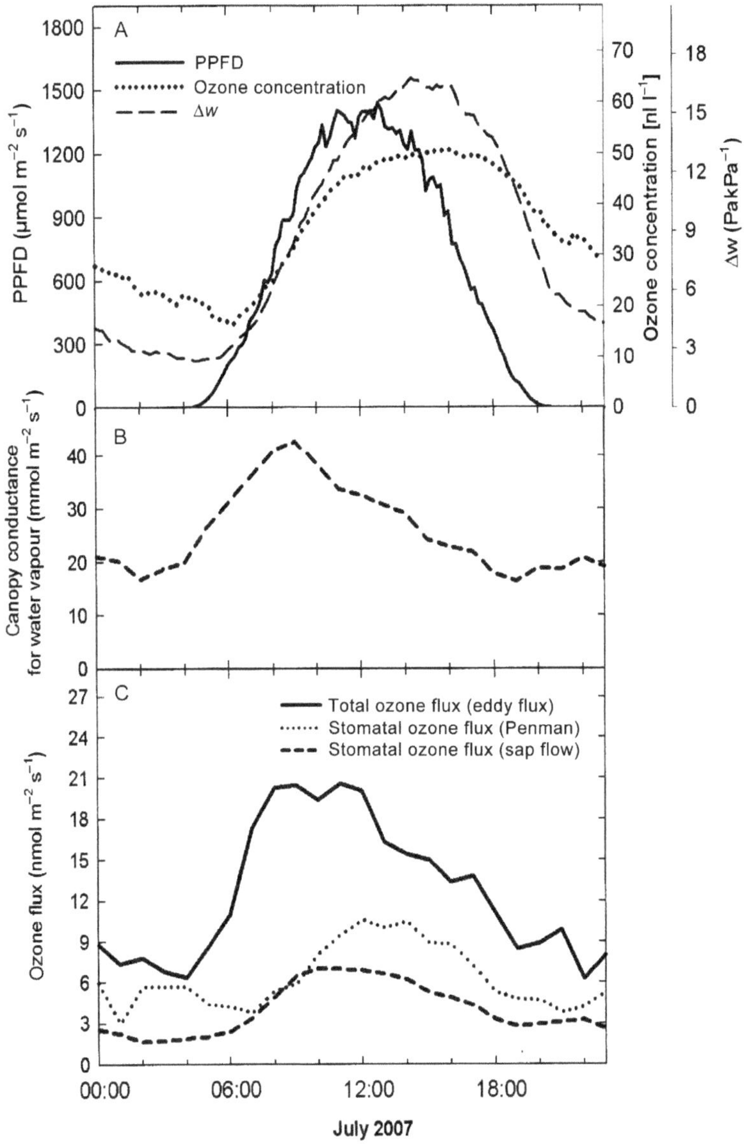

FIGURE 2.3 Diurnal course of monthly mean (July 2–29, 2007) of (A) photosynthetically active photon flux density (PPFD), leaf–air mole fraction difference of water vapour (Δw) and O_3 concentration; (B) leaf area-scaled canopy conductance for water vapour (Gc); and (C) ground surface area-scaled total stand O_3 flux, calculated stomatal stand-level O_3 flux according to the Penman–Monteith algorithm and stomatal canopy O_3 flux as derived from sap-flow assessment. *Reprinted from Nunn et al. (2010); with permission from Elsevier.*

also include evapotranspiration sources other than tree canopies, which can contribute up to 20% of total stand evapotranspiration (Köstner, 2001). In addition, sap-flow-scaled stomatal O_3 flux emphasises physiologically modulated canopy conductance (Figure 2.3B; Wieser et al. 2006, 2008), while calculations based on micrometeorological methods rather follow the diurnal course of O_3 (Figure 2.3A; Nunn et al., 2010).

2.5 INTERACTIONS WITH ATMOSPHERIC COMPOSITION AND CLIMATE

Climate change affects air pollution and its impacts on vegetation through influences on both the atmosphere and the biosphere, as well as on their interactions. Concentrating here on O_3, the climate change projected for this century can be expected to decrease the global O_3 burden, if no changes in precursor emissions are assumed (Stevenson et al., 2006). The decrease is mainly due to the warming-induced increase of atmospheric water vapour that will intensify O_3 destruction. However, this effect is not globally uniform but depends on the pollution climate: in unpolluted (low NO_x concentration) regions, the O_3 concentrations decrease, but generally the opposite is true in polluted environments. Atmospheric warming will also directly affect O_3 and other pollution through the temperature dependence of chemical reaction rates, such as those of O_3-producing VOC oxidation reactions, and the decomposition of peroxyacetyl nitrate that acts as a major reservoir of O_3 precursors (Jacob and Winner, 2009).

As discussed in the previous section, interactions with the biosphere result in substantial mass fluxes to and from the atmosphere, which significantly affect atmospheric composition and climate. Increasing CO_2 concentrations as such tend to result in less efficient stomatal exchange of gases, as plants optimise their water-use efficiency; however, the canopy-level responses of forests are poorly understood (King et al., 2013, this vol.). BVOCs, such as isoprene, monoterpenes, sesquiterpenes and oxygenated hydrocarbons, consume hydrogen radicals (OH) and contribute to the tropospheric O_3 production, as well as to the formation of secondary organic aerosol (SOA) and the growth of newly formed particles. The OH radical is the major oxidant that determines the 'self-cleansing' capacity of the atmosphere, for example, the removal of CH_4 (Lelieveld et al., 2004). As a powerful GHG, O_3 imposes a positive (warming) radiative forcing (RF) to the climate system, while SOA contributes to the negative RF of atmospheric particulate matter (Carslaw et al., 2010).

As summarised by Loreto and Fares (2013, this vol.), BVOC emissions are linked with climate change in a variety of ways. In current conditions, emissions are exponentially dependent on temperature. In future climates, they can be expected to increase as a result of the lengthening growing season and the migration of treelines. However, the long-term temperature effect is less clear, and recent studies even suggest that rising atmospheric CO_2 concentration may inhibit isoprene emissions by uncoupling production rates from photosynthesis (Peñuelas and Staudt, 2010). BVOC emissions may also be altered by O_3, drought and other biotic and abiotic factors (Loreto and Fares, 2013, this vol.); for example, emissions may be enhanced by O_3 concentration increases (Peñuelas and Staudt, 2010). Even though the climate–BVOC interactions are complex and partly unknown, current knowledge suggests increasing emissions as a result of global warming, with implications for atmospheric chemistry and climate. Importantly, the increased emissions of mono- and sesquiterpenes can produce more SOA and thus enhance the associated negative RF, in particular, locally over boreal forests (Carslaw et al., 2010).

Removal by dry deposition controls the residence time and near-surface concentrations of various gaseous and particulate compounds. For O_3, for example, deposition to vegetation and other surfaces constitutes a major term in the global tropospheric mass budget, which balances the net chemical production and stratospheric input (Stevenson et al., 2006). While the O_3 production involves BVOCs, the most reactive BVOCs, such as sesquiterpenes, also contribute, *in situ*, to the chemical within-canopy O_3 sink, thus representing part of the apparent non-stomatal deposition flux of O_3 (Fowler et al., 2009; Loreto and Fares, 2013, this vol.). As discussed above, deposition fluxes are effectively controlled by vegetation characteristics, both physical and physiological. Thus, climate-induced biome shifts, human activities such as deforestation and changes in vegetation structure and species composition will modify these fluxes, atmospheric chemistry and the related mass budgets (Ganzeveld et al., 2010).

Both the stomatal uptake and non-stomatal deposition processes discussed in the previous section are affected by climatic factors, resulting in complex interactions and feedbacks. Both elevated CO_2 and O_3 concentrations generally reduce the stomatal conductance of trees (King et al., 2013, this vol.; Matyssek et al., 2010). The coupling with climate is further complicated by the fact that, while increasing atmospheric CO_2 concentrations stimulate the primary productivity of terrestrial vegetation, the enhanced C sequestration is partly offset by reduced photosynthetic capacity due to excessive O_3 uptake (King et al., 2013, this vol.). This suppression of the global C sink results in additional accumulation of anthropogenic CO_2 emissions in the atmosphere and thus can be perceived as a source of indirect RF (Sitch et al., 2007).

The importance of the atmosphere–biosphere exchange in controlling near-surface concentrations was clearly demonstrated during the heat wave that took place in central Europe in August 2003, when the persistent anticyclonic conditions led to efficient O_3 formation in the lower troposphere (Solberg et al., 2008). This resulted in very high O_3 concentrations exceeding 200 ppb (Fiala et al., 2003). While a host of processes, including forest fire emissions (Sofiev, 2013, this vol.; Solberg et al., 2008), were involved in this O_3 build-up, a key role was played by suppressed plant uptake due to stomatal closure, which resulted from high temperatures and soil moisture deficits

(Granier et al., 2007). In addition, the positive temperature response of BVOC emissions most likely contributed significantly to the enhanced O_3 formation during this photochemical episode (Curci et al., 2009; Solberg et al., 2008).

For adverse O_3 effects on vegetation that are closely associated with stomatal uptake (Dizengremel et al., 2013, this vol.), it is noteworthy that the concentration–deposition interaction described above entails negative feedback, since the elevated ambient O_3 concentrations do not produce a proportional enhancement of uptake flux due to the reduced stomatal conductance (Löw et al., 2006). This feedback is illustrated by the eddy covariance measurements reported by Gerosa et al. (2008). These data show that the stomatal uptake flux accumulated by a Holm oak forest stand during the late summer of 2003 was about half of that observed in cooler and wetter conditions a year later, in spite of the much higher O_3 concentrations in 2003. On the other hand, in northern Europe, the stomatal O_3 deposition to boreal forests, and thus the phytotoxic O_3 dose, may in the future be enhanced by higher atmospheric temperatures (Tuovinen et al., 2013, this vol.).

From the point of view of air quality, removal of air pollutants by forests has a positive effect that can be considered an ecosystem service (Cudlin et al., 2013, this vol.; Knoke and Hahn, 2013, this vol.). For example, measurements in a treeline larch forest in Italy showed that stomatal uptake and non-stomatal sinks in total removed 53 kg of O_3 per hectare during a 3-month period (Gerosa et al., 2013). This is a significant ecosystem service, taking into account that presently O_3 has no effect on the physiological behaviour of treeline-associated conifers (Wieser et al., 2009).

2.6 CONCLUSIONS AND FURTHER RESEARCH DIRECTIONS

In conclusion, the exchange of trace gases between forest ecosystems and the atmosphere involves fundamental processes that need to be quantified when studying the ecophysiological functioning of forests and the chemical and physical environment the forests are exposed to. The magnitude of trace gas exchange between forests and the atmosphere depends on a complex interplay between ecophysiological, edaphic and meteorological factors and processes. Scaling functional processes of forest trees from leaves and compartments (soil, canopy) to stands, ecosystems and, finally, to the landscape level (Wieser et al., 2008) is fundamental for understanding the capacity of forest ecosystems to mitigate climate change and air pollution effects and to adapt to changing environmental conditions (Matyssek et al., 2012). For fostering cause-effect based understanding, eddy covariance and sap-flow measurement methods should be applied in combination for quantitatively assessing total, stomatal, and non-stomatal fluxes in forest stands. Such an approach becomes more achievable if comprehensive supersites are established for integrated research on the effects of air pollution and climate change on forest ecosystems across Europe (Mikkelsen et al., 2013, this vol.; Paoletti et al., 2013, this vol.).

ACKNOWLEDGEMENTS

This work was supported by the ÉCLAIRE project funded under the EC 7th Framework Programme of the European Commission (Grant agreement 28910) and partly by the Austrian Science Fund Project (FWF P 22206-B16) 'Transpiration of conifers in contrasting Environments'.

REFERENCES

Altimir, N., Kolari, P., Tuovinen, J.-P., Vesala, T., Bäck, J., Suni, T., Kulmala, M., Hari, P., 2006. Foliage surface ozone deposition: a role for surface moisture? Biogeosciences 3, 209–228.

Arneth, A., Harrison, S.P., Zaehle, S., Tsigaridis, K., Menon, S., Bartlein, P.J., Feichter, J., Korhola, A., Kulmala, M., O'Donnell, D., Schurgers, G., Sorvari, S., Vesala, T., 2010. Terrestrial biogeochemical feedbacks in the climate system. Nat. Geosci. 3, 525–532.

Ashmore, M.R., 2005. Assessing the future global impacts of ozone on vegetation. Plant Cell Environ. 28, 949–964.

Breuer, L., Papen, H., Butterbach-Bahl, K., 2000. N_2O emission from tropical forest soils of Australia. J. Geophys. Res. 105, 26353–26367.

Brito, P., TruJillo, J.L., Morales, D., Jimenez, M.S., Wieser, G., 2013. Soil moisture overshadows temperature control over soil CO_2 efflux in a Pinus canariensis forest at treeline in Tenerife, Canary Islands. Acta Oecologica 48, 1–6. http://dx.doi.org/10.1016/j.actao.2013.01.006.

Butterbach-Bahl, K., 2002. CH_4. In: Gasche, R., Papen, H., Rennenberg, H. (Eds.), Trace Gas Exchange in Forest Ecosystems. Kluwer Academic Publishers, The Netherlands, pp. 211–226.

Bytnerowicz, A., Arbaugh, M.J., Andersen, C., Riebau, A.R., 2009. Integrated research on wildland fires and air quality: needs and recommendations. In: Bytnerowicz, A., Arbaugh, M., Riebau, A., Andersen, C. (Eds.), Wildland Fires and Air Pollution, Development in Environmental Science, vol. 8, pp. 585–602.

Cape, J.N., 2002. Nitrogen oxides. In: Gasche, R., Papen, H., Rennenberg, H. (Eds.), Trace Gas Exchange in Forest Ecosystems. Kluwer Academic Publishers, The Netherlands, pp. 245–255.

Carslaw, K.S., Boucher, O., Spracklen, D.V., Mann, G.W., Rae, J.G.L., Woodward, S., Kulmala, M., 2010. A review of natural aerosol interactions and feedbacks within the Earth system. Atmos. Chem. Phys. 10, 1701–1737.

Cieslik, S., 2004. Ozone uptake by various surface types: a comparison between dose and exposure. Atmos. Environ. 38, 2409–2420.

Cieslik, S., 2009. Ozone fluxes over various ecosystems in Italy. Environ. Pollut. 157, 1487–1496.

Coe, H., Gallagher, M.W., Choularton, T.W., Dore, C., 1995. Canopy scale measurements of stomatal and cuticular O_3 uptake by Sitka spruce. Atmos. Environ. 29, 1413–1423.

Conrad, R., 1996. Soil microorganisms as controllers of atmospheric trace gases (H_2, CO_2, CH_4, OCS, N_2O, NO). Mircobiol. Rev. 60, 609–640.

Conrad, S.G., Solomon, A.R., 2009. Effects of wildland fire on regional and global carbon stocks in a changing environment. In: Bytnerowicz, A., Arbaugh, M., Riebau, A., Andersen, C. (Eds.), Wildland Fires and Air Pollution, Development in Environmental Science, vol. 8, pp. 109–138.

Cudlin, P., Seják, J., Pokorný, J., Albrechtová, J., Bastian, O., Marek, M., 2013. Forest ecosystem services under climate change and air pollution. Dev. Environ. Sci. 13, 521–546.

Curci, G., Beekmann, M., Vautard, R., Smiatek, G., Steinbrecher, R., Theloke, J., Friedrich, R., 2009. Modelling study of the impact of isoprene and terpene biogenic emissions on European ozone levels. Atmos. Environ. 43, 1444–1455.

Dang, Q.L., Margolis, H.A., Coyea, M.R., Sy, M., Collatz, G.J., 1997. Regulation of branch-level-gas exchange of boreal trees: roles of shoot water potential and vapour pressure difference. Tree Physiol. 17, 512–535.

Dizengremel, P., Jolivet, Y., Tuzet, A., Raineri, A., Le Thiec, D., 2013. Integrative leaf level phytotoxic ozone dose assessment for forest risk modelling. Dev. Environ. Sci. 13, 267–288.

Erickson, H., Keller, M., Davidson, E.A., 2001. Nitrogen oxide fluxes and nitrogen cycling during postagricultural succession and forest fertilization in the humid tropics. Ecosystems 4, 67–84.

Fares, S., McKay, M., Holzinger, R., Goldstein, A.H., 2010. Ozone fluxes in a *Pinus ponderosa* ecosystem are dominated by non-stomatal processes: evidence from long-term continuous measurements. Agric. Forest Meteorol. 150, 420–431.

Fiala, J., Cernikovsky, L., de Leeuw, F., Kurfuerst, P., 2003. Air Pollution by Ozone in Europe in Summer 2003. Topic Report 3/2003. European Environment Agency, Copenhagen, Denmark.

Fowler, D., Pilegaard, K., Sutton, M.A., et al., 2009. Atmospheric composition change: ecosystems-atmosphere interactions. Atmos. Environ. 43, 5193–5267.

Ganzeveld, L., Bouwman, L., Stehfest, E., van Vuuren, D.P., Eickhout, B., Lelieveld, J., 2010. Impact of future land use and land cover changes on atmospheric chemistry-climate interactions. J. Geophys. Res. 115, D23301.

Garratt, J.R., 1992. The Atmospheric Boundary Layer. Cambridge University Press, Cambridge, UK.

Gerosa, G., Finco, A., Mereu, S., Vitale, M., Manes, F., Ballarin Denti, A., 2008. Comparison of seasonal variations of ozone exposure and fluxes in a Mediterranean Holm oak forest between the exceptionally dry 2003 and the following year. Environ. Pollut. 157, 1737–1744.

Gerosa, G., Finco, A., Negri, A., Marzuoli, R., Wieser, R., 2013. Ozone fluxes to a larch forest ecosystem at the timberline in the Italian Alps. In: Cerbu, G., Hanewinkel, M., Gerosa, G., Jand, R. (Eds.), Management Strategies to Adapt Alpine Space Forests to Climate Change Risks. INTECH open science, New York, USA, in press.

Granier, A., Reichstein, M., Bréda, N., et al., 2007. Evidence for soil water control on carbon and water dynamics in European forests during the extremely dry year: 2003. Agric. Forest Meteorol. 143, 123–145.

Grünhage, L., Jäger, H.-J., 1994. Influence of the atmospheric conductivity on the ozone exposure of plants under ambient conditions: considerations for establishing ozone standards to protect vegetation. Environ. Pollut. 85, 125–129.

Havranek, W.M., Wieser, G., 1994. Design and testing of twig chambers for ozone fumigation and gas-exchange measurements in mature trees. Proc. Roy. Soc. Edinb. B Biol. Sci. 102, 541–546.

Hinckley, T.M., Lassioe, J.P., Running, S.W., 1978. Temporal and spatial variations in the water status of forest trees. Forest Sci. Monogr. 20, 27 p.

Hodnebrog, Ø., Solberg, S., Strodal, F., Svendby, T.M., Simpson, D., Gauss, M., Hilboll, A., Pfister, G.G., Turquety, S., Richter, A., Burrows, J.P., Denier van der Gon, H.A.C., 2012. Impact of forest fires, biogenic emissions and high temperatures on the elevated Eastern Mediterranean ozone levels during the hot summer of 2007. Atmos. Chem. Phys. 12, 8727–8750.

Iacobelli, A., MacCaughery, J.H., 1993. Stomatal conductance in a northern temperate deciduous forest: temporal and spatial patterns. Can. J. For. Res. 23, 245–252.

Inclan, R., Ribas, A., Penuelas, J., Gimeno, B.S., 1999. The relative sensitivity of different Mediterranean plant species to ozone exposure. Water Air Soil Pollut. 116, 273–277.

IPCC, 2007. Climate change 2007: the physical science basis. In: Solomon, S., Qin, D., Manning, M., Chen, Z., Marquis, M., Averyt, K.B., Tignor, M., Miller, H.L. (Eds.),

Contribution of Working Group I to the Fourth Assessment Report of the Intergovernmental Panel on Climate Change. Cambridge University Press, Cambridge, UK, and New York, USA.

Jacob, D.J., Winner, D.A., 2009. Effect of climate change on air quality. Atmos. Environ. 43, 51–63.

Jaffe, D.A., Wigder, N.L., 2012. Ozone production from wildfires: a critical review. Atmos. Environ. 50, 1–10.

Janssens, I.A., Lankreijer, H., Matteucci, G., Kowalski, A.S., Buchmann, N., Epron, D., Pilegaard, K., Kutsch, W., Longdoz, B., Grünwald, T., Montagnani, L., Dore, S., Rebmann, C., Moors, E.J., Grelle, Ü., Rannik, K., Morgenstern, S., Oltchev, R., Clement, J., Guðmundsson, S., Minerbi, P., Berbigier, A., Ibrom, A., Moncrieff, J., Aubinet, M., Bernhofer, C., Jensen, N.O., Vesala, T., Granier, A., Schulze, E.D., Lindroth, A., Dolman, A.J., Jarvis, P.G., Ceulemans, R., Valentini, R., 2001. Productivity overshadows temperature in determining soil and ecosystem respiration across European forests. Glob. Chang. Biol. 7, 269–278.

Janssens, I.A., Dore, S., Epron, D., Lankreijer, H., Buchmann, N., Longdoz, B., Brossaud, J., Montagnani, L., 2003. Climatic influences on seasonal and spatial differences in soil CO_2 efflux. In: Valentini, R. (Ed.), Fluxes of Energy, Water and Carbon Dioxide of European Forests, Ecological Studies. Springer, Berlin, pp. 235–256.

Karlsson, P.E., Hansson, M., Höglund, H.-O., Pleijel, H., 2006. Ozone concentration gradients and wind conditions in Norway spruce (*Picea abies*) forests in Sweden. Atmos. Environ. 40, 1610–1618.

King, J., Liu, L., Aspinwall, M., 2013. Tree and forest responses to interacting elevated atmospheric CO_2 and tropospheric O_3: a synthesis of experimental evidence. Dev. Environ. Sci. 13, 179–208.

Knoke, T., Hahn, A., 2013. Global change and the role of forests in future land-use systems. Dev. Environ. Sci. 13, 569–587.

Kolb, T.E., Fredericksen, T.S., Steiner, K.C., Skelly, J.M., 1997. Issues in scaling tree size and age responses to ozone: a review. Environ. Pollut. 98, 195–208.

Konovalov, I.B., Beekmann, M., Kuznetsova, I.N., Yurova, A., Zvyagintsev, A.M., 2011. Atmospheric impacts of the 2010 Russian wildfires: integrating modelling and measurements of an extreme air pollution episode in the Moscow region. Atmos. Chem. Phys. 11, 10031–10056.

Körner, C., Ja, Scheel, Bauer, H., 1979. Maximum leaf diffusive conductance in vascular plants. Photosynthetica 13, 45–82.

Köstner, B., 2001. Evaporation and transpiration from forests in Central Europe—relevance of patch-level studies for spatial scaling. Meteorol. Atmos. Phys. 76, 69–82.

Köstner, B., Matyssek, R., HeilMeier, H., Clausnitzer, F., Nunn, A.J., Wieser, G., 2008. Sap flow measurements as a basis for assessing trace-gas exchange of trees. Flora 203, 14–33.

Kozovits, A.R., Bustamante, M.M.C., 2013. Land use change, air pollution and climate change—vegetation response in Latin America. Dev. Environ. Sci. 13, 411–427.

Kruse, J., Simon, J., Rennenberg, H., 2013. Soil respiration and soil organic matter (SOM) decomposition in response to climate change. Dev. Environ. Sci. 13, 131–149.

Lavigne, M.B., Forster, R.J., Goodine, G., 2004. Seasonal and annual changes in soil respiration in relation to soil temperature, water potential and trenching. Tree Physiol. 24, 415–424.

Lelieveld, J., Dentener, F.J., Peters, W., Krol, M.C., 2004. On the role of hydroxyl radicals in the self-cleansing capacity of the troposphere. Atmos. Chem. Phys. 4, 2337–2344.

Leverenz, B., Deans, J.D., Ford, E.D., Jarvis, P.G., Miolne, R., Whitehead, D., 1982. Systematic spatial variation of stomatal conductance in a Sitka spruce plantation. J. Appl. Ecol. 19, 835–851.

Loreto, F., Fares, S., 2013. Biogenic volatile organic compounds and their impacts on biosphere-atmosphere interactions. Dev. Environ. Sci. 13, 57–75.

Löw, M., Herbinger, K., Nunn, A.J., Häberle, K.-H., Leuchner, M., Heerdt, C., Werner, H., Wipfler, P., Pretsch, H., Tausz, M., Matyssek, R., 2006. Extraordinary drought of 2003 overrules ozone impact on adult beech trees (*Fagus sylvatica*). Trees 20, 539–548.

Matyssek, R., Karnosky, D.F., Wieser, R., Percy, K., Oksanen, E., Grams, T.E.E., Kubiske, M., Hanke, D., Pretzsch, H., 2010. Advances in understanding ozone impact on forest trees: messages from novel phytotron and free-air fumigation studies. Environ. Pollut. 158, 1990–2006.

Matyssek, R., Wieser, G., Calfapietra, C., de Vries, W., Dizengremel, P., Ernst, D., Jolivet, Y., Mikkelsen, T.N., Mohren, G.M.J., LeThiec, D., Tuovinen, J.-P., Weatherall, A., Paoletti, E., 2012. Forests under climate change and air pollution: gaps in understanding and future directions for research. Environ. Pollut. 160, 57–65.

Mikkelsen, T.N., Ro-Poulsen, H., Hovmand, M.F., Jensen, N.O., Pilegaard, K., Egeløv, A.H., 2004. Five-year measurements of ozone fluxes to a Danish Norway spruce canopy. Atmos. Environ. 38, 2361–2371.

Mikkelsen, T.N., Clarke, N., Danielewska, A., Fischer, R., 2013. Towards supersites in forest ecosystem monitoring and research. Dev. Environ. Sci. 13, 475–495.

Monks, P.S., Granier, C., Fuzzi, S., et al., 2009. Atmospheric composition change—global and regional air quality. Atmos. Environ. 43, 5268–5350.

Neirynck, J., Gielen, B., Janssens, I.A., Ceulemans, R., 2012. Insights into ozone deposition patterns from decade-long ozone flux measurements over a mixed temperate forest. J. Environ. Monit. 14, 1684–1695.

Nunn, A.J., Cieslik, S., Metzger, U., Wieser, G., Matyssek, R., 2010. Combining sap flow and eddy covariance approaches to derive stomatal and non-stomatal O_3 fluxes in a forest stand. Environ. Pollut. 185, 2014–2022.

Paoletti, E., de Vries, W., Mikkelsen, T.N., Ibrom, A., Larsen, K.S., Tuovinen, J.-P., Serengil, Y., Yurtseven, I., Wieser, G., Matyssek, R., 2013. Key indicators of air pollution and climate change impacts at forest Supersites. Dev. Environ. Sci. 13, 497–518.

Pechony, O., Shindell, D.T., 2010. Driving forces of global wildfires over the past millennium and the forthcoming century. Proc. Natl. Acad. Sci. U.S.A. 107, 19167–19170.

Peñuelas, J., Staudt, M., 2010. BVOCs and global change. Trends Plant Sci. 15, 133–144.

Potter, C.S., Davidson, E.A., Verchot, L.V., 1996. Estimation of global biogeochemical controls and seasonality in soil methane consumption. Chemosphere 32, 2219–2246.

Raich, J.W., Potter, C.S., 1995. Global patterns of carbon dioxide emissions from soils. Global Biogeochem. Cycles 9, 23–36.

Raich, J.W., Schlesinger, W.H., 1992. The global carbon dioxide flux in soil respiration and its relationship to vegetation and climate. Tellus B 44, 81–99.

Rannik, Ü., Altimir, N., Mammarella, I., Bäck, J., Rinne, J., Ruuskanen, T.M., Hari, P., Vesala, T., Kulmala, M., 2012. Ozone deposition into a boreal forest over a decade of observations: evaluating deposition partitioning and driving variables. Atmos. Chem. Phys. 12, 12165–12182.

Reich, P.B., 1987. Quantifying plant response to ozone: a unifying theory. Tree Physiol. 3, 63–91.

Reich, P.B., Walters, M.B., Ellsoworth, D.S., 1992. Leaf life-span in relation to leaf, plant, and stand characteristics among diverse ecosystems. Ecol. Monogr. 62, 365–393.

Schimel, D.S., 1995. Terrestrial ecosystems and the carbon cycle. Glob. Change Biol. 1, 77–91.

Schindlbacher, A., Zechmeister-Boltenstern, S., Jandl, R., 2009. Carbon losses due to soil warming: do autotrophic and heterotrophic soil respiration respond equally? Glob. Change Biol. 15, 901–913.

Sitch, S., Cox, P.M., Collins, W.J., Huntingford, C., 2007. Indirect radiative forcing of climate change through ozone effects on the land-carbon sink. Nature 488, 791–794. http://dx.doi.org/10.1038/nature06059.

Smith, K.A., Ball, T., Connen, F., Dobbie, K.E., Massherder, J., Rey, A., 2003. Exchange of greenhouse gases between soil and atmosphere: interactions of soil physical factors and biological processes. Eur. J. Soil Sci. 54, 779–791.

Sofiev, M., 2013. Wild-land fires: monitoring, plume modelling, impact on atmospheric composition and climate. Dev. Environ. Sci. 13, 451–472.

Solberg, S., Hov, Ø., Søvde, A., Isaksen, I.S.A., Coddeville, P., De Backer, H., Forster, C., Orsolini, Y., Uhse, K., 2008. European surface ozone in the extreme summer 2003. J. Geophys. Res. 113, D07307.

Squire, G.R., Black, C.R., 1981. Stomatal behaviour in the field. In: Jarvis, P.G., Mansfield, T.A. (Eds.), Stomatal Physiology. Cambridge University Press, Cambridge, UK, pp. 223–245.

Stevenson, D.S., Dentener, F.J., Schulz, M.G., 2006. Multimodel ensemble simulations of present-day and near-future tropospheric ozone. J. Geophys. Res. 111, D08301.

Tuovinen, J.-P., Simpson, D., Mikkelsen, T.N., Emberson, L.D., Ashmore, M.R., Aurela, M., Cambridge, H.M., Hovmand, M.F., Jensen, N.O., Laurila, T., Pilegaard, K., Ro-Poulsen, H., 2001. Comparisons of measured and modelled ozone deposition to forests in Northern Europe. Water Air Soil Pollut. Focus 1, 263–274.

Tuovinen, J.-P., Hakola, H., Karlsson, P.E., Simpson, D., 2013. Air pollution risks to northern European forests in a changing climate. Dev. Environ. Sci. 13, 77–99.

van der Werf, G.R., Randerson, J.T., Giglio, L., Collatz, G.J., Mu, M., Kasibhatla, P.S., Morton, D.C., DeFries, R.S., Jin, Y., van Leeuwen, T.T., 2010. Global fire emissions and the contribution of deforestation, savanna, forest, agricultural, and peat fires (1997–2009). Atmos. Chem. Phys. 10, 11707–11735.

Westerling, A.L., Hidalgo, H.G., Cayan, D.R., Swetnam, T.W., 2006. Warming and earlier spring increase western U.S. forest wildfire activity. Science 313, 940–943.

Wieser, G., 2002. Exchange of trace gases at the tree—atmosphere interface: ozone. In: Gasche, R., Papen, H., Rennenberg, H. (Eds.), Trace Gas Exchange in Forest Ecosystems. Kluwer Academic Publishers, The Netherlands, pp. 211–226.

Wieser, G., 2004. Seasonal variation of soil respiration in a *Pinus cembra* forest at the upper timberline in the Central Austrian Alps. Tree Physiol. 24, 475–480.

Wieser, G., Havranek, W.M., 1993. Ozone uptake in the sun and shade crown of spruce: quantifying the physiological effects of ozone exposure. Trees 7, 227–232.

Wieser, G., Havranek, W.M., 1995. Environmental control of ozone uptake in *Larix decidua* Mill.: a comparison between different altitudes. Tree Physiol. 15, 253–258.

Wieser, G., Häsler, R., Götz, B., Koch, W., Havranek, W.M., 2000. Role of climate, crown position, tree age and altitude in calculated ozone flux into needles of *Picea abies* and *Pinus cembra*: a synthesis. Environ. Pollut. 110, 415–422.

Wieser, G., Matyssek, R., Köstner, B., Oberhuber, W., 2003. Quantifying ozone uptake at the canopy level of spruce, pine and larch trees at the alpine timberline: an approach based on sap flow measurements. Environ. Pollut. 126, 5–8.

Wieser, G., Luis, V.C., Cuevas, E., 2006. Quantification of ozone uptake at the stand level in a *Pinus canariensis* forest in Tenerife, Canary Islands: an approach based on sap flow measurements. Environ. Pollut. 140, 383–386.

Wieser, G., Matyssek, R., Cieslik, S., Paoletti, E., Ceulemans, R., 2008. Upscaling ozone flux in forests from leaf to landscape. Ital. J. Agron. 3, 35–41.

Wieser, G., Matyssek, R., Luzian, R., Zwerger, P., Pindur, P., Gruber, A., 2009. Effects of atmospheric and climate change at the timberline of the Central European Alps. Ann. For. Sci. 66, 402.

Wieser, G., Matyssek, R., Götz, B., Grünhage, L., 2012. Branch cuvettes as means of ozone risk assessment in adult forest tree crowns: combining experimental and modeling capacities. Trees 26, 1703–1712. http://dx.doi.org/10.1007/s00468-102-0715-6.

Chapter 3

Nutrients or Pollutants? Nitrogen Deposition to European Forests

Werner Eugster[1] and Matthias Haeni
Department of Environmental Systems Science, Institute of Agricultural Sciences, ETH Zürich, Zürich, Switzerland
[1]Corresponding author: e-mail: eugsterw@ethz.ch

Chapter Outline

3.1 INTRODUCTION

The effects of N deposition on forest ecosystems are described in the conceptual framework of Aber et al. (1989, 1998), which was further refined by Emmett (2007). N deposition affects all ecosystem compartments and alters

Developments in Environmental Science, Vol. 13. http://dx.doi.org/10.1016/B978-0-08-098349-3.00003-7

many biogeochemical processes. Its effect is modified by site conditions, such as nutrient availability in the soil, species composition of the forest stand and management history, in combination with the changing climate conditions and pollutant loads.

In this chapter, we will focus on N deposition loads to European forest ecosystems. The emphasis is on the most recent scientific developments aiming at quantifying N dry deposition, which tends to be the dominant pathway of N deposition in temperate climates where the majority of days do not see rainfall, and where wet deposition is hence unable to represent total ecosystem-scale N inputs. For conventional N deposition measurements, it is, however, indispensable to consult the standard protocols of ICP Forests (2010). It is not the intent of this chapter to duplicate the wealth of existing literature on non-dry deposition processes, but for the sake of completeness, a very brief summary of all relevant pathways will be given to put dry deposition into the context of total N deposition. Effects of N deposition are always a combination of several input pathways and cannot be cleanly teased apart. But with respect to dry deposition, above-ground N uptake processes must be more strongly investigated than would be required in cases where the primary pathway of anthropogenic N inputs is wet deposition.

N input must be seen in a long-term perspective and on an integrative ecosystem scale. At short timescales and low N deposition rates, the nutritional effect may be dominant, primarily increasing tree growth. At longer timescales and higher N deposition rates, this effect can become detrimental. This transition may be slow and cannot easily be generalised. Short-term studies with seedlings are best suited to study specific physiological processes, but the results obtained in this way cannot readily be transferred to mature trees and whole forest ecosystems.

N is the most important nutrient for plant growth and it is often the limiting factor for net primary production in temperate forest ecosystems (Vitousek and Field, 2001). Elaborated methods have been established to measure wet deposition within ICP Forests, and canopy balance models are used to estimate the total N deposition. At many European locations, however, dry deposition of reduced and oxidised N-containing gases is an even larger component of N deposition than wet deposition (see e.g. Eugster et al., 1998; Flechard et al., 2011). Newer developments now allow for a better quantification of dry deposition rates of gases and particles. After a short summary on effects of N deposition to forest ecosystems, this chapter not only focuses on these recent developments that may in the future complement existing N deposition measurements in Europe but also briefly addresses the other components of total N deposition measurements which must remain an essential part of long-term monitoring efforts even at sites where gaseous dry deposition is the primary source of N deposition.

FIGURE 3.1 Schematic showing the similarities and differences between nutrient and pollutant inputs to ecosystems by rainfall (left) and by fog deposition (right). *Reprinted from Eugster (2008).*

3.2 EFFECTS OF NITROGEN DEPOSITION TO FOREST ECOSYSTEMS

N uptake (Figure 3.1) is typically via the roots of the trees, but it has been shown by many studies that canopy N uptake (CNU) in forests also plays an important role (e.g. Schaefer and Reiners, 1990; Sievering et al., 2007). Earlier studies have mostly focused on foliar uptake by saplings and young plants in experimental treatments, and the results obtained in this way were then scaled up to whole canopies of more mature trees. It has been recognised that neglecting CNU in some of the widely used canopy budget models is probably a wrong assumption (ICP Forests, 2010, p. 39). This indicates where the most urgent present-day research needs are: to obtain realistic and accurate estimates of ecosystem-scale total N deposition and to assess all pathways of N uptake that may have detrimental effects on ecosystems.

It is agreed that N inputs below the critical loads are—in the long term—not detrimental to ecosystem structure and functions. Nitrogen deposition rates above a critical threshold will—again in the long term—lead to an impairment of the functions, depending on the site quality and other stressors present. Having said that, it should, however, also be recalled that the effect of

N deposition on species composition in forests starts at rather low levels, and the apparently positive effect of added nutrients may have a strong negative effect on biodiversity (e.g. Bobbink et al., 1998, 2010; Kozovits and Bustamante, 2013, this vol.; Vitousek et al., 1997). This may be a key issue to keep in mind in natural and semi-natural, extensively managed forest ecosystems in Europe. For Central Europe, Ellenberg (1990) showed that the vast majority of endangered vascular plant species growing in forests and woodlands are associated with N-poor environments (up to 40% of the species that grow in the N-poor class 2 on the nominal scale by Ellenberg et al., 1991).

Foliar uptake leads to the fact that estimates of N input via throughfall and bulk deposition collection with subsequent calculation of total deposition with models may substantially underestimate actual N inputs to forest ecosystems (Eilers et al., 1992). Nave and Curtis (2011) estimated that more than 50% of atmospheric N deposition is actually intercepted and retained in forest canopies. Moreover, Eilers et al. (1992) found that above-ground uptake of N may cause nutritional disturbances by changes in the N:Ca and N:Mg ratios in the needles, which may lead to Mg-deficiency symptoms of Norway spruce (Eilers et al., 1992). Sparks (2009) clearly showed that direct foliar uptake of N is a direct addition of N to plant metabolism and could potentially more readily influence plant growth (Figure 3.1, right) compared to soil-deposited N (Figure 3.1, left). Present day ecosystem process models, however, still do not explicitly simulate this more direct N pathway (see Sparks, 2009), and hence cannot always adequately represent the effects of N deposition on forest growth.

Nihlgard (1985) drew the attention of the scientific community to N as a possible cause for forest dieback, and since then, an ever-increasing number of studies have addressed the question how N deposition—mostly in combination with other pollutants—affects tree growth and health status. Krupa (2003) provided a concise overview over the different pathways and effects on terrestrial ecosystems, and even more detailed information on specific aspects of pollutant fluxes to different forest tree species, which can be found in the book by Elling et al. (2007).

3.2.1 What are the Typical Detrimental Effects of Nitrogen Deposition?

N deposition affects all compartments of forests. The narrowing of the carbon to nitrogen (C/N) ratio of the forest floor indicates accumulation of N and is a measure to assess the risk of nitrate release in the seepage water. In general, a C/N below 25 can be seen as the threshold for enhanced nitrate leaching via the seepage water (e.g. Aber et al., 2003; Dise et al., 2009; Gundersen et al., 1998). At the same time, the composition of the understory is altered in many regions leading to a reduced biodiversity (Bobbink et al., 2010; Nordin et al., 2006).

The N to nutrient ratios (N/Mg, N/K, N/P) in needles and leaves tend to become larger, indicating a shortage of other nutrients which will lead to an enhanced sensitivity to natural stress factors, like a lowered frost resistance and enhanced susceptibility to insects (e.g. Flückiger and Braun, 1999; Gundersen, 1998; Schulze, 1989).

There is increasing evidence that in Central Europe the phosphorous (P) nutrition is also declining (Khanna et al., 2007; Prietzel et al., 2008), leading to—at least regional—reduced growth (Braun et al., 2010). The pathways are not fully understood, but a reduction of P uptake via roots due to N-mediated inhibition of mycorrhiza is discussed (Aber et al., 1989; Braun et al., 2010; see Kraigher et al., 2013, this vol.). The linkages between C, N and P must be taken into account for the modelling of carbon sequestration (Goll et al., 2012; see also Bytnerowicz et al., 2013, this vol.).

To a certain degree, N acts as a fertiliser. If the natural N limitation is removed, the trees react with enhanced growth. This is probably one reason for the observed increased growth of trees in Europe in the past decades (e.g. Spieker et al., 1996). However, this does not imply that the effect is also positive at the ecosystem level and over longer time periods; biodiversity of a forest, for example, is normally determined primarily by the understory vegetation where many N-sensitive species may disappear, while the few tree species thrive (see review by Bobbink et al., 2010). De Schrijver et al. (2008), for example, emphasise that the effects of N deposition on soil acidification, groundwater and surface water quality, biodiversity and ecosystem services must be considered in addition to the simple nutritional effect that N deposition may exert on tree biomass increment alone.

The input of N above the demand for tree growth is termed N saturation, which means that no further increase of growth results from additional N input (De Vries et al., 2006, 2008; Laubhann et al., 2009). There are indications that the enhanced growth is today halted on many sites (Nellemann and Thomsen, 2001). The excess N cannot be taken up or stored in soil or vegetation and percolates through the soil, leading to acidification. This in turn means a depletion of base cations (Mg, K, Ca) or an increase of the amount of acidic cations (Al, Mn, H) in the soil. If the acidification potential of nitrogen is higher than the natural weathering rate of the soil forming bedrock, the soils acidify.

To integrate all these aspects, an ecosystem-specific critical threshold is defined above which ecosystem damage is likely to occur. This is taken into account in the calculation of the critical loads, which are the basis for the air pollution abatement policy in the UNECE region (UNECE, 2005). The site-specific calculated critical load for nutritional N is for most forests in the range of 10–20 kg N ha^{-1} yr^{-1} (Spranger et al., 2004, p. 125); the empirical critical load for coniferous woodland, for example, is about 5–15 kg N ha^{-1} yr^{-1} (Bobbink and Hettelingh, 2010). On the European scale, 74% of the total terrestrial area (forests and other ecosystems) of the EU27 countries (52% for all European countries) is at risk of nutrient N eutrophication (reference year 2000; Slootweg et al., 2010, p. 17).

Knowledge of the effects of N on the ecosystem level is essential. To find and verify cause–effect relations and to establish critical thresholds, the input of N in forests and the interrelations with other pollutants should be known as accurately as possible.

3.3 THE COMPONENTS OF NITROGEN DEPOSITION

The ICP Forests (2010) manual lists all relevant components of N deposition:

1. Wet-only deposition
2. Bulk deposition
3. Dry deposition
4. Occult deposition
5. Throughfall deposition
6. Stemflow deposition
7. Total atmospheric deposition

These components are non-additive and basically reflect the difficulty of separating one component clearly from another. However, this would not be that critical, if a good measurement method existed to accurately quantify total atmospheric deposition. Its definition according to ICP Forests (2010) is the sum of wet-only deposition and dry deposition, but not including occult deposition. A more integrative definition would define total deposition as wet (rainfall and snow, Section 3.3.1), occult (fog, Section 3.3.2) and dry deposition (Section 3.3.3) as we will do in this text. In areas with substantial snowfall, one could argue that snow deposition is a separate pathway for N input, but for the sake of simplicity, we will consider snow deposition a component of wet deposition, since for the ecosystem this input becomes typically only relevant when the snow melts.

Despite the difficulties in terminology and how to measure the components, it should be clear that total N input by all relevant pathways must be quantified accurately to relate measurable effects to the N deposition rates in order to specify and quantify detrimental effects (see Section 3.2) on ecosystems.

3.3.1 Wet Deposition by Rainfall and Snow

3.3.1.1 Wet-Only Deposition

Wet deposition is the most important pathway of N deposition in tropical rainforests without a dry period. In temperate forests, the contribution of wet deposition to total N deposition, however, depends (1) on the probability of rain at a given site and (2) on the concentrations of N compounds in rainfall.

At localities where rain is rare, it is clear that wet deposition may be substantially less than 50% of total deposition. For example, in western Switzerland, Eugster et al. (1998) found wet deposition rates on the order of 7–8 kg N ha^{-1} yr^{-1}, which contributed roughly 14–33% to total N deposition. Wet deposition is often a larger-scale phenomenon since very local rains are restricted to short periods in summer, and even then the ions in the rainwater are often transported over larger distances and hence small-scale variability in wet deposition is small compared to the variability of the other components.

3.3.1.2 Bulk Deposition

If bulk deposition is measured in place of wet-only deposition, then the deposition estimates tend to be higher. Even higher than bulk deposition are N deposition values determined from throughfall data in forest canopies. Throughfall is believed to reflect the composition of both incident precipitation (wet deposition) and dry deposition from N-containing components that are deposited on branches and foliage during dry periods, which are then washed off by precipitation and can be collected as throughfall (e.g. Thimonier et al., 2005). If CNU takes place, or leaching from the canopy, then some assumptions must be made to be able to derive dry deposition from the difference between throughfall deposition and bulk deposition measurements (Thimonier et al., 2005). Namely, gaseous components can be taken up via stomata (see Cieslik et al., 2013, this vol.) and hence will not get deposited on leaf surfaces and thus also will not be washed off. It may, however, hold for aerosol deposition of solid, dry, fine dust particles. The N input from aerosols is, however, a much smaller component than gaseous dry deposition (see Section 3.3.3).

3.3.1.3 Deposition by Snow

Very little attention has been given to N deposition by snow so far. In northern countries and in mountain ranges, this, however, may be a relevant additional pathway of N input. The scavenging of aerosol particles by falling snow flakes was found to be somewhat greater than that for liquid precipitation (Kyrö et al., 2009) and corresponded to rain intensities around 3 mm h^{-1} (rather intense downpour) even when snowfall remained below 0.2 mm h^{-1}. This means that snow deposition of N should be carefully considered in polluted areas where the increased efficiency for scavenging aerosols may constitute a significant pathway for N deposition. Because snow deposition occurs during the dormant season of the vegetation, it may not immediately affect the ecosystem, but during spring snow melt, a peak influx of deposited ions may occur. For lakes and streams, this was termed 'acid shock' (Jeffries et al., 1979) which in a similar way can also be expected for N inputs deposited by snow.

3.3.2 Occult Deposition by Fog

In temperate forests, water inputs by fog are rather small, but N concentrations in fogwater are one to two orders of magnitude larger than in rainwater, and hence deposition amounts can be considerable at localities with frequent fog. At the Lägeren mixed forest in Switzerland, for example, the N input by fog was quantified at 4.3 kg N ha^{-1} yr^{-1} which is 16.4% of total N deposition (Burkard et al., 2003). The tree species composition is dominated by beech, followed by Norway spruce, ash, sycamore, silver fir and eight less abundant species. This is 62% of wet deposition (6.9 kg N ha^{-1} yr^{-1}) or 28% of dry deposition (15 kg N ha^{-1} yr^{-1}). Also at the Waldstein site in the Fichtelgebirge, Germany, fog deposition contributed a similar amount of N to total deposition as was measured as wet-only deposition (Thalmann et al., 2002).

Although these are important N additions, an experimental treatment of selected trees at the Lägeren site with a mist that reflects N concentration in fogwater primarily showed an increased growth effect without clear signs of induced stress symptoms (Wortman et al., 2012). Elemental concentrations in Lägeren beech leaves showed that both N/P and N/K ratios were 6% and 19% higher, respectively, for N-amended leaves in comparison with control leaves. This supports the expectation of CNU (Figure 3.1) which is enhanced by occult deposition compared to the shorter contact time and lower N concentrations brought about by rainwater. On the other hand, occult deposition to temperate forests should not be overestimated: often the high frequencies of fog occurrence coincide with the cold period of the year when trees are dormant, and hence it remains to be investigated in more detail how important the timing of fog and its N input is to forest ecosystems.

3.3.3 Dry Deposition of Gases and Aerosol Particles

3.3.3.1 Methods for Field Measurement

Gaseous dry deposition is most directly measured by the eddy covariance (EC) method (Aubinet et al., 2012). This method uses a three-dimensional anemometer—mostly one working with ultrasonic impulses to determine wind speed and air temperature—in combination with a fast gas analyzer that allows measuring every tenth of a second or so to systematically probe turbulent transport of the gas of interest. It is then assumed that the deposition rate equals the flux rate measured with EC at a certain height above the forest canopy. Such measurements are feasible for most gases such as NO (nitric oxide), NO_2 (nitrogen dioxide), NO_x (the sum of NO and NO_2), HNO_3 (nitric acid) and peroxyacetile nitrate.

The dry deposition of aerosol particles can also be measured with EC, but it is still a challenge: while instruments exist to measure the number flux of particles, it is not yet possible to measure the exact chemical composition at a time interval suitable for EC. Also gaseous fluxes of NH_3 are not easily measured directly with EC, as Ferrara et al. (2012) have shown. Hence, many scientists chose a variant of the EC method that allows the deployment of relatively slow sensors via the relaxed eddy accumulation (REA) technique: the gas of interest or aerosol components are sampled in two reservoirs, one which collects air whenever the vertical wind speed is positive and above a given threshold value, while the other collects air from down-moving wind gusts with a wind speed above that threshold. The concentration difference between the two reservoirs can then be analysed with an analyser that requires considerably longer analysis time than that required by EC (which is on the order of fractions of a second in most cases). The deposition flux is then derived from that concentration gradient times a turbulent mixing coefficient (which relates the gradient to the corresponding flux) and a correction factor (β) to correct for the effect of the REA wind speed thresholds on the flux estimates. In the case of aerosols, the REA method was used in several combinations to obtain defensible flux estimates.

All these methods are expensive, require high scientific and technical skills to operate and can only be carried out at very few forest sites that meet all these requirements. Hence, probably the most important advance in the methodology of dry deposition flux measurements in the past decade was made during the NitroEurope project (http://www.nitroeurope.eu) with its development of a low-cost Delta denuder system for sampling gaseous NH_3, NO_2, HNO_3, HONO, SO_2, HCl and aerosol NH_4^+, NO_3^-, SO_4^{2-}, Cl^-, Na^+, Ca^{2+} and Mg^{2+} (Figure 3.2). This allows obtaining monthly mean concentrations of the most important N gases and aerosol components, from which deposition can be estimated via transfer models. In this way, an independent estimate of dry deposition can be made that does not require any assumptions on how the difference between throughfall and bulk precipitation may actually be related to dry deposition.

3.3.3.2 *Contribution to Total N Deposition*

Aerosol dry deposition was only estimated to be on the order of 2.0 to 2.3 kg N ha^{-1} yr^{-1} in parts of Switzerland (Eugster et al., 1998). Fluxes derived from Delta denuders, as presented by Flechard et al. (2011) in their Table 3, were used for additional analyses presented here. These numbers suggest that aerosol dry deposition (NH_4^+ and NO_3^-) to European forests ranges between 0.3 and 6.6 kg N ha^{-1} yr^{-1} with a median of 3.0 kg N ha^{-1} yr^{-1} and the gaseous dry deposition (NH_3, NO_2 and HNO_3) ranges between 1.2 and 25.8 kg N ha^{-1} yr^{-1}. On average, gaseous dry deposition was a factor 1.5 to 4.5 larger than aerosol deposition (median 2.6). The most challenging finding by Flechard et al. (2011), however, might be that these estimates appear rather sensitive to deposition model assumptions, as can be seen in Figure 3.3.

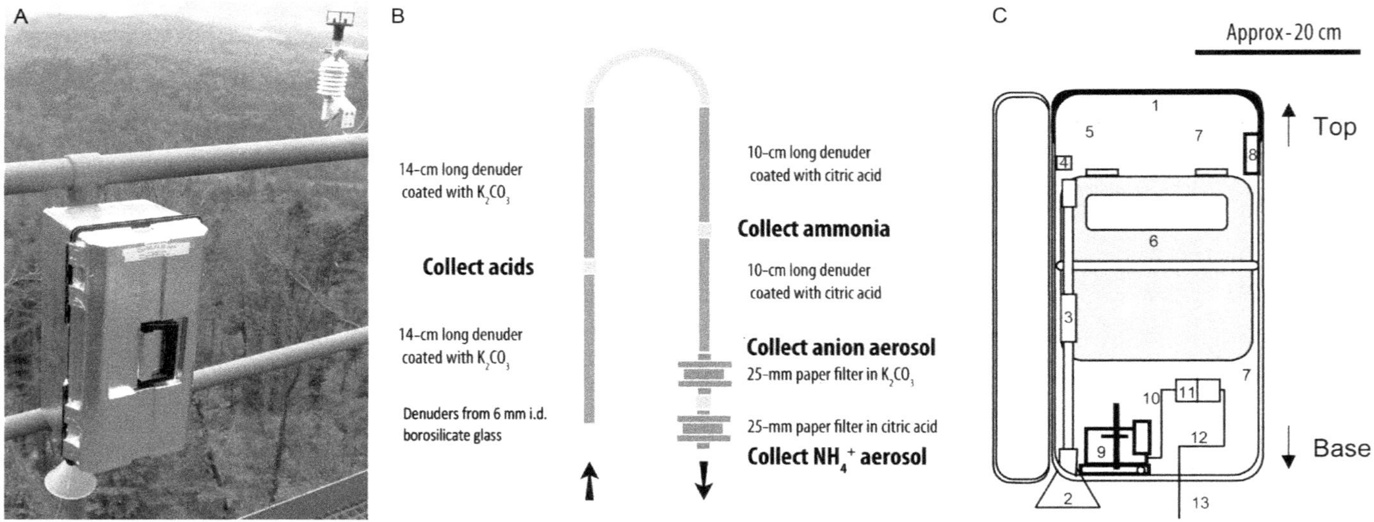

FIGURE 3.2 Delta denuder system for sampling gaseous NH_3, NO_2, HNO_3, HONC, SO_2, HCl and aerosol NH_4^+, NO_3^-, SO_4^{2-}, Cl^-, Na^+, Ca^{2+}, Mg^{2+} (A) on site operational at Lägeren mixed forest in Switzerland; (B) denuder sample train which is exchanged every month; (C) interior of housing with door open. Numbers in (C) are 1, durable casing; 2, sampling inlet (funnel); 3, 4, sampling train; 5, 7, tube connection; 6, gas volume flow meter; 8, in-line large particle filter; 9, air pump with heater; 10–13, electrical installation. *Modified after http://www.nitroeurope.eu/sites/nitroeurope.eu/files/neu_data/Dissemination-deliverables/NEU_delta_ws_tfmm_Helsinki_may2006_presentation.pdf.*

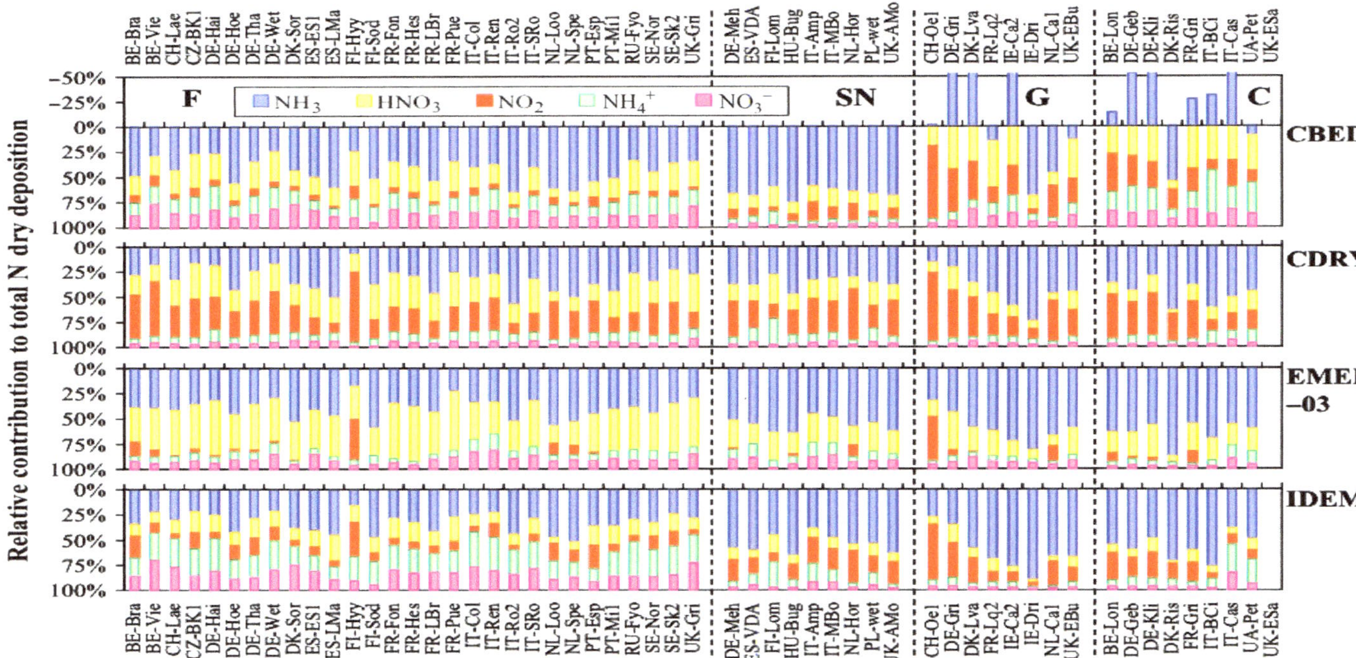

FIGURE 3.3 Relative contributions of reactive N species to total inorganic N dry deposition to 29 European forest ecosystems (F) in comparison with 9 semi-natural grasslands (SN), 8 cut or grazed grasslands (G) and 9 croplands (C). Deposition was calculated with measured monthly mean concentrations of NH_3, HNO_3, NO_2, NH_4^+ and NO_3^- using four different deposition models (CBED, CDRY, EMEP, IDEM). Location names start with an abbreviation for the country followed by a three-character location acronym. *Reprinted from Flechard et al. (2011).*

The four deposition models used to quantify dry deposition based on mean concentration measurements were CBED used in the United Kingdom, CDRY used by Environment Canada, EMEP used in Europe under the Convention on Long Range Transboundary Air Pollution (see also Tuovinen et al., 2013, this vol.) and IDEM used in The Netherlands. All four models use a surface-vegetation-atmosphere transfer scheme that quantifies a deposition velocity v_d for each gaseous and particulate component of dry deposition. The deposition rate is then computed as v_d times the mean concentration. Differences in the models to quantify v_d hence reflect differences in degree of complexity in how the canopy is modelled, and differences in how atmospheric conditions (turbulence, stability) are treated by the model and how stomatal uptake is computed (see also Cieslik et al., 2013, this vol.).

A principal component analysis of total N dry deposition and all individual components (NH_3, HNO_3, NO_2, NH_4^+ and NO_3^-) obtained from the four deposition models is shown in Figure 3.4. Despite the large differences between models in their relative contributions to total N dry deposition seen in Figure 3.3, the PCA in Figure 3.4 clearly reveals that in absolute numbers, the model differences are not that important: 96.7% of total variance in the dry deposition estimates for forest ecosystems in Europe can be explained by the first principal component. Total N deposition from the CBED model from the United Kingdom (Smith et al., 2000) almost perfectly represents the first principal component (Figure 3.4). This means that the CBED total is the best representation for the whole set of deposition estimates. It is more strongly influenced by NH_3 deposition than the other three models and assumes much lower NO_2 deposition rates than the Dutch IDEM model (Bleeker et al., 2004; Erisman et al., 1994) and—even more pronounced—the Canadian CDRY model (Zhang et al., 2003). Despite these functional differences, the overall statistical average N dry deposition as a function of altitude (Figure 3.5) is surprisingly similar for the CBED and IDEM models. Deposition rates are, however, substantially larger than what EMEP and CDRY predict from exactly the same measured concentration data obtained from low-cost Delta denuder systems.

3.4 RECENT DEVELOPMENTS TO ASSESS EFFECTS ON TREE GROWTH

At sites with long-term EC flux measurements, the net CO_2 flux can be directly related to N deposition. For other sites where EC flux measurements are not feasible, some progress has recently been made to use point-dendrometer data with high temporal resolution to monitor growth rates over the season, which in the long term should provide similarly important data to link N deposition to tree growth.

Stem radius variations of trees are a measure for biomass accumulation at the monthly, seasonal and annual timescale (Zweifel et al., 2010). Figure 3.6

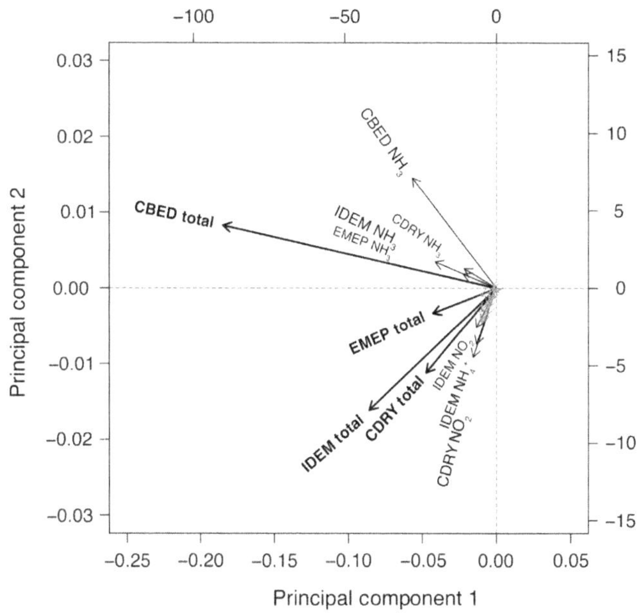

FIGURE 3.4 Principal component analysis (covariance-based PCA) of dry deposition estimates for 29 European forest ecosystems based on data published by Flechard et al. (2011). Deposition was calculated with measured monthly mean concentrations of NH_3, HNO_3, NO_2, NH_4^+ and NO_3^- using four different deposition models (CBED, CDRY, EMEP, IDEM). The graph shows the first two principal components explaining 98.4% of total variance (96.7% and 1.7% for the first and second component, respectively). This indicates that total deposition (bold vectors) strongly depends on assumptions made by the deposition model. With the British CBED gaseous deposition via NH_3 is by far the most relevant dry deposition pathway, whereas the Canadian CDRY model more strongly emphasises NO_2 dry deposition. Unnamed grey arrows represent all other deposition pathways that do not lead to substantial contributions to total dry deposition by any of the models used by Flechard et al. (2011).

shows the concept of how tree growth can be quantified from these measurements: on the annual timescale, the time traces of continuous high-resolution stem diameter measurements (Figure 3.6A) and independent cumulative net CO_2 flux measurements performed by EC (Figure 3.6B) are strikingly similar.

3.4.1 Dendrometric Measurements

The extra noise seen in the dendrometer data reflects the diurnal water cycle in the tree (see Zweifel et al., 2010) which is unrelated to tree growth at short (daily, weekly) timescales, whereas the signal at longer timescales is related to environmental factors, mostly temperature trend and water availability, including bark and the degradation of dead phloem cells, in turn affecting plant water balance and plant growth (see also Cocozza et al., 2012; Zweifel and Häsler, 2000, 2001). An important additional insight into tree

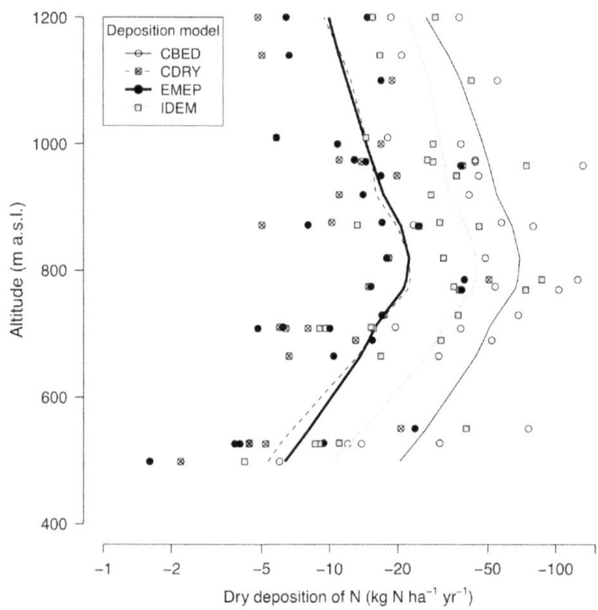

FIGURE 3.5 Altitude dependence of dry deposition (log scale) of total inorganic N to 29 European forest ecosystems determined from four deposition models that use measured monthly mean concentrations. All models show a local maximum for sites with an altitude around 800 m a.s.l., indicating the specific relevance of N deposition to montane forest ecosystems. *Data were taken from Flechard et al. (2011).*

growth is shown during the winter where the natural shrinking of the bark is similar to the respiration losses measured by EC outside the growing season. This signal adds value on top of conventional simple tree-ring width measurements which only reflect growing-season conditions.

3.4.2 Net Ecosystem Productivity Measurements

The CO_2 sources (respiration) and CO_2 sinks (assimilation) are large budget components of opposite directions that sum to a relatively small net ecosystem productivity (NEP) over a year (Barford et al., 2001; Körner et al., 2005; Zweifel et al., 2010). As the net budget of these components is directly measured with EC, it is generally assumed that the accuracy of the annual estimates of NEP should be better than the accuracy of the difference between the estimates of the two large components gross primary production and total ecosystem respiration determined separately (Buchmann and Schulze, 1999).

By installing dendrometers at EC research sites within Europe, we envisage gaining dendrometric growth data of various tree species under different climates. These will then be related to NEP measured by EC as a proof-of-concept that NEP at annual and possibly even monthly resolution (during the growing season) could be obtained from representative dendrometer measurements.

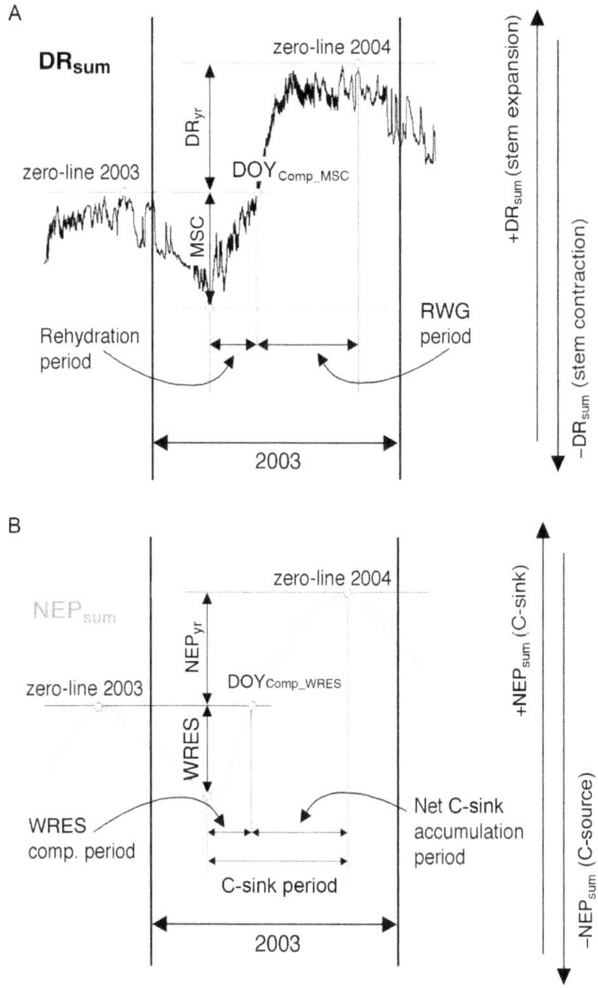

FIGURE 3.6 The time traces of dendrometer measurements DR on single trees (A) and of ecosystem-scale eddy covariance measurements of net ecosystem productivity NEP (B) are very similar in a subalpine Norway spruce forest at Davos, Switzerland. Both panels show the full year 2003 with the last months of 2002 and the first months of 2004. Using such relationships, it should become possible to see effects of pollutants and nitrogen deposition at the whole-tree scale and at time resolutions well below the annual resolution of tree-ring widths that have been used so far. Current research focuses on how the relationship between DR (A) and NEP (B) holds for other forest types in different climatic zones of Europe. *Reprinted from Zweifel et al. (2010).*

3.5 POLICY RELEVANCE OF THE KNOWLEDGE ON NITROGEN DEPOSITION

The reduction of N deposition is a matter of the UNECE Convention on Long Range Transboundary Air Pollution (CLRTAP), which came into force in 1979. The Sophia protocol on reduction of nitrous oxides from 1988 was the first based on the effect-based approach, in that it used the critical loads to calculate the required deposition reduction to reach ecosystem protection. The Gothenburg protocol (UNECE, 1999), the 'Protocol to Abate Acidification, Eutrophication and Ground-level Ozone', is also effect based and regards also the interactions between acidification, eutrophication and ozone. Since 2008, the Task Force on reactive N under the Working Group on Strategies and Review of the CLRTAP is mandated to develop a better understanding of the integrated, multi-pollutant nature of reactive N. The main aim is the development of integrated solutions for N reductions. This aim requires adequate research networks, and the development of supersites for integrated forest ecosystem research (Mikkelsen et al., 2013, this vol.) is an important step in this direction.

In Switzerland, the effects of N are monitored at nine Level II sites of ICP Forests and on cantonal monitoring sites (IAP). At more than 15 of these sites, selected to form an N- and water-deficit gradient (TreeNet-CH[1]), the long-term NEP is measured with point dendrometers. At two of these sites, NEP is measured also with the EC method. An initial study by Zweifel et al. (2010) has shown excellent agreement between EC-derived NEP and tree diameter increments (Section 3.4). This unique network started in 2011 and has since then delivered high-resolution dendrometer data and ancillary meteorological data. It is envisaged to combine the TreeNet-CH network with the Swiss Fluxnet network (Lägeren and Davos) and the regular forest inventories (National Forest Inventory, NFI) to be able to compile quarterly if not even monthly interpolated maps of actual tree growth throughout a growing season. This would allow us to see effects of N inputs in combination with climate anomalies in due time. At the annual timescale, such information is expected to provide a more robust basis to estimate the forest CO_2 budget, as is expected in the national inventory reporting under the Kyoto Protocol. This will allow estimating the mitigation potential of forests and its relationship with N inputs and other environmental factors.

ACKNOWLEDGEMENTS

We thank Dr. Sabine Augustin from the Swiss Federal Office for the Environment for her valuable contribution to this chapter, in particular, to the policy relevance aspects. The authors were supported by funds received from the Swiss Staatssekretariat für Bildung und Forschung (contract SBF Nr. C10.0101) to participate in COST Action FP0903 'Climate Change and Forest Mitigation and Adaptation in a Polluted Environment', and by the Swiss National Science Foundation (grant PDFMP3_132562). We also thank the Swiss Federal Office for the Environment for funding the Swiss TreeNet network and Dr. Roman Zweifel for scientific collaboration and technical support with dendrometer measurements.

1. http://www.treenet.info.

REFERENCES

Aber, J.D., Nadelhoffer, K.J., Steudler, P., Melillo, J.M., 1989. Nitrogen saturation in northern forest ecosystems. Bioscience 39, 378–386.

Aber, J.D., McDowell, W., Nadelhoffer, K.J., Magill, A., Berntson, G., Kamakea, M., McNulty, S., Currie, W., Rustad, L., Fernandez, I., 1998. Nitrogen saturation in northern forest ecosystems—hypotheses revisited. Bioscience 48, 921–934.

Aber, J.D., Goodale, C.L., Ollinger, S.V., Smith, M.-L., Magill, A.H., Martin, M.E., Hellett, R.A., Stoddard, J.L., 2003. Is nitrogen deposition altering the nitrogen status of northeastern forests? Bioscience 53, 375–389.

Aubinet, M., Vesala, T., Papale, D. (Eds.), 2012. Eddy Covariance—A Practical Guide to Measurement and Data Analysis. Springer, Dordrecht, Heidelberg, London, New York, 449 pp.

Barford, C., Wofsy, S., Goulden, M., Munger, J., Pyle, E., Urbanski, S., Hutyra, L., Saleska, S., Fitzjarrald, D., Moore, K., 2001. Factors controlling long-and short-term sequestration of atmospheric CO_2 in a mid-latitude forest. Science 294 (5547), 1688–1691.

Bleeker, A., Reinds, G.J., Vermeulen, A.T., de Vries, W., Erisman, J.W., 2004. Critical loads and resent deposition thresholds of nitrogen and acidity and their exceedances at the level II and level I monitoring plots in Europe. Technical report, ECN Report ECN-C-04-117, Petten, The Netherlands.

Bobbink, R., Hettelingh, J.P. (Eds.), 2010. Review and Revision of Empirical Critical Loads and Dose-Response Relationships. Coordination Centre for Effects, National Institute for Public Health and the Environment (RIVM), The Netherlands, 243 pp.

Bobbink, R., Hornung, M., Roelofs, J.G.M., 1998. The effects of air borne nitrogen pollutants on species diversity in natural and semi-natural European vegetation. J. Ecol. 86, 717–738.

Bobbink, R., Hicks, K., Galloway, J., Spranger, T., Alkemade, R., Ashmore, M., Bustamante, M., Cinderby, S., Davidson, E., Dentener, F., Emmett, B., Erisman, J.-W., Fenn, M., Gilliam, F., Nordin, A., Pardo, L., De Vries, W., 2010. Global assessment of nitrogen deposition effects on terrestrial plant diversity: a synthesis. Ecol. Appl. 20, 30–59.

Braun, S., Thomas, V.F.D., Quiring, R., Flückiger, W., 2010. Does nitrogen deposition increase forest production? The role of phosphorus. Environ. Pollut. 158, 2043–2052.

Buchmann, N., Schulze, E.D., 1999. Net CO_2 and H_2O fluxes of terrestrial ecosystems. Global Biogeochem. Cycles 13, 751–760.

Burkard, R., Bützberger, P., Eugster, W., 2003. Vertical fogwater flux measurements above an elevated forest canopy at the Lägeren research site, Switzerland. Atmos. Environ. 37 (21), 2979–2990.

Bytnerowicz, A., Fenn, M., McNulty, S., Yuan, F., Pourmokhtarian, A., Driscoll, C., Meixner, T., 2013. Interactive effects of air pollution and climate change on forest ecosystems in the United States—current understanding and future scenarios. Dev. Environ. Sci. 13, 333–369.

Cieslik, S., Tuovinen, J.-P., Baumgarten, M., Matyssek, R., Brito, P., Wieser, G., 2013. Gaseous exchange between forests and the atmosphere. Dev. Environ. Sci. 13, 19–36.

Cocozza, C., Giovannelli, A., Lasserre, B., Cantini, C., Lombardi, F., Tognetti, R., 2012. A novel mathematical procedure to interpret the stem radius variation in olive trees. Agric. For. Meteorol. 161, 80–93.

Clarke, N., Zlindra, D., Ulrich, E., Mosello, R., Derome, J., Derome, K., König, N., Lövblad, G., Draaijers, G.P.J., Hansen, K., Thimonier, A., Waldner, P., 2010. Sampling and analysis of deposition. 66 pp. Part XIV. In: Manual on Methods and Criteria for Harmonized Sampling, Assessment, Monitoring and Analysis of the Effects of Air Pollution on Forests. UNECE, ICP Forests, Hamburg. ISBN: 978-3-926301-03-1. http://www.icp-forests.org/Manual.htm.

De Schrijver, A., Verheyen, K., Mertens, J., Staelens, J., Wuyts, K., Muys, B., 2008. Nitrogen saturation and net ecosystem production. Nature 451, E1.

De Vries, W., Reinds, G.J., Gundersen, P., Sterba, H., 2006. The impact of nitrogen deposition on carbon sequestration in European forests and forest soils. Glob. Chang. Biol. 12, 1151–1173.

De Vries, W., Solberg, S., Dobbertin, M., Sterba, H., Laubhahn, D., Reinds, G.J., Nabuurs, G.-J., Gundersen, P., Sutton, M.A., 2008. Ecologically implausible carbon response? Nature 451, E1–E3.

Dise, N.B., Rothwell, J.J., Gauci, V., van der Salm, C., de Vries, W., 2009. Predicting dissolved inorganic nitrogen leaching in European forests using two independent databases. Sci. Total Environ. 407, 1798–1808.

Eilers, G., Brumme, R., Matzner, E., 1992. Above-ground N-uptake from wet deposition by Norway spruce (*Picea abies* Karst.). For. Ecol. Manage. 51, 239–249.

Ellenberg, H., 1990. Ökologische Veränderungen in Biozönosen durch Stickstoffeintrag. In: Ammoniak in der Umwelt—Kreisläufe, Wirkungen, Minderung. Kuratorium für Technik und Bauwesen in der Landwirtschaft, Darmstadt, pp. 44.1–44.24.

Ellenberg, H., Weber, H.E., Düll, R., Wirth, V., Werner, W., Paulissen, D., 1991. Zeigerwerte von Pflanzen in Mitteleuropa. In: Scripta Geobotanica. vol. 18. Erich Goltze, Göttingen, 248 pp.

Elling, W., Heber, U., Polle, A., Beese, F., 2007. Schädigung von Waldökosystemen: Auswirkungen anthropogener Umweltveränderungen und Schutzmassnahmen. Elsevier Spektrum Akademie-Verlag, München, 422 pp.

Emmett, B., 2007. Nitrogen saturation of terrestrial ecosystems: some recent findings and their implications for our conceptual framework. Water Air Soil Pollut. 7, 99–109.

Erisman, J.W., van Pul, A., Wyers, P., 1994. Parametrization of surface resistance for the quantification of atmospheric deposition of acidifying pollutants and ozone. Atmos. Environ. 28, 2595–2607.

Eugster, W., 2008. Fog research. Die Erde 139 (1–2), 1–10.

Eugster, W., Perego, S., Wanner, H., Leuenberger, A., Liechti, M., Reinhardt, M., Geissbühler, P., Gempeler, M., Schenk, J., 1998. Spatial variation in annual nitrogen deposition in a rural region in Switzerland. Environ. Pollut. 102 (S1), 327–335.

Ferrara, R., Loubet, B., Di Tommasi, P., Bertolini, T., Magliulo, V., Cellier, P., Eugster, W., Rana, G., 2012. Eddy covariance measurement of ammonia fluxes: comparison of high frequency correction methodologies. Agric. For. Meteorol. 158, 30–42.

Flechard, C.R., Nemitz, E., Smith, R.I., Fowler, D., Vermeulen, A.T., Blecker, A., Erisman, J.W., Simpson, D., Zhang, L., Tang, Y.S., Sutton, M.A., 2011. Dry deposition of reactive nitrogen to European ecosystems: a comparison of inferential models across the NitroEurope network. Atmos. Chem. Phys. 11, 2703–2728.

Flückiger, W., Braun, S., 1999. Nitrogen and its effects on growth, nutrient status and parasite attacks in beech and Norway Spruce. Water Air Soil Pollut. 116, 99–110.

Goll, D.S., Brovkin, V., Parida, B.R., Reick, C.H., Kattge, J., Reich, P.B., van Bodegom, P.M., Niinemets, U., 2012. Nutrient limitation reduces land carbon uptake in simulations with a model of combined carbon, nitrogen and phosphorus cycling. Biogeosciences 9, 3547–3569.

Gundersen, P., 1998. Effects of enhanced nitrogen deposition in a spruce forest at Klosterhede, Denmark, examined by NH_4NO_3 addition. For. Ecol. Manage. 101, 251–268.

Gundersen, P., Callesen, I., de Vries, W., 1998. Nitrate leaching is related to forest floor C/N ratios. Environ. Pollut. 102, 403–407.

Jeffries, D., Cox, C., Dillen, P., 1979. Depression of pH in lakes and streams in central Ontario during snowmelt. J. Fish. Res. Board Can. 36, 640–646.

Khanna, P.K., Bauhus, J., Meiwes, K.J., Kohler, M., Rumpf, S., Schönfelder, E., 2007. Assessment of changes in the phosphorous status of forest ecosystems in Germany—literature review and analysis of existing data. Report to the German Federal Ministry of Food, Agriculture and Consumer Protection.

Körner, C., Asshof, R., Bignucolo, O., Hattenschwiler, S., Keel, S.G., Pelaez-Riedl, S., Pepin, S., Siegwolf, R.T.W., Zotz, G., 2005. Carbon flux and growth in mature deciduous forest trees exposed to elevated CO_2. Science 309 (5739), 1360–1362.

Kozovits, A.R., Bustamante, M.M.C., 2013. Land use change, air pollution and climate change—vegetation response in Latin America. Dev. Environ. Sci. 13, 411–427.

Kraigher, H., Bajc, M., Grebenc, T., 2013. Mycorrhizosphere complexity. Dev. Environ. Sci. 13, 151–177.

Krupa, S.V., 2003. Effects of atmospheric ammonia (NH_3) on terrestrial vegetation: a review. Environ. Pollut. 124, 179–221.

Kyrö, E.-M., Grönholm, T., Vuollekoski, H., Virkkula, A., Kulmala, M., Laakso, L., 2009. Snow scavenging of ultrafine particles: field measurements and parameterization. Boreal Env. Res. 14, 527–538.

Laubhann, D., Sterba, H., Reinds, G.-J., De Vries, W., 2009. The impact of atmospheric deposition and climate on forest growth in European monitoring plots: an individual tree growth model. For. Ecol. Manage. 258, 1751–1761.

Mikkelsen, T.N., Clarke, N., Danielewska, A., Fischer, R., 2013. Towards supersites in forest ecosystem monitoring and research. Dev. Environ. Sci. 13, 475–495.

Nave, L.E., Curtis, P.S., 2011. Uptake and partitioning of simulated atmospheric N inputs in *Populus tremuloides—Pinus strobus* forest mesocosms. Botany 89, 379–386.

Nellemann, C., Thomsen, M.G., 2001. Long-term changes in forest growth: potential effects of nitrogen deposition and acidification. Water Air Soil Pollut. 128, 197–205.

Nihlgard, B., 1985. The ammonium hypothesis—an additional explanation to the forest dieback in Europe. Ambio 14, 1–8.

Nordin, A., Strengom, J., Ericson, L., 2006. Responses to ammonium and nitrate additions by boreal plants and their natural enemies. Environ. Pollut. 141, 167–174.

Prietzel, J., Rehfuess, K.E., Stetter, U., Pretzsch, H., 2008. Changes in soil chemistry, stand nutrition, and stand growth at two Scots pine (Pinus sylvestris L.) sites in Central Europe during 40 years after fertilization, liming, and lupine introduction. Eur. J. For. Res. 127, 43–61.

Schaefer, D.A., Reiners, W.A., 1990. Throughfall chemistry and canopy processing mechanisms. In: Lindberg, S.E., Page, A.L., Norton, S.E. (Eds.), Acidic Precipitation. Sources, Deposition, and Canopy Interactions, vol. 3. Springer, New York, pp. 241–284.

Schulze, E.-D., 1989. Air pollution and forest decline in a spruce (*Picea abies*) forest. Science 244, 776–783.

Sievering, H., Tomaszewski, T., Torizzo, J., 2007. Canopy uptake of atmospheric N deposition at a conifer forest: part I—canopy N budget, photosynthetic efficiency and net ecosystem exchange. Tellus 59B, 483–492.

Slootweg, J., Posch, M., Hettelingh, J.-P. (Eds.), 2010. Progress in the Modelling of Critical Thresholds and Dynamic Modelling, Including Impacts on Vegetation in Europe. CCE Status Report 2010, Coordination Centre for Effects, http://www.rivm.nl/cce.

Smith, R.I., Fowler, D., Sutton, M.A., Flechard, C., Coyle, M., 2000. Regional estimation of pollutant gas deposition in the UK: model description, sensitivity analyses and outputs. Atmos. Environ. 34, 3757–3777.

Sparks, J.P., 2009. Ecological ramifications of the direct foliar uptake of nitrogen. Oecologia 159, 1–13.

Spieker, H., Mielkäinen, K., Köhl, M., Skovsgaard, J. (Eds.), 1996. Growth trends in European forests: studies from 12 countries. EFI Research Report No. 5, Springer-Verlag, Berlin, 372 pp.

Spranger, T., Lorenz, U., Gregor, H.-D. (Eds.), 2004. Manual on Methodologies and Criteria for Modelling and Mapping Critical Loads and Levels and Air Pollution Effects, Risks and Trends. Federal Environmental Agency, Berlin, Texte, 52 (4), 266 pp.

Thalmann, E., Burkard, R., Wrzesinsky, T., Eugster, W., Klemm, O., 2002. Ion fluxes from fog and rain to an agricultural and a forest ecosystem in Europe. Atmos. Res. 64, 147–158.

Thimonier, A., Schmitt, M., Waldner, P., Rihm, B., 2005. Atmospheric deposition on Swiss long-term forest ecosystem research (LWF) plots. Environ. Monit. Assess. 104, 81–118.

Tuovinen, J.-P., Hakola, H., Karlsson, P.E., Simpson, D., 2013. Air pollution risks to northern European forests in a changing climate. Dev. Environ. Sci. 13, 77–99.

UNECE, 1999. The 1999 Gothenburg Protocol to Abate Acidification, Eutrophication and Ground-level Ozone. http://www.unece.org/fileadmin/DAM/env/lrtap/full%20text/1999%20Multi.E.Amended.2005.pdf.

UNECE, 2005. Development, Modelling and Mapping of Critical Loads and Their Input Data. http://www.unece.org/fileadmin/DAM/env/documents/2005/eb/wg1/EB.AIR.WG.1.2005.10.e.pdf.

Vitousek, P.M., Field, C.B., 2001. Input/output balances and nitrogen limitation in terrestrial ecosystems. In: Schulze, E.-D. et al., (Ed.), Global Biogeochemical Cycles in the Climate System. Academic Press, San Diego, pp. 217–225.

Vitousek, P.M., Aber, J.D., Howarth, R.W., Likens, G.E., Matson, P.A., Schindler, D.W., Schlesinger, W.H., Tilman, D.G., 1997. Human alteration of the global nitrogen cycle: sources and consequences. Ecol. Appl. 7 (3), 737–750.

Wortman, E., Tomaszewski, T., Waldner, P., Schleppi, P., Thimonier, A., Eugster, W., Buchmann, N., Sievering, H., 2012. Atmospheric nitrogen deposition and canopy retention influences on photosynthetic performance at two high nitrogen deposition Swiss forests. Tellus B 64, 1–14.

Zhang, L., Brook, J.R., Vet, R., 2003. A revised parameterization for gaseous dry deposition in air-quality models. Atmos. Chem. Phys. 3, 2067–2082.

Zweifel, R., Häsler, R., 2000. Frost-induced reversible shrinkage of bark of mature subalpine conifers. Agric. For. Meteorol. 102, 213–222.

Zweifel, R., Häsler, R., 2001. Dynamics of water storage in mature subalpine *Picea abies*: temporal and spatial patterns of change in stem radius. Tree Physiol. 21, 561–569.

Zweifel, R., Eugster, W., Etzold, S., Dobbertin, M., Buchmann, N., Häsler, R., 2010. Link between continuous stem radius changes and net ecosystem productivity of a subalpine Norway spruce forest in the Swiss Alps. New Phytol. 187, 819–830.

Chapter 4

Biogenic Volatile Organic Compounds and Their Impacts on Biosphere–Atmosphere Interactions

Francesco Loreto*[,1] and Silvano Fares[†]
*National Research Council (CNR), IPP, Florence, Italy
[†]Consiglio per la ricerca e la sperimentazione in agricoltura (CRA), Research Center for the Soil–Plant System (RPS), Rome, Italy
[1]Corresponding author: e-mail: francesco.loreto@ipp.cnr.it

Chapter Outline

4.1 GENERALITIES ON BIOGENIC VOLATILE ORGANIC COMPOUNDS

Volatile organic compounds emitted by vegetation, hereafter called biogenic volatile organic compounds (BVOC), are a large set of hydrocarbons produced by plants in various tissues above and below ground, and with a wide range of functions, ranging from plant ecology to being key actors in atmospheric chemistry, physics and climate.

Thousands of compounds are emitted from plants (Knudsen and Gershenzon, 2006), with some of the most representative compounds shown in Figure 4.1.

Developments in Environmental Science, Vol. 13. http://dx.doi.org/10.1016/B978-0-08-098349-3.00004-9

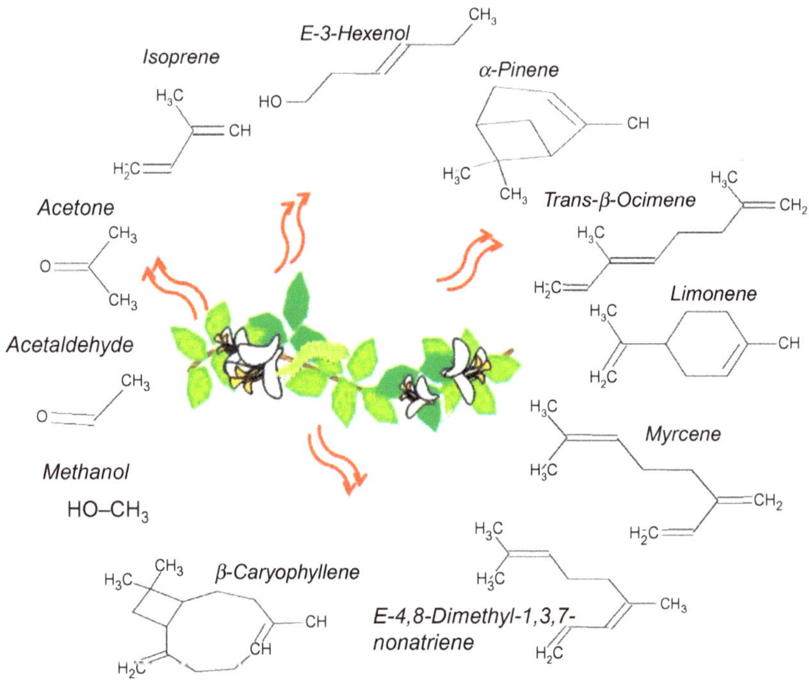

FIGURE 4.1 Molecular structure of BVOC emitted from flowers and leaves. From methanol, clockwise, the most representative BVOC are listed following a growing number of carbon atoms: oxygenated VOC (methanol, acetaldehyde, acetone), isoprene, a C-6 compound (E-3-hexenol) monoterpenes (a-pinene, *trans*-b-ocimene, limonene, myrcene), a hemiterpene ((E)-4,8-dimethyl-1,3,7-nonatriene) and a sesquiterpene (b-caryophyllene).

BVOC include the isoprenoids (isoprene, monoterpenes, sesquiterpenes) as well as alkanes, alkenes, carbonyls, alcohols, esters, ethers and acids. According to the number of C5 units, isoprenoids are subdivided into hemiterpenes (C5, e.g. isoprene, prenyl residue of citokine), monoterpenes (C10, e.g. pinene, menthol, camphor), sesquiterpenes (C15, e.g. β-caryophyllene, abscisic acid), diterpenes (C20, e.g. gibberellins, phytole, tocopherole, retinol), triterpenes (C30, e.g. sterols, saponins), tetraterpenes (C40, e.g. carotenoids) and prenols and polyterpenes (>C45, e.g. plastochinone, ubichinone, India rubber) (Kesselmeier and Staudt, 1999).

Oxygenated volatile organic compounds (hereafter referred to by the acronym OVOC) are also an important class of BVOC (Fares et al., 2011; Koppmann and Wildt, 2007; Steiner and Goldstein, 2007). Methanol, acetone and acetaldehyde are the most abundant OVOC produced by plants.

BVOC are produced by different organs; for example, leaves (Loreto and Schnitzler, 2010), flowers (Knudsen et al., 1993) and roots (Steeghs et al., 2004). Many BVOC are emitted constitutively, and the emissions can be observed throughout the plant life cycles or, more often, at specific developmental stages (e.g. leaf and needle maturation, senescence, flowering and fruit ripening). Other BVOC are not constitutively produced but are induced by biotic and abiotic stresses. Stresses may, in fact, stimulate or quench constitutive emissions or may induce *de novo* synthesis and emissions (Loreto et al., 2006).

The emission of BVOC is biosynthetically controlled by abiotic factors such as temperature and/or light, atmospheric CO_2 concentration or nutrition. For isoprenoids, the light-dependent emissions (Loreto and Sharkey, 1990; Loreto et al., 1996) are closely linked to the availability of photosynthetic intermediates (Lichtenthaler et al., 1997; Loreto et al., 1996) through the chloroplastic methylerythritol phosphate (MEP) pathway, responsible for the formation of the basic C5 units isopentenyl diphosphate (IDP) and dimethylallyl diphosphate (DMADP) (Figure 4.2). The MEP pathway provides IDP and DMADP for hemiterpene and monoterpene biosynthesis, although recent studies revealed that the MEP pathway may also contribute to sesquiterpene formation (Sallaud et al., 2009; Tholl et al., 2005) (Figure 4.2). A second biosynthetic source of isoprenoid formation is represented by the cytosolic mevalonic acid pathway from the acetyl-CoA pathway which takes place in the cytosol, generates essentially sesquiterpenes, and is mostly driven by

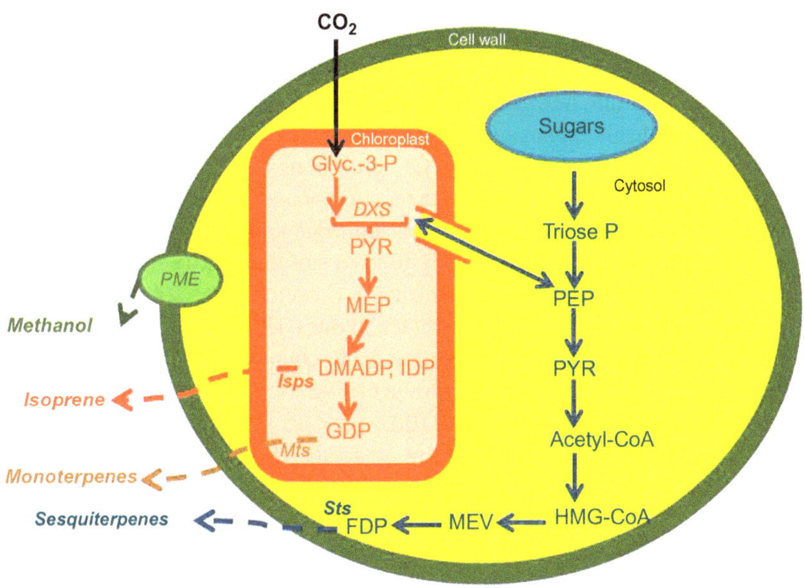

FIGURE 4.2 Scheme of the two biosynthetic pathways of isoprenoid formation: Mevalonate (MEV, inside the cytosol) and methyl erythrose phosphate (MEP, inside the chloroplast). Acetyl-coA, acetyl coenzyme A; DMADP, dimethylallyl diphosphate; DXS, 1-deoxy-D-xylulose 5-phosphate synthase; FDP, farnesyl diphosphate; GDP, geranyl diphosphate; HMG-CoA, 3-hydroxy-3-methylglutaryl-CoA reductase; IDP, isopentenyl diphosphate; ISPS, isoprene synthase; MTS, monoterpene synthase; PEP, phosphoenolpyruvate; PME, pectin methylesterase; PYR, pyruvate; STS, sesquiterpene synthase; Triose P, triose phosphate.

temperature (Figure 4.2). The strong temperature dependence of all BVOC irrespective of the pathway of formation is based not only on the temperature influencing the kinetics of reaction of the dedicated enzymes (e.g. terpene synthases, TPS) but also on the volatility of the BVOC stored in reservoir organs (e.g. resin ducts, glands) (Loreto and Schnitzler, 2010). In addition, a metabolic crosstalk between both biosynthetic routes is possible (Schuhr et al., 2003) when IDP is transported from chloroplasts to the cytosol (Laule et al., 2003).

OVOC have biosynthetic pathways different from isoprenoids. Methanol is produced by demethylation of pectins in cell walls (Obendorf, 1990), and emissions occur under phenological modification of leaf tissues during leaf expansion and senescence (Fall, 2003; Huve et al., 2007) or after mechanical stresses. Previous studies reported methanol emission in response to oxidative stress following cutting and during drying of grass (Karl et al., 2001) or following leaf wounding (Loreto et al., 2006). Acetone is released during senescence (de Gouw et al., 1999) and oxidative stress on plants (e.g. from ozone) (Cojocariu et al., 2005), but the biosynthetic pathway of acetone formation in leaves still deserves more investigation. Relevant emissions of acetaldehyde have been observed mainly under anoxic conditions in roots (Kreuzwieser et al., 1999) and leaves (Graus et al., 2004; Karl et al., 2002a,b). Acetaldehyde is also emitted by leaves in large quantities during and after abiotic stresses (Fall et al., 1999; Loreto et al., 2006), or after light to dark transitions (Karl et al., 2002a).

Advanced sampling methods in combination with gas chromatography–mass spectroscopy and new techniques for online analysis have led, in the past years, to a significant expansion of our knowledge on the occurrence and temporal and spatial distribution of BVOC emissions (Tholl et al., 2006). For instance, the PTR-MS (proton transfer reaction–mass spectrometry) technique allows for quantitative measurements of anthropogenic or BVOC (Lindinger et al., 1998) and has been successfully used to simultaneously measure fluxes of several BVOC in laboratory experiments (Loreto et al., 2006) and in the field using disjunct eddy covariance techniques (Karl et al., 2002b). The recent development of a high-resolution PTR coupled with a time-of-flight detector (PTR-TOF-MS) allows ultra-sensitive (<1 ppt) measurements of hundreds of masses simultaneously (10 Hz) and to separate compounds with the same molecular mass but different structure (Jordan et al., 2009). This instrument has been recently employed in the field to study fluxes of a multitude of BVOC using eddy covariance technique (Müller et al., 2010; Park et al., 2012) and in the laboratory to study fast response of BVOC emitted from plants in response to abiotic stress (Brilli et al., 2012).

4.2 BVOC AND THE ATMOSPHERE: FLUXES AND CONCENTRATIONS (SINKS, SOURCES)

Most BVOC are largely lipophilic and their vapour pressure is high enough to be released into the atmosphere in significant amounts. Thanks to this high volatility, BVOC are emitted to the atmosphere at an estimated global rate of 1–1.5 Pg C/year (Guenther et al., 2006). Emission sources are mainly terrestrial vegetation (leaves, in particular), while contributions from oceans and aquatic vegetation are much less known but estimated to account for less than 2% of global emissions (Shaw et al., 2010). Isoprenoid emission is estimated to account for more than half of the total emission of BVOC (Guenther et al., 2006) and is constitutively 10 times higher than other emissions, as heavy emitters can release isoprene at rates of 50–100 nmol/m^2/s, representing up to 2–5% of the photosynthetic net carbon uptake in tree species (Kesselmeier and Staudt, 1999). Moreover, volatile isoprenoid emissions are far more sustained over time than the emission of other constitutive or induced volatiles, for which emissions are transient by nature and limited to specific periods after stress or during specific developmental stages.

One important notion is that, as most BVOC are strongly reactive, they do not accumulate in the atmosphere, despite being emitted at large rates. After biosynthesis, BVOC, especially those which do not carry oxygen atoms and are rich on conjugated double bonds, can react with ozone and reactive oxygen species (ROS) in the gas phase within leaves or canopies, or just above canopies. Therefore, *in situ*, BVOC emission represents a sink for ozone, acting as anti-oxidant by removing the oxidant molecules before they reach the leaf intercellular spaces, or even after foliar acquisition but before ROS can impinge into cellular membranes. However, not all BVOC are equally reactive with ozone. Monoterpenes and, in particular, sesquiterpenes are the most reactive BVOC, with reaction time in the order of seconds as in the case of β-caryophyllene, a typical sesquiterpene emitted by many plant species (Shu and Atkinsons, 1994). This low reaction time allows within-canopy reactions with ozone, in particular in closed and stratified canopies where low vertical mixing promotes retention of air masses long enough for these reactions to occur. In plant ecosystems, a direct method to measure ozone losses by BVOC is not yet available, but micrometeorological techniques allow quantification of stomatal and total ozone deposition (Cieslik et al., 2013, this vol.), therefore calculating non-stomatal ozone deposition as the difference between these two terms. Results from experimental measurements show that in the central hours of the day, when light and temperature maximise BVOC emission, non-stomatal ozone deposition is larger than stomatal deposition. This was observed in a multitude of crop and forest ecosystems: a *Pinus ponderosa* forest (Fares et al., 2010; Kurpius and Goldstein, 2003), a Sitka spruce forest (Coe et al., 1995), a Mediterranean oak forest (Gerosa et al., 2005), a northern mixed hardwood forest (Hogg et al., 2007), in a sub-alpine ecosystem (Zeller and Nikolov, 2000) and a citrus plantation (Fares et al., 2012a,b). Although deposition on dry and wet surfaces was found to be a significant contributor to non-stomatal deposition (Altimir et al., 2006; Zhang et al., 2003), theoretical calculation of the chemical lifetime of ozone in canopies suggested that reactive BVOC can contribute up to 30% of total ozone deposition (Ciccioli et al., 1999; Fares et al., 2012a,b; Kurpius and

Goldstein, 2003). This ozone-scavenging property of BVOC is likely taking place mostly in the upper part of the canopy, where BVOC emission is higher due to direct illumination and higher temperature conditions.

Since the role of BVOC as scavengers of ozone and other ROS is now acknowledged, a few BVOC can have a dual role in atmospheric chemistry. In fact, besides directly removing ozone in the gas phase near the emission sites, it is long known that the same compounds can contribute to form ozone *ex situ*, when they are involved in photochemical reactions in the presence of sunlight and nitrogen oxides (NO_x) (Chameides et al., 1988; Papiez et al., 2009; Figure 4.3). BVOC oxidation also leads to the growth of secondary organic aerosol (SOA) and formation of atmospheric aerosol (Henze and Seinfeld, 2006; Kanakidou et al., 2005), accounting for a significant fraction of SOA produced in the atmosphere (Goldstein and Galbally, 2007). In the atmosphere, ozone acts as a potent greenhouse gas; its radiative-forcing potential is of near-equal magnitude to that of methane, making it the third largest contributor to anthropogenic warming.

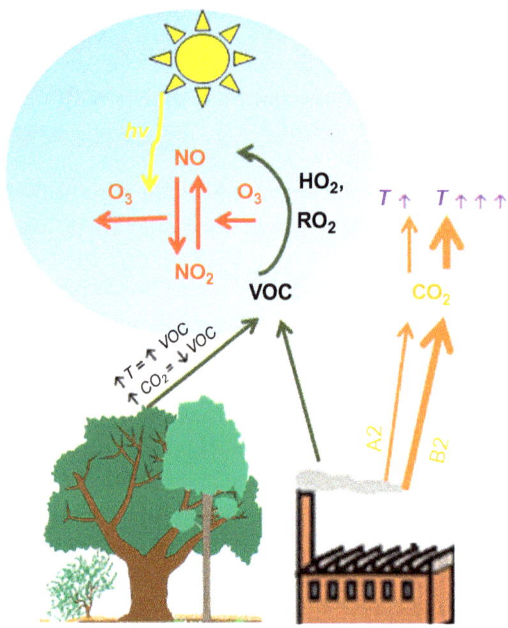

FIGURE 4.3 Simplified scheme of plant–atmosphere exchanges with emphasis on (1) VOC emission from biogenic and anthropogenic sources, (2) interaction of VOC in the photochemical reactions which lead to ozone formation, (3) CO_2 emission from anthropogenic sources which leads to increase in global temperature according to two different IPCC projections: A2 (very heterogeneous world with high population growth, slow economic development and slow technological change) and B2 (intermediate population and economic growth with more attention to social and environmental sustainability). The increase in temperature has a positive effect on VOC emission, but CO_2 increase inhibits VOC production.

This dual role of BVOC emission in ozone removal and ozone formation in the atmosphere indicates the importance of achieving a reliable estimate of BVOC fluxes over different plant ecosystems, over time and space, and in response to biotic and abiotic stress. Because of the large amount of experimental data that are being collected (Fowler et al., 2009), models are continuously updated to improve quantification of BVOC emissions on a regional and global basis (Arneth et al., 2008; Guenther et al., 2006; Tuovinen et al., 2013, this vol.). Such models are generally parameterized by integrating environmental data, plant distribution, and biomass density, and by using emission algorithms which incorporate the emission capacity of BVOC species measured experimentally under standardised conditions of 30 °C and 1000 μmol/m^2/s of photosynthetically active radiation. The actual emission is then reproduced by using algorithms that take into consideration the dependence on light and temperature of the different BVOC species (Guenther et al., 2006; Monson et al., 1992; Niinemets et al., 2004). The emissions of isoprene and some monoterpenes primarily respond to light and temperature because their chloroplast biosynthesis depends on photosynthetic intermediates, and their release takes place immediately after production (Figure 4.2; Lichtenthaler et al., 1997). Other BVOC (mainly monoterpenes and sesquiterpenes) are volatilized from storage pools, which uncouples them from photosynthetic carbon availability. These emissions are mainly temperature dependent; therefore, algorithms considering temperature responses only are used (Harley et al., 1996; Niinemets et al., 2004; Tingey et al., 1980). While the emission of monoterpenes from storage organs is widely acknowledged (Loreto and Schnitzler, 2010), sesquiterpenes have been long overlooked but are now emerging as an important class of isoprenoids (for a review, see Duhl et al., 2008). Recent results suggest the total emission rates of sesquiterpenes to be similar to monoterpenes (Ormeño et al., 2010), which raises questions about the too large amount of carbon that plants should invest on sesquiterpene biosynthesis, when accounting for the low volatility of these compounds. Emission of sesquiterpenes is also of great interest for atmospheric chemistry, as they contribute to SOA even more than monoterpenes (Lee et al., 2006; Ng et al., 2006).

Despite current efforts to improve estimates of BVOC emissions, a lot remains to be done. In particular, due to climate change and consequent more widespread exposure to biotic and abiotic stress, empirical emission algorithms may still fail to accurately predict real BVOC emissions. As the biochemical and physiological control on some BVOC (e.g. isoprene) is well known (Loreto and Schnitzler, 2010; Monson et al., 2012), it may be possible to further implement model estimation capability by also implementing process-based analyses, particularly taking into account induced emissions (Niinemets et al., 2010).

Unlike the emissions of volatile isoprenoids (with the exception noted above of sesquiterpenes), knowledge of OVOC emissions only dates from the past decade (Koppmann and Wildt, 2007; Steiner and Goldstein, 2007). Similar to isoprenoids, OVOC can notably influence the oxidising capacity and the ozone-forming potential of the atmosphere, while also increasing concentrations of HO_x and peroxyacetyl nitrates, and possibly contributing to the formation of organic aerosol (Singh et al., 2001). Methanol global emissions are now estimated at 100–240 Tg/year (Galbally and Kirstine, 2002; Jacob et al., 2005; Millet et al., 2008), and methanol emission can be comparable to isoprene emission, for example, in actively growing Mediterranean ecosystems (Davison et al., 2009). Acetone is the most abundant ketone in the atmosphere (Koppmann and Wildt, 2007), with global emissions estimated at 95 Tg/year (Jacob et al., 2002) mainly from rural areas (Ciccioli et al., 1999; Riemer et al., 1998). This compound is emitted primarily from terrestrial ecosystems and oceans but is also produced in the atmosphere in large amounts from oxidation of hydrocarbons of both anthropogenic and biogenic origin (Goldstein and Schade, 2000). Acetaldehyde is another OVOC that is directly emitted from oceanic and terrestrial sources, but is also a product of hydrocarbon oxidation in the atmosphere. An estimate of acetaldehyde emission was recently reported by Millet et al. (2009), with values which are similar to acetone.

4.3 BVOC AND PLANT PHYSIOLOGY AND ECOLOGY: MEMBRANE PROTECTION, ANTI-OXIDANTS, PLANT COMMUNICATION

The roles of BVOC and the reason why they are produced should be studied in a multi-disciplinary manner. Indeed, BVOC seem to be involved in plant defence against both abiotic and biotic stresses (Dicke and Baldwin, 2010; Loreto and Schnitzler, 2010; Vickers et al., 2009). Isoprene seems to be particularly important in protection against abiotic stresses. In several laboratory studies in which different plant species were exposed to high ozone concentration, plants emitting isoprene were more resistant to the oxidative stress than plants in which isoprene was not formed, or was chemically inhibited (Loreto and Fares, 2007; Loreto and Velikova 2001). Isoprene emitters are clearly more resistant than non-emitters also to other environmental stresses, particularly high temperatures (Singsaas et al., 1997; Velikova et al., 2011). The effect is especially clear during short, intense heat spikes, for example, experienced by plants during the warmest hours of summer days (Singsaas and Sharkey, 1998). Sharkey and colleagues attributed the isoprene effect to its ability to stabilise cellular membranes, thus protecting the photosynthetic apparatus against heat-induced denaturation of membrane lipids (Sharkey and Yeh, 2001). Perhaps a similar effect may also protect against oxidative stresses, as less ROS are formed when membrane integrity allows high rates of photosynthesis in stressed leaves (Velikova et al., 2012). However, reactive BVOC produced inside leaves are also thought to scavenge substantial amounts of ROS, thereby directly reducing oxidative damage to sensitive plant organelles and structures, or indirectly modulating ROS-induced signalling of programmed cellular death (Velikova et al., 2004). In a recent experiment, Jardine et al. (2012) tracked the reaction products between isoprene and ROS inside leaves through leaf and branch feeding with pyruvate-2-13C. These authors, indeed, observed a temperature-dependent emission of the labelled isoprene oxidation products methyl-vinyl ketone and methachrolein. Isoprene oxidation, related to increasing oxidative stresses in leaves whose photosynthesis was inhibited by high temperature, was suggested to produce the observed products (Jardine et al., 2012). By attributing a clear physiological role to isoprene, plant investment of carbon in this important compound remains justified. It is suggested that models of biogenic emissions and tropospheric chemistry should incorporate isoprene oxidation and the products of this reaction, to better understand both the oxidising power of the troposphere and plant response to climate change.

From an ecological point of view, BVOC mediate vital interactions between plants and other organisms. They are emerging as one of the most important actors in plant communication with other organisms, and as a fundamental defence of plants against biotic stress as they are able to deter pathogens and herbivores (Dicke and Baldwin, 2010). Further, pioneering research has shown that tiny emissions of BVOC at the leaf level are able to attract parasitoids or predators of the herbivorous arthropods (reviewed by Dicke and Baldwin, 2010). Thus, using BVOC emissions, plants attacked by herbivores 'call friendly organisms for help', establishing multi-trophic interactions. Recent experiments suggest that plants may even use BVOC to systemically induce the production of defensive chemicals in parts of the plant that have not been attacked yet (Brilli et al., 2009) or to inform neighbouring plants about arriving infestations (Li et al., 2012). The evolution of a BVOC-based self-defence capacity by plants has promising application for the biological control of herbivores in crops and forests through integrated pest management practices (Kos et al., 2009). Generally, constitutive expression of resistant traits in plants is accompanied by fitness costs when insect pests are absent. This may explain why isoprene, the main BVOC constitutively emitted, does not seem to have an effect in plant communication (with exceptions, see, e.g. Loivamäki et al., 2008). On the other hand, biochemical mechanisms activated after herbivore attacks (e.g. triggered by chemicals in herbivore saliva, regurgitants, faeces or even by the simple mechanical damage caused by feeding) may produce a range of 'induced' BVOC that have importance in direct or indirect plant defence (Dicke and van Loon, 2000). Of course, the cost of such attack-induced responses is, in turn, influenced by environmental conditions and by the effective presence of competitors and/or natural enemies of herbivorous arthropods at reachable distance. The impact of atmospheric pollution may have very important consequences for the distance that volatile cues released by plants travel. In a very relevant experiment, McFrederick et al. (2008) reported dramatic decreases of scent trails as a consequence of ozone and NO_x pollution because of the reaction between the pollutants and the BVOCs. Similarly, Pinto et al. (2007) observed a reduction of induced terpenes and green leaf volatiles by reaction with ozone in a tritrophic system reproduced in the laboratory.

Not all herbivore-induced BVOC have the same function. In some cases, the BVOC are stress markers or elicitors of biochemical mechanisms that eventually generate volatile (e.g. monoterpenes) or non-volatile (e.g. phenolics) defensive compounds. In other cases, the same BVOC can have multiple functions. As one example, methyl salicylate (MeSA) is traditionally considered as a pathogen-related defensive BVOC, eliciting formation of many fungistatic or fungicide plant metabolites (Dicke et al., 1990; Shulaev et al., 1997). However, MeSA has also been identified in laboratory measurements as an insect-induced plant–plant signalling compound (Kessler and Baldwin, 2001; Park et al., 2007). Moreover, if the plant is infested with herbivorous insects, the release of MeSA may aid recruiting plant bodyguards (whether predators or parasitoids) (De Boer and Dicke, 2004). Finally, evidence is now collected from field measurements, suggesting that MeSA may also respond to abiotic stresses, for example, being induced as a response to chilling events in combination with drought stress (Karl et al., 2008). Jasmonic acid and methyl jasmonate (MeJA), the latter being considered BVOC, are also produced by plants in response to many biotic and abiotic stresses (particularly herbivory and wounding), and build up in the damaged parts of the plant inducing cross-tolerance. Laboratory measurements with tomato fruit showed that MeJA and MeSA can increase the resistance to chilling injury by inducing pathogenesis-related protein genes (Ding et al., 2002). Signalling crosstalk between salicylic and jasmonic acids in response to biotic stresses is also described by Thaler et al. (2012).

4.4 BVOC AND CLIMATE CHANGE: WARMER = MORE FRAGRANT WORLD?

As temperature dependence is the main driver of BVOC emissions, future scenarios of increased global temperatures up to 4 °C above current averages (IPCC, 2007) may lead to a substantial increase of BVOC emission (Figure 4.3). Rising temperature may also expand the vegetative season, in turn, causing BVOC emissions to spread beyond the current seasonality, especially in boreal ecosystems (Peñuelas et al., 2009). On the other hand, monoterpene emissions by Mediterranean vegetation may be further stimulated by co-occurrence of pollution and high irradiance, two more global change factors that are known to elicit positive responses in this class of BVOC (Loreto et al., 1996, 2004). When alluding to the typical fragrance of many monoterpenes, this scenario may lead to a 'more fragrant' world, as discussed by Peñuelas and Staudt (2010). Increased emissions in response to short-term increases of temperature have been well documented, the total BVOC emission being predicted to increase by up to 30–45% (Peñuelas and Llusia, 2003). Short-term acclimation to high temperatures might also be predicted because of modifications in the expression levels of the terpene synthases (Blanch et al., 2011; Sharkey et al., 1999). However, experiments carried out in our laboratory have shown no acclimation of isoprene to seasonally higher temperatures, for example, during summer in southern Europe (Centritto et al., 2011). Longer-term and field studies are still missing, and are deeply needed, to assess whether acclimation to higher average temperatures will also affect future BVOC emissions, reducing their foreseen increase.

If temperature and pollution concur in stimulating the secondary metabolisms that produce BVOC, some other consequences of climate change could influence BVOC emissions negatively. For instance, enhanced drought stress in Mediterranean ecosystems could affect both primary and secondary metabolism. Isoprenoid metabolism seems to be resistant to drought, but emission of isoprene may be curbed by reduced protein and substrate levels (Fortunati et al., 2008; Rennenberg et al., 2006). Recovery from drought may transiently stimulate isoprene emission (Brilli et al., 2007; Sharkey and Loreto, 1993) but at the same time inhibit isoprene response to temperature (Fortunati et al., 2008). Rising CO_2 is another important component of global changes (King et al., 2013, this vol.). If in one side CO_2 increases the productivity of plant ecosystems (Korner, 2006), on the other side, many laboratory and field experiments have shown important isoprene inhibition under high CO_2 (Rosenstiel et al., 2003; Way et al., 2011; Figure 4.3). However, the negative effect of CO_2 does not show on all experiments with isoprene (e.g. it was not seen in field-grown poplars by Loreto et al., 2007) and does not significantly affect monoterpenes. If rising CO_2 truly inhibits isoprene emission, this effect may have important implications for future emission scenarios, especially when incorporated in regional or global emission models. Depending on the model used, the CO_2 may fully compensate for the temperature effect and rates of isoprene emissions over Europe might be equal or even lower than present (Arneth et al., 2008). Young et al. (2009) predicted a decrease in tropospheric ozone concentration in polluted source regions when accounting for a possibly reduced amount of isoprene emitted into the atmosphere, while an increase of ozone concentration was predicted for less-polluted areas due to the isoprene nitrate and peroxyacetylnitrate chemistry.

Two more factors in climate change, the increased availability of nitrogen because of over-fertilisation (see Eugster and Haeni, 2013, this vol.) and the increase of UV radiation, have been investigated with respect to their impact on BVOC emission. Higher emissions of isoprenoids were observed at increasing nitrogen fertilisation in the greenhouse (Lerdau et al., 1995; Litvak et al., 1996) and in a field experiment on citrus (Fares et al., 2012a,b). However, nitrogen input had minor effects on two Mediterranean tree species in combination with high temperature and drought, inferring the presence of other limiting factors in those conditions (Blanch et al., 2009). Interestingly, also phosphorous availability, which is again assumed to increase in eutrophic conditions, will not enhance isoprene emission and may indeed surprisingly reduce overall emissions (Fares et al., 2008). Concerning UV radiation, UV-B seemed to promote isoprene emission from a subarctic peatland (Tiiva et al., 2007), but UV-A had a negative effect on isoprene emission from poplar saplings (Pallozzi et al., 2013). Overall, we conclude that more research is needed to better understand the effects of UV radiation on BVOC emission.

Clearly, the picture of the impact of future environmental conditions on BVOC emission, and its consequences at the level of ecosystem stability, is still unclear. While more studies may further elucidate mechanisms behind single factor responses, the interactive effect of simultaneously changing factors should be studied in order to get a reasonable estimate of future BVOC levels. From an evolutionary point of view, at least two important issues deserve further investigation: (i) co-evolution of plants and other organisms have occurred so far without human interferences and within the frame of slowly changing environmental conditions. If current predictions of rising BVOC emissions in response to climate change remain confirmed, other components of the ecosystems must perhaps rapidly adapt to more intense plant cues, except when BVOC are quenched by pollutants. (ii) If the emission of volatile isoprenoids truly protects plants against oxidative stresses, then isoprenoid emitters may outcompete non-emitters or low-emitters in a more oxidative atmosphere (Lerdau, 2007). Other components of the ecosystems must then again quickly learn the changes in the plant alphabet to reconstitute trophic interactions before losing fitness and changing ecosystem stability. Changes in biodiversity may result from this BVOC-driven plant competition (more information on future forests under climate changes are provided by Knoke and Hahn, 2013, this vol.). Future actions from stakeholders and policy makers should take into account the likely essential role of isoprenoids in protecting plants, especially in polluted areas. As an example, this may influence the choice of plant species for green infrastructures for urban and periurban planning.

REFERENCES

Altimir, N., Kolari, P., Tuovinen, J.P., Vesala, T., Bäck, J., Suni, T., Kulmala, M., Hari, P., 2006. Foliage surface ozone deposition: a role for surface moisture? Biogeosciences 3, 1–20.

Arneth, A., Monson, R.K., Schurgers, G., Niinemets, Ü., Palmer, P.I., 2008. Why are estimates of global terrestrial isoprene emissions so similar (and why is this not so for monoterpenes)? Atmos. Chem. Phys. 8, 4605–4620.

Blanch, J.S., Penuelas, J., Sardans, L., Llusia, J., 2009. Drought, warming and soil fertilization effects on leaf volatile terpene concentrations in *Pinus halepensis* and *Quercus ilex*. Acta Physiol. Plant. 31, 207–218.

Blanch, J.S., Llusia, J., Niinemets, U., Noe, S.M., Penuelas, J., 2011. Instantaneous and historical temperature effects on alpha-pinene emissions in Pinus halepensis and Quercus ilex. J. Environ. Biol. 32 (1), 1–6.

Brilli, F., Barta, C., Fortunati, A., Lerdau, M., Loreto, F., Centritto, M., 2007. Response of isoprene emission and carbon metabolism to drought in white poplar saplings. New Phytol. 175, 244–254.

Brilli, F., Ciccioli, P., Frattoni, M., Prestininzi, M., Spanedda, A.F., Loreto, F., 2009. Constitutive and herbivore-induced monoterpenes emitted by *Populus* × *euroamericana* leaves are key volatiles that orient *Chrysomela populi* beetles. Plant Cell Environ. 32, 542–552.

Brilli, F., Ruuskanen, T.M., Schnitzhofer, R., Müller, M., Breitenlechner, M., Bittner, V., Wohlfahrt, G., Loreto, F., Hansel, A., 2012. Detection of plant volatiles after leaf wounding and darkening by proton transfer reaction "time-of-flight" mass spectrometry (PTR-TOF). PLoS One 6 (5), 20419.

Centritto, M., Brilli, F., Fodale, R., Loreto, F., 2011. Different sensitivity of isoprene emission, respiration and photosynthesis to high growth temperature coupled with drought stress in black poplar (*Populus nigra*) saplings. Tree Physiol. 31, 275–286.

Chameides, W.L., Lindsay, R.W., Richardson, J., Kiang, C.S., 1988. The role of biogenic hydrocarbons in urban photochemical smog: Atlanta as a case study. Science 241, 1473–1475.

Ciccioli, P., Brancaleoni, E., Frattoni, M., Di Palo, V., Valentini, R., Tirone, G., Seufert, G., Bertin, N., Hansen, U., Csiky, O., Lenz, R., Sharma, M., 1999. Emission of reactive terpene compounds from orange orchards and their removal by within-canopy processes. J. Geophys. Res. 104, 8077–8094.

Cieslik, S., Tuovinen, J.-P., Baumgarten, M., Matyssek, R., Brito, P., Wieser, G., 2013. Gaseous exchange between forests and the atmosphere. Dev. Environ. Sci. 13, 19–36.

Coe, H., Gallagher, W., Choularton, W., Dore, C., 1995. Canopy scale measurements of stomatal and cuticular O-3 uptake by Sitka spruce. Atmos. Environ. 29, 1413–1423.

Cojocariu, C., Esher, P., Heinz-Haberle, K., Matyssek, R., Rennemberg, H., Kreuzwieser, J., 2005. The effect of ozone on the emission of carbonyls from leaves of adult *Fagus sylvatica*. Plant Cell Environ. 28, 603–611.

Davison, B., Taipale, R., Langford, B., Misztal, P., Fares, S., Matteucci, G., Loreto, F., Cape, J.N., Rinne, J., Hewitt, C.N., 2009. Concentrations and fluxes of biogenic volatile organic compounds above a Mediterranean macchia ecosystem in western Italy. Biogeosciences 6, 1655–1670.

De Boer, J.G., Dicke, M., 2004. Experience with methyl salicylate affects behavioral responses of a predatory mite to blends of herbivore-induced plant volatiles. Entomol. Exp. Appl. 110, 181–189.

De Gouw, J.A., Howard, C.J., Custer, T.J., Fall, R., 1999. Emissions of volatile organic compounds from cut grass and clover are enhanced during the drying process. Geophys. Res. Lett. 26 (7), 811–814.

Dicke, M., Baldwin, I.T., 2010. The evolutionary context for herbivore-induced plant volatiles: beyond the 'cry for help'. Trends Plant Sci. 15, 167–175.

Dicke, M., van Loon, J.J.A., 2000. Multitrophic effects of herbivore-induced plant volatiles in an evolutionary context. Entomol. Exp. Appl. 97, 237–249.

Dicke, M., Sabelis, M.W., Takabayashi, J., Bruin, J., Posthumus, M.A., 1990. Plant strategies of manipulating predator-prey interactions through allelochemicals: prospects for application in pest control. J. Chem. Ecol. 16, 3091–3118.

Ding, C.K., Wang, C.Y., Gross, K.C., Smith, D.L., 2002. Jasmonate and salicylate induce the expression of pathogenesis-related-protein genes and increase resistance to chilling injury in tomato fruit. Planta 214, 895–901.

Duhl, T.R., Helmig, D., Guenther, A., 2008. Sesquiterpene emissions from vegetation: a review. Biogeosciences 5, 761–777.

Eugster, W., Haeni, M., 2013. Nutrients or pollutants? Nitrogen deposition to European forests. Dev. Environ. Sci. 13, 37–56.

Fall, R., 2003. Abundant oxygenates in the atmosphere: a biochemical perspective. Chem. Rev. 103, 4941–4951.

Fall, R., Karl, T., Hansel, A., Jordan, A., Lindinger, W., 1999. Volatile organic compounds emitted after leaf wounding: on-line analysis by proton-transfer-reaction mass spectrometry. J. Geophys. Res. 104, 15963–15974.

Fares, S., Brilli, F., Noguès, I., Velikova, V., Tsonev, T., Dagli, S., Loreto, F., 2008. Isoprene emission and primary metabolism in *Phragmites australis* grown under different phosphorous levels. Plant Biol. 10, 38–43.

Fares, S., McKay, M., Holzinger, R., Goldstein, A.H., 2010. Ozone fluxes in a Pinus ponderosa ecosystem are dominated by non-stomatal processes: evidence from long-term continuous measurements. Agr. Forest Meteorol. 150, 420–431.

Fares, S., Gentner, D.R., Park, J.H., Ormeno, E., Karlik, J., Goldstein, A.H., 2011. Biogenic 5 emissions from Citrus species in California. Atmos. Environ. 45, 4557–4568.

Fares, S., Weber, R.J., Park, J.H., Gentner, D.R., Karlik, J., Goldstein, A.H., 2012a. Ozone deposition to an orange orchard: partitioning between stomatal and non-stomatal sinks. Environ. Pollut. 169, 258–266.

Fares, S., Park, J.H., Gentner, D., Weber, R., Ormeno, E., Karlik, J., Goldstein, A.H., 2012b. Seasonal cycles of biogenic volatile organic compound fluxes and concentrations in a California citrus orchard. Atmos. Chem. Phys. 12, 9865–9880.

Fortunati, A., Barta, C., Brilli, F., Centritto, M., Zimmer, I., Schnitzler, J.P., Loreto, F., 2008. Isoprene emission is not temperature-dependent during and after severe drought-stress: a physiological and biochemical analysis. Plant J. 55, 687–697.

Fowler, D., Pilegaard, K., Sutton, M.A., Ambus, P., Raivonen, M., Duyzer, J., Simpson, D., Fagerli, H., Fuzzi, S., et al., 2009. Atmospheric composition change: ecosystems–atmosphere interactions. Atmos. Environ. 43 (33), 5193–5267.

Galbally, I.E., Kirstine, W., 2002. The production of methanol by flowering plants and the global cycle of methanol. J. Atmos. Chem. 43, 195–229.

Gerosa, G., Vitale, M., Finco, A., Manes, F., Ballarin Denti, A., Cieslik, S., 2005. Ozone uptake by an evergreen Mediterranean Forest (*Quercus ilex*). Part I. Micrometeorological flux measurements and flux partitioning. Atmos. Environ. 39, 3255–3266.

Goldstein, A.H., Schade, G.W., 2000. Quantifying biogenic and anthropogenic contributions to acetone mixing ratios in a rural environment. Atmos. Environ. 34 (29–30), 4997–5006.

Goldstein, A.H., Galbally, I.E., 2007. Known and unexplored organic constituents in the earth's atmosphere. Environ. Sci. Technol. 41 (5), 1514–1521.

Graus, M., Schnitzler, J.P., Hansel, A., Cojocariu, C., Rennenberg, H., Wisthaler, A., Kreuzwieser, J., 2004. Transient release of oxygenated volatile organic compounds during light-dark transitions in Grey poplar leaves. Plant Physiol. 135 (4), 1967–1975.

Guenther, A., Karl, T., Harley, P., Wiedinmyer, C., Palmer, P.I., Geron, C., 2006. Estimates of global terrestrial isoprene emissions using MEGAN (Model of Emissions of Gases and Aerosols from Nature). Atmos. Chem. Phys. 6, 3181–3210.

Harley, P., Guenther, A., Zimmerman, P., 1996. Effects of light, temperature and canopy position on net photosynthesis and isoprene emission from sweetgum (*Liquidambar styraciflua*) leaves. Tree Physiol. 16 (1e2), 25–32.

Henze, D.K., Seinfeld, J.H., 2006. Global secondary organic aerosol from isoprene oxidation. Geophys. Res. Lett. 33, L09812. http://dx.doi.org/10.1029/2006GL025976.

Hogg, A., Uddling, J., Ellsworth, D., Carroll, M.A., Pressley, S., Lamb, B., Vogel, C., 2007. Stomatal and non-stomatal fluxes of ozone to a northern mixed hardwood forest. Tellus 59B, 514–525.

Huve, K., Christ, M., Kleist, E., Niinemets, U., Uerlings, R., Walter, A., Wildt, J., 2007. Simultaneous growth and emission measurements demonstrate an interactive control of methanol release by leaf expansion and stomata. J. Exp. Bot. 58 (7), 1783–1793.

IPCC, 2007. The physical science basis. Contribution of working group I. In: Solomon, S. et al., (Eds.), Fourth Assessment Report of the Intergovernmental Panel on Climate Change. Cambridge University Press, Oxford, UK, pp. 1–996.

Jacob, D.J., Field, B.D., Jin, E.M., Bey, I., Li, Q., Logan, J.A., Yantosca, R.M., Singh, H.B., 2002. Atmospheric budget of acetone. J. Geophys. Res. 107 (D10), 4100.

Jacob, D.J., Field, B.D., Li, Q., Blake, D.R., de Gouw, J., Warneke, C., Hansel, A., Wisthaler, A., Singh, H.B., Guenther, A., 2005. Global budget of methanol: constraints from atmospheric observations. J. Geophys. Res. 110 (D8), D08303.

Jardine, K., Abrell, L., Jardine, A., Huxman, T., Saleska, S., Arneth, A., Monson, R., Karl, T., Fares, S., Loreto, F., Goldstein, A., 2012. Within-plant isoprene oxidation confirmed by direct emissions of oxidation products methyl vinyl ketone and methacrolein. Glob. Change Biol. 18, 973–984.

Jordan, A., Haidacher, S., Hanel, G., Hartungen, E., Märk, L., Seehauser, H., Schottkowsky, R., Sulzer, P., Märk, T.D., 2009. A high resolution and high sensitivity time-of-flight proton-transfer-reaction mass spectrometer (PTR-TOF-MS). Int. J. Mass Spectrom. 286, 122–128.

Kanakidou, M., Seinfeld, J.H., Pandis, S.N., Barnes, I., Dentener, F.J., Facchini, M.C., et al., 2005. Organic aerosol and global climate modelling: a review. Atmos. Chem. Phys. 5, 1053–1123.

Karl, T., Guenther, A., Lindinger, C., Jordan, A., Fall, R., Lindinger, W., 2001. Eddy covariance measurements of oxygenated volatile organic compound fluxes from crop harvesting using a redesigned proton-transfer-reaction mass spectrometer. J. Geophys. Res. Atmos. 106 (D20), 24157–24167.

Karl, T., Curtis, A.J., Rosenstiel, T.N., Monson, R.K., Fall, R., 2002a. Transient releases of acetaldehyde from tree leaves—products of a pyruvate overflow mechanism? Plant Cell Environ. 25, 1121–1131.

Karl, T.G., Spirig, C., Rinne, J., Stroud, C., Prevost, P., Greenberg, J., Fall, R., Guenther, A., 2002b. Virtual disjunct eddy covariance measurements of organic compound fluxes from a subalpine forest using proton transfer reaction mass spectrometry. Atmos. Chem. Phys. 2, 279–291.

Karl, T., Guenther, A., Turnipseed, A., Patton, E.G., Jardine, K., 2008. Chemical sensing of plant stress at the ecosystem scale. Biogeosciences 5, 1287–1294.

Kesselmeier, J., Staudt, M., 1999. Biogenic volatile organic compounds (VOC): an overview on emission, physiology and ecology. J. Atmos. Chem. 33, 23–88.

Kessler, A., Baldwin, I.T., 2001. Defensive function of herbivore-induced plant volatile emissions in nature. Science 291, 2141–2144.

King, J., Liu, L., Aspinwall, M., 2013. Tree and forest responses to interacting elevated atmospheric CO_2 and tropospheric O_3: a synthesis of experimental evidence. Dev. Environ. Sci. 179–208.

Knoke, T., Hahn, A., 2013. Global change and the role of forests in future land-use systems. Dev. Environ. Sci. 13, 569–587.

Knudsen, J.T., Gershenzon, J., 2006. The chemical diversity of floral scent. In: Dudareva, N., Pichersky, E. (Eds.), Biology of Floral Scent. Taylor & Francis, Oxford, UK, pp. 27–52.

Knudsen, J.T.L., Tollsten, L., Bergström, G., 1993. Floral scents—a checklist of volatile compounds isolated by head-space techniques. Phytochemistry 33, 253–280.

Koppmann, R., Wildt, J., 2007. Oxygenated volatile organic compounds. In: Koppmann, R. (Ed.), Volatile Organic Compounds in the Atmosphere. Blackwell Publishing Ltd., Oxford, UK, pp. 129–172 (Chapter 4).

Korner, C., 2006. Plant CO_2 responses: an issue of definition, time and resource supply. New Phytol. 172, 393–411.

Kos, M., van Loon, J.J.A., Dicke, M., Vet, L.E.M., 2009. Transgenic plants as vital components of integrated pest management. Trends Biotechnol. 27, 621–627.

Kreuzwieser, J., Scheerer, U., Rennenberg, H., 1999. Metabolic origin of acetaldehyde emitted by poplar (*Populus tremula* x *P. alba*) trees. J. Exp. Bot. 50, 757–765.

Kurpius, M.R., Goldstein, A.H., 2003. Gas-phase chemistry dominates O_3 loss to a forest, implying a source of aerosols and hydroxyl radicals to the atmosphere. Geophys. Res. Lett. 30 (7), 1371, 24–1, 24–4.

Laule, O., Furholz, A., Chang, H.S., Zhu, T., Wang, X., Heifetz, P.B., Gruissem, W., Lange, M., 2003. Crosstalk between cytosolic and plastidial pathways of isoprenoid biosynthesis in *Arabidopsis thaliana*. Proc. Natl. Acad. Sci. U.S.A. 100, 6866–6871.

Lee, A., Goldstein, A.H., Keywood, M.D., Gao, S., Varutbangkul, V., Bahreini, R., Ng, N.L., Flagan, R.C., Seinfeld, J.H., 2006. Gas-phase products and secondary aerosol yields from the ozonolysis of ten different terpenes. J. Geophys. Res. 111, D07302. http://dx.doi.org/10.1029/2005JD006437.

Lerdau, M., 2007. A positive feedback with negative consequences. Science 316, 212–213.

Lerdau, M., Matson, P., Fall, R., Monson, R., 1995. Ecological controls over monoterpene emissions from Douglas-fir (*Pseudotzuga menziesii*). Ecology 76 (5), 2640–2647.

Li, T., Holopainen, J.K., Kokko, H., Tervahauta, A.J., Blande, J.D., 2012. Herbivore-induced aspen volatiles temporally regulate two different indirect defences in neighbouring plants. Funct. Ecol. 26, 1176–1185.

Lichtenthaler, H.K., Schwendler, J., Disch, A., Rohmer, M., 1997. Biosynthesis of isoprenoids in higher plant chloroplasts proceeds via a mevalonate-independent pathway. FEBS Lett. 400, 271–274.

Lindinger, W., Hansel, A., Jordan, A., 1998. On-line monitoring of volatile organic compounds at pptv levels by means of proton-transfer-reaction mass spectrometry (PTR-MS). Medical applications, food control and environmental research. Int. J. Mass. Spectrom. Ion. Process. 173, 191–241.

Litvak, M.E., Loreto, F., Harley, P.C., Sharkey, T.D., Monson, R.K., 1996. The response of isoprene emission rate and photosynthetic rate to photon flux and nitrogen supply in aspen and white oak trees. Plant Cell Environ. 19, 549–559.

Loivamäki, M., Mumm, R., Dicke, M., Schnitzler, J.-P., 2008. Isoprene interferes with the attraction of bodyguards by herbaceous plants. Proc. Natl. Acad. Sci. U.S.A. 105, 17430–17435.

Loreto, F., Velikova, V., 2001. Isoprene produced by leaves protects the photosynthetic apparatus against ozone damage, quenches ozone products, and reduces lipid peroxidation of cellular membranes. Plant Physiol. 127, 1781–1787.

Loreto, F., Fares, S., 2007. Is ozone flux inside leaves only a damage indicator? Clues from volatile isoprenoid studies. Plant Physiol. 143, 1096–1100.

Loreto, F., Schnitzler, J.P., 2010. Abiotic stresses and induced BVOCs. Trends Plant Sci. 15 (3), 154–166.

Loreto, F., Sharkey, T.D., 1990. A gas exchange study of photosynthesis and isoprene emission in red oak (*Quercus rubra* L.). Planta 182, 523–531.

Loreto, F., Ciccioli, P., Cecinato, A., Brancaleoni, E., Frattoni, M., Tricoli, D., 1996. Influence of environmental factors and air composition on the emission of alpha-pinene from *Quercus ilex* leaves. Plant Physiol. 110 (1), 267–275.

Loreto, F., Pinelli, P., Manes, F., Kollist, H., 2004. Impact of ozone on monoterpene emissions and evidences for an isoprene-like antioxidant action of monoterpenes emitted by *Quercus ilex* (L.) leaves. Tree Physiol. 24, 361–367.

Loreto, F., Barta, C., Brilli, F., Nogues, I., 2006. On the induction of volatile organic compound emissions by plants as consequence of wounding or fluctuations of light and temperature. Plant Cell Environ. 29, 1820–1828.

Loreto, F., Centritto, M., Barta, C., Calfapietra, C., Fares, S., Monson, R.K., 2007. The relationship between isoprene emission rate and dark respiration rate in white poplar (*Populus alba* L.) leaves. Plant Cell Environ. 30, 662–669.

McFrederick, Q.S., Kathilankal, J.C., Fuentes, J.D., 2008. Air pollution modifies floral scent trails. Atmos. Environ. 42 (10), 2336–2348.

Millet, D.B., Jacob, D.J., Custer, T.G., de Gouw, J.A., Goldstein, A.H., Karl, T., Singh, H.B., Sive, B.C., Talbott, R.W., Warneke, C., Williams, J., 2008. New constraints on terrestrial and oceanic sources of atmospheric methanol. Atmos. Chem. Phys. 8, 6887–6905.

Millet, D.B., Atlas, E.L., Blake, D.R., Blake, N.J., Diskin, G.S., Holloway, J.S., Hudman, R.C., Meinardi, S., Ryerson, T.B., Sachse, G.W., 2009. Halocarbon emissions from the United States and Mexico and their global warming potential. Environ. Sci. Technol. 43 (4), 1055–1060.

Monson, R.K., Jaeger, C., Adams, W., Driggers, E., Silver, G., Fall, R., 1992. Relationship among isoprene emission rate, photosynthesis and isoprene synthase activity as influenced by temperature. Plant Physiol. 92, 1175–1180.

Monson, R., Grote, R., Niinemets, U., Schnitzler, J.P., 2012. Modeling the isoprene emission rate from leaves. New Phytol. 195, 541–559.

Müller, M., Graus, M., Ruuskanen, T.M., Schnitzhofer, R., Bamberger, I., Kaser, L., Titzmann, T., Hörtnagl, L., Wohlfahrt, G., Karl, T., Hansel, A., 2010. First eddy covariance flux measurements by PTR-TOF. Atmos. Meas. Tech. 3, 387–395.

Ng, N.L., Kroll, J.H., Keywood, M.D., Bahreini, R., Varutbangkul, V., Flagan, R.C., Seinfeld, J.H., Lee, A., Goldstein, A.H., 2006. Contribution of first- versus second generation products to secondary organic aerosols formed in the oxidation of biogenic hydrocarbons. Environ. Sci. Technol. 40 (7), 2283–2297.

Niinemets, U., Loreto, F., Reichstein, M., 2004. Physiological and physico-chemical controls on foliar volatile organic compound emissions. Trends Plant Sci. 9, 180–186.

Niinemets, U., Arneth, A., Kuhn, U., Monson, R.K., Penuelas, J., Staudt, M., 2010. The emission factor of volatile isoprenoids: stress, acclimation, and developmental responses. Biogeosciences 7, 2203–2223. http://dx.doi.org/10.5194/bg-7-2203-2010.

Obendorf, R.L., 1990. Methanol accumulation in maturing seeds. J. Exp. Bot. 41, 489–495.

Ormeño, E., Gentner, D.R., Fares, S., Karlik, J., Park, J.H., Goldstein, A.H., 2010. Sesquiterpenoid emissions from agricultural crops: correlations to monoterpenoid emissions and leaf terpene content. Environ. Sci. Technol. 44, 3758–3764.

Pallozzi, E., Fortunati, A., Marino, G., Loreto, F., Agati, G., Centritto, M., 2013. BVOC emission from *Populus* × *canadensis* saplings in response to acute UV-A radiation. Physiol. Plant. 148 (1), 51–61. http://dx.doi.org/10.1111/j.1399-3054.2012.01687.x.

Papiez, M.R., Potosnak, M.J., Guenther, A.B., Matsunaga, S.N., Stockwell, W.R., 2009. The impacts of reactive terpene emissions from plants on air quality in Las Vegas, Nevada. Atmos. Environ. 43 (27), 4109–4123.

Park, S.W., Kaimoyo, E., Kumar, D., Mosher, S., Klessig, D.F., 2007. Methyl salicylate is a critical mobile signal for plant systemic acquired resistance. Science 318, 113–116.

Park, J.-H., Goldstein, A.H., Timkovsky, J., Fares, S., Weber, R., Karlik, J., Holzinger, R., 2012. Eddy covariance emission and deposition flux measurements using proton transfer reaction-time of flight-mass spectrometry (PTR-TOF-MS): comparison with PTR-MS measured vertical gradients and fluxes. Atmos. Chem. Phys. Discuss. 12, 20435–20482. http://dx.doi.org/10.5194/acpd-12-20435-2012.

Peñuelas, J., Llusia, J., 2003. BVOCs: plant defense against climate warming? Trends Plant Sci. 8, 105–109.

Peñuelas, J., Staudt, M., 2010. BVOCs and global change. Trends Plant Sci. 15, 133–144.

Peñuelas, J., Rutishauser, T., Filella, I., 2009. Phenology feedbacks on climate change. Science 324, 887–888.

Pinto, D.M., Blande, J.M., Nykänen, R., Dong, W.X., Nerg, A.M., Holopainen, J.K., 2007. Ozone degrades common herbivore-induced plant volatiles: does this affect herbivore prey location by predators and parasitoids? J. Chem. Ecol. 33, 683–694.

Rennenberg, H., Loreto, F., Polle, S.A., Brilli, F., Fares, S., Beniwal, R.S., Gessler, A., 2006. Physiological responses of forest trees to heat and drought. Plant Biol. 8 (5), 556–571.

Riemer, D., Pos, W., Milne, P., Farmer, C., Zika, R., Apel, E., Olszyna, K., Kliendienst, T., Lonneman, W., Bertman, S., Shepson, P., Starn, T., 1998. Observations of nonmethane hydrocarbons and oxygenated volatile organic compounds at a rural site in the southeastern United States. J. Geophys. Res. 103 (D12), 28111–28128.

Rosenstiel, T.N., Potosnak, M.J., Griffin, K.L., Fall, R., Monson, R.K., 2003. Increased CO_2 uncouples growth from isoprene emission in an agriforest ecosystem. Nature 421, 256–259.

Sallaud, C., Rontein, D., Onillon, S., Jabes, F., Duffe, P., Giacalone, C., Thoraval, S., Escoffier, C., Herbette, G., Leonhardt, N., Causse, M., Tissier, A., 2009. A novel pathway for sesquiterpene biosynthesis from Z, Z-farnesyl pyrophosphate in the wild tomato *Solanum habrochaites*. Plant Cell 21, 301–317.

Schuhr, C.A., Radykewicz, T., Sagner, S., Latzel, C., Zenk, M.H., Arigoni, D., Bacher, A., Rohdich, F., Eisenreich, W., 2003. Quantitative assessment of crosstalk between the two isoprenoid biosynthesis pathways in plants by NMR spectroscopy. Phytochem. Rev. 2, 3–16.

Sharkey, T.D., Loreto, F., 1993. Water stress, temperature, and light effects on the capacity for isoprene emission and photosynthesis of kudzu leaves. Oecologia 95, 328–333.

Sharkey, T.D., Yeh, S., 2001. Isoprene emission from plants. Annu. Rev. Plant. Physiol. Plant. Mol. Biol. 52, 407–436.

Sharkey, T., Singsaas, E.L., Lerdau, M.T., Geron, C.D., 1999. Weather effects on isoprene emission capacity and applications in emissions algorithms. Ecol. Appl. 9, 1132–1137.

Shaw, S., Gantt, B., Meskhidze, N., 2010. Production and emissions of marine isoprene and monoterpenes: a review. Adv. Meteorol. http://dx.doi.org/10.1155/2010/408696, Article ID 408696.

Shu, Y., Atkinson, R., 1994. Rate constants for the gas-phase reactions of O_3 with a series of terpenes and OH radical formation from the O_3 reactions with sesquiterpenes at 296 K. Int. J. Chem. Kinet. 26, 1193–1205.

Shulaev, V., Silverman, P., Raskin, I., 1997. Airborne signalling by methyl salicylate in plant pathogen resistance. Nature 385, 718–721.

Singh, H., Chen, Y., Staudt, A., Jacob, D., Blake, D., Heikes, B., Snow, J., 2001. Evidence from the Pacific troposphere for large global sources of oxygenated organic compounds. Nature 410 (6832), 1078–1081.

Singsaas, E.L., Sharkey, T.D., 1998. The regulation of isoprene emission responses to rapid leaf temperature fluctuations. Plant Cell Environ. 21, 1181–1188.

Singsaas, E.L., Lerdau, M., Winter, K., Sharkey, T.D., 1997. Isoprene increases thermotolerance of isoprene-emitting species. Plant Physiol. 115, 1413–1420.

Steeghs, M., Bais, H.P., de Gouw, J., Goldan, P., Kuster, W., Northway, M., Fall, R., Vivanco, J.M., 2004. Proton-transfer-reaction mass spectrometry as a new tool for real time analysis of root-secreted volatile organic compounds in Arabidopsis. Plant Physiol. 135, 47–58.

Steiner, A., Goldstein, A.H., 2007. Biogenic volatile organic compounds. In: Koppmann, R. (Ed.), Volatile Organic Compounds in the Atmosphere. Blackwell Publishing Ltd., Oxford, UK.

Thaler, J.S., Parris, T.H., Whiteman, N.K., 2012. Evolution of jasmonate and salicylate signal crosstalk. Trends Plant Sci. 17 (5), 260–270.

Tholl, D., Chen, F., Petri, J., Gershenzon, J., Pichersky, E., 2005. Two sesquiterpene syntheses are responsible for the complex mixture of sesquiterpenes emitted from *Arabidopsis* flowers. Plant J. 42, 757–771.

Tholl, D., Boland, W., Hansel, A., Loreto, F., Rose, U.S.R., Schnitzler, J.P., 2006. Practical approaches to plant volatile analysis. Plant J. 45, 540–560.

Tiiva, P., Rinnan, R., Faubert, P., Räsänen, J., Holopainen, T., Kyrö, E., Holopainen, J.K., 2007. Isoprene emission from a subarctic peatland under enhanced UV-B radiation. New Phytol. 176, 346–355.

Tingey, D., Manning, M., Grothaus, L., Burns, W., 1980. Influence of light and temperature on monoterpene emission rates from slash pine. Plant Physiol. 65, 797–801.

Tuovinen, J.-P., Hakola, H., Karlsson, P.E., Simpson, D., 2013. Air pollution risks to northern European forests in a changing climate. Dev. Environ. Sci. 13, 77–99.

Velikova, V., Edreva, A., Loreto, F., 2004. Endogenous isoprene protects *Phragmites australis* leaves against singlet oxygen. Physiol. Plant. 122, 219–225.

Velikova, V., Várkonyi, Z., Szabó, M., Maslenkova, L., Nogues, I., Kovács, L., Peeva, V., Busheva, M., Garab, G., Sharkey, T.D., Loreto, F., 2011. Increased thermostability of thylakoid membranes in isoprene-emitting leaves probed with three biophysical techniques. Plant Physiol. 157, 905–916.

Velikova, V., Sharkey, T.D., Loreto, F., 2012. Stabilization of thylakoid membranes in isoprene-emitting plants reduces formation of reactive oxygen species. Plant Signal. Behav. 7 (1), 139–141.

Vickers, C.E., Gershenzon, J., Lerdau, M.T., Loreto, F., 2009. A unified mechanism of action for isoprenoids in plant abiotic stress. Nat. Chem. Biol. 5, 283–291.

Way, D.A., Schnitzler, J.P., Monson, R.K., Jackson, R.B., 2011. Enhanced isoprene-related tolerance of heat- and light-stressed photosynthesis at low, but not high, CO_2 concentrations. Oecologia 166, 273–282.

Young, P.J., Arneth, A., Schurgers, G., Zeng, G., Pyle, J.A., 2009. The CO_2 inhibition of terrestrial isoprene emission significantly affects future ozone projections. Atmos. Chem. Phys. 9, 2793–2803.

Zeller, K.F., Nikolov, N.T., 2000. Quantifying simultaneous fluxes of ozone, carbon dioxide and water vapor above a subalpine forest ecosystem. Environ. Pollut. 107, 1–20.

Zhang, L., Brook, J.R., Vet, R., 2003. A revised parameterization for gaseous dry deposition in air-quality models. Atmos. Chem. Phys. 3, 2067–2082.

Chapter 5

Air Pollution Risks to Northern European Forests in a Changing Climate

Juha-Pekka Tuovinen*[,1], Hannele Hakola*, Per Erik Karlsson[†] and David Simpson[‡,§]

*Finnish Meteorological Institute, Helsinki, Finland

[†]IVL Swedish Environmental Research Institute, P.O. Box 53021, Gothenburg, Sweden

[‡]EMEP MSC-W, Norwegian Meteorological Institute, P.O. Box 43, Oslo, Norway

[§]Chalmers University of Technology, Gothenburg, Sweden

[1]Corresponding author: e-mail: juha-pekka.tuovinen@fmi.fi

Chapter Outline

5.1 INTRODUCTION

The anthropogenic climate change that is driven by greenhouse gas (GHG) emissions and land-use changes poses a global-scale challenge of adaptation and mitigation. However, changes in climate characteristics, such as temperature and precipitation, do not proceed uniformly in time and space. Northern Europe is one of the regions where more intense increases in surface temperatures are predicted than for the global average (Christensen et al., 2007). Moreover, precipitation amounts are predicted to increase within this region, even if slowly, and the periods of snow and ice cover to reduce (Jylhä et al., 2008).

Developments in Environmental Science, Vol. 13. http://dx.doi.org/10.1016/B978-0-08-098349-3.00005-0

The changes in meteorological factors obviously have a direct influence on atmospheric pollution. However, there are also indirect effects, stemming from the influence of climate change on emission patterns, both anthropogenic and biogenic. While the link between energy consumption and climate is evident, generating gradually decreasing emission trends in central and northern Europe (Pilli-Sihvola et al., 2010), more abrupt developments cannot be ruled out. The observed decline of the Arctic sea ice (Comiso, 2012), in particular, is likely to significantly increase the emissions of various air pollutants and climate forcers within this area, as the melting ice leaves the Arctic Ocean increasingly open for human activities (AMAP, 2009).

In addition to anthropogenic emissions, biogenic compounds play an important role in the chemical processing in the atmosphere. Notably, the production rates of many biogenically emitted compounds that act as precursors of ozone (O_3) depend strongly on temperature (Fowler et al., 2009). As O_3 is a secondary photochemical pollutant, produced in the troposphere via oxidation of carbon monoxide (CO), methane and other volatile organic compounds (VOCs) in the presence of nitrogen oxides (NO_x), its concentrations are sensitive to meteorological variability in general. Ozone is perceived as a large-scale air pollution problem in Europe as well as worldwide due to its deleterious effects on human health, vegetation and materials (Matyssek et al., 2013, this vol.; Monks et al., 2009); it is also an efficient GHG.

The terrestrial biosphere is closely coupled with atmospheric composition and climate. Many of the related biogeochemical feedbacks, such as the interactions between O_3 phytotoxicity, carbon (C) sequestration and climate, are poorly understood (Arneth et al., 2010). Nevertheless, O_3 impacts on forest trees are widely documented and the current O_3 levels are known to reduce the biomass of northern temperate and boreal forests (Wittig et al., 2009). Detrimental effects on tree physiology and growth have also been observed in different parts of Fennoscandia (Karlsson et al., 2009; Manninen et al., 2009).

The growth of northern temperate and boreal forests is generally limited by the availability of nitrogen (N) (Tamm, 1991), and thus N fertilisers are applied to a substantial fraction of northern coniferous forests. Atmospheric N deposition explains part of the increased growth rates of European forests (Butterbach-Bahl et al., 2011). On the other hand, excess N deposition can have adverse effects including the risk of N leakage to surface waters, tree nutrient imbalances, increased tree susceptibility to pathogens and pests, and changes in vegetation species composition (Bobbink and Hettelingh, 2011; Eugster and Haeni, 2013, this vol.).

The abundance of N deposition, as well as any anthropogenic air pollution, evidently depends on the efficiency of emission control and the related abatement strategies and air quality legislation. While abatement measures have led to reductions in sulphur dioxide (SO_2) emissions in Europe by well over 70% since 1980, NO_x and ammonia emissions have only decreased by about 30%

(Tørseth et al., 2012). Correspondingly, whereas the acidifying load to ecosystems due to SO_2 emissions has been eliminated to a large extent, N deposition is still so intense as to induce a widespread risk of eutrophication (Fagerli et al., 2012). The O_3 concentration trends resulting from reductions of NO_x and other precursor emissions remain to be fully rationalised (Jonson et al., 2006; Tørseth et al., 2012).

In this chapter, we will analyse results from an atmospheric chemistry-transport model (CTM) to evaluate the risk that O_3 concentrations and N deposition pose to northern European forests. In addition, we will assess how this risk will change as a response to projected modifications to the present climatic and emission patterns. We will concentrate on the European continent north of about 55 °N with a special focus on boreal forests in the Nordic countries, in particular, the northernmost parts of the continent in relation to the developments in the Arctic. The main emphasis is on the O_3-induced risk, but we will also consider the role of N deposition.

In the following sections, we will first discuss in more detail the mechanisms and feedbacks involved in the interactions between climate change and air pollution. After this, we present model calculations of O_3 fluxes and N deposition for present and future conditions. In the final section, we discuss the challenges in assessing the air pollution risks to forests and their C stocks. We also highlight the importance of adaptive capacity of the forest sector to climate change and the synergy between the air quality and climate change policies.

5.2 INTERACTIONS AND FEEDBACKS

As mentioned in Section 5.1, climate change may also trigger abrupt changes, and in this respect, the Arctic must be considered a highly critical area. The warming rate of the Arctic (north of 60 °N) is approximately twice as high as in the Northern Hemisphere on average (Serreze and Barry, 2011). Most notably, the Arctic sea ice is retreating rapidly, the extent of 'perennial' ice (that survives the summer) at a rate exceeding 10% per decade and the thicker multi-year ice even faster (Comiso, 2012). It has been predicted that the Arctic may be nearly ice free in summer before the end of the twenty-first century, possibly as imminently as in the 2030s (Wang and Overland, 2012).

Reduced sea-ice cover is likely to significantly expand shipping operations in the Arctic Basin, potentially making it possible to establish new trade routes. Of particular interest are the trans-Arctic Northeast and Northwest Passages that both reportedly had ice-free conditions in 2008 for the first time in recorded history (AMAP, 2009). According to Corbett et al. (2010), the NO_x emissions from Arctic shipping will increase by a factor of 1.7 and 3.8 by 2030 and 2050, respectively, compared to 2004, and CO emissions correspondingly by a factor of 2.3–5.2. If new shipping routes are introduced, these increases will be considerably higher (Corbett et al., 2010). Large-scale oil and gas extraction are already taking place in the Arctic and improved accessibility to the off-shore resources will provide further potential. However, emission controls are also improving, so atmospheric emissions do not necessarily increase as a result of expanding production (Peters et al., 2011).

Terrestrial ecosystems release vast quantities of biogenic VOCs (BVOCs) and nitrogen oxide into the atmosphere (Cieslik et al., 2013, this vol.; Loreto and Fares, 2013, this vol.). Possible increases in these temperature-controlled emissions on one hand and in the NO_x and CO emissions due to Arctic shipping on the other may enhance regional-scale O_3 formation in the troposphere. Based on a global CTM and a scenario of high emissions, Granier et al. (2006) suggested that, as a consequence of intensified ship operations, surface O_3 concentrations in the North American Arctic could soar to similar levels (40–60 ppb) as those experienced in the industrialised regions in the Northern Hemisphere. In Europe, significant increases in human and ecosystem exposure to O_3 would extend to the boreal regions, thus counteracting the emission reduction commitments of European countries set by the Gothenburg Protocol of the UNECE Convention on Long-range Transboundary Air Pollution (CLRTAP) (Fagerli et al., 2012).

The changing global distributions of O_3 precursor emissions not only give rise to new local and regional air pollution hot spots (Koike et al., 2013, this vol.; Laakso et al., 2013, this vol.; Pandey et al., 2013, this vol.; Pozzer et al., 2012), but they also affect the global background O_3 concentrations (Monks et al., 2009). Atmospheric dynamics control the transport of air masses in which O_3 is formed, thereby introducing considerable natural variability in the O_3 concentrations observed in northern Europe (Colette et al., 2011), as well as in the source–receptor relationships of N compounds (Carmichael et al., 2010). For example, the systematic variation of atmospheric pressure patterns known as the North Atlantic Oscillation (NAO) has an important influence on the long-range transport across the Atlantic Ocean, the intrusion of O_3-enriched stratospheric air into the troposphere and the smaller scale atmospheric processes that control O_3 formation. Future predictions of the summertime NAO suggest an enhanced frequency of anticyclonic weather patterns particularly in northwestern Europe (Folland et al., 2009); these patterns favour positive O_3 anomalies (Pausata et al., 2012). In winter, NAO explains part of the variation in wet N deposition observed on the Norwegian coast (Hole et al., 2008).

As discussed by Cieslik et al. (2013, this vol.), in some regions climate change may partly cancel out the benefit achieved by emission control measures or impose a 'climate penalty' (Wu et al., 2008). Broadly, intensified O_3 destruction by water vapour, diminished dry deposition and enhanced BVOC emissions are expected to increase the mean O_3 concentrations in southern Europe but to decrease them north of 50 °N, although these perturbations are quite limited (Katragkou et al., 2011; Langner et al., 2012a). However, much larger impacts may be experienced on shorter timescales. This became evident during the heat wave in the summer of 2003 (Solberg et al., 2008), which bears a resemblance to warming scenarios for the latter part of the twenty-first century and therefore can be considered to represent conditions that will become more frequent in the future (Coumou and Rahmstorf, 2012; see also Cieslik et al., 2013, this vol.).

Few studies have addressed the combined effect of both climate and precursor emission projections on European near-surface O_3 concentrations. The hemispheric simulations for 2100 by Hedegaard et al. (2013) suggest that climate change may increase the annual mean O_3 concentrations in the Arctic, similar to the enhancement in southern Europe mentioned above. The projected emission reductions (scenario RCP4.5; van Vuuren et al., 2011) have an opposite effect, and thus the overall change remains limited, except for the region of highest precursor emission density. Langner et al. (2012b) estimated significant O_3 reductions across the whole Europe, on average 5 ppb per century for the daily maximum concentration in April–September in the region north of 50 °N, assuming RCP4.5 and a positive trend of 0.1 ppb yr^{-1} for the background O_3 concentration. A global multi-model ensemble indicates broadly similar reductions for the annual mean (Young et al., 2013). However, the potential for drastic changes in the Arctic emissions were not included in these studies.

5.3 RISK OF IMPACTS

5.3.1 Approach

Previous studies on potential harmful vegetation effects of future O_3 concentrations have addressed mainly crops (e.g. Avnery et al., 2011; Hollaway et al., 2012) and those that exist for European forests have typically expressed the risk of damage in terms of the AOT40 index (Colette et al., 2012; Engardt et al., 2009). AOT40 is based on ambient concentrations (exposure) only, but as the plant physiological effects depend on the O_3 uptake via the stomata (dose), the POD_Y (Phytotoxic Ozone Dose over a flux threshold of Y, $m^{-2}\,s^{-1}$) is currently considered a more tenable risk metric (Dizengremel et al., 2013, this vol.; Grünhage et al., 2013, this vol.; Karlsson et al., 2007; UNECE, 2010). This index was used by Klingberg et al. (2010), who calculated $POD_{1.6}$ for selected European locations using CTM-based data in an 'off-line' mode (rather than directly within the CTM).

For the present calculations, we used the EMEP CTM (Simpson et al., 2012), in which the stomatal uptake model constitutes an integral part of the deposition module determining the surface exchange fluxes. Hence, POD_Y is derived as a consistent component of the model dynamics. We calculated the spatial distribution of the POD_1 dose index for the boreal coniferous and deciduous trees for different climate and emission scenarios. The corresponding rates of wet and dry N deposition, both for oxidised and reduced inorganic compounds, are obtained from the same model runs. The spatial resolution of these data is about 50 km × 50 km.

In the first set of calculations, the anthropogenic emissions were kept at the present level to assess the effect of climate change on air pollution, with meteorological input data derived for the periods 2000–2009 and 2040–2049. The meteorological data are based on the SRES A1B scenario, downscaled with the Swedish Rossby Centre regional climate model RCA3, while emissions are from the RCP4.5 scenario for 2000 (van Vuuren et al., 2011). These data and calculations have been described in more detail by Langner et al. (2012a).

In a second set of calculations, we ran the EMEP CTM for the years 2005 and 2030 to investigate the impact of Arctic shipping changes, as provided by the emission inventory of Corbett et al. (2010). We adopted the high-growth scenario, in which the Arctic NO_x emissions are increased to 329 Gg NO_2 yr^{-1} in 2030 (below referred to as 'Arctic ships'). We also estimated the effect of the potential global diversion traffic through emerging routes ('Arctic ships with diversion'), which adds 410 Gg NO_2 yr^{-1}. We assumed that both Northeastern and Northwestern Passages will become navigable by 2030 and that the additional emissions are equally distributed between these two routes (Corbett et al., 2010). The reference case for 2030 ('2030 baseline'), in which no increases in the Arctic shipping are included, was calculated with the emission data from the PRIMES energy model that were used as a basis for the revised Gothenburg Protocol of the CLRTAP (Wagner et al., 2010). All model calculations with the 2030 emissions assume present-day meteorological conditions (year 2005).

It should be noted that future O_3 concentrations may be affected by changes in the tropospheric background O_3 levels (Langner et al., 2012b), for example, as a result of increasing emissions in the rapidly developing parts of Asia (Koike et al., 2013, this vol.; Pandey et al., 2013, this vol.). However, the trends in background O_3 over Europe seem to have flattened out in recent years (Monks et al., 2009). In all model calculations, we therefore adopt the simplifying assumption that the future background concentrations are the same as those for the runs representing the 2000s. Another complication that we do not allow for (for technical reasons) is the soil moisture modifications in the climate change-effect investigations. However, the EMEP CTM does account for the effects of changing temperature and humidity, which are the main factors controlling climate change responses in northern Europe.

The model results are compared with the critical levels (CLs) that for O_3 represent the dose limits above which direct adverse effects on receptors, here forest trees, may occur. The critical POD_1 level of Norway spruce determined within the CLRTAP is 8 mmol m^{-2} (projected leaf area) and that of beech and birch is 4 mmol m^{-2}, as accumulated during a growing season (Mills et al., 2011b). Exceedance of these CLs implies a risk of significant biomass loss. It has been shown previously that a flux-based metric indicates a more widespread risk of damage and a less steep south–north gradient than the concentration-based AOT40 metric (Simpson et al.,

2007). The POD_Y index also better predicts the areas of Europe where O_3-induced damage has been observed for crops and semi-natural vegetation (Mills et al., 2011a).

The estimates of N deposition are compared with the empirical CLs that are based on observed changes in the structure (species abundance, composition and/or diversity) and functioning (N leaching, decomposition or mineralisation rate) of ecosystems. For the forest ecosystems within the study area, this CL ranges from 0.5 to 2.0 g N m^{-2} yr^{-1} depending on the ecosystem type and properties (Bobbink and Hettelingh, 2011).

5.3.2 Climate Change *versus* Emission Reductions

5.3.2.1 Current Situation

The POD_1 calculated with present-time emissions and climate exceeds the CL of the deciduous species (birch and beech) across the whole modelling domain (not shown). Except for a limited area in northernmost Europe, the same is true for the CL derived for Norway spruce (Figure 5.1A). Within the EMEP CTM, the O_3 dose is calculated based on the parameterisations developed for the land cover classification incorporated into the model (Simpson et al., 2012), which slightly differs from the species-specific parameterisations required for a fully consistent comparison with the CLs of UNECE (2010). However, this discrepancy does not prevent us from concluding that, in spite of the fact that O_3 concentrations are not as high as in southern Europe, O_3 uptake is so large as to pose a significant risk to the forests growing in northern latitudes.

The large-scale geographical distribution of the total inorganic N deposition is broadly similar to that of the O_3 dose (Figure 5.1B). In northern Europe, the atmospheric N supply to ecosystems is clearly lower than in the more polluted regions of the continent. The modelled deposition in southern Fennoscandia is within 0.8–1.5 g N m^{-2} yr^{-1}, which is in accordance with the measurements made in Sweden (Pihl Karlsson et al., 2011). This N deposition exceeds the lower limit of the CL that varies within 0.5–1.0 g N m^{-2} yr^{-1} for the northern temperate pine forests as well as for boreal forests in general (Bobbink and Hettelingh, 2011).

5.3.2.2 Climate Change

As discussed above, climate change influences O_3 concentrations and deposition fluxes through a wide range of processes. The present model calculations indicate a change in the mean concentration in a range of −2 .... +2 ppb across Europe from 2000–2009 to 2040–2049 (data not shown, but see Langner et al., 2012a for similar results). In contrast to O_3 concentration showing decreases in the north, and vice versa, POD_1 increases in northern and central Europe (Figure 5.2A). In northern Fennoscandia, where POD_1 of coniferous forests in the early 2000s was less than 10 mmol m^{-2}, these

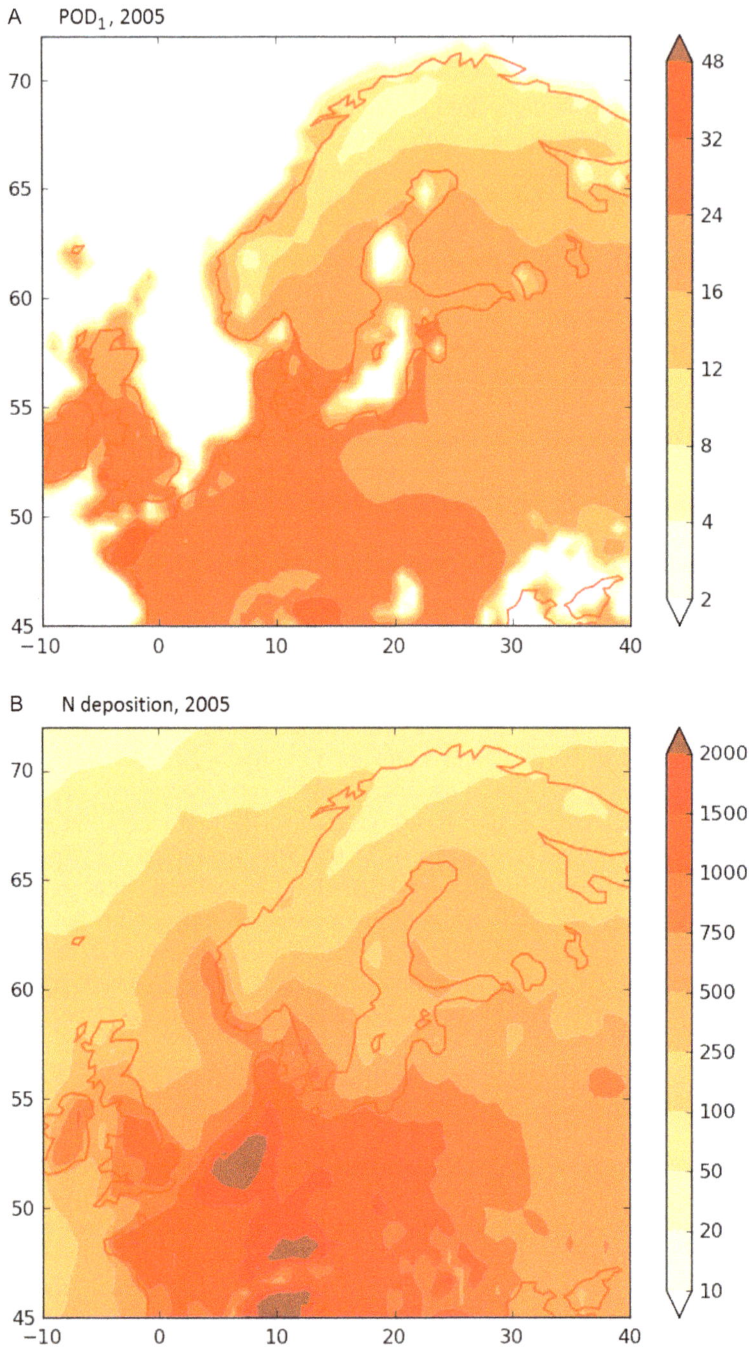

FIGURE 5.1 Current (2005) situation for (A) phytotoxic ozone dose POD_1 for coniferous forests ($mmol\ m^{-2}$) and (B) total (reduced + oxidised) annual deposition of nitrogen ($mg\ m^{-2}$).

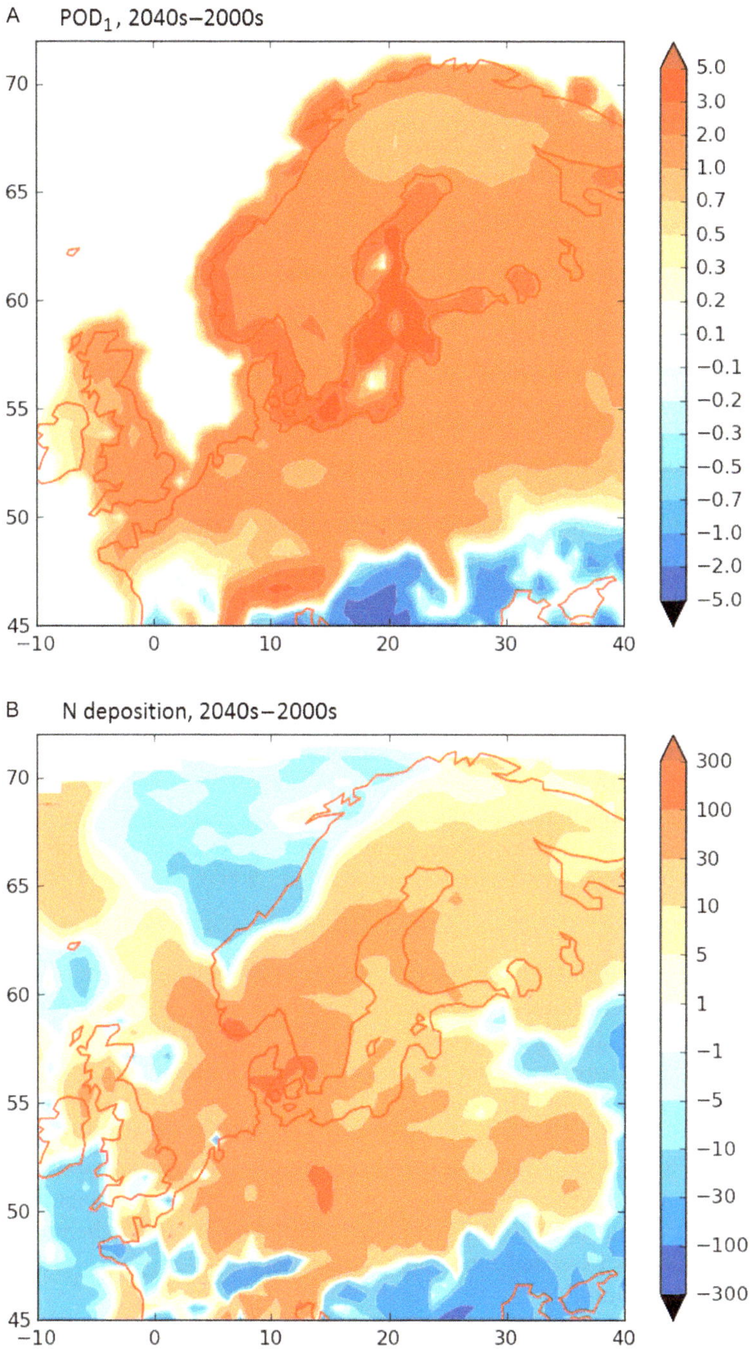

FIGURE 5.2 Effect of climate change as a difference between the periods 2000–2009 and 2040–2049 ('2040s'–'2000s') for (A) phytotoxic ozone dose POD$_1$ for coniferous forests (mmol m^{-2}) and (B) total (reduced + oxidised) annual deposition of nitrogen (mg m^{-2}).

increments exceed 10%, increasing to 20% with elevation within the Scandinavian Mountains region. These increases are associated with the higher atmospheric temperatures that in northern Europe favour rather than limit stomatal uptake.

A similar temperature effect was previously shown by Klingberg et al. (2010); however, their stomatal parameterisation had a significantly higher optimal temperature than that employed for the calculations shown in Figure 5.2A. Klingberg et al. (2010) also demonstrated that uptake rates and the corresponding risk of negative impacts will be reduced, if rising CO_2 concentrations limit stomatal uptake, but stand-level evidence of this is limited and controversial for forests (King et al., 2013, this vol.; Uddling et al., 2009).

It should be noted that the north–south difference in the climate change effect is likely to be stronger than that shown in Figure 5.2A, since the soil moisture deficits that limit stomatal exchange were not taken into account in these model calculations. Climate change scenarios typically suggest wetter conditions in northern Europe, while dry spells and droughts become longer and more frequent in the south (Christensen et al., 2007).

For N, our model calculations indicate that the total deposition in northern Europe will be substantially influenced, generally increased, by climate change (Figure 5.2B). In central Sweden, the relative increase is almost 20% in the 2040s, in relation to the 2000–2009 period. The highest absolute increase of about 100 mg N m^{-2} yr^{-1} is found in Denmark and the southernmost parts of the Scandinavian Peninsula. This increment can mostly be attributed to wet deposition, in accordance with the higher precipitation amounts projected for the future climate of northern Europe. Reduced and oxidised N compounds contribute approximately equal amounts to this increase. However, wet deposition patterns are highly sensitive to the uncertainties in precipitation scenarios (Langner et al., 2005).

5.3.2.3 *Emission Reductions*

The projected reductions of O_3 precursor emissions by 2030 will generally mitigate the O_3 stress on European vegetation, except for the areas of high NO_x emission density, where O_3 levels may be increased as a result of reduced titration by NO (Figure 5.3A). In northern Europe, the POD_1 reductions are of similar magnitude (1–2 mmol m^{-2}) as the increases due to climate change (Figure 5.2A). Hence the 'climate penalty' introduced above, that attenuates the efficiency of emission control, clearly plays a substantial role within this region in 2030. Even without the climate change increment, the critical POD_1 is exceeded on a large scale, despite the projected precursor emission reductions.

Emission reductions, as reflected in the PRIMES-based scenario for 2030, will reduce the total N deposition in central Europe by more than 300 mg m^{-2} yr^{-1}. Within our study area, the absolute reductions are smaller,

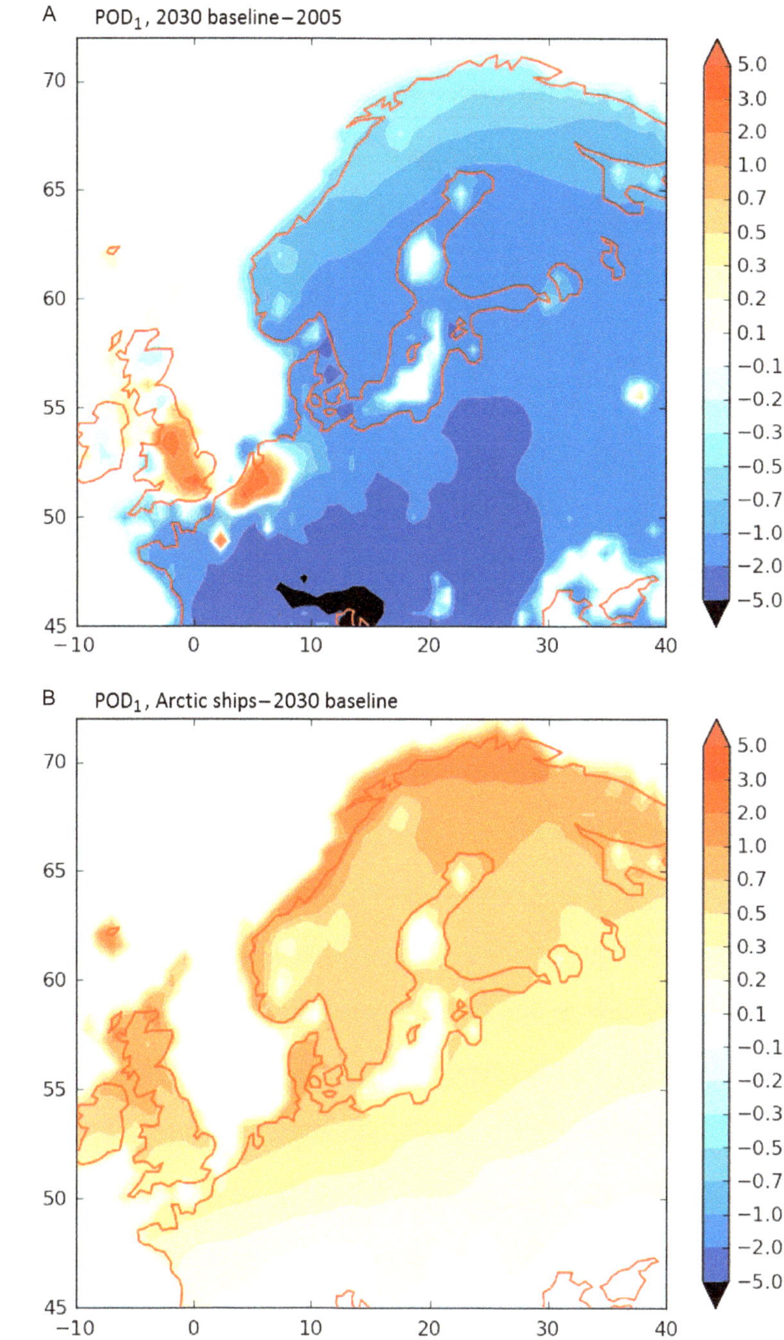

FIGURE 5.3 Changes in phytotoxic ozone dose POD_1 for coniferous forests (mmol m^{-2}) due to (A) emission changes between 2005 and 2030 without increased Arctic shipping ('2030 baseline'–'2005'), (B) increased Arctic shipping emissions in 2030 ('Arctic ships'–'2030 baseline'), (C) as (B) but emissions from diversion routes included, shown as 'Arctic ships with diversion'–'Arctic ships' and (D) as (C) but shown as 'Arctic ships with diversion'–'2030 baseline'.

(Continued)

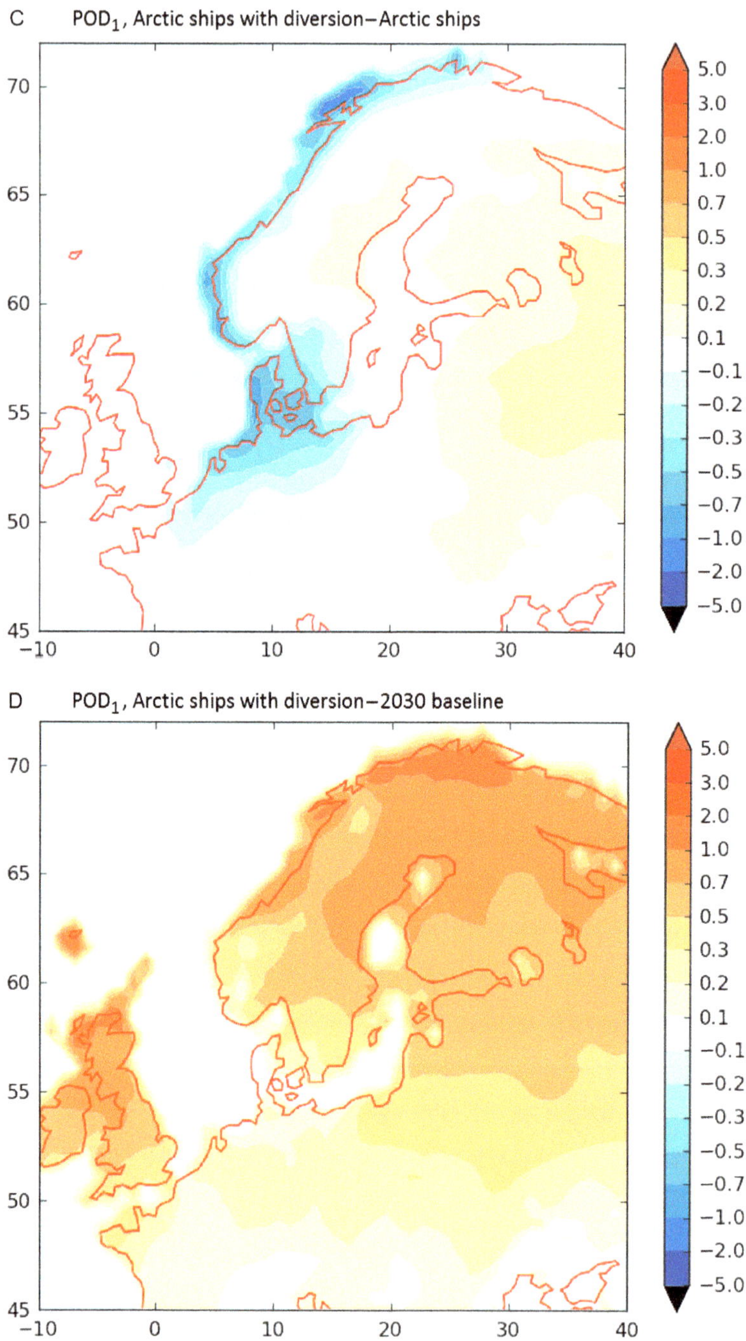

FIGURE 5.3—Cont'd

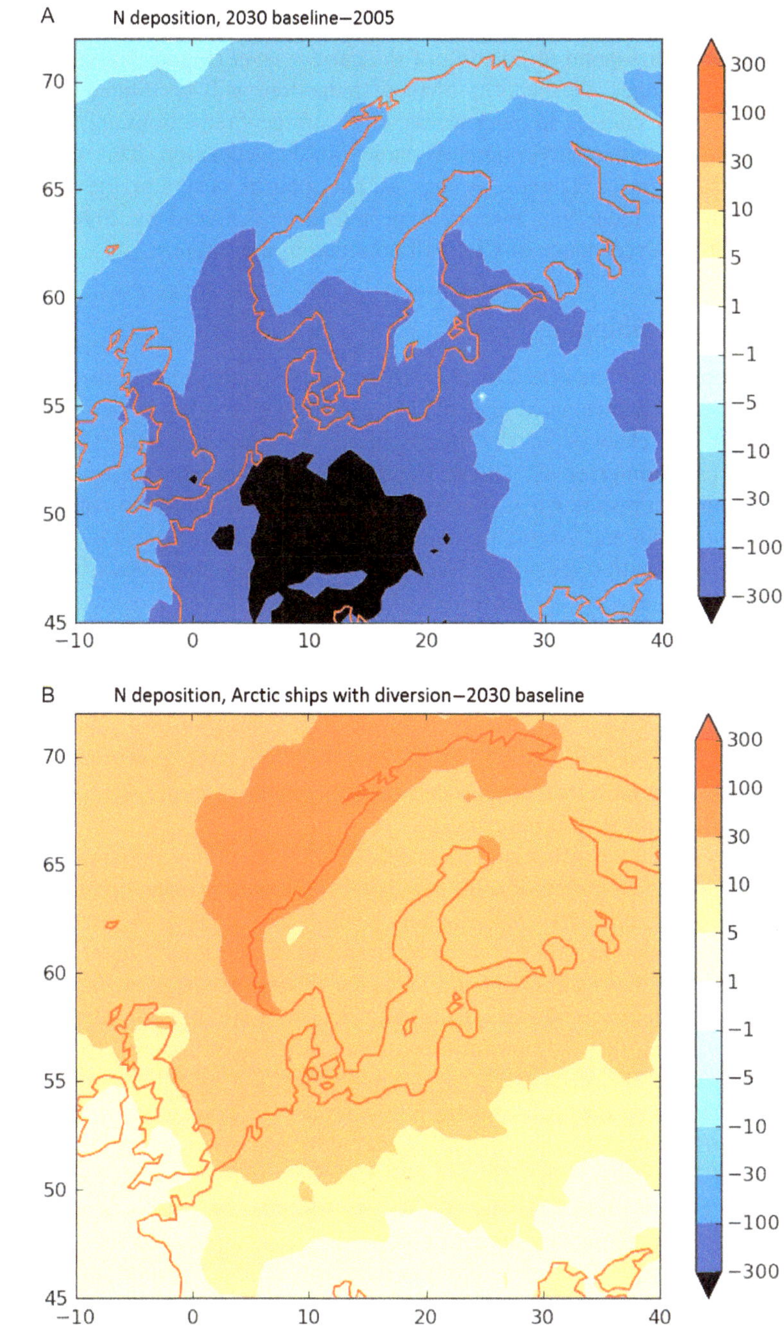

FIGURE 5.4 Changes in total (reduced + oxidised) annual deposition of nitrogen (mg m^{-2}) due to (A) emission changes between 2005 and 2030 without increased Arctic shipping ('2030 baseline'–'2005') and (B) increased Arctic shipping emissions including diversion routes in 2030 ('Arctic ships with diversion'–'2030 baseline').

but they still correspond to 20–30% of the current level (Figure 5.4A). Similarly to O_3 changes, the benefits of these reductions will be diluted by the effect of climate change; in other words, more stringent reductions are needed for meeting a certain environmental target for N deposition. This climate change penalty factor is almost as large as, but opposite in sign to, the change due to emission reductions that are expected to be achieved by 2030. This means that the exceedance of CLs will continue in the future.

5.3.3 Arctic Shipping

Previous studies on the effect of intensifying shipping activities in the Arctic have employed hemispheric-scale CTMs and shown somewhat inconsistent results on the changes in the tropospheric O_3 concentrations (Dalsøren et al., 2013; Granier et al., 2006). Figure 5.3 shows our higher resolution CTM simulation results when assuming different ship emission scenarios for 2030. If no additional Arctic ship emissions are included in the input data, the POD_1 distribution (Figure 5.3A) is rather similar to that shown for the mid-2000s (Figure 5.1A). The projected Arctic shipping will compound the O_3 stress on vegetation across the whole modelling domain. Within the region we focus on in this study, the additional contribution to POD_1 is more than 0.1 mmol m^{-2} and exceeds 1 mmol m^{-2} in the northernmost parts of the region (Figure 5.3B). In the north, 1 mmol m^{-2} constitutes a substantial (>10%) proportion of the total POD_1 accumulated during the growing season, which shows the relevance of considering future shipping in the Arctic to the O_3-induced risk of vegetation damage.

An access to the Northeastern Passage would increase maritime traffic and consequently the NO_x emissions in the vicinity of the Norwegian coast of the North Sea. These emissions result in both positive and negative O_3 concentration changes depending on the distance from this source area (Figure 5.3C). This means that these additional emissions are sufficiently large and spatially restricted to manifest nonlinear effects associated with atmospheric chemistry. Near the sources, that is, along the coastline, the NO_x emissions efficiently titrate O_3, while elsewhere the effect is smaller and can be either positive or negative. However, the O_3 increment is smaller and geographically more limited than that obtained when including the additional Arctic shipping emissions in 2030 (Figure 5.3B). The net effect of these two emission changes is that the region of highest increases is shifted towards the south-east and the peak values are slightly reduced (Figure 5.3D), as compared to the 'Arctic ships' scenario with no diversion traffic. Furthermore, the NO_x titration effect is not discernable anymore.

The Arctic shipping will potentially enhance the total N deposition also, due to increased deposition of the oxidised N originating from the NO_x emissions of shipping activities (Figure 5.4B). The northern ecosystems receive this additional N supply mainly in the form of wet deposition. Without the diversion traffic, the deposition increment amounts up to 30 mg N m^{-2} yr^{-1} (not shown), which in relative terms corresponds to an even larger increase than that in POD_1, but affects a more limited area. The maximum of this addition is located north of 65 °N.

The diversion traffic adds up to 50 mg N m^{-2} yr^{-1} to N deposition (not shown), but this is concentrated on more southern areas. As a net effect, N deposition will be substantially higher along most of the Norwegian coast (Figure 5.4B). The atmospheric processes controlling the link between NO_x emissions and N deposition do not exhibit nonlinearity similar to that related to O_3 (Figure 5.3C).

5.4 DISCUSSION AND CONCLUSIONS

The dose–response relationships behind the CLs, defined as the expected annual reduction in the whole-tree biomass of young trees related to a certain annual O_3 dose (Mills et al., 2011b), suggest that the POD_1 levels shown in Figure 5.1A could be associated with significant O_3 impacts on northern coniferous forests, with annual biomass losses of up to 7% (and even higher for deciduous forests) within the present study area. However, we emphasise that these dose–response relationships should be interpreted with great care, owing to experimental as well as theoretical limitations.

The O_3 impact experiments are generally based on the approach that young trees are exposed to O_3 in open-top chambers for a certain time period, after which the tree biomass is assessed (e.g. Karlsson et al., 2007). The percentage reduction in the biomass between the high and low O_3 treatments, estimated as an annual value, is assumed to represent the corresponding annual biomass reduction at a certain O_3 dose for all-aged trees growing in a certain region under current O_3 levels, relative to pre-industrial levels. Thus, this differential exposure approach relies on the presumptions that the dose–response relationship has reached a 'steady state' and that it is representative of that between the high and low O_3 scenarios under current, real forest-stand conditions.

Even neglecting any potential bias due to experimental setup, it is important to note that the proportional biomass loss obtained in this way is not equal to the proportional reduction in the actual growth rates of trees. Hence this estimate is not fully comparable with the data obtained from forest inventories that could be used as a reference. For instance, the current critical POD_1 level of Norway spruce has been associated with an annual 2% reduction in the total biomass (Mills et al., 2011b). However, applying this reduction rate over a typical rotation period of about 60 years results in a biomass at the time of harvest that is only 30% of that if there were no O_3. It is obvious that O_3 does not have such a dramatic effect on the growth of Norway spruce, not even in southern Sweden where POD_1 is more than twice as high as the CL (Figure 5.1A).

Quantification of the actual growth rate and its reduction due to O_3 would provide a metric for a more realistic impact assessment, but this requires data on the biomass at the start of the O_3 exposure experiment, which are seldom measured. Because of these complications, the POD_1 results reported here should be interpreted as a degree of damage risk, rather than a metric that could be translated into impacts on the roundwood supply for the forest sector industry, even though the CLRTAP methodology (Grünhage et al., 2013, this vol.; Mills et al., 2011b; UNECE, 2010) explicitly advocates such an option. Nevertheless, the widespread and in places multifold exceedance of CLs indicate that O_3 uptake is so large as to pose a risk to the forests growing in northern latitudes.

The limitations discussed above, as well as general limitations in the robustness of flux estimates (Tuovinen et al., 2009), affect not only risk mapping but also the possibilities to assess the impacts of current and future O_3 concentrations on C sequestration in forests. For example, Karlsson (2012) estimated C sequestration relying on the AOT40- rather than POD-based response functions, since the experimental data available for the former made it possible to relate O_3 to the relative growth rates of trees. According to this study, the annual increment in the living biomass C stocks of boreal and temperate forests in 10 northern and central European countries would be 10% higher in the absence of O_3 pollution. This shows that the abatement of air pollution entails climate change mitigation co-benefits. It can be expected that reducing O_3 concentrations in northern and central European would enhance C sequestration, which is a key regulating ecosystem service provided by forests (Cudlín et al., 2013, this vol.; Nijnik and Miller, 2013, this vol.), for at least some decades into the future.

The impacts of N deposition on forest ecosystems can be both positive and negative (Eugster and Haeni, 2013, this vol.). At non-fertile sites, such as most of northern Fennoscandia, N addition has the potential to increase forest growth, and hence forest C sequestration. On the other hand, forest soils in some areas of southern Sweden are already close to N saturation, with a high risk of N leakage to surface waters (Akselsson et al., 2010), while changes in the ground vegetation species composition may already have started (Nordin et al., 2005). Furthermore, it has been demonstrated that relatively small N additions affect the species richness of the epiphytic lichen communities (Johansson et al., 2012). Hence, if the future N load to northern forests will remain essentially at the current level, due to the counteracting effects of emission reductions and climate change, this may develop into a serious environmental issue in the future. This risk will be higher the higher is the additional N load due to ship traffic.

Even without any changes in air quality, the projected climatic changes will strongly affect the European forests. These effects will have both ecological and economical consequences. In the boreal zone, atmospheric warming will prolong the growing season and, in general, increase net primary productivity and forest biomass. However, as these forests are intensively managed, the forest C balance will be largely determined by the scale of timber harvests. Forest management infrastructure and technology are highly developed in northern Europe, providing large socio-economic adaptive capacity in the forest sector to climate change, as compared to the southern and eastern parts of the continent (Lindner et al., 2010). The issue of management response to climate change is further elaborated by Cudlín et al. (2013, this vol.) and Knoke and Hahn (2013, this vol.).

The air quality co-benefits of GHG emission abatement have been estimated to be of a similar order of magnitude to abatement costs, but these have not been incorporated into integrated assessments of climate policy until recently (Nemet et al., 2010). According to Rafaj et al. (2013), implementation of the measures to meet the climate change targets of the European Union would halve the annual costs of the implementation of air pollution control legislation, while significantly reducing health and environmental impacts. This kind of co-benefit analysis adopts climate change as the primary driver. In contrast, the C sequestration discussion above assumes the point of view of air pollution abatement, with the potential positive climate change interactions considered as co-benefits of the air quality policy.

However, only a fraction of air quality measures provide substantial warming mitigation (Shindell et al., 2012) and, conversely, a climate-driven policy may have negative air quality effects; for example, small-scale biomass combustion may increase emissions of particulate matter. In spite of these trade-offs, there is a clear need for pursuing the climate change and air quality policies in tandem (Pleijel, 2009), which necessitates structural changes rather than 'end-of-pipe' measures as well as consideration of the climate-related interactions highlighted above.

The results presented in this chapter demonstrate that climate change may offset the efficiency of emission reductions that alone would reduce the air pollution stress on northern European forests. Similar global-scale integration is needed to control the emissions from maritime transport that in the future will play an increasingly important role within this region, especially in the coastal and northernmost parts that are vulnerable to the developments in the Arctic.

ACKNOWLEDGEMENTS

This work was supported by the European Union (ÉCLAIRE project, no. 282910), the Nordic Council of Ministers (EnsClim project, no. KoL-10-04) and the United Nations Economical Commission for Europe (EMEP programme).

REFERENCES

Akselsson, C., Belyazid, S., Hellsten, S., Klarqvist, M., Pihl-Karlsson, G., Karlsson, P.E., Lundin, L., 2010. Assessing the risk of N leaching across a steep N deposition gradient in Swedish forests using different monitoring and modelling approaches. Environ. Pollut. 158, 3588–3595. http://dx.doi.org/10.1016/j.envpol.2010.08012.

AMAP, 2009. AMAP 2009 update on selected climate issues of concern. In: Observations, Short-Lived Climate Forcers, Arctic Carbon Cycle, and Predictive Capability. Arctic Monitoring and Assessment Program (AMAP), Oslo, Norway. http://library.arcticportal.org/1220/1/Climate_Update_2009.pdf.

Arneth, A., Harrison, S.P., Zaehle, S., Tsigaridis, K., Menon, S., Bartlein, P.J., Feichter, J., Korhola, A., Kulmala, M., O'Donnell, D., Schurgers, G., Sorvari, S., Vesala, T., 2010. Terrestrial biogeochemical feedbacks in the climate system. Nat. Geosci. 3, 525–532. http://dx.doi.org/10.1038/ngeo905.

Avnery, S., Mauzerall, D.L., Liu, J., Horowicz, L.W., 2011. Global crop yield reductions due to surface ozone exposure: 2. Year 2030 potential crop production losses and economic damage under two scenarios of O_3 pollution. Atmos. Environ. 45, 2297–2309. http://dx.doi.org/10.1016/j.atmosenv.2011.01.002.

Bobbink, R., Hettelingh, J.-P. (Eds.), 2011. Review and Revision of Empirical Critical Loads and Dose-Response Relationships. Report 680359002. National Institute for Public Health and the Environment, Bilthoven, The Netherlands. http://www.rivm.nl/bibliotheek/rapporten/680359002.pdf.

Butterbach-Bahl, K., Nemitz, E., Zaehle, S., Billen, G., Boeckx, P., Erisman, J.W., Garnier, J., Upstill-Goddard, R., Kreuzer, M., Oenema, O., Reis, S., Schaap, M., Simpson, D., de Vries, W., Winiwarter, W., Sutton, M.A., 2011. Nitrogen as a threat to the European greenhouse balance. In: Sutton, M.A., Howard, C.M., Erisman, J.W., Billen, G., Bleeker, A., Grennfelt, P., van Grinsven, H., Grizzetti, B. (Eds.), The European Nitrogen Assessment. Cambridge University Press, Cambridge, UK, pp. 434–462.

Carmichael, G., Wild, O., Fiore, A., Holloway, T., Lin, M., Doherty, R., Jonson, J.E., Fu, J., Gong, S., Chin, M., Mauzerall, D., Wang, Y., Collins, B., Schulz, M., Reidmiller, D., Liu, J., Zhang, Y., Lam, Y.F., 2010. Global and regional modelling. In: Dentener, F., Keating, T., Akimoto, H. (Eds.), Hemispheric Transport of Air Pollution 2010. Part A: Ozone and Particulate Matter. Air Pollution Studies 17, Economic Commission for Europe, Geneva, Switzerland. pp. 135–198. http://www.htap.org/activities/2010_Final_Report/HTAP%202010%20Part%20A%20110407.pdf.

Christensen, J.H., Hewitson, B., Busuioc, A., Chen, A., Gao, X., Held, I., Jones, R., Kolli, R.K., Kwon, W.-T., Laprise, R., Magaña Rueda, V., Mearns, L., Menéndez, C.G., Räisänen, J., Rinke, A., Sarr, A., Whetton, P., 2007. Regional climate projections. In: Solomon, S., Qin, D., Manning, M., Chen, Z., Marquis, M., Averyt, K.B., Tignor, M., Miller, H.L. (Eds.), Climate change 2007: The Physical Basis. Contribution of Working Group I to the Fourth Assessment Report of the Intergovernmental Panel on Climate Change. Cambridge University Press, Cambridge, UK, pp. 847–940.

Cieslik, S., Tuovinen, J.-P., Baumgarten, M., Matyssek, R., Brito, P., Wieser, G., 2013. Gaseous exchange between forests and the atmosphere. Dev. Environ. Sci. 13, 19–36.

Colette, A., Granier, C., Hodnebrog, Ø., Jakobs, H., Maurizi, A., Nyiri, A., Bessagnet, B., D'Angiola, A., D'Isidoro, M., Gauss, M., Meleux, F., Memmesheimer, M., Mieville, A., Rouïl, L., Russo, F., Solberg, S., Stordal, F., Tampieri, F., 2011. Air quality trends in Europe over the past decade: a first multi-model assessment. Atmos. Chem. Phys. 11, 11657–11678. http://dx.doi.org/10.5194/acp-11-11657-2011.

Colette, A., Granier, C., Hodnebrog, Ø., Jakobs, H., Maurizi, A., Nyiri, A., Rao, S., Amann, M., Bessagnet, B., D'Angiola, A., Gauss, M., Heyes, C., Klimont, Z., Meleux, F., Memmesheimer, M., Mieville, A., Rouïl, L., Russo, F., Schucht, S., Simpson, D., Stordal, F., Tampieri, F., Vrac, M., 2012. Future air quality in Europe: a multi-model assessment of projected exposure to ozone. Atmos. Chem. Phys. 12, 10613–10630. http://dx.doi.org/10.5194/acp-12-10163-2012.

Comiso, J.C., 2012. Large decadal decline of the Arctic multiyear ice cover. J. Climate 25, 1176–1193. http://dx.doi.org/10.1175/jcli-d-11-00113.1.

Corbett, J.J., Lack, D.A., Winebrake, J.J., Harder, S., Silberman, J.A., Gold, M., 2010. Arctic shipping emissions inventories and future scenarios. Atmos. Chem. Phys. 10, 9689–9704. http://dx.doi.org/10.5194/acp-10-9689-2010.

Coumou, D., Rahmstorf, S., 2012. A decade of weather extremes. Nat. Clim. Change 2, 491–496. http://dx.doi.org/10.1038/nclimate1452.

Cudlín, P., Seják, J., Pokorný, J., Albrechtová, J., Bastian, O., Marek, M., 2013. Forest ecosystem services under climate change and air pollution. Dev. Environ. Sci. 13, 521–546.

Dalsøren, S.B., Samset, B.G., Myhre, G., Corbett, J.J., Minjares, R., Lack, D., Fuglestvedt, J.S., 2013. Environmental impacts of shipping in 2030 with a particular focus on the Arctic region. Atmos. Chem. Phys. 13, 1941–1955. http://dx.doi.org/10.5194/apc-13-1941-2013.

Dizengremel, P., Jolivet, Y., Tuzet, A., Ranieri, A., Le Thiec, D., 2013. Integrative leaf level phytotoxic ozone dose assessment for forest risk modelling. Dev. Environ. Sci. 13, 267–288.

Engardt, M., Bergström, R., Andersson, C., 2009. Climate and emission changes contributing to changes in near-surface ozone in Europe over the coming decades: results from model studies. Ambio 38, 452–458.

Eugster, W., Haeni, M., 2013. Nutrients or pollutants?—Nitrogen deposition to European forests. Dev. Environ. Sci. 13, 37–56.

Fagerli, H., Simpson, D., Nyíri, Á., Mareckova, K., Wankmüller, R., 2012. Acidification, eutrophication and ground level ozone: from 1990 towards the revised Gothenburg Protocol target year (2020). In: Transboundary Acidification, Euthrophication and Ground Level Ozone in Europe in 2010. EMEP Report 1/2012. Norwegian Meteorological Institute, Oslo, Norway, pp. 41–57. http://emep.int/publ/reports/2012/status_report_1_2012.pdf.

Folland, C.K., Knight, J., Linderholm, H.W., Fereday, D., Ineson, S., Hurrell, J.W., 2009. The summer North Atlantic Oscillation: past, present, and future. J. Climate 22, 1082–1103. http://dx.doi.org/10.1175/2008jcli2459.1.

Fowler, D., Pilegaard, K., Sutton, M.A., et al., 2009. Atmospheric composition change: ecosystems-atmosphere interactions. Atmos. Environ. 43, 5193–5267. http://dx.doi.org/10.1016/j.atmosenv.2009.07.068.

Granier, C., Niemeier, U., Jungclaus, J.H., Emmons, L., Hess, P., Lamarque, J.-F., Walters, S., Brasseur, G.P., 2006. Ozone pollution from future ship traffic in the Arctic northern passages. Geophys. Res. Lett. 33, L13807. http://dx.doi.org/10.1029/2006GL026180.

Grünhage, L., Matyssek, R., Wieser, G., Häberle, K.-H., Leuchner, M., Menzel, A., Dieler, J., Pretzsch, H., Grimmeisen, W., Zimmermann, L., Raspe, S., Schröder, M., 2013. Flux-based ozone risk assessment for adult beech and spruce forests. Dev. Environ. Sci. 13, 251–266.

Hedegaard, G.B., Christensen, J.H., Brandt, J., 2013. The relative importance of impacts from climate change vs. emissions change on air pollution levels in the 21st century. Atmos. Chem. Phys. 13, 3569–3585. http://dx.doi.org/10.5194/acp-13-3569-2013.

Hole, L.R., de Wit, H.A., Aas, W., 2008. Influence of summer and winter climate variability on nitrogen wet deposition in Norway. Hydrol. Earth Syst. Sci. 12, 405–414. http://dx.doi.org/10.5194/hess-12-405-2008.

Hollaway, M.J., Arnold, S.R., Challinor, A.J., Emberson, L.D., 2012. Intercontinental transboundary contributions to ozone-induced crop yield losses in the Northern Hemisphere. Biogeosciences 9, 271–292. http://dx.doi.org/10.5194/bg-9-271-2012.

Johansson, O., Palmqvist, K., Olofsson, J., 2012. Nitrogen deposition drives lichen community changes through differential species responses. Glob. Chang. Biol. 18, 2626–2635. http://dx.doi.org/10.1111/j.1365-2486.2012.02723.x.

Jonson, J.E., Simpson, D., Fagerli, H., Solberg, S., 2006. Can we explain the trends in European ozone levels? Atmos. Chem. Phys. 6, 51–66. http://www.atmos-chem-phys.org/acp/6/51/.

Jylhä, K., Fronzek, S., Tuomenvirta, H., Carter, T.R., Ruosteenoja, K., 2008. Changes in frost, snow and Baltic sea ice by the end of the twenty-first century based on climate model projections for Europe. Clim. Change 86, 441–462. http://dx.doi.org/10.1007/s10584-007-9310-z.

Karlsson, P.E., 2012. Ozone impacts on carbon sequestration in northern and central European forests. IVL Report B 2065. IVL Swedish Environmental Research Institute, Gothenburg, Sweden. http://www.ivl.se/download/18.14817b28139c413dc751ad/1347976887032/B2065+.pdf.

Karlsson, P.E., Braun, S., Broadmeadow, M., Elvira, S., Emberson, L., Gimeno, B.S., Le Thiec, D., Novak, K., Oksanen, E., Schaup, M., Uddling, J., Wilkinson, M., 2007. Risk assessments for forest trees: the performance of the ozone flux versus the AOT concepts. Environ. Pollut. 146, 608–616. http://dx.doi.org/10.1016/j.envpol.2006.06.012.

Karlsson, P.E., Pleijel, H., Danielsson, H., Pihl Karlsson, G., Piikki, K., Uddling, J., 2009. Evidence for impacts of near-ambient ozone concentrations on vegetation in southern Sweden. Ambio 38, 425–431.

Katragkou, E., Zanis, P., Kioutsioukis, I., Tegoulias, I., Melas, D., Krüger, B.C., Coppola, E., 2011. Future climate change impacts on summer surface ozone from regional climate-air quality simulations over Europe. J. Geophys. Res. 116, D22307. http://dx.doi.org/10.1029/2011JD015899.

King, J., Liu, L., Aspinwall, M., 2013. Tree and forest responses to interacting elevated atmospheric CO_2 and tropospheric O_3: a synthesis of experimental evidence. Dev. Environ. Sci. 13, 179–208.

Klingberg, J., Engardt, M., Uddling, J., Karlsson, P.E., Pleijel, H., 2010. Ozone risk for vegetation in the future climate of Europe based on stomatal ozone uptake calculations. Tellus A 63, 174–187. http://dx.doi.org/10.1111/j.1600-0870.2010.00465.x.

Knoke, T., Hahn, A., 2013. Global change and the role of forests in future land-use systems. Dev. Environ. Sci. 13, 569–587.

Koike, T., Watanabe, M., Hoshika, Y., Kitao, M., Matsumura, H., Funada, R., Izuta, T., 2013. Effects of ozone on forest ecosystems in East and Southeast Asia. Dev. Environ. Sci. 13, 371–390.

Laakso, L., Beukes, J.P., Van Zyl, P.G., Pienaar, J., Josipovic, M., Venter, A., Jaars, K., Vakkari, V., Labuschagne, C., Chiloane, K., Tuovinen, J.-P., 2013. Ozone concentrations and their potential impacts on vegetation in southern Africa. Dev. Environ. Sci. 13, 429–450.

Langner, J., Bergström, R., Foltescu, V., 2005. Impact of climate change on surface ozone and deposition of sulphur and nitrogen in Europe. Atmos. Environ. 39, 1129–1141. http://dx.doi.org/10.1016/j.atmosenv.204.09.082.

Langner, J., Engardt, M., Baklanov, A., Christensen, J.H., Gauss, M., Geels, C., Hedegaard, G.B., Nuterman, R., Simpson, D., Soares, J., Sofiev, M., Wind, P., Zakey, A., 2012a. A multi-model study of impacts of climate change on surface ozone in Europe. Atmos. Chem. Phys. 10, 10423–10440. http://dx.doi.org/10.5194/acp-12-10423-2012.

Langner, J., Engardt, M., Andersson, C., 2012b. European summer surface ozone 1990–2100. Atmos. Chem. Phys. 12, 10097–10105. http://dx.doi.org/10.5194/acp-12-10097-2012.

Lindner, M., Maroschek, M., Netherer, S., Kremer, A., Barbati, A., Garzia-Gonzalo, J., Seidl, R., Delzon, S., Corona, P., Kolström, M., Lexer, M.J., Marchetti, M., 2010. Climate change impacts, adaptive capacity, and vulnerability of European forest ecosystems. For. Ecol. Manage. 259, 698–709. http://dx.doi.org/10.1016/j.foreco.2009.09.023.

Loreto, F., Fares, S., 2013. Biogenic Volatile Organic Compounds and their impacts on biosphere-atmosphere interactions. Dev. Environ. Sci. 13, 57–75.

Manninen, S., Huttunen, S., Tømmervik, H., Hole, L.R., Solberg, S., 2009. Northern plants and ozone. Ambio 38, 406–412.

Matyssek, R., Wieser, G., Fleischmann, F., Grünhage, L., 2013. Ozone Research, quo vadis? Lessons from the free–air canopy fumigation experiment at Kranzberg Forest. Dev. Environ. Sci. 13, 103–129.

Mills, G., Hayes, F., Simpson, D., Emberson, L., Norris, D., Harmens, H., Büker, P., 2011a. Evidence of widespread effects of ozone on crops and (semi-)natural vegetation in Europe (1990–2006) in relation to AOT40- and flux-based risk maps. Glob. Chang. Biol. 17, 592–613. http://dx.doi.org/10.1111/j.1365-2486.2010.02217.x.

Mills, G., Pleijel, H., Braun, S., Büker, P., Bermejo, V., Calvo, E., Danielsson, H., Emberson, L., González Fernández, I., Grünhage, L., Harmens, H., Hayes, F., Karlsson, P.-E., Simpson, D., 2011b. New stomatal flux-based critical levels for ozone effects on vegetation. Atmos. Environ. 45, 5064–5068. http://dx.doi.org/10.1016/j.atmosenv.2011.06.009.

Monks, P.S., Granier, C., Fuzzi, S., et al., 2009. Atmospheric composition change—global and regional air quality. Atmos. Environ. 43, 5268–5350. http://dx.doi.org/10.1016/j.atmenv.2009.08.021.

Nemet, G.F., Holloway, T., Meier, P., 2010. Implications of incorporating air-quality co-benefits into climate change policymaking. Environ. Res. Lett. 5, 014007. http://dx.doi.org/10.1088/1748-9326/5/1/014007.

Nijnik, M., Miller, D., 2013. Targeting sustainable provision of forest ecosystem services with special focus on carbon sequestration. Dev. Environ. Sci. 13, 547–568.

Nordin, A., Strengbom, J., Witzell, J., Näsholm, T., Ericsson, L., 2005. Nitrogen deposition and the biodiversity of boreal forests: implications for the nitrogen critical load. Ambio 34, 20–24.

Pandey, V., Oksanen, E., Singh, N., Sharma, C., 2013. Impacts of Air Pollution and Climate Change on plants: implications for India. Dev. Environ. Sci. 13, 391–409.

Pausata, F.S.R., Pozzoli, L., Vignati, E., Dentener, F.J., 2012. North Atlantic Oscillation and tropospheric ozone variability in Europe: model analysis and measurements intercomparison. Atmos. Chem. Phys. 12, 6357–6376. http://dx.doi.org/10.5194/acp-12-6357-2012.

Peters, G.P., Nilssen, T.B., Lindholt, L., Eide, M.S., Glomsrød, S., Eide, L.I., Fuglestvedt, J.S., 2011. Future emissions from shipping and petroleum activities in the Arctic. Atmos. Chem. Phys. 11, 5305–5320. http://dx.doi.org/10.5194/acp-11-5305-2011.

Pihl Karlsson, G., Akselsson, C., Hellsten, S., Karlsson, P.E., 2011. Reduced European emissions of S and N—Effects on air concentrations, deposition and soil water chemistry in Swedish forests. Environ. Pollut. 159, 3571–3582. http://dx.doi.org/10.1016/j.envpol.2011.08.007.

Pilli-Sihvola, K., Aatola, P., Ollikainen, M., Tuomenvirta, H., 2010. Climate change and electricity consumption—Witnessing increasing or decreasing use and costs? Energy Policy 38, 2409–2419. http://dx.doi.org/10.1016/j.enpol.2009.12.033.

Pleijel, H. (Ed.), 2009. Air Pollution and Climate Change—Two Sides of the Same Coin? Swedish Environmental Protection Agency, Stockholm, Sweden.

Pozzer, A., Zimmermann, P., Doering, U.M., van Aardenne, J., Tost, H., Dentener, F., Janssens-Maenhout, G., Lelieveld, J., 2012. Effects of business-as-usual anthropogenic emissions on air quality. Atmos. Chem. Phys. 12, 6915–6937. http://dx.doi.org/10.5194/acp-12-6915-2012.

Rafaj, P., Schöpp, W., Russ, P., Heyes, C., Amann, M., 2013. Co-benefits of post-2012 global climate mitigation policies. Mitig. Adapt. Strategies Glob. Change 18, 801–824. http://dx.doi.org/10.1007/s11027-012-9390-6.

Serreze, M.C., Barry, R.G., 2011. Processes and impacts of Arctic amplification: a research synthesis. Global Planet. Change 77, 85–96. http://dx.doi.org/10.1016/j.gloplacha.2011.03.004.

Shindell, D., Kuylenstierna, J.C.I., Vignati, E., van Dingenen, R., Amann, M., Klimont, Z., Anenberg, S.C., Muller, N., Janssens-Maenhout, G., Raes, F., Schwartz, J., Faluvegi, G., Pozzoli, L., Kupiainen, K., Höglund-Isaksson, L., Emberson, L., Streets, D., Ramanathan, V., Hicks, K., Oanh, N.T.K., Milly, G., Williams, M., Demkine, V., Fowler, D., 2012. Simultaneously mitigating near-term climate change and improving human health and food security. Science 335, 183–189. http://dx.doi.org/10.1126/science.1210026.

Simpson, D., Ashmore, M.R., Emberson, L., Tuovinen, J.-P., 2007. A comparison of two different approaches for mapping potential ozone damage to vegetation. A model study. Environ. Pollut. 146, 715–725. http://dx.doi.org/10.1016/j.envpol.2006.04.013.

Simpson, D., Benedictow, A., Berge, H., Bergström, R., Emberson, L.D., Fagerli, H., Flechard, C.R., Hayman, G.D., Gauss, M., Jonson, J.E., Jenkin, M.E., Nyíri, A., Richter, C., Semeena, V.S., Tsyro, S., Tuovinen, J.-P., Valdebenito, Á., Wind, P., 2012. The EMEP MSC-W chemical transport model—technical description. Atmos. Chem. Phys. 12, 7825–7865. http://dx.doi.org/10.5194/acp-12-7825-2012.

Solberg, S., Hov, Ø., Søvde, A., Isaksen, I.S.A., Coddeville, P., De Backer, H., Forster, C., Orsolini, Y., Uhse, K., 2008. European surface ozone in the extreme summer 2003. J. Geophys. Res. 113, D07307. http://dx.doi.org/10.1029/2007JD009098.

Tamm, C.O., 1991. Nitrogen in terrestrial ecosystems. Ecological Studies 81, Springer-Verlag, Berlin, Germany.

Tørseth, K., Aas, W., Breivik, K., Fjæraa, A.M., Fiebig, M., Hjellbrekke, A.G., Lund Myhre, C., Solberg, S., Yttri, K.E., 2012. Introduction to the European Monitoring and Evaluation Programme (EMEP) and observed atmospheric composition change during 1972–2009. Atmos. Chem. Phys. 12, 5447–5481. http://dx.doi.org/10.5194/acp-12-5447-2012.

Tuovinen, J.-P., Emberson, L., Simpson, D., 2009. Modelling ozone fluxes to forests for risk assessment: status and prospects. Ann. For. Sci. 66, 401. http://dx.doi.org/10.1051/forest/2009024.

Uddling, J., Teclaw, R.W., Pregitzer, K.S., Ellsworth, D.S., 2009. Leaf and canopy conductance in aspen and aspen-birch forests under free-air enrichment of carbon dioxide and ozone. Tree Physiol. 29, 1367–1380. http://dx.doi.org/10.1093/treephys/tpp070.

UNECE, 2010. Mapping critical levels for vegetation. Manual on methodologies and criteria for modelling and mapping critical loads and levels and air pollution effects, risks and trends. United Nations Economic Commission for Europe (UNECE) Convention on Long-range Transboundary Air Pollution, Geneva, Switzerland. http://icpvegetation.ceh.ac.uk/manuals/mapping_manual.html.

van Vuuren, D.P., Edmonds, J., Kainuma, M., Riahi, K., Thomson, A., Hibbard, K., Hurtt, G.C., Kram, T., Krey, V., Lamarque, J.-F., Masui, T., Meinshausen, M., Nakicenovic, N., Smith, S.J., Rose, S.K., 2011. The representative concentration pathways: an overview. Clim. Change 109, 5–31. http://dx.doi.org/10.1007/s10584-011-0148-z.

Wagner, F., Amann, M., Bertok, I., Cofala, J., Heyes, C., Klimont, Z., Rafaj, P., Schöpp, W., 2010. Baseline emission projections and further cost-effective reductions of air pollution impacts in Europe—a 2010 perspective. NEC Scenario Analysis Report 7. International

Institute of Applied Systems Analysis (IIASA), Laxenburg, Austria. http://gains.iiasa.ac.at/images/stories/reports/NEC7_Interim_report_20100827.pdf.

Wang, M., Overland, J.E.E., 2012. A sea ice free summer Arctic within 30 years – and update from CMIP5 models. Geophys. Res. Lett. 39, L18501. http://dx.doi.org/10.1029/2012GL052868.

Wittig, V.E., Ainsworth, E.A., Naidu, S.L., Karnosky, D.F., Long, S.P., 2009. Quantifying the impact of current and future tropospheric ozone on tree biomass, growth, physiology and biochemistry: a quantitative meta-analysis. Glob. Chang. Biol. 15, 396–424. http://dx.doi.org/10.1111/j.1365-2486.2008.01774.x.

Wu, S., Mickley, L.J., Leibensperger, E.M., Jacob, D.J., Rind, D., Streets, D.G., 2008. Effects of 2000–2050 global change on ozone air quality in the United States. J. Geophys. Res. 113, D06302. http://dx.doi.org/10.1029/2007JD008917.

Young, P.J., Archibald, A.T., Bowman, K.W., et al., 2013. Pre-industrial to end 21st century projections of tropospheric ozone from the Atmospheric Chemistry and Climate Model Intercomparison Project (ACCMIP). Atmos. Chem. Phys. 13, 2063–2090. http://dx.doi.org/10.5194/acp-13-2063-2013.

Part III

Significance of Biotic Processes in Forest Ecosystem Response

Chapter 6

Ozone Research, *Quo Vadis*? Lessons from the Free-Air Canopy Fumigation Experiment at Kranzberg Forest

Rainer Matyssek*[,1], Gerhard Wieser[†], Frank Fleischmann[‡] and Ludger Grünhage[§]

[*]*Technische Universität München, Ecophysiology of Plants, Hans-Carl-von-Carlowitz-Platz 2, D-85354 Freising-Weihenstephan, Germany*

[†]*Department of Alpine Timberline Ecophysiology, Federal Office and Research Centre for Forests, Innsbruck, Austria*

[‡]*Pathology of Woody Plants, Technische Universität München, Freising, Germany*

[§]*Department of Plant Ecology, Justus-Liebig University, Giessen, Germany*

[1]*Corresponding author: e-mail: matyssek@wzw.tum.de*

Chapter Outline

Developments in Environmental Science, Vol. 13. http://dx.doi.org/10.1016/B978-0-08-098349-3.00006-2

6.1 INTRODUCTION

Why must one still raise the question of future ozone (O_3) research on forest trees and ecosystems? Ground-level O_3 as acting on trees and forests through anthropogenically enhanced regimes is of concern in this chapter, along with arising physiological, ecological and socio-economic consequences. As the latter aspect will be dealt with in detail in Part VII of this book, this question is perhaps surprising, given the plethora of relevant physiological and ecological research that has been conducted since the 1940s (Karnosky et al., 2007; Klingberg et al., 2010; Lefohn, 1992; Matyssek and Sandermann, 2003; Matyssek et al., 2010a). As our knowledge increased, judgement about the injurious potential of ozone on vegetation has progressed during the intervening decades (Keating et al., 2004). Ozone was a concern as a local agent of high toxicity during the middle of the twentieth century, and it was debated as a regionally acting driver of forest decline during later decades (Miller and McBride, 1999; Sandermann et al., 1997). It was recognized at the beginning of the twenty-first century as a component of climate change of global dimensions (Fowler et al., 2008; Matyssek et al., 2013). Evidence of the global scope of O_3 pollution is the long-range transport of O_3 plumes across and between continents (Derwent et al., 2004; Newell and Evans, 2000) and projections of continuing high or further increasing background O_3 concentrations (Dentener et al., 2006; Vingarzan, 2004). New regional "hot spots" of anthropogenic O_3 formation have emerged across the northern and southern hemispheres (Fowler et al., 1999, 2008; Sitch et al., 2007; see Part V). Further, the scientific consensus continues to be that ground-level O_3 at enhanced concentrations is potentially the most detrimental air pollutant to vegetation (Fowler et al., 2008; Matyssek and Sandermann, 2003; Wittig et al., 2009).

Attention has moved from regarding ozone as a primary causative factor of tree death towards a stressor that in interaction with other factors affects tree performance (e.g. King et al., 2005; Watanabe et al., 2012; Zak et al., 2011). Such modified O_3 impact may weaken stress resistance, competitiveness, fitness and, ultimately, carbon sink strength (Edwards and Zak, 2011; Karnosky et al., 2007; Miller and McBride, 1999). As a result, radiative forcing can be promoted in the atmosphere through reduced photosynthetic CO_2 withdrawal due to the chronic O_3 impact (Ren et al., 2011; Sitch et al., 2007). The intrinsic link between atmospheric CO_2 enhancement and continued O_3 formation is mediated, in addition, through increased irradiance, if in mesic regions cloudiness is lowered upon climate warming, although one should be aware of reported uncertainties in such a context (Wuebbles et al., 1999). In parallel, O_3 formation is driven by the emission of anthropogenic O_3 precursors from industries, traffic and forest burning (Grulke et al., 2009; Martins et al., 2012; Stockwell et al., 1997). Such processes together foster atmospheric CO_2 enrichment (Figure 6.1).

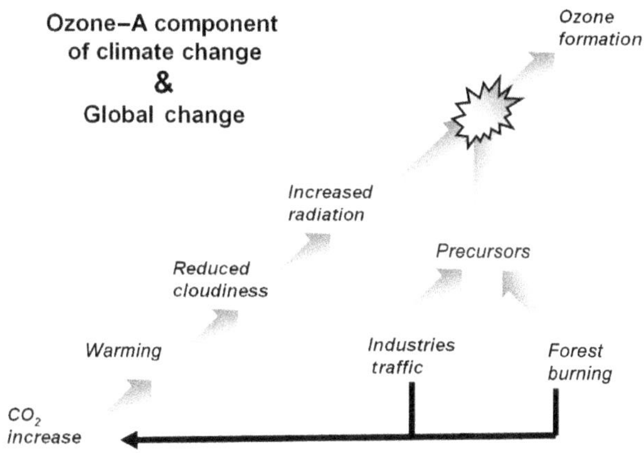

FIGURE 6.1 Physico-chemical pathways within the atmosphere and in interaction with terrestrial anthropogenic surface processes, illustrating enhancement of tropospheric ground-level O_3 regimes as a component of climate change and global change, when including land-use changes, for example, as imposed by forest burning (see text for details).

As forests are crucial globally for carbon (C) storage, decline of forest C sink strength due to elevated ground-level ozone must continue to be an active area of research. Loss of global primary productivity is expected to become substantial according to modelling predictions, particularly in view of the limited annual amount of additional C fixation by ecosystems under increased atmospheric CO_2 (Fowler et al., 2008; King et al., 2013, this vol.; Sitch et al., 2007). However, regarding forest ecosystems, modelling is limited by shortcomings in parameterization and validation. There is a severe shortage of suitable databases, given that research throughout the decades on empirical evidence of O_3 responses predominantly derived from chamber studies with juvenile trees rather than controlled stand-level experiments with mature individuals (Kolb and Matyssek, 2001). It is the stand situation, however, which is decisive for ecologically meaningful O_3 risk assessment (Matyssek and Innes, 1999). The evidential deficit is socio-economically crucial, given the lack of reliable decision tools in silviculture for mitigating atmospheric CO_2 increase and fostering efficient fast-growing tree plantations within the scope of the Kyoto Protocol ("clean development mechanisms"; Oberthür and Ott, 1999). Gaps in knowledge have been compensated, to some extent at least, through controlled, long-term field experimentation (up to one decade) on rapidly growing tree plantations (Aspen Face Experiment: Karnosky et al., 2003) and adult forest trees under silvicultural stand conditions (Kranzberg Forest Experiment (KFE): Matyssek et al., 2007, 2010a,b). Both studies made use of novel free-air O_3 fumigation methodology (Karnosky et al., 2001, 2007; Nunn et al., 2002; Werner and Fabian, 2002), allowing canopy-level O_3 exposure of whole trees while avoiding micro-climatic artefacts with consequences for physiological responsiveness (cf. Matyssek et al., 1997).

Here, we focus on the study of adult trees, being aware that the outcome from both free-air O_3 experiments has been described elsewhere (Matyssek et al., 2010a). Major results from the study of adult trees will be summarized and taken as a starting point for locating shortcomings of the existing knowledge base (*sensu* "*lessons learnt*"), new questions emerging and conceptual and methodological advancements available for envisioning "next-generation" O_3 research (*sensu* "*quo vadis*?"), based on controlled experiments in ageing closed-canopy forests. The significance of biotic processes in forest tree response to O_3 impact will be emphasized (according to Part III of this book).

6.2 OZONE AS PART OF FACTORIAL COMPLEXES

Introducing lessons from the KFE and about the future of O_3 research on forest trees, it is conducive to envision tropospheric O_3 impact as nested into multiple abiotic–biotic interrelationships (Figure 6.2). As nutrition, in particular high nitrogen supply, can affect tree sensitivity to CO_2 by favouring primary productivity (cf. Körner, 2006), the sensitivity to ozone can also increase as reflected by enhanced leaf turnover or reduced productivity (Matyssek et al., 2012a; Watanabe et al., 2012). O_3-induced limitation of primary productivity, in turn, may reduce nutrient demand and uptake from soils (Weigt et al., 2012). Nevertheless, O_3 defence can raise the demand (Luedemann et al., 2005), and high nutrient supply may ameliorate O_3 effects (Takemoto et al., 2001) or promote them (Yamaguchi et al., 2010). O_3 and CO_2 can interact in tree response, with elevated CO_2 ameliorating O_3 effects

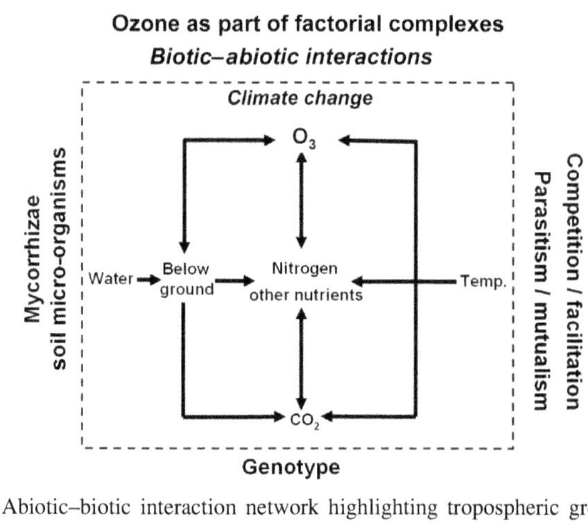

FIGURE 6.2 Abiotic–biotic interaction network highlighting tropospheric ground-level ozone and its effects on vegetation as subjected to multiple factorial influences (arrows denote cause–effect relationships).

or inversely, enhanced O_3 counteracting stimulation by CO_2 (King et al., 2013, this vol.; Kubiske et al., 2007; Zak et al., 2011).

The balance of the above interactions depends on abiotic site factors (like nutrition) and/or the biotic components of the genotype or age of the trees (Karnosky et al., 1998; Kolb and Matyssek, 2001; McGrath et al., 2010). In view of climate change, enhanced temperature may drive nutrient availability through mineralization (influencing effects by CO_2 and O_3), as long as water supply stays non-limiting (Figure 6.2). Drought, however, can hinder stomatal CO_2 and O_3 uptake, curtailing productivity or preventing decline, respectively (Löw et al., 2006), whereas the impact of low soil moisture on root systems may reduce tree C sink strength and mineralization. Conversely, above-ground O_3 stress on trees can impede below-ground C allocation (as commonly observed: Matyssek and Sandermann, 2003) and exacerbate drought effects via limited root growth, although stimulation may occur upon O_3 impact (see below and Nikolova et al., 2010). Tree response can become most complex if nesting of O_3 action, in addition to interference with other abiotic factors, comprises biotic interactions at the site, beyond such by tree-associated features (like genotype, age). The effects of competition or facilitation by neighbouring plants need to be considered (Cahill, 2012; Grams et al., 2012; Kozovits et al., 2005), along with host–parasite relationships (Karnosky et al., 2002; Osswald et al., 2012) and/or rhizospheric interactions (Agerer et al., 2012). Figure 6.2 illustrates that tree performance under O_3 stress can hardly be generalized, but strongly depends on prevailing biotic–abiotic conditions. In particular, the mostly neglected biotic component, both internally (see Section 6.4.1) and externally (see Section 6.4.2) to the tree, represents an important but under-studied driver of ecosystem response to elevated O_3. Awareness of the respective factors will guide the lessons learnt (see Section 6.3) and identification of unanswered questions—challenging upcoming experimental O_3 research at representative forest sites (Sections 6.5 and 6.6).

6.3 THE KRANZBERG FOREST EXPERIMENT AS A STARTING POINT

6.3.1 Motivation of the Experiment

Prior to the year 2000, adult-tree O_3 response was not available from controlled experimentation under forest conditions (Matyssek and Innes, 1999). O_3 exposure experiments using individual branch bags or cuvettes had left behind debates about canopy representativeness, micro-climatic bias in branch response and branch-level C autarchy in O_3 defence relative to the non-fumigated canopy parts (Sandermann et al., 1997; Wieser et al., 2012). Experimental free-air O_3 exposure systems allowing whole-tree approaches without micro-climatic bias appeared, therefore, as an ultimate choice (Karnosky et al., 2001; Musselman and Hale, 1997). The only experiment of such kind had been conducted within tree crowns on clusters of sun and shade foliage (Tjoelker et al., 1995). The novelty of KFE was the employment of free-air O_3 exposure methodology to up to 27 m high and about 60-year-old European beech (*Fagus sylvatica*) and Norway spruce trees (*Picea abies*) throughout the growing seasons of 8 consecutive years in a mixed forest in central Europe under closed-canopy conditions (see details in Karnosky et al., 2007; Nunn et al., 2002; Pretzsch et al., 2010a; Werner and Fabian, 2002). Five neighbouring trees of each species under the given O_3 regime at the site ($1 \times O_3$) served as control, whereas five other trees were exposed to $2 \times O_3$, which followed $1 \times O_3$ at experimental enhancement by a factor of two, allowing 150 nL L^{-1} at maximum (tree individual being the statistical unit, Matyssek et al., 2007). The aim of the study was the cause–effect clarification of O_3 response at the cell-, organ- and whole-tree level (Nunn et al., 2005a; Sandermann and Matyssek, 2004) relative to $1 \times O_3$ (Matyssek et al., 2007, 2010b).

6.3.2 Outcome

A key finding was a decline in annual stem productivity of beech under $2 \times O_3$, representing a 44% loss (about 10 m^3 ha^{-1} $year^{-1}$), on average, relative to $1 \times O_3$ during the 8-year study period (Pretzsch et al., 2010a). The effect became apparent from whole-stem diameter–height relationships, being undetectable through conventional breast height analysis (Wipfler et al., 2005). The major O_3 effect on beech leaves was stomatal narrowing across entire crowns rather than decline in photosynthesis or canopy-level gross primary productivity (Kitao et al., 2009, 2012). The above-ground O_3 impact was internally signalled below ground, as reflected by enhanced soil respiration around stems of $2 \times O_3$ trees and increased annual fine-root productivity and fungal diversity of mycorrhizae (Grebenč and Kraigher, 2007; Haberer et al., 2007; Nikolova et al., 2010). The increased C sink strength below ground was mediated in leaves by O_3-caused decline of cytokinins and their drain, via xylem transport, to the foliage (Winwood et al., 2007), which relieved cytokinin-associated inhibition of fine-root production (Matyssek et al., 2012a; Riefler et al., 2006), in the presence of annually varying molecular and biochemical O_3 responses, incipient macroscopic leaf injury and accelerated leaf senescence (Matyssek et al., 2010b; Nunn et al., 2005a). $2 \times O_3$ hardened beech against the fungal leaf endophyte *Apiognomonia errabunda* during summer (Bahnweg et al., 2005; Olbrich et al., 2010). $2 \times O_3$ did not inhibit whole-stem productivity in spruce, although soil respiration was enhanced around $2 \times O_3$ trees (Nikolova et al., 2010). Drought rather than O_3 stress limited stem productivity of beech and spruce during the exceptionally warm and dry year of 2003 (cf. Ciais et al., 2005; Löw et al., 2006), when also O_3-driven below-ground stimulations were nullified in both species (Nikolova et al., 2009, 2010). Drought suspended the endophytic infection (see above; Bahnweg et al., 2005).

6.3.3 Remaining Uncertainties

Apart from restrictions in the experimental design of this case study, this being unique to date, biotic factors could not be weighted in the O_3 response of the adult forest trees. Uncovered drivers were genotype (Karnosky et al., 1998; Kolb and Matyssek, 2001; McGrath et al., 2010), although setting the stage for metabolic defence of O_3 stress and acclimation, and inter-organismic relationships expressed through competition or facilitation amongst trees, or through micro-organisms (bacteria and fungi) and animals (e.g. insects: Hillstrom et al., 2010; Lindroth, 2010) via parasitism, mutualism or herbivory. In the following Section 6.4, we will highlight the significance of biotic factors for the current frontiers in understanding tree susceptibility to ozone before outlining requirements for experimental concepts and designs as well as feasibilities in Section 6.5.

6.4 THE ECOLOGICAL SIGNIFICANCE OF BIOTIC FACTORS FOR DEVELOPING NEW O_3 RESEARCH

6.4.1 Biological Relevance of Tree Features for O_3 Response

6.4.1.1 Actual and Effective O_3 Dose

O_3 effects on trees depend on a three-step chain of events: external O_3 exposure, stomatal O_3 uptake and metabolic effects which, depending on the defence capacity (Calatayud et al., 2010; Matyssek et al., 2008; Wieser and Matyssek, 2007), reflect the effective O_3 dose. The latter represents the integrated flux (i.e. actual O_3 uptake) over time as causing, *sensu* medical definitions, metabolic alteration (Wieser and Tausz, 2006). O_3 uptake inherently reflects stress avoidance as stomatal regulation may counteract O_3 influx, whereas defence is an expression of stress tolerance.

Upon O_3 entering the intercellular air space, reactive oxygen species are formed (Ernst et al., 2012). In response, antioxidant fluctuations may buffer the O_3 influx (Tausz et al., 2007). Based on the conceptual model by Massman et al. (2000) evaluating O_3 injury in plants, Wieser et al. (2002a) suggested linking O_3 influx to area-based antioxidant concentrations in the leaves (Figure 6.3). Similarly, Kolb and Matyssek (2001) suggested relating O_3 influx to net photosynthetic CO_2 uptake rate, with high ratios indicating O_3 sensitivity. Such criteria were conducive to interpreting differential O_3 susceptibility between juvenile and mature forest trees (Wieser et al., 2002a), tree species (Wieser et al., 2012) and trees under chamber and free-air conditions (Nunn et al., 2005b). As O_3/CO_2 uptake ratios positively correlate with specific leaf area (SLA) (cf. Kolb and Matyssek, 2001, presuming high SLA under shading: Kitao et al., 2009), this leaf feature may serve as a proxy for weighing the risk of O_3 influx, given that antioxidant levels decrease towards high SLA (Figure 6.3) in parallel to lowered photosynthetic capacity. Wieser

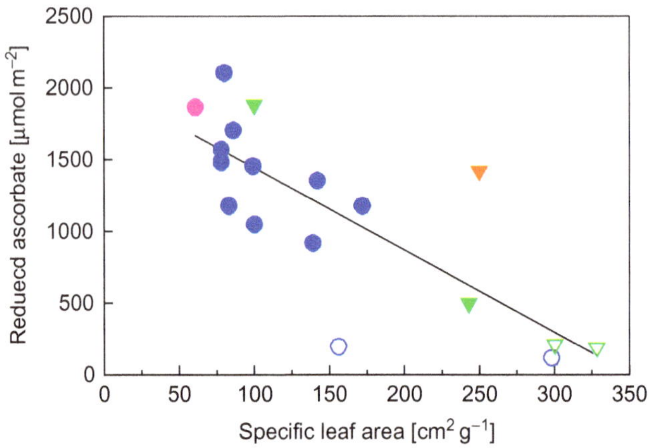

FIGURE 6.3 The correlation between specific leaf area and leaf surface area-based reduced ascorbate concentration in sun (closed symbols) and shade (open symbols) leaves of differently aged trees of *Picea abies* (blue circles), *Pinus cembra* (magenta circles), *Larix decidua* (orange triangles) and *Fagus sylvatica* (green triangles). Fit represents linear regression: $y = -5.73x + 2016$, $R^2 = 0.79$, $P < 0.001$. *Compiled after data from Wieser et al. (2002a,b, 2003, 2012) and Herbinger et al. (2007).*

and Matyssek (2007) postulated that induction of O_3 injury occurs if defence costs are not met by assimilate supply.

Variability in O_3 sensitivity relates to genotype (see above). For example, high-altitude provenances of *P. abies* (Havranek et al., 1990) and *Larix decidua* (Wieser, 1995) were less sensitive to O_3 than provenances from low altitude (see also Cieslik et al., 2013, this vol.). High variability among genotypes in O_3 response was also found in *Populus* (Coleman et al., 1995; Karnosky et al., 2007) during the Aspen FACE Experiment (see Pregitzer and Talhelm, 2013, this vol.). Canopy features determine avoidance *versus* tolerance of O_3 stress through their source–sink capacities in O_3–BVOC (biogenic volatile organic compounds) interaction (see Cieslik et al., 2013, this vol.; Loreto and Fares, 2013, this vol.).

6.4.2 Role of Biotic Interactions

One lesson from KFE is the recognition of dealing with a limited number of individual trees (see above). In adult beech and spruce trees, different genotypes perhaps were reflected in the high metabolic variation observed (Nunn et al., 2005a). KFE, in addition, was restricted to one site, that is one specific scenario of biotic interactions. Beyond genotype, tree O_3 susceptibility is substantially driven by competition with neighbours (Kubiske et al., 2007; Matyssek et al., 2010a; McDonald et al., 2002), although mostly this has been shown for juvenile trees. Hence, external biotic determinants of O_3 sensitivity are highlighted here, namely, influences of neighbouring trees and mutualistic or parasitic micro-organisms (for effects of herbivory, see Couture and Lindroth, 2013, this vol.).

6.4.2.1 *Competition Between Trees*

Competition as a modifier of O_3 sensitivity needs to be clarified for late-successional beech and spruce at advanced tree age, as managed mixed forests of these species are of great importance in central Europe (Scherer-Lorenzen et al., 2005). Such near-natural systems are of higher benefit for warranting ecological and socio-economic services than the widely prevalent artificial spruce monocultures (Olsthoorn et al., 1999). Beech/spruce mixtures are known for over- and under-yielding in stem productivity relative to respective monocultures (Pretzsch et al., 2010b). Competition may turn into facilitation, enabling over-yielding along edaphic gradients, in particular if they become growth limiting (cf. Callaway and Walker, 1997). Such change may be mediated through altered sensitivity of species/genotype mixtures to O_3 stress.

Understanding O_3 stress as one component of climate change (see Section 6.1), association with variable water availability is of particular relevance for tree mixtures, as drought and O_3 impact can ultimately curtail carbon gain, depending on the extent of stomatal opening, and eventually cause injury (Matyssek et al., 2006). Beech and spruce are possibly characterized by contrasting drought-related mechanisms of decline (Lyr et al., 1992). Spruce is supposed to behave "isohydrically", that is closing stomata early under water limitation and stabilizing water relations, so that starvation may result from the prolonged exclusion of CO_2 (cf. McDowell et al., 2008; Sala, 2009). In comparison, "anisohydric" beech may sustain high transpiration and CO_2 uptake rates until hydraulic collapse leads to extensive cavitation in the xylem and disrupted phloem transport (cf. Hartmann, 2011). In mixtures, "risk-taking" beech may profit, therefore, from the soil water not being used by "conservative" spruce, adding complexity to O_3–drought interaction at the stand level. During the exceptionally warm and dry growing season of 2003 in Central Europe (Ciais et al., 2005), spruce reflected dormancy by ceasing fine-root production as opposed to beech (Nikolova et al., 2009). Clarification beyond episodic observation is mandatory for lowering the silvicultural risk in tending mixed forests under the enhanced O_3 regimes of climate change (Pretzsch et al., 2010b).

6.4.2.2 *Below-Ground Processes Affecting Competition*

Agerer et al. (2012) summarize the indirect belowground effects of the $2 \times O_3$ regime of KFE. O_3 interference at the tree canopy level led to increased soil respiration and fine-root production (Nikolova et al., 2010), accompanied by increased density but altered community structure of ectomycorrhizae (ECM) on beech (Haberer et al., 2007). Specific N uptake was reduced as possibly an ECM-mediated O_3 effect (Haberer et al., 2007) and corroborated by Weigt et al. (2012) upon ^{15}N soil labelling. In spruce, shifts occurred from "medium" and "long-distance" towards "short-distance" and compact "contact" ectomycorrhizal exploration types after 8 years of $2 \times O_3$ exposure. Loss in mycelial below-ground soil volume occupation amounted to about 80% relative to $1 \times O_3$ (Agerer et al., 2012). Below-ground loss in mycelial exploration capacity upon above-ground O_3 impact is likely to affect tree competitiveness. Unanswered questions remain regarding potential influences by O_3-stressed trees on the interaction between their ECM and associated bacteria and on the bacterial organization within biofilms on mycorrhizae. As EMC-associated bacteria can enhance nutrient mobilization (Calvaruso et al., 2007; Smith and Read, 2008), and given their physiological capacities in general ("helper bacteria", Duponnois and Garbaye, 1991), holobiontic understanding (Matyssek and Lüttge, 2013) of O_3-inflicted interference within beneficial below-ground tree, fungal and bacterial interactions (cf. Werner and Zadworny, 2003) poses a challenge for research. This holds even more so, if mutualistic interaction networks become parasitic, with ozone being a stressor of such capacity.

6.4.2.3 *Parasitic Interactions*

These were only marginally covered at KFE (Bahnweg et al., 2005). Direct impact of O_3 on fungal development can be neglected at concentrations <150–$200\ nL\ L^{-1}$ (Manning and Tiedemann, 1995) so that effects of the prevalent chronic O_3 stress on host-pathogen interactions result from changes within the host plant. Differentiation in upcoming research is required between the action of necrotrophic and biotrophic pathogens, which require destroyed host tissue before invasion or living host cells, respectively. Typically, disease intensity by necrotrophic fungi is high, as they benefit from the preceding injury by ozone (James et al., 1980a,b), whereas injury by biotrophic pathogens stays restricted in O_3-exposed plants, either upon lowered nutritional quality or pre-activation of defence under O_3 stress. Complexity in such differentiation was indicated during KFE, given the infestation of beech leaves by *A. errabunda* (Bahnweg et al., 2005; Olbrich et al., 2010), a fungal biotrophic leaf endophyte that can become necrotrophic by fall to cause the "beech blight disease". O_3 stress confined the biotrophic phase, but significantly promoted the nectrotrophic phase, although the underlying mechanisms are not well understood.

Root exudates are crucial in shaping microbial community composition in the rhizosphere, and its productivity and turnover. Relevance exists for the complex O_3-driven interactions between pathogens and antagonistic microorganisms, complicating the predictability of pathogenesis, especially of root diseases. As above-ground O_3 affects the carbon allocation belowground (see above and Kruse et al., 2013, this vol.), intensity and chemical composition of root exudation are likely to change (McCrady and Andersen, 2000). Related evidence, contrasting with herbaceous plants (Doornbos et al., 2012), is missing for trees. Research is needed, therefore, to clarify global change impact on tree diseases under field conditions, with O_3 being one driver. To this end, fostering of the above mentioned "holobiontic" understanding of plant–microbe interactions is crucial for advancing our understanding and ability to predict forest responses to air pollution and climate change. Respective approaches are compelling as they represent the natural plant life, ecology and evolution.

6.4.3 Biotic Interactions and Controlled Field Experimentation

The challenge of experimentation under forest conditions on tree responses to ozone while being subject to biotic interactions is a matter of the appropriate control. It is conceivable to unravel effects of competition or facilitation amongst neighbouring trees, while being exposed to experimental O_3 stress. Assuming employment of free-air fumigation methodology with unchanged ambient air as the control, competitive or facilitative effects on O_3 response may be unveiled by comparing plots in forests with trees in mono *versus* multi-specific association, either naturally or silviculturally established.

The situation is more complex, unveiling impact of interaction with other organisms than trees on the tree O_3 responses, since controlled introduction (e.g. of pathogens) or withdrawal (e.g. of mycorrhizal fungi or rhizospheric bacteria) is technically not feasible under forest conditions. Nevertheless, ways exist to circumvent such limitation. One way is to analyse system response, that is of trees and interacting other organisms, as naturally occurring at the stand scale, but experimentally exposed in neighbouring plots to differing free-air O_3 regimes. Although system responsiveness can be discerned in such way, mechanistic grounds remain obscured, owing to the limitation in varying the presence of biotic associates. Such ultimate limitation can be compensated by conceptually linking field research with controlled multi-organismic chamber studies. Matyssek et al. (2010a) proposed such concepts (Figure 6.4), in particular, if phytotrons allow for advanced ecological experimentation. Complementarity may be created between free-air and chamber approaches, as both require validation by scenarios that prevail at experimentally untouched forest sites as a reference. In such ways, mechanisms can be approached that may improve interpretation of tree O_3 responses at forest sites, under the "bias" of biotic interactions.

6.5 GUIDING O_3 RESEARCH INTO THE FUTURE

Future research on O_3 action in forest trees and ecosystems needs to integrate three pathways of evidential advancement: experimentation, process monitoring and modelling. Gain in cause–effect-based knowledge is the aim of

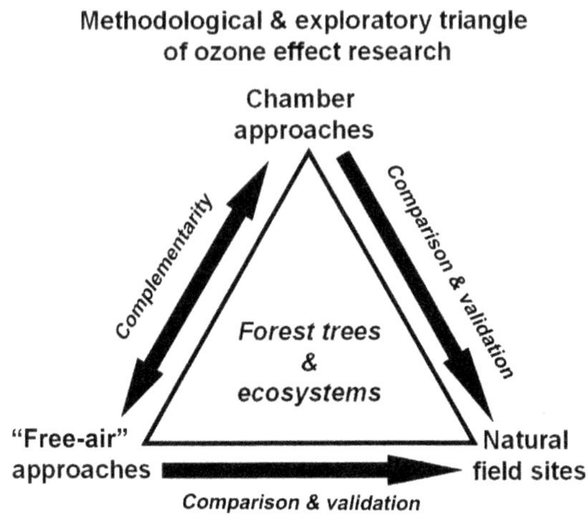

FIGURE. 6.4 Conceptual triangle for directing research on O_3 effects on forest trees and ecosystems, highlighting the complementarity of experimental free-air exposure and chamber/phytotron approaches towards ecologically relevant consolidation of knowledge (as mediated through scenario comparisons and validation with independent field data; see text for details, from Matyssek et al., 2010a).

experimentation, which creates the mechanistic grounds of understanding O_3 action within and between atmosphere and vegetation. Advancing from restricted case studies to hierarchically arranged spatio-temporal scales is the subsequent step, while examining available evidence for ecological representativeness (see Cieslik et al., 2013, this vol.). The latter task is to be ensured through cross-referencing between experimental outcomes and monitoring databases on ecosystem development (see Part VI). This kind of evaluation fosters and validates numeric modelling (see Part IV), promoting cause–effect process scaling between trees, ecosystems and the globe with O_3 stress as a key component of climate change. Eventually, modelling enables ecologically reliable tool development for risk assessment. In the following section, conceptual and methodological requirements will be highlighted for experimentation in relation to monitoring and modelling with a global perspective.

6.5.1 Experimentation

6.5.1.1 *Setting the Stage*

Future experimental demands for clarifying the action of enhanced tropospheric O_3 levels in forest trees and ecosystems are analogous to postulations by Calfapietra et al. (2009) on the effects of elevated CO_2 concentrations. In both cases, the scope of mechanistic understanding is to be advanced towards widened spatio-temporal scales. One means of consolidating ecologically relevant knowledge is field experiments, which imminently need to approach the large-scale and long-term dimensions of forest ecosystem functioning. The perspective represents cause–effect-based understanding of biotic–abiotic interactions in forest ecosystems under O_3 impacts (cf. Figure 6.2), including feedbacks with atmospheric processes, while functionally linking tree with ecosystem level and beyond at appropriate temporal scales (Sandermann and Matyssek, 2004). Ecosystems of mature closed-canopy forests have not yet been the subject of such integrated O_3 research.

6.5.1.2 *Conceptual Requirements*

Striving for the forest ecosystem-scale means attaining evidence of O_3 impact in interaction with other crucial drivers of system performance. These together with the biotic stressors are mainly temperature, precipitation and nutrition (cf. Figure 6.2), all being under anthropogenic influences (e.g. through high N deposition and/or increasing atmospheric CO_2 concentration). The interplay between the climatic and nutritional drivers is crucial in governing the sink strength of C sequestration, although the interference of ozone with the factorial complex underlying sequestration is hardly understood at the tree and ecosystem levels. Evidently, whole-system C balances need to be quantified under O_3 stress, implying particular focus on factorial dependences of below-ground C fluxes and allocation (Agerer et al., 2012; see Kruse et al., 2013, this vol.). This is substantial for fathoming transitions between ecosystem C sink and source capacities under O_3 stress (cf. Matyssek et al., 2010b; Nikolova et al., 2010). The long-term challenge is clarification of the extent to which ecosystem productivity and source/sink capacities in C storage may vary under chronic O_3 impact.

Informed forest management requires the analysis of O_3-related ecosystem–atmosphere feedbacks in relation to prevalent tree species associations and progressive climate change. A major deficit in comprehending such feedbacks is the limited understanding of BVOC action, with its capacity for mitigating or exacerbating canopy-level O_3 regimes, depending on the particular physico-chemical conditions in the atmosphere and on the forest type (Matyssek et al., 2012a; Stockwell et al., 1997). Tree species and forest type relate to another rudimentary understanding, that is of mechanisms that underlie variable productivity. An area of research that has been neglected is the eventuality that elevated O_3 may enable some O_3-tolerant tree species to gain facilitative advantage (cf. Callaway, 2012) from growing in multi-species conditions. Hence, clarification of tree and ecosystem responsiveness to ozone is needed for multi-species associations. Such research will unveil functional grounds of genotype-specific and tree community-dependent ecosystem behaviour under O_3 stress, that is mechanisms of functional plasticity as the crucial biotic component in ecosystem response (Matyssek et al., 2012b). This information will improve silvicultural practises to minimize O_3 impacts on woody-plant systems.

6.5.1.3 Methodological Means

Free-air fumigation methodology has reached such a technological stage from experience gained throughout the past several decades that spatio-temporal extensions in experimentation now become feasible. This approach can reach the hectare scale (i.e. ecosystem scale) and, temporally, the scale of decades. In addition, more than one enhanced gas regime of O_3 and/or CO_2 are manageable for locating threshold ranges towards altered system behaviour or injury initiation (Karnosky et al., 2003; King et al., 2013, this vol.). The shortcoming that levels $<1 \times O_3$ cannot be accomplished is outweighed by whole-tree approaches even with tall individuals in the absence of micro-climatic artefacts (Karnosky et al., 2007; Werner and Fabian, 2002). For experimentation in tall forests with dense canopies, in particular, the free-air fumigation methodology developed for KFE is advantageous (Werner and Fabian, 2002) and performs promisingly also for FACE approaches, that is experiments with elevated CO_2 levels (Grams et al., 2011). The employed tubing system allows release of enhanced gas regimes throughout the stand canopy at close distances to leaves, not requiring wind direction to be accounted for, while preventing high over-dosing at gas outlet holes.

In agreement with Calfapietra et al. (2009), plots within ecosystems also during free-air O_3 experiments might be 30 m in diameter or larger, depending on pre-experimental analysis of stand heterogeneity. Plot design must account for variation in soil nutrition, which can dictate experimental outcome (cf. Figure 6.2). As for CO_2 studies, most proficient among them would be suites of large-scale, ecosystem-level O_3 experiments along climatic and/or edaphic gradients of continental or even hemispherical dimension. Structural and genetic diversity along such gradients, however, will hardly be met to the extent desired.

Fewer O_3 free-air fumigation experiments have been conducted in woody-plant systems as compared with free-air CO_2 experiments; but both kinds have focused on the temperate climate zone (Calfapietra et al., 2009). Hence, emphasis should be shifted towards the tropics and subtropics. Also neglected has been the boreal zonobiome with its coniferous ecosystems. Nevertheless, the overall scarcity of O_3 experiments in forest ecosystems necessitates research on broad-leafed temperate forests, given their ecological diversity across the northern hemisphere. In general, awareness should be directed to such forest ecosystems, which are crucial in global C cycling and storage both in the (sub)tropical and temperate zonobiomes.

6.5.2 Monitoring and Modelling

Monitoring O_3 flux into leaves is the primary stage for up-scaling from the tree to the stand/ecosystem and landscape level (see Cieslik et al., 2013, this vol.; Wieser et al., 2008). Since transpiration and O_3 influx are coupled through stomatal regulation, sap flow assessment of crown transpiration can be used to derive tree-level O_3 uptake (Matyssek et al., 2008; Wieser et al., 2003). Hence, the sap flow approach for monitoring stomatal O_3 influx as related to climatic factors and soil moisture enables validation of models used for risk assessment. The acceptance of sap flow-derived whole-tree O_3 uptake as a seasonal monitoring database for risk assessment will depend on the ease of model applications (Emberson et al., 2000; Matyssek et al., 2007). One step towards simplification is the finding that canopy conductance (G_C) is similarly related to leaf to air vapour mole fraction difference across contrasting tree species and site conditions (see Cieslik et al., 2013, this vol.; Wieser et al., 2008). Advancement in mechanistically modelling O_3 risks for forest trees is demonstrated by Grünhage et al. (2013, this vol.), so that focus in the following will stay within the scope of this chapter.

A challenge to modelling is that living organisms, after perturbation, tend to return to homeostasis by repair and/or compensatory mechanisms. Decline in productivity results, therefore, from the inability to repair or compensate for changes of cellular functions and structures (Tingey and Andersen, 1991; Tingey and Taylor, 1982), according to

$$\begin{aligned} \text{injury} &= f(\text{effective dose, repair/compensatory capacity}) \\ &= f(\text{stomatal uptake, detoxification capacity, repair/compensatory capacity}). \end{aligned}$$

Hence, mechanistic modelling must cover three parameterization steps for realistic risk assessment: (i) stomatal uptake, (ii) detoxification and (iii) repair/compensation (cf. Figure 6.2 in Matyssek et al., 2012a). To date, however, such parameterization of the fluctuating detoxification capacity and of repair and/or compensatory mechanisms has not become available.

As shown by Grünhage et al. (2012, 2013, this vol.), the LRTAP Convention's response functions for beech and spruce approximate realistic biomass losses for adult trees. Up-scaled response functions together with sap flow-based O_3 flux assessment will become a practical tool for tree- or stand-level based O_3 risk assessments to be employed regionally (see Cieslik et al., 2013, this vol.; Grünhage et al., 2013, this vol.). Nevertheless, without new free-air O_3 exposure experiments, thoroughly mechanistic and validated risk assessment based on stomatal flux and fluctuating detoxification and repair/compensatory capacities will not be achieved.

6.5.3 Exemplifying Integrated Research

6.5.3.1 The Option of "Supersites"

A promising approach to transposing conceptual and experimental needs is research "supersites" for mechanistically clarifying effects of air pollution and climate change interaction on forest tree and ecosystem functioning, as introduced by Fischer et al. (2011) and extended by Matyssek et al.

(2012a,b). As supersite capacities will be explored in Part VI of this volume, here we focus on the core of the concept, which strives to integrate experimental research, monitoring activities and modelling under a process-based and spatio-temporally cross-scale perspective. Such sites are to catalyze interdisciplinary collaboration across transnational research networks that can be spread across ecological gradients at the transcontinental, hemispherical or even global scales. O_3 research will intrinsically be integrated at supersites into the overall flux analysis at the ecosystem–atmosphere interface. Such integration will allow us to weight O_3 impacts on forest ecosystem services, such as maintaining species or genotype diversity and controlling water and carbon fluxes, and feedbacks to climate.

6.5.3.2 *Designing Transcontinental O_3 Research*

Complementary to research at supersites, how might transcontinental O_3 experiments on forest trees and ecosystems be designed? An integrated approach might rest on five interrelated processes: (i) controlled field experimentation with sensitive tree genotypes in plantations under enhanced O_3 regimes, (ii) assessment of O_3 uptake of maturing forest stands under prevalent site scenarios as affecting productivity, extended by experimental O_3 enhancements at the ecosystem ("hectare") scale, (iii) cross-referencing outcomes from (i) and (ii) with databases from monitoring networks (e.g. ICP Forests) for assessing ecological representativeness and general validity, (iv) process-based modelling of tree and ecosystem performance on the basis of (i) through (iii) across spatio-temporal scales and (v) transfer of knowledge (i–iv) to policy making and risk assessment within socio-economic contexts.

Regarding (i), the ultimate experimental methodology to be chosen is free-air O_3 canopy fumigation, employing more than one enhanced regime to enable derivation of threshold ranges towards injury development. Sensitive genotypes are preferred to foster mechanistic understanding of the "effective O_3 dose" through integrated molecular, biochemical and ecophysiological analysis (Matyssek et al., 2012b) upon assessment of the actual O_3 dose through the sap flow approach. The analysis must comprise the tree–soil system. Respective study sites should be arranged by continental climate zones: in the case of Europe, at least one site each for Fennoscandia, Central Europe and the Mediterranean region. Extension to forest sites with late-successional tree species is provided by (ii), including silvicultural scenarios. Additional edaphic and ecological characteristics or economically important forest sites should widen the N–S transect of (i), climatically, into the Atlantic-continental W–E dimension. Performance of maturing forests under their prevalent site scenarios, including monitoring of ambient O_3 impact, is the focus here, relating sap flow-derived O_3 uptake to productivity and providing ecological contexts for mechanistic evidence from (i). Such contexts should be strengthened by large-scale experimental free-air O_3 canopy fumigations (hectare basis) at some of the forest sites, extended by detailed mechanistic above- and below-ground analysis. The evidence from (i) and (ii) will be widened through (iii) to the zonobiome (e.g. pan-European scale) by cross-evaluation with monitoring databases and examination of large-scale representativeness and validity. Modelling (iv) integrates novel evidence across spatio-temporal scales into process-based numerical frameworks for ascertaining predictive reliability and suggesting advanced hypotheses for empirical research (i–iii). The spin-off from such conceptual interaction is tool development for practice related to risk assessment, socio-economic evaluation of O_3 effects and policy making. The inclusion of the ecosystem level in (i) through (iv) allows the coverage of multi-factorial settings.

6.5.3.3 *Global-Scale Perspective*

Due to the high mobility in the troposphere, plumes of anthropogenic O_3 and its precursors can expand over large geographical scales, so that regions distant from the pollutant source can suffer from enhanced O_3 background levels or episodes (e.g. Ireland within the lee of North America, Derwent et al., 2004). In parallel, new "hot spots" of O_3 exposure are increasing in severity and area around the globe (Sitch et al., 2007). Contrasting with the northern hemisphere, few studies have been undertaken in the southern hemisphere regarding the impact of O_3 on vegetation (e.g. Bulbovas et al., 2010; Furlan et al., 2008; Pina and Moraes, 2010). During the twenty-first century, the global gross primary production, however, is expected to become reduced by 14–23% due to O_3 impact, corresponding to a significant suppression of the global land C sink strength (Sitch et al., 2007). Due to a deficiency of databases on plant sensitivity to O_3 in tropical ecosystems and other regions (e.g. east Asia, Latin America, Indian sub-continent; see Koike et al., 2013, this vol.; Kozovits and Bustamante, 2013, this vol.; Laakso et al., 2013, this vol.; Pandey et al., 2013, this vol.), however, modelling suffers from significant uncertainties in extrapolating from North American and European tree species and ecosystems (see Part VI). Such uncertainties demand integrating study sites on both hemispheres in joint experimental research networks (see Section 6.5.1).

6.6 *QUO VADIS*? CONCLUSIONS, PERSPECTIVES AND POLICY IMPLICATIONS

Empirical evidence obtained during recent decades has led to the scientific recognition that enhanced ground-level O_3 regimes are undoubtedly important components of global change. However, this recognition has resulted in integrated research efforts towards a mechanistic understanding of the underlying interactions in a few forest ecosystems that are not representative globally and do not provide extrapolation through model development. This chapter has demonstrated the need to assess enhanced O_3 impacts on forest trees and ecosystems as part of multi-factorial interaction networks. Although their principles of action become understandable in qualitative terms, capacities are restricted in understanding the plasticity and, hence, in reliably predicting quantitative system response (cf. Matyssek et al., 2012b). This is a crucial shortcoming, in particular for forest ecosystems in view of their paramount importance globally for C sequestration and storage and, hence, confinement of CO_2-driven radiative forcing in the atmosphere. Given such outlined interaction networks between ground-level ozone and other relevant environmental factors related to global change, integrated concepts are needed when asking "*quo vadis?*" in view of O_3 research on forest trees and ecosystems. The need is affirmed by the outcome from KFE as one recent case study on adult forest trees that elucidated stand-level O_3 responsiveness while identifying a range of unanswered but crucial research questions about ecosystem response to O_3 stress. The following integrated research concepts are proposed to fill important gaps in knowledge:

- Analysis of tropospheric ozone in interaction with other chemical constituents and climatic drivers of the atmosphere as well as an agent of physicochemical feedback cycles with forest ecosystems;
- Clarification of the role of ozone within forest ecosystems as part of abiotic and biotic interaction networks, both during the process of physicochemical impact and in system response;
- Clarification of O_3-related cause–effect relationships in response as constituents of concurrently prevailing competitive/facilitative interactions, as well as host–parasite and mycorrhizal interaction networks;
- Ascertaining the ecological significance of tree genotype and ontogenetic and phenological stage, as well as tree and stand structure during the spatio-functional up-scaling process of O_3 responsiveness to the ecosystem level;
- Widening assessments from the ecosystem at the regional and hemispheric scales through cross-linked experimental, monitoring and modelling approaches, based on biome-level edaphic and climatic gradients, which should be covered across the tropical/sub-tropical and temperate/boreal zonobiomes.

As the need for each of these five concepts is fundamental *per se*, the order of priority appears to be secondary in view of the evident progress resulting from any of them, if actually executed.

It was stressed that the conceptual and methodological means needed to cope with the above-listed integrations have become available during recent decades. A crucial tool in experimentation will be ecosystem-scale free-air O_3 canopy exposure systems similar to the one developed for KFE. This approach can be linked with advanced phytotron studies for a mechanistic basic understanding, but needs to be linked with transcontinental monitoring databases and research on "supersites" for ecological validation, and with modelling for extrapolation. The latter needs to serve hypothesis development in testing response threshold ranges towards ecosystem decline. Integrated approaches of this kind envision "next-generation" O_3 research on ageing forest ecosystems and tree plantations beyond the sapling stage. However, to what extent such a vision will be realized depends on farsightedness at the level of policy making.

Therefore, relevance for the socio-economic context in terms of related societal benefits of the proposed research must be explored. It is important to be aware of the prevalent ecological and economic deficiencies, namely, that predictions of the mitigation of the global C sink strength still need to rely on modelling, which is parameterized mostly through agricultural datasets from crop plants and restricted in validation, given the current lack of appropriate forest ecosystem-level databases. This deficit is crucial, given the dramatic decline predicted in the global C sink strength due to the chronically enhanced O_3 impact on forest ecosystems. Any amount of CO_2 not being assimilated and stored by woody-plant systems, including tree plantations for renewable bioenergy production, will reinforce atmospheric heating (i.e. drive radiative forcing). This global effect exacerbates the economic aspect of the yield loss, restricts the substitution of fossil energy sources and jeopardizes ecosystem services related to human well-being and health. It also has political implications in view of international carbon trading, given the O_3-caused constraint on Clean-Development Mechanisms as initiated by the Kyoto Protocol. The above outline of next-generation O_3 research will decrease uncertainty through development of reliable cause–effect empirical tools for environmental policy making and risk assessment within prevailing socio-economic contexts. The multi-factorial rationale will prevent inappropriate mono-causal conclusions about ecosystem-level O_3 action, still influencing ecologically meaningful O_3 risk assessment today.

REFERENCES

Agerer, R., Hartmann, A., Pritsch, K., Raidl, S., Schloter, M., Verma, R., Weigt, R., 2012. Plants and their ectomycorrhizosphere: cost and benefit of symbiotic soil organisms. Ecol. Stud. 220, 213–242.

Bahnweg, G., Heller, W., Stich, S., Knappe, C., Betz, G., Heerdt, C., Kehr, R.D., Ernst, D., Langebartels, C., Nunn, A.J., Rothenburger, J., Schubert, R., Wallis, P., Muller-Starck, G., Werner, H., Matyssek, R., Sandermann, H., 2005. Beech leaf colonization by the endophyte *Apiognomonia errabunda* dramatically depends on light exposure and climatic conditions. Plant Biol. 7, 659–669.

Bulbovas, P., Moraes, R.M., Rinaldi, M.C.S., Cunha, A.L., Delitti, W.B.C., Domingos, M., 2010. Leaf antioxidant fluctuations and growth responses in saplings of *Caesalpinia echinata* Lam. (brazilwood) under an urban stressing environment. Ecotoxicol. Environ. Saf. 73, 664–670.

Cahill Jr., J.F., 2012. Plant competition: can understanding trait-behavior linkages offer a new perspective on very old questions? Nova Acta Leopoldina 114, 115–126.

Calatayud, V., Marco, F., Cervero, J., Sánchez-Peña, G., Sanz, M.J., 2010. Contrasting ozone sensitivity in related evergreen and deciduous shrubs. Environ. Pollut. 158, 3580–3587.

Calfapietra, C., Fares, S., Loreto, F., 2009. Volatile organic compounds from Italian vegetation and their interaction with ozone. Environ. Pollut. 157, 1478–1486.

Callaway, R.M., 2012. Facilitation, competition and the organization of plant communities. Nova Acta Leopoldina 114, 147–158.

Callaway, R.M., Walker, L.R., 1997. Competition and facilitation: a synthetic approach to interactions in plant communities. Ecology 78, 1958–1965.

Calvaruso, C., Turpault, M.-P., Leclerc, E., Frey-Klett, P., 2007. Impact of ectomycorrhizosphere on the functional diversity of soil bacterial and fungal communities from a forest stand in relation to nutrient mobilization processes. Microb. Ecol. 54, 567–577.

Ciais, Ph., Reichstein, M., Viovy, N., Granier, A., Ogée, J., Allard, V., Aubinet, M., Buchmann, N., Bernhofer, Chr., Carrara, A., Chevallier, F., De Noblet, N., Friend, A.D., Friedlingstein, P., Grünwald, T., Heinesch, B., Keronen, P., Knohl, A., Krinner, G., Loustau, D., Manca, G., Matteucci, G., Miglietta, F., Ourcival, J.M., Papale, D., Pilegaard, K., Rambal, S., Seufert, G., Soussana, J.F., Sanz, M.J., Schulze, E.D., Vesala, T., Valentini, R., 2005. Europe-wide reduction in primary productivity caused by the heat and drought in 2003. Nature 437 (7058), 529–534.

Cieslik, S., Tuovinen, J.-P., Baumgarten, M., Matyssek, R., Brito, P., Wieser, G., 2013. Gaseous exchange between forests and the atmosphere. Dev. Environ. Sci. 13, 19–36.

Coleman, M.D., Dickson, R.E., Isebrands, J.G., Karnosky, D.F., 1995. Carbon allocation and partitioning in aspen clones varying in sensitivity to tropospheric ozone. Tree Physiol. 15, 721–726.

Couture, J.J., Lindroth, R.L., 2013. Impacts of atmospheric change on tree-arthropod interactions. Dev. Environ. Sci. 13, 227–248.

Dentener, F., Stevenson, D., Ellingsen, K., van Noije, T., Schultz, M., Amann, M., Atherton, C., Bell, N., Bergmann, D., Bey, I., Bouwman, L., Butler, T., Cofala, J., Collins, B., Drevet, J., Doherty, R., Eickhout, B., Eskes, H., Fiore, A., Gauss, M., Hauglustaine, D., Horowitz, L., Isaksen, I.S.A., Josse, B., Lawrence, M., Krol, M., Lamarque, J.F., Montanaro, V., Muller, J.F., Peuch, V.H., Pitari, G., Pyle, J., Rast, S., Rodriguez, J., Sanderson, M., Savage, N.H., Shindell, D., Strahan, S., Szopa, S., Sudo, K., Van Dingenen, R., Wild, O., Zeng, G., 2006. The global atmospheric environment for the next generation. Environ. Sci. Technol. 40, 3586–3594.

Derwent, R.G., Stevenson, D.S., Collins, W.J., Johnson, C.E., 2004. Intercontinental transport and the origins of the ozone observed at surface sites in Europe. Atmos. Environ. 38, 1891–1901.

Doornbos, R.F., van Loon, L.C., Bakker, P., 2012. Impact of root exudates and plant defense signaling on bacterial communities in the rhizosphere. A review. Agron. Sustain. Dev. 32, 227–243.

Duponnois, R., Garbaye, J., 1991. Mycorrhiza helper bacteria associated with Douglas fir—*Laccaria laccata* symbiosis: effects in aseptic and in glasshouse conditions. Ann. Forest Sci. 48, 239–251.

Edwards, I.P., ZAK, D.R., 2011. Fungal community composition and function after long-term exposure of northern forests to elevated atmospheric CO_2 and tropospheric O_3. Glob. Chang. Biol. 17, 2184–2195.

Emberson, L.D., Ashmore, M.R., Cambridge, H.M., Simpson, D., Tuovinen, J.P., 2000. Modelling stomatal ozone flux across Europe. Environ. Pollut. 109, 403–413.

Ernst, D., Jürgensen, M., Bahnweg, G., Heller, W., Müller-Starck, G., 2012. Common links of molecular biology with biochemistry and physiology in plants under ozone and pathogen attack. In: Matyssek, R., Schnyder, H., Oßwald, W., Ernst, D., Munch, J.C., Pretzsch, H.

(Eds.), Growth and Defence in Plants—Resource Allocation at Multiple Scales. Ecological Studies, vol. 220. Springer, Heidelberg, pp. 29–52.

Fischer, R., Aas, W., De Vries, W., Clarke, N., Cudlin, P., Leaver, D., Lundin, L., Matteucci, G., Matyssek, R., Mikkelsen, T.N., Mirtl, M., Öztürk, Y., Papale, D., Potocic, N., Simpson, D., Tuovinen, J.P., Vesala, T., Wieser, G., Paoletti, E., 2011. Towards a transnational system of supersites for forest monitoring and research in Europe—an overview on present state and future recommendations. iForest 4, 167–171.

Fowler, D., Cape, J.N., Coyle, M., Flechard, C., Kuylenstienra, J., Hicks, K., Derwent, D., Johnson, C., Stevenson, D., 1999. The global exposure of forests to air pollutants. Water Air Soil Pollut. 116, 5–32.

Fowler, D., Amann, M., Anderson, R., Ashmore, M., Cox, P., Depledge, M., Derwent, D., Grennfelt, P., Hewitt, N., Hov, O., Jenkin, M., Kelly, F., Liss, P., Pilling, M., Pyle, J., Slingo, J., Stefenson, D., 2008. Ground-level Ozone in the 21st Century: Future Trends, Impacts and Policy Implications. The Royal Society Policy Document 15/08, London, p. 132.

Furlan, C.M., Moraes, R.M., Bulbovas, P., Sanz, M.J., Domingos, M., Salatino, A., 2008. *Tibouchina pulchra* (Cham.) Cogn., a native Atlantic Forest species, as a bio-indicator of ozone: visible injury. Environ. Pollut. 152, 361–365.

Grams, T.E.E., Werner, H., Kuptz, D., Ritter, W., Fleischmann, F., Andersen, C., Matyssek, R., 2011. A free-air system for long-term stable carbon isotope labeling of adult forest trees. Trees 25, 187–198.

Grams, T.E.E., Daigo, M.J., Winkler, J.B., Gayler, S., Matyssek, R., 2012. Growth and space use in competitive interactions between juvenile trees. Ecol. Stud. 220, 273–286.

Grebenč, T., Kraigher, H., 2007. Changes in the community of ectomycorrhizal fungi and increased fine root number under adult beech trees chronically fumigated with double ambient ozone. Plant Biol. 9, 279–287.

Grulke, N.E., Minnich, R.A., Paine, T.D., Seybold, S.J., Chavez, D.J., Fenn, M.E., Riggan, P.J., Dunn, A., 2009. Air pollution increases forest susceptibility to wildfires: a case study in the San Bernardino mountains in Southern California. Dev. Environ. Sci. 8, 365–403.

Grünhage, L., Matyssek, R., Häberle, K.H., Wieser, G., Metzger, U., Leuchner, M., Menzel, A., Dieler, J., Pretzsch, H., Grimmeisen, W., Zimmermann, L., Raspe, S., 2012. Flux-based ozone risk assessment for adult beech forests. Trees. http://dx.doi.org/10.1007/s00468-012-0716-5.

Grünhage, L., Matyssek, R., Wieser, G., Häberle, K.H., Leuchner, M., Menzel, A., Dieler, J., Pretzsch, H., Grimmeisen, W., Zimmermann, L., Raspe, S., Schröder, M., 2013. Flux-based ozone risk assessment for adult beech and spruce forests. Dev. Environ. Sci. 13, 251–266.

Haberer, K., Grebenc, T., Alexou, M., Gessler, A., Kraigher, H., Rennenberg, H., 2007. Effects of long-term free-air ozone fumigation on d15N and total N in *Fagus sylvatica* and associated mycorrhizal fungi. Plant Biol. 9, 242–252.

Hartmann, H., 2011. Will a 385 million year-struggle for light become a struggle for water and for carbon?—how trees may cope with more frequent climate change-type drought events. Glob. Chang. Biol. 17 (1), 642–655.

Havranek, W.M., Pfeifhofer, H., Grill, G., 1990. Pigmentgehalte und Gaswechsel von Tief- und Hochlagenfichten nach chronischer Ozonbelastung. Forstwissenschaftiches Centralblatt 109, 200–209.

Herbinger, K., Then, C., Haberer, K., Alexou, M., Löw, M., Remele, K., Rennenberg, H., Matyssek, R., Grill, D., Wieser, G., Tausz, M., 2007. Gas exchange and antioxidative compounds in young beech trees under free-air ozone exposure and comparisons to adult trees. Plant Biol. 9, 288–297.

Hillstrom, M., Meehan, T.D., Kelly, K., Lindroth, R.L., 2010. Soil carbon and nitrogen mineralization following deposition of insect frass and greenfall from forests under elevated CO_2 and O_3. Plant Soil 336, 75–85.

James, R.L., Cobb, F.W., Miller, P.R., Parmeter, J.R., 1980a. Effects of oxidant air-pollution on susceptibility of pine roots to *Fomes annosus*. Phytopathology 70, 560–563.

James, R.L., Cobb, F.W., Wilcox, W.W., Rowney, D.L., 1980b. Effects of photochemical oxidant injury of Ponderosa and Jeffrey pines on susceptibility of sapwood and freshly cut stumps to *Fomes annosus*. Phytopathology 70, 704–708.

Karnosky, D.F., Podila, G.K., Gagnon, Z., Pechter, P., Akkapeddi, A., Coleman, M., Dickso, R.E., Isebrands, J.G., 1998. Genetic control of responses to interacting O_3 and CO_2 in *Populus tremuloides*. Chemosphere 36, 807–812.

Karnosky, D.F., Gielen, B., Ceulemans, R., Schlesinger, W.H., Norby, R.J., Oksanen, E., Matyssek, R., Hendrey, G.R., 2001. FACE systems for studying the impacts of greenhouse gases on forest ecosystems. In: Karnosky, D.F., Scarascia-Mugnozza, G., Ceulemans, R., Innes, J.L. (Eds.), The Impacts of Carbon dioxide and Other Greenhouse Gases on Forest Ecosystems. CABI Press, New York, pp. 297–324.

Karnosky, D.F., Percy, K.E., Xiang, B.X., Callan, B., Noormets, A., Mankovska, B., Hopkin, A., Sober, J., Jones, W., Dickson, R.E., Isebrands, J.G., 2002. Interacting elevated CO_2 and tropospheric O_3 predisposes aspen (*Populus tremuloides* Michx.) to infection by rust (*Melampsora medusae f. sp tremuloidae*). Glob. Chang. Biol. 8, 329–338.

Karnosky, D.F., Zak, D., Pregnitzer, K., Awmack, C., Bockheim, J., Dickson, R., Hendrey, G., Host, G., King, J., Kopper, B., Kruger, E., Kubiske, M., Lindroth, R., Mattson, W., McDonald, E., Noormets, A., Oksanen, E., Parsons, W., Percy, K., Podila, G., Riemenschneider, D., Sharma, P., Thakur, R., Sober, A., Sober, J., Jones, W., Anttonen, S., Vapaavuori, E., Manskovska, B., Heilman, W., Isebrands, J., 2003. Tropospheric O_3 moderates responses of temperate hardwood forests to elevated CO_2: a synthesis of molecular to ecosystem results from the Aspen FACE project. Funct. Ecol. 17, 289–304.

Karnosky, D.F., Werner, H., Holopainen, T., Percy, K., Oksanen, T., Oksanen, E., Heerdt, C., Fabian, P., Nagy, J., Heilman, W., Cox, R., Nelson, N., Matyssek, R., 2007. Free-air exposure systems to scale up ozone research to mature trees. Plant Biol. 9, 181–190.

Keating, T.W., West, J.J., Farrell, A.E., 2004. Prospects for international management of intercontinental air pollution transport. In: Stohl, A. (Ed.), The Intercontinental Transport of Air Pollution. Springer, Berlin, pp. 295–320.

King, J.S., Pregitzer, K.S., Kubiske, M.E., Hendrey, G.R., Giardina, C.P., McDonald, E.P., Karnosky, D.F., 2005. Tropospheric O_3 compromises net primary production in young stands of trembling aspen, paper birch, and sugar maple in response to elevated atmospheric CO_2. New Phytol. 168, 623–636.

King, J., Liu, L., Aspinwall, M., 2013. Tree and forest responses to interacting elevated atmospheric CO_2 and tropospheric O_3: a synthesis of experimental evidence. Dev. Environ. Sci. 13, 179–208.

Kitao, M., Löw, M., Heerdt, C., Grams, T.E.E., Häberle, K.-H., Matyssek, R., 2009. Effects of chronic elevated ozone exposure on gas exchange responses of adult beech trees (*Fagus sylvatica*) as related to the within-canopy light gradient. Environ. Pollut. 157, 537–544.

Kitao, M., Winkler, J.B., Loew, M., et al., 2012. How closely does stem growth of adult beech (*Fagus sylvatica*) relate to net carbon gain under experimentally enhanced ozone stress? Environ. Pollut. 166, 108–115.

Klingberg, J.M., Enghardt, M., Uddling, J., Karlsson, P.E., Pleijel, H., 2010. Ozone risk for vegetation in the future climate of Europe based on stomatal ozone uptake calculations. Tellus 63, 174–187.

Koike, T., Watanabe, M., Hoshika, Y., Kitao, M., Matsumura, H., Funada, R., Izuta, T., 2013. Effects of ozone on forest ecosystems in East and Southeast Asia. Dev. Environ. Sci. 13, 371–390.

Kolb, T.E., Matyssek, R., 2001. Limitations and perspectives about scaling ozone impact in trees. Environ. Pollut. 115, 373–393.

Körner, C., 2006. Plant CO_2 responses: an issue of definition, time and resource supply. New Phytol. 172, 393–411.

Kozovits, A.R., Bustamante, M.M.C., 2013. Land use change, air pollution and climate change—vegetation response in Latin America. Dev. Environ. Sci. 13, 411–427.

Kozovits, A.R., Matyssek, R., Blaschke, H., Göttlein, A., Grams, T.E.E., 2005. Competition increasingly dominates the responsiveness of juvenile beech and spruce to elevated CO_2 and/or O_3 concentrations throughout two subsequent growing seasons. Glob. Chang. Biol. 11, 1387–1401.

Kruse, J., Simon, J., Rennenberg, H., 2013. Soil respiration and soil organic matter (SOM) decomposition in response to climate change. Dev. Environ. Sci. 13, 131–149.

Kubiske, M.E., Quinn, V.S., Marquardt, P.E., Karnosky, D.F., 2007. Effects of elevated atmospheric CO_2 and/or O_3 on intra- and interspecific competitive ability of aspen. Plant Biol. 9, 342–355.

Laakso, L., Beukes, J.P., Van Zyl, P.G., Pienaar, J., Josipovic, M., Venter, A., Jaars, K., Vakkari, V., Labuschagne, C., Chiloane, K., Tuovinen, J.-P., 2013. Ozone concentrations and their potential impacts on vegetation in southern Africa. Dev. Environ. Sci. 13, 429–450.

Lefohn, A.S., 1992. Surface Level Ozone Exposure and their Effects on Vegetation. Lewis Publishers, Chelsea, MI, p. 366.

Lindroth, R.L., 2010. Impacts of elevated atmospheric CO_2 and O_3 on forests: phytochemistry, trophic interactions, and ecosystem dynamics. J. Chem. Ecol. 36, 2–21.

Loreto, F., Fares, S., 2013. Biogenic volatile organic compounds and their impacts on biosphere-atmosphere interactions. Dev. Environ. Sci. 13, 57–75.

Löw, M., Herbinger, K., Nunn, A.J., Häberle, K.-H., Leuchner, M., Heerdt, C., Werner, H., Wipfler, P., Pretzsch, H., Tausz, M., Matyssek, R., 2006. Extraordinary drought of 2003 overrules ozone impact on adult beech trees (*Fagus sylvatica*). Trees 20, 539–548.

Luedemann, G., Matyssek, R., Fleischmann, F., Grams, T.E.E., 2005. Acclimation to ozone affects host/pathogen interaction and competitiveness for nitrogen on juvenile *Fagus sylvatica* and *Picea abies* trees infected with *Phytophthora citricola*. Plant Biol. 7, 640–649.

Lyr, H., Fiedler, H.J., Tranquillini, W., 1992. Physiologie und Ökologie der Gehölze. G. Fischer Verlag, Jena.

Manning, W.J., V. Tiedemann, A., 1995. Climate change: potential effects of increased atmospheric Carbon dioxide (CO_2), ozone (O_3), and ultraviolet-B (UV-B) radiation on plant diseases. Environ. Pollut. 88, 219–245.

Martins, V., Miranda, A.I., Carvalho, A., Schaap, M., Borrego, C., Sa, E., 2012. Impact of forest fires on particulate matter and ozone levels during the 2003, 2004 and 2005 fire seasons in Portugal. Sci. Total. Environ. 414, 53–62.

Massman, W.J., Musselman, R.C., Lefohn, A.S., 2000. A conceptual ozone-dose response model to develop a standard to protect vegetation. Atmos. Environ. 34, 45–759.

Matyssek, R., Innes, J.L., 1999. Ozone—a risk factor for trees and forests in Europe? Water Air Soil Pollut. 116, 199–226.

Matyssek, R., Lüttge, U., 2013. Gaia: the planet Holobiont. Nova Acta Leopoldina 114, 325–344.

Matyssek, R., Sandermann, H., 2003. Impact of Ozone on Trees: An Ecophysiological Perspective. In: Progress in Botany, vol. 64. Springer Verlag, Heidelberg, pp. 49–404.

Matyssek, R., Havranek, W.M., Wieser, G., Innes, J.L., Matyssek, R., Havranek, W.M., Wieser, G., Innes, J.L., 1997. Ozone and the forests in Austria and Switzerland. In: Sandermann Jr., H., Wellburn, A.R., Heath, R.L. (Eds.), Forest Decline and Ozone: A Comparison of Controlled Chamber and Field Experiments. Ecological Studies, vol. 127. Springer-Verlag, Berlin, pp. 95–134.

Matyssek, R., Le Thiec, D., Löw, M., Dizengremel, P., Nunn, A.J., Häberle, K.H., 2006. Interaction between drought stress and O_3 stress in forest trees. Plant Biol. 8, 11–17.

Matyssek, R., Bytnerowicz, A., Karlsson, P.E., Paoletti, E., Sanz, M., Schaub, M., Wieser, G., 2007. Promoting the O_3 flux concept for European forest trees. Environ. Pollut. 146, 587–607.

Matyssek, R., Sandermann, H., Wieser, G., Booker, F., Cieslik, S., Musselman, R., Ernst, D., 2008. The challenge of making ozone risk assessment for forest trees more mechanistic. Environ. Pollut. 156, 567–582.

Matyssek, R., Karnosky, D.F., Wieser, G., Percy, K., Oksanen, E., Grams, T.E.E., Kubiske, M., Hanke, D., Pretzsch, H., 2010a. Advances in understanding ozone impact on forest trees: messages from novel phytotron and free-air fumigation studies. Environ. Pollut. 158 (6), 1990–2006.

Matyssek, R., Wieser, G., Ceulemans, R., Rennenberg, H., Pretzsch, K., Häberle, K.H., Löw, M., Nunn, A.J., Werner, H., Wipfler, P., Oßwald, W., Nikolova, P., Hanke, D.E., Kraigher, H., Tausz, M., Bahnweg, G., Kitao, M., Dieler, J., Sandermann, H., Herbinger, K., Grebenc, T., Blumenröther, M., Deckmyn, G., Grams, T.E.E., Heerdt, C., Leuchner, M., Fabian, P., Häberle, K.H., 2010b. Enhanced ozone strongly reduces carbon sink strength of adult beech (*Fagus sylvatica*)—resume from the free-air fumigation study at Kranzberg Forest. Environ. Pollut. 158, 2527–2532.

Matyssek, R., Wieser, G., Calfapietra, C., de Vries, W., Dizengremel, P., Ernst, D., Jolivet, Y., Mikkelsen, T.N., Mohren, G.M.J., le Thiec, D., Tuovinen, J.-P., Weatherall, A., Paoletti, E., 2012a. Forests under climate change and air pollution: gaps in understanding and future directions for research. Environ. Pollut. 160, 57–65.

Matyssek, R., Gayler, S., zu Castell, W., Oßwald, W., Ernst, D., Pretzsch, H., Schnyder, H., Munch, J.C., 2012b. Predictability of plant resource allocation: new theory needed? In: Matyssek, R., Schnyder, H., Osswald, W., Ernst, D., Munch, J.C., Pretzsch, H. (Eds.), Growth and Defence in Plants. Ecological Studies, vol. 220. Springer, Heidelberg, pp. 433–449.

Matyssek, R., Kosovits, A.R., Schnitzler, J.-P., Pretzsch, H., Dieler, J., Wieser, G., 2013. Forest trees under air pollution as a factor of climate change. In: Tausz, M., Grulke, N. (Eds.), Trees in a Changing Environment: Ecophysiology, Adaptation and Future Survival. Series Plant Ecophysiology, Springer, Heidelberg, in press.

McCrady, J.K., Andersen, C.P., 2000. The effect of ozone on below-ground carbon allocation in wheat. Environ. Pollut. 107, 465–472.

McDonald, E.P., Kruger, E.L., Riemenschneider, D.E., Isebrands, J.G., 2002. Competitive status influences tree-growth responses to elevated CO_2 and O_3 in aggrading stands. Funct. Ecol. 16, 792–801.

McDowell, N.G., Pockman, W.T., Allen, C.D., Breshears, D.D., Cobb, N., Kolb, T., Plaut, J., Sperry, J., West, A., Williams, D.G., Yepez, E.A., 2008. Mechanisms of plant survival and mortality during drought: why do some plants survive while others succumb to drought? New Phytol. 178 (4), 719–739.

McGrath, J.M., Karnosky, D.F., Ainsworth, A.E., 2010. Spring leaf flush in aspen *(Populus tremuloides)* clones is altered by long-term growth at elevated carbon dioxide and elevated ozone concentration. Environ. Pollut. 158, 1023–1028.

Miller, P.R., McBride, J.M., 1999. Oxidant Air Pollution Impacts in the Montane Forest of Southern California. Ecological Studies. vol. 134. Springer, Heidelberg, pp. 317–336.

Musselman, R.C., Hale, B.A., 1997. Methods for controlled and field ozone exposures of forest tree species in North America. In: Sandermann Jr., H., Wellburn, A.R., Heath, R.L. (Eds.), Forest Decline and Ozone: A Comparison of Controlled Chamber and Field Experiments. Ecological Studies, vol. 127. Springer-Verlag, Berlin, pp. 277–315.

Newell, R.E., Evans, M.J., 2000. Seasonal changes in pollutant transport to the North Pacific: the relative importance of Asian and European sources. Geophys. Res. Lett. 27, 2509–2512.

Nikolova, P., Raspe, S., Andersen, C., Mainiero, R., Blaschke, H., Matyssek, R., Häberle, K.H., 2009. Effects of the extreme drought in 2003 on soil respiration in a mixed forest. Eur. J. Forest Res. 128, 87–98.

Nikolova, P.S., Andersen, C.P., Blaschke, H., Matyssek, R., Häberle, K.H., 2010. Below-ground effects of enhanced tropospheric ozone and drought in a beech/spruce forest (*Fagus sylvatica* L./*Picea abies* [L.] Karst). Environ. Pollut. 158 (4), 1071–1078.

Nunn, A.J., Reiter, I.M., Häberle, K.-H., Werner, H., Langebartels, C., Sandermann, H., Heerdt, C., Fabian, P., Matyssek, R., 2002. "Free-air" ozone canopy fumigation in an old-growth mixed forest: concept and observations in beech. Phyton 42, 105–119.

Nunn, A.J., Reiter, I.M., Häberle, K.-H., Langebartels, C., Bahnweg, G., Pretzsch, H., Sandermann, H., Matyssek, R., 2005a. Response pattern in adult forest trees to chronic ozone stress: identification of variations and consistencies. Environ. Pollut. 136, 365–369.

Nunn, A.J., Kozovits, A.R., Reiter, I.M., Heerdt, C., Leuchner, M., Lütz, C., Liu, X., Winkler, J.B., Grams, T.E.E., Häberle, K.-H., Werner, H., Fabian, P., Rennenberg, H., Matyssek, R., 2005b. Comparison of ozone uptake and responsiveness between phytotron study with young and field grown adult beech (*Fagus sylvatica*). Environ. Pollut. 137, 494–506.

Oberthür, S., Ott, H.E., 1999. The Kyoto Protocol: International Climate Policy for the 21st Century. Springer, Heidelberg, p. 441.

Olbrich, M., Knappe, C., Wenig, M., Gerstner, E., Haberle, K.-H., Kitao, M., Matyssek, R., Stich, S., Leuchner, M., Werner, H., Schlink, K., Muller-Starck, G., Welzl, G., Scherb, H., Ernst, D., Heller, W., Bahnweg, G., 2010. Ozone fumigation (twice ambient) reduces leaf infestation following natural and artificial inoculation by the endophytic fungus *Apiognomonia errabunda* of adult European beech trees. Environ. Pollut. 158, 1043–1050.

Olsthoorn, A.F.M., Bartelink, H.H., Gardiner, J.J., Pretzsch, H., Hekhuis, H.J., Franc, A., 1999. Management of Mixed-species Forest: Silviculture and Economics. IBN Scientific Contributions.

Osswald, W., Fleischmann, F., Treutter, D., 2012. Host–parasite interactions and trade-offs between growth- and defence-related metabolism under changing environments. Ecol. Stud. 220, 53–84.

Pandey, V., Oksanen, E., Singh, N., Sharma, C., 2013. Impacts of air pollution and climate change on plants: implications for India. Dev. Environ. Sci. 13, 391–409.

Pina, J.M., Moraes, R.M., 2010. Gas exchange, antioxidants and foliar injuries in saplings of a tropical woody species exposed to ozone. Ecotoxicol. Environ. Saf. 73, 685–691.

Pregitzer, K.S., Talhelm, A.F., 2013. Belowground carbon cycling at aspen FACE: dynamic responses to CO_2 and O_3 in developing forests. Dev. Environ. Sci. 13, 209–226.

Pretzsch, H., Dieler, J., Matyssek, R., Wipfler, P., 2010a. Tree and stand growth of mature Norway spruce and European beech under long-term ozone fumigation. Environ. Pollut. 158, 1061–1070.

Pretzsch, H., Block, J., Dieler, J., Dong, P.H., Kohnle, U., Nagel, J., Spellmann, H., Zingg, A., 2010b. Comparison between the productivity of pure and mixed stands of Norway spruce and European beech along an ecological gradient. Ann. Forest Sci. 67, 712.

Ren, W., Tian, H., Tao, B., Chappelka, A., Sun, G., Lu, C., Liu, M., Chen, G., Xu, X., 2011. Impacts of tropospheric ozone and climate change on net primary productivity and net carbon exchange of China's forest ecosystems. Global Ecol. Biogeogr. 20, 391–406.

Riefler, M., Novak, O., Strnad, M., Schmülling, T., 2006. *Arabidopsis* cytokinin receptor mutants reveal functions in shoot growth, leaf senescence, seed size, germination, root development, and cytokinin metabolism. Plant Cell 18, 40–54.

Sala, A., 2009. Lack of direct evidence for the carbon-starvation hypothesis to explain drought-induced mortality in trees. Proc. Natl. Acad. Sci. U. S. A. 106, E68.

Sandermann Jr., H., Matyssek, R., 2004. Scaling up from molecular to ecological processes. In: Sandermann, H. (Ed.), Molecular Ecotoxicology of Plants. Ecological Studies, vol. 170. Springer Verlag, Heidelberg, pp. 207–226.

Sandermann Jr., H., Wellburn, A.R., Heath, R.L., 1997. Forest decline and ozone: synopsis. In: Sandermann Jr., H., Wellburn, A.R., Heath, R.L. (Eds.), Forest Decline and Ozone: A Comparison of Controlled Chamber and Field Experiments. Ecological Studies, vol. 127. Springer-Verlag, Berlin, pp. 369–377.

Scherer-Lorenzen, M., Körner, C., Schulze, E.-D., 2005. Forest Diversity and Function: Temperate and Boreal Systems. Springer, Heidelberg.

Sitch, S., Cox, P.M., Collins, W.J., Huntingford, C., 2007. Indirect radiative forcing of climate change through ozone effects on the land-carbon sink. Nature 448, 791–794.

Smith, S.E., Read, D.J., 2008. Mycorrhizal Symbiosis, 33rd ed. Academic Press, San Diego, London.

Stockwell, W.R., Kramm, G., Scheel, H.-E., Mohnen, V.A., Seiler, W., 1997. Ozone formation, destruction and exposure in Europe and the United States. In: Sandermann Jr., H., Wellburn, A.R., Heath, R.L. (Eds.), Forest Decline and Ozone, A Comparison of Controlled Chamber and Field Experiments. Ecological Studies, vol. 127. Springer, Berlin, pp. 1–38.

Takemoto, B.K., Bytnerowicz, A., Fenn, M.E., 2001. Current and future effects of ozone and atmospheric nitrogen deposition on California's mixed conifer forests. Forest Ecol. Manage. 144, 159–173.

Tausz, M., Grulke, N.E., Wieser, G., 2007. Defence and avoidance of ozone under global change. Environ. Pollut. 147, 525–531.

Tingey, D.T., Andersen, C.P., 1991. The physiological basis of differential plant sensitivity to changes in atmospheric quality. In: Taylor, G.E., Pitelka, L.F., Clegg, M.T. (Eds.), Ecological Genetics and Air Pollution. Springer, New York, pp. 209–235.

Tingey, D.T., Taylor, G.E., 1982. Variation in plant response to ozone: a conceptual model of physiological events. In: Unsworth, M.H., Ormrod, D.P. (Eds.), Effects of Gaseous Air Pollution in Agriculture and Horticulture. Butterworth Scientific, London, pp. 113–138.

Tjoelker, M.G., Volin, J.C., Oleksyn, J., Reich, P.B., 1995. Interaction of ozone pollution and light effects on photosynthesis in a forest canopy experiment. Plant Cell Environ. 18, 895–905.

Vingarzan, R., 2004. A review of surface O_3 background levels and trends. Atmos. Environ. 38, 3431–3442.

Watanabe, M., Yamaguchi, M., Matsumura, H., Kohno, Y., Izuta, T., 2012. Risk assessment of ozone impact on *Fagus crenata* in Japan: consideration of atmospheric nitrogen deposition. Eur. J. Forest Res. 131, 475–484.

Weigt, R.B., Häberle, K.H., Millard, P., Metzger, U., Ritter, W., Blaschke, H., Göttlein, A., Matyssek, R., 2012. Ground-level ozone differentially affects nitrogen acquisition and allocation in mature European beech (*Fagus sylvatica*) and Norway spruce (*Picea abies*) trees. Tree Physiol. 32, 1259–1273. http://dx.doi.org/10.1093/treephys/tps087.

Werner, H., Fabian, P., 2002. Free-air fumigation of mature trees. Environ. Sci. Pollut. Res. 9, 117–121.

Werner, A., Zadworny, M., 2003. In vitro evidence of mycoparasitism of the ectomycorrhizal fungus *Laccaria laccata* against *Mucor hiemalis* in the rhizosphere of *Pinus sylvestris*. Mycorrhiza 13, 41–47.

Wieser, H., 1995. Auswirkungern chronicher Ozonbelastung auf die gaswechselphysiologie junger Hoch- und Tieflagenherkünfte der Lärchce (*Larix decidua* Mill.) in Klimakammern. Diploma thesis, Botany University, Innsbruck.

Wieser, G., Matyssek, R., 2007. Linking ozone uptake and defense towards a mechanistic risk assessment for forest trees. New Phytol. 174, 7–9.

Wieser, G., Tausz, M., 2006. Critical levels for ozone: further applying and developing the flux-based concept. UNECE workshop report. BFW-Wien.

Wieser, G., Tegischer, K., Tausz, M., Häberle, K.-H., Grams, T.E.E., Matyssek, R., 2002a. Age effects on Norway spruce (*Picea abies*) susceptibility to ozone uptake: a novel approach relating stress avoidance to defense. Tree Physiol. 22, 583–590.

Wieser, G., Hecke, K., Tausz, M., Häberle, K.-H., Grams, T.E.E., Matyssek, R., 2002b. The role of antioxidative defense in determining ozone sensitivity of Norway spruce (*Picea abies* (L.) Karst.) across tree age: implications for the sun and shade crown. Phyton 42, 245–253.

Wieser, G., Matyssek, R., Köstner, B., Oberhuber, W., 2003. Quantifying ozone uptake at the canopy level of spruce, pine and larch trees at the alpine timberline: an approach based on sap flow measurements. Environ. Pollut. 126, 5–8.

Wieser, G., Matyssek, R., Cieslik, S., Paoletti, E., Ceulemans, R., 2008. Upscaling ozone flux in forests from leaf to landscape. Ital. J. Agron. 3, 35–41.

Wieser, G., Matyssek, R., Götz, B., Grünhage, L., 2012. Branch cuvettes as means of ozone risk assessment in adult forest tree crowns: combining experiential and modeling capacities. Trees 26, 1703–1712. http://dx.doi.org/10.1007/s00468-102-0715-6.

Winwood, J., Pate, A.E., Price, A.J., Hanke, D.E., 2007. Effects of long-term, free-air ozone fumigation on the cytokinin content of mature beech trees. Plant Biol. 9, 265–278.

Wipfler, P., Seifert, T., Heerdt, C., Werner, H., Pretzsch, H., 2005. Growth of adult Norway spruce (*Picea abies* [L.] Karst.) and European beech (*Fagus sylvatica* L.) under free-air ozone fumigation. Plant Biol. 7, 611–618.

Wittig, V.E., Ainsworth, E.A., Naidu, S.L., Karnosky, D.F., Long, S.P., 2009. Quantifying the impact of current and future tropospheric ozone on tree biomass, growth, physiology and biochemistry: a quantitative meta-analysis. Glob. Chang. Biol. 15, 396–424.

Wuebbles, D.J., Jain, A., Edmonds, J., Harvey, D., Hayhoc, K., 1999. Global change: state of the science. Environ. Pollut. 100, 57–86.

Yamaguchi, M., Watanabe, M., Matsumura, H., Kohno, Y., Izuta, T., 2010. Effects of ozone on nitrogen metabolism in the leaves of *Fagus crenata* seedlings under different soil nitrogen loads. Trees 24, 175–184.

Zak, D.R., Pregitzer, K.S., Kubiske, M.E., Burton, A.J., 2011. Forest productivity under elevated CO_2 and O_3: positive feedbacks to soil N cycling sustain decade-Iong net primary productivity enhancement by CO_2. Ecol. Lett. 14, 1220–1226.

Chapter 7

Soil Respiration and Soil Organic Matter Decomposition in Response to Climate Change

Jörg Kruse[1], Judy Simon and Heinz Rennenberg
Chair of Tree Physiology, Institute of Forest Sciences, University of Freiburg, Freiburg, Germany
[1]Corresponding author: e-mail: joerg.kruse@ctp.uni-freiburg.de

Chapter Outline

7.1 INTRODUCTION

On a global scale, terrestrial soils store substantial amounts of organic carbon. Estimates of total organic carbon converge on 1500 Gt C (Kirschbaum, 2000), considerably more than that present in above-ground vegetation (ca. 560 Gt C; Matyssek et al., 2010a). The soil organic carbon (SOC) reservoir is roughly twice as large as the atmospheric CO_2–carbon reservoir. In simple terms, organic carbon enters the soil via autotrophic photosynthesis and is returned to the atmosphere via soil CO_2 respiration. Each year, roughly one-sixth of the atmospheric CO_2 passes through ecosystems (Trumbore, 2006), with soil respiration accounting for 70–80% of total terrestrial respiration (Ryan and Law, 2005). While carbon exchanges between atmosphere and biosphere are almost balanced, slight departures from 'steady-state' conditions can have

Developments in Environmental Science, Vol. 13. http://dx.doi.org/10.1016/B978-0-08-098349-3.00007-4

significant consequences for existing carbon pools. The postindustrial increase of atmospheric CO_2 originating from anthropogenic sources indicates insufficient sink strength to keep atmospheric CO_2 concentrations in balance (see Matyssek et al., 2013a, this vol.). In the long term, carbon transfer into soils might attenuate rates of atmospheric CO_2 increase, mediated by 'CO_2 fertilization' of the land vegetation (Iverson et al., 2008; Schimel, 1995; but also see Houghton, 2007). On the other hand, global warming has the potential to accelerate decomposition of soil organic matter (SOM) and, hence, soil respiration (see King et al., 2013, this vol.). After the last glacial maximum (ca. 10,000 years BC), for example, large soil carbon stocks developed in northern latitudes under a cool and comparatively stable climate (Hobbie et al., 2000; Schultz, 2008). In general, soil carbon stocks are inversely related to latitudinal and altitudinal gradients of average temperature (Jones et al., 2005). Further, turnover rates of 'young' soil carbon seem largely controlled by mean annual temperature and are faster in warm as compared to cold biomes. This was demonstrated by tracing ^{14}C-carbon, liberated during thermonuclear weapon testing in the early 1960s, through different SOM reservoirs (Trumbore et al., 1996). Such results imply that climate warming, predicted to be most pronounced in northern latitudes, might alter the balance between soil carbon input and decomposition (see King et al., 2013, this vol.). The fate of recalcitrant carbon stocks in a warmer world is of particular concern. If enhanced temperatures trigger fractional losses of old soil carbon, the resultant increase of atmospheric CO_2 might reinforce global warming (Davidson et al., 2000).

Yet there is considerable uncertainty about the importance of temperature for decomposition of old SOM (Bond-Lamberty and Thomson, 2010; Davidson et al., 2000; Giardina and Ryan, 2000; Kirschbaum, 2006; Liski et al., 1999). The temperature sensitivity of SOM decomposition is highly variable, depending on the history of soil cultivation and other site-specific factors like vegetation type, litter quality (see Kraigher et al., 2013, this vol.; Pregitzer and Talhelm, 2013, this vol.) and soil developmental stage (Neff and Hooper, 2002). For example, atmospheric deposition of nutrients (e.g. nitrate, ammonium, sulphate) and protons may be beneficial for microbial communities by increasing nutrient availability, or detrimental because of imbalanced nutrition with the consequences of impaired growth, leaching of pollutants into the groundwater and soil acidification (Greaver et al., 2012; Rennenberg, 1984; Rennenberg et al., 1998). Controversies also arose because different approaches to assessing the temperature sensitivity of SOM decomposition led to different conclusions (Conant et al., 2011; Lützow and Kögel-Knabner, 2009). Attempts to obtain information about SOM decomposition frequently rely on temperature-induced shifts of bulk soil respiration. Bulk CO_2 efflux is a valuable, non-invasive measure of physiological activity, integrating all processes relevant to respiration in undisturbed soils. On the other hand, bulk soil respiration does not distinguish between autotrophic root respiration and heterotrophic microbial respiration, and hence cannot be equated with SOM decomposition. For example, air pollutants such as SO_2 (Randewig et al., 2012; Rennenberg, 1984) or ozone (Löw et al., 2012) can cause stomatal closure thereby reducing photosynthetic CO_2 fixation (see Matyssek et al., 2013b, this vol.; Cieslik et al., 2013, this vol.; Couture and Lindroth, 2013, this vol.; King et al., 2013, this vol.) and, hence, assimilate transport from the leaves to the roots in the phloem. Ozone may also have an immediate effect on sucrose unloading into the phloem (Rennenberg et al., 1996) with the consequence of enhanced sugar levels in the leaves at reduced sugar transport in the phloem. Reduced sugar transport from the leaves to the roots will not only affect root growth, but also sugar exudation into the rhizosphere and, hence, mycorrhizal fungi (see Kraigher et al., 2013, this vol.; Pregitzer and Talhelm, 2013, this vol.) and microbial biomass (Dannemann et al., 2009; Pena et al., 2010; see Matyssek et al., 2013b, this vol.). Nevertheless, phytohormonal perturbances were reported to stimulate belowground C flux in adult forest trees (Matyssek et al., 2010b; Winwood et al., 2007). Air pollution significantly contributes to regional and temporal variation in soil respiration (i.e. Nikolova et al., 2010).

One might therefore question whether it is appropriate to relate short-term effects on soil respiration to altered decomposition of old SOM. SOM undergoes chemical degradation at different rates, producing a connected series of progressively more recalcitrant SOM and turnover times that range from days to millennia. Only about 5–15% of total soil respiration (R_s) is fuelled by decomposition of the more recalcitrant, old carbon fraction with comparatively slow rates of turnover (i.e. decades to centuries). The majority of soil carbon belongs to this fraction, but the fate of old carbon reservoirs cannot be predicted on the basis of soil respiration measurements alone (Subke and Bahn, 2010). Increased soil respiration observed during field-warming experiments returned, within a few years, to pre-warming levels (Hartley et al., 2007; Melillo et al., 2002). Such observations have been related to swift depletion of 'labile' carbon compounds, respired during the initial period following soil warming (Eliasson et al., 2005; Hartley et al., 2007; Kirschbaum, 2004). Whatever the reason, the temperature sensitivity of soil respiration appears to be flexible and can adjust to prolonged soil warming. Temperature response functions obtained from short-term warming experiments are thus not adequate to predict long-term effects of climate warming on soil respiration and on behaviour of old carbon stocks. Many carbon balance models simplistically assume fixed temperature sensitivity of respiration (e.g. setting $Q_{10}=2$; Cox et al., 2000), glossing over warming-induced shifts in substrate supply or soil water status that may alter the temperature sensitivity of soil respiration (Reichstein et al., 2005).

Davidson et al. (2006) stressed the lack of mechanistic understanding underlying the apparent flexibility of respiratory sensitivity to soil temperature. In this review, we summarize temperature effects on processes controlling soil carbon fluxes and their significance for the temperature sensitivity of bulk soil respiration. In doing so, we try to distinguish between immediate temperature effects on soil respiration and secondary effects that follow temperature changes with some delay—for example, changes in soil water potential. We will highlight central concepts which recently emerged to represent soil respiration more realistically in carbon balance models and to analyze the behaviour of soil carbon stocks in response to climate warming. These concepts are similar to those developed for plants, but we will also draw attention to significant differences between temperature effects on plant and microbial respiration. In this way, we also aim to improve the exchange of ideas and theoretical approaches prevalent in the plant and soil sciences.

7.2 THE INSTANTANEOUS TEMPERATURE RESPONSE OF SOIL RESPIRATION

When the temperature sensitivity of soil respiration is determined via short-term incubation studies, the temperature response of microbial respiration can be well approximated by Arrhenius-type and Q_{10} models (i.e. Jenkins and Adams, 2010; Reichstein et al., 2005). In this respect, soil microbial respiration does not differ from the instantaneous temperature response of plants:

$$R_h = R_{REF} \times \exp\left[\frac{E_o}{rT} \times \left(\frac{T - T_{REF}}{T \times T_{REF}}\right)\right], \tag{7.1}$$

where R_h is the rate of heterotrophic CO_2 respiration (i.e. CO_2 efflux per unit soil area or volume), R_{REF} is the respiration rate at some reference temperature (T_{REF}), T is the temperature (K), r is the universal gas constant (8.314 J mol^{-1} K^{-1}) and E_o is the overall activation barrier to soil CO_2 loss (kJ mol^{-1}; Kruse et al., 2011; Lloyd and Taylor, 1994). Q_{10} values seem more prominent in soil as compared to plant studies, determining the temperature sensitivity of respiration:

$$R_h = R_{REF} \times Q_{10}^{\left[\frac{T - T_{REF}}{10}\right]}, \tag{7.2}$$

where Q_{10} is the proportional increase of respiration per 10 °C increment of temperature (Kätterer et al., 1998). To some extent, it is surprising that these simple equations, originating in the late nineteenth century to describe the temperature dependency of a single chemical reaction, can be applied to such complex processes as microbial and plant respiration (Clarke, 2006). Nonetheless, functional representations of the instantaneous temperature response of soil respiration by Equations (7.1) and (7.2) remain approximations. These equations neglect deviations from strict exponential behaviour, which become apparent in 'dynamic responses' of soil and plant respiration to differing measurement temperature (Atkin and Tjoelker, 2003; Kirschbaum, 1995; Kruse et al., 2011; Lloyd and Taylor, 1994). In phenomenological terms, high Q_{10} values ranging from 4 to 8 were observed at low soil temperatures (5–10 °C), decreasing to low Q_{10} values ranging from 1 to 3 at high soil temperature (30–35 °C; Kirschbaum, 1995). Although Q_{10} values must be interpreted with some caution (as Q_{10} depends on the temperature interval chosen for calculation; Kruse et al., 2011), available evidence suggests that dynamically changing temperature sensitivity of soil respiration is a phenomenon that cannot be neglected (also cf. Lloyd and Taylor, 1994). The mechanisms that give rise to this dynamic response, however, are hardly understood. Explanation requires distinguishing between 'direct', or 'immediate' temperature effects, and 'indirect' temperature effects on the temperature sensitivity of soil respiration. As an example for 'indirect' effects, enhanced temperatures are usually associated with soil drying (which is amplified by large water loss of transpiring plants in field studies; Rennenberg et al., 2006). Changed soil water potential significantly alters the 'dynamic response' of respiration, which was demonstrated by Kirschbaum (2000). Davidson et al. (2006) provided a theoretical framework relevant for both immediate and secondary temperature effects on the temperature sensitivity of microbial respiration.

Microbes produce enzymes to degrade polymeric organic compounds into soluble monomers that can be assimilated (amino acids, organic acids, diverse sugar monomers, phenols, etc.) (see Kraigher et al., 2013, this vol.). De-polymerization is widely regarded as the bottleneck controlling availability of soluble compounds, and hence microbial activity and soil respiration (Bengtson and Bengtson, 2007). The activation barrier to degradation of 'labile' carbon compounds such as proteins, starch and cellulose seems lower than that of more recalcitrant compounds (i.e. chitin and lignin, tannin, waxes, or other polyphenolic compounds, Conant et al., 2011; Davidson and Janssens, 2006; Karhu et al., 2010). 'Stable' compounds undergo a chain of reactions catalyzed by different enzymes, partly explaining their greater 'intrinsic' temperature sensitivity for decomposition. The set of enzymes required for degradation of organic polymers can therefore have substantial influence on the availability of dissolved organic carbon, and hence on the availability of assimilable substrates. Temperature has a direct influence on enzyme activity. Davidson et al. (2006) pointed out that the temperature sensitivities of maximum enzyme activity (V_{max}) and the half-saturation constant that reflects affinity of any specific enzyme for its substrate (K_m) can cancel each other to produce a dynamic temperature response of the catalyzed reaction (deviating from Arrhenius kinetics):

$$E_R = \frac{V_{max} \times Q_{10(V_{max})}^{(T-T_{REF})/10} \times C}{\left[K_m \times Q_{10(K_m)}^{(T-T_{REF})/10}\right] + C} \tag{7.3}$$

where E_R is the reaction rate catalyzed by any specific enzyme following Michaelis–Menten kinetics and C is the substrate used by this enzyme (Davidson et al., 2006). Temperature-dependent reaction rates only follow Arrhenius kinetics when substrate concentrations are much greater than K_m.

Under these circumstances, V_{max} attained at any given temperature is mainly controlled by the amount of active enzymes present in the soil solution. The validity of this concept was recently confirmed by German et al. (2012), who measured the temperature sensitivities of extracellular cellobiohydrolase, β-glucosidase, β-xylosidase, α-glucosidase and *N*-acetyl-β-D-glucosaminidase. The Q_{10} for V_{max} of these enzymes ranged from 1.53 to 2.27, and that for K_m ranged from 0.90 to 1.57. Some enzymes exhibited greater temperature sensitivity in boreal as compared to tropical biomes (e.g. β-glucosidase), indicating adaptation of enzyme-binding ability, and thereby functionality, to prevailing temperature regimes (German et al., 2012). In arctic tundra soils, extracellular enzymes were more sensitive to temperature in early spring than during the summer (Wallenstein et al., 2009). Further, enzyme activity at any given temperature was low in summer, possibly due to N-limitation of enzyme production (Wallenstein et al., 2009). Microbial production and release of extracellular enzymes seem thus regulated by substrate availability and quality (see Sections 7.3 and 7.4; see Kraigher et al., 2013, this vol.), which, in turn, depend on vegetation phenology and soil water status. In temperate and Mediterranean climates, for example, desiccation of soils during the summer impedes free diffusion of exoenzymes and substrates—and hence reaction rates (also see Öquist et al., 2009). Taken together, immediate temperature effects on Q_{10} of heterotrophic respiration depend on the quality of organic polymers and the kinetic properties of enzymes that degrade them (Malcolm et al., 2009b), whereas 'indirect' effects become more relevant in the long term. In addition to temperature-induced effects on soil water potential, temperature may alter the balance between sorption and desorption of physicochemically protected SOM (discussed in Section 7.3).

Diffusion limitation of enzyme activity that may sporadically occur in the extracellular soil medium is less important in live organisms, although temperature can affect cellular membrane fluidity (Kruse et al., 2011). While the temperature sensitivity of microbial respiration crucially depends on the diversity and quality of organic polymers (Davidson and Janssens, 2006), autotrophic plants mainly metabolize starch and sugars that were previously produced in photosynthesis. The flexibility of Q_{10} observed in plants is therefore more closely related to 'acclimation' of metabolism to changed environmental conditions (Atkin et al., 2005). Similar rates of respiration observed in plants that were exposed to different growth temperatures are diagnostic for temperature acclimation of respiration. Kruse et al. (2012) recently showed that respiratory acclimation in *Eucalyptus* spp. helps adjust metabolism to fluctuating temperature, so that growth rates remain stable and potentially harmful substrate depletion is avoided. The temperature sensitivity of plant root respiration is an important component of bulk soil respiration (Boone et al., 1998), and root respiration was shown to acclimate swiftly to temperature changes (Atkin et al., 2000). By contrast, there is no clear-cut evidence for *active* adjustment of microbial physiology and respiration (Bradford et al., 2008; Hartley et al., 2008, 2009; Malcolm et al., 2009a), although temperature might significantly affect microbial community composition (see Section 7.4). Temperature acclimation of root respiration therefore interferes with interpretation of the temperature sensitivity of heterotrophic respiration (Luo et al., 2001), the latter reflecting substrate supply and quality (Gershenson et al., 2009). Autotrophic and heterotrophic processes are notoriously hard to study separately (Trumbore, 2006). Soil incubation studies in the laboratory or 'trenching' of soil areas in field studies are biased, and cannot inform about microbial, heterotrophic respiration in undisturbed soils. For example, root exudation of soluble organic compounds promotes the decomposition of SOM by microbes (Bahn et al., 2010); an effect known as 'priming' (Kuzyakov et al., 2000). In recent years, it has become increasingly apparent that vegetation and soil processes are intimately linked (see Pregitzer and Talhelm, 2013, this vol.), and roughly 50% of bulk soil respiration is attributable to the transfer of recent assimilate into the soil (Hartley et al., 2006; Högberg and Read, 2006). While transfer of recent assimilates supports microbial decomposition of SOM, soil scientists are confronted with the dilemma that variable supply of low-molecular weight compounds can mask the degradation of more recalcitrant compounds, which matters most in the long term.

7.3 SHORT-TERM FLUCTUATION OF SUBSTRATE SUPPLY WITH POSSIBLE LONG-TERM EFFECTS ON SOIL RESPIRATION

Reliable prediction of climate change effects on soil carbon stocks necessitates the detection of comparatively small carbon fluxes related to degradation (or build-up) of old SOM, against the background of large temporal variations in soil CO_2 efflux associated with degradation of new assimilates. The input of readily degradable organic compounds, including release from new leaf litter and decaying fine roots (Fig. 7.1), can show wide fluctuation on timescales ranging from days to years. The transfer of new photosynthate dominates current soil respiration, which was impressively demonstrated by stem-girdling experiments (Dannemann et al., 2009; Högberg et al., 2001), and was termed the 'high-speed link' between canopy and soil processes (Högberg and Read, 2006). Similarly, the carbon flux anomaly that was observed during the exceptionally hot and dry European summer of 2003 is thought to have been a consequence of a strong photosynthesis reduction through inhibition of stomatal gas exchange, which translated into a massive drop of assimilate transport and, hence, soil respiration (Ciais et al., 2005; Granier et al., 2007). However, the fact that some forest ecosystems turned into a CO_2 *source* during the 2003 summer implies that soil respiration did not decrease as strongly as canopy photosynthesis (Vetter et al., 2008). Apparently, soil respiration

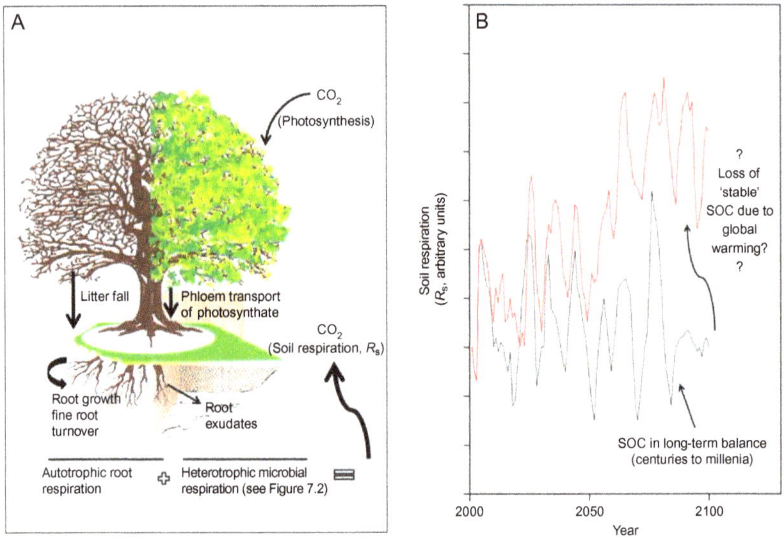

FIGURE 7.1 Soil respiration as affected by substrate availability. (A) Soil respiration is the sum of autotrophic and heterotrophic respiration, both ultimately driven by carbon substrate supply via photosynthesis. On timescales ranging from days to weeks, soil respiration heavily depends on current rates of photosynthesis, linking phloem carbohydrate transport to respiration of roots and mycorrhizae, and root exudation of labile carbon compounds to respiration of free-living microbes. On timescales ranging from weeks to months, input of soil organic carbon (SOC) can be affected by altered fine root turnover (and microbial turnover; see Fig. 7.2), responding to fluctuating resource availability (water, nitrogen). Input of SOC may also vary seasonally and inter-annually, depending on leaf litter fall and litter quality. (B) On timescales ranging from decades to centuries (to millennia), climate change may favour the breakdown of recalcitrant SOC into assimilable compounds (cf. Fig. 7.2), hence increasing 'background' soil respiration. It is currently debated whether global warming has the potential to cause degradation of soil carbon stocks, with possible feedback on atmospheric CO_2.

was partly fuelled by older and more recalcitrant carbon reservoirs under these conditions.

Rapidly changing availability of plant-derived labile assimilates exerts strong influence on both the capacity and the temperature sensitivity of microbial respiration. So far, we have discussed immediate temperature effects on microbial respiration, derived from soil incubation studies (Section 7.2). Such studies can provide 'snapshots' of the temperature sensitivity of heterotrophic respiration, which may, however, be biased and cannot be projected into the future. On the other hand, interpretation of temperature effects on respiration of intact soil is also problematic. Rates of soil respiration observed under field conditions are not solely controlled by temperature (Yuste et al., 2007). Field studies rather record variation in respiration capacity, as affected by fluctuating substrate supply and, hence, microbial growth and population size.

Free-living soil bacteria are sometimes categorized as *r*-strategists (Allison, 2005; Paterson et al., 2009), which continue to propagate unless environmental conditions deteriorate and substrate supply wanes. Bacterial communities prefer readily degradable substrates with low activation barriers of degradation (see Section 7.2). By contrast, mycorrhizal and saprophytic fungi are often grouped as *K*-strategists, releasing enzymes that make more recalcitrant compounds accessible (Allison, 2005; Fontaine and Barot, 2005; cf. Section 7.4; see Kraigher et al., 2013, this vol.). The turnover of bacterial populations (of *r*-strategists) can be quick and co-varies with the capacity for heterotrophic respiration. Drought stress, for instance, which is indirectly related to high temperatures, causes bacterial death and soil respiration rates to drop (Schindelbacher et al., 2012). In spite of generally low respiration rates, the temperature sensitivity of SOM decomposition might actually increase, as the shortage of aboveground assimilates compels soil microbes to use a larger fraction of more recalcitrant carbon compounds. When dry spells are interrupted by precipitation events, soil microbes that survived resume proliferation, rapidly increasing the capacity for heterotrophic soil respiration (Schindelbacher et al., 2012). This respiratory burst is initially fuelled by dead microbes and fine roots that accumulated during the drought period (Borken and Matzner, 2009), closing the 'microbial loop' that largely controls variation of the soil CO_2 efflux on timescales ranging from weeks to months (cf. Fig. 7.2). The 'instantaneous' temperature response functions presented in Section 7.2 are too static. They do not capture the temporal variation of respiratory capacity and temperature sensitivity, as affected by the speed of microbial turnover and temporal fluctuation of substrate supply and quality. These processes can affect soil carbon fluxes on timescales ranging from years to decades. Frequent drying and wetting can diminish the wetting pulse of soil respiration (Schindelbacher et al., 2012), because repeated deposition of organic substances increases hydrophobicity ('water repellency') of soil surfaces, thereby restricting accessibility by extracellular enzymes (Borken and Matzner, 2009).

A large fraction of SOC is protected from enzymatic degradation. Organic carbon that might otherwise be decomposable can be bound to soil minerals or physically occluded within soil aggregates (Six et al., 2002). Physicochemical reactions can be expected to respond differently to soil warming than enzyme-mediated microbial reactions (Thornley and Cannell, 2001), which may have strong repercussions on substrate accessibility (cf. Kemmitt et al., 2008). Organic carbon compounds bind with minerals via covalent bonds and several kinds of non-covalent bonds (van der Waals forces, charge interactions and hydrogen bonding; Conant et al., 2011). The latter may be divided into fast, high-affinity reactions regulated by equilibrium thermodynamics and slow, low-affinity reactions regulated by diffusion. While the net effect of soil warming on covalent bonding and aggregate turnover is probably small, sorption–desorption processes are strongly temperature-dependent (Conant

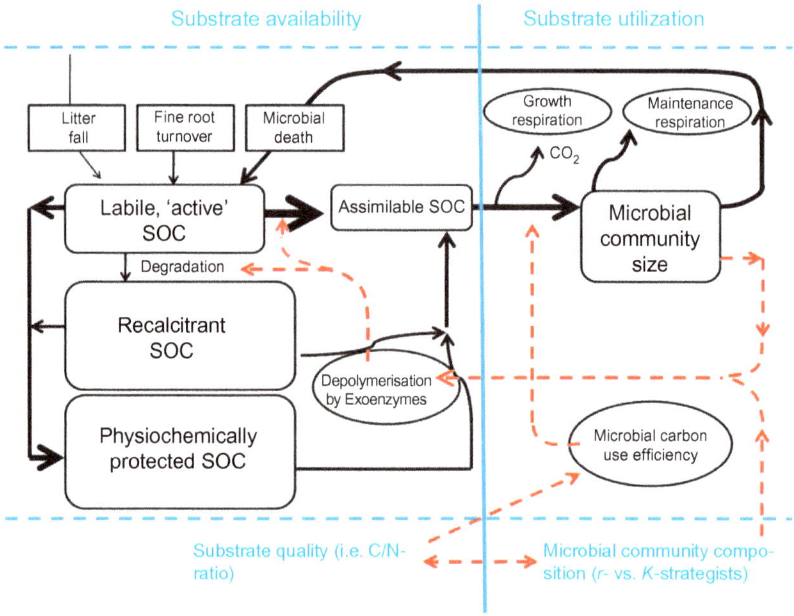

FIGURE 7.2 Effects of substrate availability and substrate utilization on heterotrophic soil respiration. Current rates of soil respiration are driven by the availability of assimilable soil organic carbon (SOC). Labile SOC that can easily be degraded by bacterial *r*-strategists contributes most to the pool of assimilable substrates. Rates of heterotrophic respiration depend on microbial turnover, affecting growth respiration, maintenance respiration, and the re-cycling of microbial SOC. Short-term turnover of microbial biomass may mask the effects of increased decomposition of existing, old carbon stocks in soil respiration studies. These compounds are more resistant to degradation, which is mainly catalyzed by exoenzymes produced by *K*-strategists. Also note that a large fraction of this carbon is physicochemically protected from enzymatic degradation. The efficiency of substrate assimilation and incorporation into microbial biomass (i.e. microbial carbon use efficiency, CUE_m) depends on substrate quality and many environmental variables, including temperature. Black arrows indicate soil carbon fluxes and red arrows indicate significant feedbacks on distinct carbon fluxes.

et al., 2011). Theoretical considerations and experimental results reviewed by Conant et al. (2011) suggest that increased soil temperature favours desorption relative to adsorption of organic compounds with high affinity (but no covalent link) to soil minerals, making them accessible to enzymatic degradation. This effect depends on bonding strength and is most pronounced for molecules that have the greatest affinity for soil mineral surfaces—equivalent to a large activation barrier to desorption. On the other hand, warmer temperatures can favour the *adsorption* of SOM during slowly forming, low-affinity interactions with soil minerals dominated by diffusion processes (Conant et al., 2011), which, in turn are confounded by variable soil moisture. Conant et al. (2011) concluded that it is necessary to refocus on individual processes controlling the decomposability of SOM. In this way, the complexity underlying short- and long-term variation in decomposition of SOM might eventually be resolved by process-based models. Experimentally, stable isotope approaches seem most promising to assist the resolution of this complexity, and the interpretation of soil respiratory sensitivity to temperature (Paterson et al., 2009). These approaches rely on distinguishable differences in the isotopic signatures of organic substrates consumed in respiration (Trumbore, 2006). Pulse-labelling experiments using $^{13}CO_2$ inform about the speed of assimilate transfer into soils, and the emergence of the label in different carbon reservoirs (i.e. mycorrhiza, soil microbes, etc.; Högberg et al., 2007). On longer timescales, researchers have taken advantage of past vegetation shifts from C_4 to C_3 plants (or vice versa) to quantify the contribution of 'old' heterotrophically respired CO_2 to total soil respiration (Hanson et al., 2000). Nonetheless, we are not aware of any study that combined vegetation shift approaches with experimental soil warming, which might provide the most reliable estimate about the decomposition of existing carbon stocks in a warmer world.

The amount of carbon that enters the resistant pool of SOC, or which is returned to the atmosphere, ultimately depends on both organic matter protection (discussed earlier) and how effectively the decomposer community converts carbon to biomass (discussed next).

7.4 MICROBIAL CARBON USE EFFICIENCY AS AFFECTED BY TEMPERATURE

The effective microbial carbon use efficiency (CUE_m) is defined as the ratio of growth over carbon uptake. By definition, high CUE_m promotes growth, while low CUE_m favours respiration (Manzoni et al., 2012):

$$\mathrm{CUE_m} = \frac{G}{U} = \frac{U - R_\mathrm{G} - R_\mathrm{M} - R_\mathrm{E} - R_\mathrm{O} - \mathrm{EX}}{U} \tag{7.4}$$

where U is the microbial C-uptake, G is the net microbial growth, EX represents the excretion of extracellular enzymes and metabolites, R_G, R_M, R_E and R_O are, respectively, respiration for growth, maintenance, extracellular enzyme production and overflow respiration (Manzoni et al., 2012). Overflow respiration is thought to be associated with substrate oxidation, while the resultant energy (i.e. ATP) may not support growth or maintenance processes, but rather be dissipated as heat. The applicability of overflow respiration concepts to plant metabolism is discussed by Kruse et al. (2011). Growth respiration is usually assumed to be proportional to the uptake flux, whereas maintenance respiration is a function of the amount of microbial biomass (Manzoni et al., 2012). The concept of carbon conversion efficiency also plays a central role in plant physiology (Kruse and Adams, 2008). For plants, however, the usefulness of the distinction between growth and maintenance respiration is debated. While this distinction makes sense at the level of entire organisms (like microbes), it is often applied to separate plant organs (Kruse et al., 2011). In fact, the very concepts of growth and maintenance respiration originally stem from microbial studies (Amthor, 2000), and ought not to be transmitted uncritically to plant respiration studies.

For microbial communities, the mass balance of microbial carbon (C_B) can be written as (Manzoni et al., 2012):

$$\frac{dC_B(t)}{dt} = G - BD = CUE_m \times U - BD \tag{7.5}$$

where BD is the microbial mortality and the microbial growth is $G = CUE_m \times U$ (cf. Equation 7.4). It is apparent from Equation (7.5) that the balance between microbial growth and death crucially depends on CUE_m. If, for example, microbial carbon is withdrawn from the 'microbial loop' by physicochemical stabilization processes, SOC stocks tend to increase (cf. Fig. 7.2).

Most carbon balance models include a constant CUE_m, differing widely between individual models (Manzoni et al., 2012). However, sensitivity analyzes have shown that CUE_m is one of the parameters to which model output is most sensitive (Wetterstedt and Ågren, 2011). There is mounting experimental evidence that CUE_m is variable, depending on substrate quality, microbial community composition and temperature (Manzoni et al., 2012). Experimental evidence suggests that CUE_m of aquatic microbes declines with increasing temperature, which might also be true for terrestrial ecosystems (Manzoni et al., 2012). On the other hand, CUE_m is also sensitive to possible, temperature-induced changes in nutrient availability and microbial community composition, dominating ecosystem responses in the long term. First, the degradability of organic compounds is tightly linked to the C/N-ratio of detritus. Soil microbes have conservative stoichiometries relative to the elemental heterogeneity of their environment (Sistla and Schimel, 2012), and fungi have higher C/N than bacteria (Six et al., 2006). As detritus C/N approaches that of its decomposers, microbes shift from nitrogen immobilization to mineralization, thereby regulating plant-available nitrogen resources. The growth of heterotrophic microbes is generally C-limited, but when net nitrogen immobilization occurs, excess carbon might increase 'overflow respiration' or be excreted, causing CUE_m to decline (see Equation 7.4; Manzoni et al., 2012). Greater fungal to bacterial dominance is often associated with lower N-availability and increased oxidative enzyme activities, mediating the degradation of resistant carbon compounds (cf. Section 7.2). By contrast, higher N-availability is correlated with hydrolytic enzyme activities that are associated with degradation of more labile compounds (Sistla and Schimel, 2012). There is some evidence that zymogenous (fast-growing and opportunistic) microbial communities use carbon less efficiently than autochthonous (slow-growing and humus-degrading) communities (Manzoni et al., 2012). Zymogenous populations are prevalent in soils with ample access to glucose, where they achieve rapid growth, albeit at low CUE_m. Such a trade-off between rates of growth and respiratory efficiency has also been observed in plants (Kruse and Adams, 2008; Kruse et al., 2012). While enhanced C/N of detritus might decrease CUE_m, this effect can be counteracted by the slow growth of microbial communities decomposing more resistant SOM.

Soil-warming experiments may initially affect microbial priorities of substrate use, favouring consumption of labile over resistant compounds. Prolonged warming, however, can result in decreased microbial growth (Rousk et al., 2012), reductions in microbial biomass (Bradford et al., 2008), or altered microbial community composition (Balser and Wixon, 2009). Frequently, the optimum temperature of bacterial growth is adjusted to prevailing temperature (Bárcenas-Moreno et al., 2009). Ecosystem perturbation induced by climate warming will most likely alter CUE_m and, concomitantly, soil CO_2 efflux. However, these changes are not easily predicted, because they involve ecosystem-level equilibrations occurring over different timescales. Thus, prediction of organic matter decomposition and soil respiration in a warmer world remains a challenging task.

7.5 SCIENTIFIC CONCLUSIONS

Generalizations about the impact of climate change on soil respiration and SOM decomposition are hardly possible, because such effects will vary between biomes and between soil and vegetation types. Climate change will presumably affect seasonal and diurnal fluctuations of soil respiration and SOM decomposition, because temperature increases are predicted to be most pronounced in winter and at night-time, at least in the mid- and high latitudes. Some ecosystems may more frequently experience heat waves during the summer. Such temperature effects on soil processes cannot be analyzed separately from associated drought effects. This review highlighted key processes that are pivotal for understanding the spatial and temporal variation of soil respiration, in order to help predict SOM decomposition under climate change. Substrate availability and quality are central to both autotrophic root and heterotrophic microbial respiration, dominating both the capacity and temperature sensitivity of soil respiration. Degradation of labile SOM has lower sensitivity to soil temperature than degradation of more recalcitrant compounds, which requires a different set of extra-cellular enzymes. Availability of labile and new compounds is largely driven by aboveground processes, including input of plant litter and root exudation of carbohydrates, whereas accessibility of old and recalcitrant compounds more crucially depends on physicochemical processes. Sensitivities of biotic and physicochemical processes to temperature and soil moisture differ and cause fluctuation in substrate availability and hence substrate utilization. The efficiency of substrate utilization by microbes (CUE_m) determines CO_2 loss to the atmosphere versus retention of organic carbon in the soil system. CUE_m largely depends on the composition of the decomposer community, encompassing microbial *r*- and *K*-strategists. Prolonged soil warming frequently causes a drop in labile compounds, accompanied by adjustments of the soil microbial community that have to rely on older SOM. Together with increased desorption of recalcitrant SOM from soil minerals, microbial adjustments to enhanced temperature may facilitate degradation of existing 'old' soil carbon stocks. On the other hand, enhanced speed of microbial turnover that is driven by higher frequency of drying/wetting cycles may accelerate the transfer of labile compounds to the pool of physicochemically protected soil carbon. It is presently not known how altered temperature regimes and precipitation/evapotranspiration dynamics ('climate change') translate into long-term adjustments of microbial respiration, carbon turnover and SOM decomposition.

7.6 POLITICAL IMPLICATIONS

Some politicians call for concerted actions to confine average global warming to 2 °C, since such a 'moderate' extent of warming within the twenty-first century is thought to be still 'manageable' (whatever that means). If a consensus emerges on this point, political decision making will require a sound information basis. Model predictions to evaluate alternative courses of action are indispensable to guide such decision making. This review identified some crucial feedbacks between biological processes and climate change which are not yet understood. The complexity of atmosphere–biosphere–pedosphere interactions will presumably necessitate the refinement of existing process-based models to enhance our confidence in model predictions. Available climate models reflect the present state of scientific knowledge and politicians ought to be aware that model predictions are necessarily afflicted with some uncertainty. We are, however, confident that ecosystem research will continue to narrow this gap of uncertainty and clarify the consequences of desirable/undesirable socioeconomic trajectories for terrestrial ecosystems and the future climate (Matyssek et al., 2013c, this vol.).

REFERENCES

Allison, S.D., 2005. Cheaters, diffusion and nutrients constrain decomposition by microbial enzymes in spatially structured environments. Ecol. Lett. 8, 626–635.

Amthor, J.S., 2000. The McCree-de Wit-Penning de Vries-Thornley respiration paradigms: 30 years later. Ann. Bot. 86, 1–20.

Atkin, O.K., Tjoelker, M.G., 2003. Thermal acclimation and the dynamic response of plant respiration to temperature. Trends Plant Sci. 8, 343–351.

Atkin, O.K., Edwards, E.J., Loveys, B.R., 2000. Response of root respiration to changes in temperature and its relevance to global warming. New Phytol. 147, 141–154.

Atkin, O.K., Bruhn, D., Hurry, V.M., Tjoelker, M.G., 2005. The hot and the cold: unraveling the variable response of plant respiration to temperature. Funct. Plant Biol. 32, 87–105.

Bahn, M., Janssens, I.A., Reichstein, M., Smith, P., Trumbore, S.E., 2010. Soil respiration across scales: towards an integration of patterns and processes. New Phytol. 186, 292–296.

Balser, T.C., Wixon, D.L., 2009. Investigating biological control over soil carbon temperature sensitivity. Glob. Chang. Biol. 15, 2935–2949.

Bárcenas-Moreno, G., Gómez-Brandón, M., Rousk, J., Bååth, E., 2009. Adaptation of soil microbial communities to temperature: comparison of fungi and bacteria in a laboratory experiment. Glob. Chang. Biol. 15, 2950–2957.

Bengtson, P., Bengtson, G., 2007. Rapid turnover of DOC in temperature forests accounts for increased CO_2-production at elevated temperatures. Ecol. Lett. 10, 783–790.

Bond-Lamberty, B., Thomson, A., 2010. Temperature-associated increases in the global soil respiration record. Nature 464, 579–582.

Boone, R.D., Nadelhoffer, K.J., Canary, J.D., Kaye, J.P., 1998. Roots exert a strong influence on the temperature sensitivity of soil respiration. Nature 396, 570–572.

Borken, W., Matzner, E., 2009. Reappraisal of drying and wetting effects on C and N mineralization and fluxes in soils. Glob. Chang. Biol. 15, 808–824.

Bradford, M.A., Davies, C.A., Frey, S.D., Maddox, T.R., Mellilo, J.M., Mohan, J.E., Reynolds, J.F., Treseder, K.K., Wallenstein, M.D., 2008. Thermal adaptation of soil microbial respiration to elevated temperature. Ecol. Lett. 11, 1316–1327.

Ciais, Ph., Reichstein, M., Viovy, N., Granier, A., et al., 2005. Europe-wide reduction in primary productivity caused by the heat and drought in 2003. Nature 437, 529–533.

Cieslik, S., Tuovinen, J.-P., Baumgarten, M., Matyssek, R., Brito, P., Wieser, G., 2013. Gaseous exchange between forests and the atmosphere. Dev. Environ. Sci. 13, 19–36.

Clarke, A., 2006. Temperature and the metabolic theory of ecology. Funct. Ecol. 20, 405–412.

Conant, R.T., Ryan, M.G., Ågren, G.I., Birge, H., Davidson, W.A., Eliasson, P.E., Evans, S.E., Frey, S.D., Giardina, C.P., Hopkins, F.M., Hyvönen, R., Kirschbaum, M.U.F., Lavallee, J.M., Leifeld, J., Parton, W.J., Steinweg, J.M., Wallenstein, M.D., Wetterstedt, J.M., Bradford, M.A., 2011. Temperature and soil organic matter decomposition rates—synthesis of current knowledge and a way forward. Glob. Chang. Biol. 17, 3392–3404.

Couture, J.J., Lindroth, R.L., 2013. Impacts of atmospheric change on tree-arthropod interactions. Dev. Environ. Sci. 13, 227–248.

Cox, P.M., Betts, R.A., Jones, C.D., Spall, S.A., Totterdell, I.J., 2000. Acceleration of global warming due to carbon-cycle feedbacks in a coupled climate model. Nature 408, 184–187.

Dannemann, M., Simon, J., Gasche, R., et al., 2009. Tree girdling provides insight on the role of labile carbon in nitrogen partitioning between soil microorganisms and adult European beech. Soil Biol. Biochem. 41, 1622–1631.

Davidson, E.A., Janssens, I.A., 2006. Temperature sensitivity of soil carbon decomposition and feedbacks to climate change. Nature 440, 165–173.

Davidson, E.A., Trumbore, S.E., Amundson, R., 2000. Soil warming and organic carbon content. Nature 408, 789–790.

Davidson, E.A., Janssens, I.A., Luo, Y., 2006. On the variability of respiration in terrestrial ecosystems: moving beyond Q_{10}. Glob. Chang. Biol. 12, 154–164.

Eliasson, P.E., McMurtrie, R.E., Pepper, D.A., Strömgren, M., Linder, S., Ågren, G.I., 2005. The response of heterotrophic CO_2 flux to soil warming. Glob. Chang. Biol. 11, 167–181.

Fontaine, S., Barot, S., 2005. Size and functional diversity of microbe populations control plant persistence and long-term soil carbon accumulation. Ecol. Lett. 8, 1075–1087.

German, D.P., Marcelo, K.R.B., Stone, M.M., Allison, S.T., 2012. The Michaelis-Menten kinetics of soil extracellular enzymes in response to temperature: a cross-latitudinal study. Glob. Chang. Biol. 18, 1468–1479.

Gershenson, A., Bader, N.E., Cheng, W., 2009. Effects of substrate availability on the temperature sensitivity of soil organic matter decomposition. Glob. Chang. Biol. 15, 176–183.

Giardina, C.P., Ryan, M.G., 2000. Evidence that decomposition rates of organic carbon in mineral soil do not vary with temperature. Nature 404, 858–861.

Granier, A., Reichstein, M., Breda, N., et al., 2007. Evidence for soil water control on carbon and water dynamics in European forests during the extremely dry year: 2003. Agric. Forest. Meteorol. 143, 123–145.

Greaver, T.L., Sullivan, T.J., Herrick, J.D., et al., 2012. Ecological effects of nitrogen and sulphur air pollution in the US: what do we know? Front. Ecol. Environ. 10 (7), 365–372.

Hanson, P.J., Edwards, N.T., Garten, C.T., 2000. Separating root and soil microbial contributions to soil respiration: a review of methods and observations. Biogeochemistry 48, 115–146.

Hartley, I.P., Armstrong, A.F., Murthy, R., Barren-Gafford, G., Ineson, P., Atkin, O.K., 2006. The dependence of respiration on photosynthesic substrate supply and temperature: integrating leaf, soil and ecosystem measurements. Glob. Chang. Biol. 12, 1954–1968.

Hartley, I.P., Heinemeyer, A., Ineson, P., 2007. Effects of three years of soil warming and shading on the rate of soil respiration: substrate availability and not thermal acclimation mediates observed response. Glob. Chang. Biol. 13, 1761–1770.

Hartley, I.P., Hopkins, D.W., Garnett, M.H., Sommerkorn, M., Wookey, P.A., 2008. Soil microbial respiration in arctic soil does not acclimate to temperature. Ecol. Lett. 11, 1092–1100.

Hartley, I.P., Hopkins, D.W., Garnett, M.H., Sommerkorn, M., Wookey, P.A., 2009. No evidence for compensatory thermal adaptation of soil microbial respiration in the study of Bradford et al. (2008). Ecol. Lett. 12, 12–14.

Hobbie, S.E., Schimel, J.P., Trumbore, S.E., Randerson, J.R., 2000. Controls over carbon storage and turnover in high-latitude soils. Glob. Chang. Biol. 6, 196–210.

Högberg, P., Read, D.J., 2006. Towards a more plant physiological perspective on soil ecology. Trends Ecol. Evol. 21, 548–554.

Högberg, P., Nordgren, A., Buchmann, N., Taylor, A.F.S., Ekblad, A., Högberg, M.N., Nyberg, G., Ottosson-Löfvenius, M., Read, D.J., 2001. Large-scale forest girdling shows that current photosynthesis drives soil respiration. Nature 411, 789–792.

Högberg, P., Högberg, M.N., Göttlicher, S.G., Betson, N.R., Keel, S.G., Metcalfe, D.B., Campbell, C., Schindelbacher, A., Hurry, V., Lundmark, T., Linder, S., Näsholm, T., 2007. High temporal resolution tracing of photosynthate carbon from the tree canopy to forest soil microorganisms. New Phytol. 177, 220–228.

Houghton, R.A., 2007. Balancing the global carbon budget. Annu. Rev. Earth Planet. Sci. 35, 313–347.

Iverson, C.M., Ledford, J., Norby, R.J., 2008. CO_2 enrichment increases carbon and nitrogen input from fine roots in a deciduous forest. New Phytol. 179, 837–847.

Jenkins, M., Adams, M.A., 2010. Vegetation type determines heterotrophic respiration in subalpine Australian ecosystems. Glob. Chang. Biol. 16, 209–219.

Jones, R.J.A., Hiederer, R., Rusco, E., Montanarella, E., 2005. Estimating organic carbon in the soils of Europe for policy support. Eur. J. Soil Sci. 56, 655–671.

Karhu, K., Fritze, H., Tuomi, M., Vanhala, P., Spetz, P., Kitunen, V., Liski, J., 2010. Temperature sensitivity of organic matter decomposition in two boreal forest soils. Soil Biol. Biochem. 42, 72–82.

Kätterer, T., Reichstein, M., Andrén, O., Lomander, A., 1998. Temperature dependence of organic matter decomposition: a critical review using literature data analyzed with different models. Biol. Fert. Soils 27, 258–262.

Kemmitt, S.J., Lanyon, C.V., Waite, I.S., Wen, Q., Addiscott, T.M., Bird, N.R.A., O'Donnell, A.G., Brookes, P.C., 2008. Mineralization of native soil organic matter is not regulated by size, activity or composition of the soil microbial biomass—a new perspective. Soil Biol. Biochem. 40, 61–73.

King, J., Liu, L., Aspinwall, M., 2013. Tree and forest responses to interacting elevated atmospheric CO_2 and tropospheric O_3: a synthesis of experimental evidence. Dev. Environ. Sci. 13, 179–208.

Kirschbaum, M.U.F., 1995. The temperature dependence of soil organic matter decomposition, and the effect of global warming on soil organic C storage. Soil Biol. Biochem. 27, 753–760.

Kirschbaum, M.U.F., 2000. Will changes in soil organic carbon act as a positive or negative feedback on global warming? Biogeochemistry 48, 21–51.

Kirschbaum, M.U.F., 2004. Soil respiration under prolonged soil warming: are rate reductions caused by acclimation or substrate loss? Glob. Chang. Biol. 10, 1870–1877.

Kirschbaum, M.U.F., 2006. The temperature dependence of organic-matter decomposition—still a topic of debate. Soil Biol. Biochem. 38, 2510–2518.

Kraigher, H., Bajc, M., Grebenc, T., 2013. Mycorrhizosphere complexity. Dev. Environ. Sci. 13, 151–177.

Kruse, J., Adams, M.A., 2008. Sensitivity of respiratory metabolism and efficiency to foliar nitrogen during growth and maintenance. Glob. Chang. Biol. 14, 1233–1251.

Kruse, J., Rennenberg, H., Adams, M.A., 2011. Steps towards a mechanistic understanding of respiratory temperature responses. New Phytol. 189, 659–677.

Kruse, J., Turnbull, T., Adams, M.A., 2012. Disentangling respiratory acclimation and adaptation to growth temperature by *Eucalyptus* spp. New Phytol. 195, 149–163.

Kuzyakov, Y., Friedel, J.K., Stahr, K., 2000. Review of mechanisms and quantification of priming effects. Soil Biol. Biochem. 38, 3267–3278.

Liski, J., Ilvesniemi, H., Mäkelä, A., Westmann, C.J., 1999. CO_2 emissions from soil in response to climate warming are overestimated—the decomposition of old soil organic matter is tolerant to temperature. Ambio 29, 171–174.

Lloyd, J., Taylor, J.A., 1994. On the temperature dependence of soil respiration. Funct. Ecol. 8, 315–323.

Löw, M., Deckmyn, G., Op de Beeck, M., Blumenröther, M., Oßwald, W., Alexou, M., Jehnes, S., Haberer, K., Rennenberg, H., Herbinger, K., Häberle, K.-H., Bahnweg, G., Winwood, J., Pate, A., Hanke, D., Ceulemans, R., Matyssek, R., Tausz, M., 2012. Multivariate analysis of physiological parameters reveals a consistent O_3 response pattern in leaves of adult beech (*Fagus sylvatica*). New Phytol. 196, 162–172.

Luo, Y., Wan, S., Hui, D., Wallace, L.L., 2001. Acclimatization of soil respiration to warming in a tall grass prairie. Nature 413, 622–625.

Lützow, M., Kögel-Knabner, I., 2009. Temperature sensitivity of soil organic matter decomposition—what do we know? Biol. Fertil. Soils 46, 1–15.

Malcolm, G.M., López-Gutiérrez, J.C., Koide, R.T., 2009a. Little evidence for respiratory acclimation by microbial communities to short-term shifts in temperature in red pine (*Pinus resinosa*) litter. Glob. Chang. Biol. 15, 2485–2492.

Malcolm, G.M., López-Gutiérrez, J.C., Koide, R.T., 2009b. Temperature sensitivity of respiration differs among forest floor layers in a *Pinus resinosa* plantation. Soil Biol. Biochem. 41, 1075–1079.

Manzoni, S., Taylor, P., Richter, A., Porporato, A., Ågren, G.I., 2012. Environmental and stoichiometric controls on microbial carbon-use efficiency in soils. New Phytol. 196, 79–91.

Matyssek, R., Fromm, J., Rennenberg, H., Roloff, A., 2010a. Biologie der Bäume. Eugen Ulmer KG, Stuttgart, Germany.

Matyssek, R., Wieser, G., Ceulemans, R., Rennenberg, H., Pretzsch, H., Haberer, K., Löw, M., Nunn, J.J., Werner, H., Wipfler, P., Oßwald, W., Nikolova, P., Hanke, D., Kraigher, H., Tausz, M., Bahnweg, G., Kitao, M., Dieler, J., Sandermann, H., Herbinger, K., Grebenc, T., Blumenröther, M., Deckmyn, G., Grams, T.E.E., Heerdt, C., Leuchner, M., Fabian, P., Häberle, K.H., 2010b. Enhanced ozone strongly reduces carbon sink strength of adult beech (*Fagus sylvatica*)—resumé from the free-air fumigation study at Kranzberg Forest. Environ. Pollut. 158, 2527–2532.

Matyssek, R., Clarke, N., Cudlin, P., Mikkelsen, T.N., Tuovinen, J.-P., Wieser, G., Paoletti, E., 2013a. Climate change, air pollution and global challenges: understanding and perspectives from forest research. Dev. Environ. Sci. 13, 3–16.

Matyssek, R., Wieser, G., Fleischmann, F., Grünhage, L., 2013b. Ozone research, quovadis? Lessons from the free-air canopy fumigation experiment at Kranzberg forest. Dev. Environ. Sci. 13, 103–129.

Matyssek, R., Knoke, T., Clarke, N., Cudlin, P., Mikkelsen, T.N., Tuovinen, J.-P., Wieser, G., Paoletti, E., 2013c. Conclusions and perspectives. Dev. Environ. Sci. 13, 591–607.

Melillo, J.M., Steudler, P.A., Aber, J.D., Newkirk, K., Lux, H., Bowles, F.P., et al., 2002. Soil warming and carbon-cycle feedbacks to the climate system. Science 298, 2173–2176.

Neff, J.C., Hooper, D.U., 2002. Vegetation and climate controls on potential CO_2, DOC and DON production in northern latitude soils. Glob. Chang. Biol. 8, 872–884.

Nikolova, P.S., Andersen, C.P., Blaschke, H., Matyssek, R., Häberle, K.H., 2010. Belowground effects of enhanced tropospheric ozone and drought in a beech/spruce forest (*Fagus sylvatica* L./*Picea abies* [L.] Karst). Environ. Pollut. 158, 1071–1078.

Öquist, M.G., Sparrmann, T., Klemedtsson, L., Harrysson, D., Grip, H., Schleuchers, J., Nilsson, M., 2009. Water availability controls microbial temperature responses in frozen soil CO_2 production. Glob. Chang. Biol. 15, 2715–2722.

Paterson, E., Midwood, A.J., Millard, P., 2009. Through the eye of the needle: a review of isotope approaches to quantify microbial processes mediating soil carbon balance. New Phytol. 184, 19–33.

Pena, R., Offermann, C., Simon, J., Naumann, P., Gessler, A., Holst, J., Dannenmann, M., Mayer, H., Kögel-Knabner, I., Rennenberg, H., Polle, A., 2010. Girdling affects ectomycorrhizal diversity and reveals functional differences of EM community composition in a beech forest. Appl. Environ. Microbiol. 76, 1831–1841.

Pregitzer, K.S., Talhelm, A.F., 2013. Belowground carbon cycling at aspen FACE: dynamic responses to CO_2 and O_3 in developing forests. Dev. Environ. Sci. 13, 209–226.

Randewig, D., Hamisch, D., Herschbach, C., Eiblmeier, M., Gehl, C., Jurgeleit, J., Skerra, J., Mendel, R.R., Rennenberg, H., Hänsch, R., 2012. Sulfite oxidase controls sulfur metabolism under SO_2-exposure in *Arabidopsis thaliana*. Plant Cell Environ. 35, 100–115.

Reichstein, M., Subke, J.A., Angeli, A.C., Tenhunen, J.D., 2005. Does the temperature sensitivity of decomposition of soil organic matter depend upon water content, soil horizon, or incubation time? Glob. Chang. Biol. 11, 1754–1767.

Rennenberg, H., 1984. The fate of excess sulfur in higher plants. Annu. Rev. Plant Physiol. 35, 121–153.

Rennenberg, H., Herschbach, C., Polle, A., 1996. Consequences of air pollution on shoot–root interactions. J. Plant Physiol. 148, 296–301.

Rennenberg, H., Kreutzer, K., Papen, H., Weber, P., 1998. Consequences of high loads of nitrogen for spruce (*Picea abies* L.) and beech (*Fagus sylvatica* L.) forests. New Phytol. 139, 71–86.

Rennenberg, H., Loreto, F., Polle, A., 2006. Physiological responses of forest trees to heat and drought. Plant Biol. 8, 556–571.

Rousk, J., Frey, S., Bååth, E., 2012. Temperature adaptation of bacterial communities in experimentally warmed forest soils. Glob. Chang. Biol. 18, 3252–3259.

Ryan, M.G., Law, B.E., 2005. Interpreting, measuring, and modeling soil respiration. Biogeochemistry 73, 3–27.

Schimel, D.S., 1995. Terrestrial ecosystems and the carbon cycle. Glob. Chang. Biol. 1, 77–91.

Schindelbacher, A., Wunderlich, S., Borken, W., Kitzler, B., Zechmeister-Boltenstern, S., Jandl, R., 2012. Soil respiration under climate change: prolonged summer drought offsets soil warming effects. Glob. Chang. Biol. 18, 2270–2279.

Schultz, J., 2008. Die Ökozonen der Erde. Verlag Eugen Ulmer, Stuttgart, Germany.

Sistla, S.A., Schimel, J.P., 2012. Stoichiometric flexibility as a regulator of carbon and nutrient cycling in terrestrial ecosystems under change. New Phytol. 196, 68–78.

Six, J., Conant, R.T., Paul, E.A., Paustian, K., 2002. Stabilization mechanisms of soil organic matter: implications for C-saturation of soils. Plant Soil 241, 155–176.

Six, J., Frey, S.D., Thiet, R.K., Batten, K.M., 2006. Bacterial and fungal contributions to carbon sequestration in agroecosystems. Soil Sci. Soc. Am. J. 70, 555–569.

Subke, J.A., Bahn, M., 2010. On the 'temperature sensitivity' of soil respiration: can we use the immeasurable to predict the unknown? Soil Biol. Biochem. 42, 1653–1656.

Thornley, J.H.M., Cannell, M.G.R., 2001. Soil carbon storage response to temperature: a hypothesis. Ann. Bot. 87, 591–598.

Trumbore, S., 2006. Carbon respired by terrestrial ecosystems—recent progress and challenges. Glob. Chang. Biol. 12, 141–153.

Trumbore, S.E., Chadwick, O.A., Amundson, R., 1996. Rapid exchange between soil carbon and atmospheric carbon dioxide driven by temperature change. Science 272, 393–396.

Vetter, M., Churkina, G., Jung, M., et al., 2008. Analyzing the causes and spatial pattern of the European 2003 carbon flux anomaly using seven models. Biogeosciences 5, 561–583.

Wallenstein, M.D., McMahon, S.K., Schimel, J.P., 2009. Seasonal variation in enzyme activities and temperature sensitivities in Arctic tundra soils. Glob. Chang. Biol. 15, 1631–1639.

Wetterstedt, J.Å.M., Ågren, G.I., 2011. Quality or decomposer efficiency—which is most important in the temperature response of litter decomposition? A modeling study using the GLUE methodology. Biogeosciences 8, 477–487.

Winwood, J., Pate, A.E., Price, J., Hanke, D.E., 2007. Effects of long-term, free-air ozone fumigation on the cytokinin content of mature beech trees. Plant Biol. 9, 265–278.

Yuste, J.C., Baldocchi, D.D., Gershenson, A., Goldstein, A., Misson, L., Wong, S., 2007. Microbial soil respiration and its dependency on carbon inputs, soil temperature and moisture. Glob. Chang. Biol. 13, 2018–2035.

Chapter 8

Mycorrhizosphere Complexity

Hojka Kraigher[1], Marko Bajc and Tine Grebenc
Slovenian Forestry Institute, Ljubljana, Slovenia
[1]Corresponding author: e-mail: hojka.kraigher@gozdis.si

Chapter Outline

Developments in Environmental Science, Vol. 13. http://dx.doi.org/10.1016/B978-0-08-098349-3.00008-6

8.1 INTRODUCTION: THE ROLE OF MYCORRHIZAE IN ECOSYSTEM FUNCTIONS AND PROCESSES

The evolution of terrestrial organisms is linked to overcoming harsh environmental conditions, which has lead to the development of a variety of mechanisms and mutual relationships (Plett et al., 2012). The exchange of water and minerals for carbohydrates has linked two kingdoms, plants and fungi, to form the mutualistic symbiosis called 'mycorrhiza' (Frank in 1886), enabling the partners to survive and scavenge in stressful conditions with limited supplies in time and space. In northern boreal and temperate forests, fine roots of the majority of stand-forming trees are ectomycorrhizal (Kraigher, 1999), while the understory plants can share the ectomycorrhizal fungi, or have any of the other forms of mycorrhizal symbiosis, such as arbuscular and ericoid mycorrhizae (ERM) (Read, 1991).

Ectomycorrhizae (ECM; Figure 8.1) are characterized by the Hartig net, a labyrinth of hyphae around the primary cortex cells, the fungal mantle ensheathing the fine roots and the extramatrical mycelium (EMM) designating all mycelial structures emanating from the hyphal mantle into the soil; (Agerer, 2001). This EMM is formed by single differently structured hyphae, or by differently complex hyphal-mycelial strands, forming rhizomorphs, which can extend over significant distances in forest ecosystems (Agerer, 2001; Allen, 2011; Smith and Read, 2008).

The activity of roots influences the soils around them, forming the so-called rhizosphere (Curl and Truelove, 1986); the one influenced by mycorrhizal roots is the mycorhizosphere and the soil around fungal mycelium is the hyphosphere (Marschner 1995, cit. in Kraigher et al., 2007; Frey-Klett et al., 2007). The rhizosphere differs from bulk soil by its lower pH, lower water potential, lower partial pressure of O_2 and higher partial pressure of CO_2, and higher concentrations of soluble carbohydrates. This leads to a higher concentration of micro-organisms, and a different community structure of these organisms (Curl and Truelove, 1986). Soil bacteria are also able to multiply inside the roots as endophytes and modulate plant growth and development, with implications for plant productivity and phytoremediation (Bulgarelli et al., 2012). Rhizosphere bacterial communities were also found to be determined by soil type and host genetics (Lundberg et al., 2012).

Within the fungal kingdom, with presumably 1.5 million species, about 200 fungal species of the *Glomeromycota* form arbuscular mycorrhizae (AM) (Schüßler et al., 2001), and more than 6000 species, mainly from the *Ascomycota* and *Basidiomycota*, can form ectomycorrhizae and ERM (Tedersoo et al., 2010). However, only about 500 ectomycorrhizal species have been characterized by their morphological and anatomical traits (Agerer and Rambold, 2004–2011, www.deemy.de) and further classified according to their extramatrical mycelium into five exploration types (ET; Agerer, 2001, Figure 8.1, Right). Even less studied is the physiology of

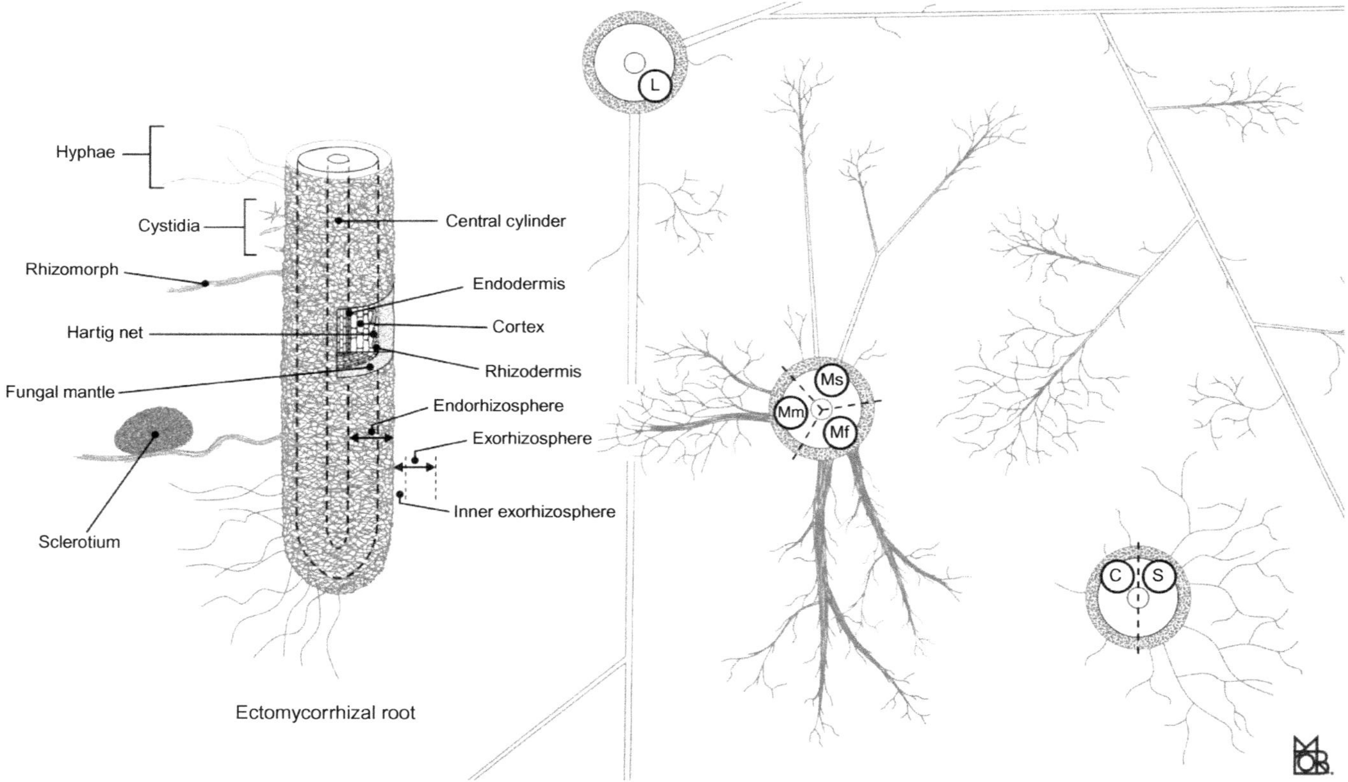

FIGURE 8.1 Left: An ectomycorrhizal root, showing different mycorrhizal structures, and the rhizosphere terminology, where exo- and endo-rhizosphere include the fungal mantle. Right: A schematic presentation of four exploration types (ET), based on characteristics of the EMM. C, contact ET; S, short distance ET; M, medium distance ET (Mf, fringe, Mm, mat, Ms, smooth); L, long distance ET. *Left panel: Modified after Curl and Truelove (1986). Right panel: Modified and re-drawn after Agerer (2001).*

ectomycorrhizae as cultivation of only a few mycorrhizal fungi has been successful, while *in situ* methods in the field are limited.

In this chapter, mycorrhizal forms and ecological strategies are presented, with the main emphasis on ectomycorrhizal fungi and their role in forest ecosystems and in belowground processes.

8.1.1 Mycorrhizal Life Strategies

Mycorrhizae differing in anatomical structures and enzyme systems interact differently with the environment and sources of nutrients (Table 8.1), which is reflected by their distribution in the world biomes (Figure 8.2). Global climate change effects are expected to alter this distribution.

8.1.2 Identification and Characterisation of Types of Ectomycorrhizae

Ectomycorrhizae are formed predominantly by *Basidio*- and *Ascomycota* and can be characterised by their morphological and anatomical traits (Agerer and Rambold, 2004–2011, www.deemy.de). Plant and fungal partners can be identified by DNA-sequence based comparison analyses, using mainly the ITS

TABLE 8.1 Interactions of Three Dominant Mycorrhizal Forms of Woody Plants with the Environment

Mycorrhizae/ characteristics	Ericoid mycorrhizae	Ectomycorrhizae	Arbuscular mycorrhizae
Predominant climate zone	Tundra	Boreal and temperate zones	(Sub)tropical savannas and forests
Nutrient availability	Poor with mineral N and P	Seasonally available N and P	High N, P limited
pH	Below 3–4	Below 4–7	5–9
C:N ratio	120:100	110:20	30:10
Decomposition	5 years or more	2–5 years	1–2 years
Humus	Raw humus	Three humus forms	Mull
Plant forms and understory	Ericaceous bushes	Forest trees with AM, ericaceous or arbutoid understory	Herbaceous plants and tropical trees
Fungal enzymes	Proteases, ligninases, phenol oxidases	Polyphenol oxidases, proteases, cellulases, phosphatases	Alkaline phosphatases

Adapted from Read (1991).

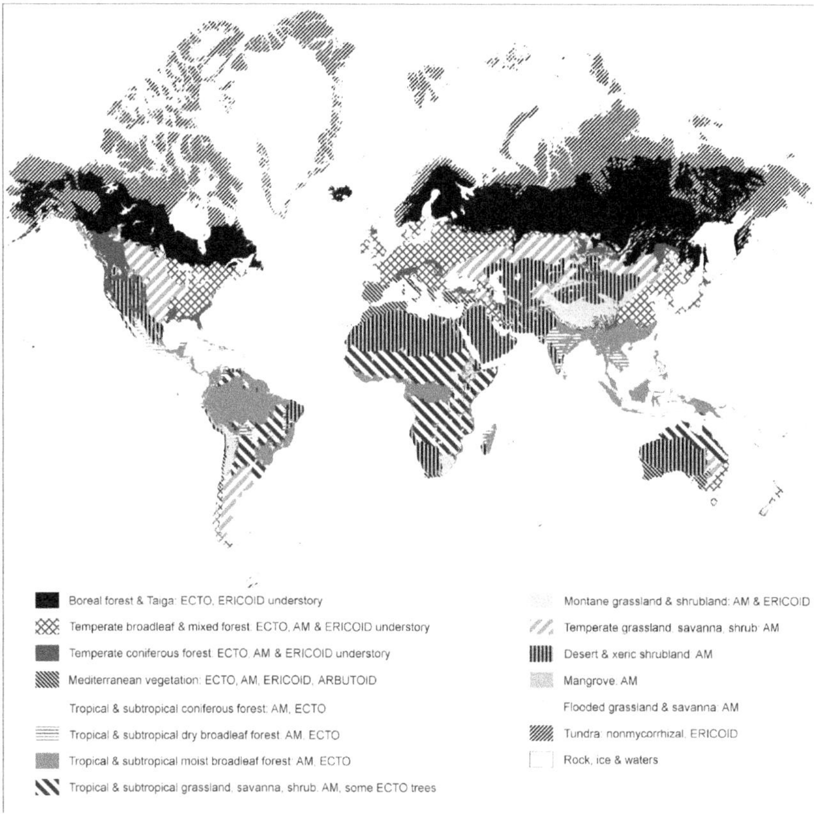

FIGURE 8.2 Distribution of different forms of mycorrhizae in the world biomes (most information on mycorrhizal distribution based on Read (1991), biomes distribution map by Olson et al. (2001)).

internal transcribed spacer (Abarenkov et al., 2010; Eberhardt, 2010; White et al., 1990). Together, these methods provide insights into belowground fungal diversity and are the basis for further understanding the physiological functions and ecological roles of ectomycorrhizal fungi and their mycelial networks in ecosystems. Furthermore, these data reveal the functional compatibility between the fungal partner and the plant (Gianninnazzi-Pearson, 1984, cit. in Kraigher, 1996) and allow identification of the fungal community structure in belowground environments as reviewed by Wallander et al. (2013).

8.1.3 Plant and Ecosystem Services Provided by Mycorrhizal Fungi

Mycorrhizae enable the plant (macrosymbiont) a better growth in natural and anthropogenically disturbed environments with mechanisms that can be defined at the morphological, physiological and ecological level (Allen, 2011;

Kraigher, 1996, 1999; Smith and Read, 2008), which are highlighted below, and provide a number of ecosystem as well as remediation services.

Morphological mechanisms include (i) changed fine root growth pattern and architecture, (ii) colonisation of a bigger soil volume compared to non-mycorrhizal fine roots, (iii) colonisation of smaller soil compartments due to smaller diameter of mycorrhizal hyphae (considering the scales: the diameter of fine roots is measured in millimeter, of root hairs in 0.1 mm, of mycorrhizal hyphae in micrometer), (iv) higher overall nutrient absorption per area unit (as seen by enhanced uptake of N-sources in mycorrhizal seedlings), (v) reduced distance of transport between the soil and the root in exploration types with simple emanating mycelium and an increased long-range transport by mass flow through highly differentiated rhizomorphs, (vi) changes in soil structure induced by fungal mycelium, such as formation of soil aggregates or hydrolisation of complex organic residues and (vii) fungal mantle functioning as a mechanical barrier for invasion of pathogenic organisms and toxic elements in soils.

Physiological mechanisms include the following: (i) mobilisation of poorly accessible macronutrients by specific transporters, (ii) release of macronutrients from organic compounds using similar enzymatic systems as saprotrophic fungi, (iii) storage of phosphates and other nutrients in the fungal mantle or vesicles of arbuscular fungi, (iv) immobilisation or selective uptake of toxic (heavy) metal ions, (v) tolerance to or immobilisation of pathogenic organisms in the fungal mantle, (vi) exudation of antibiotics into the mycorrhizosphere, (vii) modifications of pH in the mycorrhizosphere and (viii) exudation of organic acids and siderophores that enable uptake of some micronutrients from forest soils.

These mechanisms allow mycorrhizal plants to grow on nutrient-poor substrates and even tolerate mining residues or dry ash, enabling utilisation of mycorrhizal plants in bioremediation technologies and for recultivation of polluted substrates or elimination of pollutants from substrates, for example, in phytostabilisation or in phytoextraction technologies.

Ecological mechanisms include (i) inter- and intra-specific networks that link sources and sinks of C within the ecosystem, (ii) water and nutrients redistributed in time and space within the ecosystem, (iii) a range of interactions with animal and micro-organism communities in the mycorrhizosphere occur, and (iv) diversity of the bio-component in forest soils and that of the whole ecosystem is altered and supported.

8.1.4 Mycorrhizae and Soil Processes

Mycorrhizal fungi influence the formation or stabilisation of soil at the macro aggregate (53–250 μm) and micro aggregate (<53 μm) scales with mechanisms that Rillig and Mummey (2006) differentiate into physical (such as altered water regimes and translocation), biochemical and biological processes.

Additionally, ECM fungal species differ in their effects on root hydraulic conductivity, their influence on root biomass and their potential to colonise soil due to differences in mycelium architecture (Rillig and Mummey, 2006). Allen (2007, cit. in Allen, 2011) discussed the fungal kingdom from an ecological perspective: though the prevalent classification is that fungi are 'microorganisms' with respect to their size, they are 'macro-organisms packaged in microscopic units', which may extend across many hectares and form complex networks.

The potential to exploit nutrients contained in soil organic matter (SOM) has been additionally attributed to extracellular enzymes secreted by ectomycorrhizal and ERM fungi (Courty et al., 2010; Cullings and Courty, 2009; Pritsch and Garbaye, 2011; Rineau and Courty, 2011). Using enzyme activity measurements on individual mycorrhizal roots, potential enzyme activities of numerous ECM species demonstrated a widely distributed enzymatic potential for phosphatase, various plant cell-wall degrading, and chitinolytic enzymes, while enzyme activities such as phenol oxidase or proteolytic enzymes were restricted to a few taxa. In some cases, the interactions between saprotrophic soil fungal communities led to the decrease of litter decay in the presence of ECM fungi, known as the Gadgill effect. However, interactions of mycorrhizal communities with other soil microbial components are still understudied (Finlay, 2008).

8.1.5 Mycorrhizosphere Interactions

The community structure of soil organisms can be defined at different scales and trophic levels (Koide, 2012).

According to the classical view of mycorrhizal symbiosis, the mycorrhizal fungus facilitates the uptake of water and nutrients to the plant, while the plant provides the fungus with carbohydrates. The evidence that mycorrhizal fungal mycelia can link the whole belowground part of the ecosystem in a network has prompted scientists to name them 'common mycelial networks' or 'common mycorrhizal networks' (CMN) (Read, 1998; Simard et al., 1997). The importance of mycorrhizal diversity has been emphasised even more in studies on biodiversity and ecosystem productivity (such as Kraigher, 1999; Read, 1998; van der Heijden et al., 1998) and carbon dynamics (Simard et al., 2012; van der Heijden and Horton, 2009). The mycorrhizal networks facilitate transfer of resources through different constituents of a forest ecosystem, including plants of the same or different species, and compartmentalise valuable resources away from potential disruption, such as competition from other soil microbes, fauna, chemical adsorption of nutrients to soil particles or physical disturbance of the soil structure (Simard et al., 2012). The CMN have been proposed to fulfil different roles in ecosystems in which they can act as facilitators of plant growth. In these, the plant has a net benefit for growth, or the conditions for growth are either ameliorated or worsened and lead towards competition. Similarly, water can also move through multiple belowground pathways with greatest transfer efficiency through root grafts and CMN (Allen, 2011).

Mycorrhizal networks are ubiquitous in nature; therefore, facilitation via networks may be as important as competition in the structuring, functioning and sustainability of plant and fungal communities (Simard et al., 2012). Furthermore, considering net transfer of C from assimilating broadleaves to shaded conifers (Simard et al., 1997) and from assimilating conifers to broadleaves just initiating their spring leaf burst (Simard et al., 2012), the role of mycorrhizal mycelium can be interpreted as the basic belowground network connecting the whole ecosystem into a complex functioning unit (Kraigher, 1999), providing 'belowground collaboration' between organisms. It suggests also that mycorrhizal networks play an important role in the functioning of ecosystems as complex adaptive systems, in their ecological stability and evolution (Levin, 2005).

A number of organisms living in forest soils from different kingdoms are linked through a variety of interactions. Plant growth promoting bacteria and mycorrhization helper bacteria can be considered as multiple symbioses in soils (Frey-Klett et al., 2007), while belowground diversity has been shown to support biodiversity and productivity of whole ecosystems (Read, 1998; van der Heijden et al., 1998). From the classical differentiation of interactions among organisms, such as mutualistic, predator, parasitic, and competitive interactions, the terms socialism and facilitation were introduced in the characterisation of the roles of mycorrhizal fungal networks in forest soils (Read, 1998; van der Heijden and Horton, 2009).

8.1.6 Mycorrhizal Exploration Types: Functional Considerations of Differing Structures

Ectomycorrhizae represent almost 100% of all fine roots of the majority of stand-forming broadleaf and coniferous tree species in northern temperate and boreal forests (Kraigher, 1999). The fungal partner is primarily responsible for ectomycorrhizal system architecture, morphological and anatomical characteristics, and the extent of emanating elements—that is, extramatrical mycelia (EMM) growing from the root into the mycorrhizosphere. The plant symbiont determines the growth pattern (such as dichotomous fine mycorrhizal root growth in pines), the extent to which the Hartig net develops around primary cortex cells (a single layer in angiosperms or more layers in conifers) and any additional layers of tannin cells, which characterise several conifer ectomycorrhizal types (Kraigher, 1996).

Among about 6000 fungal species that can form ectomycorrhizae, a great differentiation in growth characteristics, structure, and function has been established, yet all physiological studies and understanding of the functioning of ectomycorrhizae have been hampered so far by limited growth in laboratory conditions and complexity of interactions in the mycorrhizosphere

(Wallander et al., 2013). Therefore, Agerer (2001) has proposed a differentiation of types of ectomycorrhizae, based on species characteristics, into five exploration types (ET), based on the amounts of emanating hyphae and the presence and differentiation of rhizomorphs, suggesting their differentiation and resemblances in ecological characteristics and function (Figure 8.1, Right). These types range from so-called contact exploration types with smooth mycorrhizal tips having only a few short emanating hyphae to long distance exploration types with highly differentiated rhizomorphs. Exploration types have been differentiated based on about 400 different morphotypes of ectomycorrhizae (www.deemy.de), representing about 7% of known fungi that can form ectomycorrhizae (Ekblad et al., 2013, modified). Characteristics of extramatrical mycelium length, space occupation, and biomass for different exploration types have been considered and related to carbon costs, niche occupation and nutrient mobilization, competition, and interaction with other organisms (Agerer et al., 2012). However, more physiological and ecological studies should support the proposed functional diversity and belowground successions in changing mycorrhizal communities as simplified by differentiation into ET.

8.2 MYCORRHIZAE UNDER STRESS AND DISTURBANCE

Stress is considered to be a significant deviation from the conditions optimal for life, eliciting changes and responses at all functional levels of the organism (Larcher, 1995 cit. in Kraigher and Al Sayegh-Petkovšek, 2011). Natural or anthropogenic *disturbances* are traditionally viewed as events that initiate primary or secondary succession, and *succession* is an orderly unidirectional process of community change in which communities replace each other sequentially until a stable (self-reproducing) community is reached (Johnson and Miyanishi, 2007). Therefore, stress and disturbance can both lead to successions; however, disturbance is better viewed with respect to ground cover changes (such as by fire, erosion, or direct human-induced changes), while stress is better viewed with respect to the changes of different environmental conditions (e.g. due to climate change and air pollution).

The belowground mycorrhizal communities change differently due to stress and disturbance effects on their niches and survival functions, which forms the basis for *mycobioindication*—a method to detect stress and disturbance in forest soils (For a recent review see Kraigher and Al Sayegh-Petkovšek, 2011).

8.2.1 Pollutants

A slow acidification of soils is a natural process in most coniferous forests, but due to high atmospheric inputs of *acidifying pollutants*, such as S and N compounds, the percentage of forests in Europe with a soil pH below 4.2

increased in the twentieth century (Arnolds, 1991). *N deposition* has increased across a large number of temperate conifer and deciduous ecosystems and was found to correlate with changes in the number of ectomycorrhizal (ECM) community attributes. The response to *N fertilisation* is also species dependent. A reduction of the biodiversity of types of ECM in polluted areas was shown in spruce forests (Kraigher, 1999), whereas in beech forests, no negative effects on biodiversity were determined (Kraigher et al., 2007). A difference between coniferous and deciduous nutrient acquisition strategy was proposed (Kalliokoski et al., 2010). Also, atmospheric pollution, leading to soil acidification and *elevated concentrations of trace elements* in the soil, affect soil microorganisms and microbial processes. Effects of pollutants on forests and mycorrhiza are summarized in Table 8.2.

In conclusion, a high species-specific differentiation in survival and functioning was found but has been only studied in relatively few ectomycorrhizal fungi so far. Considering the pollution-induced changes in fungal community structure (including changes in exploration types), which lead to changes in the functional ecology of trees, such studies should be further supported. Also, methodological approaches for better understanding the belowground regulators of processes in forest ecosystems should be further developed.

8.2.2 Selected Anthropogenic Disturbance Regimes

Forest fires differ according to their impact on the ecosystem through direct or indirect effects such as burned vegetation, burning, loss or change in organic matter aboveground and belowground, and prompt release of high amounts of nutrients, which affect soil erosion and vegetation regeneration, and modify the restoration of community structure, including the belowground compartment (Buscardo et al., 2011). Physical, chemical and biological changes at burned sites always cause losses of or changes in organic matter aboveground and belowground.

Forest fires are likely to become more frequent as a consequence of predicted climate changes, as already shown in the Mediterranean area (Moriondo et al., 2006). The drift in soil microbial community after fire also has an important application in reforestation of burned areas, where mycorrhizal seedlings have a better survival rate compared to non-mycorrhizal (*ibid.*).

Effects of selected anthropogenic disturbances on forests and mycorrhiza are summarized in Table 8.3.

Forest management is a term concerning technical and scientific aspects in forestry, providing a framework of tools for decision making, including a range of silvicultural operations that influence the development of a stand over time. Management intensity can vary from completely natural development without human intervention up to highly intensive silvicultural approaches with the key aim to achieve target economic or ecological demands (Duncker et al., 2012). The changes in quality and diversity of the

TABLE 8.2 Effects of Pollution on Forests and Mycorrhiza

Pollutant	Effect on environment	Effect on mycorrhiza	References
Acidifying pollutants	• Decreased soil pH • Increased release of heavy metals (Al) • Manipulated pH and P	• Reduced ECM abundance • Reduced number of morphotypes • Different sensitivity of ECM types, therefore, shift in community structure • Shift to *Ascomycota* and shift from AM to ECM fungi	Brunner (2001), Cudlin et al. (2007), Read (1991), Kraigher (1999) and Kluber et al. (2012)
N fertilization and depostion		• Decrease in fractional colonisation by ECM fungi—decrease in number of ECM roots • Decrease in ECM sporocarp production, rhizomorph formation and EMM • Decline in abundance and diversity of ECM fungi • Shift in ECM community composition with an increase of nitrophillic spp. • In polluted spruce forests *Tylospora* spp. and *Lactarius rufus* dominated the ECM community (75%) • Species from genera *Cortinarius, Tricholoma* and *Russula* sensitive to N enrichment resulting in reduced abundance	Kraigher and Al Sayegh-Petkovšek (2011), Kraigher (1999), Kraigher et al. (2007), Taylor et al. (2000), Nilsson and Wallander (2003), Kalliokoski et al. (2010), Deslippe et al. (2011) and Lilleskov (2005)
Trace elements	• Elevated concentrations of trace elements due to soil acidification	• Abundance of tolerant spp., for example *Amanita, Albatrellus ovinus, Leccinum scabrum* likely to increase • Accumulation of metals in sporocarps and EMM	Kraigher and Al Sayegh-Petkovšek (2011)

AM, arbuscular mycorrhiza; ECM, ectomycorrhiza; EMM, extramatrical mycelium.

TABLE 8.3 Effects of Anthropogenic Disturbance on Forests and Mycorrhiza

Anthropogenic disturbance	Effect on environment	Effect on mycorrhiza, soil microorganisms and roots	References
Forest fire	• Reduction of C:N ratio • Soil erosion • Change of community structure	• Reduction of microbial activity • Shift to mycorrhizal spp. better adapted to high flushes of nutrients and early colonizers of seedlings • Reduction of fine root biomass and ECM density • More ECM in deep horizons, less in litter, organic and shallow mineral soil	Dumm et al. (2008)
Forest management: felling and thinning	• Gap formation • Reduced above- and belowground diversity	• Complex influence on above- and belowground processes • Reduced numbers of fine roots and their dynamics, diversity indices and ECM composition even in small canopy gaps	Jones et al. (2003), Grebenc et al. (2009) and Buée et al. (2005)
Forest management: cultivation practices	• Reduced above- and belowground diversity	• Reduced mycorrhizal frequency in cultivated vs. natural or naturalised stands • Reduced natural mycorrhization on previously agricultural lands • High N treatment may increase ECM and the proportion of fine roots in spruce and pine seedlings grown in pots, increasing seedling competitiveness	Hrynkiewicz et al. (2012), Beniwal et al. (2011) and Gruffman et al. (2012)
Forest management: liming in conifer forests or fertilization in plantations	• Changed soil nutrient composition and pH	• Shift in ECM community structure toward an increase in the relative abundance of dominant and stress tolerant mycorrhiza species (e.g. *Cenococcum geophilum*)	Bakker et al. (2000), Fransson et al. (2000) and Benucci et al. (2011)

TABLE 8.3 Effects of Anthropogenic Disturbance on Forests and Mycorrhiza—Cont'd

Anthropogenic disturbance	Effect on environment	Effect on mycorrhiza, soil microorganisms and roots	References
		• High pH and low N regime applied in truffle plantations to suppress competing ECM species	
Soil compaction	• Changes in soil structure and soil removal • Physical damage to seedlings	• Decrease in ECM abundance and diversity • Physical damage to seedlings' mycorrhiza (roots and fungi)	Wiensczyk et al. (2002) and Waltert et al. (2002)

AM, arbuscular mycorrhiza; ECM, ectomycorrhiza.

ECM fungal inoculum in gaps in comparison to closed canopy stands indicates a significant effect of forest management practice on the mycorrhizosphere community with a long-term effect on the regeneration success (Cline et al., 2007).

Soil compaction in forests occurs due to recreational human activities (hiking, mushroom picking, etc.) or during forest management practices and harvests. To avoid compaction effects, proper recreation measures and limits should be applied in frequently visited forest sites (Waltert et al., 2002) and particular care should be taken to retain the forest floor during wood harvest.

Predicting the effects of forest management practices and soil compaction on forest ecosystem functioning, in particular to the belowground compartment, is complex and rarely applied at the level of mycorrhizosphere parameters. Several attempts to model mycorrhizosphere processes related to ECM and their changes under management practices were published (Deckmyn et al., 2011; Meyer et al., 2010; Orwin et al., 2011). Compared to AM models, the modelling of ECM fungi and their position in mycorrhizosphere processes remains understudied and in need of additional knowledge and principles.

8.2.3 Climate Change

By definition, *climate change* means a change of climate that is attributed directly or indirectly to human activity, which alters the composition of the global atmosphere and which is observed in addition to natural climate variability over comparable time periods. Its main anthropogenic cause is production of greenhouse gases and its most evident consequences are changing temperatures, precipitation regimes, elevated CO_2 and tropospheric ozone.

C balances have been considered in minimising anthropogenic climate change processes and its consequences. Since about half of the terrestrial C sink is located in forests, and the average ratio of the C pool in soils relative to vegetation ranges from 5:1 in boreal, 2:1 in temperate and 1:1 in tropical forests (Jarvis et al., 2005), the role of forest soils and the belowground regulators of C dynamics have been identified as important topics for future research (Matyssek et al., 2010).

Global climate change and other stresses and disturbances may weaken C uptake by forest ecosystems and render forests into a C source leading to a positive feedback on global climate. On the other hand, enhancing and strengthening C fluxes into stable forest C pools can offset anthropogenic CO_2 emissions and minimize risks of abrupt climate change. A selection of the most evident consequences influencing mycorrhizal diversity and functioning is presented (Table 8.4).

Elevated temperature effects on the mycorrhizosphere constitute a sum of major direct effects on ectomycorrhizal fungi in soil and indirect effects through changes in the physiology of symbiotic plant partners. The results of direct temperature effects on ectomycorrhizal fungi activity/respiration are inconsistent (Kruse et al., 2013, this vol.).

Water regimes in forest ecosystems are likely to change with changing climate. Water uptake in plants is partially mediated by ectomycorrhizal fungi and dependent on osmotic potential- (Duddridge et al., 1980). The transportation of water (hydraulic lift and redistribution) is coupled with the allocation of key nutrients and supported by CMN, thereby providing stability for forest ecosystems (Simard et al., 1997). This has led to describing the mycorrhizal fungi as acting as 'highways for water and nutrients in arid soils' (Allen, 2011).

The reduction of soil moisture, that is, *drought*, has a significant effect on the functioning of the CMN. A number of rare ectomycorrhizal fungi species might provide a means for the forest to adapt to changing environmental conditions, or facilitate the establishment of the future generations of trees in a forest ecosystem (Pena et al., 2010). For the other extreme water regime, *flooding and soil water saturation*, laboratory experiments have revealed various degrees of intolerance of different ectomycorrhizal fungi.

Any drastic change in water regime in forest ecosystems tends to lead to similar effects in the mycorrhizosphere, specifically a drift in the mycorrhiza community toward more resistant species and species with specific morphologic structures and a dieback of intolerant species and strains.

Under *elevated* CO_2 the increased carbon allocation to ectomycorrhizal roots did not automatically translate into an increased uptake of limiting nutrients and hence to increased forest productivity, as the increase in ECM roots and mycelia would also increase nutrient limitation of the system. This phenomenon has been described as the progressive nitrogen limitation hypothesis (Hu et al., 2006; Luo et al., 2004).

TABLE 8.4 Effects of Climate Change on Forests and Mycorrhiza

Climate change	Effect on environment	Effect on mycorrhiza	References
Elevated temperature	• Usually in combination with drought, nutrients and CO_2 changes • Most pronounced impact on edge and marginal populations	• Dominance of indirect temperature effect through the availability of carbohydrates as the primary trigger for ECM responses • Increase in fungal biomass and colonization rates at stand level • Selection of higher temperature tolerant ECM species at ecosystem level • Increasing potential for nutrient acquisition in arctic mycorrhizal species	Staddon et al. (2002), Ekblad et al. (2013), Pietkäinen et al. (2005), Heinemeyer et al. (2007), Hawkes et al. (2008) and Kytoviita and Ruotsalainen (2007)
Drought	• Reduced water availability	• Different day-night water distribution and re-distribution • Higher mycorrhizal colonization rates under mild drought • Reduced mycorrhizal colonization rates under severe drought • Seasonal droughts shift the ECM community towards species with higher capability of long-distance water transportation (rhizomorph formation) or to generally stress tolerant species (e.g. *Cenococcum geophilum*)	Querejeta et al. (2003) cit. in Allen (2011), Shi et al. (2002) and Swaty et al. (2004)
Flooding and soil water saturation	• Excess of water, dilution of nutrients and temporary anoxia	• Shift in the mycorrhizal fungi community toward more flooding resistant species • Growth of flooded trees better when inoculated with a combination of ECM and AM fungi, which is common in species adapted to occasional flooding (e.g. poplars)	Zhong et al. (1997)

Continued

TABLE 8.4 Effects of Climate Change on Forests and Mycorrhiza—Cont'd

Climate change	Effect on environment	Effect on mycorrhiza	References
Elevated CO_2	• Progressive nitrogen limitation (PNL) hypothesis	• Increased carbon allocation to ectomycorrhizal roots • Effects on mycorrhizal roots and mycelia not consistent: reported effects include increased root tip abundance and changes in root:shoot ratio; opposite or no effects also reported • Changes of ECM fungal communities with a possible increase in abundance of more carbon demanding species (long range exploration types)	Luo et al. (2004), Hu et al. (2006), Alberton et al. (2005) cit. in Cudlin et al. (2007), Weigt et al. (2011) cit. in Ekblad et al. (2013) and Ekblad et al. (2013)
Tropospheric O_3	• Reduced carbon sink strength in forest ecosystems • Reduced and chemically altered carbon flow to the mycorrhizosphere	• ECM diversity of beech seedlings in lysimeters not affected • Exposed beech seedlings in pots exhibited significantly reduced ECM diversity • Small changes in ECM diversity of adult beech trees, but significant changes in ECM community composition • Change in interspecies interactions predicted	Pritsch et al. (2009), Železnik et al. (2007); Grebenc and Kraigher (2007a,b) and Haberer et al. (2007)
Elevated temperature	• Usually in combination with drought, nutrients and CO_2 changes • Most pronounced impact on edge and marginal populations	• Physiologic responses of ECM fungi described by square-root functions • Dominance of indirect temperature effect through the availability of carbohydrates as the primary trigger for ECM responses • Increase in fungal biomass and colonization rates at stand level • Selection of higher temperature tolerant ECM species at ecosystem level • Increasing potential for nutrient acquisition in arctic mycorrhizal species	Staddon et al. (2002), Ekblad et al. (2013), Pietkäinen et al. (2005), Heinemeyer et al. (2007), Hawkes et al. (2008) and Kytoviita and Ruotsalainen (2007)

TABLE 8.4 Effects of Climate Change on Forests and Mycorrhiza—Cont'd

Climate change	Effect on environment	Effect on mycorrhiza	References
Drought	• Reduced water availability	• Different day-night water distribution: during the day, uptake of deep water through oak tap-roots, during the night, re-distribution of this water into the mycorrhizal mycelium enabling N uptake through water droplets • Higher mycorrhizal colonization rates under mild drought • Reduced mycorrhizal colonization rates under severe drought • Seasonal droughts shift the ectomycorrhizal community toward species with higher capability of long-distance water transportation (rhizomorph formation) or to generally stress tolerant species (*Cenococcum geophilum*)	Querejeta et al. (2003) cit. in Allen (2011), Shi et al. (2002) and Swaty et al. (2004)
Flooding and soil water saturation	• Excess of water, dilution of nutrients and temporary anoxia	• Shift in the mycorrhizal fungi community toward more flooding-resistant species • Growth of flooded trees better when inoculated with a combination of ecto- and arbuscular mycorrhizal fungi, which is common in species adapted to occasional flooding (e.g. poplars)	Zhong et al. (1997)
Elevated CO_2	• Progressive nitrogen limitation (PNL) hypothesis	• Increased carbon allocation to ectomycorrhizal roots • Effects on mycorrhizal roots and mycelia not consistent: reported effects include increased root tip abundance and changes in root:shoot ratio; opposite or no effects also reported	Luo et al. (2004), Hu et al. (2006), Alberton et al (2005) cit. in Cudlin et al. (2007); Weigt et al. (2011) cit. in

Continued

TABLE 8.4 Effects of Climate Change on Forests and Mycorrhiza—Cont'd

Climate change	Effect on environment	Effect on mycorrhiza	References
		• Changes of ectomycorrhizal fungal communities with a possible increase in abundance of more carbon demanding species (long range exploration types)	Ekblad et al. (2013) and Ekblad et al. (2013)
Tropospheric O_3	• Reduced carbon sink strength in forest ecosystems • Reduced and chemically altered carbon flow to the mycorrhizosphere	• ECM diversity of beech seedlings in lysimeters not affected • Exposed beech seedlings in pots exhibited significantly reduced ECM diversity • Small changes in ECM diversity of adult beech trees, but significant changes in ECM community composition • Change in interspecies interactions predicted	Pritsch et al. (2009), Železnik et al. (2007), Grebenc and Kraigher (2007a,b) and Haberer et al. (2007)

AM, arbuscular mycorrhiza; ECM, ectomycorrhiza.

Tropospheric ozone at predicted future background concentrations of over 40 ppb is supposed to strongly reduce carbon sink strength in forest ecosystems, thus diminishing offsetting for a key carbon sink for current CO_2 emissions (Wittig et al., 2009). As mycorrhizal fungi depend on plant carbon, ozone has been proposed as an important factor impairing soil nutrient cycling and forest tree growth and nutrition in an indirect manner via reduced and chemically altered carbon flow to the mycorrhizosphere (Andersen, 2003). Ozone effects on mycorrhizal communities differed between young and old trees and were not consistent across several studies. A change in interspecies interactions has been predicted (Matyssek et al., 2013, this vol.).

A detailed overview of elevated O_3 and CO_2 is provided by King et al. (2013, this vol.).

8.3 MYCORRHIZAL INFLUENCE ON CARBON STORES AND BIODIVERSITY: THE FACILITATION CONCEPT

8.3.1 Carbon Stores and Mycorrhizal Networks

The total belowground flux of carbon (C) represents between 25% and 63% of gross primary production and has a large influence on the physical, chemical and biological properties of the soil. The fine roots, although contributing less than 2% of tree biomass in mature forests, significantly influence belowground C stocks due to their rapid turnover (Brunner and Godbold, 2007). Fine roots of trees are always associated with mycorrhizal fungi, and approximately 20–30% of current assimilates are allocated to them. It is assumed that at least 30% of the microbial biomass and 80% of the fungal biomass belong to the extramatrical mycelium of the mycorrhizal fungi. Mycorrhizal symbioses increase the sink strength of roots for carbon, by increasing metabolic activity of the root system and by increasing the size of the root-hyphal sink through facilitation of interactions among the complex belowground community. The C flux through EMM is estimated to be large, it may be important for formation of SOM and it may directly or indirectly affect decomposition of SOM (Ekblad et al., 2013). The complexity of belowground interactions is shown in Figure 8.3, including in part also interactions between other than mycorhizal organisms in forest soils (Couture and Lindroth, 2013, this vol.).

Several models of C dynamics and forest growth have been applied in different climate zones and forest types, yet all lack any specification on C fluxes derived from fine root and mycorrhizal mycelia turnover. There is an internationally recognized need to improve the knowledge about, and develop high-quality tools for estimating belowground biomass/C-equivalents and C-fluxes through different compartments in the soil and (myco)rhizosphere. Due to the species dependent structural and functional differences among different ectomycorrhizal fungal systems, any modifications, shifts or successions in the ectomycorrhizal community may lead to important differences in C dynamics in forest ecosystems. Furthermore, other aspects of species-specific functional diversity should be considered, such as the role of ECM in temporal and spatial distribution and re-distribution of N and P, as well as trace elements in forest ecosystems.

8.3.2 Resource Partitioning and the Facilitation: Collaboration Concepts

Resource partitioning through *competition*, resource sharing through CMN and a variation in host responses to microbes (including fungi) and soil community feedbacks (Bever et al., 2010) have so far been largely biased by a plant-centric approach. Also, discussions on low-diversity plant communities mainly focus on competitive dominance, inhibition, or positive feedbacks that plant community development might be based on (*ibid.*). Belowground biotic complexity was found to drive aboveground dynamics through a soil community feedback (Pendergast et al., 2013). *Facilitative interactions*, that is, beneficial interactions among different organisms, were placed in focus several years ago (van der Heijden and Horton, 2009) and have raised interest in including a new conceptual model in different areas of ecology (Brooker

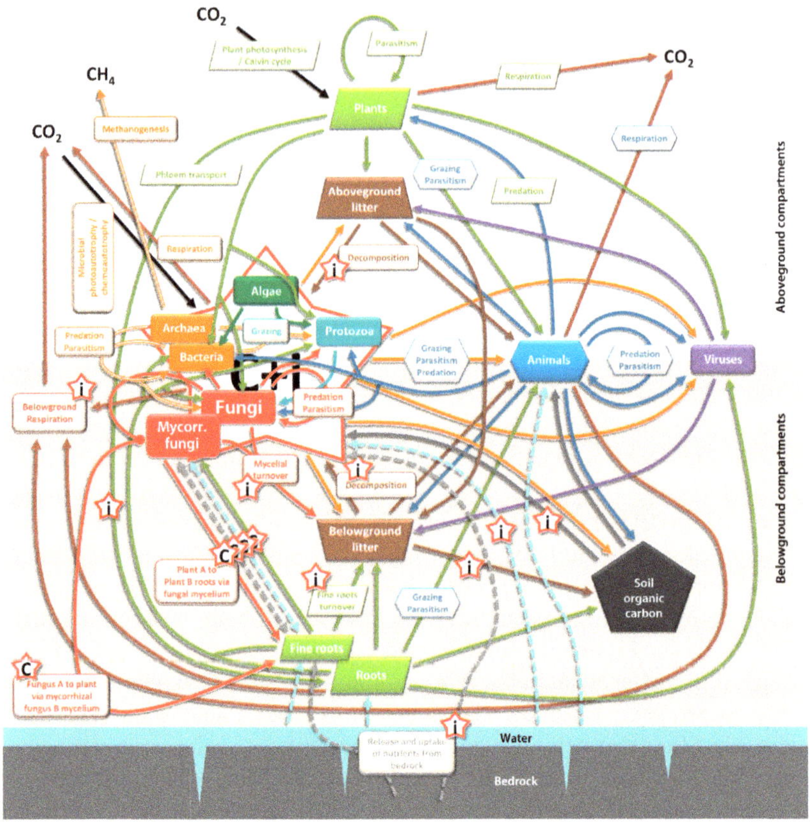

FIGURE 8.3 A simplified schematic representation of complexity of biota, biological interactions and carbon translocation in land ecosystems. Arrows indicate the direction of the carbon transfer. Full arrow lines represent translocations associated with common metabolic, nutrient uptake and transport pathways; dashed arrow lines represent nutrient release and uptake from bedrock in grey and water uptake and transport in light blue. Individual groups of organisms with associated activities, processes and carbon compounds originating from them are presented in different colours and shapes: light green parallelogram for plants; dark blue hexagon for animals; rounded rectangles for micro biota including algae in dark green, bacteria and archaea in orange with associated carbon translocations in white arrow lines in orange outline, protozoa in light blue and fungi in light red (solid orange arrow lines represent carbon translocations common to all groups within micro biota); purple rounded rectangles for viruses including viroids, all other virus-like particles and satellites; brown trapezoid for combined litter and translocation of litter-originating carbon; dark grey pentagon for soil organic carbon; dark red arrow lines represent release of carbon dioxide as a result of aerobic respiration of all groups of organisms; autotrophic assimilation of carbon dioxide by black arrow lines. Processes and translocations that are affected by facilitative-collaborative interactions involving fungi or impacted by fungal activities are marked by 'C' and 'i' in a star-shaped outline, respectively. Stress and disturbance, natural or anthropogenic, elicit changes and responses at all functional levels of organisms and often lead to changes in the composition of the biota, from the level of the gene to the level of populations, communities and entire ecosystems.

and Callaway, 2009). Mutualism and facilitation have been subsumed under a common concept called '*beneficial interactions*' (*ibid.*), which we here propose to name '*belowground collaboration*', and includes carbon, water and nutrient transfer among organisms of the same and different taxons and taxonomic groups, thus supporting their survival and influencing sustainable development of the ecosystem.

Facilitation has been discussed as paving the way for ecological speciation, enhancing local adaptation, and allowing long-term adaptation to environmental extremes in marginal habitats and in contact zones (Liancourt et al., 2012). Also, a supporting role of the existing diversity both above- and belowground has been found to promote persistence of organisms which, according to current research methods, have been named 'low-quality fungal partners' (Hart et al., 2013). Therefore, the distribution of different forms of mycorrhiza in the world's biomes (Figure 8.2) is subject to change and will determine future distribution, adaptation and survival of large terrestrial biomes.

8.4 CONCLUSIONS AND PROSPECTS FOR FURTHER RESEARCH AND MONITORING

Potential mechanisms of mycorrhizae to enhance plant tolerance to stressed or disturbed environments can be defined at the morphological, physiological and ecological level. They allow water, assimilates and mineral nutrients to re-allocate in time and space. As species and populations differ in their functional traits, the fungal community structure and its changes can provide key information for any interaction, bioindication or disturbance studies.

Pollutants in forest ecosystems influence the mycorrhizosphere by acidification, altered toxicity or imbalance of ions and nutrients. Natural and anthropogenic physical disturbances such as forest fires or forest management practices, or factors connected to climate change such as drought, flooding, increased soil temperature, increased CO_2 and tropospheric ozone, change the mycorrhizal community. As all of these factors cause community shifts with unknown effects on the complex mycorrhizosphere network, altered carbon storage capacities and reduced stress resilience may lead to increased mortality in forests and reverse the role of forests in climate change scenarios.

The large C flux through EMM is an important factor of SOM formation and decomposition. Moreover, the mycorrhizosphere biotic complexity was found to drive aboveground dynamics through a soil community feedback. Facilitation among different organisms is merging into a common concept of 'beneficial interactions' and 'belowground collaboration'. Although CMN provide an important component for long-term functioning of terrestrial ecosystems and their disruption may lead to changes in C dynamics and degradation of ecosystems, they are still an understudied field.

The conservation of the belowground component in terrestrial ecosystems should be a priority area of future research and monitoring programmes. The system of monitoring of belowground biodiversity, supported by studies of diversity at different levels, should be better incorporated in future common European strategies and national legislative measures for protection of forests and soils that are closely connected to socio-economic and demographic relationships in rural environments.

A *model of collaboration* among different organisms with the supporting role of ectomycorrhizal networks, and a whole ecosystem diversity, with the highly diverse belowground component and its functions, should be included in future models on C dynamics in terrestrial ecosystems.

ACKNOWLEDGEMENTS

The text was prepared for the common COST action FP 0903, and supported by research programme P4-0104 and project L4-4319, financed by the Slovenian Research Agency, co-financed by the Ministry of agriculture and the environment of the Republic of Slovenia. The text profited by invaluable comments and corrections made by Dr. Karin Pritsch and two reviewers.

REFERENCES

Abarenkov, K., Henrik Nilsson, R., Larsson, K.H., Alexander, I.J., Eberhardt, U., Erland, S., Høiland, K., Kjøller, R., Larsson, E., Pennanen, T., Sen, R., Taylor, A.F.S., Tedersoo, L., Ursing, B.M., Vrålstad, T., Liimatainen, K., Peintner, U., Kõljalg, U., 2010. The UNITE database for molecular identification of fungi—recent updates and future perspectives. New Phytol. 186 (2), 281–285.

Agerer, R., 2001. Exploration types of ectomycorrhizae. A proposal to classify ectomycorrhizal mycelial systems according to their patterns of differentiation and putative ecological importance. Mycorrhiza 11, 107–114.

Agerer, R., Rambold, G., 2004–2011. DEEMY—An Information System for Characterization and Determination of Ectomycorrhizae. In: München, Germany [first posted on 2004-06-01; most recent update: 2011-01-10], www.deemy.de.

Agerer, R., Christan, J., Mayr, C., Hobbie, E., 2012. Isotopic signatures and trophic status of Ramaria. Mycol. Prog. 11, 47–59.

Allen, M.F., 2011. Linking water and nutrients through the vadose zone: a fungal interface between the soil and plant systems. J. Arid Land 3 (3), 155–163.

Andersen, C.P., 2003. Source-sink balance and carbon allocation below ground in plants exposed to ozone (Tansley review). New Phytol. 157, 213–228.

Arnolds, E., 1991. Decline of ectomycorrhizal fungi in Europe. Agric. Ecosyst. Environ. 35, 209–244.

Bakker, M.R., Garbaye, J., Nys, C., 2000. Effect of liming on the ectomycorrhizal status of oak. For. Ecol. Manage. 126, 121–131.

Beniwal, R.S., Hooda, M.S., Polle, A., 2011. Amelioration of planting stress by soil amendment with a hydrogel-mycorrhiza mixture for early establishment of beech (Fagus sylvatica L.) seedlings. Ann. For. Sci. 68 (4), 803–810.

Benucci, G.M.N., Raggi, L., Albertini, E., Grebenc, T., Bencivenga, M., Falcinelli, M., Di Massimo, G., 2011. Ectomycorrhizal communities in a productive Tuber aestivum orchard: composition, host influence and species replacement. FEMS Microbiol. Ecol. 76, 170–184.

Bever, J.D., Dickie, I.A., Facelli, E., Facelli, J.M., Klironomos, J., Moora, M., Rillig, M.C., Stock, W.D., Tibbett, M., Zobel, M., 2010. Rooting theories of plant community ecology in microbial interactions. Trends Ecol. Evol. 25, 466–476.

Brooker, R.W., Callaway, R.M., 2009. Facilitation in the conceptual melting pot. J. Ecol. 97 (6), 1117–1120.

Brunner, I., 2001. Ectomycorrhizas: their role in forest ecosystems under the impact of acidifying pollutants. Perspect. Plant Ecol. Evol. Syst. 4 (1), 13–27.

Brunner, I., Godbold, D.L., 2007. Tree roots in a changing world. J. For. Res. 12, 78–82.

Buée, M., Vairelles, D., Garbaye, J., 2005. Year-round monitoring of diversity and potential metabolic activity of the ectomycorrhizal community in a beech (Fagus sylvatica) forest subjected to two thinning regimes. Mycorrhiza 15, 235–243.

Bulgarelli, D., Rott, M., Schlaeppi, K., Ver Loren van Themaat, E., Ahmadinejad, N., Assenza, F., Rauf, P., Huettel, B., Reinhardt, R., Schmeltzer, E., Peplies, J., Gloeckner, F.O., Amann, R., Eickhorst, T., Schulze-Lefert, P., 2012. Revealing structure and assembly cues for Arabidopsis root-inhabiting bacterial microbiota. Nature 488, 91–95.

Buscardo, E., Freitas, H., Pereira, J.S., de Angelis, P., 2011. Common environmental factors explain both ectomycorrhizal species diversity and pine regeneration variability in a post-fire Mediterranean forest. Mycorrhiza 21 (6), 549–558.

Cline, E., Vinyard, B., Edmonds, R., 2007. Spatial effects of retention trees on mycorrhizas and biomass of Douglas-fir seedlings. Can. J. For. Res. 37 (2), 430–438.

Courty, P.E., Buée, M., Diedhiou, A.G., Frey-Klett, P., Le Tacon, F., Rineau, F., Turpault, M.P., Uroz, S., Garbaye, J., 2010. The role of ectomycorrhizal communities in forest ecosystem processes: new perspectives and emerging concepts. Soil Biol. Biochem. 42, 679–698.

Couture, J.J., Lindroth, R.L., 2013. Impacts of atmospheric change on tree-arthropod interactions. Dev. Environm. Sci. 13, 227–248.

Cudlin, P., Kieliszewska-Rokicka, B., Rudawska, M., Grebenc, T., Alberton, O., Lehto, T., Bakker, M.R., Börja, I., Konopka, B., Leski, T., Kraigher, H., Kuyper, T.W., 2007. Fine roots and ectomycorrhizas as indicators of environmental change. Plant Biosyst. 141 (3), 406–525.

Cullings, K., Courty, P.-E., 2009. Saprotrophic capabilities as functional traits to study functional diversity and resilience of ectomycorrhizal community. Oecologia 161 (4), 661–664.

Curl, E.A., Truelove, B., 1986. The Rhizosphere. Springer-Verlag, Berlin.

Deckmyn, G., Campioli, M., Muys, B., Kraigher, H., 2011. Simulating C and N cycles in forest soils: including the active role of micro-organisms in the ANAFORE forest model. Ecol. Model. 222, 1972–1985.

Deslippe, J.R., Hartmann, M., Mohn, W.W., Simard, S.W., 2011. Long-term experimental manipulation of climate alters the ectomycorrhizal community of Betula nana in Arctic tundra. Glob. Chang. Biol. 17, 1625–1636.

Duddridge, J.A., Malibari, A., Read, D.J., 1980. Structure and function of mycorrhizal rhizomorphs with special reference to their role in water transport. Nature 287, 834–836.

Dumm, G., Fins, L., Graham, R.T., Jain, T.B., 2008. Distribution of fine roots of Ponderosa pine and Douglas-fir in a Central Idaho forest. West. J. Appl. For. 23 (4), 202–205.

Duncker, P.S., Barreiro, S.M., Hengeveld, G.M., Lind, T., Mason, W.L., Ambrozy, S., Spiecker, H., 2012. Classification of forest management approaches: a new conceptual framework and its applicability to European forestry. Ecol. Soc. 17 (4), 51.

Eberhardt, U., 2010. A constructive step towards selecting a DNA barcode for fungi. New Phytol. 187 (2), 265–268.

Ekblad, A., Wallander, H., Godbold, D.L., Cruz, C., Johnson, D., Baldrian, P., Björk, R.G., Epron, D., Kieliszewska-Rokicka, B., Kjöller, R., Kraigher, H., Matzner, E., Neumann, J., Plassard, C., 2013. The production and turnover of extramatrical mycelium of ectomycorrhizal fungi in forest soils: role in carbon cycling. Plant Soil 366 (1–2), 1–27.

Finlay, R.D., 2008. Ecological aspects of mycorrhizal symbiosis: with special emphasis on the functional diversity of interactions involving the extraradical mycelium. J. Exp. Bot. 59, 1115–1126.

Fransson, M.A., Taylor, A.F.S., Finlay, R.D., 2000. Effects of continuous optimal fertilization on belowground ectomycorrhizal community structure in a Norway spruce forest. Tree Physiol. (Oxford, UK) 20, 599–606.

Frey-Klett, P., Garbaye, J., Tarkka, M., 2007. The mycorrhiza helper bacteria revisited. New Phytol. 176, 22–36.

Grebenc, T., Kraigher, H., 2007a. Changes in the community of ectomycorrhizal fungi and increased fine root number under adult beech trees chronically fumigated with double ambient ozone concentration. Plant Biol. 9 (2), 279–287.

Grebenc, T., Kraigher, H., 2007b. Types of ectomycorrhiza of mature beech and spruce at ozone-fumigated and control forest plots. Environ. Monit. Assess. 128 (1), 47–59.

Grebenc, T., Christensen, M., Vilhar, U., Čater, M., Martin, M.P., Simončič, P., Kraigher, H., 2009. Response of ectomycorrhizal community structure to gap opening in natural and managed temperate beech-dominated forests. Can. J. For. Res. 39, 1375–1386.

Gruffman, L., Ishida, T., Nordin, A., Näsholm, T., 2012. Cultivation of Norway spruce and Scots pine on organic nitrogen improves seedling morphology and field performance. For. Ecol. Manage. 276, 118–124.

Haberer, K., Grebenc, T., Alexou, M., Gessler, A., Kraigher, H., Rennenberg, H., 2007. Effects of long-term free-air ozone fumigation on δ15N and total N in Fagus sylvatica and associated mycorrhizal fungi. Plant Biol. 9 (2), 242–252.

Hart, M.M., Forsythe, J., Oshowski, B., Bucking, H., Jansa, J., Kiers, E.T., 2013. Hiding in a crowd—does diversity facilitate persistence of a low-quality fungal partner in the mycorrhizal symbiosis? Symbiosis 59 (1), 47–56. http://dx.doi.org/10.1007/s13199-012-0197-8.

Hawkes, C., Hartley, I., Ineson, P., Fitter, A., 2008. Soil temperature affects allocation within arbuscular mycorrhizal networks and carbon transport from plant to fungus. Glob. Chang. Biol. 14, 1–10.

Heinemeyer, A., Hartley, I., Evans, S., Carreira de la Fuentes, J., Ineson, P., 2007. Forest soil CO_2 flux: uncovering the contribution and environmental responses of ectomycorrhizas. Glob. Chang. Biol. 13, 1786–1797.

Hrynkiewicz, K., Toljander, Y.K., Baum, C., Fransson, P.M.A., Taylor, A.F.S., Weih, M., 2012. Correspondence of ectomycorrhizal diversity and colonisation of willows (Salix spp.) grown in short rotation coppice on arable sites and adjacent natural stands. Mycorrhiza 22, 603–613.

Hu, S., Tu, C., Chen, X., Gruver, J.B., 2006. Progressive N limitation of plant response to elevated CO_2: a microbiological perspective. Plant Soil 289, 47–58.

Jarvis, P.G., Ibrom, A., Linder, S., 2005. Carbon forestry—managing forests to conserve carbon. In: Griffiths, H., Jarvis, P.G. (Eds.), The Carbon Balance of Forest Biomes. Taylor and Francis Group, United Kingdom, pp. 331–349.

Johnson, A., Miyanishi, K., 2007. Disturbance and succession. In: Johnson, E.A., Miyanishi, K. (Eds.), Plant Disturbance Ecology. The Process and the Response. Elsevier, Amsterdam, pp. 1–14.

Jones, R.H., Mitchell, R.J., Stevens, G.N., Pecot, S.D., 2003. Controls of fine root dynamics across a gradient of gap sizes in a pine woodland. Oecologia 134, 132–143.

Kalliokoski, T., Pennanen, T., Nygren, P., Sievanen, R., Helmisaari, H.S., 2010. Belowground interspecific competition in mixed boreal forests: fine root and ectomycorrhiza characteristics along stand developmental stage and soil fertility gradients. Plant Soil 330 (1/2), 73–89.

King, J., Liu, L., Aspinwall, M., 2013. Tree and forest responses to interacting elevated atmospheric CO_2 and tropospheric O_3: a synthesis of experimental evidence. Dev. Environm. Sci. 13, 179–208.

Kluber, L.A., Carrino-Kyker, S.R., Coyle, K.P., DeForest, J.L., Hewins, C.R., Shaw, A.N., Smemo, K.A., Burke, D.J., 2012. Mycorrhizal response to experimental pH and P manipulation in acidic hardwood forests. PLoS One 7 (11), e48946, 10 pp.

Koide, R.T., 2012. Context-dependent interaction hierarchies and the organization of ectomycorrhizal fungal communities. In: Hock, B. (Ed.), The Mycota, Volume IX: Fungal Associations. Springer-Verlag, Berlin, pp. 181–196.

Kraigher, H., 1996. Types of ectomycorrhizae - their taxonomy, role and application. Zbornik gozdarstva in lesarstva 49, 33–66.

Kraigher, H., 1999. Diversity of types of ectomycorrhizae on Norway spruce in Slovenia. Phyton (Horn, Austria) 39 (3), 199–202.

Kraigher, H., Al Sayegh-Petkovšek, S., 2011. Mycobioindication of stress in forest ecosystems. In: Rai, M., Varma, A. (Eds.), Diversity and Biotechnology of Ectomycorrhizae. Soil Biology, vol. 25. Springer-Verlag, Berlin, pp. 301–322.

Kraigher, H., Al Sayegh-Petkovšek, S., Grebenc, T., Simončič, P., 2007. Types of ectomycorrhiza as pollution stress indicators: case studies in Slovenia. Environ. Monit. Assess. 128, 31–45.

Kruse, J., Simon, J., Rennenberg, H., 2013. Soil respiration and soil organic matter (SOM) decomposition in response to climate change. Dev. Environm. Sci. 13, 131–149.

Kytoviita, M.M., Ruotsalainen, A.L., 2007. Mycorrhizal benefit in two low arctic herbs increases with increasing temperature. Am. J. Bot. 94 (8), 1309–1315.

Levin, S.A., 2005. Self-organization and the emergence of complexity in ecological systems. Bioscience 55, 1075–1079.

Liancourt, P., Choler, P., Gross, N., Thibert-Plante, X., Tielbörger, K., 2012. How facilitation may interfere with ecological speciation. Int. J. Ecol. 2012, 11. http://dx.doi.org/10.1155/2012/725487.

Lilleskov, E.A., 2005. How do composition, structure, and function of mycorrhizal fungal communities respond to nitrogen deposition and ozone exposure? In: Dighton, J., White, J., Oudemans, P. (Eds.), The Fungal Community: Its Organization and Role in the Ecosystem. Taylor and Francis, Boca Raton, pp. 769–801.

Lundberg, D.S., Lebeis, S.L., Paredes, S.H., Yourstone, S., Gehring, J., Malfatti, S., Tremblay, J., Engelbrektson, A., Kunin, V., Glavina del Rio, T., Edgar, R.C., Eickhorst, T., Ley, R.E., Hugenholtz, P., Green Tringe, S., Dang, J.L., 2012. Defining the core Arabidopsis thaliana root microbiome. Nature 488, 86–90.

Luo, Y., Su, B., Currie, W.S., Dukes, J.S., Finzi, A., Hartwig, U., Hungate, B., McMurtrie, R.S., Oren, R., Parton, W.J., Pataki, D.E., Shaw, R., Zak, D.R., Field, C.B., 2004. Progressive nitrogen limitation of ecosystem responses to rising atmospheric carbon dioxide. Bioscience 54, 731–739.

Matyssek, R., Wieser, G., Ceulemans, R., Rennenberg, H., Pretzsch, H., Haberer, K., Löw, M., Nunn, A.J., Werner, H., Wipfler, P., Oßwald, W., Nikolova, P., Hanke, D.E., Kraigher, H., Tausz, M., Bahnweg, G., Kitao, M., Dieler, J., Sandermann, H., Herbinger, K., Grebenc, T., Blumenröther, M., Deckmyn, G., Grams, T.E.E., Heerdt, C., Leuchner, M.,

Fabian, P., Haeberle, K.H., 2010. Enhanced ozone strongly reduces carbon sink strength of adult beech (Fagus sylvatica): resume from the free-air fumigation study at Kranzberg Forest. Environ. Pollut. 158, 2527–2532.

Matyssek, R., Wieser, G., Fleischmann, F., Grünhage, L., 2013. Ozone Research, quo vadis ? Lessons from the free-air canopy fumigation experiment at Kranzberg Forest. Dev. Environm. Sci. 13, 103–129.

Meyer, A., Grote, R., Polle, A., Butterbach-Bahl, K., 2010. Simulating mycorrhiza contribution to forest C- and N-cycling—the MYCOFON model. Plant Soil 327, 493–517.

Moriondo, M., Good, P., Durao, R., Bindi, M., Giannakopoulos, C., Corte-Real, J., 2006. Potential impact of climate change on fire risk in the Mediterranean area. Climate Res. 31, 85–95.

Nilsson, L.O., Wallander, H., 2003. Production of external mycelium by ectomycorrhizal fungi in a Norway spruce forest was reduced in response to nitrogen fertilization. New Phytol. 158, 409–416.

Olson, D.M., Dinerstein, E., Wikramanayake, E.D., Burgess, N.D., Powell, G.V.N., Underwood, E.C., D'amico, J.A., Itoua, I., Strand, H.E., Morrison, J.C., Colby, J.L., Allnut, T.F., Ricketts, T.H., Kura, Y., Lamoreux, J.F., Wettengel, W.W., Hedao, P., Kassem, K.R., 2001. Terrestrial ecoregions of the World: a new map of life on Earth. Bioscience 51, 933–938.

Orwin, K.H., Kirschbaum, M.U., St John, M.G., Dickie, I.A., 2011. Organic nutrient uptake by mycorrhizal fungi enhances ecosystem carbon storage: a model-based assessment. Ecol. Lett. 14 (5), 493–502.

Pena, R., Offermann, C., Simon, J., Naumann, P.S., Geßler, A., Holst, J., Mayer, H., Kögel-Knabner, I., Rennenberg, H., Polle, A., 2010. Girdling affects ectomycorrhizal diversity and reveals functional differences of EM community composition in a mature beech forest (*Fagus sylvatica*). Appl. Environ. Microbiol. 76, 1831–1841.

Pendergast, T.H., Burke, D.J., Carson, W.P., 2013. Belowground biotic complexity drives aboveground dynamics: a test of the soil community feedback model. New Phytol. 197 (4), 1300–1310. http://dx.doi.org/10.1111/nph.12105, published online: 11 Jan 2013.

Pietkäinen, J., Pettersson, M., Bååth, E., 2005. Comparison of temperature effects on soil respiration and bacterial and fungal growth rates. FEMS Microbiol. Ecol. 52, 49–58.

Plett, J.M., Kohler, A., Martin, F., 2012. De-constructing a mutualist: how the molecular blueprints of model symbiotic fungi are changing our understanding of mutualism. In: Hock, B. (Ed.), The Mycota, Volume IX: Fungal Associations. Springer-Verlag, Berlin, pp. 93–118.

Pritsch, K., Garbaye, J., 2011. Enzyme secretion by ECM fungi and exploitation of mineral nutrients from soil organic matter. Ann. For. Sci. 68, 25–33.

Pritsch, K., Esperschuetz, J., Haesler, F., Raidl, S., Winkler, B., Schloter, M., 2009. Structure and activities of ectomycorrhizal and microbial communities in the rhizosphere of *Fagus sylvatica* under ozone and pathogen stress in a lysimeter study. Plant Soil 323 (1), 97–109.

Read, D.J., 1991. Mycorrhizas in ecosystems. Experientia 47 (4), 376–390.

Read, D.J., 1998. Plants on the web. Nature 396, 22–23.

Rillig, M.C., Mummey, D.L., 2006. Mycorrhizas and soil structure. New Phytol. 171 (1), 41–53.

Rineau, F., Courty, P.-E., 2011. Secreted enzymatic activities of ectomycorrhizal fungi as a case study of functional diversity and functional redundancy. Ann. For. Sci. 68 (1), 69–80. http://dx.doi.org/10.1007/s13595-010-0008-4.

Schüßler, A., Schwarzott, D., Walker, C., 2001. A new fungal phylum, the Glomeromycota: phylogeny and evolution. Mycol. Res. 105 (12), 1413–1421.

Shi, L.B., Guttenberger, M., Kottke, I., Hampp, R., 2002. The effect of drought on mycorrhizas of beech (*Fagus sylvatica* L.): changes in community structure, and the content of carbohydrates and nitrogen storage bodies of the fungi. Mycorrhiza 12 (6), 303–311.

Simard, S.W., Jones, M.D., Durall, D.M., Perry, D.A., Myrold, D.D., 1997. Reciprocal transfer of carbon isotopes between ectomycorrhizal Betula papyrifera and Pseudotsuga menziesii. New Phytol. 137, 529–542.

Simard, S.W., Beiler, K.J., Bingham, M.A., Deslippe, J.R., Philip, L.J., Teste, F.P., 2012. Mycorrhizal networks: Mechanisms, ecology and modelling. Fungal Biol. Rev. 26, 39–60, Review.

Smith, S.E., Read, D.J., 2008. Mycorrhizal Symbiosis. Academic Press, New York, 800 pp.

Staddon, P.L., Heinemeyer, A., Fitter, A.H., 2002. Mycorrhizas and global environmental change: research at different scales. Plant Soil 244, 253–261.

Swaty, R.L., Deckert, R.J., Whitham, T.G., Gehring, C.A., 2004. Ectomycorrhizal abundance and community composition shifts with drought: predictions from tree rings. Ecology 85 (4), 1072–1084.

Taylor, A.F.S., Martin, F., Read, D.J., 2000. Fungal diversity in ectomycorrhizal communities of Norway spruce [Picea abies (L.) Karst.] and beech (Fagus sylvatica L.) along north–south transects in Europe. In: Detlef Schulze, E.D. (Ed.), Carbon and Nitrogen Cycling in European Forest Ecosystems. Ecological Studies, vol. 142. Springer-Verlag, Heidelberg, pp. 343–365.

Tedersoo, L., May, T., Smith, M., 2010. Ectomycorrhizal lifestyle in fungi: global diversity, distribution, and evolution of phylogenetic lineages. Mycorrhiza 20 (4), 217–263.

van der Heijden, M.G.A., Horton, T.R., 2009. Socialism in soil? The importance of mycorrhizal fungal networks for facilitation in natural ecosystems. J. Ecol. 97 (6), 1139–1150.

van der Heijden, M.G.A., Klironomos, J.N., Ursic, M., Moutoglis, P., Streitwolf-Engel, R., Boller, T., Wiemken, A., Sanders, I.R., 1998. Mycorrhizal fungal diversity determines plant biodiversity, ecosystem variability and productivity. Nature 396, 69–72.

Wallander, H., Ekblad, A., Godbold, D.L., Johnson, D., Bahr, A., Baldrian, P., Björk, R.G., Kieliszewska-Rokicka, B., Kjøller, R., Kraigher, H., Plassard, C., Rudawska, M., 2013. Evaluation of methods to estimate production, biomass and turnover of ectomycorrhizal mycelium in forests soils—a review. Soil Biol. Biochem. 57, 1034–1042.

Waltert, B., Wiemken, V., Rusterholz, H.P., Boller, T., Baur, B., 2002. Disturbance of forest by trampling: effects on mycorrhizal roots of seedlings and mature trees of *Fagus sy*lvatica. Plant Soil 243, 143–154.

White, T.J., Bruns, T., Lee, S., Taylor, J., 1990. Amplification and direct sequencing of fungal ribosomal RNA genes for phylogenetics. In: Innis, M.A., Gelfand, D.H., Sninsky, J.J., White, T.J. (Eds.), PCR Protocols: A Guide to Methods and Applications. Academic Press, New York, pp. 315–322.

Wiensczyk, A.M., Gamiet, S., Durall, D.M., Jones, M.D., Simard, S.W., 2002. Ectomycorrhizae and forestry in British Columbia: a summary of current research and conservation strategies. BC J. Ecosyst. Manage. 2 (1), 1–20.

Wittig, V.E., Ainsworth, E.A., Naidu, S.L., Karnosky, D.F., Long, S.P., 2009. Quantifying the impact of current and future tropospheric ozone on tree biomass, growth, physiology and biochemistry: a quantitative meta-analysis. Glob. Chang. Biol. 15 (2), 396–424.

Železnik, P., Hrenko, M., Then, C., Koch, N., Grebenc, T., Levanič, T., Kraigher, H., 2007. CASIROZ: root parameters and types of ectomycorrhiza of young beech plants exposed to different ozone and light regimes. Plant Biol. 9, 298–308.

Zhong, Z., Zhenhui, W., Xiping, L., 1997. Influence of flooding on effects of mixed inoculation with VA- and ectomycorrhizal fungi of Populus tomentosa. J. Northwest For. Coll. 12 (4), 7–12.

Chapter 9

Tree and Forest Responses to Interacting Elevated Atmospheric CO_2 and Tropospheric O_3: A Synthesis of Experimental Evidence

John King*[,1], Lingli Liu[†] and Michael Aspinwall[‡]

*Department of Forestry and Environmental Resources, North Carolina State University, Raleigh, North Carolina, USA

†State Key Laboratory of Vegetation and Environmental Change, Institute of Botany, Chinese Academy of Sciences, Beijing, China

‡Hawkesbury Institute for the Environment, University of Western Sydney, Richmond, New South Wales, Australia

[1]Corresponding author: e-mail: john_king@ncsu.edu

Chapter Outline

Dedication: We dedicate this chapter to the memory of our friend and colleague, Dr. Fitzgerald Booker. Fitz made many lasting contributions to air pollution science through his primary research and leadership in organizing informal and formal research collaborations, such as the Air Pollution Workshop and NE-1013. He also contributed to the career advancement of many by unselfishly sharing his lab, expertise and time. He was a meticulous editor and author, careful researcher and patient teacher. His life is a reminder to appreciate each day, be helpful to those around us, and do good work.

Developments in Environmental Science, Vol. 13. http://dx.doi.org/10.1016/B978-0-08-098349-3.00009-8

9.1 INTRODUCTION

It has long been recognized that development of ecosystem models to project the effects of global change on terrestrial carbon, water and nutrient cycling must consider how multiple environmental factors *interact* to influence ecosystem composition and function (Mooney et al., 1991; Niinemets, 2010). Recent studies have shown explicitly that responses to single environmental stress factors are often (usually) not additive, suggesting that conclusions from single factor experiments be interpreted with caution (Dieleman et al., 2012). Given the limitations of our ability to understand higher-order interactions of more than three or four factors at a time, and the practical limitations of conducting such experiments on large scales, the question is: Which factors should be studied? Because of its long mean-residence time in the atmosphere, continued increase in emissions and direct effects on plant physiological processes, it has been widely concluded that elevated atmospheric CO_2 (eCO_2), and especially how it interacts with rising global temperature, is of particular importance (Calfapietra et al., 2010; Dieleman et al., 2012). Similarly, both nitrogen (N) limitation and enrichment continue to be problems around the world, and how N availability interacts with eCO_2 continues to be an active area of research (Calfapietra et al., 2010; Reich et al., 2006). However, the interaction of eCO_2 and elevated tropospheric O_3 (eO_3) is uniquely compelling in that both pollutants are assimilated via gas exchange of the leaf, directly influence leaf area development and stomatal conductance and alter plant physiology in generally antagonistic ways, with potential to cascade through ecosystem processes (Karnosky et al., 2003).

The transition to a new energy infrastructure with significantly decreased fossil CO_2 emissions is still decades away and there is rapid economic growth but lagging environmental policy in many parts of the world; therefore, the interaction of $eCO_2 \times eO_3$ will continue to impact terrestrial ecosystems over a planning horizon beyond 2100. Although air pollution policy in Europe and North America has largely succeeded in decreasing peak O_3 concentrations in those regions (see Chapter 5), economic modelling studies and documented transoceanic O_3 transport suggest increasing background concentrations worldwide for decades to come (Fowler et al., 1999; Li et al., 2002). Globally, production of eO_3 exceeds its consumption. Regionally, rapid economic development in Central Europe, Asia and South America is adding new sources to the global eO_3 burden (see also Koike et al., Pandey et al., Kozovits and Bustamante, Laasko et al., this volume). It is possible these new source regions will experience a similar co-evolution of economic development and degradation of air quality, followed by improved scientific understanding and development of enlightened environmental policy, as occurred elsewhere (e.g. the Clean Air Act of the U.S. dates to 1970). Lessons learned from the decades of research on eCO_2 and eO_3 could benefit social development in ways that preserve environmental quality, locally and globally. In particular, development of metrics for eO_3 dose–response relationships is still a work in progress, and widespread adoption of air quality standards would be beneficial.

In order to improve our predictive understanding of terrestrial ecosystem responses to climate change and to protect current environmental quality, there is a pressing need to summarize the available data on interacting eCO_2 and eO_3, identify gaps in knowledge and make recommendations for future research. We sought to accomplish these goals by surveying the peer-reviewed literature with the following questions in mind:

a. Do the available data support the hypothesis of increased leaf-level photosynthesis and decreased stomatal conductance under eCO_2, and the inverse under eO_3?
b. Do the available data support the hypothesis that forest productivity increases under eCO_2 and decreases under eO_3?
c. Do the available data support the hypothesis that litter quality and decomposition rates will be affected by eCO_2 and eO_3, resulting in changes in soil organic carbon (SOC) pools?
d. What is the net effect of the *interaction* of eCO_2 and eO_3 on all of the above and do responses change with tree/forest age or length of exposure?

9.2 LITERATURE SURVEY METHODS

To focus on $eCO_2 \times eO_3$ interaction experiments, we surveyed the peer-reviewed scientific literature from 1990 to 2012 using academic Internet search engines with access to all major databases in the natural sciences (e.g. AGRICOLA, CAB, Plant Science, Ecology Abstracts, BIOSIS, Environmental Science & Pollution Management, JSTOR, Web of Science). We restricted the review to studies that simultaneously applied fumigation of eCO_2 and eO_3 and used only widely accessible primary sources (no proceedings, reviews/meta-analyses, edited book volumes, textbooks, abstracts, etc.). For each variable (listed in Tables 9.1–9.3), we searched the keywords: variable name, 'elevated CO_2', 'carbon dioxide', 'tropospheric O_3', 'ozone', 'forest', 'tree' and 'interaction'. If a paper contained values for multiple species or communities, they were included in the database as individual observations. If other treatments were part of the experiment, we used the values for the control in the experiment; however this occurred only on a few occasions. Trees were grouped as 'young', if the plants were 5 years of age or less,

TABLE 9.1 Effects of Elevated Atmospheric CO_2 (eCO_2) and Elevated Tropospheric O_3 (eO_3) on Selected Physiology and Water Use Parameters for Young (5 Years or Less) and Intermediate (6–12 years) Age Trees and Forest Stands

Parameter	Observations GC:GH:OT: FC	Age (years)	A:E CO_2 exp. (ppm)	A:E O_3 exp. (ppb)	Control (mean, *n*, CV)	eCO_2 (%, *n*, CV)	eO_3 (%, *n*, CV)	$eCO_2 \times eO_3$ (%, *n*, CV)	Sources
Young: 5 years or less									
A_{CO_2}	10:0:2:6	1.7 (1.0)	358:605 (8:67)	30:73 (19:18)	10.0, 18, 0.58	22.6, 18, 0.50	−12.5, 18, 0.49	13.0, 18, 0.47	Grams et al. (1999), Kull et al. (1996), Loats and Rebbeck (1999), Noormets (2001), Noormets et al. (2010), Olszyk et al. (2002), Sharma et al. (2003), and Volin et al. (1998)
g_s	10:0:4:3	1.7 (1.8)	359:629 (9:69)	30:77 (23:18)	194.4, 17, 0.58	−26.1, 17, 0.61	−9.6, 17, 0.58	−26.4, 17, 0.60	Gardner et al. (2005), Grams et al. (1999), Kull et al. (1996), Loats and Rebbeck (1999), Noormets (2001), Noormets et al. (2001, 2010), Olszyk et al. (2002), Sharma et al. (2003), Volin and Reich (1996) and Volin et al. (1998)
iWUE	9:0:2:2	1.4 (0.9)	358:618 (9:67)	25:74 (20:18)	0.053, 13, 0.36	83.6, 13, 0.27	2.5, 13, 0.33	65.4, 13, 0.32	Kull et al. (1996), Loats and Rebbeck (1999), Olszyk et al. (2002), Sharma et al. (2003) and Volin et al. (1998)

LAI	0:0:0:3	3.7 (1.2)	353:554 (6:10)	36:53 (NA)	2.1, 3, 0.17	14.9, 3, 0.21	−17.6, 3, 0.14	−11.5, 3, 0.19	Isebrands et al. (2001), Karnosky et al. (2005) and Noormets (2001)
Intermediate: 6–12 years									
A_{CO_2}	0:0:3:12	8.6 (1.4)	362:588 (4:60)	31:65 (16:16)	12.5, 15, 0.13	45.0, 15, 0.15	−14.1, 15, 0.11	35.6, 15, 0.16	Darbah et al. (2010), Kets et al. (2010) and Riikonen et al. (2005)
g_s	0:0:0:2	7.5 (0.7)	360:560 (NA)	20:48 (NA)	300.0, 2, 0.19	−19.2, 2, 0.19	−6.8, 2, 0.11	−24.3, 2, 0.09	Kets et al. (2010)
iWUE	0:0:0:2	7.5 (0.7)	360:560 (NA)	20:48 (NA)	0.044, 2, 0.13	69.1, 2, 0.17	−0.7, 2, 0.17	83.6, 2, 0.09	Kets et al. (2010)
LAI	0:0:0:2	7.5 (0.7)	369:526 (1:1)	35:47 (3:6)	3.9, 2, 0.04	32.4, 2, 0.16	−25.6, 2, 0	6.4, 2, 0.05	Uddling et al. (2008)

No data were available from mature trees. 'Observations' refers to the number of observations from facilities of the following types: GC, growth chamber; GH, green house; OT, open-top chamber; FC, FACE. Age, A:E CO_2 and O_3 exposures consist of the mean for ambient:elevated levels of CO_2 and O_3 and respective standard deviations.

TABLE 9.2 Effects of Elevated Atmospheric CO_2 and Tropospheric O_3 on Selected Biomass and Net Primary Production (NPP) Parameters for Young (5 Years or Less) and Intermediate (6–12 Years) Age Trees and Forest Stands

Parameter	Observations GC:GH:OT: FC	Age (years)	A:E CO_2 exp. (ppm)	A:E O_3 exp. (ppb)	eCO_2 (%, *n*, CV)	eO_3 (%, *n*, CV)	$eCO_2 \times eO_3$ (%, *n*, CV)	Sources
Young: 5 years or less								
Foliage	10:4:12:3	2.9 (2.1)	368:634 (19:89)	27:88 (16:46)	18.4, 26, 1.24	−15.8, 28, 1.07	18.3, 28, 3.92	Broadmeadow and Jackson (2000), Dickson et al. (1998, 2001), Gardner et al. (2005), Gaucher et al. (2003), King et al. (2005), Liu et al. (2004), Loats and Rebbeck (1999), Rebbeck and Scherzer (2002), Richet et al. (2012), Rodenkirchen et al. (2009) and Watanabe et al. (2010)
Stem/wood	11:8:9:3	3.2 (2.2)	374:637 (20:94)	27:84 (16:39)	22.8, 28, 1.05	−10.2, 31, 1.82	10.5, 31, 2.47	Dickson et al. (1998, 2001), Gardner et al. (2005), Gaucher et al. (2003), King et al. (2005), Kozovits et al. (2005), Liu et al. (2004), Loats and Rebbeck (1999), Polle et al. (1993), Rebbeck and Scherzer (2002), Richet et al. (2012), Rodenkirchen et al. (2009) and Watanabe et al. (2010)
Coarse root	5:4:1:3	4.6 (1.4)	376:663 (23:70)	33:86 (9:46)	34.0, 11, 0.77	−2.7, 12, 4.01	23.3, 12, 1.35	King et al. (2005), Liu et al. (2004), Rebbeck and Scherzer (2002), Rodenkirchen et al. (2009) and Watanabe et al. (2010)

Fine root	6:4:1:3	4.5 (1.4)	379:665 (25:68)	33:86 (8:44)	41.8, 11, 0.93	−8.2, 12, 2.16	37.1, 11, 1.2	King et al. (2005), Liu et al. (2004), Phillips et al. (2009), Rebbeck and Scherzer (2002) and Watanabe et al. (2010)
FR turnover	1:0:0:0	3.0 (NA)	420:690 (NA:NA)	36:80 (NA: NA)	26.0, 1, NA	5.0, 1, NA	0.0, 1, NA	Phillips et al. (2009)
Total AG	8:0:5:3	3.4 (1.9)	361:641 (16:73)	34:79 (10:43)	26.8, 16, 0.54	−9.7, 16, 1.28	16.7, 16, 1.27	Broadmeadow and Jackson (2000), Gardner et al. (2005), King et al. (2005), Loats and Rebbeck (1999), Polle et al. (1993), Riikonen et al. (2004) and Utriainen et al. (2000)
Total BG	9:0:11:3	2.5 (1.8)	365:609 (21:85)	27:88 (17:44)	26.5, 21, 0.82	−19.1, 23, 1.10	3.6, 22, 3.87	Broadmeadow and Jackson (2000), Dickson et al. (1998, 2001), Gaucher et al. (2003), King et al. (2005), Loats and Rebbeck (1999), Olszyk et al. (2001), Rodenkirchen et al. (2009) and Utriainen et al. (2000)
Total AGBG	9:0:11:3	2.5 (1.8)	362:608 (14:84)	26:89 (18:44)	29.6, 21, 0.63	−18.3, 23, 1.04	14.5, 23, 2.25	Broadmeadow and Jackson (2000), Dickson et al. (1998, 2001), Gaucher et al. (2003), King et al. (2005), Loats and Rebbeck (1999), Rodenkirchen et al. (2009), Utriainen et al. (2000) and Watanabe et al. (2010)
ANPP	0:0:0:3	3.0 (NA)	348:540 (NA:NA)	37:52 (NA: NA)	40.9, 3, 0.43	−15.9, 3, 0.72	12.1, 3, 0.97	King et al. (2005)

Continued

TABLE 9.2 Effects of Elevated Atmospheric CO_2 and Tropospheric O_3 on Selected Biomass and Net Primary Production (NPP) Parameters for Young (5 Years or Less) and Intermediate (6–12 Years) Age Trees and Forest Stands—Cont'd

Parameter	Observations GC:GH:OT: FC	Age (years)	A:E CO_2 exp. (ppm)	A:E O_3 exp. (ppb)	eCO_2 (%, *n*, CV)	eO_3 (%, *n*, CV)	eCO_2 × eO_3 (%, *n*, CV)	Sources
Intermediate: 6–12 years								
Foliage	0:0:4:6	8.7 (2.1)	362:610 (13:94)	46:67 (12:21)	38.0, 10, 0.75	−13.41, 10, 0.61	12.2, 10, 0.75	Kasurinen et al. (2005), King et al. (2005), Riikonen et al. (2004) and Talhelm et al. (2012)
Stem/wood	0:0:2:3	7.2 (1.6)	360:612 (16:98)	46:68 (12:22)	42.9, 5, 0.59	−13.0, 5, 0.81	22.5, 5, 0.68	King et al. (2005) and Riikonen et al. (2004)
Coarse root	0:0:0:3	6.0 (NA)	348:540 (NA:NA)	37:52 (NA: NA)	49.0, 3, 0.38	−19.35, 3, 0.25	13.9, 3, 0.84	King et al. (2005)
Fine root	0:0:0:11	7.4 (0.9)	354:536 (4:3)	36:50 (0.5:0.9)	53.7, 11, 1.65	1.82, 11, 8.08	47.8, 11, 0.72	King et al. (2005), Pregitzer et al. (2008) and Rhea and King (2012)
FR turnover	0:0:0:1	8.0 (NA)	348:540 (NA:NA)	37:52 (NA: NA)	2.8, 1, NA	1.4, 1, NA	17.6, 1, NA	Pregitzer et al. (2008)
Total AG	0:0:2:3	7.2 (1.6)	360:612 (16:99)	46:68 (12:22)	41.3, 5, 0.63	−13.7, 5, 0.83	22.4, 5, 0.73	King et al. (2005) and Riikonen et al. (2004)

Total BG	0:0:2:3	7.2 (1.6)	360:612 (16:99)	46:68 (12:22)	36.8, 5, 0.60	−19.8, 5, 0.57	21.4, 5, 0.92	King et al. (2005) and Riikonen et al. (2004)
Total AGBG	0:0:2:3	7.2 (1.6)	360:612 (16:99)	46:68 (12:22)	40.4, 5, 0.62	−15.0, 5, 0.76	22.1, 5, 0.80	King et al. (2005) and Riikonen et al. (2004)
ANPP	0:0:0:3	6 (NA)	348:540 (NA:NA)	37:52 (NA: NA)	50.8, 3, 0.35	−11.1, 3, 0.97	20.7, 3, 1.12	King et al. (2005)
Total NPP	0:0:0:1	12 (NA)	348:560 (NA:NA)	35:55 (NA: NA)	12.0, 1, NA	8, 1, NA	0, 1, NA	Zak et al. (2011)

No data were available from mature trees. 'Observations' refers to the number of observations from facilities of the following types: GC, growth chamber; GH, green house; OT, open-top chamber; FC, FACE. Age, A:E CO_2 and O_3 exposures consist of the mean for ambient:elevated levels of CO_2 and O_3 and respective standard deviations.

TABLE 9.3 Effects of Elevated Atmospheric CO_2 and Tropospheric O_3 on Selected Litter Biochemical and Decomposition-Related Soil Parameters for Young (5 Years or Less) and Intermediate (6–12 Years) Age Trees and Forest Stands

Parameter	Observations GC:GH:OT:FC	Age (years)	A:E CO_2 exp. (ppm)	A:E O_3 exp. (ppb)	eCO_2 (%, *n*, CV)	eO_3 (%, *n*, CV)	$eCO_2 \times eO_3$ (%, *n*, CV)	Sources
Young: 5 years or less								
Leaf condensed tannins	0:0:0:9	3.8 (0.8)	360:560 (0:0)	41:62 (5:8)	15.5, 9, 1	2.6, 9, 0.6	35.3, 9, 0.9	Holton et al. (2003), Kopper and Lindroth (2003), Parsons et al. (2004, 2008) and Talhelm et al. (2009)
Leaf phenolics	0:0:2:3	2.8 (0.5)	358:600 (4:69)	39:62 (6:8)	8.9, 5, 0.79	−9.9, 5, 0.7	0.5, 5, 1.1	Holton et al. (2003), Kainulainen et al. (2003), Parsons et al. (2004) and Sallas et al. (2001)
Leaf lignin	0:1:0:1	3.5 (2)	360:610 (0:70)	41:63 (7:10)	5.0, 1, NA	−15.4, 2, 0.3	22.5, 2, 1.3	Boerner and Rebbeck (1995) and Parsons et al. (2004)
Leaf C	0:2:0:0	5 (0)	350:750 (0:0)	20:75 (0:0)	−1.0, 2, 1	−1.8, 2, 0.9	−1.9, 2, 1.1	Pfirrmann et al. (1996)
Leaf N	0:3:10:7	3.2 (1.2)	364:657 (9:77)	29:57 (9:12)	−14.2, 16, 0.7	−11.5, 20, 1.1	−27.6, 20, 0.46	Boerner and Rebbeck (1995), Broadmeadow and Jackson (2000), Holton et al. (2003), Kopper and Lindroth (2003), Olszyk et al. (2001), Parsons et al. (2004, 2008), Pfirrmann et al. (1996) and Scherzer et al. (1998)
Leaf C/N	0:0:0:2	3.0 (0)	360:560 (0:0)	45:70 (0:0)	14.9, 2, 0.5	−5.4, 2, 0.7	32.3, 2, 0.6	Lindroth et al. (2001)

Fine root N	0:2:2:0	3.5 (1.7)	355:695 (6:64)	28:70 (9:6)	−2.6, 4, 0.5	4.6, 4, 0.5	−1.6, 4, 1.0	Olszyk et al. (2001) and Pfirrmann et al. (1996)
Litter decomposition (MRT)	0:1:0:5	1.2 (0.4)	360:576 (0:41)	36:58 (0:7)	2.9, 6, 0.9	−2.9, 6, 0.7	47.1, 6, 1.1	Boerner and Rebbeck (1995) and Parsons et al. (2004, 2008)
Soil carbon	0:0:0:1	4 (0)	360:560 (0:0)	46:70 (0:0)	−2.8, 1, NA	5.5, 1, NA	−5.0, 1, NA	Loya et al. (2003)
Intermediate: 6–12 years								
Leaf condensed tannins	0:0:5:10	8.9 (1.6)	363:603 (11:72)	38:58 (6:14)	19.6, 15, 0.7	34.1, 15, 1	46.1, 15, 1.0	Agrell et al. (2005), Couture et al. (2012), Hillstrom et al. (2010), Kasurinen et al. (2007), Liu et al. (2005, 2009) and Peltonen et al. (2005)
Leaf phenolics	0:0:6:4	9.5 (1.6)	360:633 (8:76)	34:53 (6:13)	14.6, 9, 0.6	26.3, 9, 0.9	42.0, 9, 0.9	Couture et al. (2012), Kasurinen et al. (2006, 2007) and Liu et al. (2005, 2009)
Leaf lignin	0:0:1:6	8.6 (2.0)	368:572 (12:57)	38:56 (6:12)	7.9, 7, 0.9	9.3, 7, 0.9	−9.1, 7, 0.6	Couture et al. (2012), Liu et al. (2005, 2009) and Oksanen et al. (2005)
Litter C	0:0:0:3	7.7 (1.2)	357:557 (6:6)	42:63 (6:12)	−1.6, 3, 0.4	0.9, 3, 0.3	−1.1, 3, 0.7	Hillstrom et al. (2010) and Liu et al. (2005)
Leaf N	0:0:3:11	10.0 (2.0)	372:595 (12:88)	35:51 (9:15)	−8.4, 14, 0.7	−6.5, 14, 0.7	−10.9, 14, 0.7	Agrell et al. (2005), Couture et al. (2012), Hillstrom et al. (2010), Kasurinen et al. (2006), Liu et al. (2005), Oksanen et al. (2005) and Talhelm et al. (2012)

Continued

TABLE 9.3 Effects of Elevated Atmospheric CO_2 and Tropospheric O_3 on Selected Litter Biochemical and Decomposition-Related Soil Parameters for Young (5 Years or Less) and Intermediate (6–12 Years) Age Trees and Forest Stands—Cont'd

Parameter	Observations GC:GH:OT:FC	Age (years)	A:E CO_2 exp. (ppm)	A:E O_3 exp. (ppb)	eCO_2 (%, *n*, CV)	eO_3 (%, *n*, CV)	$eCO_2 \times eO_3$ (%, *n*, CV)	Sources
Leaf C/N	0:0:2:4	9.0 (1.8)	378:633 (15:113)	32:57 (14:24)	12.5, 6, 0.4	14.5, 6, 0.6	15.7, 6, 0.6	Couture et al. (2012), Kasurinen et al. (2007) and Liu et al. (2005)
Fine root condensed tannins	0:0:0:3	6.5 (0)	360:560 (0:0)	45:70 (0:0)	−34.0, 3, 0.4	−16.8, 3, 0.2	−49.2, 3, 0.7	Chapman et al. (2005)
Fine root lignin	0:0:0:3	6.5 (0)	360:560 (0:0)	46:70 (0:0)	58.5, 3, 0.9	21.0, 3, 0.4	53.6, 3, 1.0	Chapman et al. (2005)
Fine root N	0:0:0:3	6.5 (0)	360:560 (0:0)	46:70 (0:0)	−5.6, 3, 0.9	−5.9, 3, 0.8	−12.1, 3, 0.5	Chapman et al. (2005)
Fine root C/N	0:0:0:3	6.5 (0)	360:560 (0:0)	46:70 (0:0)	11.7, 3, 0.2	−4.1, 3, 0.8	15.7, 3, 0.6	Chapman et al. (2005)
Litter decomposition (MRT)	0:0:0:2	6 (0)	360:560 (0:0)	46:70 (0:0)	0.7, 2, 0.4	12.3, 2, 0.1	5.0, 2, 0.3	Liu et al. (2009)
Soil carbon	0:0:0:3	11 (0)	360:560 (0:0)	46:70 (0:0)	1.5, 3, 0.5	5.4, 3, 0.9	−11.3, 3, 1	Talhelm et al. (2009)

No data were available from mature trees. 'Observations' refers to the number of observations from facilities of the following types: GC, growth chamber; GH, green house; OT, open-top chamber; FC, FACE. Age, A:E CO_2 and O_3 exposures consist of the mean for ambient:elevated levels of CO_2 and O_3 and respective standard deviations.

and as 'intermediate', for 6–20 years of age. We searched the literature for studies of 'mature' trees, 20 years of age or older, but found none. The longest running study was 12 years. The literature was searched until no new sources were identified; then summary statistics (mean, *n*, SD, CV) were calculated from statistically significant relative responses reported for each parameter across studies (Tables 9.1–9.3). Linear regression was used to test correlations between parameter relative responses and tree age and eCO_2/eO_3 exposure level. Formal meta-analysis was not performed due to a paucity of data for many variables.

9.3 FOREST RESPONSES TO INTERACTING eCO_2 AND eO_3

9.3.1 Experimental Design, Duration and Exposure Regimes

The availability of data varied widely, from a low of 1 observation for parameters such as fine root turnover (FRT) and fine root longevity, to a high of 31 for stem/wood biomass (Tables 9.1–9.3). The total number of observations was greatest from FACE experiments (176), followed by open-top chambers (104), growth chambers (88) and then greenhouses (29) (Tables 9.1–9.3). At first glance, one might expect the smaller-scale less-expensive experimental platforms to be the most widely reported, but the abundance of FACE/OTC observations speaks of the importance and efficiency of such large-scale, multi-investigator, multi-year experiments. The average length of experiments for young systems was 3.3 years, but this was highly variable, with some studies as short as 0.2 years and the longest over 5.0 years. For intermediate systems, the mean age was 7.7 years, but again this was variable with the shortest studies running for 6.0 years and the longest single experiment, the Aspen FACE Project, running for 12.0 years. Although there have been several experiments of eCO_2 and eO_3 alone on mature forest trees (Körner et al., 2005; Matyssek et al., 2006, 2010; Pretzsch et al., 2010), there have been no $eCO_2 \times eO_3$ experiments, leaving a corresponding gap in our knowledge on the effects of this interaction on advanced-age ecosystems. The mean (SD) reported ambient CO_2 concentration was 347 (17) ppm, and that for eCO_2 was 603 (81) ppm, with the variance generally being very low. The mean (SD) ambient O_3 concentration was 34.8 (12) ppb and that for eO_3 was 70.1 (29) ppb, with proportionally much larger variance, perhaps indicating the difficulty of maintaining uniform O_3 target concentrations relative to eCO_2, or more variable background conditions. Although seasonal mean treatment concentrations may not be the most meaningful estimate of exposure or dose, they were the most reliably reported metric and therefore the best basis for comparison between experiments.

9.3.2 Tree Physiology (Gas Exchange and Water Relations)

We focused on the following traits, which were most widely reported in the literature, and are important determinants of leaf- and canopy-level carbon and water fluxes: net photosynthesis (A_{CO_2}, μmol CO_2 m^{-2} s^{-1}), stomatal conductance (g_s, mmol H_2O m^{-2} s^{-1}), instantaneous water use efficiency (iWUE, μmol CO_2 $mmol^{-1}$ H_2O) and leaf area index (LAI, m^{-2} m^{-2}). We attempted to find information on stand-level transpiration (E, mm $year^{-1}$) and stand-level WUE (mg biomass produced m^{-3} water transpired $year^{-1}$), but were unsuccessful. A_{CO_2} and g_s were the most abundant data types (Table 9.1).

In young trees, eCO_2 stimulated A_{CO_2} (+23%) while eO_3 reduced it (−13%) (Table 9.1). Ozone decreased A_{CO_2} by reducing carboxylation efficiency and increasing respiratory costs associated with detoxification and repair processes (Kolb and Matyssek, 2001). Under $eCO_2 \times eO_3$, A_{CO_2} increased by 13%. Therefore, while eO_3 compromised the response of A_{CO_2} to eCO_2, the negative effects of eO_3 were largely ameliorated by eCO_2. Importantly, the absolute response of A_{CO_2} to each treatment varied by about 50%, illustrating the somewhat broad range of responses of A_{CO_2} to eCO_2 and eO_3 in young trees. The effects of eCO_2, eO_3 and $eCO_2 \times eO_3$ on A_{CO_2} were more pronounced in intermediate-aged trees than in young trees, counter to our expectation, with eCO_2 alone stimulating A_{CO_2} by nearly 45% (Table 9.1), which may be an indication that the larger trees, often grown in unrestricted soil in the field, experienced less co-limitation of eCO_2 responses by other factors (Thomas and Strain, 1991). Elevated O_3 reduced A_{CO_2} by a similar magnitude to that observed in young trees (−14%). Interestingly, in intermediate-aged trees growing under $eCO_2 \times eO_3$, A_{CO_2} remained 35% higher than control trees, which demonstrated that eCO_2 more than counteracted the negative effects of eO_3 on leaf carbon gain. Variation in the absolute response of A_{CO_2} to each treatment was small, with CV between 0.10 and 0.20. We found no studies which measured A_{CO_2} under $eCO_2 \times eO_3$ in mature trees. Regardless of tree size or age, it is more likely that trees (i.e. species, genotypes) with higher g_s will have the highest negative response to eO_3 (Herbinger et al., 2007; Kolb and Matyssek, 2001; Reich, 1987; Volin et al., 1998).

Young trees showed consistent reductions in g_s under all treatments, that is, eCO_2, eO_3 and $eCO_2 \times eO_3$ (Table 9.1). The reduction in g_s under $eCO_2 \times eO_3$ was effectively equivalent to the reduction in g_s under eCO_2 alone. Similarly, Medlyn et al. (2001) found a reduction in g_s under eCO_2 (−21%) across different aged trees and species. Despite the consistent reductions in g_s across treatments, variation in g_s was somewhat high (CV~0.60), again demonstrating the wide range of stomatal responses among species and studies. Remarkably, we found only two data points for g_s measured on intermediate-aged trees growing under $eCO_2 \times eO_3$ (Kets et al., 2010). Although the CVs for g_s were low across treatments, sparse data limit our ability to make broad inferences. In any case, g_s decreased by 24% under $eCO_2 \times eO_3$ in intermediate-age trees, similar to the g_s reduction in young trees under that treatment. Again, we found no g_s data for mature trees growing under $eCO_2 \times eO_3$. Interestingly, these data show that reductions in g_s under eCO_2 and eO_3 are relatively similar across tree ages, with g_s more affected by eCO_2 than eO_3 (Table 9.1). Moreover, these data generally match the 21% reduction in g_s under eCO_2, reported by meta-analysis (Medlyn et al., 2001).

From studies which report both A_{CO_2} and g_s, we estimated iWUE (A_{CO_2}/g_s). The standard model suggests that eCO_2 increases iWUE (Ellsworth et al., 1995; Long et al., 2004) and eO_3 reduces iWUE (Kolb and Matyssek, 2001). Data from the current literature survey show that eCO_2 increased iWUE similarly across tree ages (+69–88%), and eO_3 had little to no effect, regardless of tree age (Table 9.1); however, there is evidence that mature trees may respond differently (Kitao et al., 2009; Löw et al., 2006). It was seen that iWUE increased substantially under eCO_2 × eO_3 in young and intermediate-aged trees. Compared to A_{CO_2} and g_s, iWUE showed the lowest CVs suggesting a more consistent pattern of responses to eCO_2. Overall, these results do not suggest that eCO_2 ameliorates the negative effects of eO_3 on iWUE because eO_3 had little effect on iWUE in the studies reviewed here (but see Matyssek and Sandermann, 2003).

Interestingly, we found only three data points reporting LAI of young trees growing under eCO_2 × eO_3. Although the data are minimal, they suggest that eCO_2 stimulates LAI (+15%), eO_3 reduced LAI (−18%) and eCO_2 does not ameliorate the negative effects of eO_3 on LAI in young trees (Table 9.1). Intermediate-aged trees show similar responses to eCO_2 and eO_3, but this was again more pronounced; LAI increased more with eCO_2, and decreased more with eO_3. Moreover, although the data were limited to one study (Uddling et al., 2008), LAI increased by 6% under eCO_2 × eO_3. The results from young and intermediate-age trees suggest that eO_3 severely limits increased LAI under eCO_2.

Data on stand-level E and WUE under eCO_2 × eO_3 were non-existent and there is generally little information on canopy-level processes in older stands exposed to either eCO_2 or eO_3. This gap in the data makes it difficult to examine how ontogeny influences *canopy-level* responses to eCO_2 and eO_3. Studies by Uddling et al. (2008, 2009) provide the most insight into stand-level water use under eCO_2 × eO_3. Their results suggest that eO_3 does not significantly affect tree water use and canopy conductance, but rather, eCO_2 increases tree water use because of increased growth and leaf area. Similarly, Tricker et al. (2009) found that eCO_2 did not reduce stand water use because eCO_2 increases the amount of transpiring leaf area. Thus, water use at the leaf and canopy level may show very different responses to eCO_2. However, in agreement with leaf-level responses, Ellsworth et al. (1995), Wullscheger and Norby (2001) and Wullscheger et al. (2002) found reduced stand water use under eCO_2, although not of the same magnitude of leaf-level responses, most likely due to issues of scale and environmental variability. Notably, it is hypothesized that eCO_2 increases stand-level WUE, yet we found only one study which estimated stand-level WUE under eCO_2 (Wullscheger and Norby, 2001). Wullscheger and Norby (2001) found that eCO_2 increased stand-level WUE. Clearly, the data on canopy-level responses to eCO_2 and

eO_3 alone, and more so as $eCO_2 \times eO_3$, are extremely limited and sometimes contradictory, making it difficult to identify response trends and predict forest ecosystem responses to interacting eCO_2 and eO_3.

Lastly, it has been noted that ontogeny be considered when examining the effects of eCO_2 and eO_3 on forest ecosystems (Chappelka and Samuelson, 1998; Matyssek et al., 2010; Norby et al., 2005). Mature and young trees are structurally different and have different rates of water and nutrient uptake, which ultimately determine their response to eCO_2 and eO_3. As reviewed by Chappelka and Samuelson (1998), seedlings often have greater g_s, which results in higher O_3 uptake and greater reductions in A_{CO_2}, compared to mature trees. For many of the same reasons, young trees may be more responsive to eCO_2 (Medlyn et al., 2001). Furthermore, under $eCO_2 \times eO_3$, responses that were significant at young ages (Noormets et al., 2001) may no longer be significant as trees mature (Uddling et al., 2009). Based on our analysis, under interacting eCO_2 and eO_3, A_{CO_2}, iWUE and LAI were less compromised by eO_3 in intermediately aged trees than young trees, and g_s showed similar responses to $eCO_2 \times eO_3$ across tree age groups. While there were no studies on mature trees in this survey, this trend suggests that the positive effects of eCO_2 become more pronounced with age while the negative effects of eO_3 may become less. However, much of the data on physiological processes has been collected on young trees in growth chamber experiments with biotic and abiotic stress agents excluded (Ainsworth and Long 2005; Matyssek et al., 2010). Thus, in terms of physiological processes, important questions remain with regard to how mature forests growing in natural conditions respond to interacting $eCO_2 \times eO_3$, and much is yet to be learned regarding stand-level E and WUE being affected by $eCO_2 \times eO_3$ (Karnosky et al., 2003, 2005; Matyssek et al., 2010; Pretzsch et al., 2010). More extensive, long-term FACE-type experiments in mature forest ecosystems will provide the most insight into forest ecosystem responses to atmospheric change.

9.3.3 Biomass and Net Primary Production

Biomass and net primary production (NPP) data for both young and intermediate-age trees consisted of foliage, stem/wood, coarse root (CR), fine root (FR), FRT or life span, total aboveground biomass (AG), total belowground biomass (BG), total above- and belowground biomass (AGBG), aboveground NPP (ANPP) and total NPP (Table 9.2). Data were more abundant for young ecosystems (mean $n=15$) compared to intermediate (mean $n=5$), and in both groups the order of data abundance was generally foliage = stem wood > AG = BG = AGBG > FR > CR > ANPP > FRT = NPP. The exception to this was that fine root biomass data were much more abundant ($n=11$) than the overall mean in intermediate-age systems. Only one study, the Aspen FACE Project, provided ecosystem-level estimates of ANPP and NPP (King et al., 2005; Zak et al., 2011; Pregitzer and Talhelm, this volume).

As with the broader ecological literature, data on FRT were very limited, restricted to only two studies but using a total of 5 years of root observation data from minirhizotrons (Phillips et al., 2009; Pregitzer et al., 2008). Therefore, our knowledge of these key components of terrestrial C cycling as affected by the $eCO_2 \times eO_3$ interaction remains very basic.

Elevated CO_2 was generally a highly significant factor that stimulated biomass production of all plant parts in both young and intermediate-age trees in the studies surveyed. The mean stimulation due to eCO_2 across all parameters in young trees was 29.6% relative to the control, with a low of 18.4% for foliage and a high of 41.8% for fine roots. The variance of these responses, as estimated by the coefficient of variation, ranged from 0.43 for ANPP to 1.24 for foliage, reflecting the fact that the former was estimated from a single study using uniform methods while the latter came from many different studies. The mean stimulation due to eCO_2 relative to the control in intermediate-age trees was 36.7%, with a low of 2.8% for FRT and a high of 53.7% for fine root biomass. This was contrary to our expectation that younger trees or forest ecosystems would have higher responsiveness to eCO_2. Although based on fewer observations, the CV for intermediate-age tree responses was less than for young trees (mean of 0.7 compared to 0.8), which again could be related to the smaller sample size, but also may suggest that responses become more consistent through time. The magnitude of the stimulation due to eCO_2 overall, and the relatively high responsiveness of fine roots in particular are consistent with the broader eCO_2 literature (Curtis and Wang, 1998; Dieleman et al., 2012; Norby et al., 2005).

Elevated O_3, also generally highly significant, decreased biomass production almost universally across parameters, although exceptions occurred related to fine roots in both young and intermediate-age trees (Table 9.2). The mean decrease in biomass production in young trees relative to the control was −12.5%, with the most responsive parameter being total belowground biomass (−19.1%) and the least being CRs (−2.7%). The CV of these responses averaged 1.65, more than double that for eCO_2. For intermediate-age trees, the decrease in biomass production due to eO_3 relative to the control averaged −15.0%, which as for eCO_2, was greater than the mean response of young trees. The CV of the mean decrease in biomass production (0.65) was much lower than for young trees. Again, this could be a function of fewer observations or less variable responses in older trees. Interestingly, fine root biomass in intermediate-age trees showed a mean stimulation due to eO_3 of 1.82% relative to the control, but this was accompanied by a CV of 8.08. The extremely large variance occurred because of the 11 observations, 6 were negative (average of −20.4%) and 5 were positive (average of 28.5%). FRT was stimulated by eO_3 relative to the control in both young (5.0%) and intermediate (2.8%) age trees, but the data were from a single study in each case. Clearly, fine roots are highly responsive to eO_3 but the direction of response is variable across ecosystems, age classes and environmental conditions. At the Aspen FACE project, the pure aspen community showed a consistent stimulation in fine root biomass under eO_3 over time (King et al., 2005; Pregitzer et al., 2008; Rhea and King, 2012), whereas the aspen-birch and aspen maple communities did not, suggesting a strong genetic effect on ecosystem function. A later analysis (Zak et al., 2011) incorporated estimates of turnover with the fine root biomass and estimated a non-significant 8% *increase* in total NPP in the aspen community in the 12th year of the experiment. That fine root dynamics can have such a large impact on total NPP, and potentially net ecosystem productivity (NEP), in response to air pollution illustrates the continuing need for a much better understanding of belowground processes (see also Kraigher et al., this volume).

Finally, in the experiments surveyed, the $eCO_2 \times eO_3$ interaction was generally not statistically significant and resulted in positive stimulation of all biomass production parameters in both young and intermediate-age trees relative to the control (Table 9.2). This suggests that the effects of eCO_2 and eO_3, when occurring together, are generally additive, and that the positive effects of eCO_2 more than offset the negative effects of eO_3 on biomass production. This is consistent with the magnitude of the relative responses for eCO_2 and eO_3 treatments when applied alone. The mean stimulation in biomass in the $eCO_2 \times eO_3$ treatment, relative to the control across parameters, was 15.1% in young trees, with a mean CV of 2.16, which is a higher variance than for eCO_2 and eO_3. In intermediate-age trees, the mean increase across biomass production parameters in the $eCO_2 \times eO_3$ plots relative to the control was 20.1%, with a mean CV of 0.82. Again, a larger relative treatment response in intermediate compared to young trees is counter to our expectation, and the lower variance suggests a smoothing out of responses through time or a narrower range of variation in responses due to fewer studies. The most responsive parameter was again fine root biomass with increases under $eCO_2 \times eO_3$ of 37.1% and 47.8% in young and intermediate-age trees, respectively. The least responsive parameters were FRT (0%) and total belowground biomass (3.6%) relative to control in young trees, and total NPP (0%) in intermediate-age trees, but the ability to make inferences is limited by the small number of studies.

9.3.4 Litter Decomposition and Soil Relations

The parameters for litter biochemistry surveyed include condensed tannins (mg/g), total soluble phenolics (mg/g), lignin (mg/g), C (mg/g), N (mg/g) and C/N for leaf and fine root litter. The parameters for soil carbon cycling include litter mean resident time (MRT, year), which represents litter decomposition rate and total soil carbon (g/m^2). Most of the data were from studies conducted in young tree systems with ages ranging from 1 to 5 years, and intermediate tree systems with ages ranging from 6 to 12 years. As with physiology and biomass production, no data were available for mature (>20 years of age) tree/forest systems (Figure 9.1).

FIGURE 9.1 Researchers collect leaf litter for biochemical analysis and to estimate litter biomass production using mesh-lined baskets (left) and recover field incubation bags of foliage and fine root litter (right) to estimate decomposition dynamics as affected by the $eCO_2 \times eO_3$ interaction at the Aspen FACE Experiment in Rhinelander, Wisconsin, USA. *Photo credit: John King.*

Growth of trees under eCO_2 generally increased the concentrations of secondary metabolites in leaves. In leaf litter of young and intermediate-age trees, eCO_2 increased the concentrations of condensed tannins by a mean of 15.5% and 19.6%, respectively, and increased the concentration of total soluble phenolics by a mean of 8.9% and 14.6%, respectively. Although most of the individual studies included in this review did not detect a significant effect, the overall mean response of lignin to eCO_2 was positive in leaf and fine root litter. Elevated CO_2 increased lignin concentration by 5% and 7.9% in leaf litter of young and intermediate trees, respectively, and by 58.5% in fine root litter of intermediate-age trees. The increases in the concentrations of secondary metabolites under eCO_2 can be explained on the basis of growth-differentiation balance hypothesis (Herms and Mattson, 1992). Elevated CO_2 increases production of carbohydrates but plant growth does not increase proportionally because of limited soil nutrient supply (Luo et al., 2004). Elevated CO_2 therefore leads to more carbohydrates allocated to secondary metabolic pathways.

The responses of condensed tannins, phenolics and lignin to eO_3 were different for young and intermediated age trees. In young trees, eO_3 had a small impact on the concentration of condensed tannins, but decreased phenolics (−9.9%) and lignin (−15.4%) concentration in leaf litter. In intermediate-age trees, eO_3 increased the concentrations of condensed tannins by 34.1%, phenolics by 26.3% and lignin by 7.9% in leaf litter. The different response between young and intermediate-age trees was probably due to shifts in carbon allocation strategy from growth to defence in older trees. It has been shown that eO_3 depresses photosynthesis but triggers anti-oxidant defence responses (Kangasjarvi et al., 1994). In young trees, as lower carbohydrate supply constrains allocation to secondary metabolism, eO_3 had small or negative impacts on the synthesis of secondary metabolites. With tree growth, the up-regulation of the shikimic acid pathway leads to an increase in anti-oxidants under elevated O_3 (Lindroth, 2010), that may explain the higher condensed tannin and phenolic concentrations in leaf litter (Lindroth, 2010; Matyssek et al., 2013; Couture and Lindroth, 2013; Pregitzer and Talhlem, 2013, this vol.).

In addition to altering the concentrations of secondary metabolites, eCO_2 also tended to decrease N concentration in leaf and fine root litter, with leaf litter showing a greater response (−14.2% and −8.4% for young and intermediate trees, respectively) than fine root litter (−2.6% and −5.6% for young and intermediate trees, respectively). Lower N concentration was accompanied by higher C/N ratio in leaf and fine root litter for both young and intermediate-age trees. It has been shown that eCO_2 often leads to an accumulation of carbohydrates, especially starch, in leaves (Lindroth, 2010). The dilution effect of increased carbohydrates therefore resulted in a lower N concentration and higher C/N ratio in leaf litter. Elevated O_3 showed a similar reduction in N concentration. Leaf and fine root litter grown under eO_3 had lower N concentration and higher C/N ratio, except for fine root litter of young trees, which increased by a mean of 4.6%.

Most studies included in this review found that the interaction of eCO_2 and eO_3 was not statistically significant for leaf chemistry changes (Chapman et al., 2005; Holton et al., 2003; Kasurinen et al., 2006; Liu et al., 2005; see also Chapters 11 and 15). Litter chemistry under the $eCO_2 \times eO_3$ treatment was therefore often influenced by the additive effects of eCO_2 and eO_3. Compared to eCO_2 or eO_3 only treatments, $eCO_2 \times eO_3$ treatment resulted in greater increases in the concentrations of condensed tannins, phenolics and C/N, and a greater decrease in N concentration (Table 9.3).

The changes in litter quality under eCO_2 and eO_3, such as increases in condensed tannins, phenolics and C/N, are expected to lead to lower decomposition rates. Mean responses found here suggest that eCO_2 decreased litter decomposition in both young and intermediate tree systems, while eO_3 stimulated litter decomposition in young tree systems but reduced it in intermediate tree systems (Table 9.3). However, the changes in litter decomposition were usually not significant in individual experiments, or the impacts were often transient. For example, Liu et al. (2009) found eO_3 decreased litter mass loss only in the first year, but not in the second year. In an OTC experiment, Kasurinen et al. (2006) found that litter decomposition was not affected by eO_3 exposure in the first 2 years, but a slower decomposition rate was found in the third year of incubation.

All data on changes in soil C were from only two studies conducted at the Aspen FACE Project at different stages of stand development (Loya et al., 2003; Talhelm et al., 2009). Although the mean responses showed eCO_2 and eO_3 altered total soil carbon in young and intermediate tree systems (Table 9.3), the changes were not statistically significant (Talhelm et al., 2009). The 11-year study at Aspen FACE indicated that the effects of eCO_2 and eO_3 on soil C varied with stand development. After the first 4 years of fumigation at the Aspen FACE, Loya et al. (2003) found that the formation of new soil carbon under $eCO_2 \times eO_3$ treatment was 51% less than that under eCO_2-only treatment. However, a later study by Talhelm et al. (2009) suggested that the initial reduction in new C accumulation in eO_3 under eCO_2 was only a temporary effect. The amount of new soil C under the eCO_2 and $eCO_2 \times eO_3$ appears to have converged in the fifth year of the experiment. After 11 years of fumigation, neither eCO_2 nor eO_3 induced significant impacts on new soil carbon accumulation or total soil carbon content. As with estimates of NPP, however, inferences on the effects of the $eCO_2 \times eO_3$ interaction on soil C formation are necessarily limited by the lack of data.

9.4 SUMMARY OF PHYSIOLOGY, BIOMASS PRODUCTION AND SOC CYCLING RESPONSES TO $eCO_2 \times eO_3$

We were able to locate 58 studies published between 1990 and 2012 reporting primary data on forest tree physiology, biomass production, litter quality and decomposition responses from $eCO_2 \times eO_3$ interaction experiments. Studies were abundant for young trees (<6 years) from growth chamber, greenhouse and OTC experiments. As might be expected, longer-term studies on intermediate-age trees (6–12 years) were less abundant and primarily from OTC and FACE experiments. We found that no $eCO_2 \times eO_3$ studies have been conducted on mature forest trees (>20 years). This represents a large gap in our knowledge of forest response to air pollution and climate change because we know that physiological and ecosystem functional processes change through stand development, leaving this important interaction unrepresented for much of the world's forests. Data on leaf-level photosynthesis, stomatal conductance, foliage, fine root and aboveground biomass, and leaf/litter N concentration were most abundant, but were severely lacking for many other parameters important to the storage and cycling of C, water and nutrients in forest ecosystems. In particular, our knowledge of canopy-scale processes (evapotranspiration, gross primary production, net ecosystem exchange), belowground C inputs to soil (fine root production and turnover) and changes in SOC formation is particularly limited, yet these are some of the most important processes to understand.

There were commonalities in how eCO_2 and eO_3, alone and in combination, affected physiology, biomass production and litter decomposition, and the net effect of the $eCO_2 \times eO_3$ interaction was fairly consistent. In terms of physiology, there was a strong positive response of A_{CO_2} to eCO_2 and a negative response to eO_3 that was more than compensated for under $eCO_2 \times eO_3$ (i.e. a net increase relative to control). Stomatal conductance was strongly decreased by eCO_2 but not greatly affected by eO_3; hence, the $eCO_2 \times eO_3$ treatment was similar to eCO_2. Instantaneous WUE therefore increased under eCO_2 with and without simultaneous exposure to eO_3. Leaf area index increased under eCO_2, decreased with eO_3 and showed slight increase with $eCO_2 \times eO_3$. For most physiological parameters, responses of intermediate-age trees were greater than in young trees, which was counter to expectation. In terms of biomass production, there were generally strong positive and strong negative responses to eCO_2 and eO_3, respectively. As with physiology, the net effect of the $eCO_2 \times eO_3$ interaction was generally not statistically significant, but still important to quantify because relative responses (relative to control) were generally intermediate compared to eCO_2 and eO_3 when applied alone. Few studies reported significant effects of eCO_2 or eO_3 on biomass partitioning (data not shown), and as for physiology, ontogenetic effects were counter to expectation: that is, relative responses were greater and less variable for intermediate-age compared to young trees. Linear regression did not reveal any significant correlations of relative responses versus age or exposure level, which could be due to limited data available.

Finally, in terms of SOC cycling, concentrations of secondary metabolites (condensed tannins, soluble phenolics and lignin) generally increased in response to eCO_2, and responses were again larger in intermediate-age compared to young trees. Interestingly, eO_3 decreased concentrations of secondary metabolites in young trees but increased them in intermediate-age trees, perhaps indicating a shift in C allocation from growth to defence with age. Litter N concentration decreased with both eCO_2 and eO_3, except for young fine root litter, in which it increased with eO_3. Changes in litter quality may be expected to change decomposition dynamics, and the data of our survey suggest that eCO_2 decreased litter decomposition in both young and intermediate tree systems, while eO_3 stimulated litter decomposition in young tree systems but decreased it in intermediate-age systems. However, the changes in litter decomposition were usually not significant in individual experiments, or the impacts were often small or transient. Data on SOC from long-term studies were extremely limited (only one study longer than 5 years), but suggest little significant effect due to eCO_2 or O_3, alone or in combination. The above conclusions should be tempered by the realization that most studies reviewed here were performed on young plants under highly controlled conditions, often with abundant water and nutrients. More work is needed to determine long-term forest responses to eCO_2 and eO_3, especially in older age classes, and as affected by interacting biotic and abiotic environmental stresses.

9.5 MOVING FORWARD

In closing, we may ask: What are the most pressing scientific questions remaining on forest responses to interacting eCO_2 and eO_3? and How can science best be used to inform policy to enhance social welfare? A good illustration of science and policy working together is research that showed the potential for eO_3 to decrease forest productivity in response to eCO_2 in young north-temperate forests (Isebrands et al., 2001; Karnosky et al., 2003; King et al., 2005), which if representative, could imply air pollution limitation of the capacity of the world's forests to counteract the build-up of atmospheric CO_2. This science supported the decision to decrease the secondary air quality standard for O_3 in the United States (U.S. Federal Register, 27 March 2008), with resulting health benefits for humans and vegetation. However, how widespread this productivity response is, whether the key processes governing C storage act uniformly across forests and stages of stand development and how they interact with other environmental stresses (e.g. drought) is still very poorly understood. Clearly, this survey has demonstrated that data for most important C cycling processes, especially for later stages of forest development, are very limited. There simply have not been enough long-term, large-scale experiments, which incidentally are also some of the most efficient in terms of scientific output for the resources invested. Critical data gaps remain for canopy-level processes affecting C assimilation and water cycling, belowground processes affecting SOC formation and nutrient dynamics and long-term net ecosystem CO_2 exchange with the atmosphere. Several long-term experiments have successfully demonstrated that exposure of mature forest ecosystems to eCO_2 and eO_3 (individually) is possible using a system of pressurized tubing to distribute the fumigation gas throughout the forest canopy (Körner et al., 2005; Matyssek et al., 2010; Chapters 1 and 6). This technology promises to provide a lower cost alternative to the more technically challenging forced-air FACE approach to forest fumigation that could facilitate future $eCO_2 \times eO_3$ experiments. Such studies in natural ecosystems are especially important for process-level understanding and model parameterization, and they are technically feasible.

Moving forward, when thinking about future research on interacting air pollution and climate change, and potential development of a distributed network of long-term research 'supersites' (e.g. see Mikkelsen et al., 2013, this vol.), we recommend considering a subset of projects that examine the interaction of eCO_2 and eO_3 over the major stages of stand development in both coniferous and deciduous forests. This review suggests that more information on intermediate-age and mature ecosystems is most urgently needed. Empirical studies should work iteratively with development of process-based ecosystem models to refine mechanistic understanding of ecosystem function and test hypotheses related to future forest responses to climate and air pollution. The improved ecosystem models can then be linked to larger-scale models to extrapolate the interactive effects of $eCO_2 \times eO_3$ on carbon, water and nutrient cycling at the region-to-global scale, and forward in time.

Finally, an argument can be made that current environmental policy lags behind the science we already have at hand, and a goal for the future should be scientifically literate public officials who implement legislation based on sound science in a timely manner. For example, we know well that negative health consequences arise for humans and vegetation when background tropospheric O_3 concentration exceeds 60 ppbv for any length of time, yet the primary and secondary air quality standards for tropospheric O_3 in the United States remain fixed at 75 ppbv. Similarly, there is strong scientific consensus that the rising atmospheric CO_2 concentration is directly linked to the changing climate, yet global CO_2 emissions continue to increase at an increasing rate and development of alternative energy sources proceeds at a snail's pace. This feeds back to our responsibility as educators to develop an educated electorate who will elect public officials who embrace science as the basis of sound policy, and who will hold them accountable for failing to do so. An informed leadership must possess the foresight and courage to invest in the science that will guide development of environmental policy that balances near-term economic welfare and long-term climate stability, environmental quality and sustainability to enhance the quality of life for a rapidly rising population.

ACKNOWLEDGEMENTS

During preparation of this manuscript, J. K. was supported by an International Francqui Foundation Visiting Professorship at the University of Antwerp and by a research scholarship from the Commission for International Exchange (U.S. State Department) Fulbright program. Additional support for J. K. was provided by USDA NIFA grant

2011-67009-20089, USDA CSREES grant 2009-10001-05113, US Forest Service Eastern Forest Environmental Threat Assessment Center grant 08-JV-11330147-038 and DOE–BER Terrestrial Ecosystem Sciences program grants 11-DE-SC-0006700 ER65189.

REFERENCES

Agrell, J., Kopper, B., McDonald, E.P., et al., 2005. CO_2 and O_3 effects on host plant preferences of the forest tent caterpillar (*Malacosoma disstria*). Glob. Chang. Biol. 11, 588–599.

Ainsworth, E.A., Long, S.P., 2005. What have we learned from 15 years of free-air CO_2 enrichment (FACE)? A meta-analytic review of the responses of photosynthesis, canopy properties and plant production to rising CO_2. New Phytol. 165, 351–372.

Boerner, R.E.J., Rebbeck, J., 1995. Decomposition and nitrogen release from leaves of three hardwood species grown under elevated O_3 and or CO_2. Plant Soil 170, 149–157.

Broadmeadow, M.S.J., Jackson, S.B., 2000. Growth responses of *Quercus petraea*, *Fraxinus excelsior* and *Pinus sylvestris* to elevated carbon dioxide, ozone and water supply. New Phytol. 146, 437–451.

Calfapietra, C., Ainsworth, E.A., Beier, C., et al., 2010. Challenges in elevated CO_2 experiments on forests. Trends Plant Sci. 15, 5–10.

Chapman, J.A., King, J.S., Pregitzer, K.S., et al., 2005. Effects of elevated concentrations of atmospheric CO_2 and tropospheric O_3 on decomposition of fine roots. Tree Physiol. 25, 1501–1510.

Chappelka, A.H., Samuelson, L.J., 1998. Ambient ozone effects on forest trees of the eastern United States: a review. New Phytol. 139, 91–108.

Couture, J.J., Meehan, T.D., Lindroth, R.L., 2012. Atmospheric change alters foliar quality of host trees and performance of two outbreak insect species. Oecologia 168, 863–876.

Couture, J.J., Lindroth, R.L., 2013. Impacts of atmospheric change on tree-arthropod interactions. Dev. Environm. Sci. 13, 227–248.

Curtis, P.S., Wang, X.Z., 1998. A meta-analysis of elevated CO_2 effects on woody plant mass, form, and physiology. Oecologia 113, 299–313.

Darbah, J.N.T., Kubiske, M.E., Nelson, N., Kets, K., Riikonen, J., Sober, A., Rouse, L., Karnosky, D.F., 2010. Will photosynthetic capacity of aspen trees acclimate after long-term exposure to elevated CO_2 and O_3? Environ. Pollut. 158, 983–991.

Dickson, R.E., Coleman, M.D., Riemenschneider, D.E., Isebrands, J.G., Hogan, G.D., Karnosky, D.F., 1998. Growth of five hybrid poplar genotypes exposed to interacting elevated CO_2 and O_3. Can. J. Forest Res. 28, 1706–1716.

Dickson, R.E., Coleman, M.D., Pechter, P., Karnosky, D., 2001. Growth and crown architecture of two aspen genotypes exposed to interacting ozone and carbon dioxide. Environ. Pollut. 115, 319–334.

Dieleman, W., Vicca, S., Dijkstra, F.A., Hagedorn, F., Hovenden, M.J., Larsen, K., Morgan, J.A., Volder, A., Beier, C., Dukes, J.S., King, J., Leuzinger, S., Linder, S., Luo, Y., Oren, R., De Angelis, P., Tingey, D., Hoosbeek, M.R., Janssens, I.A., 2012. Simple additive effects are rare: a quantitative review of plant biomass and soil processes responses to combined manipulations of CO_2 and temperature. Glob. Chang. Biol. 18, 2681–2693.

Ellsworth, D.S., Oren, R., Huang, C., et al., 1995. Leaf and canopy responses to elevated CO_2 in a pine forest under free-air CO_2 enrichment. Oecologia 104, 139–146.

Fowler, D., Cape, J.N., Coyle, M., Flechard, C., Kuylenstierna, J., Hicks, K., Derwent, R.G., Johnson, C.E., 1999. The global exposure of forests to air pollutants. Water Air Soil Pollut. 116, 5–32.

Gardner, S.D.L., Freer-Smith, P.H., Tucker, J., Taylor, G., 2005. Elevated CO_2 protects poplar (*Poplus trichocarpa* x *P. deltoides*) from damage induced by O_3: identification of mechanisms. Funct. Plant Biol. 32, 221–235.

Gaucher, C., Costanzo, N., Afif, D., Maufette, Y., Chevrier, N., Dizengremel, P., 2003. The impact of elevated ozone and carbon dioxide on young *Acer saccharum* seedlings. Physiol. Plant. 117, 392–402.

Grams, T.E.E., Anegg, S., Häberle, K.-H., Langebartels, C., Matyssek, R., 1999. Interactions of chronic exposure to elevated CO_2 and O_3 levels in the photosynthetic light and dark reactions of European beech (*Fagus sylvatica*). New Phytol. 144, 95–107.

Herbinger, K., Then, C., Haberer, K., Alexou, M., Löw, M., Remele, K., Rennenberg, H., Matyssek, R., Grill, D., Wieser, G., Tausz, M., 2007. Gas exchange and antioxidative compounds in young beech trees under free-air ozone exposure and comparisons to adult trees. Plant Biol. 9, 288–297.

Herms, D.A., Mattson, W.J., 1992. The dilemma of plants: to grow or defend. Q. Rev. Biol. 67, 283–335.

Hillstrom, M., Meehan, T.D., Kelly, K., et al., 2010. Soil carbon and nitrogen mineralization following deposition of insect frass and greenfall from forests under elevated CO_2 and O_3. Plant Soil 336, 75–85.

Holton, M.K., Lindroth, R.L., Nordheim, E.V., 2003. Foliar quality influences tree-herbivore-parasitoid interactions: effects of elevated CO_2, O_3, and plant genotype. Oecologia 137, 233–244.

Isebrands, J.G., McDonald, E.P., Kruger, E., Hendrey, G., Percy, K., Pregitzer, K., Sober, J., Karnosky, D.F., 2001. Growth responses of *Populus tremuloides* clones to interacting elevated carbon dioxide and tropospheric ozone. Environ. Pollut. 115, 359–371.

Kainulainen, P., Holopainen, T., Holopainen, J.K., 2003. Decomposition of secondary compounds from needle litter of Scots pine grown under elevated CO_2 and O_3. Glob. Chang. Biol. 9, 295–304.

Kangasjarvi, J., Talvinen, J., Utriainen, M., et al., 1994. Plant defense systems induced by ozone. Plant Cell Environ. 17, 783–794.

Karnosky, D.F., Zak, D.R., Pregtizer, K.S., et al., 2003. Tropospheric O_3 moderates responses of temperate hardwood forests to elevated CO_2: a synthesis of molecular to ecosystem results from the Aspen FACE project. Funct. Ecol. 17, 289–304.

Karnosky, D.F., Pregitzer, K.S., Zak, D.R., Kubiske, M.E., Hendrey, G.R., Weinstein, D., Nosal, M., Percy, K.E., 2005. Scaling ozone responses of forest trees to the ecosystem level in a changing environment. Plant Cell Environ. 28, 965–981.

Kasurinen, A., Keinanen, M.M., Kaipainen, S., Nilsson, L.-O., Vapaavuori, E., Kontro, M.H., Holopainen, T., 2005. Below-ground responses of silver birch trees exposed to elevated CO_2 and O_3 levels during three growing seasons. Glob. Chang. Biol. 11, 1167–1179.

Kasurinen, A., Riikonen, J., Oksanen, E., et al., 2006. Chemical composition and decomposition of silver birch leaf litter produced under elevated CO_2 and O_3. Plant Soil 282, 261–280.

Kasurinen, A., Peltonen, P.A., Julkenen-Titto, R., et al., 2007. Effects of elevated CO_2 and O_3 on leaf litter phenolics and subsequent performance of litter-feeding soil macrofauna. Plant Soil 292, 25–43.

Kets, K., Darbah, N.T., Sober, A., Riikonen, J., Sober, J., Karnosky, D.F., 2010. Diurnal changes in photosynthetic parameters of *Populus tremuloides*, modulated by elevated concentrations of CO_2 and/or O_3 and daily climatic variation. Environ. Pollut. 158, 1000–1007.

King, J.S., Kubiske, M.E., Pregitzer, K.S., Hendrey, G.R., McDonald, E.P., Giardina, G.P., Quinn, V.S., Karnosky, D.F., 2005. Tropospheric O_3 compromises net primary production in young stands of trembling aspen, paper birch and sugar maple in response to elevated atmospheric CO_2. New Phytol. 168, 623–636.

Kitao, M., Löw, M., Heerdt, C., Grams, T.E.E., Häberle, K.H., Matyssek, R., 2009. Effects of chronic elevated ozone exposure on gas exchange responses of adult beech trees (*Fagus sylvatica*) as related to the within-canopy light gradient. Environ. Pollut. 157, 537–544.

Kolb, T.E., Matyssek, R., 2001. Limitations and perspectives about scaling ozone impacts in trees. Environ. Pollut. 115, 373–393.

Kopper, B.J., Lindroth, R.L., 2003. Effects of elevated carbon dioxide and ozone on the phytochemistry of aspen and performance of an herbivore. Oecologia 134, 95–103.

Körner, C., Roman, A., Bignucolo, O., et al., 2005. Carbon flux and growth in mature deciduous forest trees exposed to elevated CO_2. Science 309, 1360–1362.

Kozovits, A.R., Matyssek, R., Blaschke, H., Gottlein, A., Grams, T.E.E., 2005. Competition increasingly dominates the responsiveness of juvenile beech and spruce to elevated CO_2 and/or O_3 concentrations throughout two subsequent growing seasons. Glob. Chang. Biol. 11, 1387–1401.

Kull, O., Sober, M.D., Dickson, R.E., Isebrands, J.G., Gagnon, Z., Karnosky, D.F., 1996. Photosynthesis responses of aspen clones to simultaneous exposures of ozone and CO_2. Can. J. Forest Res. 26, 639–648.

Li, Q., Jacob, D.J., Bey, I., et al., 2002. Transatlantic transport of pollution and its effects on surface ozone in Europe and North America. J. Geophys. Res. 107, 4166–4187.

Lindroth, R.L., 2010. Impacts of elevated atmospheric CO_2 and O_3 on forests: phytochemistry, trophic interactions, and ecosystem dynamics. J. Chem. Ecol. 36, 2–21.

Lindroth, R.L., Kopper, B.J., Parsons, W.F.J., et al., 2001. Consequences of elevated carbons dioxide and ozone for foliar chemical composition and dynamics in trembling aspen (*Populus tremuloides*) and paper birch (*Betula papyrifera*). Environ. Pollut. 115, 395–404.

Liu, X., Kozovits, A.R., Grams, T.E.E., Blaschke, H., Rennenberg, H., Matyssek, R., 2004. Competition modifies effects of enhanced ozone/carbon dioxide concentrations on carbohydrate and biomass accumulation in juvenile Norway spruce and European beech. Tree Physiol. 24, 1045–1055.

Liu, L.L., King, J.S., Giardina, C.P., 2005. Effects of elevated concentrations of atmospheric CO_2 and tropospheric O_3 on leaf litter production and chemistry in trembling aspen and paper birch communities. Tree Physiol. 25, 1511–1522.

Liu, L., King, J.S., Giardina, C.P., et al., 2009. The influence of chemistry, production and community composition on leaf litter decomposition under elevated atmospheric CO_2 and tropospheric O_3 in a northern hardwood ecosystem. Ecosystems 12, 401–416.

Loats, K.V., Rebbeck, J., 1999. Interactive effects of ozone and elevated carbon dioxide on the growth and physiology of black cherry, green ash, and yellow-poplar seedlings. Environ. Pollut. 106, 237–248.

Long, S.P., Ainsworth, E.A., Rogers, A., et al., 2004. Rising carbon dioxide: plants FACE the future. Annu. Rev. Plant Biol. 55, 591–628.

Löw, M., Herbinger, K., Nunn, A.J., Häberle, K.-H., Leuchner, M., Heerdt, C., Werner, H., Wipfler, P., Pretzsch, H., Tausz, M., Matyssek, R., 2006. Extraordinary drought of 2003 overrules ozone impact on adult beech trees (*Fagus sylvatica*). Trees 20, 539–548.

Loya, W.M., Pregitzer, K.S., Karberg, N.J., et al., 2003. Reduction of soil carbon formation by tropospheric ozone under increased carbon dioxide levels. Nature 425, 705–707.

Luo, Y., Su, B., Currie, W.S., et al., 2004. Progressive nitrogen limitation of ecosystem responses to rising atmospheric carbon dioxide. Bioscience 54, 731–739.

Matyssek, R., Sandermann, H., 2003. Impact of Ozone on Trees: An Ecophysiological Perspective. In: Progress in Botany, vol. 64. Springer Verlag, Heidelberg, pp. 349–404.

Matyssek, R., Le Thiec, D., Low, M., Dizengremel, P., Nunn, A.J., Haberle, K.-H., 2006. Interactions between drought and O_3 stress in forest trees. Plant Biol. 8, 11–17.

Matyssek, R., Wieser, G., Ceulemans, R., et al., 2010. Enhances ozone strongly reduces carbon sink strength of adult beech (*Fagus sylvatica*)—resume from the free-air fumigation study at Kranzberg Forest. Environ. Pollut. 158, 2527–2532.

Matyssek, R., Schnyder, H., Oßwald, W., Ernst, D., Munch, J.C., Pretzsch, H., 2013. Growth and Defence in Plants: Resource Allocation at Multiple Scales. In: Ecological Studies, vol. 220. Springer, Berlin, p. 467.

Medlyn, B.E., Laitat, E., Rey, A., et al., 2001. Stomatal conductance of forest species after long-term exposure to elevated CO_2 concentrations: a synthesis. New Phytol. 149, 247–264.

Mikkelsen, T.N., Clarke, N., Danielewska, A., Fischer, R., 2013. Towards supersites in forest ecosystem monitoring and research. Dev. Environ. Sci. 13, 475–495.

Mooney, H.A., Winner, W.E., Pell, E.J., 1991. Responses of Plants to Multiple Stresses. Academic Press, San Diego, 422.

Niinemets, U., 2010. Responses of forest trees to single and multiple environmental stresses from seedlings to mature plants: past stress history, stress interactions, tolerance and acclimation. For. Ecol. Manage. 260, 1623–1639.

Noormets, A., 2001. The effects of interacting CO_2 and O_3 on the carbon balance of aspen (Populus tremuloides Michx.) grown in an open-air CO_2 and O_3 enrichment system. PhD thesis. Michigan Technological University.

Noormets, A., Sober, A., Pell, E.J., Dickson, R.E., Podila, G.K., Sober, J., Isebrands, J.G., Karnosky, D.F., 2001. Stomatal and non-stomatal limitation to photosynthesis in two trembling aspen (*Populus tremuloides* Michx.) clones exposed to elevated CO_2 and/or O_3. Plant Cell Environ. 24, 327–336.

Noormets, A., Kull, O., Kubiske, M.E., Karnosky, D.F., 2010. Elevated CO_2 response of photosynthesis depends on ozone concentration in aspen. Environ. Pollut. 158, 992–999.

Norby, R.J., Ceulemans, R., DeAngelis, P., et al., 2005. Forest response to elevated CO_2 is conserved across a broad range of productivity. Proc. Natl. Acad. Sci. 102, 18052–18056.

Oksanen, E., Riikonen, J., Kaakinen, S., et al., 2005. Structural characteristics and chemical composition of birch (*Betula pendula*) leaves are modified by increasing CO_2 and ozone. Glob. Chang. Biol. 11, 732–748.

Olszyk, D.M., Johson, M.G., Phillips, D.L., Seidler, R.J., Tingey, D.T., Watrud, L.S., 2001. Interactive effects of CO_2 and O_3 on a ponderosa pine plant/litter/soil mesocosm. Environ. Pollut. 115, 447–462.

Olszyk, D., Tingey, D., Wise, C., Davis, E., 2002. CO_2 and O_3 alter photosynthesis and water vapor exchange for *Pinus ponderosa* needles. Phyton 42, 121–134.

Parsons, W.F.J., Lindroth, R.L., Bockheim, J.G., 2004. Decomposition of *Betula papyrifera* leaf litter under the independent and interactive effects of elevated CO_2 and O_3. Glob. Chang. Biol. 10, 1666–1677.

Parsons, W.F.J., Bockheim, J.G., Lindroth, R.L., 2008. Independent, interactive, and species-specific responses of leaf litter decomposition to elevated CO_2 and O_3 in a northern hardwood forest. Ecosystems 11, 505–519.

Peltonen, P.A., Vapaavuori, E., Julkunen-Tiitto, R., 2005. Accumulation of phenolic compounds in birch leaves is changed by elevated carbon dioxide and ozone. Glob. Chang. Biol. 11, 1305–1324.

Pfirrmann, T., Barnes, J.D., Steiner, K., et al., 1996. Effects of elevated CO_2, O_3 and K deficiency on Norway spruce (*Picea abies*): nutrient supply, content and leaching. New Phytol. 134, 267–278.

Phillips, D.L., Johnson, M.G., Tingey, D.T., Storm, M.J., 2009. Elevated CO_2 and O_3 effects on fine-root survivorship in ponderosa pine mesocosms. Oecologia 160, 827–837.

Polle, A., Pfirrmann, T., Chakrabarti, S., Rennenberg, H., 1993. The effects of enhanced ozone and enhanced carbon dioxide concentrations on biomass, pigments and antioxidative enzymes in spruce needles. Plant Cell Environ. 16, 311–316.

Pregitzer, K.S., Burton, A.J., King, J.S., Zak, D.R., 2008. Soil respiration, root biomass, and root turnover following long-term exposure of northern forests to elevated atmospheric CO_2 and tropospheric O_3. New Phytol. 180, 153–161.

Pregitzer, K.S., Talhlem, A.F., 2013. Belowground carbon cycling at aspen FACE: dynamic responses to CO_2 and O_3 in developing forests. Dev. Environ. Sci. 13, 209–226.

Pretzsch, H., Dieler, J., Matyssek, R., Wipfler, P., 2010. Tree and stand growth of mature Norway spruce and European beech under long-term ozone fumigation. Environ. Pollut. 158, 1061–1070.

Rebbeck, J., Scherzer, A.J., 2002. Growth responses of yellow-poplar (Liridendron tulipifera L.) exposed to 5 years of O_3 alone or combined with elevated CO_2. Plant Cell Environ. 25, 1527–1537.

Reich, P.B., 1987. Quantifying plant response to ozone: a unifying theory. Tree Physiol. 3, 63–91.

Reich, P.B., Hobbie, S.E., Lee, T., et al., 2006. Nitrogen limitation constrains sustainability of ecosystem response to CO_2. Nature 440, 922–925.

Rhea, L.K., King, J.S., 2012. Depth-dependency of trembling aspen and paper birch small-root responses to eCO_2 and eO_3. Plant Soil 355, 215–229.

Richet, N., Afif, D., Tozo, K., Pollet, B., Maillard, P., Huber, P., Huber, F., Priault, P., Banvoy, J., Gross, P., Dizengremel, P., Lapierre, C., Perre, P., Cabane, M., 2012. Elevated CO_2 and/or ozone modify lignification in the wood of poplars (*Populus tremula* x *alba*). J. Exp. Bot. 63, 4291–4301.

Riikonen, J., Lindsberg, M.-M., Holopainen, T., Oksanen, E., Lappi, J., Peltonen, P., Vapaavuori, E., 2004. Silver birch and climate change: variable growth and carbon allocation responses to elevated concentrations of carbon dioxide and ozone. Tree Physiol. 24, 1227–1237.

Riikonen, J., Holopainen, T., Oksanen, E., Vapaavuori, E., 2005. Leaf photosynthetic characteristics of silver birch during three years of exposure to elevated concentrations of CO_2 and O_3 in the field. Tree Physiol. 25, 621–632.

Rodenkirchen, H., Gottlein, A., Kozovits, A.R., Matyssek, R., Grams, T.E.E., 2009. Nutrient contents and efficiencies of beech and spruce saplings as influenced by competition and O_3/CO_2 regime. Eur. J. For. Res. 128, 117–128.

Sallas, L., Kainulainen, P., Utriainen, J., et al., 2001. The influence of elevated O_3 and CO_2 concentrations on secondary metabolites of Scots pine (*Pinus sylvestris* L.) seedlings. Glob. Chang. Biol. 7, 303–311.

Scherzer, A.J., Rebbeck, J., Boerner, R.E.J., 1998. Foliar nitrogen dynamics and decomposition of yellow-poplar and eastern white pine during four seasons of exposure to elevated ozone and carbon dioxide. For. Ecol. Manage. 109, 355–366.

Sharma, P., Sober, A., Sober, J., Podila, G.K., Kubiske, M.E., Mattson, W.J., Isebrands, J.G., Karnosky, D.F., 2003. Moderation of [CO_2]-induced gas exchange responses by elevated tropospheric O_3 in trembling aspen and sugar maple. Ekológia 22, 318–331.

Talhelm, A.F., Pregitzer, K.S., Zak, D.R., 2009. Species-specific responses to atmospheric carbon dioxide and tropospheric ozone mediate changes in soil carbon. Ecol. Lett. 12, 1219–1228.

Talhelm, A.F., Pregitzer, K.S., Giardina, C.P., 2012. Long-term leaf production response to elevated atmospheric carbon dioxide and tropospheric ozone. Ecosystems 15, 71–82.

Thomas, R.B., Strain, B.R., 1991. Root restriction as a factor in photosynthetic acclimation of cotton seedlings grown in elevated carbon dioxide. Plant Physiol. 96, 627–634.

Tricker, P.J., Pecchiari, M., Bunn, S.M., et al., 2009. Water use of a bioenergy plantation increases in a future high CO_2 world. Biomass Bioenergy 33, 200–208.

Uddling, J., Teclaw, R.M., Kubiske, M.E., Pregitzer, K.S., Ellsworth, D.S., 2008. Sap flux in pure aspen and mixed-aspen birch forests exposed to elevated concentrations of carbon dioxide and ozone. Tree Physiol. 28, 1231–1243.

Uddling, J., Teclaw, R.M., Pregitzer, K.S., et al., 2009. Leaf and canopy conductance in aspen and aspen-birch forests under free-air enrichment of carbon dioxide and ozone. Tree Physiol. 29, 1367–1380.

Utriainen, J., Janhunen, S., Helmisaari, H.-S., Holopainen, T., 2000. Biomass allocation, needle structural characteristics and nutrient composition in Scots pine seedlings exposed to elevated CO_2 and O_3 concentrations. Trees 14, 475–484.

Volin, J.C., Reich, P.B., 1996. Interaction of elevated CO_2 and O_3 on growth, photosynthesis and respiration of three perennial species grown in low and high nitrogen. Physiol. Plant. 97, 674–684.

Volin, J.C., Reich, P.B., Givnish, T.J., 1998. Elevated carbon dioxide ameliorates the effects of ozone on photosynthesis and growth: species respond similarly regardless of photosynthetic pathway or plant functional group. New Phytol. 138, 315–325.

Watanabe, M., Umemoto-Yamaguchi, M., Koike, T., Izuta, T., 2010. Growth and photosynthetic response of Fagus crenata seedlings to ozone and/or elevated carbon dioxide. Landsc. Ecol. Eng. 6, 181–190.

Wullscheger, S.D., Norby, R.J., 2001. Sap velocity and canopy transpiration in a sweetgum stand exposed to free-air CO_2 enrichment (FACE). New Phytol. 150, 489–498.

Wullscheger, S.D., Gunderson, C.A., Hanson, P.J., et al., 2002. Sensitivity of stomatal and canopy conductance to elevated CO_2 concentration-interacting variables and perspectives of scale. New Phytol. 153, 485–496.

Zak, D.R., Pregitzer, K.S., Kubiske, M.E., Burton, A.J., 2011. Forest productivity under elevated CO_2 and O_3: positive feedbacks to soil N cycling sustain decade-long net primary productivity enhancement by CO_2. Ecol. Lett. 14, 1220–1226.

Chapter 10

Belowground Carbon Cycling at Aspen FACE: Dynamic Responses to CO_2 and O_3 in Developing Forests

Kurt S. Pregitzer[1] and Alan F. Talhelm
College of Natural Resources, University of Idaho, Moscow, Idaho, USA
[1]*Corresponding author: e-mail: kpregitzer@uidaho.edu*

Chapter Outline

10.1 INTRODUCTION

Belowground processes such as root growth, soil respiration, and soil carbon (C) formation are fundamental components of forest C cycling. Soils are often the largest pool of C within forests (Pregitzer and Euskirchen, 2004) and soil respiration is the second largest terrestrial C flux (Bond-Lamberty and Thomson, 2010). On average, more than one-third of forest gross primary productivity is allocated belowground in order to sustain processes such as root respiration and the production of fine roots and mycorrhizae (Litton et al., 2007). However, although the negative impacts of tropospheric ozone (O_3)

Developments in Environmental Science, Vol. 13. http://dx.doi.org/10.1016/B978-0-08-098349-3.00010-4

exposure on plant growth and the potential for rising atmospheric carbon dioxide (CO_2) concentrations to increase plant productivity have been studied for decades (Dunn, 1959; Rich, 1964; Siegenthaler and Oeschger, 1978), the effects of these gases on belowground processes such as root growth, soil C storage, and mycorrhizal productivity remain poorly resolved (Anderson, 2003; Hungate et al., 2009; Nikolova et al., 2010; Norby and Zak, 2011; King et al., 2013, this vol.; Kraigher et al., 2013, this vol.; Matyssek et al., 2013, this vol.). In part, this is because many belowground processes remain at the frontier of scientific research (Brantley et al., 2011) and some fundamental properties of belowground systems, such as the turnover rate of fine roots (Gaudinski et al., 2010) and the composition of soil organic matter (Schmidt et al., 2011), are still actively debated.

Elevated CO_2 and elevated O_3 are each thought to affect belowground processes indirectly, with direct effects on leaf-level processes creating changes in plant physiology and productivity and shifting in belowground C inputs (e.g. Anderson, 2003; Zak et al., 2000; King et al., 2013, this vol.). This cascade

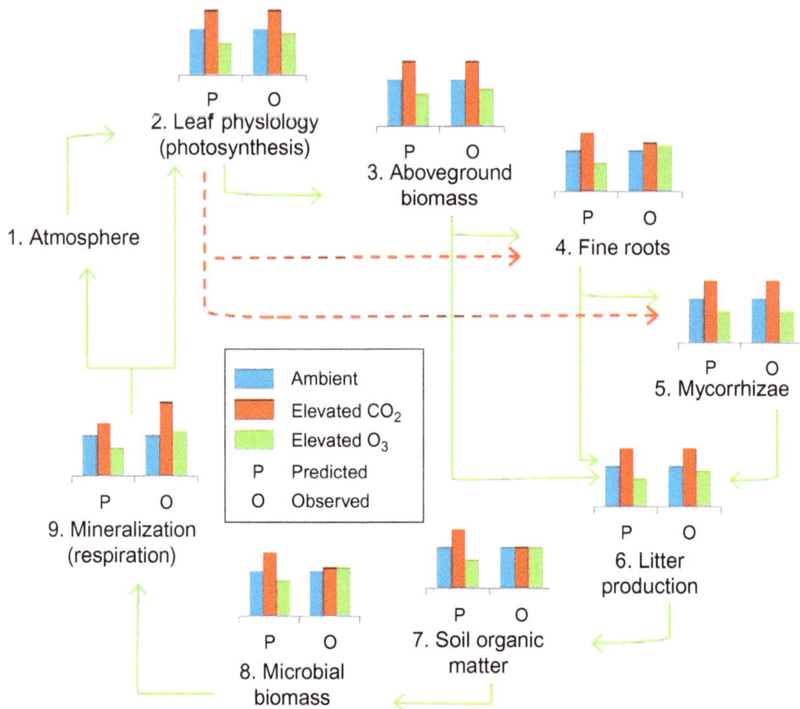

FIGURE 10.1 Conceptual diagram of primary belowground processes. Bar graphs are qualitative and illustrate differences between ambient and the main effects of CO_2 and O_3. Arrows represent directional connections among pools and fluxes, with solid arrows representing predicted connections and the dashed arrows representing potential new connections based on the results from Aspen FACE.

of effects is shown in Figure 10.1 and begins with changes in atmospheric composition (1) that impact leaf-level physiological processes, particularly photosynthesis (2). These changes in photosynthesis often affect aboveground growth (3), which creates parallel impacts on fine roots (4) and mycorrhizae (5). Fine roots, mycorrhizae, and aboveground growth contribute to the flux of litter into the soil (6), both through mortality and exudation. Litter inputs form soil organic matter (7), which is the substrate for heterotrophic microorganisms (8). Fine roots, mycorrhizae, and soil heterotrophs contribute to the mineralization (9) of organic compounds, releasing CO_2 through respiration and making nitrogen available for plant uptake.

Many of the early studies on the effects of elevated CO_2 or O_3 on plant growth were conducted in greenhouses or growth chambers (Ceulemans and Mousseau, 1994; Karnosky et al., 2005) and consequently, belowground processes such as root growth were often constrained by soil volume (Ceulemans and Mousseau, 1994). Scaling results from tree-level chamber studies to the ecosystem is also difficult (Ceulemans and Mousseau, 1994; Pye, 1988). For instance, there is wide inter-specific variation in growth in response to CO_2 (Bazzaz et al., 1990; Tolley and Strain, 1984) and O_3 (e.g. Dunn, 1959; Karnosky and Steiner, 1981; Tingey and Reinert, 1975), but the influence of this variation on ecosystem function is difficult to predict (Bradley and Pregitzer, 2007).

The invention of Free-Air CO_2 Enrichment (FACE) technology in the early 1990s (Hendrey et al., 1999) made it possible to expose whole forest stands to elevated concentrations of CO_2, reducing limitations on experimental scale and allowing more accurate characterizations of soil–atmosphere linkages. However, insights into competitive interactions under elevated CO_2 were limited because FACE technology was predominately applied to established forests, and particularly to single-species plantations (Norby and Zak, 2011). In addition, although FACE technology has been adapted to O_3 research (Matyssek et al., 2010; Matyssek et al., 2013, this vol.), ecosystem-scale experiments quantifying the effects of O_3 or the interactive effects of CO_2 and O_3 on forests are rare (Ainsworth et al., 2012; King et al., 2013, this vol.).

10.2 THE ASPEN FACE EXPERIMENT

In this context, the Aspen FACE experiment was established in the north-central United States (45.61°N, 89.51°W). This experiment was designed to test the influence of competitive interactions on the response of developing stands of trembling aspen (*Populus tremuloides* Michaux) to CO_2 and/or O_3 in a factorial randomized complete block design. This species of aspen is the most widely distributed tree in North America and common in several regions subject to high O_3 exposures (Karnosky et al., 2003). At Aspen FACE, aspen was grown in either mixed species communities at equal densities with birch (*Betula papyrifera* Marsh.) or maple (*Acer saccharum* Marsh.)

TABLE 10.1 Pre-Treatment (1997) Soil Physical and Chemical Properties Within the Top 10 cm of Mineral Soil

	Mean	SE
Soil texture		
Sand (%)	55.6	0.9
Silt (%)	36.6	0.8
Clay (%)	7.8	0.4
Gravimetric moisture content (θ at water holding capacity)	0.11	0.00
Bulk density (mg m^{-2})	1.33	0.03
Total C (mg g^{-1})	15.30	0.84
Total N (mg g^{-1})	1.19	0.07
C:N	12.9	0.2
Extractable P (μg g^{-1})	136.8	6.0
Cation exchange capacity (cmol(+)kg^{-1})	0.54	0.06
Base saturation (%)	57.5	5.5

Values are means with standard error (SE).

or in mixed genotype communities (five clones varying in sensitivity to CO_2 or O_3), all representing common forest types in the region. All communities were planted with 1 m spacing in July 1997; fumigation began shortly thereafter and continued until mid-2009. During the 11 full growing seasons of this experiment, average fumigation concentrations were 532 μl l^{-1} for CO_2 (+44% relative to ambient) and 46 nl l^{-1} for O_3 (+30%; Zak et al., 2011) and the forest communities grew from small trees (<0.25 m tall) to dense stands (>8 m tall). Soils (Table 10.1) are Alfic Haplorthods with a sandy loam Ap horizon overlaying a sandy clay loam Bt horizon (Dickson et al., 2000).

Here, we use the Aspen FACE experiment as a comprehensive case study to test the conceptual framework of how CO_2 and O_3 affect belowground C cycling. We briefly summarize the changes in aboveground productivity and then follow the path of C belowground, from fine roots and mycorrhizae to free-living soil microorganisms and soil respiration. We conclude by returning to the conceptual framework and putting the results into a broader scientific context.

10.2.1 Net Primary Productivity and Community Composition

Over the first 3 years of the experiment, these forest communities responded to CO_2 and O_3 as predicted: elevated CO_2 increased net primary productivity

(NPP), while elevated O_3 reduced NPP (King et al., 2005). These changes were consistent with leaf-level measurements of light-saturated photosynthesis, which were higher under elevated CO_2 and lower under elevated O_3 (Darbah et al., 2010). However, the effects of elevated O_3 were smaller and less consistent than the effects of elevated CO_2 (Kets et al., 2010). Interestingly, the negative effect of O_3 on NPP began to diminish after 3–5 years (depending on the community; King et al., 2005) and during the final 3 years of the experiment, elevated O_3 no longer reduced NPP (Zak et al., 2011). The recovery in NPP may be a consequence of compensatory growth of O_3-tolerant plants in response to the suppression of O_3-sensitive species and genotypes (Zak et al., 2011). For instance, exposure to elevated O_3 and ambient CO_2 increased the relative importance of birch and maple over aspen (Kubiske et al., 2007). However, the effect of elevated O_3 on leaf litter production (g m^{-2}) remained relatively constant over the last 7 years of the experiment, averaging −13% (Talhelm et al., 2012). In contrast, elevated CO_2 had a sustained positive effect on NPP (+25% in 2008; Zak et al., 2011) and leaf litter production (+32% in 2008; Talhelm et al., 2012).

10.2.2 Fine Roots Dynamics

In these developing forests, fine root (<1 mm in diameter) biomass increased quickly during the first half of the experiment before stabilizing over the last 4 years of the experiment (Figure 10.2). Within this overall pattern, the response of fine root biomass to the treatments was highly dynamic. In the

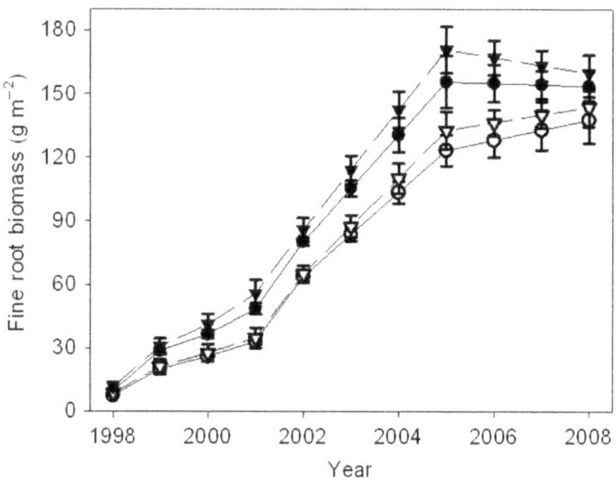

FIGURE 10.2 Fine root (<1 mm in diameter) biomass during the Aspen FACE experiment. Data from estimates in King et al. (2005), Pregitzer et al. (2006, 2008), and Zak et al. (2011) for the top 25 cm of mineral soil. Circles represent ambient O_3, triangles represent elevated O_3, empty symbols represent ambient CO_2, and filled symbols represent elevated CO_2. Error bars are ±1 standard error.

initial sampling (1999), elevated CO_2 nearly doubled fine root biomass in the top 10 cm of mineral soil, but elevated O_3 did not have a significant effect (King et al., 2001). In subsequent samplings in 2000 and 2002 (to 25 cm in depth; Figure 10.2), elevated CO_2 created increases in fine root biomass that were smaller but consistent across all community types (averaging +37%), while the response to elevated O_3 was community-specific (King et al., 2005; Pregitzer et al., 2008; Figure 10.3). Elevated O_3 decreased fine root biomass within the aspen-birch and aspen-maple communities in these two samplings, but O_3 unexpectedly stimulated fine root biomass by more than 30% within the aspen-only community (King et al., 2005; Pregitzer et al., 2008). The response of fine root biomass to elevated CO_2 was still positive in 2005 and 2008 (Zak et al., 2011), but this stimulation shrank further to 27% in 2005 and 11% in 2008. The effect of elevated O_3 was positive across all communities in 2008 (+4%), but the stimulation of fine root biomass in the aspen-only community was smaller (+9%) than in previous estimates. Less is known about the response of roots deeper in the soil. However, Rhea and King (2012) sampled fine roots in the aspen-only and aspen-birch communities to 1 m depth in 2005 and found that neither treatment significantly affected fine root biomass below the top 30 cm of soil.

It is interesting to contrast leaves and fine roots, the two most ephemeral major elements of plant biomass. Like roots, leaves initially displayed both rapid growth and large treatment effects, followed by a period of more subtle growth and smaller treatment effects (King et al., 2005; Talhelm et al., 2012). Leaf and fine root biomass both responded positively to elevated CO_2 in all three communities and negatively to O_3 in the aspen-birch and aspen-maple communities (Figure 10.2), but the positive response of fine roots to elevated O_3 in the aspen community was not observed for leaves (Figure 10.2). Although this divergent response was unique among the three communities, the negative response to elevated O_3 was consistently stronger for leaf biomass than for fine root biomass in all three communities (Figure 10.3). Similarly, the ratio of leaf area to fine root area had a positive, though less consistent, response to O_3 when measured in 2005 (Rhea and King, 2012).

The production and mortality of fine roots in the aspen community was estimated using minirhizotron observations from 2002 to 2005 (Pregitzer et al., 2008). Production rates (mm mm^{-1}) were not affected by CO_2 or O_3, but elevated O_3 significantly increased root mortality rates in 1 year (2003). Consequently, changes in production and mortality biomass (g m^{-2} $year^{-1}$) were mostly a function of changes in standing biomass (Pregitzer et al., 2008). These responses contrast with observations of leaf life span in the aspen-only community, in which leaf abscission was delayed by elevated CO_2 and unaffected by O_3 (Riikonen et al., 2008; Taylor et al., 2008).

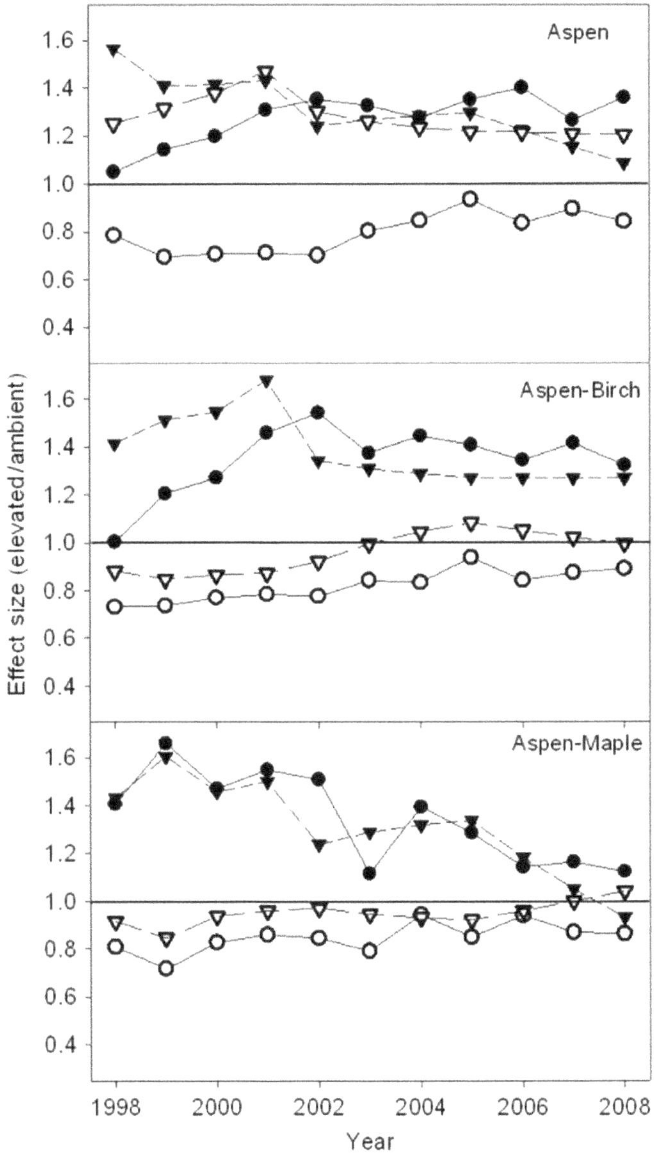

FIGURE 10.3 Relative treatment effect size (elevated/ambient) for leaf biomass and fine root (<1 mm in diameter) biomass for each community. Leaf biomass effects from King et al. (2005) and Talhelm et al. (2012); fine root data sources as in Figure 10.2. Triangles represent fine roots, circles represent leaves, empty symbols represent O_3 effects, and filled symbols represent CO_2 effects. Horizontal lines represent no treatment response.

10.2.3 Mycorrhizal Fungi

There are no estimates of the mycorrhizal biomass or hyphal turnover rates. However, insight into the influence of CO_2 and O_3 on mycorrhizae is available from observations of ectomycorrhizal sporocarps (2003–2006) within the aspen-only and aspen-birch communities (Andrew and Lilleskov, 2009) and analyses of fungal ribosomal DNA within the mineral soil in 2002 (Lesaulnier et al., 2008) and 2007 (Edwards and Zak, 2011). Elevated CO_2 increased overall sporocarp production by an average of 96% in the aspen-only community and 70% in the aspen-birch community, principally by increasing the abundance of late-successional ectomycorrhizal taxa. Elevated CO_2 also increased the species richness of these sporocarps. Notably, elevated CO_2 increased the abundance of the early-successional ectomycorrhizal taxa *Inocybe* in the 2002 molecular analyses (Lesaulnier et al., 2008), but this change was not evident in 2007 (Edwards and Zak, 2011).

Elevated O_3 did not affect sporocarp species richness and only affected sporocarp production and community composition under ambient CO_2 (Andrew and Lilleskov, 2009). Here, elevated O_3 decreased sporocarp production by an average of 12% in the aspen-birch community and, in contrast to the response of fine root biomass, decreased sporocarp production by 48% in the aspen-only community. Elevated O_3 affected sporocarp community composition under ambient CO_2 by increasing the abundance of early-successional taxa such as *Inocybe* and *Paxillus*. Assessments of fungal community composition in 2007 based on ribosomal DNA abundance provided more evidence that elevated O_3 had increased the abundance of *Inocybe*, but in general, O_3 effects on ectomycorrhizal community structure in these molecular analyses were small and independent of CO_2 (Edwards and Zak, 2011). In addition to *Inocybe*, ectomycorrhizal communities within the soil under elevated O_3 had relatively more species of *Cortinarius* and fewer *Laccaria* and *Tomentella* species (Edwards and Zak, 2011). Although Andrew and Lilleskov (2009) found more early-successional species under elevated O_3 and more late-successional species under elevated CO_2, there were clear differences in sporocarp community composition between the ontogenetically similar ambient and $+CO_2+O_3$ treatments.

10.2.4 Soil Respiration

For the first 7 years of the experiment, soil respiration was measured only in the two most productive communities, aspen-birch and aspen-only. Elevated CO_2 stimulated soil respiration in these two communities throughout the first seven growing seasons of the experiment (1998–2005), but the size of this stimulation varied by community, year, and season (King et al., 2001, 2004; Pregitzer et al., 2006). For instance, the annual CO_2 effect during the first 4 years ranged from +13% to +41% within the aspen-only community and from +43% to +77% within the aspen-birch community (King et al., 2004). Elevated CO_2 continued to have a smaller effect on soil respiration in the aspen-only community than the aspen-birch community throughout the first decade of the experiment (Pregitzer et al., 2006, 2008) and effect of elevated CO_2 was not significant within the aspen-only community in 2002. Here, soil respiration rates were similar in the aspen-only and aspen-birch communities under elevated CO_2, but soil respiration tended to be higher under ambient CO_2 in the aspen-only community (Pregitzer et al., 2006, 2008). Later in the experiment (2005–2007), soil respiration was also quantified within the aspen-maple community. Patterns in growing season soil respiration in this community were similar to those in the aspen-birch community in terms of the size of the overall soil CO_2 flux and the relative effects of elevated CO_2. During these 3 years, the annual stimulation across all three communities of growing seasonal soil respiration by elevated CO_2 ranged from 25% to 31% (Pregitzer et al., 2008). In general, the magnitude of the CO_2 stimulation varied seasonally; it was strongest in the middle of the growing season when soil respiration was greatest and weaker during the spring and fall when soil temperatures were low (King et al., 2001; Pregitzer et al., 2006).

Elevated O_3 had much less consistent effects on soil respiration. There was no significant O_3 effect on soil respiration in 1998, but in 1999 elevated O_3 caused a 20% decrease in soil respiration late in the growing season (August to November; King et al., 2001). Elevated O_3 significantly decreased soil respiration within the aspen-only community in 2002, but in the two subsequent years there was a significant $CO_2 \times O_3$ interaction ($P < 0.01$) in both communities and the effect of elevated O_3 became positive under elevated CO_2 (Pregitzer et al., 2006). Rates of soil respiration were consistently 5–10% greater in the $+CO_2+O_3$ treatment than in the $+CO_2$ treatment from 2005 to 2007, but the $CO_2 \times O_3$ interaction was not significant because elevated O_3 often only slightly decreased soil respiration under ambient CO_2. This slight positive effect of O_3 was most consistent in the aspen-only community, matching the response of fine root biomass (Pregitzer et al., 2008). In fact, soil respiration rates in 2005 were significantly correlated with fine root biomass across treatments and communities ($r = 0.72$; $P = 0.008$; Pregitzer et al., 2008).

In addition to increasing the flux of CO_2 from the soil, elevated CO_2 also increased the partial pressure of CO_2 within the soil (Karberg et al., 2005; King et al., 2001). The increase appears to have resulted primarily from greater rhizosphere C fluxes rather than changes in soil moisture or temperature that would reduce the diffusion of CO_2 (Karberg et al., 2005; King et al., 2001). Additional CO_2 within the soil is important because it can dissolve into the soil solution and increase the rate of two key processes: soil weathering and long-term C sequestration within oceanic carbonates (Kump et al., 2000). In measurements throughout the 2002 growing season, Karberg et al.

(2005) observed that increases in the partial pressure of soil CO_2 resulted in a 22% increase in dissolved inorganic carbon (DIC) and a 14% increase in the concentration of carbonic acid in soil solution. Soil solution alkalinity also increased by 210% (Karberg et al., 2005), likely due to an enhancement of primary mineral weathering by carbonic acid. Although elevated O_3 did not affect either the soil partial pressure of CO_2 or the concentration of DIC, we note that DIC concentrations in 2002 followed the same rank order as in soil respiration measurements made in subsequent years (2003–2004; Pregitzer et al., 2006): $+CO_2+O_3 > +CO_2 >$ ambient $> +O_3$. In contrast to the changes in DIC, measurements of dissolved organic carbon during the first 2 years of the experiments found no clear effect of elevated CO_2 or O_3 (King et al., 2001).

10.2.5 Soil Organic Matter

In addition to changing the amount of fine root and leaf litter, CO_2 and O_3 also influenced the chemical composition of plant litter and low molecular-weight organic compounds in the surface mineral soil. In the mineral soil, exposure to elevated CO_2 and ambient O_3 created a marginally significant ($P=0.08$) increase in sugar concentrations (approximately +10%) and had a larger effect on soluble phenolics ($P=0.07$, approximately +40%), but there were no other treatment effects on sugar or soluble phenolic concentrations (Johnson and Pregitzer, 2007). These differences contrast with changes in plant litter chemistry. Concentrations of non-structural carbohydrates (sugars and starch) and structural carbohydrates (cellulose and hemi-cellulose) in fine roots collected in 2002 were not affected by the treatments (Chapman et al., 2005). Concentrations of sugars and soluble phenolics in leaf litter increased in response to CO_2 (Liu et al., 2005; Parsons et al., 2008), but the greatest increases tended to be in the interaction ($+CO_2+O_3$) treatment rather than with exposure to elevated CO_2 alone ($+CO_2$). Soil amino acid concentrations responded individualistically to CO_2 and to O_3. Elevated CO_2 increased aspartic acid concentrations, while elevated O_3 decreased arginine and methionine concentrations. Elevated O_3 decreased valine concentrations under ambient CO_2, but had no effect under elevated CO_2. The 11 other amino acids assayed were unaffected by the treatments (Johnson and Pregitzer, 2007).

Averaged across all communities and treatments, soil C content within the top 20 cm of mineral soil increased from 3350 ± 230 g m^{-2} when first measured in 2001 to 4010 ± 160 g m^{-2} in 2008 (Talhelm et al., 2009). Although elevated CO_2 consistently increased NPP (King et al., 2005; Zak et al., 2011), fine root production, and leaf litter inputs (Talhelm et al., 2012), there was no increase in soil C content (Figure 10.3). Soil C content was not significantly affected by elevated CO_2 in the aspen-birch and aspen-maple communities. In the aspen-only community, soil C accrued more slowly under elevated CO_2 than under ambient CO_2 during the last several years of the experiment, creating a large and significant negative effect ($-1740\ g\ m^{-2}$) of elevated CO_2 during the final year of the experiment ($P=0.003$; Figure 10.3). Likewise, although elevated O_3 reduced NPP early in the experiment, it did not affect soil C content. The mass of the soil organic horizon (forest floor) was somewhat greater under elevated CO_2 and smaller under elevated O_3 in both 2004 and 2008, but these differences were not significant (Zak et al., 2011). The lack of soil C accumulation under elevated CO_2 contrasts with studies of leaf litter and fine root decomposition (Chapman et al., 2005; Liu et al., 2009; Parsons et al., 2008), wherein elevated CO_2 either slowed decomposition (Chapman et al., 2005; Parsons et al., 2008) or had no significant effect (Liu et al., 2009). For elevated O_3, litter decomposition studies had mixed results: increased decomposition (Parsons et al., 2008), decreased decomposition (Liu et al., 2009; Parsons et al., 2008), or no effect (Chapman et al., 2005).

Insight into the changes in C cycling within the mineral soil can be gained by isolating and quantifying different fractions of soil organic matter. Hofmockel et al. (2011) partitioned soil organic matter sampled from the top 20 cm of mineral soil in 2003, 2004, and 2007 into several physical fractions: coarse particulate ($>250\ \mu m$), fine particulate (250–53 μm), and mineral-associated ($<53\ \mu m$). The majority of soil C was found in the mineral-associated fraction, nearly $3000\ g\ m^{-2}$. The pools of coarse particulate and fine particulate organic matter were considerably smaller, with each accounting for approximately $500\ g\ m^{-2}$ of C. Although elevated CO_2 did not have significant effects on any of these fractions overall, it did affect how the pools of coarse particulate and mineral-associated soil C changed through time. Coarse particulate C pools increased faster under elevated CO_2 and were 19% greater under elevated CO_2 in 2007. However, this increase was matched by a trend towards decreased mineral-associated C as the effect of elevated CO_2 changed from +7% in 2003 to −1% in 2007. Together, these trends suggest that while there is no significant overall effect on soil C content, the processing of soil C has been altered by elevated CO_2 (Hofmockel et al., 2011). The effects of elevated CO_2 on coarse particulate soil C also varied significantly by community, with a 6% decrease within the aspen-only community and a 23% increase in the aspen-birch community. There was a similar pattern in the pool of mineral-associated soil C, but these differences were not significant. Elevated O_3 did not significantly affect the amount of C in these three soil fractions, but did increase the C/N ratio and decrease the pool of soil N within the coarse particulate and fine particulate fractions.

The fossil-fuel derived CO_2 used to create the elevated CO_2 treatment is heavily depleted in ^{13}C. This depletion is reflected in the $\delta^{13}C$ of plant tissue and creates an opportunity to distinguish 'new' C added to the soil since the beginning of the fumigation and 'old' pre-existing C. In the initial survey of these pools within the aspen-only and aspen-birch communities in 2001, the amount of total new soil C formed under elevated CO_2 was 51% lower under elevated O_3 ($+CO_2+O_3$ vs. $+CO_2$; Loya et al., 2003). Using a chemical fractionation technique, Loya et al. (2003) also found that approximately one-third of this new C was recalcitrant acid-insoluble organic matter and that elevated O_3 reduced the formation of new acid-insoluble C by 48%. When these soils were incubated, more C was respired in the interaction ($+CO_2+O_3$) treatment than under elevated CO_2 alone, suggesting that the changes in soil C pools were a product of increased C losses rather than decreased C inputs (Loya et al., 2003). However, the negative effect of elevated O_3 on soil C formation under elevated CO_2 was considerably weaker within a wider analysis that included all three communities and measurements from 2001 to 2008 (Talhelm et al., 2009). Elevated O_3 did not significantly decrease the amount of new soil C formed under elevated CO_2 in the aspen-maple community in 2001 and the negative effects of O_3 within the aspen-birch and aspen-maple communities weakened in subsequent measurements and were no longer significant. The input of new soil C under elevated CO_2 was evident within each of the three physical soil C fractions, but inputs of new C to the fine particulate and mineral-associated soil C were offset by loss of old soil C (Hofmockel et al., 2011). Elevated O_3 reduced the fraction of new C somewhat within each of the three fractions, but not significantly. However, elevated O_3 slightly, but significantly, increased the amount of old C within the coarse particulate and fine particulate soil C pools (Hofmockel et al., 2011). Although this isotopic technique cannot be used to determine why elevated CO_2 reduced total soil C content within the aspen-only community, it is notable that the largest loss of pre-existing ‘old’ soil C under elevated CO_2 occurred within this community (Talhelm et al., 2009).

10.2.6 Soil Microorganisms and Extracellular Enzymes

The lack of soil C accumulation under elevated CO_2 was surprising, but consistent with observations of soil extracellular enzymes. Cellobiohydrolase and *N*-acetylglucosaminidase are extracellular enzymes important in the microbial breakdown of cellulose and chitin, respectively, which are two of the most abundant biomolecules within soil. Early in the experiment (1999), when the relative effects of elevated CO_2 on fine roots and leaf litter were large, elevated CO_2 stimulated the activity of cellobiohydrolase and *N*-acetylglucosaminidase by approximately 50% (Larson et al., 2002). A similar stimulation of these enzymes by elevated CO_2 was also observed in 2001 and 2002 (Chung et al., 2006). These observations were consistent with laboratory incubations in which elevated CO_2 increased the respiration of added ^{13}C-labelled substrates for these enzymes (Phillips et al., 2002). Interestingly, the effect of elevated CO_2 on cellobiohydrolase was considerably stronger under ambient O_3 than under elevated O_3 in both the field and the laboratory incubation, but this was not apparent for *N*-acetylglucosaminidase

(Edwards and Zak, 2011; Phillips et al., 2002). The activities of other soil extracellular enzymes used to breakdown soil organic matter, including phosphatase, peroxidase, 1,4-α-glucosidase, 1,4-β-glucosidase, 1,4-β-xylosidase, and1,4-β-*N*-acetylglucose-aminidase also tended to be higher under elevated CO_2 early in the experiment, but these differences were sometimes not significant (Chung et al., 2006; Larson et al., 2002).

As the relative effects of elevated CO_2 on fine roots and leaf litter declined, the stimulation of cellobiohydrolase and *N*-acetylglucosaminidase by elevated CO_2 also declined and was no longer significant by 2008 (Edwards and Zak, 2011). The shrinking effect of elevated CO_2 on *N*-acetylglucosaminidase resulted from an increase in enzyme activity under ambient CO_2 rather than a decrease under elevated CO_2, but the changes through time in cellobiohydrolase were less consistent (Edwards and Zak, 2011). Elevated O_3 had a smaller impact on extracellular enzymes. Elevated O_3 reduced cellobiohydrolase activity within the Oe horizon by 17% in 2008 (Edwards and Zak, 2011) and reduced the activity of 1,4-β-glucosidase by 25% in the surface mineral soil in 2001–2002 (Chung et al., 2006). Matching the inconsistent positive response of some enzymes to CO_2, a similar group of enzymes tended to decrease under elevated O_3.

Elevated CO_2 and elevated O_3 each increased microbial (Larson et al., 2002) and fungal biomass (Chung et al., 2006), but not significantly. Although plant community type and soil horizon appear to be the dominant controls, elevated CO_2 and O_3 did influence microbial community composition within the mineral and organic soil (Chung et al., 2006; Edwards and Zak, 2011). For instance, elevated CO_2 appears to increase the abundance of *Sistotrema* fungi, while elevated O_3 shifted the composition of agaricomycete fungi (Edwards and Zak, 2011). Elevated CO_2 also increased bacterial taxonomic richness (Dunbar et al., 2012; Lesaulnier et al., 2008), but decreased archaeal diversity and reduced the abundance of archaeal DNA by 50% (Lesaulnier et al., 2008). There is evidence that elevated CO_2 altered the abundance of a number of individual bacterial and archaeal taxa (Dunbar et al., 2012; Lesaulnier et al., 2008). For instance, of 1313 individual bacterial taxa identified at Aspen FACE, 75 taxa significantly increased or decreased in abundance in response to elevated CO_2 (Dunbar et al., 2012). However, the functional roles of individual taxa are poorly defined (Dunbar et al., 2012).

10.3 CONCLUSIONS AND IMPLICATIONS

Belowground C cycling responded dynamically to the CO_2 and O_3 treatments within the developing forests at Aspen FACE. Often, these processes were consistent with changes in aboveground productivity (Figure 10.3). For instance, elevated CO_2 increased photosynthesis and aboveground productivity, resulting in concomitant increases in fine root biomass, soil respiration, the partial pressure of soil CO_2 and DIC. However, some responses diverged from expectations in several important ways, particularly for O_3 (Figure 10.1). First, elevated O_3 reduced leaf biomass more than fine root biomass in all communities, contrasting with reviews that have found decreased root:shoot ratios under elevated O_3 (Anderson, 2003; Grantz et al., 2006; Wittig et al., 2009). Further, elevated O_3 stimulated fine root biomass in the aspen-only community and stimulated soil respiration in all communities during the latter part of the experiment. A positive response of both fine root biomass and soil respiration was found in another free-air forest O_3 experiment (Nikolova et al., 2010; Matyssek et al., 2013, this vol.) and, in general, the response of fine root biomass to O_3 is highly variable in developing forests (King et al., this vol.). Likewise, mycorrhizal sporocarp production responded differently to CO_2 or O_3 than did fine roots, sometimes in the opposite direction. Finally, changes in productivity were not reflected in soil C content (Figure 10.4), as extracellular enzyme activity changed in parallel with litter inputs.

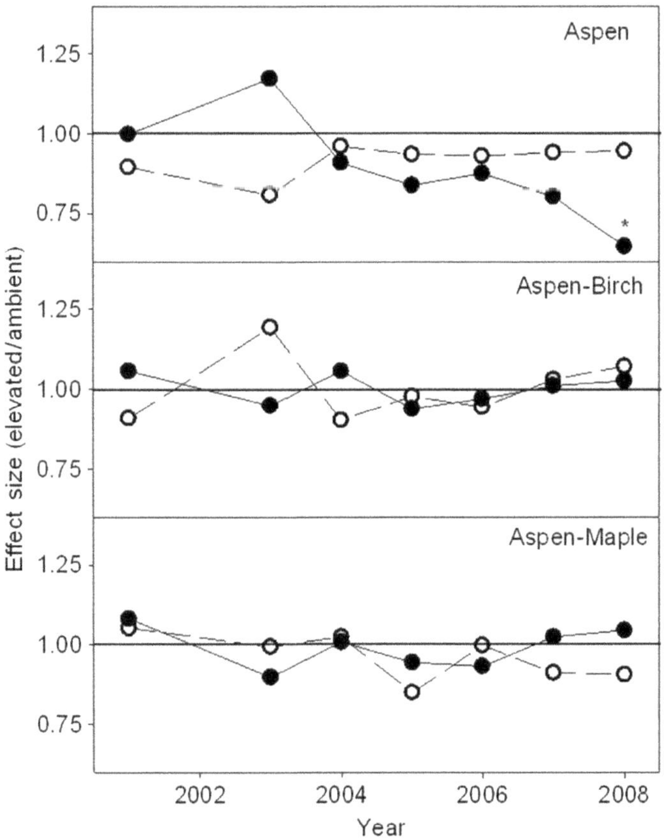

FIGURE 10.4 Relative treatment effect sizes for soil carbon content for each community. Empty symbols represent O_3 effects and filled symbols represent CO_2 effects. Significant ($P < 0.05$) CO_2 effect denoted with asterisk.

Overall, the differences between the predicted and observed effects on belowground processes illustrate the gaps in our knowledge of these processes and the value in long-term field experiments. Unfortunately, similar experiments (in scale and duration) are too few in number to know if the responses of belowground processes at Aspen FACE are universal or exceptional. Because of the large role of belowground processes in terrestrial C cycling (Bond-Lamberty and Thomson, 2010; Litton et al., 2007; Pregitzer and Euskirchen, 2004), accurate predictions of future terrestrial productivity and C sequestration will require further long-term experiments to achieve a more sophisticated understanding of how these processes respond to air pollution, climate change, and other global change factors. Without accurate predictions of future forest C cycling, the ability to design effective climate policy will be limited.

ACKNOWLEDGEMENTS

The Aspen FACE experiment was supported by funding from the U.S. Department of Energy and by the USDA Forest Service. We sincerely thank Donald Zak, Dave Karnosky, Andrew Burton, Jud Isebrands, George Hendrey, Richard Dickson, John Nagy, and Keith Lewin for their invaluable contributions.

REFERENCES

Ainsworth, E.A., Yendrek, C.R., Sitch, S., Collins, W.J., Emberson, L.D., 2012. The effects of tropospheric ozone on net primary productivity and implications for climate change. Annu. Rev. Plant Biol. 63, 637–661.

Anderson, C.P., 2003. Source–sink balance and carbon allocation below ground in plants exposed to ozone. New Phytol. 157, 213–228.

Andrew, C.A., Lilleskov, E.A., 2009. Productivity and community structure of ectomycorrhizal fungal sporocarps under increased atmospheric CO_2 and O_3. Ecol. Lett. 12, 813–822.

Bazzaz, F.A., Coleman, J.S., Morse, S.R., 1990. Growth responses of seven major co-occurring tree species of the northeastern United States to elevated CO_2. Can. J. Forest Res. 20, 1479–1484.

Bond-Lamberty, B., Thomson, A., 2010. Temperature-associated increases in the global soil respiration record. Nature 464, 579–582.

Bradley, K.L., Pregitzer, K.S., 2007. Ecosystem assembly and terrestrial carbon balance under elevated CO_2. Trends Ecol. Evol. 22, 538–547.

Brantley, S.L., Megonigal, J.P., Scatena, F.N., Balogh-Brunstad, Z., Barnes, R.T., et al., 2011. Twelve testable hypotheses on the geobiology of weathering. Geobiology 9, 140–165.

Ceulemans, R., Mousseau, M., 1994. Effects of elevated atmospheric CO_2 on woody plants. New Phytol. 127, 425–446.

Chapman, J.A., King, J.S., Pregitzer, K.S., Zak, D.R., 2005. Effects of elevated concentrations of atmospheric CO_2 and tropospheric O_3 on decomposition of fine roots. Tree Physiol. 25, 1501–1510.

Chung, H., Zak, D.R., Lilleskov, E.A., 2006. Fungal community composition and metabolism under CO_2 and O_3. Oecologia 147, 143–154.

Darbah, J.N.T., Kubiske, M.E., Nelson, N., Kets, K., Riikonen, J., et al., 2010. Will photosynthetic capacity of forest trees acclimate after long-term exposure to elevated CO_2 and O_3? Environ. Pollut. 158, 983–991.

Dickson, R.E., Lewin, K.F., Isebrands, J.G., Coleman, M.D., Heilman, W.E., et al., 2000. Forest atmosphere carbon transfer and storages (FACTS-II): the aspen free-air CO_2 and O_3 enrichment (FACE) project: an overview. General Technical Report NC-214. USDA Forest Service North Central Research Station, St. Paul, MN, USA.

Dunbar, J., Eichorst, S.A., Gallegos-Graves, L.V., Silva, S., Xie, G., et al., 2012. Common bacterial responses in six ecosystems exposed to 10 years of elevated atmospheric carbon dioxide. Environ. Microbiol. 14, 1145–1158.

Dunn, D.B., 1959. Some effects of air pollution on Lupinus in the Los Angeles area. Ecology 40, 621–625.

Edwards, I.P., Zak, D.R., 2011. Fungal community composition and function after long-term exposure of northern forests to elevated atmospheric CO_2 and tropospheric O_3. Glob. Chang. Biol. 17, 2184–2195.

Gaudinski, J.B., Torn, M.S., Riley, W.J., Dawson, T.E., Joslin, J.D., Majdi, H., 2010. Measuring and modeling the spectrum of fine-root turnover times in three forests using isotopes, minirhizotrons, and the Radix model. Global Biogeochem. Cycles 24, GB3029.

Grantz, D.A., Gunn, S., Vu, H.-B., 2006. O_3 impacts on plant development: a meta-analysis of root/shoot allocation and growth. Plant Cell Environ. 29, 1193–1209.

Hendrey, G.R., Ellsworth, D.S., Lewin, K.F., Nagy, J., 1999. A free-air enrichment system for exposing tall forest vegetation to elevated atmospheric CO_2. Glob. Chang. Biol. 5, 293–309.

Hofmockel, K.S., Zak, D.R., Moran, K.K., Jastrow, J.D., 2011. Changes in forest soil organic matter pools after a decade of elevated CO_2 and O_3. Soil Biol. Biochem. 43, 1518–1527.

Hungate, B.A., Van Groenigen, K.J., Six, J., Jastrow, J.D., Luo, Y., et al., 2009. Assessing the effect of elevated carbon dioxide on soil carbon: a comparison of four meta-analyses. Glob. Chang. Biol. 15, 2020–2034.

Johnson, R.M., Pregitzer, K.S., 2007. Concentration of sugars, phenolic acids, and amino acids in forest soils exposed to elevated atmospheric CO_2 and O_3. Soil Biol. Biochem. 39, 3159–3166.

Karberg, N., Pregitzer, K.S., King, J.S., Friend, A.L., Wood, J.R., 2005. Soil carbon dioxide partial pressure and dissolved inorganic carbonate chemistry under elevated carbon dioxide and ozone. Oecologia 142, 296–306.

Karnosky, D.F., Steiner, K.C., 1981. Provenance and family variation in response of *Fraxinus americana* and *Fraxinus pennsylvanica* to ozone and sulfur dioxide. Phytopathology 71, 804–807.

Karnosky, D.F., Zak, D.R., Pregitzer, K.S., Awmack, C.S., Bockheim, J.G., et al., 2003. Tropospheric O_3 moderates responses of temperate hardwood forests to elevated CO_2: a synthesis of molecular to ecosystem results from the Aspen FACE project. Funct. Ecol. 17, 289–304.

Karnosky, D.F., Pregitzer, K.S., Zak, D.R., Kubiske, M.E., Hendrey, G.R., Weinstein, D., Nosal, M., Percy, K.E., 2005. Scaling ozone responses of forest trees to the ecosystem level in a changing climate. Plant Cell Environ. 28, 965–981.

Kets, K., Darbah, J.N.T., Sober, A., Riikonen, J., Sober, J., Karnosky, D.F., 2010. Diurnal changes in photosynthetic parameters of Populus tremuloides, modulated by elevated concentrations of CO_2 and/or O_3 and daily climatic variation. Environ. Pollut. 158, 1000–1007.

King, J., Liu, L., Aspinwall, M., 2013. Forest ecosystem response to interacting elevated atmospheric CO_2 and tropospheric O_3: a synthesis of experimental evidence. Dev. Environ. Sci. 13, 179–208.

King, J.S., Pregitzer, K.S., Zak, D.R., Sober, J., Isebrands, J.G., Dickson, R.E., Hendrey, G.R., Karnosky, D.F., 2001. Fine-root biomass and fluxes of soil carbon in young stands of paper birch and trembling aspen as affect by elevated atmospheric CO_2 and tropospheric O_3. Oecologia 128, 237–250.

King, J.S., Hanson, P.J., Bernhardt, E., DeAngelis, P., Norby, R.J., Pregitzer, K.S., 2004. A multi-year synthesis of soil respiration responses to elevated atmospheric CO_2 from four forest FACE experiments. Glob. Chang. Biol. 10, 1027–1042.

King, J.S., Kubiske, M.E., Pregitzer, K.S., Hendrey, G.R., McDonald, E.P., Giardina, C.P., Quinn, V.S., Karnosky, D.F., 2005. Tropospheric O_3 compromises net primary production in young stands of trembling aspen, paper birch and sugar maple in response to elevated atmospheric CO_2. New Phytol. 168, 623–636.

Kraigher, H., Bajc, M., Grebanc, T., Pritsch, K., 2013. Mycorrhizosphere complexity. Dev. Environ. Sci. 13, 151–177.

Kubiske, M.E., Quinn, V.S., Marquardt, P.E., Karnosky, D.F., 2007. Effects of elevated atmospheric CO_2 and/or O_3 on intra-and interspecific competitive ability of aspen. Plant Biol. 9, 342–355.

Kump, L.R., Brantley, S.L., Arthur, M.A., 2000. Chemical weathering, atmospheric CO_2, and climate. Annu. Rev. Earth Planet. Sci. 28, 611–667.

Larson, J.L., Zak, D.R., Sinsabaugh, R.L., 2002. Microbial activity beneath temperate trees growing under elevated CO_2 and O_3. Soil Sci. Soc. Am. J. 66, 1848–1856.

Lesaulnier, C., Papamichail, D., McCorkle, S., Ollivier, B., Skiena, S., Taghavi, S., Zak, D.R., van der Lelie, D., 2008. Elevated atmospheric CO_2 affects soil microbial diversity associated with trembling aspen. Environ. Microbiol. 10, 926–941.

Litton, C.M., Raich, J.W., Ryan, M.G., 2007. Carbon allocation in forest ecosystems. Glob. Chang. Biol. 13, 2089–2109.

Liu, L., King, J.S., Giardina, C.P., 2005. Effects of elevated atmospheric CO_2 and tropospheric O_3 on leaf litter production and chemistry in trembling aspen and paper birch ecosystems. Tree Physiol. 15, 1511–1522.

Liu, L., King, J.S., Giardina, C.P., Booker, F.L., 2009. The influence of chemistry, production and community composition on leaf litter decomposition under elevated atmospheric CO_2 and tropospheric O_3 in a northern hardwood ecosystem. Ecosystems 12, 401–416.

Loya, W.M., Pregitzer, K.S., Karberg, N.J., King, J.S., Giardina, C.P., 2003. Reduction of soil carbon formation by tropospheric ozone under increased carbon dioxide levels. Nature 425, 705–707.

Matyssek, R., Wieser, G., Ceulemans, R., Rennenberg, H., Pretzsch, H., et al., 2010. Enhanced ozone strongly reduces carbon sink strength of adult beech (*Fagus sylvatica*)—resume from the free-air fumigation study at Kranzberg Forest. Environ. Pollut. 158, 2527–2532.

Matyssek, R., Wieser, G., Fleischmann, F., Grunhage, L., 2013. Ozone Research, quo vadis? Lessons from the Kranzberg Forest Experiment. Dev. Environ. Sci. 13, 103–129.

Nikolova, P.S., Andersen, C.P., Blaschke, H., Matyssek, R., Haberle, K.-H., 2010. Belowground effects of enhanced tropospheric ozone and drought in a beech/spruce forest (*Fagus sylvatica* L./*Picea abies* [L.] Karst). Environ. Pollut. 158, 1071–1078.

Norby, R.J., Zak, D.R., 2011. Ecological lessons from free-air CO_2 enrichment (FACE) experiments. Annu. Rev. Ecol. Evol. Syst. 42, 181–203.

Parsons, W.F.J., Bockheim, J.G., Lindroth, R.L., 2008. Independent, interactive, and species-specific responses of leaf litter decomposition to elevated CO_2 and O_3 in a northern hardwood forest. Ecosystems 11, 505–519.

Phillips, R.L., Zak, D.R., Holmes, W.E., White, D.C., 2002. Microbial community composition and function beneath temperate trees exposed to elevated atmospheric carbon dioxide and ozone. Oecologia 131 (2), 236–244.

Pregitzer, K.S., Euskirchen, E.E., 2004. Carbon cycling and storage in world forests: biome patterns related to forest age. Glob. Chang. Biol. 10, 2052–2077.

Pregitzer, K., Loya, W., Kubiske, M., Zak, D., 2006. Soil respiration in northern forests exposed to elevated carbon dioxide and ozone. Oecologia 148, 503–516.

Pregitzer, K., Burton, A., King, J., Zak, D., 2008. Soil respiration, root biomass, and root turnover following long-term exposure of northern forests to elevated atmospheric CO_2 and tropospheric O_3. New Phytol. 180, 153–161.

Pye, J.M., 1988. Impact of ozone on the growth and yield of trees: a review. J. Environ. Qual. 17, 347–360.

Rhea, L.K., King, J.S., 2012. Depth-dependency of trembling aspen and paper birch small-root responses to eCO_2 and eO_3. Plant Soil 355, 215–229.

Rich, S., 1964. Ozone damage to plants. Annu. Rev. Phytopathol. 2, 253–266.

Riikonen, J., Kets, K., Darbah, J., Oksanen, E., Sober, A., et al., 2008. Carbon gain and bud physiology in *Populus tremuloides* and *Betula papyrifera* under long-term exposure to elevated concentrations of CO_2 and O_3. Tree Physiol. 28, 243–254.

Schmidt, M.W., Torn, M.S., Abiven, S., Dittmar, T., Guggenberger, G., et al., 2011. Persistence of soil organic matter as an ecosystem property. Nature 478, 49–56.

Siegenthaler, U., Oeschger, H., 1978. Predicting future atmospheric carbon dioxide levels. Science 199, 388–395.

Talhelm, A.F., Pregitzer, K.S., Zak, D.R., 2009. Species-specific responses to atmospheric carbon dioxide and tropospheric ozone mediate changes in soil carbon. Ecol. Lett. 12, 1219–1228.

Talhelm, A.F., Pregitzer, K.S., Giardina, C.P., 2012. Long-term leaf production response to elevated atmospheric carbon dioxide and tropospheric ozone. Ecosystems 15, 71–82. http://dx.doi.org/10.1007/s10021-011-9493-z.

Taylor, G., Tallis, M.J., Giardina, C.P., Percy, K.E., Miglietta, F., et al., 2008. Future atmospheric, CO_2 leads to delayed autumnal senescence. Glob. Chang. Biol. 14, 264–275.

Tingey, D.T., Reinert, R.A., 1975. The effect of ozone and sulphur dioxide singly and in combination on plant growth. Environ. Pollut. 9, 117–125.

Tolley, L.C., Strain, B.R., 1984. Effects of CO_2 enrichment and water stress on growth of *Liquidambar styraciflua* and *Pinus taeda* seedlings. Can. J. Bot. 62, 2135–2139.

Wittig, V.E., Ainsworth, E.A., Naidu, S.L., Karnosky, D.F., Long, S.P., 2009. Quantifying the impact of current and future tropospheric ozone on tree biomass, growth, physiology and biochemistry: a quantitative meta-analysis of the ozone literature. Glob. Chang. Biol. 15, 396–424.

Zak, D.R., Pregitzer, K.S., King, J.S., Holmes, W.E., 2000. Elevated atmospheric CO_2, fine roots and the response of soil microorganisms: a review and hypothesis. New Phytol. 147, 201–222.

Zak, D.R., Pregitzer, K.S., Kubiske, M.E., Burton, A.J., 2011. Forest productivity under elevated CO_2 and O_3: positive feedbacks to soil N cycling sustain decade-long net primary productivity enhancement. Ecol. Lett. 14, 1220–1226.

Chapter 11

Impacts of Atmospheric Change on Tree–Arthropod Interactions

John J. Couture*[,1] and Richard L. Lindroth[†]

*Department of Forest and Wildlife Ecology, University of Wisconsin–Madison, Madison, Wisconsin, USA

†Department of Entomology, University of Wisconsin–Madison, Madison, Wisconsin, USA

[1]Corresponding author: e-mail: jjcouture@wisc.edu

Chapter Outline

Developments in Environmental Science, Vol. 13. http://dx.doi.org/10.1016/B978-0-08-098349-3.00011-6

11.1 INTRODUCTION

Forests face an uncertain future. Activities such as deforestation and fossil fuel combustion have dramatically altered the Earth's atmospheric composition. Concentrations of atmospheric carbon dioxide (CO_2) are at the highest levels estimated over the last 26 million years and are currently 40% greater than prior to the industrial revolution. Levels of another important greenhouse gas, tropospheric ozone (O_3), have also increased by approximately 40% since the beginning of the industrial age. These two gases play significant roles in the functioning of forest ecosystems, as both CO_2 and O_3 alter tree metabolism and growth. Typically, enriched CO_2 enhances tree productivity, whereas elevated O_3 decreases tree growth. The effects of CO_2 and O_3 on trees are also represented in changes in tree chemical composition, which can cascade through trophic levels to influence ecosystem processes important for forest health. The effects of atmospheric change on forest processes are critically important, as forests provide a suite of ecosystem services that are central to terrestrial biome functioning. Forests are the largest terrestrial carbon sink, play a dominant role in global carbon and hydrologic cycling, and serve as reservoirs of biodiversity. If we are to understand and predict responses of forests to future environments, improved understanding of the influence of atmospheric change on forest processes is necessary.

Central to numerous aspects of forest functioning is one of the most dominant taxa of animals on Earth: arthropods. Arthropods are the most abundant and

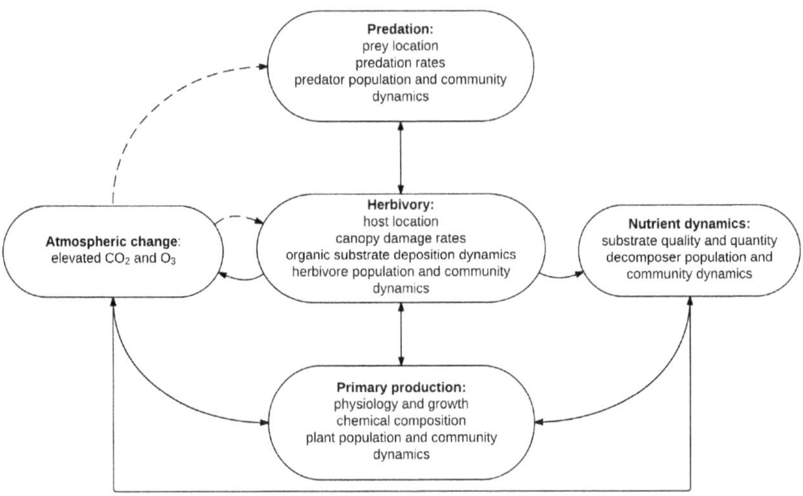

FIGURE 11.1 Schematic outline of the effects of CO_2 and O_3 on forest ecosystem processes. Solid lines represent direct effects for adjoining boxes, and indirect effects for non-adjoining boxes, for which strong experimental confirmation exists. Dashed lines represent direct effects for adjoining boxes, for which developing evidence exists. Double-ended arrows indicate that interactions proceed in both directions.

speciose animals in forest ecosystems, influencing processes such as herbivory, predation, and nutrient cycling (Figure 11.1). These actions ultimately affect forest primary productivity and the capacity of forests to act as carbon sinks. Advancing our knowledge of how arthropods will respond to global change and how they will influence forest ecosystem functioning will enhance our ability to understand how forest health will be affected in future environments.

In this chapter, we review and summarize the current state of knowledge regarding the direct effects of CO_2 and O_3 on tree foliar chemistry and responses by forest arthropods. We will highlight limitations and knowledge gaps in current methodologies that evaluate the ecosystem-level responses by arthropods to environmental change. This review will focus on the effects of elevated CO_2 and O_3 on tree–arthropod interactions mediated through phytochemistry, not effects mediated through climate (e.g. warming). In addition, this chapter will emphasize the ecosystem-level consequences of tree–arthropod interactions in future atmospheres, with a special emphasis on forest systems.

11.2 EFFECTS OF CO_2 AND O_3 ON TREE GROWTH AND CHEMISTRY

Phytochemical responses to elevated CO_2 and O_3 play important roles in plant–arthropod interactions and ecosystem functioning, and have been extensively reviewed (Ainsworth et al., 2012; Bezemer and Jones, 1998; Bidart-Bouzat and Imeh-Nathaniel, 2008; Coviella and Trumble, 1999; Lindroth, 2010, 2011; Robinson et al., 2012; Stiling and Cornelissen, 2007; Valkama et al., 2007; Wittig et al., 2009; Zvereva and Kozlov, 2006). Below, we briefly summarize and expand the findings of those reviews.

11.2.1 Elevated CO_2

As the substrate for photosynthesis, CO_2 dramatically influences plant growth and development. Enriched CO_2 environments generally increase plant growth by enhancing photosynthetic efficiency (Ainsworth and Long, 2005; King et al., 2013, this vol.; Körner, 2006; Leakey et al., 2009; Long et al., 2004; Norby and Zak, 2011; Norby et al., 2005). The resulting increase in plant growth has fostered the notion that accelerated forest productivity may offset anthropogenic carbon emissions by acting as a carbon sink. However, the magnitude of increase in growth experienced by forests under elevated CO_2 levels is strongly influenced by a number of factors, including nutrient availability, stand age and composition, O_3 levels, altered nutrient cycling, and herbivory (Couture J.J., Meehan T.D., Kruger E.L., and Lindroth R.L., unpublished data; Drake et al., 2011; Finzi et al., 2007; King et al., 2005; Körner, 2006; Kubiske et al., 2007; Lou et al., 2004; McDonald et al., 2002; Norby and Zak, 2011; Zak et al., 2011).

As plant growth is modified under elevated CO_2, so is plant chemical composition. Elevated CO_2 generally reduces foliar nitrogen, thereby increasing the ratio of carbon to nitrogen (C:N). Carbohydrate responses to enhanced CO_2 are somewhat variable, with simple sugar levels either increasing or not changing, while starch levels generally increase. The most commonly studied secondary metabolites in tree foliage have been the carbon-based compounds, primarily phenolics and terpenoids. Enriched CO_2 generally increases foliar tannin concentrations, and, to a lesser degree, simple phenolic compounds and either increases or has no effect on levels of lignin. Responses by terpenoids to elevated CO_2 have generally been nominal, with most studies finding no change.

While some phytochemicals respond in a consistent manner to elevated CO_2 levels, a number of other foliar constituents do not, making generalities difficult. These inconsistencies are likely due to the suite of abiotic and biotic factors that affect phytochemical responses to enriched CO_2. The influence of multiple variables acting in concert on phytochemistry is an important, yet understudied, factor governing the effect of global change on forest systems.

A specific example of an understudied factor that can influence tree chemical response to atmospheric change is ontogeny. Little is known about how enriched CO_2 will influence foliar chemistry of long-lived tree species through multiple developmental stages. Examination of phytochemical data from aspen trees collected from over a decade of exposure to elevated CO_2 at the Aspen FACE (Free Air Carbon dioxide and ozone Enrichment) research facility revealed that the magnitude of chemical responses to elevated CO_2 varied among tree developmental stages, with nitrogen and phenolic glycoside concentrations having more pronounced responses in younger than older trees (Figure 11.2).

Tree stand age and ontogeny, increased soil nutrient availability, improved tree–microbe relationships, altered tree competitive interactions, and accelerated nutrient cycling are all suggested to influence long-term responses by forests to elevated CO_2 (Drake et al., 2011; Körner, 2006; Kubiske et al., 2007; Zak et al., 2011, 2012). These factors also influence phytochemical responses to atmospheric change that can, in turn, cascade to affect interactions across multiple trophic levels. Further research is needed to understand the influence of multiple factors on forest responses to environmental change.

11.2.2 Elevated O_3

Generated via the oxidation of hydrocarbons by OH radicals in the presence of sunlight, O_3 is arguably the phytotoxin most damaging to plants. While cuticular absorption of ozone can occur, the main conduit into the leaf is the stomata. Once inside the plant, O_3 elicits oxidative stress, leading to a cascade of physiological responses including stomatal closure, cell death, and upregulation of a number of defence-related genes resulting in enhanced synthesis of reactive oxygen scavengers, flavonoids, volatile terpenoids, and phenolic compounds (Dizengremel et al., 2013, this vol.; Heath, 2008). The

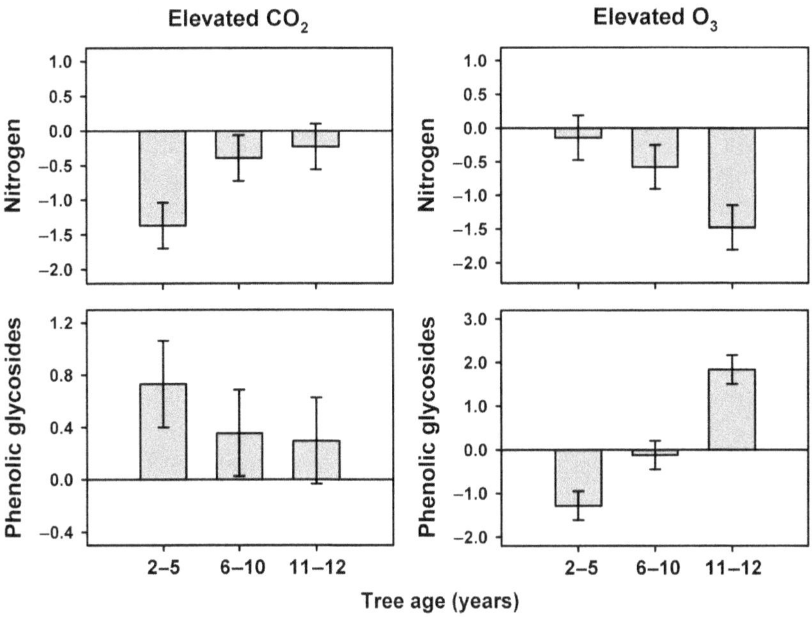

FIGURE 11.2 Mean effect size of foliar nitrogen and phenolic glycosides in response to elevated CO_2 (left panel) and O_3 (right panel) at three developmental stages of aspen at Aspen FACE. Error bars represent 95% confidence intervals.

combination of these responses generally decreases photosynthesis, increases photorespiration, and alters whole plant carbon balance, ultimately reducing overall plant growth (Ainsworth et al., 2012; Matyssek et al., 2013, this vol.; Wittig et al., 2009).

The effects of O_3 on primary and secondary metabolism directly alter plant chemical composition. Plant nutrient levels may increase, decrease, or remain the same. Simple sugars either decrease or are unaffected, while starch levels often decline. Similar to CO_2 experiments, the secondary metabolites examined in trees exposed to elevated O_3 have largely been limited to phenolics and terpenoids. Usually, levels of phenolics and flavonoids increase, while lignin concentrations either increase or remain the same. Terpenoid levels can increase, decrease, or remain unaffected.

Three major factors influencing tree chemical responses to elevated O_3 are duration of exposure (i.e. transient vs. chronic), tree sensitivity, and developmental stage (Ainsworth et al., 2012; Gupta et al., 2005; Karnosky et al., 1998; Oksanen, 2003a,b). The limited work that has tracked tree responses through relatively long-term exposure to damaging levels of O_3 has found that the negative effects are exacerbated as exposure duration increases (Oksanen, 2003a,b). Trembling aspen is a genetically diverse model tree species that exhibits a wide range of sensitivity to O_3, resulting in genotype-specific gene regulation, chemical, and growth responses to enhanced O_3 levels (Gupta et al., 2005; Karnosky et al., 1998). Little work, however, has examined the responses of long-lived tree species through multiple developmental stages. Phytochemical analysis of aspen trees exposed to elevated O_3 for over a decade at Aspen FACE revealed that foliar nitrogen concentrations were generally higher and phenolic levels were generally lower in younger trees exposed to elevated O_3 and that the opposite patterns were true for older trees (Couture J.J., Holeski L.M., and Lindroth R.L., unpublished data; Figure 11.2). These findings suggest complex interactions among tree physiology and growth, developmental stage, and exposure duration that were borne out only through long-term experimental research.

11.2.3 Interactions Between CO_2 and O_3

Because CO_2 and tropospheric O_3 have widely contrasting effects on plant growth and chemistry, and levels of both gases are regulated by anthropogenic activities, knowledge of their independent and interactive influences on plants is critical. One major question concerning the co-occurring effects of these gases on forest productivity is the balance between enhanced growth of trees exposed to enriched CO_2 and reduced growth of trees exposed to elevated O_3. At the Aspen FACE research site, elevated O_3 offset enhanced growth under enriched CO_2 in younger forest stands (Karnosky et al., 2005; King et al., 2005). However, by the time of final harvest (12 years after onset of treatments), O_3 had no detectable effect on the growth increase under elevated CO_2 because of enhanced productivity of O_3-tolerant trees (Zak et al., 2011).

When combined, elevated CO_2 and O_3 can produce either additive or interactive effects on foliar chemistry. Elevated O_3 can exacerbate or ameliorate the decline in foliar nutrient levels and the increase in C:N observed under elevated CO_2, depending on tree species and O_3 sensitivity. Responses of secondary metabolites in leaves of plants exposed to co-occurring levels of elevated CO_2 and O_3 vary among tree species and even within metabolite classes (Lindroth, 2010, 2011; Peltonen et al., 2005, 2010; Valkama et al., 2007). In short, primary and secondary phytochemical responses to both elevated CO_2 and O_3 will be variable and will depend on atmospheric composition and the tree species and chemical constituents examined.

11.3 EFFECTS OF CO_2 AND O_3 ON CANOPY AND SOIL ARTHROPODS

Arthropods account for the majority of all known animal species and when combined with plants, comprise a substantial proportion of terrestrial biodiversity. Because of their inherent value to global biodiversity and their influence on ecosystem functioning, it is no surprise that arthropods dominate studies examining the effects of environmental change on plant–animal interactions.

A number of recent reviews have documented the effects of atmospheric change on plant–arthropod interactions (Bezemer and Jones, 1998; Bidart-Bouzat and Imeh-Nathaniel, 2008; Coviella and Trumble, 1999; Lindroth, 2010, 2011; Robinson et al., 2012; Stiling and Cornelissen, 2007; Valkama et al., 2007; Wittig et al., 2009; Zvereva and Kozlov, 2006). Below, we summarize and expand those findings with special emphasis on forest systems.

11.3.1 Herbivores: Individuals, Populations, and Communities

Herbivores play influential roles in forest ecosystems, directly and indirectly affecting forest productivity (Figure 11.1). Little evidence exists to suggest that elevated CO_2 will *directly* affect herbivorous arthropods; rather, the major influence of elevated CO_2 on arthropod herbivores will be through altered plant quality.

Most studies reporting the effects of atmospheric change on plant–arthropod interactions have focused on responses by individual arthropods. These studies have examined host location and preference, performance (i.e. growth, consumption, etc.), and reproductive success. Elevated CO_2 and O_3 alter host plant location and preference, but the responses are highly variable, making generalizations difficult (Agrell et al., 2005; Coleman and Jones, 1988; Freiwald et al., 2008; Kopper and Lindroth, 2003; Peltonen et al., 2006; Stange, 1997; Traw et al., 1996; Wang et al., 2009). Arthropod growth and feeding performance are also affected when feeding on foliage exposed to elevated CO_2 and O_3. Herbivorous arthropods generally exhibit reduced growth and food processing efficiencies, and increased consumption rates and development times, under elevated CO_2. The increase in consumption under elevated CO_2 is likely a compensatory response to decreased host nutritional quality (i.e. the need to increase consumption because of reduced nutrient concentrations). Compensatory consumption, however, can increase exposure to a wide variety of secondary compounds that can further exacerbate declines in growth and food processing efficiency. Elevated O_3 generally, but not always, enhances the performance of arthropod herbivores, increasing pupal masses and decreasing development times, but not affecting their growth or consumption.

Reproduction rates by forest arthropods will also likely be altered in future environments. Hillstrom et al. (2010) observed that fecundity of the green weevil decreased under elevated CO_2, but was not affected under elevated O_3. Awmack et al. (2004) reported that aphid fecundity decreased under both elevated CO_2 and O_3. And while Lindroth et al. (1997) found that elevated CO_2 had no effect on gypsy moth fecundity, and Couture and Lindroth (2012) reported that gypsy moth egg production slightly increased under elevated CO_2 but dramatically decreased under elevated O_3.

Individual herbivore responses to atmospheric change have been highly divergent among feeding guilds, as highlighted by Bezemer and Jones

(1998). To date, a disproportionate amount of studies have been conducted on chewing insects, compared with other feeding guilds, and these studies have mostly focused on insects from the order Lepidoptera. Robinson et al. (2012) further confirmed variation among feeding guilds in responses to elevated CO_2, reporting that chewing herbivores generally increase consumption and development time and decrease growth, conversion efficiencies, and fecundity under elevated CO_2. Phloem feeders, however, generally display opposite patterns. These findings highlight that a wider range of feeding guilds should be examined to fully understand herbivore responses to environmental change.

Compared with the large number of studies examining performance of individual arthropods, relatively few studies have investigated arthropod population responses to atmospheric change. Population sizes of tree-feeding aphids have either increased or remained the same under elevated CO_2 levels (Awmack et al., 2004; Dochery et al., 1997; Percy et al., 2002) and have increased under elevated O_3 (Awmack et al., 2004). Over a period of 11 years, Stiling et al. (2009) found that in Florida scrub oak forest, elevated CO_2 decreased densities of leafminers and leaftiers in the first 5 years, but increased densities of these herbivores over the subsequent 6 years. The reduction in herbivore densities in the first 5 years was attributed to an increase in plant-mediated mortality and increased attack rate by natural enemies (Rossi et al. 2004; Stiling et al., 2003), while the increase in herbivore densities over the next 6 years was due to the positive growth effect of elevated CO_2 on tree productivity (Stiling et al., 2009).

In addition to herbivore populations, complex herbivore communities, such as those found in forest ecosystems, are also likely to change in future environments (Lindroth, 2010; Tylianakis et al., 2008). At Aspen FACE, Hillstrom and Lindroth (2008), using pan traps, reported that while elevated CO_2 had little effect on total herbivore abundance, it reduced numbers of phloem-feeding arthropods and tended to increase numbers of chewing herbivores. Similarly, Hamilton et al. (2012), using stick traps, found that elevated CO_2 did not affect total abundance of insects collected, but did decrease numbers of chewing herbivores, thereby altering community composition. Elevated O_3 also alters herbivore community composition, but results have been somewhat inconsistent. For example, while Hillstrom and Lindroth (2008) observed that elevated O_3 decreased numbers of chewing herbivores and increased numbers of phloem feeders, Jones and Paine (2006) found exposure to enriched O_3 shifted community composition from phloem-feeding to chewing insects. In the most exhaustive study to date, Hillstrom M.L. and Lindroth R.L. (unpublished data) visually censused insect communities on aspen and paper birch for 3 years at Aspen FACE. Recording over 36,000 insects from almost 300 species, they found that, while the responses were highly variable, elevated CO_2 tended to increase the abundance of phloem feeders and decrease the abundance of chewing herbivores on aspen, but had little effect on insects on paper birch. They also reported that elevated O_3 tended to increase the number of chewing herbivores and decrease the number of phloem feeders on aspen, but had little impact on herbivores on paper birch.

In summary, herbivores will likely be influenced by elevated CO_2 and O_3, but the direction and magnitude of responses will be difficult to predict. The variety of herbivore responses is likely due to interactions among abiotic and biotic factors (Lindroth, 2010; Robinson et al., 2012). Future research should focus on expanding the experimental conditions examined in global change research to more completely understand the complex of factors influencing herbivore responses to future environments. Given the important roles that arthropods play in regulating forest productivity and structure, understanding responses by complex arthropod communities to global change should be a research priority.

11.3.2 Predators: Individuals, Populations, and Communities

While less studied than herbivores, arthropod predators are important regulators of herbivore populations, and thus indirectly involved in influencing forest ecosystem processes (Figure 11.1). Atmospheric change may influence predator–prey interactions by altering prey quality (e.g. nutritional- or defence-related chemical composition), modifying the defensive behaviour of prey, affecting predator location of prey, or by shifting the prey community structure and increasing predator competition.

Examining the influence of prey quality on parasitoid performance, Roth and Lindroth (1995) and Holton et al. (2003) observed that parasitoid survivorship and growth were largely unaffected when caterpillars were fed foliage from elevated CO_2, but were reduced when larvae were fed foliage from O_3. Prey-defensive behaviours are also influenced by atmospheric change, as Mondor et al. (2004) found that the pheromone-mediated defence escape response of a tree-feeding aphid was diminished under elevated CO_2, but was enhanced under elevated O_3. Location of prey can be affected by atmospheric pollution, specifically as O_3 can interact with complex volatile organic compounds to degrade cues used by predators to locate prey (Gate et al., 1995; Pinto et al., 2007; Vuorinen et al., 2004). A recent review by Butler et al. (2009) reported that exposure to elevated CO_2 either increased, decreased, or had no effect, while elevated O_3 had a predominantly negative effect, on predator performance and prey location in forest ecosystems. In summary, the combination of these findings suggests that atmospheric change will influence predator–prey interactions and may alter the role predators play in regulating forest ecosystem processes.

Similar to herbivores, predator populations and communities are also responsive to atmospheric change. Hillstrom and Lindroth (2008) reported that elevated CO_2 increased the abundance of figitid and ichneumonid parasitoids, but not numbers of braconid or chalcid parasitoids, while elevated O_3 reduced total numbers of parasitoids at Aspen FACE. Sanders et al. (2004) found that predators were more abundant in ambient, compared with elevated, CO_2 environments in a grass-dominated understory of a sweetgum monoculture. Conversely, in the understory of a loblolly pine forest exposed to elevated CO_2, Hamilton et al. (2012) reported that numbers of predators increased dramatically, ultimately shifting arthropod community structure. We are aware of no studies examining predator reproductive responses across multiple generations in forest ecosystems. However, Holton et al. (2003) found that while survivorship was reduced under elevated O_3, pupal masses of a parasitic fly were unaffected when developing in caterpillars fed foliage from either elevated CO_2 or O_3. While not a direct measure of fecundity, insect pupal masses are generally correlated with reproductive success, and the findings of Holton et al. (2003) suggest that population levels responses by some predators to atmospheric change may be driven more by changes in larval survival than by altered reproductive success.

11.3.3 Decomposers: Individuals, Populations, and Communities

Arthropod decomposers govern nutrient dynamics in forest ecosystems (Figure 11.1). Detritivores break down organic materials, influence the microbial community, and alter the soil quality by depositing faecal matter and their carcasses and by inoculating litter substrate with microbes. These actions facilitate conversion of organically bound nutrients into mineralized forms and can substantially influence rates of nutrient turnover in forests (Hättenschwiler and Gasser, 2005). The effects of decomposers on forest processes are regulated by the quality and quantity of organic substrates deposited to the forest floor, which in turn are influenced by atmospheric change.

Isopods exhibit a variety of responses in terms of preference and performance on leaf litter grown under elevated CO_2. Cotrufo et al. (1998) found that isopod consumption decreased on ash litter from elevated, compared with ambient, CO_2 environments, while Hättenschwiler et al. (1999) reported that isopod consumption increased on beech, but not spruce, leaf litter under elevated CO_2. And Cotrufo et al. (2005) observed that isopod litter preference was unchanged for five of six tree species grown under either ambient or elevated CO_2.

Additional studies have addressed the effects of litter from different atmospheric environments on the individual performance, population growth, and community structure of detritivores. Litter from elevated CO_2 and O_3 generally has little effect on isopod growth and mortality (Hättenschwiler and Bretscher, 2001; Kasurien et al., 2007). Alternatively, Kasurien et al. (2007) reported that young earthworms exhibited reduced growth rates when fed silver birch litter exposed to enhanced CO_2 and O_3, with reduced litter quality being implicated as the primary factor for earthworm responses. Meehan et al. (2010) also observed that earthworm growth and processing efficiencies decreased when fed aspen litter from elevated CO_2, but found no detectable effects when fed litter from elevated O_3. Additionally, Meehan et al. (2010) reported that the population growth of collembolans decreased when fed aspen litter from elevated CO_2, but increased when fed litter from elevated O_3.

The community structure of soil arthropods is also affected by atmospheric change. A recent meta-analysis of the effects of global change on soil arthropods found that abundance of detritivores increased in elevated CO_2 environments (Blankinship et al., 2011). The increase in abundance of soil arthropods under enhanced CO_2, however, diminished (and in cases, decreased) as increasingly realistic field studies were employed. For example, Hansen et al. (2001) reported a 33% reduction in soil arthropod abundance in enhanced CO_2 plots at the Duke Forest FACE site. Loranger et al. (2004) observed that soil arthropod abundance was reduced under independent treatments of elevated CO_2 and O_3 at Aspen FACE, although the combination of elevated CO_2 and O_3 had no effect on soil arthropod abundance. Contrary to the previous two studies, Haimi et al. (2005) found little difference in soil microfauna abundance in open-top chambers comparing elevated and ambient CO_2 environments, with the exception that acaridid mite numbers were generally lower under elevated, compared with ambient, CO_2.

While evidence exists that atmospheric change will affect decomposer dynamics in forest ecosystems and that the role they play in nutrient cycling will be altered, there is no clear indication of a general pattern. Very similar to responses by other arthropod guilds outlined in this review, responses by detritivores to environmental change will likely be species-specific and influenced by a host of interacting factors, such as substrate quality and climate shifts.

11.4 EFFECTS OF CO_2 AND O_3 ON ARTHROPOD-MEDIATED ECOSYSTEM PROCESSES

Arthropods influence biomass production, succession patterns, nutrient cycling, and energy flow in ecosystems (Fajvan and Wood, 1996; Hunter, 2001; Lindroth, 2010; Lovett et al., 2002, 2006; Mattson and Addy, 1975). The influence of arthropods on ecosystem processes will be determined through the responses by and interactions among plants, herbivores, predators, and detritivores and may occur at the level of individuals, populations, and communities (Figure 11.1). Growing evidence suggests that elevated levels of CO_2 and O_3 will alter the influence of arthropods on canopy damage, organic substrate deposition, and nutrient dynamics, and those changes will in turn alter forest productivity.

11.4.1 Canopy Defoliation

At endemic population levels, herbivorous arthropods defoliate approximately 5–15% of the canopy in temperate forests (Cebrian, 1999; Cyr and Pace, 1993). Damage increases considerably during outbreak years and nearly all of the foliage in an area can be removed by herbivores, shifting forests from acting as carbon sinks to carbon sources. Canopy damage is a combination of herbivore population size, community composition, and individual consumption rates, all of which have been shown to respond to atmospheric change.

Few studies to date have examined the influence of elevated CO_2, and no published studies have examined the effects of elevated O_3, on canopy defoliation in forested ecosystems. Stiling et al. (2002) found that myrtle oak trees exposed to elevated CO_2 concentrations experienced reduced chewing and mining damage. They attributed this decrease to increased plant- and predator-mediated herbivore mortality. Hamilton et al. (2004) reported reduced herbivore damage in one of three tree species in an understory of a loblolly pine forest exposed to elevated, compared with ambient, levels of CO_2. The authors were unable to resolve damage rates with variation in leaf quality characteristics, as leaf chemistry did not respond to enhanced levels of CO_2. Knepp et al. (2005) observed that elevated CO_2 reduced average foliar damage among 12 potted hardwood saplings in only one of 3 years measured. The authors suggested that changes in foliar quality (i.e. decreased nitrogen concentrations and specific leaf area and increased ratios of carbon to nitrogen) in some of the tree species may have been responsible for the variation in herbivory; however, foliar chemical responses among the tree species examined were inconsistent. Recently, in the most exhaustive study to date, Couture J.J., Meehan T.D., and Lindroth R.L. (unpublished data) examined the effects of elevated CO_2 and O_3 on canopy damage rates in trembling aspen and paper birch over 3 years at the Aspen FACE research facility. They found that elevated CO_2 dramatically increased and elevated O_3 generally decreased canopy damage of trembling aspen and paper birch over the 3 years. Some, but not all, of the changes in canopy damage were associated with phytochemical variation in response to elevated CO_2 and O_3, suggesting that multiple factors interact to influence canopy damage responses to environmental change.

While accumulating evidence suggests that canopy defoliation rates will change under future atmospheric conditions, little attention has focused on the impact that changes in damage rates will have on forest productivity. Over long periods of time, an increase in chronic, endemic levels of damage can substantially reduce individual tree growth, and may outweigh the effects of periodic, outbreak damage (Zvereva et al., 2012). Moreover, plant photosynthetic efficiency is generally reduced in the undamaged leaf tissue surrounding the damaged area (Aldea et al., 2006), suggesting that a reduction in productivity based on the area of tissue damaged will underestimate actual reductions in productivity, although this reduction may be ameliorated in leaves exposed to elevated CO_2 (Nabity et al., 2012).

In recent work at Aspen FACE, we sought to determine the impact of canopy defoliation on forest productivity (Couture J.J., Meehan T.D., Kruger E.L., and Lindroth R.L., unpublished data). We used a light use efficiency model to quantify the impact of canopy defoliation, compared with no canopy damage, on stand productivity in aspen and aspen-birch forest communities exposed to elevated CO_2 and O_3. We found that the negative effect of herbivory on potential productivity (i.e. productivity in the absence of canopy defoliation) more than doubled in communities exposed to elevated, compared with ambient, CO_2. In contrast, the negative effect of herbivory on forest productivity was offset by approximately one-quarter in trees exposed to elevated, compared with ambient, O_3 (Figure 11.3). While forest productivity increased under enhanced CO_2 at Aspen FACE (Zak et al., 2011), this work suggests that the negative impact of herbivores on the potential of forest stands to sequester anthropogenically produced carbon will increase under elevated levels of CO_2.

11.4.2 Substrate Deposition and Nutrient Dynamics

Arthropods transfer organic substrate from the canopy to the forest floor via deposition of frass, greenfall, and arthropod biomass. Organic input by arthropods can alter the timing and quality of organic deposition during the growing season and can contribute to the overall retention or loss of nutrients from ecosystems (Lovett et al., 2002; Townsend et al., 2004). To date, the effects of environmental change on arthropod-mediated substrate deposition have not been well characterized.

Foliar chemistry is a strong determinant of frass chemistry and many foliar traits important for nutrient cycling are present in arthropod frass. For example, C:N and tannin levels are correlated between foliage and frass (Kagata and Ohgushi, 2012; Madritch et al., 2007). At Aspen FACE, Hillstrom et al. (2010) found that key chemical characteristics of frass collected from white-marked tussock moths were largely similar to those in foliage the caterpillars consumed: frass exhibited higher tannin concentrations when larvae were fed foliage from elevated CO_2 levels and lower nitrogen concentrations when fed foliage from elevated O_3. While changes in frass quality under the different treatments had little impact on microbial respiration or nitrate leaching in soil microcosm experiments, the quantity of frass dramatically altered microbial respiration and nitrate export. Examining a wider range of folivores, we recently observed species-level variation in the frass quality of multiple herbivores fed foliage exposed to elevated CO_2 and O_3 (Couture J.J., and Lindroth R.L., unpublished data). Notably, in this study, we found that two periodic defoliators of northern temperate hardwoods in North America, gypsy moths and forest tent caterpillars, produced frass of different chemical compositions when fed foliage exposed to elevated CO_2 and O_3. This finding suggests that nutrient dynamics may be different during outbreak events specific to these defoliators

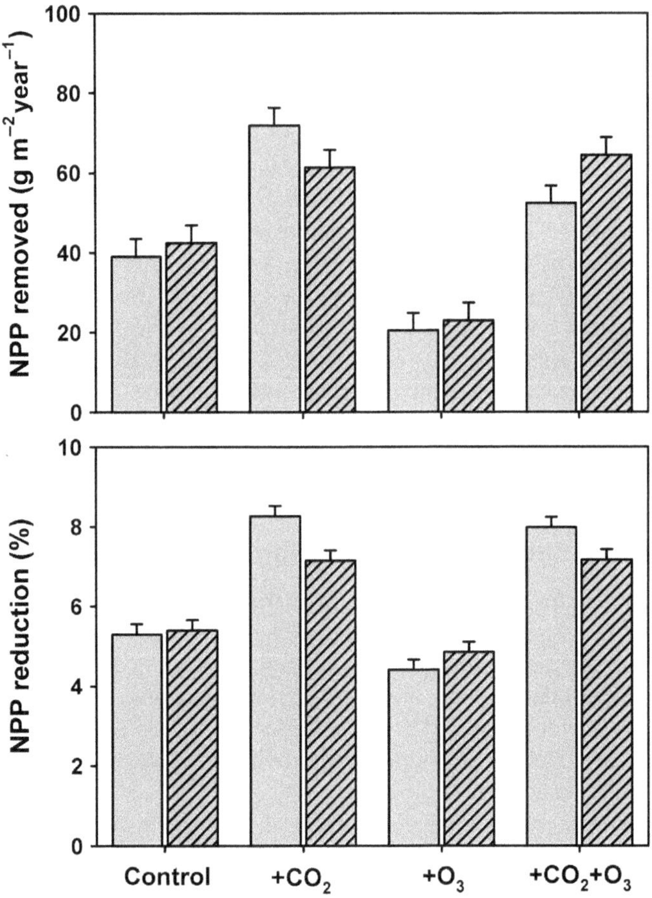

FIGURE 11.3 Independent and interactive effects of elevated CO_2 and O_3 on the amount of carbon removed annually and the percent reduction in net primary production (NPP) due to insect-related canopy damage in aspen-only (open bars) and aspen-birch (hatched bars) communities at Aspen FACE over 3 years. Error bars represent ±1 SE. Control represents ambient CO_2 and O_3; $+CO_2$ is elevated CO_2, ambient O_3; $+O_3$ is ambient CO_2, elevated O_3; $+CO_2+O_3$ is elevated CO_2 and O_3.

under future atmospheric conditions. In the most comprehensive study to date, Meehan T.D., Couture J.J., and Lindroth R.L. (unpublished data) examined the effects of elevated CO_2 and O_3 on arthropod-mediated organic deposition at Aspen FACE. They documented that, on average, the combined flux of frass and greenfall over 3 years increased by 35% under elevated, compared with ambient, CO_2 and decreased by 12% in elevated, compared with ambient, O_3. Moreover, the altered input rates affected the total amount of carbon, nitrogen, and tannins transferred from the canopy to the forest floor.

The combination of findings highlighted above suggests that changes in both herbivore community composition and the timing, quantity, and quality of organic material deposited by herbivores will alter nutrient dynamics in forests of the future. Continued research in this area is imperative if we are to understand the extent to which environmental change influences arthropod-mediated forest processes and, ultimately, forest productivity.

11.5 CONCLUSIONS AND FUTURE DIRECTIONS

Anthropogenic activities are substantially altering the composition of the Earth's atmosphere and these changes will likely have dramatic effects on forest health. Forests are historically undervalued natural resources that provide a wide variety of ecosystem services to the benefit of society, including carbon storage, carbon and hydrologic cycling, biodiversity maintenance, timber production, and aesthetic, recreational, and spiritual value. Arthropods are influential to all of the services forests provide. In addition to being integral components of forest ecosystems, arthropods also have the ability to substantially modify forest functioning. Consequently, understanding responses by arthropods to global change is requisite to improve our ability to predict forest functioning in future environments.

Our review of research on forest ecosystems reveals that elevated levels of atmospheric CO_2 and tropospheric O_3 alter the quantity and quality of primary production and these changes in turn alter processes important for forest functioning. As noted by others (Lindroth, 2010; Robinson et al., 2012; Tylianakis et al., 2008), a consistent theme emerging from research to date is that responses to environmental change are highly variable, and both species- and context-specific.

While our understanding of the relationships among arthropod-mediated forest processes and environmental change is advancing, critical knowledge gaps still exist. Below, we highlight several knowledge gaps and provide recommendations (extended from Lindroth, 2010) for future research that will enhance our understanding of the effects of atmospheric change on forest systems.

11.5.1 Knowledge Gaps

First is the general lack of knowledge of how the numerous components that comprise 'global change' will interact to affect forest ecosystems. A number of environmental factors (e.g. CO_2, O_3, precipitation, temperature) interact to produce consequences that may differ in magnitude and even direction from responses elicited by each factor operating independently. Understanding the interactive effects of multiple environmental factors on forests will be central to accurate predictions of forest health in future climates. Second is the need to continue conducting experimental research in ecologically realistic environments. Research examining complex biotic responses to multiple environmental variables in ecologically relevant settings produces more realistic outcomes and thus affords greater confidence in the results. Third is the lack of long-term, experimental research examining biotic responses to environmental change. Emerging findings from the longest studies conducted in the most environmentally realistic settings to date suggest that biotic responses to environmental change may shift over relatively short (i.e. years) time scales. Longitudinal studies experimentally manipulating environmental change variables are necessary to fully understand the responses by complex, long-lived ecosystems to global change.

11.5.2 Future Research Directions

1. Expand the range of species and biomes studied. Research should include both multiple forest types (e.g. tropical and temperate) and different forest communities within particular biomes. Most research to date has focused on deciduous and coniferous forests in temperate regions. Tropical and boreal forests are critically important to global carbon cycling and biodiversity; a focus of future research should be to understand the effects of environmental change on those forest types.
2. Include insect and plant biodiversity as experimental variables in global change research. A large body of research over the last two decades has demonstrated the linkage of biodiversity to ecosystem function. We now need to address how those relationships will be altered by global environmental change.
3. Construct studies examining forest responses to the multiple environmental variables that comprise global change. The influences of individual climate drivers are often mediated by the presence of another. Simultaneous exposure to multiple climactic drivers will undoubtedly provide the most accurate information about biotic responses to environmental change factors and provide a range of responses depending on treatment combinations.
4. Finally, long-term, experimentally driven studies needs to remain an important focus of climate change research. Accumulating evidence suggests that a number of ecological processes may vary over long-term (at least decadal) time periods. While examining responses over the entire life span of trees is not logistically practical, future studies should focus on incorporating multiple developmental stages of long-lived trees species to fully address the dynamics of forest ecosystems.

ACKNOWLEDGEMENTS

We are grateful for the contributions of multiple undergraduate, graduate, postdoctoral, and technical researchers to over 20 years of global change research by the Lindroth Lab. That work has been supported by Grants from the National Science Foundation, U.S. Department of Energy (DOE), U.S. Department of Agriculture (USDA), and the University of Wisconsin–Madison. We specifically acknowledge U.S. DOE (Office of Science, BER) Grant DE-FG02-95ER62125, USDA NIFA AFRI Grant 2011-67013-30147, and University of Wisconsin–Madison Hatch Grant WIS04898 to R. L. L., and USDA NIFA AFRI Fellowship Grant 2012-67012-19900 to J. J. C.

REFERENCES

Agrell, J., Kopper, B.J., McDonald, E.P., Lindroth, R.L., 2005. CO_2 and O_3 effects on host plant preferences of the forest tent caterpillar (*Malacosoma disstria*). Glob. Chang. Biol. 11, 588–599.

Ainsworth, E.A., Long, S.P., 2005. What have we learned from 15 years of free-air CO_2 enrichment (FACE)? A metaanalytic review of the responses of photosynthesis, canopy properties and plant production to rising CO_2. New Phytol. 165, 351–372.

Ainsworth, E.A., Yendrek, C.R., Stich, S., Collins, W.J., Emberson, L.D., 2012. The effects of tropospheric ozone on net primary productivity and implications for climate change. Annu. Rev. Plant Biol. 63, 637–661.

Aldea, M., Hamilton, J.G., Resti, J.P., Zangerl, A.R., Berenbaum, M.R., Frank, T.D., DeLucia, E.H., 2006. Comparison of photosynthetic damage from arthropod herbivory and pathogen infection in understory hardwood saplings. Oecologia 149, 221–232.

Awmack, C.S., Harringtion, R., Lindroth, R.L., 2004. Aphid individual performance may not predict population responses under elevated CO_2 or O_3. Glob. Chang. Biol. 10, 1414–1423.

Bezemer, T.M., Jones, T.H., 1998. Plant-insect herbivore interactions in elevated CO_2: quantitative analysis and guild effects. Oikos 82, 212–222.

Bidart-Bouzat, M.G., Imeh-Nathaniel, A., 2008. Global change effects on plant chemical defenses against insect herbivores. J. Integr. Plant Biol. 50, 1339–1354.

Blankinship, J.C., Niklaus, P.A., Hungate, B.A., 2011. A meta-analysis of responses of soil biota to global change. Oecologia 165, 553–565.

Butler, C.D., Becklage, N.E., Trumble, J.T., 2009. Effects of terrestrial pollutants on insect parasitoids. Environ. Toxicol. Chem. 28, 1111–1119.

Cebrian, J., 1999. Patterns in the fate and production in plant communities. Am. Nat. 154, 449–468.

Coleman, J.S., Jones, C.G., 1988. Plant stress and insect performance: cottonwood, ozone and a leaf beetle. Oecologia 76, 57–61.

Cotrufo, M.F., Briones, M.J.I., Ineson, P., 1998. Elevated CO_2 affects field decomposition rate and palatability of tree leaf litter: importance of changes in substrate quality. Soil Biol. Biochem. 30, 1565–1571.

Cotrufo, M.F., Drake, B., Ehleringer, J.R., 2005. Palatability trials on hardwood leaf litter grown under elevated CO_2: a stable carbon isotope study. Soil Biol. Biochem. 37, 1105–1112.

Couture, J.J., Lindroth, R.L., 2012. Atmospheric change alters performance of an invasive forest insect. Glob. Chang. Biol. 18, 3543–3557.

Coviella, C.E., Trumble, J.T., 1999. Effects of elevated atmospheric carbon dioxide on insect-plant interactions. Conserv. Biol. 13, 700–712.

Cyr, H., Pace, M.L., 1993. Magnitude and patterns of herbivory in aquatic and terrestrial ecosystems. Nature 361, 148–150.

Dizengremel, P., Jolivet, Y., Tuzet, A., Ranieri, A., Thiec, D., 2013. Integrative leaf-level phytotoxic ozone dose assessment for forest risk modelling. Dev. Environ. Sci. 13, 267–288.

Dochery, M., Wade, F.A., Hurst, D.K., Whittaker, J.B., Lea, P.J., 1997. Responses of tree sap-feeding herbivores to elevated CO_2. Glob. Chang. Biol. 3, 51–59.

Drake, J.E., Gallet-Budynek, A., Hofmockel, K.S., Drake, J.E., Gallet-Budynek, A., Hofmockel, K.S., Bernhardt, E.S., Bilings, S.A., Jackson, R.B., Johnsen, K.S., Lichter, J., McCarthy, H.R., McCormack, M.L., Moore, D.J.P., Oren, R., Palmroth, S., Philips, R.P., Pippen, J.S., Pritchard, S.G., Treseder, K.K., Schlesinger, W.H., DeLucia, E.H., Finzi, A.C., 2011. Increases in the flux of carbon belowground stimulate nitrogen uptake and sustain the long-term enhancement of forest productivity under elevated CO_2. Ecol. Lett. 14, 349–357.

Fajvan, M.A., Wood, J.M., 1996. Stand structure and development after gypsy moth defoliation in the Appalachian Plateau. For. Ecol. Manage. 89, 79–88.

Finzi, A.C., Norby, R.J., Calfapietra, C., Gallet-Budyneka, A., Gielene, B., Holmes, W.E., Hoosbeek, M.R., Iversen, C.M., Jackson, R.B., Kubiske, M.E., Ledford, J., Liberloo, M., Oren, R., Pollek, A., Pritchard, S., Zak, D.R., Schlesingerb, W.H., Ceulemans, R., 2007. Increases in nitrogen uptake rather than nitrogen-use efficiency support higher rates of temperate forest productivity under elevated CO_2. Proc. Natl. Acad. Sci. U.S.A. 104, 14014–14019.

Freiwald, V., Häikiö, E., Julkunen-Tiitto, R., Holopainen, J.K., Oksanen, E., 2008. Elevated ozone modifies the feeding behaviour of the common leaf weevil on hybrid aspen through shifts in developmental, chemical, and structural properties of leaves. Entomol. Exp. Appl. 128, 66–72.

Gate, I.M., McNeill, S., Ashmore, M.R., 1995. Effects of air pollution on the searching behavior of an insect parasitoid. Water Air Soil Pollut. 85, 1425–1430.

Gupta, P., Duplessis, S., White, H., Karnosky, D.F., Martin, F., Podila, G.K., 2005. Gene expression patterns of trembling aspen trees following long-term exposure to interacting elevated CO_2 and tropospheric O_3. New Phytol. 167, 129–141.

Haimi, J., Laamanen, J., Penttinen, R., Räty, M., Koponen, S., Kellomäki, S., Niemelä, P., 2005. Impacts of elevated CO_2 and temperature on the soil fauna of boreal forests. Appl. Soil Ecol. 30, 104–112.

Hamilton, J.G., Zangerl, A.R., Berenbaum, M.R., Pippen, J., Aldea, M., DeLucia, E.H., 2004. Insect herbivory on an intact forest understory under experimental CO_2 enrichment. Oecologia 138, 566–573.

Hamilton, J.G., Zangerl, A.R., Berenbaum, M.R., Sparks, J.P., Elich, L., Eisenstein, A., DeLucia, E.H., 2012. Elevated atmospheric CO_2 alters the arthropod community in a forest understory. Acta Oecol. 43, 80–85.

Hansen, R.A., Williams, R.S., Degenhardt, D.C., Lincoln, D.E., 2001. Non-litter effects of elevated CO_2 on forest floor microarthropod abundances. Plant Soil 236, 139–144.

Hättenschwiler, S., Bühler, S., Körner, C., 1999. Decomposition of tree litter produced under elevated CO_2. Oikos 85, 271–281.

Hättenschwiler, S., Bretscher, D., 2001. Isopod effects on decomposition of litter produced under elevated CO_2, N deposition and different soil types. Glob. Chang. Biol. 7, 565–579.

Hättenschwiler, S., Gasser, P., 2005. Soil animals alter plant litter diversity effects on decomposition. Proc. Natl. Acad. Sci. U.S.A. 102, 1519–1524.

Heath, R.L., 2008. Modification of the biochemical pathways of plants induced by ozone: what are the varied routes to change? Environ. Pollut. 155, 453–463.

Hillstrom, M.L., Lindroth, R.L., 2008. Elevated atmospheric carbon dioxide and ozone alter forest insect abundance and community composition. Insect Conserv. Diver. 1, 233–241.

Hillstrom, M.L., Vigue, L.M., Coyle, D.R., Raffa, K.F., Lindroth, R.L., 2010. Performance of an invasive weevil *Polydrusus sericeus* is influenced by atmospheric CO_2 and host species. Agric. For. Entomol. 12, 285–292.

Holton, M.K., Lindroth, R.L., Nordheim, E.V., 2003. Foliar quality influences tree-herbivore-parasitoid interactions: effects of elevated CO_2 and O_3. Oecologia 137, 233–244.

Hunter, M.D., 2001. Effects of elevated atmospheric carbon dioxide on insect-plant interactions. Agric. For. Entomol. 3, 153–159.

Jones, M.E., Paine, T.D., 2006. Detecting changes in insect herbivore communities along a pollution gradient. Environ. Pollut. 143, 377–387.

Kagata, H., Ohgushi, T., 2012. Carbon to nitrogen excretion ratio in lepidopteran larvae: relative importance of ecological stoichiometry and metabolic scaling. Oikos 121, 1869–1877.

Karnosky, D.F., Mankovska, B., Percy, K., Dickson, R.E., Podila, G.K., Sober, J., Noormets, A., Hendrey, G., Coleman, M.D., Kubiske, M., Pregitzer, K.S., Isebrands, J.G., 1998. Genetic control of responses to interacting tropospheric ozone and CO_2 in *Populus tremuloides*. Chemosphere 36, 807–812.

Karnosky, D.F., Pregitzer, K.S., Zak, D.R., Kubiske, M.E., Hendrey, G.R., Weinstein, D., Nosal, M., Percy, K.E., 2005. Scaling ozone responses of forest trees to the ecosystem level in a changing climate. Plant Cell Environ. 28, 965–981.

Kasurien, A., Peltonen, P.A., Julkunen-Tiitto, R., Vapaavuori, E., Nuutinen, V., Holopainen, J.K., 2007. Effects of elevated CO_2 and O_3 leaf litter phenolics and subsequent performance of litter-feeding soil macrofauna. Plant Soil 292, 25–43.

King, J.S., Kubiske, M.E., Pregitzer, K.S., Hendrey, G.R., McDonald, E.P., Giardina, C.P., Quinn, V.S., Karnosky, D.F., 2005. Tropospheric O_3 compromises net primary production in young stands of trembling aspen, paper birch and sugar maple in response to elevated atmospheric CO_2. New Phytol. 168, 623–636.

King, J., Liu, L., Aspinwall, M., 2013. Forest ecosystem response to interacting elevated atmospheric CO_2 and tropospheric O_3: a synthesis of experimental evidence. Dev. Environ. Sci. 13, 179–208.

Knepp, R.G., Hamilton, J.G., Mohan, J.E., Zangerl, A.R., Berenbaum, M.R., DeLucia, E.H., 2005. Elevated CO_2 reduces leaf damage by insect herbivores in a forest community. New Phytol. 167, 207–218.

Kopper, B.J., Lindroth, R.L., 2003. Responses of trembling aspen (*Populus tremuloides*) phytochemistry and aspen blotch leafminer (*Phyllonorycter tremuloidiella*) performance to elevated level of CO_2 and O_3. Agric. For. Entomol. 5, 17–26.

Körner, C., 2006. Plant CO_2 responses: an issue of definition, time, and resources supply. New Phytol. 172, 393–411.

Kubiske, M.E., Quinn, V.S., Marquardt, P.E., Karnosky, D.F., 2007. Effects of elevated atmospheric CO_2 and/or O_3 on intra- and interspecific competitive ability of aspen. Plant Biol. 9, 342–355.

Leakey, A.D.B., Ainsworth, E.A., Bernacchi, C.J., Rogers, A., Long, S.P., Ort, D.R., 2009. Elevated CO_2 effects on plant carbon, nitrogen, and water relations: six important lessons from FACE. J. Exp. Bot. 60, 2859–2876.

Lindroth, R.L., 2010. Impacts of elevated atmospheric CO_2 and O_3 on forests: phytochemistry, trophic interactions, and ecosystem dynamics. J. Chem. Ecol. 36, 2–21.

Lindroth, R.L., 2011. Atmospheric change, plant secondary metabolites and ecological interactions. In: Iason, G.R., Dicke, M., Hartley, S.E. (Eds.), The Ecology of Plant Secondary Metabolites: From Genes to Global Processes. Cambridge University Press, UK, pp. 120–153.

Lindroth, R.L., Roth, S., Kruger, E.L., Volin, J.C., 1997. CO_2-mediated changes in aspen chemistry: effects on gypsy moth performance and susceptibility to virus. Glob. Chang. Biol. 3, 279–289.

Long, S.P., Ainsworth, E.A., Rogers, A., Ort, D.R., 2004. Rising atmospheric carbon dioxide: plants FACE the future. Annu. Rev. Plant Biol. 55, 591–628.

Loranger, G.I., Pregitzer, K.S., King, J.S., 2004. Elevated CO_2 and O_3 concentrations differentially affect selected groups of the fauna in temperate forest soils. Soil Biol. Biochem. 36, 1521–1524.

Lou, Y., Su, B., Currie, W.S., Dukes, J.S., Finzi, A., Hartwig, U., Hungate, B., McMurtrie, R.E., Oren, R., Parton, W.J., Patanki, D.E., Shaw, M.R., Zak, D.R., Field, C.B., 2004. Progressive nitrogen limitation of ecosystem responses to rising atmospheric carbon dioxide. Bioscience 54, 731–739.

Lovett, G.M., Christenson, L.M., Groffman, P.M., Jones, C.G., Hart, J.E., Mitchell, M.J., 2002. Insect defoliation and nitrogen cycling in forests. Bioscience 54, 335–341.

Lovett, G.M., Canham, C.D., Arthur, M.A., Weathers, K.C., Fitzhugh, R.D., 2006. Forest ecosystem response to exotic pests and pathogens in eastern North America. Bioscience 56, 395–405.

Madritch, M.D., Donaldson, J.R., Lindroth, R.L., 2007. Canopy herbivory can mediate the influence of plant genotype on soil processes through frass deposition. Soil Biol. Biochem. 39, 1192–1201.

Mattson, W.J., Addy, N.D., 1975. Phytophagous insects as regulators of forest primary production. Science 190, 515–522.

Matyssek, R., Wieser, G., Fleischmann, F., Grünhage, L., 2013. Ozone research, quo vadis? Lessons from the Kranzberg forest experiment. Dev. Environ. Sci. 13, 103–129.

McDonald, E.P., Kruger, E.L., Riemenschneider, D.E., Isebrands, J.G., 2002. Competitive status influences tree-growth responses to elevated CO_2 and O_3 in aggrading aspen stands. Funct. Ecol. 16, 792–801.

Meehan, T.D., Crossley, M.S., Lindroth, R.L., 2010. Impacts of elevated CO_2 and O_3 on aspen leaf litter chemistry and earthworm and springtail productivity. Soil Biol. Biochem. 42, 1132–1137.

Mondor, E.B., Tremblay, M.N., Awmack, C.S., Lindroth, R.L., 2004. Divergent pheromone-mediated insect behaviour under global atmospheric change. Glob. Chang. Biol. 10, 1820–1824.

Nabity, P.D., Hillstrom, M.L., Lindroth, R.L., DeLucia, E.H., 2012. Elevated CO_2 interacts with herbivory to alter chlorophyll fluorescence and leaf temperature in *Betula papyrifera* and *Populus tremuloides*. Oecologia 169, 905–913.

Norby, R.J., Zak, D.R., 2011. Ecological lessons from free-air CO_2 enrichment (FACE) experiments. Ann. Rev. Ecol. Evol. Syst. 42, 181–203.

Norby, R.J., DeLucia, E.H., Gielen, B., Calfapietra, C., Giardina, C.P., King, J.S., Ledford, J., McCarthy, H.R., Moore, D.J.P., Ceulmans, R., De Angelis, P., Finzi, A.C., Karnosky, D.F., Kubiske, M.E., Lukac, M., Pregitzer, K.S., Scarascia-Mugnozza, G.E., Schlesinger, W.H., Oren, R., 2005. Forest response to elevated CO_2 is conserved across a broad range of productivity. Proc. Natl. Acad. Sci. U.S.A. 102, 18052–18056.

Oksanen, E., 2003a. Physiological responses of birch (*Betula pendula*) to ozone: a comparison between open-soil grown trees exposed for six growing seasons and potted seedlings exposed for one season. Tree Physiol. 23, 603–614.

Oksanen, E., 2003b. Responses of selected birch (*Betula pendula* Roth) clones to ozone change over time. Plant Cell Environ. 26, 875–886.

Peltonen, P., Vapaavuori, E., Julkunen-Tiitto, R., 2005. Accumulation of phenolic compounds in birch leaves is changed by elevated carbon dioxide and ozone. Glob. Chang. Biol. 11, 1305–1324.

Peltonen, P.A., Julkunen-Tiitto, R., Vapaavuori, E., Holpainen, J.K., 2006. Effects of elevated carbon dioxide and ozone on aphid oviposition preference and birch bud exudates phenolics. Glob. Chang. Biol. 12, 1670–1679.

Peltonen, P.A., Vapaavuori, E., Heinonen, J., Julkunen-Tiitto, R., Holopainen, J.K., 2010. Do elevated atmospheric CO_2 and O_3 affect food quality and performance of folivorous insects on silver birch? Glob. Chang. Biol. 16, 918–935.

Percy, K.E., Awmack, C.S., Lindroth, R.L., Kubiske, M.E., Kopper, B.J., Isebrands, J.G., Pregitzer, K.S., Hendrey, G.R., Dickson, R.E., Zak, D.R., Oksanen, E., Sober, J., Harrington, R., Karnosky, D.F., 2002. Altered performance of forest pests under atmospheres enriched by CO_2 and O_3. Nature 420, 403–407.

Pinto, D.M., Blande, J.D., Nykänen, R., Dong, W.-X., Nerg, A.M., Holopainen, J.K., 2007. Ozone degrades common herbivore-induced plant volatiles: does this affect herbivore prey location by predators and parasitoids? J. Chem. Ecol. 4, 683–694.

Robinson, E.A., Ryan, G.D., Newman, J.A., 2012. A meta-analytical review of the effects of elevated CO_2 on plant-arthropod interactions highlights the importance of interacting environmental and biological variables. New Phytol. 194, 321–336.

Rossi, A.M., Stiling, P., Moon, D.C., Cattell, M.V., Drake, B.G., 2004. Induced defensive response of myrtle oak to foliar insect herbivory in ambient and elevated CO_2. J. Chem. Ecol. 30, 1143–1152.

Roth, S.K., Lindroth, R.L., 1995. Elevated atmospheric CO_2: effects on phytochemistry, insect performance and insect-parasitoid interactions. Glob. Chang. Biol. 1, 173–182.

Sanders, N.J., Belote, R.T., Weltzin, J.F., 2004. Multitrophic effects of elevated atmospheric CO_2 on understory plant and arthropod communities. Environ. Entomol. 33, 1609–1616.

Stange, G., 1997. Effects of changes in atmospheric carbon dioxide on the location of hosts by the moth, *Cactoblastis cactorum*. Oecologia 110, 539–545.

Stiling, P., Cornelissen, T., 2007. How does elevated carbon dioxide (CO_2) affect plant-herbivore interactions? A field experiment and meta-analysis of CO_2-mediated changes on plant chemistry and herbivore performance. Glob. Chang. Biol. 13, 1–20.

Stiling, P., Cattell, M., Moon, D.C., Rossi, A.M., Hungate, B.A., Hymus, G., Drakes, B.G., 2002. Elevated atmospheric CO_2 lowers herbivore abundance, but increases leaf abscission rate. Glob. Chang. Biol. 8, 658–667.

Stiling, P., Moon, D.C., Hunter, M.D., Colson, J., Rossi, A.M., Hymus, G.J., Drake, B.G., 2003. Elevated CO_2 lowers relative and absolute herbivore density across all species of a scrub-oak forest. Oecologia 134, 82–87.

Stiling, P., Moon, D., Rossi, A., Hungate, B.A., Drake, B., 2009. Seeing the forest for the trees: long-term exposure to elevated CO_2 increases some herbivore densities. Glob. Chang. Biol. 15, 1895–1902.

Townsend, P.A., Eshleman, K.N., Welcker, C., 2004. Remote sensing of gypsy moth defoliation to assess variation in stream nitrogen concentrations. Ecol. Appl. 14, 504–516.

Traw, M.B., Lindroth, R.L., Bazzaz, F.A., 1996. Decline in gypsy moth (*Lymantria dispar*) performance in an elevated CO_2 atmosphere depends upon host plant species. Oecologia 108, 113–120.

Tylianakis, J.M., Didham, R.K., Bascompte, J.W., 2008. Global change and species interactions in terrestrial ecosystems. Ecol. Lett. 11, 1351–1363.

Valkama, E., Koricheva, J., Oksanen, E., 2007. Effects of elevated O_3, alone and in combination with elevated CO_2, on tree leaf chemistry and insect herbivore performance: a meta-analysis. Glob. Chang. Biol. 13, 184–201.

Vuorinen, T., Nerg, A., Holopainen, J.K., 2004. Ozone exposure triggers the emission of herbivore-induced plant volatiles, but does not disturb tritrophic signaling. Environ. Pollut. 131, 305–311.

Wang, X.-W., Ji, L.-Z., Zhang, Q.-H., Liu, Y., Wang, G.-Q., 2009. Effects of elevated CO_2 on feeding preference and performance of the gypsy moth (*Lymantria dispar*) larvae. J. Appl. Entomol. 133, 47–57.

Wittig, V.E., Ainsworth, E.A., Naidu, S.L., Karnosky, D.F., Long, S.P., 2009. Quantifying the impact of current and future tropospheric ozone on tree biomass, growth, physiology and biochemistry: a quantitative metaanalysis. Glob. Chang. Biol. 15, 396–424.

Zak, D.R., Pregitzer, K.S., Kubiske, M.E., Burton, A.J., 2011. Forest productivity under elevated CO_2 and O_3: positive feedbacks to soil N cycling sustain decade-long net primary productivity enhancement by CO_2. Ecol. Lett. 14, 1220–1226.

Zak, D.R., Kubiske, M.E., Pregitzer, K.S., Burton, A.J., 2012. Atmospheric CO_2 and O_3 alter competition for soil nitrogen in developing forests. Glob. Chang. Biol. 18, 1480–1488.

Zvereva, E.L., Kozlov, M.V., 2006. Consequences of simultaneous elevation of carbon dioxide and temperature for plant-herbivore interactions: a metaanalysis. Glob. Chang. Biol. 12, 27–41.

Zvereva, E.L., Zverev, V., Kozlov, M., 2012. Little strokes fell great oaks: minor but chronic herbivory substantially reduces birch growth. Oikos 121, 2036–2043.

Part IV

Mechanistic and Diagnostic Understanding for Risk Assessment and Up-Scaling

Chapter 12

Flux-Based Ozone Risk Assessment for Adult Beech and Spruce Forests

Ludger Grünhage[*,1], Rainer Matyssek[†], Gerhard Wieser[‡], Karl-Heinz Häberle[†], Michael Leuchner[§,2], Annette Menzel[§], Jochen Dieler[¶], Hans Pretzsch[¶], Winfried Grimmeisen[||], Lothar Zimmermann[||], Stephan Raspe[||] and Matthias Schröder[#]

[*]*Department of Plant Ecology, Justus-Liebig University, Giessen, Germany*

[†]*Technische Universität München, Ecophysiology of Plants, Hans-Carl-von-Carlowitz-Platz 2, D-85354 Freising-Weihenstephan, Germany*

[‡]*Department of Alpine Timberline Ecophysiology, Federal Office and Research Centre for Forests, Innsbruck, Austria*

[§]*Ecoclimatology, Technische Universität München, Wissenschaftszentrum Weihenstephan, Hans-Carl-von-Carlowitz-Platz 2, D-85354 Freising-Weihenstephan, Germany*

[¶]*Forest Growth and Yield Science, Wissenschaftszentrum Weihenstephan, Technische Universität München, Hans-Carl-von-Carlowitz-Platz 2, D-85354 Freising-Weihenstephan, Germany*

[||]*Bayerische Landesanstalt für Wald und Forstwirtschaft, Hans-Carl-von-Carlowitz-Platz 1, D-85354 Freising-Weihenstephan, Germany*

[#]*Department of Plant Ecology, Justus-Liebig University, D-35392 Giessen, Germany*

[1]*Corresponding author: e-mail: Ludger.Gruenhage@bot2.bio.uni-giessen.de*

Chapter Outline

2. Current address: Physical Geography/Landscape Ecology, Katholische Universität Eichstätt-Ingolstadt, Ostenstr. 18, D-85072 Eichstätt, Germany.

Developments in Environmental Science, Vol. 13. http://dx.doi.org/10.1016/B978-0-08-098349-3.00012-8

12.1 INTRODUCTION

Tropospheric ozone (O_3) poses a critical hazard and problem to present and future forest ecosystem services such as fibre and timber production or carbon storage capacity. As a consequence of the discussion about the reasons for the so-called neuartige Waldschäden (forest dieback; cf. Grennfelt and Schjoldager, 1984; Skärby and Selldén, 1984), ground-level ozone and its impacts on human health and vegetation have increasingly come into focus since the mid-1980s within the UNECE (United Nations Economic Commission of Europe) in the context of the Convention on Long-Range Transboundary Air Pollution (LRTAP) and the European Union (EU). A recent data synthesis, published by the International Cooperative Programme on Effects of Air Pollution on Natural Vegetation and Crops (ICP Vegetation) Coordination Centre (http://icpvegetation.ceh.ac.uk; Hayes et al., 2007; Mills et al., 2011a), suggested widespread occurrence of O_3 effects on vegetation at ambient concentrations in Europe over the period of 1990 through 2006.

A first European workshop on critical levels for O_3 to protect vegetation[1] was held in 1988 in Bad Harzburg, Germany (UNECE, 1988), followed by a second one in 1992 in Egham, United Kingdom (Ashmore and Wilson, 1992). While the 1988 long-term critical level for O_3 was defined as a 7-h mean of 25 $nl\,l^{-1}$ over the growing season, at the Egham workshop a change to critical levels based on O_3 concentrations at the top of the canopy (LRTAP Convention, 2010) *A*ccumulated *O*ver a *T*hreshold of 40 $nl\,l^{-1}$, AOT40, was recommended, assuming that plants have adapted to low, pre-industrial, naturally occurring O_3 concentrations (Paoletti and Manning, 2007). At UNECE workshops in Bern, Switzerland, 1993, and Kuopio, Finland, 1996, AOT40-based critical levels for O_3 to protect crops, semi-natural vegetation and forest trees were defined (Fuhrer and Achermann, 1994; Kärenlampi and Skärby, 1996; UBA, 1996). A more detailed description of the early phase of the critical-level approach is given in Fuhrer et al. (1997).

The AOT40-based critical levels are used as risk indicators for the LRTAP Convention's protocol to Abate Acidification, Eutrophication and Ground-level O_3 of 1999 (LRTAP Convention, 1999) as well as the European Directive on Ground-level O_3 (EU, 2002, 2008). The concentration-based critical levels underlying exposure response functions are mainly derived from experimental studies in open-top chambers. Paoletti and Manning (2007) summarized the following main reasons, cited here, why the concentration-based

1. LRTAP Convention (2010): Currently, critical levels are defined as the 'concentration, cumulative exposure or cumulative stomatal flux of atmospheric pollutants above which direct adverse effects on sensitive vegetation may occur according to present knowledge'.

critical levels are inadequate: (1) inadequacy of the scientific background, (2) insufficient database for the derivation of the standards, (3) insufficient database to select the effect to be evaluated in the field, (4) inappropriate grouping into categories (e.g. one critical level for deciduous and evergreen species), and (5) no field validation. As stated in the scientific background documents (Fuhrer and Achermann, 1994; Kärenlampi and Skärby, 1996) as well as in the LRTAP Convention's Modelling and Mapping Manual versions (LRTAP Convention, 2004, 2010; UBA, 1996), the 'concentration-based critical levels are suitable for estimating the risk of damage' only, that is, quantitative, 'economic losses should not be estimated using AOTx-based critical levels and associated response functions' (LRTAP Convention, 2010).

One of the basic rules of toxicology is that dose–response relationships can only be established if the effective dose at the target site (e.g. membranes) or at least the absorbed dose of the stressor is known (e.g. Dämmgen and Grünhage, 1998; Dämmgen et al., 1993; Massman et al., 2000; Matyssek et al., 2008).

As a toxicologically more sound concept, the stomatal flux approach was introduced into the UNECE critical-level community at the workshops in Bern and Kuopio, in 1993 and 1996, respectively (e.g. Grünhage and Jäger, 1994a,b, 1996). Progress in deriving cumulative stomatal flux-based critical levels was reviewed at LRTAP Convention workshops held in Gothenburg, Sweden, 2002 (Karlsson et al., 2003), in Obergurgl, Austria, 2005 (Wieser and Tausz, 2006), in Ispra, Italy, 2009, and as part of the LRTAP Convention's ICP Vegetation Task Force Meeting in Tervuren, Belgium, 2010. As summarized by Mills et al. (2011b), eight flux-based critical levels for effects on crops, forest trees and (semi-)natural vegetation were improved for inclusion in the current LRTAP Convention's Modelling and Mapping Manual version (LRTAP Convention, 2010) after discussions at meetings of the LRTAP Convention's ICP Forests, ICP Modelling and Mapping, Working Group on Effects and Executive Body. As stated in the LRTAP Convention's Modelling and Mapping Manual, 'flux-based critical levels and associated response functions are suitable for mapping and quantifying impacts at the local and regional scale, including effects on food security (crops), roundwood supply for the forest sector industry and loss of carbon storage capacity and other beneficial ecosystem services (forest trees), and impacts on the vitality of fodder–pasture and natural grassland species ((semi-)natural vegetation)' and in contrast to the concentration-based critical levels 'they could be used for assessing economic losses' in quantitative terms.

12.2 THE LRTAP CONVENTION'S STOMATAL O_3 FLUX APPROACH FOR FOREST TREES

The LRTAP Convention's stomatal O_3 flux approach is based on the assumption that the stomatal O_3 uptake by plants, that is, the flux of O_3 from the atmosphere through the stomata F_{stom, O_3} (nmol m^{-2} PLA s^{-1}), above a statistically deduced constant flux threshold Y (nmol m^{-2} PLA s^{-1})[2] provides an estimate of the critical amount of O_3 entering through the stomata and reaching the sites of action inside the plant (target sites, e.g. membranes). It has been shown that the cumulative stomatal O_3 uptake above a constant threshold flux of Y accumulated over a stated time period during daylight hours (global radiation > 50 W m^{-2}), the so-called phytotoxic ozone dose POD_Y (mmol m^{-2} PLA), gives stronger relationships with effects than external concentration-based exposure indices.

$$POD_Y = \sum_{i=1}^{n} \left[\max\left(F_{\text{sunlit leaf, stom}, O_3} - Y, 0\right) \cdot \Delta t\right]_i$$

POD_Y is interpreted as an estimate of the effective dose at the target sites and is calculated from hourly values of stomatal O_3 uptake by the sunlit upper canopy leaves $F_{\text{sunlit leaf, stom}, O_3}$ (nmol m^{-2} s^{-1}); n denotes the number of hours to be included in the calculation period and $\Delta t = 1$ h (LRTAP Convention, 2010).

The statistically derived constant flux threshold Y is interpreted as a provisional estimate of a detoxification threshold, below which it is assumed that any O_3 molecule absorbed by the plant will be detoxified in the apoplast before reaching a target site. To reflect this detoxification (cf. Dizengremel et al., 2013, this vol.) for beech/birch and Norway spruce, Y is set to 1 nmol O_3 m^{-2} PLA s^{-1} due to expert judgement (cf. Mills et al., 2011b) because flux–response relationships were strongest when either no or a small flux threshold were applied.

For the evaluation of O_3-related risk for reduction in annual growth rates (whole-tree biomass) of Norway spruce and beech/birch, response functions were derived from fumigation experiments mainly in open-top chambers with young trees of up to 10 years of age (Figure 12.1; LRTAP Convention, 2010). For a detailed description of the data sets and experimental conditions, see Braun and Flückiger (1995), Karlsson et al. (2004) and Uddling et al. (2004).

The stomatal O_3 uptake-response functions are based on the assumption that the O_3 fumigation concentrations measured at canopy top in the respective exposure systems reflect the O_3 concentrations at the upper surface boundary of the sunlit leaves' laminar layer. Additionally, it is assumed that under ambient conditions in the field (e.g. forest stands), the O_3 concentration at the top of the canopy is a reasonable estimate of the O_3 concentration at the upper surface boundary of the laminar layer of the sunlit upper canopy leaves, unless the roughness sub-layer near the canopy is taken into account in the deposition modelling approach (cf. LRTAP Convention, 2010; Tuovinen and

---

2. LRTAP Convention (2010): 'The projected leaf area (*PLA*, m^2) is the total area of the sides of the leaves that are projected towards the sun. *PLA* is in contrast to the total leaf area, which considers both sides of the leaves. For horizontal leaves the total leaf area is simply 2**PLA*'.

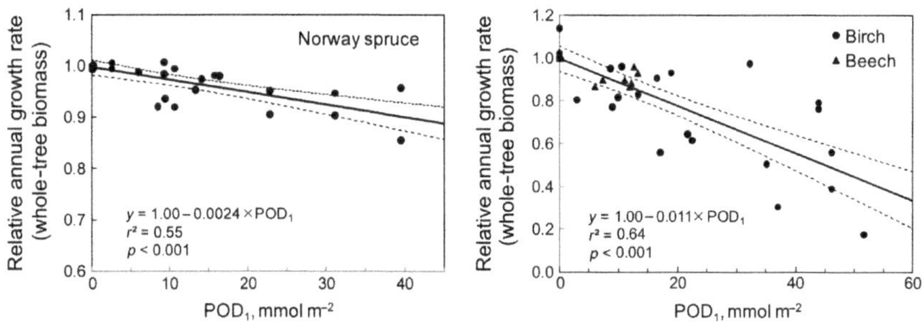

FIGURE 12.1 Relationships between relative annual growth rates (whole-tree biomass) and POD_1 for sunlit leaves of Norway spruce (left) and beech/birch (right). (POD_1 = phytotoxic ozone dose (mmol m^{-2} PLA) above a constant threshold flux of 1 nmol m^{-2} PLA s^{-1} accumulated over the respective growing season during daylight hours; PLA = projected leaf area). *Source: LRTAP Convention (2010) modified.*

Simpson, 2008). Considering both assumptions, stomatal O_3 uptake of the sunlit upper canopy leaves $F_{\text{sunlit leaf, stom}, O_3}$ (nmol m^{-2} s^{-1}) can be approximated by

$$F_{\text{sunlit leaf, stom}, O_3} = c_{O_3}(z_h) \cdot g_{\text{sunlit leaf, stom}, O_3} \cdot \frac{R_{\text{sunlit leaf, total}, O_3}}{R_{\text{sunlit leaf, laminar layer}, O_3} + R_{\text{sunlit leaf, total}, O_3}}$$

with $c_{O_3}(z_h)$ the O_3 concentration at canopy top z_h (nmol m^{-3}), $g_{\text{sunlit leaf, stom}, O_3}$ the stomatal conductance for O_3 (m s^{-1}), $R_{\text{sunlit leaf, total}, O_3}$ the total leaf resistance for O_3 (s m^{-1}) and $R_{\text{sunlit leaf, laminar layer}, O_3}$ the resistance of the sunlit-leaf laminar layer for O_3 (s m^{-1}). The total leaf resistance is calculated according to

$$R_{\text{sunlit leaf, total}, O_3} = \frac{1}{g_{\text{sunlit leaf, stom}, O_3} + g_{\text{sunlit leaf, external leaf surface}, O_3}}$$

where $g_{\text{sunlit leaf, external leaf surface}, O_3}$ is the conductance of the external leaf surface for O_3 (m s^{-1}) which currently is set constant to 0.0004 m s^{-1} according to LRTAP Convention (2010). Future research concepts should consider the derivation of a more realistic parameterization of non-stomatal O_3 deposition on the external leaf surface. The influence of the leaf laminar boundary layer on stomatal O_3 gas exchange is approximated according to McNaughton and van den Hurk (1995):

$$R_{\text{sunlit leaf, laminar layer}, O_3} = 1.3 \cdot 150 \cdot \sqrt{\frac{L_{\text{leaf}}}{u(z_h)}}$$

with the characteristic crosswind leaf dimensions $L_{\text{leaf, Norway spruce}} = 0.008$ m and $L_{\text{leaf, beech}} = 0.07$ m. $u(z_h)$ is the horizontal wind velocity at canopy height z_h (m s^{-1}). 150 as a constant exhibits the dimension $s^{0.5}$ m^{-1}, while the factor 1.3 accounts for the differences in diffusivity between sensible heat and O_3.

Stomatal O_3 uptake is parameterized for a sunlit upper canopy leaf according to a multiplicative Jarvis–Stewart approach (Jarvis, 1976; Stewart, 1988):

$$g_{\text{sunlit leaf, stom, } O_3} = g_{\text{sunlit leaf, stom, max, } O_3} \cdot \left[\min\left(f_{\text{phen}}, f_{O_3}\right)\right] \cdot f_{\text{light}} \cdot \max\left\{f_{\min}, \left(f_{\text{temp}} \cdot f_{\text{VPD}} \cdot f_{\text{SM}}\right)\right\}$$

where $g_{\text{sunlit leaf, stom, max}, O_3}$ represents the maximum level of the stomatal conductance for O_3 $g_{\text{sunlit leaf, stom}, O_3}$ per unit projected leaf area (mmol O_3 m^{-2} PLA s^{-1}), that is, the stomatal conductance under optimal environmental conditions, and f_i weighting factors expressed in relative terms. The weighting factors f_i take values between 0 and 1 as a proportion of $g_{\text{sunlit leaf, stom}, O_3}$ and incorporate the effects of plant phenology (f_{phen}), ozone load (f_{O_3}), radiation (f_{light}), air temperature (f_{temp}), water vapour pressure deficit of the air surrounding the leaves (f_{VPD}) and soil moisture (f_{SM}) on $g_{\text{sunlit leaf, stom, max}, O_3}$. $f_{\min}$ is based on the analysis of published data and set to, for example, 0.16 for Norway spruce and 0.13 for beech in continental Central Europe (LRTAP Convention, 2010). Currently, no data are available to deduce a parameterization for f_{O_3}, so this factor is set to unity. For detailed parameterization of weighting factors f_i, see LRTAP Convention (2010) and Grünhage et al. (2012b).

While for Norway spruce, it is assumed that needles are physiologically active above an air temperature of 0 °C, for beech, the physiologically active growth period is the time between leaf unfolding and the onset of dormancy (i.e. autumnal leaf shedding). Leaf discolouration is assumed to occur 20 days prior to dormancy (LRTAP Convention, 2010). If no phenological observation data are available, start and end of the beech physiologically active growth period is parameterized according to a latitude model (cf. Grünhage et al., 2012b; LRTAP Convention, 2010).

As described earlier, stomatal O_3 uptake estimations depend on $g_{\text{sunlit leaf, stom, max}, O_3}$ which cannot be measured directly. Generally, stomatal conductance for O_3 is related to that of water vapour or carbon dioxide by the ratio of the respective molecular diffusivities D. According to the values cited in LRTAP Convention (2010) for beech in continental Central Europe, the mean maximum stomatal conductance for water vapour of the sunlit leaves is 245 mmol H_2O m^{-2} PLA s^{-1}, a value identical with the maximum stomatal water vapour conductance for beech leaves of the sun crown in Kranzberg Forest as assessed by Nunn et al. (2005). For Norway spruce, $g_{\text{sunlit leaf, stom, max}, H_2O}$ is 204 mmol H_2O m^{-2} PLA s^{-1}. Since the molecular diffusivity of O_3 in air has never been measured, it must be derived from known molecular diffusivities of another gas or its characteristic properties. A critical literature review performed for the 2010 revision of the LRTAP Convention's Modelling and Mapping Manual (cf. Grünhage et al., 2012a) leads to change from the diffusivity ratio $D_{O_3}/D_{H_2O} = 0.613$ used in the previous versions of the LRTAP Convention's Modelling and Mapping Manual to the more sound diffusivity ratio $D_{O_3}/D_{H_2O} = 0.663$. Nevertheless, the $g_{\max}$ values as well as the response functions for beech and Norway spruce given in LRTAP

Convention (2010) are based on the previously used diffusivity ratio $D_{O_3}/D_{H_2O}=0.613$. The application of the more sound diffusivity ratio of $D_{O_3}/D_{H_2O}=0.663$ leads to a mean maximum stomatal O_3 conductance of 162 mmol O_3 PLA m^{-2} s^{-1} for beech and of 135 mmol O_3 PLA m^{-2} s^{-1} for Norway spruce. Based on these g_{max} values, provisionally, that is, until now not included in the LRTAP Convention's Modelling and Mapping Manual, corrected response functions for European beech

$$\text{relative total biomass (loss in annual growth rate)}_{\text{European beech}} = 1.00 - 0.0102 \cdot \text{POD}_1$$

and Norway spruce

$$\text{relative total biomass (loss in annual growth rate)}_{\text{Norway spruce}} = 1.00 - 0.0022 \cdot \text{POD}_1$$

were derived and will be used here.

It must be emphasized (i) that the estimated stomatal O_3 fluxes should not be interpreted as the actual, site-specific ones and therefore (ii) the described LRTAP Convention's stomatal O_3 flux approach should be viewed as an indication of potential risk for whole-tree losses in biomass production due to O_3 (LRTAP Convention, 2010). The estimation of 'realistic' site-related losses requires site-specific parameterizations. A risk evaluation for a situation with no soil water limitation on stomatal behaviour, that is, $f_{SM}=1$, can be interpreted as a worst-case assessment, while a risk evaluation considering the 'actual' soil water content gives an indication of the lower range of potential biomass losses due to O_3. Because the 'actual' risk for vegetation due to O_3 in a specific year and at a specific site depends, in particular, on intensity and temporal variation of precipitation, worst-case risk evaluations should be preferred if the focus lies on a general protection of vegetation.

12.3 THE KRANZBERG FOREST EXPERIMENT: A VALIDATION EXPERIMENT FOR THE LRTAP CONVENTION'S STOMATAL FLUX APPROACH FOR FOREST TREES

The 8-year canopy-level free-air O_3 enrichment experiment at Kranzberg Forest near Freising/Germany, performed during 2000–2007, is worldwide the only fumigation experiment with adult Norway spruce and European beech trees (Matyssek et al., 2010). For a detailed description, see Nunn et al. (2002), Werner and Fabian (2002), Karnosky et al. (2007), Pretzsch et al. (2010), and Matyssek et al. (2007, 2010). The study focused on 10 representative Norway spruce and 10 European beech trees, which were about 60 years old and up to 28 m high (Pretzsch et al., 2010). Five of the 10 study trees of each tree species were exposed to a twice-ambient O_3 regime ($2 \times O_3$), where the individual tree was considered as the replication unit (i.e. as a case study each on its own).

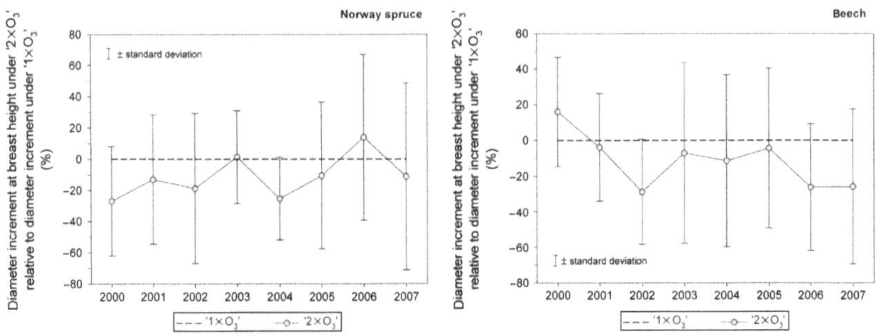

FIGURE 12.2 Year-to-year deviation in observed loss of diameter increment at breast height for Norway spruce (left) and beech (right) under twice-ambient O_3 exposure ($2 \times O_3$) relative to diameter increment under ambient O_3 (=0%).

To prevent acute O_3 injury, the O_3 levels of this regime were restricted to a maximum of 150 nl l^{-1} (Matyssek and Sandermann, 2003). Five adjacent trees under the unchanged ambient regime ($1 \times O_3$) prevailing at the site served as control. Ambient O_3 concentrations were recorded at $z = 28$ m, while the elevated O_3 concentrations were measured at $z = 20$ m inside the crown at the transition between shaded and sunlit crown area. Air temperature and air humidity were measured at $z = 24$ m and horizontal wind velocity at $z = 36$ m. Soil water potential was modelled with BROOK90 (Hammel and Kennel, 2001). Statistical analyses were performed using repeated measurement ANOVA procedures.

Based on measurements of diameter increment at breast height, a statistically not significant mean annual loss by 11.4% for Norway spruce and 11.5% for beech was observed under $2 \times O_3$ relative to $1 \times O_3$ (Figure 12.2).

As detailed in Pretzsch et al. (2010), the exposure of the tree cohorts during 8 years to $2 \times O_3$ induced some shift in resource allocation into height growth but reduction of diameter growth in both tree species. In the case of Norway spruce, this change in allometry leads to cone-shaped stem forms, a stem form which may result eventually in reduced stem stability. In the case of beech, the allometric changes led to neiloidal stem forms. Taking into account the effect of O_3 on stem shape, the 8-year exposure to $2 \times O_3$ caused, on average, a statistically non-significant increase by 2.4% in annual volume growth in Norway spruce and a decrease by 43.5% in beech in relation to the annual growth rate occurring under $1 \times O_3$ (Figure 12.3). While the O_3 effect on stem productivity of beech was significant, the year-to-year variation in losses in biomass formation was not.

As stated by Pretzsch et al. (2010), neglect of such ozone-induced changes in stem shape may lead to a flawed estimation of volume growth. Therefore, derivation of dose–effect relationships by epidemiological studies under ambient air restricted to increment measurements at breast height appears to be questionable (cf. Figures 12.2 and 12.3).

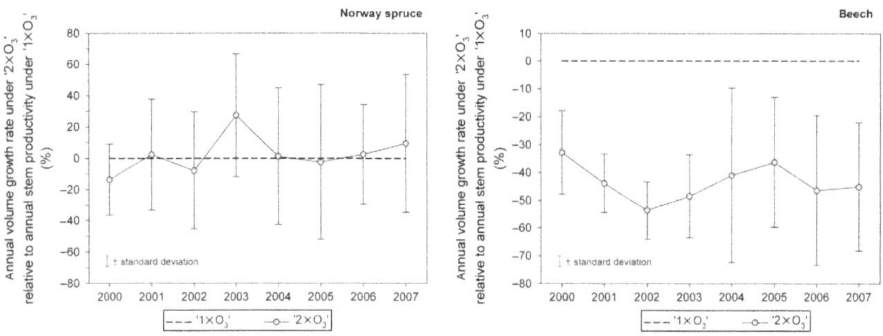

FIGURE 12.3 Year-to-year deviation in observed loss in whole-stem productivity, that is, loss in annual volume growth rate, for Norway spruce (left) and beech (right) under twice-ambient O_3 exposure ($2 \times O_3$) relative to stem productivity under ambient O_3 (=0%).

As described briefly in Grünhage et al. (2012b), for the O_3-related risk evaluation of adult Norway spruce and beech stands at local scale, the Soil–Vegetation–Atmosphere-Transfer (SVAT) model FO_3REST is under development. The leaf-level risk-assessment module of FO_3REST, used in this study, is parameterized according to LRTAP Convention (2010)—applying the site-specific f_{temp}-parameterization for beech—as described in Grünhage et al. (2012b). Due to the experimental setup of the Kranzberg Forest experiment, we assume (i) that the fumigation concentration measured at $z = 20$ m reflects the O_3 concentration at the upper surface boundary of the laminar layer of the sunlit leaves in treatment $2 \times O_3$ and (ii) that the ambient O_3 concentration measured at $z = 28$ m provides a reasonable estimate of the O_3 concentration at the upper surface boundary of the laminar layer of the sunlit leaves in treatment $1 \times O_3$. Since we assume that the horizontal wind velocity measured at $z = 36$ m is more or less identical to the wind velocity at canopy top, the actual leaf surface boundary layer resistance will be underestimated, that is, stomatal O_3 uptake will be slightly overestimated. The model result uncertainties given in Figure 12.4 are based on the achievable measurement uncertainties of the meteorological data needed as input variables for stomatal O_3 uptake calculation, which are deduced from several sources (e.g. WMO, 2008) for measurement devices of high quality (O_3 concentration: ±2%, global radiation: ±3%, air temperature: ±0.15 K, air humidity: ±3%, air pressure: ±2 hPa, horizontal wind velocity (2D sonic anemometer): $\pm 0.1 \text{ m s}^{-1}$ if wind velocity $\leq 5 \text{ m s}^{-1}$ else ±2%). It must be emphasized that the model result uncertainty due to the measurement uncertainties, that is, the precision of the devices, of the model input variables represents an important but not the total uncertainty of POD_Y estimates.

Other uncertainties can be associated with the choice of the values of internal model parameters such as, for example, $g_{\text{sunlit leaf, stom, max}, O_3}$ (cf. Tuovinen et al., 2009). But the latter uncertainties are associated with the derivation of

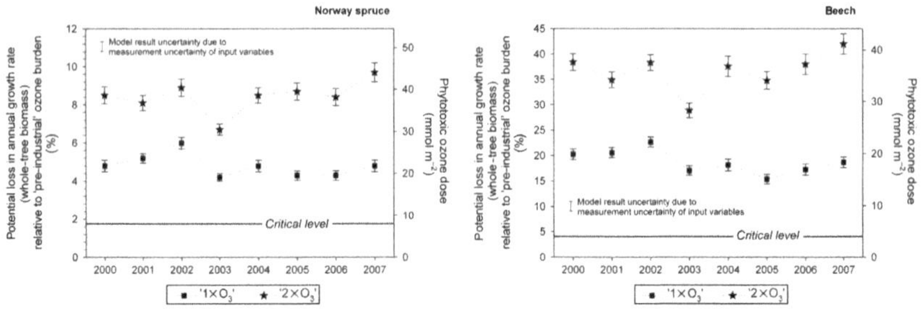

FIGURE 12.4 Year-to-year deviation in potential loss in annual growth rate (whole-tree biomass), for Norway spruce (left) and beech (right) relative to 'pre-industrial' O_3 burden under ambient ($1 \times O_3$) and twice-ambient O_3 exposure ($2 \times O_3$).

dose–effect relationships and, therefore, should not be considered in risk-assessment studies.

From the information given in LRTAP Convention (2010), it can be deduced that the estimated potential losses in biomass formation (Figure 12.4) are expressed, in a strict sense of their meaning, relative to the 'pre-industrial' O_3 burden. The range of estimated potential loss in annual growth rate (whole-tree biomass) at Kranzberg Forest is from 4.2% to 6.0% under $1 \times O_3$ and 6.7% to 9.7% under $2 \times O_3$ for Norway spruce; for beech, the ranges are 15.4% to 22.7% and 28.9% to 42.0%, respectively. During the hot and dry summer of 2003, there was a strong influence of soil moisture conditions on stomatal O_3 uptake. The LRTAP Convention's critical levels were exceeded significantly under ambient O_3 exposure for both tree species. As mentioned above, the flux–effect relations for Norway spruce and beech/birch (Figure 12.1) were derived from fumigation experiments mainly in open-top chambers with young, well-watered trees of up to 10 years of age. From the response function and the observed losses in biomass formation of juvenile Norway spruce and beech in unfiltered air (cf. Figure 12.1; Braun and Flückiger, 1995; Karlsson et al., 2004; Uddling et al., 2004), it can be concluded that the O_3 burden during the 1980s/1990s led for spruce to a 2–3% and for beech to an approximately 10% reduction in annual whole-plant biomass growth rates in ambient air relative to a 'pre-industrial' situation. As shown in Figure 12.4, the O_3 burden during the Kranzberg Forest experiment in $1 \times O3$ led to an additional potential annual loss in biomass formation in comparison to the 1980s/1990s, which corresponded with the observed increase in ambient O_3 concentrations during the past decades in Germany (cf. Figure 7 in Grünhage et al., 2012a). The estimated mean potential losses in whole-tree productivity for spruce of approximately 5% relative to the 'pre-industrial' O_3 burden is in the order of magnitude of the potential losses described by Tuovinen et al. (2013, this vol.). Obviously, the Kranzberg Forest is at high risk.

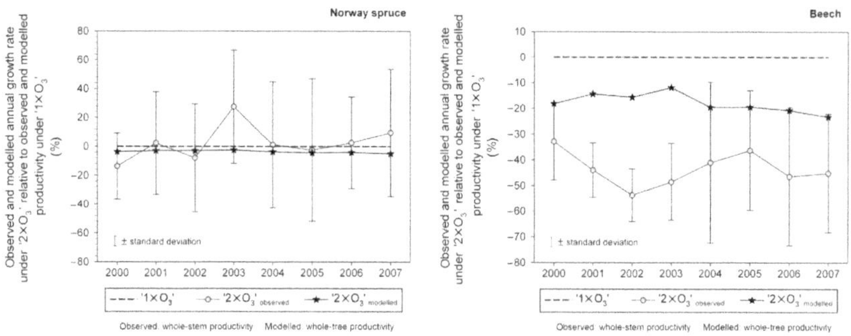

FIGURE 12.5 Year-to-year deviation in observed loss in whole-stem productivity and modelled loss in annual growth rate (whole-tree biomass) for Norway spruce (left) and beech (right) under twice-ambient O_3 exposure ($2 \times O_3$) relative to whole-stem/whole-tree productivity under ambient O_3 (=0%).

Potential losses in whole-tree biomass formation due to O_3 for treatment $2 \times O_3$ relative to the modelled potential losses under $1 \times O_3$ (i.e. the differences of the modelled losses between the two treatments) are compared in Figure 12.5 with the actually observed losses in whole-stem productivity under $2 \times O_3$ (i.e. differences between observed whole-stem productivity under $1 \times O_3$ and $2 \times O_3$). While, on average, in Norway spruce, a slight potential loss in biomass production of 3.6% was estimated, loss was significant in beech, amounting to 17.8%. Since the year-to-year variation in observed losses in whole-stem biomass formation of beech was not significant, the modelled whole-tree biomass losses for beech can be interpreted as losses in the lower range of the maximum standard deviation of the observed ones. Thus, the LRTAP Convention's risk assessment for beech based on potential stomatal O_3 uptake of sunlit leaves must be seen as conservative, indicating (i) European beech forests to be at risk and (ii) that risk may be underestimated.

12.4 CONCLUSIONS AND PERSPECTIVES FOR FUTURE O_3 RISK ASSESSMENTS AT STAND LEVEL

The LRTAP Convention's risk-assessment approach for Norway spruce and beech briefly described here is based on fumigation experiments with young trees of up to 10 years while the Kranzberg Forest canopy-level free-air O_3 fumigation study was performed with adult trees. Comparison of observed and modelled loss in annual growth under twice-ambient O_3 exposure ($2 \times O_3$) relative to whole-stem productivity under ambient O_3 ($1 \times O_3$) seems to confirm the LRTAP Convention's leaf-level stomatal flux approach and the associated response function for Norway spruce up to twice-ambient O_3 exposure. For European beech, it must be emphasized that the LRTAP Convention's methodology at leaf level may underestimate the risk for loss in whole-stem productivity due to O_3, because the estimated productivity losses are in the lower range of the standard deviation of the observed ones. When accounting for the observed increase in below-ground productivity (Nikolova et al., 2010; H. Pretzsch, personal communication), the apparent offset between the observed and modelled losses (cf. Figure 12.5) becomes smaller due to a conversion of whole-stem into whole-tree productivity. The observed impact of twice-ambient O_3 exposure on whole-stem productivity in Norway spruce and European beech at Kranzberg Forest indicates that adult beech trees are more sensitive than adult spruce ones and that O_3 impacts on C sequestration in beech forests will be an important issue for future carbon balance.

The potential underestimation of modelled biomass losses for adult beech trees highlights the need for effect–flux relationships deduced from experiments with adult trees and therefore for new experimental concepts such as 'supersites' (Mikkelsen et al., 2013, this vol.). From the comparison of the observed losses in diameter increment at breast height (Figure 12.2) and in annual volume growth rate (Figure 12.3), it can be concluded that the derivation of dose–effect relationships by epidemiological studies under ambient air based on increment measurements at breast height only is questionable.

As mentioned above, the LRTAP Convention's stomatal flux parameterization is based on O_3 concentrations at canopy top. Because O_3 concentrations and the other meteorological input variables are not measured at canopy height by the European air quality monitoring networks and the ICP Forests monitoring stations, the model input variables measured at a reference height (e.g. 3.5 m above ground for O_3 or 10 m for horizontal wind velocity) must be transformed to those at canopy top in local risk-assessment studies. Such a conversion has to be performed with an appropriate methodology, that is, a flux-gradient model (Cieslik et al., 2013, this vol.). The model input variable transformation methodology and local scale risk-assessment approach for forest stands can be split into three steps: (1) micrometeorologically correct transformation of O_3 concentrations as well as the other model input variables needed for $g_{\mathrm{sunlit\,leaf,\,stom,}O_3}$ modelling from near ground measurement height to the forest canopy top (cf. Tuovinen et al., 2009), (2) calculation of sunlit-leaf stomatal uptake and phytotoxic ozone dose (POD_1), and (3) risk evaluation. For routine applications of the risk-assessment methodology at local scale, simplifications, especially in step (1), may be appropriate, but the model result uncertainties due to such a simplification should be given. Additionally, the model result uncertainties due to the respective measurement uncertainties should be considered.

For practical use in local risk assessments, tree- and/or stand-level flux–response relationships need to be developed. Prior to new free-air experiments of adult trees at stand level for mechanistic clarification (Matyssek et al., 2013, this vol.), scaling the current leaf-level parameterizations to tree and stand level seems to be the next logical step (cf. Wieser et al., 2008). As mentioned above, the application of the current stomatal uptake–biomass loss relationships for beech—considering the observed increase in below-ground biomass—and Norway spruce seems to reflect more or less the observed losses in annual volume growth rate under doubled O_3 exposure relative to stem productivity under ambient O_3. Therefore, an up-scaling of the leaf-level stomatal O_3 uptake-based response functions to tree level seems to be appropriate and may be possible by comparison of leaf-level stomatal fluxes and tree-level ones taking into account a scaling approach such as proposed by Grünhage et al. (2011) and Matyssek et al. (2013, this vol.). These up-scaled response functions together with sap-flow-based O_3 flux estimates can be practical tools for tree/stand-level-based O_3 risk assessments at local scale.

ACKNOWLEDGEMENTS

The modelling work of Matthias Schröder was supported by the German Federal Environment Agency (Umweltbundesamt, FKZ: 3711 63 235). The free-air O_3 enrichment experiment at Kranzberg Forest was funded by 'Bayerisches Staatsministerium für Umwelt, Gesundheit und Verbraucherschutz', by 'Deutsche Forschungsgemeinschaft' (DFG) through SFB 607 'Growth and Parasite defence—Competition for Resources in Economic Plants from Agronomy and Forestry', and through the Project 'CASIROZ—The carbon sink strength of beech in a changing environment: Experimental risk assessment by mitigation of chronic ozone impact', supported by the European Commission's Directorate-General for Research, Environment Programme, 'Natural Resources Management and Services' (EVK2-2002-00165, Ecosystem Vulnerability).

REFERENCES

Ashmore, M.R., Wilson, R.B., 1992. Critical levels of air pollutants for Europe. In: Background Papers Prepared for the United Nations Economic Commission for Europe workshop on Critical Levels, Egham, United Kingdom, March 23–26. Air Quality Division, Department of the Environment, London.

Braun, S., Flückiger, W., 1995. Effects of ambient ozone on seedlings of *Fagus sylvatica* L. and *Picea abies* (L.) Karst. New Phytol. 129, 33–44.

Cieslik, S., Tuovinen, J.-P., Baumgarten, M., Matyssek, R., Brito, P., Wieser, G., 2013. Gaseous exchange between forests and the atmosphere. Dev. Environ. Sci. 13, 19–36.

Dämmgen, U., Grünhage, L., 1998. Response of a grassland ecosystem to air pollutants. V. A toxicological model for the assessment of dose-response relationships for air pollutants and ecosystems. Environ. Pollut. 101, 375–380.

Dämmgen, U., Grünhage, L., Haenel, H.-D., Jäger, H.-J., 1993. Climate and stress in ecotoxicology. A coherent system of definitions and terms. Angew. Bot. (now J. Appl. Bot.) 67, 157–162.

Dizengremel, P., Jolivet, Y., Tuzet, A., Ranieri, A., Le Thiec, D., 2013. Integrative Leaf-Level Phytotoxic Ozone Dose Assessment for Forest Risk Modelling. Dev. Environ. Sci. 13, 267–288.

EU, 2002. Directive 2002/3/EC of the European Parliament and of the Council of 12 February 2002 relating to ozone in ambient air. Official Journal of the European Communities L 067, 09/03/2002, 14–30.

EU, 2008. Directive 2008/50/EC of the European Parliament and of the Council of 21 May 2008 on ambient air quality and cleaner air for Europe. Official Journal of the European Union L 152, 1–44.

Fuhrer, J., Achermann, B., 1994. Critical levels for ozone—a UN-ECE workshop report. FAC Schriftenreihe (Eidgenössische Forschungsanstalt für Agrikulturchemie und Umwelthygiene, Bern) 16, Bern.

Fuhrer, J., Skärby, L., Ashmore, M.R., 1997. Critical levels for ozone effects on vegetation in Europe. Environ. Pollut. 97, 91–106.

Grennfelt, P., Schjoldager, J., 1984. Photochemical oxidants in the troposphere: a mounting menace. Ambio 13, 61–67.

Grünhage, L., Jäger, H.-J., 1994a. Atmospheric ozone exposure-potential for vegetation: how suitable are critical levels? In: Fuhrer, J., Achermann, B. (Eds.), Critical Levels for Ozone—a UN-ECE Workshop Report. FAC Schriftenreihe, vol. 16. Eidgenössische Forschungsanstalt für Agrikulturchemie und Umwelthygiene, Bern, pp. 222–230.

Grünhage, L., Jäger, H.-J., 1994b. Influence of the atmospheric conductivity on the ozone exposure of plants under ambient conditions: considerations for establishing ozone standards to protect vegetation. Environ. Pollut. 85, 125–129.

Grünhage, L., Jäger, H.-J., 1996. Critical levels for ozone, ozone exposure potentials of the atmosphere or critical absorbed doses for ozone: a general discussion. In: Kärenlampi, L., Skärby, L. (Eds.), Critical Levels for Ozone in Europe: Testing and Finalizing the Concepts. UN-ECE Workshop Report. Department of Ecology and Environmental Science, University of Kuopio, Kuopio, pp. 151–168.

Grünhage, L., Braden, H., Bender, J., Burkart, S., Lehmann, Y., Schröder, M., 2011. Evaluation of the ozone-related risk for winter wheat at local scale with the CRO_3PS model. Gefahrst. Reinhalt. L. 71, 90–97, available at http://www.uni-giessen.de/cms/CRO3PS.

Grünhage, L., Pleijel, H., Mills, G., Bender, J., Danielsson, H., Lehmann, Y., Castell, J.-F., Bethenod, O., 2012a. Updated stomatal flux and flux-effect models for wheat for quantifying effects of ozone on grain yield, grain mass and protein yield. Environ. Pollut. 165, 147–157.

Grünhage, L., Matyssek, R., Häberle, K.-H., Wieser, G., Metzger, U., Leuchner, M., Menzel, A., Dieler, J., Pretzsch, H., Grimmeisen, W., Zimmermann, L., Raspe, S., 2012b. Flux-based ozone risk assessment for adult beech forests. Trees 26, 1713–1721.

Hammel, K., Kennel, M., 2001. Charakterisierung und Analyse der Wasserverfügbarkeit und des Wasserhaushalts von Waldstandorten in Bayern mit dem Simulationsmodell BROOK90. Forstliche Forschungsberichte No. 185, München.

Hayes, F., Mills, G., Harmens, H., Norris, D., 2007. Evidence of Widespread Ozone Damage to Vegetation in Europe. ICP Vegetation Programme Coordination Centre, Centre for Ecology & Hydrology, Bangor, UK, 58 pp. (available at http://icpvegetation.ceh.ac.uk).

Jarvis, P.G., 1976. The interpretation of the variations in leaf water potential and stomatal conductance found in canopies in the field. Philos. Trans. R. Soc. B 273, 593–610.

Kärenlampi, L., Skärby, L., 1996. Critical Levels for Ozone in Europe: Testing and Finalizing the Concepts. UN-ECE Workshop Report. Department of Ecology and Environmental Science, University of Kuopio, Kuopio.

Karlsson, P.E., Selldén, G., Pleijel, H., 2003. Establishing Ozone Critical Levels II. UNECE Workshop Report. IVL Report B 1523, IVL Swedish Environmental Research Institute, Gothenburg.

Karlsson, P.E., Uddling, J., Braun, S., Broadmeadow, M., Elvira, S., Gimeno, B.S., le Thiec, D., Oksanen, E., Vandermeiren, K., Wilkinson, M., Emberson, L., 2004. New critical levels for ozone effects on young trees based on AOT40 and simulated cumulative leaf uptake of ozone. Atmos. Environ. 38, 2283–2294.

Karnosky, D.F., Werner, H., Holopainen, T., Percy, K., Oksanen, T., Oksanen, E., Heerdt, C., Fabian, P., Nagy, J., Heilman, W., Cox, R., Nelson, N., Matyssek, R., 2007. Free-air exposure systems to scale up ozone research to mature trees. Plant Biol. 9, 181–190.

LRTAP Convention, 1999. Protocol to the 1979 Convention on Long-Range Transboundary Air Pollution to Abate Acidification, Eutrophication and Ground-Level Ozone.

LRTAP Convention, 2004. Mapping Manual 2004. Manual on methodologies and criteria for modelling and mapping critical loads & levels and air pollution effects, risk and trends. Mapping critical levels for vegetation (Chapter 3). Texte 52/04. Umweltbundesamt, Berlin (available at http://www.icpmapping.org/Mapping_Manual).

LRTAP Convention, 2010. Chapter 3 Mapping Manual 2004. Manual on methodologies and criteria for modelling and mapping critical loads & levels and air pollution effects, risk and trends. Mapping critical levels for vegetation. 2010 revision, minor text changes made June 2011 (available at http://icpvegetation.ceh.ac.uk).

Massman, W.J., Musselman, R.C., Lefohn, A.S., 2000. A conceptual ozone dose-response model to develop a standard to protect vegetation. Atmos. Environ. 34, 745–759.

Matyssek, R., Sandermann, H., 2003. Impact of ozone on trees: an ecophysiological perspective. Prog. Bot. 64, 349–404.

Matyssek, R., Bahnweg, G., Ceulemans, R., Fabian, P., Grill, D., Hanke, D.E., Kraigher, H., Oßwald, W., Rennenberg, H., Sandermann, H., Tausz, M., Wieser, G., 2007. Synopsis of the CASIROZ case study: carbon sink strength of *Fagus sylvatica* L. in a changing environment—experimental risk assessment of mitigation by chronic ozone impact. Plant Biol. 9, 163–180.

Matyssek, R., Sandermann, H., Wieser, G., Booker, F., Cieslik, S., Musselman, R., Ernst, D., 2008. The challenge of making ozone risk assessment for forest trees more mechanistic. Environ. Pollut. 156, 567–582.

Matyssek, R., Wieser, G., Ceulemans, R., Rennenberg, H., Pretzsch, H., Haberer, K., Löw, M., Nunn, A.J., Werner, H., Wipfler, P., Oßwald, W., Nikolova, P., Hanke, D.E., Kraigher, H., Tausz, M., Bahnweg, G., Kitao, M., Dieler, J., Sandermann, H., Herbinger, K., Grebenc, T., Blumenröther, M., Deckmyn, G., Grams, T.E.E., Heerdt, C., Leuchner, M., Fabian, P., Häberle, K.-H., 2010. Enhanced ozone strongly reduces carbon sink strength of adult beech (*Fagus sylvatica*)—resume from the free-air fumigation study at Kranzberg Forest. Environ. Pollut. 158, 2527–2532.

Matyssek, R., Wieser, G., Fleischmann, F., Grünhage, L., 2013. Ozone Research, quo vadis? Lessons from the Kranzberg Forest Experiment. Dev. Environ. Sci. 13, 103–129.

McNaughton, K.G., van den Hurk, B.J.J.M., 1995. A 'Lagrangian' revision of the resistors in the two-layer model for calculating the energy budget of a plant canopy. Bound. Lay. Meteorol. 74, 261–288.

Mikkelsen, T.N., Clarke, N., Danielewska, A., Fischer, R., 2013. Towards supersites in forest ecosystem monitoring and research. Dev. Environ. Sci. 13, 475–495.

Mills, G., Hayes, F., Simpson, D., Emberson, L., Norris, D., Harmens, H., Büker, P., 2011a. Evidence of widespread effects of ozone on crops and (semi-)natural vegetation in Europe (1990-2006) in relation to AOT40- and flux-based risk maps. Glob. Chang. Biol. 17, 592–613.

Mills, G., Pleijel, H., Braun, S., Büker, P., Bermejo, V., Calvo, E., Danielsson, H., Emberson, L., Gonzáles Fernández, I., Grünhage, L., Harmens, H., Hayes, F., Karlsson, P.-E., Simpson, D., 2011b. New stomatal flux-based critical levels for ozone effects on vegetation. Atmos. Environ. 45, 5064–5068.

Nikolova, P.S., Andersen, C.P., Blaschke, H., Matyssek, R., Häberle, K.-H., 2010. Belowground effects of enhanced tropospheric ozone and drought in a beech/spruce forest (*Fagus sylvatica* L./*Picea abies* [L.] Karst). Environ. Pollut. 158, 1071–1078.

Nunn, A.J., Reiter, I.M., Häberle, K.-H., Werner, H., Langebartels, C., Sandermann, H., Heerdt, C., Fabian, P., Matyssek, R., 2002. "Free-air" ozone canopy fumigation in an old-growth mixed forest: concept and observations in beech. Phyton (Austria) Special issue: "Global change" 42, 105–119.

Nunn, A.J., Kozovits, A.R., Reiter, I.M., Heerdt, C., Leuchner, M., Lütz, C., Liu, X., Löw, M., Winkler, J.B., Grams, T.E.E., Häberle, K.-H., Werner, H., Fabian, P., Rennenberg, H., Matyssek, R., 2005. Comparison of ozone uptake and sensitivity between a phytotron study with young beech and a field experiment with adult beech (*Fagus sylvatica*). Environ. Pollut. 137, 494–506.

Paoletti, E., Manning, W.J., 2007. Toward a biologically significant and usable standard for ozone that will also protect plants. Environ. Pollut. 150, 85–95.

Pretzsch, H., Dieler, J., Matyssek, R., Wipfler, P., 2010. Tree and stand growth of mature Norway spruce and European beech under long-term ozone fumigation. Environ. Pollut. 158, 1061–1070.

Skärby, L., Selldén, G., 1984. The effects of ozone on crops and forests. Ambio 13, 68–72.

Stewart, J.B., 1988. Modelling surface conductance of pine forest. Agric. For. Meteorol. 43, 19–35.

Tuovinen, J.-P., Simpson, D., 2008. An aerodynamic correction for the European ozone risk assessment methodology. Atmos. Environ. 42, 8371–8381.

Tuovinen, J.-P., Emberson, L., Simpson, D., 2009. Modelling ozone fluxes to forests for risk assessment: status and prospects. Ann. For. Sci. 66, 401. http://dx.doi.org/10.1051/forest/2009024.

Tuovinen, J.-P., Hakola, H., Karlsson, P.E., Simpson, D., 2013. Air pollution risks to northern European forests in a changing climate. Dev. Environ. Sci. 13, 77–99.

UBA–Umweltbundesamt, 1996. Manual on methodologies and criteria for mapping critical levels/loads and geographical areas where they are exceeded. Texte 71/96, Umweltbundesamt, Berlin.

Uddling, J., Günthardt-Goerg, M.S., Matyssek, R., Oksanen, E., Pleijel, H., Selldén, G., Karlsson, P.E., 2004. Biomass reduction of juvenile birch is more strongly related to stomatal uptake of ozone than to indices based on external exposure. Atmos. Environ. 38, 4709–4719.

UNECE, 1988. ECE Critical Levels Workshop Report. Bad Harzburg, Germany, 14–18 March 1988 (Final Draft Report). United Nations Economic Commission for Europe.

Werner, H., Fabian, P., 2002. Free-air fumigation of mature trees. A novel system for the controlled ozone enrichment in grown-up beech and spruce canopies. Environ. Sci. Pollut. R. 9, 117–121.

Wieser, G., Tausz, M., 2006. Proceedings on the workshop "Critical Levels of Ozone: Further Applying and Developing the Flux-based Concept". Obergurgl, Tyrol, Austria, 15-19 November 2005. Federal Research and Training Centre for Forests, Natural Hazards and Landscape (BFW), Vienna.

Wieser, G., Matyssek, R., Then, C., Cieslik, S., Paoletti, E., Ceulemans, R., 2008. Upscaling ozone flux in forests from leaf to landscape. Ital. J. Agronomy (Rivista di Agronomia) 1, 35–41.

WMO, 2008. Guide to meteorological instruments and methods of observations. WMO-No. 8, seventh ed. World Meteorological Organisation, Geneva, Switzerland.

Chapter 13

Integrative Leaf-Level Phytotoxic Ozone Dose Assessment for Forest Risk Modelling

Pierre Dizengremel[*,†,1], Yves Jolivet[*,†], Andrée Tuzet[‡], Annamaria Ranieri[§] and Didier Le Thiec[*,†]

[*]*Université de Lorraine, UMR 1137 EEF, Champenoux, France*

[†]*INRA, UMR 1137 EEF, Champenoux, France*

[‡]*UMR INRA/AgroParisTech EGC, Thiverval-Grignon, France*

[§]*Department of Agriculture, Food and Environment, University of Pisa, Pisa, Italy*

[1]*Corresponding author: e-mail: pierre.dizengremel@univ-lorraine.fr*

Chapter Outline

Developments in Environmental Science, Vol. 13. http://dx.doi.org/10.1016/B978-0-08-098349-3.00013-X

13.1 INTRODUCTION

The harmful impact of the phytotoxic air pollutant ozone on natural plant communities, crops and forests is widely recognized (Ashmore, 2005; Matyssek and Innes, 1999; Matyssek et al., 2012a). Plants exposed to current ozone concentrations during prolonged periods of time (chronic ozone exposure) show a range of effects, including visible foliar injury and growth reduction, combined with premature senescence (Matyssek and Sandermann, 2003). In trees, meta-analyses suggest that current ozone concentration is decreasing photosynthesis by around 10% (Wittig et al., 2007). Reduction of photosynthesis is the indication of early-stage effects by ozone on cellular metabolism (Dizengremel, 2001; Heath, 2008; Renaut et al., 2009). The intracellular impact of ozone in leaves will thus depend on both the uptake of the pollutant through stomata and the detoxification capacity of the cells (Matyssek et al., 2008). The detoxification capacity must be maintained by regeneration of the antioxidant compounds, and the rate of regeneration would control the effective ozone flux which corresponds to the non-detoxified ozone molecules (Dizengremel et al., 2008; Matyssek et al., 2008; Musselman et al., 2006; Paoletti and Manning, 2007). Even though ascorbate (ASC) is currently proposed as the main actor in counteracting ozone and reactive oxygen species (ROS), recent results show that the mechanisms controlling plant sensitivity to ozone still remain obscure (Booker et al., 2012; D'Haese et al., 2005). This chapter summarizes the main knowledge on the subject and will propose pathways for overcoming gaps towards scientifically reliable, process-based risk assessment.

13.2 OZONE AND CARBON METABOLISM

13.2.1 Primary Metabolism

The biochemical events caused by ozone exposure precede macroscopic foliar visible symptoms and growth reduction. Decrease of photosynthesis was early recognized to be paralleled by declining activity and quantity of the chloroplast CO_2-fixing enzyme, ribulose-1,5-bis phosphate carboxylase/oxygenase (RuBisCO) (Degl'Innocenti et al., 2002; Dizengremel et al., 1994; Pell et al., 1992). The abundance of other enzymes of the photosynthetic carbon reduction cycle (Calvin cycle) was also shown to be decreased under ozone impact (Renaut et al., 2009). Paralleling the reduction in photosynthesis, respiratory processes are enhanced by ozone (Bagard et al., 2008; Reich, 1983). At the level of metabolic pathways, the progressive decreases in RuBisCO activity and quantity are accompanied by a strong increase in the activity and quantity of the cytosolic phosphoenolpyruvate carboxylase (PEPC), which normally plays a minor role in carbon fixation by unstressed C_3 plants (Dizengremel, 2001; Fontaine et al., 1999; Renaut et al., 2009; Saurer et al., 1995; Sehmer et al., 1998).

Under ozone stress, the RuBisCO/PEPC ratio decreases by a factor of 3–20 (Dizengremel et al., 2009). This contrasting behaviour of the two carboxylases was also observed in the field where needles of declining trees in ozone-polluted areas showed reduced RuBisCO/PEPC ratios (Dalstein et al., 2002; Wild and Schmitt, 1995). The stimulation of PEPC activity, usually linked to mitochondrial respiration and to the anaplerotic pathway of organic acid-dependent synthesis of amino acids (Dizengremel et al., 2012; Sweetlove et al., 2010) (Figure 13.1) is associated with enhanced enzymatic activities of glycolysis and the pentose phosphate pathway (Dizengremel et al., 2009; Sehmer et al., 1998). The increased involvement of those enzymes may foster the metabolic reducing power in support of detoxification mechanisms (see Section 13.3.3).

An overall integrated biochemical and physiological view of the responses of primary carbon metabolism to ozone therefore emerges, with an increase in carbohydrate breakdown at the expense of photosynthate production. In addition, part of the carbon pool is used for the synthesis of a range of secondary compounds (Figure 13.1; see also Section 13.2.2 and Figure 13.2). A body of facts indicates phosphoenolpyruvate (PEP) to be situated at the central junction of four metabolic routes towards O_3 response (Figure 13.1; Dizengremel et al., 2012): (1) mitochondrial respiration; (2) isoprene formation in chloroplasts through the methylerythritol 4-phosphate (MEP) pathway; (3) the shikimic-phenylpropanoid pathway producing aromatic

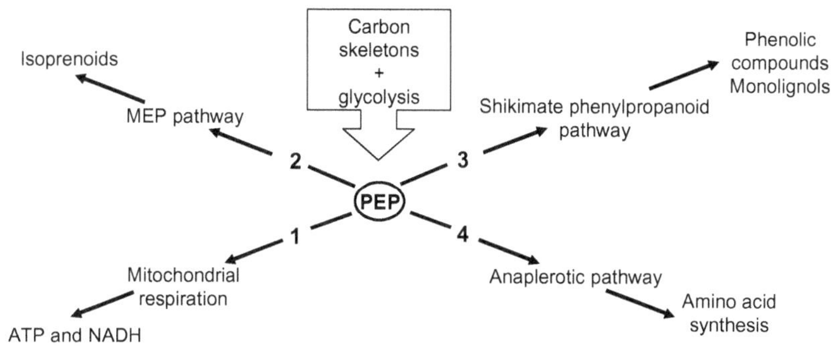

FIGURE 13.1 Schematic diagram showing the central role of phosphoenolpyruvate distributing carbon in different metabolic pathways. PEP, phosphoenolpyruvate; MEP, methylerythritol 4-phosphate (Dizengremel et al., 2012).

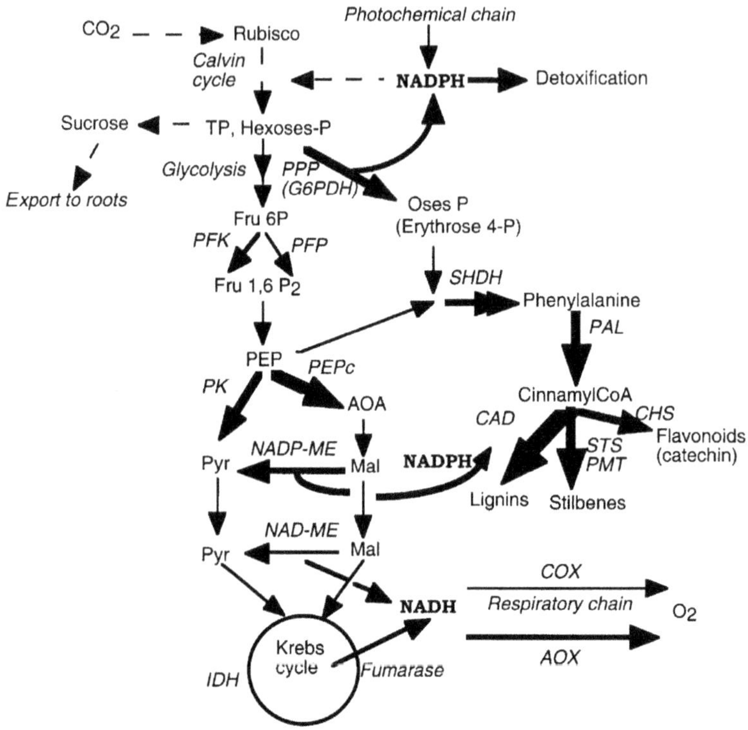

FIGURE 13.2 Implication of carbon metabolism pathways in trees exposed to ozone. AOX, alternative oxidase; CAD, cinnamyl alcohol dehydrogenase; CHS, chalcone synthase; COX, cytochrome oxidase; G6PDH, glucose 6-phosphate dehydrogenase; IDH, isocitrate dehydrogenase; ME, malic enzyme; PAL, phenylalanine ammonia lyase; PEPC, phosphoenolpyruvate carboxylase; PFK, phosphofructokinase; PFP, pyrophosphate fructose 6-phosphate phosphotransferase; PK, pyruvate kinase; PPP, pentose phosphate pathway; PMT, pinosylvin methyl transferase; SHDH, shikimate dehydrogenase; STS, stilbene synthase. Dashed arrows: inhibition; continuous arrows: stimulation, more and more marked according to the thickness of the arrow (Dizengremel, 2001).

amino acids, precursors of phenolic compounds and monolignols; and (4) the anaplerotic pathway starting with PEPC activity. This latter pathway serves to replenish the tricarboxylic acid (TCA) cycle allowing TCAs to be used for further amino acid synthesis.

13.2.2 Secondary Metabolism

The initial part of the phenylpropanoid pathway is formed by the shikimic acid pathway which links the primary to the secondary carbon metabolism (Figures 13.1 and 13.2). The first step leads to the amino acid phenylalanine. Phenylalanine is further transformed into cinnamoyl-CoA, a precursor marking the starting point of three important pathways leading to the production of stilbenes, catechin and lignins. As a result, the enhanced phenolic synthesis requires re-adjustment between primary and secondary metabolism (Castagna and Ranieri, 2009; Dizengremel, 2001) (Figure 13.2).

Ozone triggers the upregulation of antioxidant defence systems linked to the shikimic pathway (Valkama et al., 2007) and induces an accumulation of polyphenolic compounds, flavonoids and cell wall-bound polyphenols such as lignins, in various tree species (Langebartels et al., 2002). The phenylpropanoid pathway is one of the most affected targets of O_3, which induces genes, enzymes and biosynthetic products (Cabané et al., 2004; Castagna and Ranieri, 2009; Di Baccio et al., 2008; Eckey-Kaltenbach et al., 1994; Fares et al., 2010; Saleem et al., 2001).

A protective role provided by phenolic metabolites has been proposed, due to their effectiveness as radicals and ROS scavengers (Grace, 2005; Langebartels et al., 2002), probably enabling stress tolerance to the pollutant. However, according to Booker et al. (2012), free phenolics in the apoplast space would be ineffective in limiting O_3 diffusion to the plasmalemma, depending on both their low concentration and reactivity with O_3. The role of phenolic compounds under ozone stress is widely discussed in Oksanen et al., 2013, this vol.

A role in O_3 tolerance was attributed to condensed tannins and catechin because they were enriched in the tolerant aspen clones relative to the sensitive ones (Häikiö et al., 2009). A change in the phenolic pattern of total leaf extract was also observed, especially quercetin glycoside, known for high antioxidant capacity (Kontunen-Soppela et al., 2007).

An O_3-dependent induction of enzyme activities involved in the phenylpropanoid biosynthetic pathway (SHDH, PAL and CAD) was observed in poplar leaves following O_3 exposure (Cabané et al., 2004; Di Baccio et al., 2008). The activities of the biosynthetic enzymes, stilbene synthase, pinosylvin methyl transferase and chalcone synthase, were stimulated by ozone in Scots pine (Zinser et al., 1998). Similar to results reported by Koch et al. (1998) in poplar, increased PAL activity was associated with enhanced tolerance to O_3 as shown by the clone I-214 compared to the sensitive clone Eridano (Di Baccio et al., 2008). Ozone also induced leaf phenolic accumulation in both poplar clones, the increase being higher in Eridano. However, specific phenolics displayed differential trends in the two clones, with free and esterified forms of caffeic and ferulic acids changing in opposite ways in the two clones, which suggests modifications in the antioxidant capacity as well as in lignin composition (Ranieri, A., et al., unpublished; Table 13.1), according to previous observations (Cabané et al., 2004).

Salicylic acid (SA), a central molecule in the plant signalling route activated in response to O_3, is also synthesized by the shikimate pathway and frequently increases following O_3 exposure (Diara et al., 2005; Pasqualini et al., 2002). SA is believed to play a central role, in association with ethylene, in O_3-induced lesion initiation and progressive response to O_3 (Rao et al., 2002). Therefore,

TABLE 13.1 Leaf Phenolic Acid Concentration in I-214 and Eridano Poplar Clones Exposed to Filtered Air (C) or to 60 ppb O_3, 5 h a Day for 15 Days (O_3)

		I-214			Eridano		
		C	O_3	% *Variation*	C	O_3	% *Variation*
Gallic acid	Free	n.d.	n.d.		n.d.	91 (35)	+
	Glycosylated	n.d.	n.d.		n.d.	171 (65)	+
	Esterified	n.d.	n.d.		n.d.	n.d.	
	Total					262	+
Caffeic acid	Free	49 (3)	93 (10)	90	177 (7)	114 (3)	−36
	Glycosylated	817 (48)	718 (78)	−12	1377 (55)	1352 (32)	−2
	Esterified	854 (59)	112 (12)	−87	937 (38)	2757 (65)	194
	Total	1720	923	−46	2492	4223	70
Coumaric acid	Free	52 (2)	26 (2)	−50	547 (13)	860 (18)	57
	Glycosylated	581 (18)	29 (2)	−95	927 (21)	948 (20)	2
	Esterified	2586 (80)	1595 (96)	−38	2853 (66)	2997 (62)	3
	Total	3219	1650	−49	4327	4805	11
Ferulic acid	Free	354 (28)	287 (21)	−19	142 (10)	174 (21)	23
	Glycosylated	220 (17)	231 (17)	5	198 (14)	189 (22)	−5
	Esterified	694 (55)	838 (62)	21	1074 (76)	481 (57)	−55
	Total	1268	1536	−7	1414	844	−40
Cinnamic acid	Free	120 (25)	396 (38)	230	n.d.	n.d.	
	Glycosylated	n.d.	500 (49)	+	n.d.	72 (23)	+
	Esterified	367 (75)	131 (13)	−64	105 (100)	242 (77)	130
	Total	487	1027	111	105	314	199

Data (mean value, $n=3$) are expressed as mmol/gfw. Numbers in parentheses refer to percentage of free, glycosylated and esterified forms with respect to the total concentration of the individual acid. For each clone, percentage of variation of O_3-exposed samples with respect to control is reported (Ranieri et al., unpublished).

the O_3-dependent re-adjustment of the phenylpropanoid metabolism may have multiple effects, ranging from O_3/ROS scavenging to physically impeding (e.g. through lignin deposition) to signalling, altogether determining plant response to O_3. A complementary overview of all the changes in metabolism as well as in processes of defence against oxidative stress is provided by a set of technologies including transciptomics, proteomics and metabolomics (Renaut et al., 2009), as discussed in Ernst, 2013, this vol.; Oksanen et al., 2013, this vol.

In addition to the above mentioned changes in carbon metabolism at leaf cell level, an impairment of the sucrose translocation was demonstrated in the whole plant (Einig et al., 1997). This could result in a modified partitioning of carbon above and below ground, with root growth more affected than shoot growth (Wittig et al., 2009). The most probable risk for a forest tree will thus be a lower resource acquisition and a disturbed carbon allocation with a probable reduction of tree growth (Matyssek and Sandermann, 2003), although root growth may be stimulated in adult forest trees under ozone stress (Matyssek et al., 2010).

13.3 OXIDATIVE STRESS AND CARBON METABOLISM

13.3.1 Leaf Cell Response to Ozone Uptake

A crucial point is the means to determine the degree of phytotoxicity of ozone, through an index of risk for trees. The usual European index relies on the concentration of the pollutant in the external atmosphere surrounding the leaves, AOT 40. This concentration, that is, exposure-based concept, however, ignores the actual flux of ozone through stomata (which is also influenced by deposition of ozone on various surfaces including leaf cuticles; Tuzet et al., 2011). The stomatal conductance is thus the main point of control of ozone uptake within the leaf (Point 1 in Figure 13.3). It is usually admitted that ozone-tolerant species close their stomata as an avoidance mechanism in response to the pollutant (Castagna and Ranieri, 2009; Castagna et al., 2001; Degl'Innocenti et al., 2003). It has also been shown, in *Arabidopsis thaliana*, that the most O_3-sensitive genotype constitutively has a high stomatal conductance, leading to a high initial O_3 uptake rate (Brosché et al., 2010). It is, however, still poorly understood how ozone affects the stomatal function (Paoletti and Grulke, 2005), not to mention that other environmental factors (quality of light, CO_2, vapour pressure deficit) can interfere (Dumont et al., 2013). In addition, the degree of stomatal opening is not necessarily the only factor conferring sensitivity or tolerance to ozone stress (Castagna and Ranieri, 2009; Dizengremel et al., 2009; Matyssek et al., 2008). After entering the leaf through stomata, a large part of ozone quickly reacts with the apoplast, producing ROS (Laisk et al., 1989) and triggering a series of signalling cascades and plant defence responses (Kangasjärvi et al., 2005; Matyssek et al., 2012a;

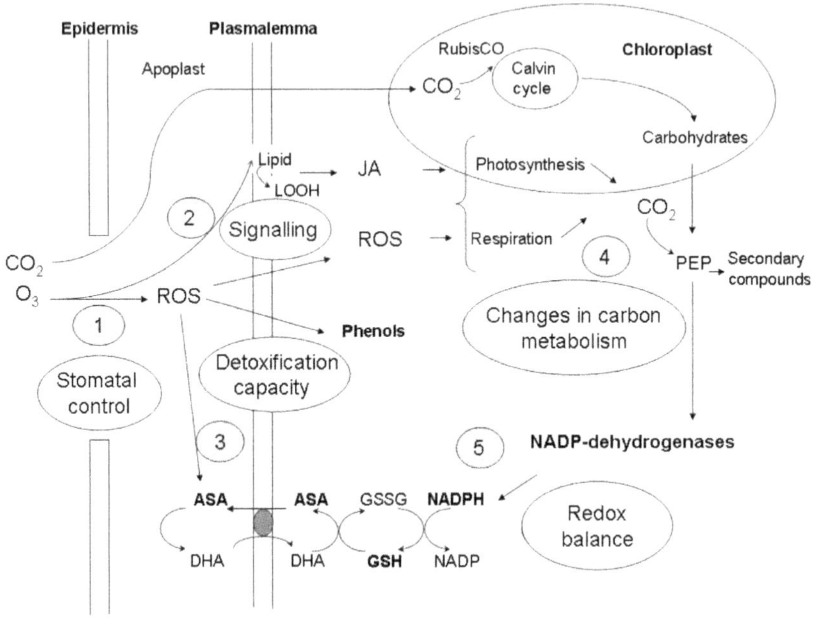

FIGURE 13.3 Model of possible O_3-dependent signalling pathways and parameters determining O_3 sensitivity. In the apoplast, ozone can induce enzymatic or non-enzymatic oxidation of unsaturated lipid to peroxides, initiating lipid-derived signalling affecting metabolic activities and gene expression. Ozone can also form ROS which can directly act as signalling messengers in the cytosol. Apoplastic ascorbate is thought to provide protection against ROS. The resulting oxidized ascorbate is regenerated in the cytosol at the expense of glutathione and NADPH. Several steps thus exist, contributing to the degree of sensitivity to ozone: (1) stomatal control, (2) constitutive detoxification capacity, (3) ratio between carboxylases (RubisCO and PEPCase) and (4) antioxidant regeneration based on redox power (i.e. NADPH availability; see text) provided through catabolic pathway. *From Baier et al. (2005), Castagna and Ranieri (2009), Dizengremel et al. (2008) and Matyssek et al. (2012a).*

Points 2 and 3, Figure 13.3). Two ways of defence are thus used to cope with oxidative stress caused by ozone, namely by preventing ozone influx through the regulation of stomatal conductance and by detoxifying ROS within the cells once ozone has been taken up (Castagna and Ranieri, 2009; Dizengremel et al., 2008; Matyssek et al., 2008, 2012a).

13.3.2 Ozone Flux and Detoxification Capacity

Conceiving the critical level of ozone in terms of injury induction to vegetation, the phytotoxic ozone dose (POD), representing the non-detoxified ozone molecules, needs to be determined (Dizengremel et al., 2008; Musselman et al., 2006). The POD concept integrates the stomatal conductance, and hence, the amount of ozone entering the leaf, as discussed above (see also Figure 13.3). Additionally, the capacity of cells is considered to neutralise

ROS, which can be inferred from antioxidant levels and enzyme activity in NADPH formation needed to regenerate ASC and glutathione (Dizengremel et al., 2009; Point 5 in Figure 13.3). Due to the mode of O_3 action, the first line of defence is represented by apoplast antioxidants, in particular by ASC (Dizengremel et al., 2008; Polle et al., 1995). However, the actual importance of apoplastic ASC in conferring O_3 tolerance is still unclear, since its rapid depletion can induce the subsequent oxidative burst, mediated through ROS production inside the foliar cells (D'Haese et al., 2005; Zechmann, 2011). Ranieri et al. (1999), by applying the model calculation developed by Chameides (1989) to estimate the ability of apoplast ASC in limiting the flux of O_3 to the plasma membrane, reported neutralization of O_3 by direct reaction with extracellular ASC in poplars under acute stress to be low. As demonstrated with two poplar clones, apoplastic ASC concentration was not related to O_3 sensitivity. The modification of antioxidant concentration by O_3 flux will provide a better estimation of the potential detoxification capacity, as suggested by Wieser et al. (2002) and Tausz et al. (2007). When whole-cell rather than apoplastic ASC concentration was taken into account, and corrected for O_3 flux, the correlation with the differential O_3 sensitivity of the two poplar clones was indeed improved (Di Baccio et al., 2008).

In the dynamics of apoplast chemistry, the role and importance of phenolic compounds (concentration and profiling) as antioxidants should be stressed. Moreover, phenolics, being incorporated in cell wall as lignin components, may contribute to thicken the cell wall, thus constituting a physical barrier to O_3 impact. Similar to what was observed for ASC, the phenolic accumulation measured in the O_3-sensitive poplar clone following exposure to the pollutant and the related antioxidant ability was restricted when data were corrected by stomatal O_3 flux, confirming the necessity to consider both avoidance and defence mechanisms in determining O_3 sensitivity (Di Baccio et al., 2008).

13.3.3 Detoxification and Carbon Metabolism

The maintenance of high antioxidant activity of ASC in the apoplast depends on a continuous supply of reduced ASC from the cytosol (Point 5 in Figure 13.3). The Halliwell-Asada pathway ensures the reduction of ASC by a sequence of redox reactions involving glutathione and NADPH. GSH is the reducing co-factor for several enzymes involved in ROS detoxification and plays a central role as an antioxidant and also in cell signalling (Noctor et al., 2012). The glutathione redox ratio remained stable in response to ozone, preserving cell signalling and suggesting that the activity of glutathione reductase (GR), catalyzing the regeneration of GSH from GSSG, is sufficient to maintain the glutathione pool in O_3-treated leaves. The enhanced availability of glutathione in the tolerant poplar genotype may be an important tolerance trait (Di Baccio et al., 2008). The regeneration of the reduced forms of ASC and glutathione in the cell can be provided for by using the reducing power of soluble redox molecules (Dizengremel et al., 2008; Noctor, 2006). It was recently demonstrated that NADPH is delivered by a series of enzymes linked to catabolic processes which are increased under ozone exposure, as mentioned above (Section 13.2.1; Dghim et al., 2013; Dizengremel et al., 2009).

In summary, high apoplastic ASC levels for counteracting O_3-caused toxicity need to be regenerated by the cell metabolism, controlling the effective O_3 flux upon O_3 uptake through stomata. Regeneration of ASC requires the reducing power of NADPH, while O_3 impact on photosynthesis is typically associated with decreasing and increasing activities of RuBisCO and PEPC, respectively. Related stimulation of catabolic pathways keeps NADPH high (Point 4 in Figure 13.3). Decline in photosynthesis due to increasing demand for detoxification and repair will curtail growth and may affect whole-tree C allocation.

13.4 IDENTIFICATION OF THE GAPS

13.4.1 The Prominent but Ambiguous Role of ASC in Counteracting Oxidative Stress

13.4.1.1 Modelling Ozone Deposition and Detoxification

As discussed above, the apoplast has been proposed to play a fundamental role in attenuating the flux of O_3 (and/or ROS) reaching the plasmalemma (Chameides, 1989; Plöchl et al., 2000). Within the apoplast, ozone can react with numerous antioxidants although ASC seems to be the most important scavenger molecule. By linking the diurnal cycling of apoplastic ASC with the uptake patterns of O_3 and its half-hourly concentrations, a first modelling attempt has resulted in MODD (Model of Ozone Deposition and Detoxification; Tuzet et al., 2011). The model accounts for the diurnal variability of detoxification processes and O_3 deposition on cuticular waxes and the soil surface (Figure 13.4). The mechanistic modelling approach to plant defence reactions is based upon the direct reaction of ozone with ASC located in the apoplast (Plöchl et al., 2000). The reactive and diffusive fluxes of O_3 through the cell wall are calculated by solving a diffusion/reaction equation. The final product of ASC oxidation, which is dehydroascorbic acid (DHA), can easily cross the plasma membrane and enter into the cytosol, where it is recycled via dehydroascorbate reductase at the expense of reduced glutathione (GSH) and NADPH. By considering the diffusion/reaction processes, through parameterization of cuticular and soil surface resistances, O_3 reaction probability is obtained. MODD is fully integrated into a soil–plant–atmosphere continuum model so that complete coupling between stomatal conductance and O_3 exchanges between leaves and the atmosphere is ensured. The observed diurnal variations in stomatal conductance, which largely control the influx of O_3 into the leaf, are well reproduced for wheat (Figure 13.5A). The patterns for

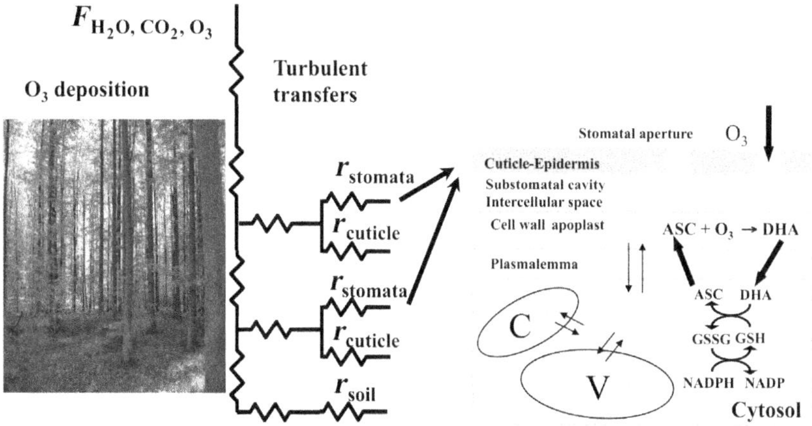

FIGURE 13.4 Scheme of the main processes considered in the Model of Ozone Deposition and Detoxification (MODD). The resistance network represents mass exchanges (water–CO_2–O_3) between forest and the atmosphere. A model flow diagram is also shown for ozone and detoxification: from the intercellular space, ozone rapidly diffuses into the aqueous media of cell walls where it reacts with apoplastic ascorbate (ASC). The final product of ASC oxidation, which is dehydroascorbic acid (DHA), can easily cross the plasma membrane and enter into the cytosol, where it is recycled via dehydroascorbate reductase in the ascorbate-glutathione cycle and returned to the apoplast.

ASC concentration, $[ASC]_{cw}$ and the ASC redox ratio, $ASC/(ASC + DHA)_{cw}$, show a close correlation with a mid-day depression when the stomata are fully open (Figure 13.5B and C). MODD also quantifies the partition of the total ozone canopy flux into the fluxes reaching the stomatal cavities, the cuticular plant surfaces and the soil beneath the canopy (Figure 13.5D–F). Sensitivity analysis underlines the significant role played by the pool of ASC located in the mesophyll cell wall in the detoxification of O_3. Modelling indicates, during the period prior to senescence, that 57% of the O_3 flux is deposited onto the cuticular surfaces and 4% on soil, whereas only 37% is absorbed by stomata. The deposition of ozone on leaf surface is thus important and this point is particularly developed in Oksanen et al., 2013, this vol.

MODD represents a novel approach for studying plant defence mechanisms against oxidative stress that can be used to predict the response of plants to other stress conditions. In addition, questions can be explored as to whether the apoplast (i) represents the site where O_3-induced ROS are first produced and perceived and (ii) acts as a fundamental barrier against ROS diffusion. However, contrasting with the cytosol, the redox homeostasis is less efficient in the apoplast, as the latter is deficient in glutathione and NAD(P)H. Accordingly, ASC is much more oxidized in the apoplast than within the cell. The lower buffering capacity of the apoplast may facilitate the signalling of the redox state as oxidative signals may readily accumulate beyond a signalling threshold. In such terms, the reported lack of correlation between O_3 tolerance

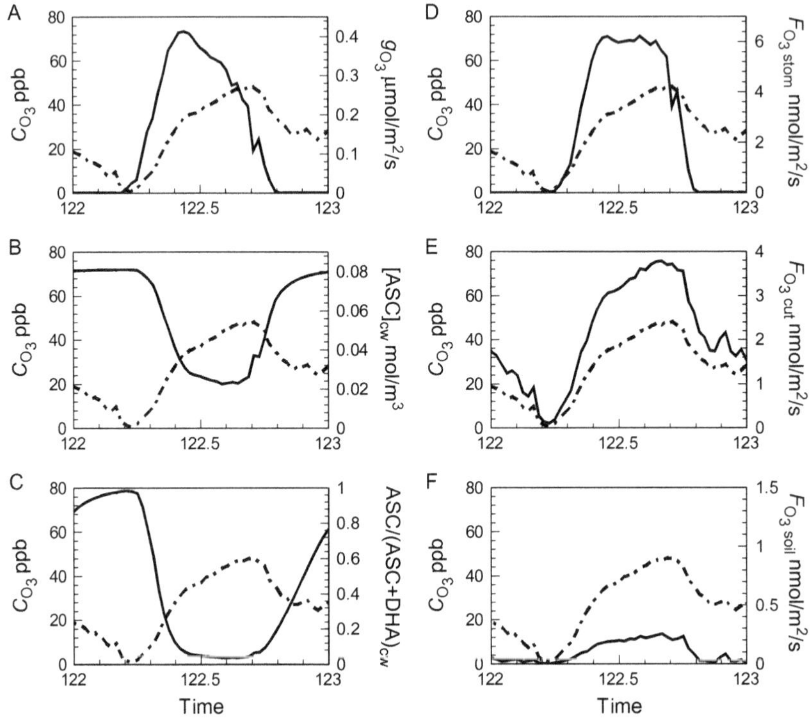

FIGURE 13.5 Daily time course (day 122) of calculated O_3 stomatal conductance. g_{O_3} (A), apoplastic ASC concentration, $[ASC]_{cw}$ (B), ascorbate redox ratio, $ASC/(ASC+DHA)_{cw}$ (C), stomatal deposition flux (D), cuticular deposition flux (E) and soil deposition flux (F) for a wheat canopy (solid line). Also plotted on these figures is the daily cycle of atmospheric measured ozone concentration (double dotted line). *After Tuzet et al. (2011).*

and extracellular levels of ASC is not surprising, since ASC, rather than simply acting as an antioxidant molecule, would participate in the redox signalling pathway by providing information on the cellular redox state (Foyer and Noctor, 2005). On such grounds, the ASC dynamics in the apoplastic space and cytosol should be integrated into the overall metabolic regenerating machinery to properly acknowledge the role of ASC in warranting O_3 tolerance.

13.4.1.2 Relative Importance of ASC, Glutathione and Other Scavengers in Detoxification

ASC rather than glutathione appears to be an effective scavenger of singlet oxygen, whereas glutathione may be superior regarding tackling of superoxide (Foyer and Noctor, 2005; Polle, 2001). However, the mean concentrations reported for ASC were about four- to fivefold higher than for glutathione

(Foyer and Noctor, 2005) with the latter being absent in the apoplast, which underlines the higher scavenging efficiency by ASC. For acting as antioxidant, ASC must be maintained by the cell in a reduced state. Such a prerequisite is usually warranted, at the expense of reduced glutathione, by monodehydroascorbate reductase or dehydroascorbate reductase. Therefore, either in the extracellular environment or within the cell, the detoxification ability of ASC is strictly linked to glutathione regeneration.

Enhanced ASC and glutathione *de novo* synthesis might be linked to ozone tolerance in poplar. Dumont, J., et al. (unpublished) found *de novo* production of ASC under ozone exposure, along with ample ASC regeneration, in the ozone-tolerant genotype. In contrast, even though the sensitive genotype showed similar ASA levels in ambient conditions in response to ozone, total ASC concentration increased due to an increase of DHA concentration, while ASA concentration remained stable. The sensitive genotype produced *de novo* ASC in response to ozone but was not able to regenerate it, which caused DHA to increase. The glutathione redox ratio remained stable in response to ozone, preserving cell signalling and suggesting, as described in Di Baccio et al. (2008), that the activity of GR, catalyzing the regeneration of GSH from GSSG, was sufficient to maintain the glutathione pool in O_3-treated leaves. The inability of the sensitive genotype to regenerate its ASA pool in response to ozone was not due to a decrease of GSH, but might be linked to a decrease of NADP(H) or due to an insufficient activity of MDHAR and DHAR. The tolerant genotype possessed higher glutathione levels, both regarding reduced and oxidized forms, and in response to ozone, the two forms increased more strongly than in the other genotype. The higher *de novo* synthesis of glutathione, and therefore, the higher availability of glutathione in the tolerant genotype, may thus be an important tolerance trait, as found in another tolerant poplar genotype (Di Baccio et al., 2008).

The limited importance of apoplastic ASC in O_3 scavenging by direct reaction under a range of O_3 concentrations has already been mentioned (Booker et al., 2012). Moreover, the same authors excluded the contribution of extracellular soluble flavonoids and sinapoyl glycosides as effective antioxidants, based on their too low concentration, despite the O_3-induced increased transcription of many phenylpropanoid genes and concentration of sinapoyl malate.

It is therefore evident that the current understanding of the protective mechanisms occurring in the apoplast and of the connections to the signalling pathway is still incomplete. Moreover, it is reasonable to hypothesize that the antioxidant and, in a wider sense, the defence potential of the apoplast may depend on a pool of different metabolites rather than on a single molecule. In addition, other defence processes within the cell, such as the xanthophyll cycle, thioredoxins, superoxidases, catalases and peroxidases, could participate, reinforcing the complexity.

13.4.2 Metabolism-Dependent Redox Power Impeded in Detoxification?

NADPH has been claimed as a key factor for allowing the regeneration of reduced ASC in the cytosol (Section 13.2.1; Dghim et al., 2013; Dizengremel et al., 2009). However, the presence of a respiratory burst oxidase homolog using NADPH was mentioned as a component of the plasma membrane (Jaspers and Kangasjärvi, 2010). This NADPH oxidase, which delivers ROS, may divert some part of the reducing power arising from the enhanced respiratory pathways. Recent results obtained in poplar nevertheless show enhanced capacity of NADP-dehydrogenases of the poplar clone *Carpaccio* under ozone exposure, accompanied by a capacity for maintaining NADPH at a high level. Conversely, NADPH decreased in the ozone-treated sensitive clone *Robusta* (Dghim et al., 2013). Thus, the capacity of plant cells for regenerating the reducing power, especially through NADPH formation, appears to contribute to enhancing O_3 tolerance.

13.5 CONCLUSIONS

13.5.1 Biochemical Markers Identifiable for Fostering Sensitivity Thresholds for Ozone?

In the light of the above-reported exchanges of ASC pools between apoplast and cytosol, it is difficult to use extracellular ASC alone as an effective marker of O_3 sensitivity. The extracellular/intracellular ASC ratio, as well as its redox state within the cytosol, better relate to the cell's ability for sustaining adequate antioxidant efficiency. As a result, the role of glutathione in keeping ASC in reduced form would become plausible, even though glutathione is involved in functions other than that of an antioxidant (e.g. transferring molecules to storage compartments). Glutathione concentration, therefore, cannot contribute exclusively to ROS scavenging. Maintenance of high redox power seems to better correspond to leaf cell performance under O_3 impact. However, ROS scavenging capacity implies a range of compounds. In particular, the complex network between the changes in metabolic pathways, namely the increase in the pathways devoted to the synthesis of amino acids, phenolics and isoprenoids, as well as the detoxification processes, must be unveiled.

Finally, since O_3 tolerance depends on mechanisms of both stress avoidance and defence, any identified marker should be corrected for O_3 influx to assess the plant detoxification capacity. Overall, proposal of a new simple index for ozone risk assessment thus seems premature. It could nevertheless be possible in the near future to add the most appropriate detoxification parameter to cellular, leaf and even field models to improve the predicted effects of ozone on higher plants: this could be considered an application procedure useful for policy makers.

13.5.2 Combination of Ozone with Other Stresses (CO_2, Drought, Temperature)

Within the global context of climatic changes, the combination of the impact of ozone with other environmental factors that are known to interact with this atmospheric pollutant must be highlighted. Several studies have been performed, essentially viewing ozone in combination with only one other factor, mostly drought, CO_2 or, less frequently, temperature. The interactive effects of ozone, CO_2 and temperature, mainly at the level of phenolic compounds, are also presented in Oksanen et al., 2013, this vol.

Individually, elevated O_3 and CO_2 have contrasting effects on plant physiology and growth, O_3 being clearly detrimental whereas CO_2 is often considered, at least at the short-term scale, as being beneficial. The impact of a combined exposure of plants to both O_3 and CO_2 is still debated since various results have been obtained, from a mitigation by CO_2 of the negative effects of O_3 (Gaucher et al., 2003; Grams et al., 1999) to the absence of any protective effect (Paoletti and Grulke, 2005; Uddling et al., 2010). Concerning more specifically the defence processes, a field experiment on aspen showed that CO_2 did not ameliorate the harmful effects of O_3 (Wustman et al., 2001). Moreover, the competition resulting from mixed plantations can modify the responses of trees when exposed to individual gases or their combinations (Kozovits et al., 2005; Liu et al., 2004). Temperature increase and the modification of precipitation regimes are the most important additional factors to be evaluated in the context of effects on forest trees by enhanced ground-level O_3 and CO_2 concentrations in the atmosphere.

Both O_3 and drought may curtail the capacity for carbon fixation in forest trees in a CO_2-enriched atmosphere. The interactions between O_3 and drought in biochemical and ecophysiological responses of trees appear to vary, depending on the genotype and factorial scenarios (Matyssek et al., 2006). During the seasonal course, the kind of effect also depends on the instances of occurrence, duration, and intensities of the two stresses relative to each other. If O_3 perturbs stomatal regulation first, tolerance to both drought and persisting O_3 exposure may be weakened, whereas drought preceding O_3 uptake may protect plants against O_3 impact through increased stomatal closure. However, even in the latter case, defence may be constrained due to limited carbon fixation resulting from the stomatal closure. In fact, drought may cause a stronger reduction in stem growth than does ozone on an annual basis (Löw et al., 2006). Drought episodes can induce carry-over effects on growth during succeeding years, subsequently having a quality similar to chronic O_3 effects. In turn, chronic O_3 effects provoke further risks, predisposing trees to other (episodic) stresses such as severe drought (or frosts or pests), which produce a sequence of events that can eventually cause a breakdown, thus imposing a burden on tree performance in a changing environment (Matyssek et al., 2006). The predicted increase in temperature is thought to slightly improve plant growth although antagonistic effects exist under enhanced tropospheric O_3 regimes (Kasurinen et al., 2012). Riikonen et al. (2009) recently demonstrated that the effects of slightly elevated temperature and O_3 on the antioxidant capacity of birch were counteractive.

13.5.3 Research Needs in the Near Future

Concerning phytotoxic ozone, the complex dynamics of O_3-driven changes need to be investigated through multi-level, interdisciplinary approaches, up-scaling O_3 response from cell to ecosystem, in order to develop models for risk assessment, which enable mechanistic description and prediction of O_3 impact on vegetation. Past thresholds for O_3 sensitivity were indeed derived mainly from studies under controlled conditions, by exposing plants (seedlings in most cases) to acute, short-term, stress. The challenge is to move from exposure-based to flux-based indices to achieve more mechanistic models for O_3 risk assessment (Matyssek et al., 2007). To this end, improved knowledge of physiological processes at the basis of O_3 uptake and stress tolerance is necessary, requiring a new evaluation of the largely arbitrary separation between primary and secondary metabolism (Dizengremel, 2001; Dizengremel et al., 2012; Matyssek et al., 2013). Moreover, O_3 must be regarded more realistically as a component of global changes (Matyssek et al., 2012b; Renaut et al., 2009), thus considering its interaction with increasing CO_2 levels, as well as with other environmental agents such as light, drought, temperature and anthropogenic land-use changes, especially in view of upcoming post-Kyoto discussions.

Terrestrial ecosystems are threatened with an irreversible shift from the present state as a result of human activities (Barnosky et al., 2012). Most of the predicted impacts of climatic changes on forest ecosystems focus on expected increases in CO_2 and temperature as well as changes in precipitation regime, with subsequent risks of insect and pathogen attacks. Clearly, the detrimental effect of ozone was, up until recently, neglected. However, the antagonistic impact of ozone in relation to elevated CO_2 on the carbon sink strength of vegetation has become emphasized (Sitch et al., 2007). The effects of elevated CO_2 and O_3 on plant physiology and ecological interactions are highly context- and species-specific, rendering attempts to identify general, global patterns in the health and sustainability of forest ecosystems frustrating (Lindroth, 2010). More generally, it is not always possible to predict the impacts of environmental change on the magnitude and direction of ecosystem carbon exchange based on manipulation of individual factors due to additive and synergistic effects (Templer and Reinmann, 2011). Future research should seek to increase the diversity of species and biomes studied, incorporate long-term, multi-factor experiments, and employ a comprehensive 'genes to ecosystems' perspective that couples genetic/genomic tools with the approaches of evolutionary and ecosystem ecology (Matyssek et al., 2012c).

In such a context, it is also important to do up-scaling from individual trees to forest ecosystems, integrating experimental results obtained with the above environmental factors into model simulations (Hanson et al., 2005; Matyssek et al., 2012c; McLaughlin et al., 2007; Sun et al., 2012). In particular, the interactive effects of ozone and water availability seem to be of prominent importance: increase in evapotranspiration in response to ambient ozone exposures is expected to episodically increase the frequency and severity of drought in forest stands (Sun et al., 2012). The model-based analyses identify critical areas of uncertainty for multivariate predictions of future ecosystem response and underscore the importance of long-term field experiments for the evaluation of acclimation and growth under complex environmental scenarios (Hanson et al., 2005; see Matyssek et al., 2013, this vol.). In brief, the integration in regional models of a pertinent ozone damage factor, identified at the cellular level, could improve risk assessment and help policy makers in making related socio-economic decisions.

ACKNOWLEDGEMENTS

This work was partly supported by the French National Research Agency through the project VMCS VULNOZ (ANR-08-VULN-012) and the Laboratory of Excellence ARBRE (ANR-12-LABXARBRE-01).

REFERENCES

Ashmore, M.R., 2005. Assessing the future global impacts of ozone on vegetation. Plant Cell Environ. 28, 949–964.

Bagard, M., Le Thiec, D., Delacôte, E., Hasenfratz-Sauder, M.P., Banvoy, J., Gérard, J., Dizengremel, P., Jolivet, Y., 2008. Ozone-induced changes in photosynthesis and photorespiration of hybrid poplar in relation to the developmental stage of the leaves. Physiol. Plant. 134, 559–574.

Baier, M., Kandlbinder, A., Golldack, D., Dietz, K.J., 2005. Oxidative stress and ozone: perception, signalling and response. Plant Cell Environ. 28, 1012–1020.

Barnosky, A.D., Hadly, E.A., Bascompte, J., Berlow, E.L., Brown, J.H., Fortelius, M., Getz, W.M., Harte, J., Hastings, A., Marquet, P.A., Martinez, N.D., Mooers, A., Roopnarine, P., Vermeij, G., Williams, J.W., Gillespie, R., Kitzes, J., Marshall, C., Matzke, N., Mindell, D.P., Revilla, E., Smith, A.B., 2012. Approaching a state shift in Earth's biosphere. Nature 486, 52–58.

Booker, F.L., Burkey, K.O., Jones, A.M., 2012. Re-evaluating the role of ascorbic acid and phenolic glycosides in ozone scavenging in the leaf apoplast of *Arabidopsis thaliana* L. Plant Cell Environ. 35, 1456–1466.

Brosché, M., Merilo, E., Mayer, F., Pechter, P., Puzorjova, I., Brader, G., Kangasjarvi, J., Kollist, H., 2010. Natural variation in ozone sensitivity among *Arabidopsis thaliana* accessions and its relation to stomatal conductance. Plant Cell Environ. 33, 914–925.

Cabané, M., Pireaux, J.-C., Léger, E., Weber, E., Dizengremel, P., Pollet, B., Lapierre, C., 2004. Condensed lignins are synthesized in poplar leaves exposed to ozone. Plant Physiol. 134, 586–594.

Castagna, A., Ranieri, A., 2009. Detoxification and repair process of ozone injury: from O_3 uptake to gene expression adjustment. Environ. Pollut. 157, 1461–1469.

Castagna, A., Nali, C., Ciompi, S., Lorenzini, G., Soldatini, G.F., Ranieri, A., 2001. Ozone exposure affects photosynthesis of pumpkin (*Cucurbita pepo*) plants. New Phytol. 152, 223–229.

Chameides, W.L., 1989. The chemistry of ozone deposition to plant leaves: role of ascorbic acid. Environ. Sci. Technol. 23, 595–600.

Dalstein, L., Torti, X., Le Thiec, D., Dizengremel, P., 2002. Physiological study of declining Pinus cembra (L.) trees in southern France. Trees 16, 299–305.

Degl'Innocenti, E., Guidi, L., Soldatini, G.F., 2002. Effect of chronic O_3 fumigation on the activity of some Calvin cycle enzymes in two poplar clones. Photosynthetica 40, 121–126.

Degl'Innocenti, E., Vaccà, C., Guidi, L., Soldatini, G.F., 2003. CO_2 photoassimilation and chlorophyll fluorescence in two clover species showing different responses to O_3. Plant Physiol. Biochem. 41, 485–493.

Dghim, A.A., Dumont, J., Hasenfratz-Sauder, M.P., Dizengremel, P., Le Thiec, D., Jolivet, Y., 2013. Capacity for NADPH regeneration in the leaves of two poplar genotypes differing in ozone sensitivity. Physiol. Plant 148, 36–50.

D'Haese, D., Vandermeiren, K., Asard, H., Horemans, N., 2005. Other factors than apoplastic ascorbate contribute to the differential ozone tolerance of two clones of *Trifolium repens* L. Plant Cell Environ. 28, 623–632.

Diara, C., Castagna, A., Baldan, B., Sodi, A.M., Sahr, T., Langebartels, C., Sebastiani, L., Ranieri, A., 2005. Differences in the kinetics and scale of signalling molecule production modulate the ozone sensitivity of hybrid poplar clones: the roles of H_2O_2, ethylene and salicylic acid. New Phytol. 168, 351–364.

Di Baccio, D., Castagna, A., Paoletti, E., Sebastiani, L. And, Ranieri, A., 2008. Could the differences in O_3 sensitivity between two poplar clones be related to a difference in antioxidant defense and secondary metabolic response to O_3 influx? Tree Physiol. 28, 1761–1772.

Dizengremel, P., 2001. Effects of ozone on the carbon metabolism of forest trees. Plant Physiol. Biochem. 39, 729–742.

Dizengremel, P., Sasek, T., Brown, K.J., Richardson, C.J., 1994. Ozone-induced changes in primary carbon metabolism enzymes of loblolly pine needles. J. Plant Physiol. 144, 300–306.

Dizengremel, P., Le Thiec, D., Bagard, M., Jolivet, Y., 2008. Ozone risk assessment for plants: central role of metabolism-dependent changes in reducing power. Environ. Pollut. 156, 11–15.

Dizengremel, P., Le Thiec, D., Hasenfratz-Sauder, M.P., Vaultier, M.N., Bagard, M., Jolivet, Y., 2009. Metabolic-dependent changes in plant cell redox power after ozone exposure. Plant Biol. 11, 35–42.

Dizengremel, P., Vaultier, M.N., Le Thiec, D., Cabané, M., Bagard, M., Gérant, D., Gérard, J., Dghim, A.A., Richet, N., Afif, D., Pireaux, J.C., Hasenfratz-Sauder, M.P., Jolivet, Y., 2012. Phosphoenolpyruvate is at the crossroads of leaf metabolic responses to ozone stress. New Phytol. 195, 512–517.

Dumont, J., Spicher, F., Montpied, P., Dizengremel, P., Jolivet, Y., Le Thiec, D., (unpublished). Effects of ozone on stomatal responses to environmental parameters (blue light, red light, CO_2 and vapour pressure deficit) in three *Populus deltoides* x *nigra* genotypes. Environ. Pollut. 173, 85–96.

Eckey-Kaltenbach, H., Ernst, D., Heller, W., Sandermann, H., 1994. Biochemical plant responses to ozone: IV. Cross-induction of defensive pathways in parsley (*Petroselinum crispum* L.) plants. Plant Physiol. 104, 67–74.

Einig, W., Lauxmann, U., Hauch, B., Hampp, R., Landolt, W., Maurer, S., Matyssek, R., 1997. Ozone-induced accumulation of carbohydrates changes enzyme activities of carbohydrate metabolism in birch leaves. New Phytol. 137, 673–680.

Ernst, D., 2013. Integrated studies on abiotic stress defence in trees: the case of ozone. Dev. Environ. Sci. 13, 289–307.

Fares, S., Oksanen, E., Lännenpää, M., Julkunen-Tiito, R., Loreto, F., 2010. Volatile emissions and phenolic compound concentrations along a vertical profile of *Populus nigra* leaves exposed to realistic ozone concentrations. Photosynth. Res. 104, 61–74.

Fontaine, V., Pelloux, J., Podor, M., Afif, D., Gérant, D., Grieu, P., Dizengremel, P., 1999. Carbon fixation in *Pinus halepensis* submitted to ozone. Opposite response of ribulose-1,5-bisphosphate carboxylase:oxygenase and phosphoenolpyruvate carboxylase. Physiol. Plant. 105, 187–192.

Foyer, C.H., Noctor, G., 2005. Redox homeostasis and antioxidant signaling: a metabolic interface between stress perception and physiological responses. Plant Cell 17, 1866–1875.

Gaucher, C., Costanzo, N., Afif, D., Mauffette, Y., Chevrier, N., Dizengremel, P., 2003. The impact of elevated ozone and carbon dioxide on young *Acer saccharum* seedlings. Physiol. Plant. 117, 392–402.

Grace, S.C., 2005. Phenolics as antioxidants. In: Smirnoff, N. (Ed.), Antioxidants and Reactive Oxygen Species in Plants. Blackwell Publishing Ltd, Oxford, UK, pp. 141–168.

Grams, T.E.E., Anegg, S., Haberle, K.H., Langebartels, C., Matyssek, R., 1999. Interactions of chronic exposure to elevated CO_2 and O_3 levels in the photosynthetic light and dark reactions of European beech (*Fagus sylvatica*). New Phytol. 144, 95–107.

Häikiö, E., Freiwald, V., Julkunen-Tiito, R., Beuker, E., Holopainen, T., Oksanen, E., 2009. Differences in leaf characteristics between ozone-sensitive and ozone-tolerant hybrid aspen (*Populus tremula* x *Populus tremuloides*) clones. Tree Physiol. 29, 53–66.

Hanson, P.J., Wullschleger, S.T., Norby, R.J., Tschaplinski, T.J., Gunderson, C.A., 2005. Importance of changing CO_2, temperature, precipitation and ozone on carbon and water cycles of an upland-oak forest: incorporating experimental results into model simulations. Glob. Chang. Biol. 11, 1402–1423.

Heath, R., 2008. Modification of the biochemical pathways of plants induced by ozone: what are the various routes of change? Environ. Pollut. 155, 453–463.

Jaspers, P., Kangasjärvi, J., 2010. Reactive oxygen species in abiotic stress signaling. Physiol. Plant. 138, 405–413.

Kangasjärvi, J., Jaspers, P., Kollist, H., 2005. Signalling and cell death in ozone-exposed plants. Plant Cell Environ. 28, 1021–1036.

Kasurinen, A., Biasi, C., Holopainen, T., Ronsi, M., Mäenpää, M., Oksanen, E., 2012. Interactive effects of elevated ozone and temperature on carbon allocation of silver birch (*Betula pendula*) genotypes in an open-air field exposure. Tree Physiol. 32, 737–751.

Koch, J.R., Scherzer, A.J., Eshita, S.M., Davis, K.R., 1998. Ozone sensitivity in hybrid poplar is correlated with a lack of defense-gene activation. Plant Physiol. 118, 1243–1252.

Kontunen-Soppela, S., Ossipov, V., Ossipova, S., Oksanen, E., 2007. Shift in birch leaf metabolome and carbon allocation during long-term open-field ozone exposure. Glob. Chang. Biol. 13, 1053–1067.

Kozovits, A.R., Matyssek, R., Blaschke, H., Göttlein, A., Grams, T.E.E., 2005. Competition increasingly dominates the responsiveness of juvenile beech and spruce to elevated CO_2 and/or O_3 concentrations throughout two subsequent growing seasons. Glob. Chang. Biol. 11, 1387–1401.

Laisk, A., Kull, O., Moldau, H., 1989. Ozone concentration in leaf intercellular air spaces is close to zero. Plant Physiol. 90, 1163–1167.

Langebartels, C., Schraudner, M., Heller, W., Ernst, D., Sandermann, H., 2002. Oxidative stress and defense reactions in plants exposed to air pollutants and UV-B radiation. In: Inzé, D.,

Van Montagu, M. (Eds.), Oxidative Stress in Plants. Taylor & Francis, London, UK, pp. 105–135.

Lindroth, R.L., 2010. Impacts of elevated atmospheric CO_2 and O_3 on forests: phytochemistry, trophic interactions, and ecosystem dynamics. J. Chem. Ecol. 36, 2–21.

Liu, X., Kozovits, A.R., Grams, T.E.E., Blaschke, H., Rennenberg, H., Matyssek, R., 2004. Competition modifies effects of enhanced ozone/carbon dioxide concentrations on carbohydrate and biomass accumulation in juvenile Norway spruce and European beech. Tree Physiol. 24, 1045–1055.

Löw, M., Herbinger, K., Nunn, A.J., Häberle, K.-H., Leuchner, M., Heerdt, C., Werner, H., Wipfler, P., Pretzsch, H., Tausz, M., Matyssek, R., 2006. Extraordinary drought of 2003 overrules ozone impact on adult beech trees (*Fagus sylvatica*). Trees 20, 539–548.

Matyssek, R., Innes, J.L., 1999. Ozone—a risk factor for trees and forests in Europe? Water Air Soil Pollut. 116, 199–226.

Matyssek, R., Sandermann, H., 2003. Impact of ozone on trees: an ecophysiological perspective. Prog. Bot. 64, 349–404.

Matyssek, R., Le Thiec, D., Löw, M., Dizengremel, P., Nunn, A.J., Häberle, K.H., 2006. Interactions between drought and O_3 stress in forest trees. Plant Biol. 8, 11–17.

Matyssek, R., Bytnerowicz, A., Karlsson, P.-E., Paoletti, E., Sanz, M., Schaub, M., Wieser, G., 2007. Promoting the O_3 flux concept for European forest trees. Environ. Pollut. 146, 587–607.

Matyssek, R., Sandermann, H., Wieser, G., Booker, F., Cieslik, S., Musselman, R., Ernst, D., 2008. The challenge of making ozone risk assessment for forest trees more mechanistic. Environ. Pollut. 156, 567–582.

Matyssek, R., Wieser, G., Ceulemans, R., Rennenberg, H., Pretzsch, K., Löw, M., Nunn, A.J., Werner, H., Wipfler, P., Osswald, W., Nikolova, P., Hanke, D.E., Kraigher, H., Tausz, M., Bahnweg, G., Kitao, M., Dieler, J., Sandermann, H., Herbinger, K., Grebenc, T., Blumenröther, M., Deckmyn, G., Grams, T.E.E., Heerdt, C., Leuchner, M., Fabian, P., Häberle, K.H., 2010. Enhanced ozone strongly reduces carbon sink strength of adult beech (*Fagus sylvatica*)—resume from the free air fumigation study at Kranzberg Forest. Environ. Pollut. 158, 2527–2532.

Matyssek, R., Wieser, G., Calfapietra, C., de Vries, W., Dizengremel, P., Ernst, D., Jolivet, Y., Mikkelsen, T.T., Mohren, G.M.J., Le Thiec, D., Tuovinen, J.P., Weatherall, A., Paoletti, E., 2012a. Forests under climate change and air pollution: gaps in understanding and future directions for research. Environ. Pollut. 160, 57–65.

Matyssek, R., Koricheva, J., Schnyder, H., Ernst, D., Munch, J.C., Osswald, W., Pretzsch, H., 2012b. The balance between resource sequestration and retention: a challenge in plant science. In: Matyssek, R., Schnyder, H., Osswald, W., Ernst, D., Munch, J.C., Pretzsch, H. (Eds.), Growth and Defence in Plants—Resource Allocation at Multiple Scales. Ecological Studies, vol. 220. Springer, Berlin, pp. 3–24.

Matyssek, R., Schnyder, H., Osswald, W., Ernst, D., Munch, J.C., Pretzsch, H., 2012c. Growth and defence in plants—resource allocation at multiple scales. In: Ecological Studies, vol. 220. Springer, Berlin, p. 467.

Matyssek, R., Wieser, G., Fleischmann, F., Grünhage, L., 2013. Ozone research, quo vadis? Lessons from the Kranzberg forest experiment. Dev. Environ. Sci. 13, 103–129.

McLaughlin, S.B., Nosal, M., Wullschleger, S.D., Sun, G., 2007. Interactive effects of ozone and climate on tree growth and water use in a southern Appalachian forest in the USA. New Phytol. 174, 109–124.

Musselman, R.C., Lefohn, A.S., Massman, W.J., Heath, R.L., 2006. A critical review and analysis of the use of exposure- and flux-based ozone indices for predicting vegetation effects. Atmos. Environ. 40, 1869–1888.

Noctor, G., 2006. Metabolic signalling in defence and stress: the central roles of soluble redox couples. Plant Cell Environ. 29, 409–425.

Noctor, G., Mhamdi, A., Chaouch, S., Han, Y., Neukermans, J., Queval, G., Foyer, C.H., 2012. Glutathione in plants: an integrated overview. Plant Cell Environ. 35, 454–484.

Oksanen, E., Keski-Saari, S., Kontunen-Soppela, S., Markku Keinänen, M., 2013. Metabolomics and transcriptomics increase our understanding about defence responses and genotypic differences of northern deciduous trees to elevating ozone, CO_2 and climate warming. Dev. Environ. Sci. 13, 309–329.

Paoletti, E., Grulke, N.E., 2005. Does living in elevated CO_2 ameliorate tree response to ozone? A review on stomatal responses. Environ. Pollut. 137, 483–493.

Paoletti, E., Manning, W.J., 2007. Toward a biologically significant and usable standard for ozone that will also protect plants. Environ. Pollut. 150, 85–95.

Pasqualini, S., Della Torre, G., Ferranti, F., Ederli, L., Piccioni, C., Reale, L., Antonielli, M., 2002. Salicylic acid modulates ozone-induced hypersensitive cell death in tobacco plants. Physiol. Plant. 115, 204–212.

Pell, E.J., Eckardt, N., Enyedi, A.J., 1992. Timing of ozone stress and resulting status of ribulose bisphosphate carboxylase/oxygenase and associated net photosynthesis. New Phytol. 120, 397–405.

Plöchl, M., Lyons, T., Ollerenshaw, J., Barnes, J., 2000. Simulating ozone detoxification in the leaf apoplast through the direct reaction with ascorbate. Planta 210, 454–457.

Polle, A., 2001. Dissecting the superoxide dismutase-ascorbate-glutathione pathway in chloroplasts by metabolic modeling. Computer simulations as a step towards flux analysis. Plant Physiol. 126, 445–462.

Polle, A., Wieser, G., Havranek, W.M., 1995. Quantification of ozone influx and apoplastic ascorbate content in needles of Norway spruce trees (*Picea abies* L. Karst.) at high altitude. Plant Physiol. 94, 312–319.

Ranieri, A., Castagna, A., Padu, E., Moldau, H., Rahi, M., Soldatini, G.F., 1999. The decay of O_3 through direct reaction with cell wall ascorbate is not sufficient to explain the different degrees of O_3-sensitivity in two polar clones. J. Plant Physiol. 150, 250–255.

Rao, M.V., Lee, H.I., Davis, K.R., 2002. Ozone-induced ethylene production is dependent on salicylic acid, and both salicylic acid and ethylene act in concert to regulate ozone-induced cell death. Plant J. 32, 447–456.

Reich, P.B., 1983. Effects of low concentration of O_3 on net photosynthesis, dark respiration and chlorophyll contents in aging hybrid poplar leaves. Plant Physiol. 73, 291–296.

Renaut, J., Bohler, S., Hausman, J.F., Hoffmann, L., Sergeant, K., Ashan, N., Jolivet, Y., Dizengremel, P., 2009. The impact of atmospheric composition on plants: a case study of ozone on poplar. Mass Spectrom. Rev. 28, 495–516.

Riikonen, J., Mäenpää, M., Alavillamo, M., Silfver, T., Oksanen, E., 2009. Interactive effect of elevated temperature and O_3 on antioxidant capacity and gas exchange in *Betula pendula* saplings. Planta 230, 419–427.

Saleem, A., Loponen, J., Pihlaja, K., Oksanen, E., 2001. Effects of long-term open-field ozone exposure on leaf phenolics of European silver birch (*Betula pendula* Roth.). J. Chem. Ecol. 27, 1049–1062.

Saurer, M., Maurer, S., Matyssek, R., Landolt, W., Günthardt-Goerg, M.S., Siegenthaler, U., 1995. The influence of ozone and nutrition on $\delta^{13}C$ in *Betula pendula*. Oecologia 103, 397–406.

Sehmer, L., Fontaine, V., Antoni, F., Dizengremel, P., 1998. Effects of ozone and elevated atmospheric carbon dioxide on carbohydrate metabolism of spruce needles. Catabolic and detoxification pathways. Physiol. Plant. 102, 605–611.

Sitch, S., Cox, P.M., Collins, W.J., Huntingford, C., 2007. Indirect oxidative forcing of climate change through ozone effects on the land-carbon sink. Nature 448, 791–795.

Sun, G., McLaughlin, S.B., Porter, J.H., Uddling, J., Mulholland, P.J., Adams, M.B., Pederson, N., 2012. Interactive influences of ozone and climate on streamflow of forested watersheds. Glob. Chang. Biol. 18, 3395–3409.

Sweetlove, L.J., Beard, K.F.M., Nunes-Nesi, A., Fernie, A.R., Ratcliffe, R.G., 2010. Not just a circle: flux modes in the plant TCA cycle. Trends Plant Sci. 15, 462–470.

Tausz, M., Grulke, N.E., Wieser, G., 2007. Defense and avoidance of ozone under global change. Environ. Pollut. 147, 525–531.

Templer, P.H., Reinmann, A.B., 2011. Multi-factor global change experiments: what have we learned about terrestrial carbon storage and exchange? New Phytol. 192, 797–800.

Tuzet, A., Perrier, A., Loubet, B., Cellier, P., 2011. Modelling ozone deposition fluxes: the relative roles of deposition and detoxification processes. Agric. Forest Meteorol. 151, 480–492.

Uddling, J., Hogg, A.J., Teclaw, R.M., Carroll, M.A., Ellsworth, D.S., 2010. Stomatal uptake of O_3 in aspen and aspen-birch forests under free-air CO_2 and O_3 enrichment. Environ. Pollut. 158, 2023–2031.

Valkama, E., Koricheva, J., Oksanen, E., 2007. Effects of elevated O_3, alone and in combination with elevated CO_2, on tree leaf chemistry and insect herbivore performance: a meta-analysis. Glob. Chang. Biol. 13, 184–201.

Wieser, G., Tegischer, K., Tausz, M., Häberle, K.H., Grams, T.E.E., Matyssek, R., 2002. Age effects on Norway spruce (*Picea abies*) susceptibility to ozone uptake: a novel approach relating stress avoidance to defense. Tree Physiol. 22, 583–590.

Wild, A., Schmitt, V., 1995. Diagnosis of damage to Norway spruce (*Picea abies*) through biochemical criteria. Physiol. Plant. 93, 375–382.

Wittig, V.E., Ainsworth, E.A., Long, S.P., 2007. To what extent do current and projected increases in surface ozone affect photosynthesis and stomatal conductance of trees? A meta-analytic review of the last 3 decades of experiments. Plant Cell Environ. 30, 1150–1162.

Wittig, V.E., Ainsworth, E.A., Naidu, S.L., Karnosky, D.F., Long, S.P., 2009. Quantifying the impact of current and future tropospheric ozone on tree biomass, growth, physiology and biochemistry: a quantitative meta-analysis. Glob. Chang. Biol. 15, 396–424.

Wustman, B.A., Oksanen, E., Karnosky, D.F., Noormets, A., Isebrands, J.G., Pregitzer, K.S., Hendrey, G.R., Sober, J., Podila, G.K., 2001. Effects of elevated CO_2 and O_3 on aspen clones varying in O_3 sensitivity: can CO_2 ameliorate the harmful effects of O_3? Environ. Pollut. 115, 473–481.

Zechmann, B., 2011. Subcellular distribution of ascorbate in plants. Plant Signal. Behav. 6, 360–363.

Zinser, C., Ernst, D., Sandermann, H., 1998. Induction of stilbene synthase and cinnamyl alcohol dehydrogenase mRNA in Scots pine (Pinus sylvestris L.) seedlings. Planta 204, 169–176.

Chapter 14

Integrated Studies on Abiotic Stress Defence in Trees: The Case of Ozone

Dieter Ernst[1]
Helmholtz Zentrum München, Neuherberg, Germany
[1]*Corresponding author: e-mail: ernst@helmholtz-muenchen.de*

Chapter Outline

14.1 INTRODUCTION

In recent decades, the effects of tropospheric ozone, a ubiquitous phytotoxic air pollutant, have been broadly studied at the physiological, biochemical and, to a minor extent, molecular biological levels (Karnosky et al., 2005; Matyssek and Sandermann, 2003) (Dizengremel et al., 2013, this vol.; Oksanen et al., 2013, this vol.). The mechanistic responses of plants at the transcriptional level have been elucidated (Heath, 2008; Langebartels et al., 2002; Matyssek et al., 2008) (see also Dizengremel et al., 2013, this vol.; Oksanen et al., 2013, this vol.), and it has also been shown that transcript patterns after acute ozone exposure or pathogen attack are comparable and appear to have several commonalities. Thus, ozone has been described as an abiotic elicitor of plant defence reactions (Sandermann et al., 1998).

Acute ozone fumigation (200–300 ppb for several hours) results in transcriptional changes in ozone-responsive genes. Microarray analyses have led to the characterisation of such genes and the links to various functional groups and metabolic pathways, including primary and secondary metabolism, cell growth and structure, energy, transcription, protein synthesis, signal transduction, disease and defence (Matyssek et al., 2008; Puckette et al., 2008; Tamaoki et al., 2003; Tosti et al., 2006). Chronic ozone fumigation experiments (up to twice the ambient level for several days, weeks or years) have been performed under controlled chamber (Bohler et al., 2010; D'Haese et al., 2006; Langebartels et al., 1997; Miyazaki et al., 2004) and free-air conditions (Karnosky et al., 2007). It has been shown in model plants such as *Arabidopsis thaliana* and in trees including beech, aspen and birch that several genes are similarly affected by chronic and acute ozone exposure, though to a lesser extent by chronic ozone fumigation (Gupta et al., 2005; Kontunen-Soppela et al., 2010a; Li et al., 2006; Miyazaki et al., 2004; Olbrich et al., 2005, 2009) (Table 14.1). Nevertheless, great differences in transcript abundances between controlled chamber and field experiments have been described, indicating additional stress factors under field conditions (Miyazaki et al., 2004; Olbrich et al., 2010b).

Forest free-air ozone fumigation experiments have contributed to a better molecular ecophysiological understanding of trees under such challenges. To date, transcriptomics in free-air forest tree research have mainly been performed using custom-made cDNA microarrays and quantitative real-time RT-PCR (Gupta et al., 2005; Jehnes et al., 2007; Kontunen-Soppela et al., 2010a,b; Olbrich et al., 2009, 2010b) (Table 14.1); however, these data are not sufficient to link molecular and ecophysiological research. Analyses of stress-affected transcriptomes in trees can provide significant information regarding the transcribed genes that are involved in stress responses and the adaptation of trees to changing environments. However, the results obtained must be considered carefully, as transcript and protein levels are not necessarily and/or clearly correlated (Gygi et al., 1999). In addition, post-translational protein modification and reactive oxygen species (ROS)/reactive nitrogen species (RNS), which influence protein function, have to be taken into consideration (Bachi et al., 2013; Beltrao et al., 2012; http://www.geneinfinity.org/sp/sp_proteinptmodifs.html).

Regarding proteomics, new protocols, platforms and workflow stages have been developed in the last 10 years, including qualitative and quantitative proteomics (e.g. two-dimensional difference in gel electrophoresis (2D-DIGE), LC-MS/MS, MALDI-MS/MS, ICPL, SILAC and label-free protocols). An excellent review on plant proteome changes under abiotic stress has recently been published (Kosova et al., 2011). Regarding forest trees, climate change and air pollution, drought-responsive proteins have been found in oak and

TABLE 14.1 List of Ozone-Affected Transcripts, Proteins and Metabolites in Trees

Species	Age	Ozone treatment	Analyses	References
Norway spruce	Seedlings	Chamber: 300, 600 ppb; 8 h/day (24 days)	Cinnamyl alcohol dehydrogenase mRNA	Galliano et al. (1993)
Norway spruce, Scots pine	Seedlings	Chamber: 200 ppb; 10 h/day (18 days)	Extensin mRNA	Schneiderbauer et al. (1995)
European beech	Saplings	Chamber: 60 ppb; 10 h/day (7 weeks)	Extensin mRNA	Schneiderbauer et al. (1995)
Norway spruce	Saplings	Chamber: 80 ppb; 24 h/day (6 weeks)	Extensin mRNA	Schneiderbauer et al. (1995)
Scots pine	Saplings	Chamber: 150, 300 ppb; 10 h/day (2 days)	Stilbene biosynthesis transcripts	Chiron et al. (2000)
Hybrid poplar (tolerant, sensitive clone)	Cuttings	Chamber: 300 ppb; 6 h/day (4 days)	Defence-related transcripts	Koch et al. (1998)
Birch (tolerant, sensitive clone)	Saplings	Chamber: 100 ppb; 12 h/day (43 days)	Defence-related transcripts	Pääkkönen et al. (1998)
Hybrid aspen (tolerant, sensitive clone)	Ramets	Chamber: 75, 112.5, 150 ppb; 8 h/day (2 days)	Ethylene biosynthesis: transcripts, ethylene	Vahala et al. (2003)
Poplar (clone)	Saplings	Chamber: 60, 120 ppb; 14 h/day (31 days)	Lignin biosynthesis: transcripts, enzyme activities, lignin content	Cabané et al. (2004)
Poplar (clones)	Cuttings	Chamber: 120 ppb; 13 h/day (35 days)	Proteome	Bohler et al. (2007, 2010)
European beech	Saplings	Greenhouse: 150–190 ppb; 8 h/day (83 days)	Ethylene biosynthesis: transcripts, metabolites	Betz et al. (2009a)
European beech	Saplings	Greenhouse: 150–190 ppb (83 days)	Shikimate pathway: transcripts, proteins, metabolites	Betz et al. (2009b)

Continued

TABLE 14.1 List of Ozone-Affected Transcripts, Proteins and Metabolites in Trees—Cont'd

Species	Age	Ozone treatment	Analyses	References
European beech	Saplings	Greenhouse: (1) 300 ppb; 8 h/day (30 days); (2) 200 ppb; 8 h/day (46 days)	Shikimate pathway: transcripts, lignin content	Betz et al. (2009c)
European beech	Saplings	Greenhouse: 150–190 ppb; 8 h/day (69 days)	Transcriptome	Olbrich et al. (2005)
Trembling aspen (tolerant clone)	Saplings	Aspen FACE: 1.5 × ambient (4 and 5 years)	Transcriptome	Gupta et al. (2005)
Paper birch	Saplings	Aspen FACE: 1.5 × ambient (7 years)	Transcriptome	Kontunen-Soppela et al. (2010a)
European beech	Mature trees	Kranzberg: 2 × ambient (4 and 5 years)	Ethylene biosynthesis, shikimate pathway, defence-related, ABA-synthesis transcripts	Betz et al. (2009c), Jehnes et al. (2007) and Matyssek et al. (2010b)
European beech	Mature trees	Kranzberg: 2 × ambient (6 and 7 years)	Transcriptome	Olbrich et al. (2010a,b)
European beech	Saplings	Lysimeter: 2 × ambient (3 and 4 years)	Transcriptome, proteome	Abril et al. (2011), Kerner et al. (2011) and Olbrich et al. (2009)
Silver birch (tolerant, sensitive clone)	Saplings	Suonenjoki, open-top-chamber: 2 × ambient (2 years)	Transcriptome	Kontunen-Soppela et al. (2010b)
Silver birch (clone)	Saplings	Kuopio, open-field: 1.5 × ambient (7 years)	Metabolome	Kontunen-Soppela et al. (2007)

poplar (Plomion et al., 2006; Sergeant et al., 2011); an altered proteomic profile was found in poplar and European beech leaves upon fumigation with realistic ozone concentrations (Bohler et al., 2007, 2010; Kerner et al., 2011) (Table 14.1).

For metabolomics, highly sophisticated methods have been developed, and an overview of existing technologies has recently been published (Weckwerth, 2010). Concerning global change and air pollution, metabolite profiling in *Populus euphratica* clearly showed differences when the trees were grown on different salt-containing soils (Brosché et al., 2005), and metabolite differences between salt-sensitive and salt-tolerant poplar were found (Janz et al., 2010). In *Eucalyptus*, divergent metabolite responses of two closely related species to water stress were found (Warren et al., 2012), and changes in the birch leaf metabolome were observed in long-term open-field ozone fumigation experiments (Kontunen-Soppela et al., 2007) (Table 14.1; Oksanen et al., 2013, this vol.). Thus, the integration of transcript, protein and metabolite data is important to understand tree responses to environmental stresses (Renaut et al., 2009) (Figure 14.1) (see also Matyssek et al., 2013a, this vol.).

14.2 OZONE EXPOSURE UNDER CONTROLLED CHAMBER/ GREENHOUSE CONDITIONS

Transcriptional changes upon acute ozone exposure (up to 200 ppb) were first reported for young spruce, pine and beech seedlings grown under controlled chamber conditions (Langebartels et al., 1997) (Figure 14.2). The ozone-induced upregulation of transcripts involved in cell wall biosynthesis and secondary metabolism has been described (Chiron et al., 2000; Galliano et al., 1993; Schneiderbauer et al., 1995) (Table 14.1; see also Dizengremel et al., 2013, this vol.; Oksanen et al., 2013, this vol.). In a subsequent study, poplar and birch saplings were also analysed under controlled ozone fumigation, and it was shown that transcripts belonging to secondary metabolism and pathogen defence were upregulated, similar to the results for herbaceous plants (Koch et al., 1998; Pääkkönen et al., 1998) (Table 14.1). The plant hormone ethylene was shown to play an important role in signalling ozone damage in herbaceous plants through an upregulation of transcripts involved in the biosynthesis of ethylene (Jaspers et al., 2005; Moeder et al., 2002; Tuomainen et al., 1997). Transcripts for 1-aminocyclopropane-1-carboxylate (ACC) synthase and oxidase were also upregulated in birch and beech saplings (Betz et al., 2009a; Nunn et al., 2005; Vahala et al., 2003) (Table 14.1), again demonstrating the close similarity of herbaceous and woody plants in response to ozone fumigation. These transcriptional changes in the leaves of European beech could then be linked to the formation of ACC and, ultimately, ethylene (Betz et al., 2009a; Nunn et al., 2005) (Table 14.1). The increased accumulation of ethylene was therefore in accordance with the cell death and leaf loss reported for beech saplings (Nunn et al., 2005). The shikimate pathway, an intermediate pathway between primary and secondary metabolisms, plays an important role in the production of

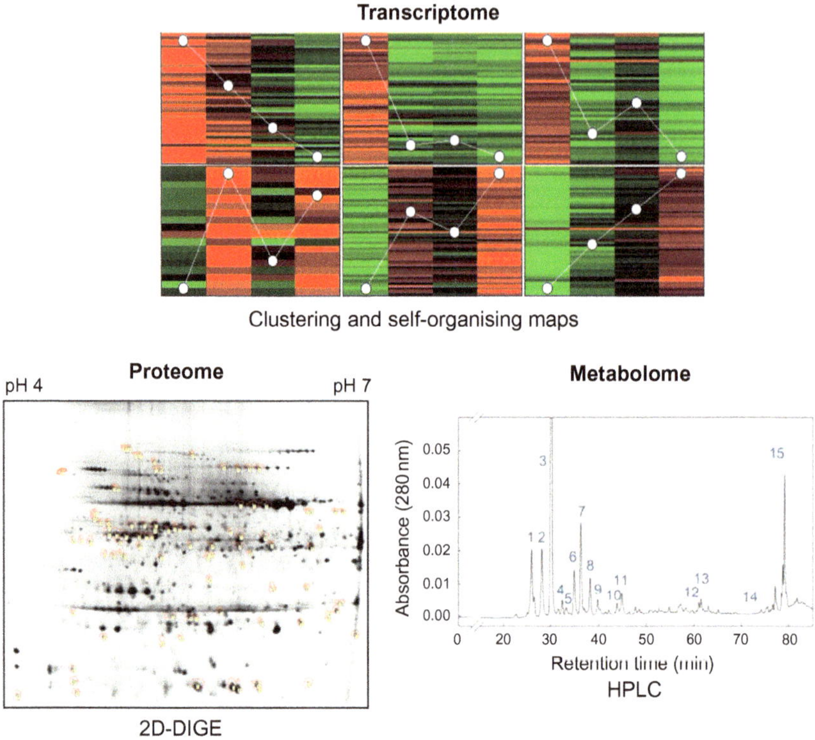

FIGURE 14.1 Examples of transcript, protein and metabolite analyses. Transcript analyses of UV-B- and cold-treated *Arabidopsis thalina* were carried out by microarrays and only strongly regulated identical transcripts are given in seven self-organising maps (red, upregulated; green, downregulated); the proteome of ozone-treated European beech leaves was analysed by 2D-DIGE (differentially expressed protein spots are circled) (taken from Kerner et al., 2011); HPLC analysis indicates soluble phenolic compounds of ozone-treated European beech leaves (taken from Betz, 2006).

flavonoids, stilbenes and lignin. All shikimate genes were upregulated in the leaves of beech saplings upon ozone treatment (Betz et al., 2009b,c) (Table 14.1), a response that has also been described in tobacco (Janzik et al., 2005). The increase of shikimate pathway transcripts in European beech saplings was in accordance with the increased protein level of 3-deoxy-D-arabino-heptulosonate-7-phosphate synthase 3 and 3-dehydroquinate dehydratase/shikimate dehydrogenase (Betz et al., 2009b) (Table 14.1). Moreover, the levels of conjugated salicylic and gentisic acid that are synthesised via the shikimate pathway were elevated as well. Interestingly, an accumulation of stress lignin was also found (Betz et al., 2009c) (Table 14.1). A similar correlation was also described for ozone-treated poplar leaves, with an increase of shikimate dehydrogenase, phenylalanine ammonia-lyase and cinnamyl alcohol dehydrogenase activity, followed by an accumulation of lignin (Cabané et al., 2004) (Table 14.1). An overall

FIGURE 14.2 Ozone exposure systems. Climate chamber with Norway spruce saplings, greenhouse with European beech saplings, lysimeter with young European beech trees, Kranzberg Forest with old European beech and Norway spruce trees.

transcriptome analysis of long-term ozone-treated beech saplings under greenhouse conditions (Olbrich et al., 2005) (Figure 14.2; Table 14.1) resulted in a gene grouping similar to that described for *Arabidopsis thaliana* (D'Haese et al., 2006; Ludwikow et al., 2004; Tamaoki et al., 2003). Functional grouping according to the Munich Information Center for Protein Sequences classification has revealed genes involved in primary metabolism, energy, cell growth/division, protein synthesis, protein destination and storage, secondary metabolism, cell structure, disease/defence, transporters, intracellular traffic, signal transduction, transcription and transposons (Olbrich et al., 2005) (Table 14.1). Taken together, the ozone treatment of herbaceous and juvenile woody plants under controlled chamber or greenhouse conditions resulted in a similar transcriptional response.

14.3 FREE-AIR EXPOSURE SYSTEMS

Over the past 20 years, chronic ozone exposure has increasingly become the focus of forest ecosystem research, and scaling up from molecular processes to the ecosystem level has resulted in complex interactions between different hierarchical levels and environmental factors (Sandermann and Matyssek, 2004). In particular, molecular biological research must be linked to forest ecophysiological research (Matyssek et al., 2012). To address these questions, free-air ozone exposure systems were established in the United States and Europe (Karnosky et al., 2007) (Figure 14.2) (Matyssek et al., 2013b, this vol.; Pregitzer et al., 2013, this vol.).

14.3.1 The Aspen Free-Air CO_2 Enrichment Experiment

In the Aspen Free-Air CO_2 Enrichment (FACE) experiment (Karnosky et al., 2007), an ozone-tolerant poplar clone was exposed for a period of 5 years to an elevated ozone concentration (1.5 × the ambient level), and the ensuing study was conducted over 2 years (for more details on the Aspen FACE experiment, Pregitzer et al., 2013, this vol.). A total of 4600 expressed sequence tags (ESTs) were evaluated using a custom-made cDNA microarray: 238 genes showed similar expression in at least one experiment, and 185 were significantly ozone regulated for the two replication years (Gupta et al., 2005) (Table 14.1). A higher expression was observed for signalling, defence-related, senescence-associated, cytoskeleton and flavonoid biosynthesis genes, whereas photosynthesis and energy related genes were downregulated. In addition, interacting effects of elevated CO_2 and ozone on the gene expression profile were observed that were not observed with the single stress treatments (Gupta et al., 2005) (Table 14.1). The transcript pattern found could be correlated to the overall patterns of physiology and poplar growth (Wustman et al., 2001). In another experiment with paper birch, the effects of ozone and CO_2 were also analysed using cDNA microarrays (Kontunen-Soppela et al., 2010a) (Table 14.1; see also Oksanen et al., 2013, this vol.)., and most of the alterations in transcript abundance were caused by ozone alone or in combination with CO_2. Similar to those found for trembling aspen, the expression profiles revealed the upregulation of genes involved in defence responses to oxidative stress and senescence and the downregulation of photosynthesis and carbon fixation-related genes (Kontunen-Soppela et al., 2010a) (Table 14.1; see also Oksanen et al., 2013, this vol.). The combined gas treatment also resulted in a differential transcript abundance not observed in the individual gas treatments. Both experiments demonstrated that CO_2 cannot completely counteract the harmful effects of ozone, at least at the transcriptional level (responses to interactions of CO_2 and ozone are also given in King et al., 2013, this vol.; Pregitzer et al., 2013, this vol.; Dizengremel et al., 2013, this vol.; Oksanen et al., 2013, this vol.).

14.3.2 The Kranzberg Ozone Fumigation Experiment

In the Kranzberg Ozone Fumigation Experiment, 65-year-old beech trees were fumigated with twice the level of ambient ozone concentrations from 2000 to 2007 (Karnosky et al., 2007) (Figure 14.2; see also Matyssek et al., 2013b, this vol.). Transcriptional analyses of all shikimate pathway genes revealed only a weak upregulation in contrast to the greenhouse study with young trees (Betz et al., 2009b,c). These differences are likely caused by the tree ontogeny, the ozone regimes used or the local additional stress conditions (Matyssek et al., 2008). Similarly, genes connected with the shikimate pathway and lignin formation were not significantly affected in another study at the same site (Jehnes et al., 2007). Transcript analyses of RNA isolated from both sun and shade leaves were performed in 2005 and 2006 using a custom-made cDNA microarray (Olbrich et al., 2005). However, changes in gene expression were only observed for approximately 0.2% of all the ESTs analysed (Olbrich et al., 2010b) (Table 14.1), which was much lower compared to the 5% of genes changed in the Aspen FACE experiment using poplar (Gupta et al., 2005) (Table 14.1). Interestingly, differences in the transcript patterns between shade and sun leaves were observed in this study (Olbrich et al., 2010a,b). Moreover, the leaves of beech trees altered their transcript levels in response to ozone. In two consecutive years, different transcription patterns were observed during the vegetation period. At the same site, differences were also observed in the transcript pattern over two consecutive years (Jehnes et al., 2007; Matyssek et al., 2010b). These results indicate that unpredictable factors, such as weather conditions, pathogens and nutrient supply, may strongly influence gene expression. Furthermore, a comparison of the effects of ozone and leaf type on gene expression revealed a stronger influence of the leaf type than that of the ozone treatment (Olbrich et al., 2010a) (Table 14.1), indicating a different photo-protective capacity of the sun versus shade leaves. A direct link between gene expression and the physiological level was demonstrated by Matyssek et al., (2010b). It was shown that the transcript level of 9-*cis*-epoxycarotenoid dioxygenase, which encodes an enzyme involved in abscisic biosynthesis, was increased under elevated ozone conditions, a result in accordance with an increase of abscisic acid, stomatal closure and net photosynthesis rates and decreased stem volume growth (Matyssek et al., 2010b).

14.3.3 Field Lysimeters

In a free-air lysimeter study, beech saplings were grown under ambient ozone levels and twice the ambient ozone levels (Figure 14.2) (Winkler et al., 2009). Transcript analyses were performed using a custom-made cDNA microarray consisting of 1246 ESTs (Olbrich et al., 2005) (Table 14.1). Transcripts belonging to disease/defence, secondary metabolism and signal transduction mainly appeared to be upregulated, whereas transcripts belonging to photosynthesis, chloroplast cell structure and the Calvin cycle were predominantly downregulated (Olbrich et al., 2009) (Table 14.1). Although the expression profiles for 2005 and 2006 were not identical, a common trend of transcriptional changes was evident. Interestingly, these differences were more pronounced in 2006, which may be explained by a higher stomatal conductance in 2006 compared to 2005 and, as a result, enhanced ozone-flux

(Olbrich et al., 2009). Proteomic analyses by 2D-DIGE showed that approximately 9% of the total proteins were differentially regulated upon ozone fumigation (Kerner et al., 2011) (Table 14.1). Comparing the transcriptomic and proteomic data, only three proteins showed a direct overlap with the corresponding transcript (Kerner et al., 2011), clearly indicating that post-translational processing and gene silencing must be taken into account in such comparative analyses (Beltrao et al., 2012; Perco et al., 2010; Sánchez-Pons et al., 2011). Moreover, physiological low amounts of ROS and RNS will result in important protein modifications (redox proteomics), which can lead to functional changes (Bachi et al., 2013). However, upon the functional classification of protein groups, a higher correlation was observed for the transcripts and proteins related to disease/defence, photosynthesis and the Calvin cycle (Kerner et al., 2011; Olbrich et al., 2009) (Table 14.1).

14.3.4 Open-Field Ozone Exposure Systems in Kuopio, Finland

An open-top chamber system was established in Finland using twice the ambient ozone concentrations and two silver birch genotypes (Vapaavuori et al., 2002). Transcriptome analyses were performed using a custom-made cDNA microarray of *Populus* (Kontunen-Soppela et al., 2010b) (Table 14.1; see also Oksanen et al., 2013, this vol.). Under these conditions, many senescence-associated genes were upregulated, which correlated with earlier leaf senescence and abscission (Kontunen-Soppela et al., 2010b). However, this transcript pattern did not necessarily reflect the corresponding protein activity; for example, the downregulation of genes involved in photosynthesis was not reflected in the net photosynthesis (Kontunen-Soppela et al., 2010b). This result is similar to the findings for beech saplings in the lysimeter study mentioned earlier, indicating the need for additional regulation processes, post-translational modifications or protein stability (Kontunen-Soppela et al., 2010b). Differences in transcript patterns were also observed between different sampling points and the two clones used, and these differences suggested that additional factors are involved in ozone tolerance and gene expression during the entire period of vegetative growth (Kontunen-Soppela et al., 2010b; Matyssek et al., 2010b; Olbrich et al., 2010b) (Table 14.1). A metabolomic study was performed during a long-term open-field ozone exposure experiment (Karnosky et al., 2007). In white birch, elevated ozone resulted in a decrease in the compounds related to carbon metabolism and the function of chloroplast membranes and pigments, whereas compounds related to the cuticular wax layer and phenolics were increased (Kontunen-Soppela et al., 2007) (Table 14.1). However, genotypic differences caused more variability in metabolite concentrations than ozone, reflecting the basic biochemical differences between genotypes (Kontunen-Soppela et al., 2007). Similarly, the differences between single trees and the field site were shown to increase the phenotypic variation in silver birch metabolites (Ossipov et al., 2008).

The metabolite profiles of different genotypes of birch trees were different, and their biochemical reactions to the same environmental factors could be quite different too. These authors conclude that the application of metabolomics to study the biochemical response of trees to environmental stress is unlikely to produce reliable results, as individual trees in the forest are of different genotypes (Ossipov et al., 2008).

14.4 NEXT-GENERATION TECHNOLOGIES

With regard to genomics and forest trees, more genomes are being analysed by next-generation sequencing (e.g. 454 sequencing, Illumina sequencing, chip sequencing), which will contribute to a better understanding of the genomic resources in forest trees. The complete genome sequence of poplar was published a few years ago (Tuskan et al., 2006), as well as the genomes of several other tree species, including conifers, Fagaceae and *Eucalyptus* (Grattapaglia et al., 2012; Kremer et al., 2012; Rigault et al., 2011). In addition, genomes of forest and fruit trees have been recently summarised (Abbott, 2012; and articles in this special issue).

For the analysis of transcriptomics, next-generation sequencing will dramatically increase our knowledge regarding the transcriptome of forest trees, and such analyses are important to understand the responses of trees to both elevated air pollution and to the entire natural environment, including mycorrhiza, pathogens and insects. Further regulation of gene expression covering non-protein-coding RNAs, small RNAs (Charon et al., 2010; Chitwood and Timmermans, 2010) and stress-responsive microRNAs for cold, heat, dehydration and UV-B radiation were found in *Populus* (Jia et al., 2009; Lu et al., 2005, 2008). In addition, epigenetic gene regulation is important (Chinnusamy and Zhu, 2009), and an epigenetic variation in trees upon environmental disturbances has been described (Johnson et al., 2009; Lira-Medeiros et al., 2010). Interestingly, memory and carryover have also been found in ozone-treated conifers (Langebartels et al., 1998; Oksanen, 2003; Sasek et al., 1991).

The next 'omic' step, proteomics, will significantly contribute to a better understanding of protein abundance and plant stress acclimation. A recent review on proteomics and forest trees summarises the state-of-the-art techniques for *Populus*, *Pinus*, *Picea*, *Fagus*, *Eucalyptus* and *Quercus* (Abril et al., 2011). However, information regarding post-translational processes and the coverage of low-abundance proteins remains limited.

The last 'omic' technology is the analysis of metabolites: metabolomics. Regarding the millions of putative structures, the great challenge will be to cover the entire metabolome using only a few protocols (Weckwerth, 2011). Moreover, as indicated earlier, metabolic studies of forest trees and environmental stress must be interpreted carefully (Ossipov et al., 2008), and species-specific responses to environmental changes must be considered (Scherling et al., 2010).

These 'omic' technologies facilitate the study of plant systems in a holistic manner, up to the ecosystem level. The integration of genomics, proteomics, transcriptomics, metabolomics, genotypes and phenotypes will generate novel insight into the responses of plants to an altered environment (Weckwerth, 2011). Although these technologies are not yet fully developed in forest trees, forthcoming data will facilitate tree systems biology. Nevertheless, a recent review focuses on the complexity, challenges and pitfalls of the integration of such a large amount of data (zu Castell and Ernst, 2012).

14.5 CONCLUSIONS

As indicated from the experiments described, controlled chamber or greenhouse conditions are valuable tools for the mechanistic and diagnostic understanding of the action of ozone on tree seedlings, saplings or young trees. Transcriptional changes could be linked to the proteome and, in part, to the metabolite level (Betz et al., 2009a,b) (Table 14.1). Similarly, this association could also be demonstrated in a long-term free-air model fumigation site with young beech trees by comparing the transcriptome with the proteome (Kerner et al., 2011; Olbrich et al., 2009). However, no consistent gene expression was observed over 2 years; only a common trend of gene expression was evident (Olbrich et al., 2009). This trend was also evident for adult beech trees at the Kranzberg forest site (Matyssek et al., 2010b; Olbrich et al., 2010b). This result clearly indicates that free-air ozone experiments are influenced by additional factors, for example, weather, pathogens, herbivores or nutrient supply (see also Oksanen et al., 2013, this vol.). What will be an additional impact of drought or temperature, even as an extreme event? Moreover, it is important to note that individual beech trees of different genotypes will also influence gene expression, as indicated for metabolite analyses (Ossipov et al., 2008). Therefore, it is important to link well-defined multifactorial controlled chamber experiments with free-air supersites research (see also Matyssek et al., 2013, this vol., and for supersites Mikkelsen et al., 2013, this vol.; Paoletti et al., 2013, this vol.), which will result in a better understanding of forest ecosystems in silviculture as well as in plantations. However, this great challenge is lab- and cost-intensive and can only be handled by large consortiums. In addition, modelling will be a useful tool to evaluate long-term field studies under complex scenarios (Gayler et al., 2012; Rötzer et al., 2012; zu Castell et al., 2012) (see also Dizengremel et al., 2013, this vol.). Ozone risk assessment will have an impact on silviculture, as well as on fast-growing clonal trees in plantations that are important for biogenic energy production. According to the Kranzberg forest studies, ozone resulted in a dramatic decrease in the productivity of European beech and in a slight increase for Norway spruce (Matyssek et al., 2010a; Pretzsch et al., 2010) (for more details, see Grünhage et al., 2013, this vol.). In addition to this economic aspect, forest ecosystems are also important for human health and recreation, thus contributing to human well-being. Molecular approaches are the basis to understanding the detrimental effects of ozone on tree growth and are thus important in a socio-economic context (details on socio-economic aspects are given in Cudlin et al., 2013, this vol.; Nijnik and Miller, 2013, this vol.; Knoke and Hahn, 2013, this vol.). An understanding of molecular forest ecology, proceeding finally to forest physiology and up to tree growth, will help stakeholders and policy-makers perform their duties and make decisions.

ACKNOWLEDGEMENTS

This work was supported by the Deutsche Forschungsgemeinschaft (SFB 607) and, in part, by the European Community (Evoltree, 6th Framework Programme; COST Action E52). The valuable feedback and input of two reviewers is gratefully acknowledged.

REFERENCES

Abbott, A., 2012. The genomes of the giants: a walk through the forest of tree genomes. Tree Genet. Genomes 8, 443, Preface.

Abril, N., Gion, J.-M., Kerner, R., Müller-Starck, G., Cerrillo, R.M.N., Plomion, C., Renaut, J., Valledor, L., Jorrin-Novo, J.V., 2011. Proteomics research on forest trees, the most recalcitrant and orphan plant species. Phytochemistry 72, 1219–1242.

Bachi, A., Dalle-Donne, I., Scaloni, A., 2013. Redox proteomics: chemical principles, methodological approaches and biological/biomedical promises. Chem. Rev. 113, 596–698.

Beltrao, P., Albanèse, V., Kenner, L.R., Swaney, D.L., Burlingame, A., Villén, J., Lim, W.A., Fraser, J.S., Frydman, J., Krogan, N.J., 2012. Systematic functional prioritization of protein posttranslational modifications. Cell 150, 413–425.

Betz, G.A., 2006. Regulation des Shikimatstoffwechsels der europäischen Buche (*Fagus sylvatica* L.) unter dem Einfluß von Ozon. Thesis, Technische Universität München. http://nbn-resolving.de/urn/resolver.pl?urn:nbn:de:bvb:91-diss20060504-0931190040.

Betz, G.A., Gerstner, E., Olbrich, M., Winkler, J.B., Langebartels, C., Heller, W., Sandermann, H., Ernst, D., 2009a. Effects of abiotic stress on gene transcription in European beech: ozone affects ethylene biosynthesis in saplings of *Fagus sylvatica* L. iForest 2, 114–118.

Betz, G.A., Gerstner, E., Stich, S., Winkler, B., Welzl, G., Kremmer, E., Langebartels, C., Heller, W., Sandermann, H., Ernst, D., 2009b. Ozone affects shikimate pathway genes and secondary metabolites in saplings of European beech (*Fagus sylvatica* L.) grown under greenhouse conditions. Trees 23, 539–553.

Betz, G.A., Knappe, C., Lapierre, C., Olbrich, M., Welzl, G., Langebartels, C., Heller, W., Sandermann, H., Ernst, D., 2009c. Ozone affects shikimate pathway transcripts and monomeric lignin composition in European beech (*Fagus sylvatica* L.). Eur. J. For. Res. 128, 109–116.

Bohler, S., Bagard, M., Oufir, M., Planchon, S., Hoffmann, L., Jolivet, Y., Hausman, J.-F., Dizengremel, P., Renaut, J., 2007. A DIGE analysis of developing poplar leaves subjected to ozone reveals major changes in carbon metabolism. Proteomics 7, 1584–1599.

Bohler, S., Sergeant, K., Lefèvre, I., Jolivet, Y., Hoffmann, L., Renaut, J., Dizengremel, P., Hausman, J.-F., 2010. Differential impact of chronic ozone exposure on expanding and fully expanded poplar leaves. Tree Physiol. 30, 1415–1432.

Brosché, M., Vinocur, B., Alatalo, E., Lamminmäki, A., Teichmann, T., Ottow, E.A., Djilianov, D., Afif, D., Bogeat-Triboulot, M.-B., Altman, A., Polle, A., Dreyer, E., Rudd, S., Paulin, L., Auvinen, P., Kangasjärvi, J., 2005. Gene expression and metabolite profiling of *Populus euphratica* growing in the Negev desert. Genome Biol. 6, R101.

Cabané, M., Pireaux, J.-C., Léger, E., Weber, E., Dizengremel, P., Pollet, B., Lapierre, C., 2004. Condensed lignins are synthesized in poplar leaves exposed to ozone. Plant Physiol. 134, 586–594.

Charon, C., Moreno, A.B., Bardou, F., Crespi, M., 2010. Non-protein-coding RNAs and their interacting RNA-binding proteins in the plant cell nucleus. Mol. Plant 3, 729–739.

Chinnusamy, V., Zhu, J.-K., 2009. Epigenetic regulation of stress responses in plants. Curr. Opin. Plant Biol. 12, 133–139.

Chiron, H., Drouet, A., Lieutier, F., Payer, H.-D., Ernst, D., Sandermann, H., 2000. Gene induction of stilbene biosynthesis in Scots pine in response to ozone treatment, wounding, and fungal infection. Plant Physiol. 124, 865–872.

Chitwood, D.H., Timmermans, M.C.P., 2010. Small RNAs are on the move. Nature 467, 415–419.

Cudlín, P., Seják, J., Pokorný, J., Albrechtová, J., Bastian, O., Marek, M., 2013. Forest ecosystem services under climate change and air pollution. In: Matyssek, R., Clarke, N., Cudlin, P., Mikkelsen, T., Tuovinen, J.P., Weiser, G., Paoletti, E. (Eds.), Climate Change, Air Pollution and Global Challenges: Understanding and Solutions from Forest Research. Dev. Environ. Sci. 13. Elsevier, Netherlands, pp. 521–546.

D'Haese, D., Horemans, N., De Coen, W., Guisez, Y., 2006. Identification of late O_3-responsive genes in *Arabidopsis thaliana* by cDNA microarray analysis. Physiol. Plant. 128, 70–79.

Dizengremel, P., Jolivet, Y., Tuzet, A., Ranieri, A., Thiec, D., 2013. Integrative leaf-level phytotoxic ozone dose assessment for forest risk modelling. Dev. Environ. Sci. 13, 267–288.

Galliano, H., Cabané, M., Eckerskorn, C., Lottspeich, F., Sandermann, H., Ernst, D., 1993. Molecular cloning, sequence analysis and elicitor-/ozone induced accumulation of cinnamyl alcohol dehydrogenase from spruce (*Picea abies* L.). Plant Mol. Biol. 23, 145–156.

Gayler, S., Priesack, E., Fleischmann, F., Heller, W., Rötzer, T., Seifert, T., Matyssek, R., 2012. Modelling the defensive potential of plants. In: Matyssek, R., Schnyder, H., Oßwald, W., Ernst, D., Munch, J.C., Pretzsch, H. (Eds.), Growth and Defence in Plants. Ecological Studies, vol. 220. Springer-Verlag, Berlin, pp. 375–399.

Grattapaglia, D., Vaillancourt, R.E., Shepherd, M., Thumma, B.R., Foley, W., Külheim, C., Potts, B.M., Myburg, A.A., 2012. Progress in Myrtaceae genetics and genomics: *Eucalyptus* as the pivotal genus. Tree Genet. Genomes 8, 463–508.

Grünhage, L., Matyssek, R., Wieser, G., Häberle, K.-H., Leuchner, M., Menzel, A., Dieler, J., Pretzsch, H., Grimmeisen, W., Zimmermann, L., Raspe, S., Schröder, M., 2013. Flux-based ozone risk assessment for adult beech and spruce forests. Dev. Environ. Sci. 13, 251–266.

Gupta, P., Duplessis, S., White, H., Karnosky, D.F., Martin, F., Podila, G.K., 2005. Gene expression patterns of trembling aspen trees following long-term exposure to interacting elevated CO_2 and tropospheric O_3. New Phytol. 167, 129–142.

Gygi, S.P., Rochon, Y., Franza, B.R., Aebersold, R., 1999. Correlation between protein and mRNA abundance in yeast. Mol. Cell. Biol. 19, 1720–1730.

Heath, R.L., 2008. Modification of the biochemical pathways of plants induced by ozone: what are the varied routes to change? Environ. Pollut. 155, 453–463.

Janz, D., Behnke, K., Schnitzler, J.-P., Kanawati, B., Schmitt-Kopplin, P., Polle, A., 2010. Pathway analysis of the transcriptome and metabolome of salt sensitive and tolerant poplar species reveals evolutionary adaption of stress tolerance mechanisms. BMC Plant Biol. 10, 150.

Janzik, I., Preiskowski, S., Kneifel, H., 2005. Ozone has dramatic effects on the regulation of the prechorismate pathway in tobacco (*Nicotiana tabacum* L. cv. Bel W3). Planta 223, 20–27.

Jaspers, P., Kollist, H., Langebartels, C., Kangasjärvi, J., 2005. Plant responses to ozone. In: Smirnoff, N. (Ed.), Antioxidative and Reactive Oxygen Species in Plants. Blackwell Publishing, Oxford, pp. 268–292.

Jehnes, S., Betz, G., Bahnweg, G., Haberer, G., Sandermann, H., Rennenberg, H., 2007. Tree internal signalling and defence reactions under ozone exposure in sun and shade leaves of European beech (*Fagus sylvatica* L.) trees. Plant Biol. 9, 253–264.

Jia, X., Ren, L., Chen, Q.-J., Li, R., Tang, G., 2009. UV-B-responsive microRNAs in *Populus tremula*. J. Plant Physiol. 166, 2046–2057.

Johnson, Ø., Kvaalen, H., Yakovlev, I., Dæhlen, O.G., Fossdal, C.G., Skrøppa, T., 2009. An epigenetic memory from time of embryo development affects climatic adaptation in Norway spruce. In: Gusta, L., Wisniewski, M., Tanino, K. (Eds.), Plant Cold Hardiness: From the Laboratory to the Field. CAB International, Wallingford, pp. 99–107.

Karnosky, D.F., Pregitzer, K.S., Zak, D.R., Kubiske, M.E., Hendrey, G.R., Weinstein, D., Nosal, M., Percy, K.E., 2005. Scaling ozone responses of forest trees to the ecosystem level in a changing climate. Plant Cell Environ. 28, 965–981.

Karnosky, D.F., Werner, H., Holopainen, T., Percy, K., Oksanen, T., Oksanen, E., Heerdt, C., Fabian, P., Nagy, J., Heilman, W., Cox, R., Nelson, N., Matyssek, R., 2007. Free-air exposure systems to scale up ozone research to mature trees. Plant Biol. 9, 181–190.

Kerner, R., Winkler, J.B., Dupuy, J.W., Jürgensen, M., Lindermayr, C., Ernst, D., Müller-Starck, G., 2011. Changes in the proteome of juvenile European beech following three years exposure to free-air elevated ozone. iForest 4, 69–76.

King, J., Liu, L., Aspinwall, M., 2013. Forest ecosystem response to interacting elevated atmospheric CO_2 and tropospheric O_3: a synthesis of experimental evidence. Dev. Environ. Sci. 13, 179–208.

Knoke, T., Hahn, A., 2013. Global change and the role of forests in future land-use systems. In: Matyssek, R., Clarke, N., Cudlin, P., Mikkelsen, T., Tuovinen, J.-P., Weiser, G., Paoletti, E. (Eds.), Climate Change, Air Pollution and Global Challenges: Understanding and Solutions from Forest Research. Dev. Environ. Sci. 13. Elsevier, Netherlands, pp. 569–587.

Koch, J.R., Schwerzer, A.J., Eshita, S.M., Davis, K.R., 1998. Ozone sensitivity in hybrid poplar is correlated with a lack of defense-gene activation. Plant Physiol. 118, 1243–1252.

Kontunen-Soppela, S., Ossipov, V., Ossipova, S., Oksanen, E., 2007. Shift in birch leaf metabolome and carbon allocation during long-term open-field ozone exposure. Glob. Chang. Biol. 13, 1053–1067.

Kontunen-Soppela, S., Parviainen, J., Ruhanen, H., Brosché, M., Keinänen, M., Thakur, R.C., Kolehmainen, M., Kangasjärvi, J., Oksanen, E., Karnosky, D.F., Vapaavuori, E., 2010a. Gene expression responses of paper birch (*Betula papyrifera*) to elevated CO_2 and O_3 during leaf maturation and senescence. Environ. Pollut. 158, 959–968.

Kontunen-Soppela, S., Riikonen, J., Ruhanen, H., Brosché, M., Somervuo, P., Peltonen, P., Kangasjävi, J., Auvinen, P., Paulin, L., Keinänen, M., Oksanen, E., Vapaavuori, E., 2010b. Differential gene expression in senescing leaves of two silver birch genotypes in response to elevated CO_2 and tropospheric ozone. Plant Cell Environ. 33, 1016–1028.

Kosova, K., Vítámvás, P., Prášil, I.T., Renaut, J., 2011. Plant proteome changes under abiotic stress—contribution of proteomics studies to understanding plant stress response. J. Proteomics 74, 1301–1322.

Kremer, A., Abbott, A.G., Carlson, J.E., Manos, P.S., Plomion, C., Sisco, P., Staton, M.E., Ueno, S., Vendramin, G.G., 2012. Genomics of Fagaceae. Tree Genet. Genomes 8, 583–610.

Langebartels, C., Ernst, D., Heller, W., Lütz, C., Payer, H.-D., Sandermann, H., 1997. Ozone responses of trees: results from controlled chamber exposures at the GSF phytotron. In: Sandermann, H., Wellburn, A., Heath, R. (Eds.), Forest Decline and Ozone: A Comparison of Controlled Chamber and Field Experiments. Ecological Studies, vol. 127. Springer-Verlag, Berlin, pp. 163–200.

Langebartels, C., Heller, W., Führer, G., Lippert, M., Simons, S., Sandermann, H., 1998. Memory effects in the action of ozone on conifers. Ecotoxicol. Environ. Saf. 41, 62–72.

Langebartels, C., Schraudner, M., Heller, W., Ernst, D., Sandermann, H., 2002. Oxidative stress and defense reactions in plants exposed to air pollutants and UV-B radiation. In: Inzé, D., Van Montagu, M. (Eds.), Oxidative Stress in Plants. Taylor & Francis, London, pp. 105–135.

Li, P., Mane, S.P., Sioson, A.A., Robinet, C.V., Heath, L.S., Bohnert, H.J., Grene, R., 2006. Effects of chronic ozone exposure on gene expression in *Arabidopsis thaliana* ecotypes and in *Thellungiella halophila*. Plant Cell Environ. 29, 854–868.

Lira-Medeiros, C.F., Parisod, C., Fernandes, R.A., Mata, C.S., Cardoso, M.A., Ferreira, P.C.G., 2010. Epigenetic variation in mangrove plants occurring in contrasting natural environment. PLoS One 5, e10326.

Lu, S., Sun, Y.-H., Shi, R., Clark, C., Li, L., Chiang, V.L., 2005. Novel and mechanical stress-responsive microRNAs in *Populus trichocarpa* that are absent from Arabidopsis. Plant Cell 17, 2186–2203.

Lu, S., Sun, Y.-H., Chiang, V.L., 2008. Stress-responsive microRNAs in *Populus*. Plant J. 55, 131–151.

Ludwikow, A., Gallois, P., Sadowski, J., 2004. Ozone-induced oxidative stress response in *Arabidopsis*: transcription profiling by microarray approach. Cell. Mol. Biol. Lett. 9, 829–842.

Matyssek, R., Sandermann, H., 2003. Impact of ozone on trees: an ecophysiological perspective. In: Esser, K., Lüttge, U., Beyschlag, W., Hellwig, F. (Eds.), Progress in Botany, vol. 64. Springer, Berlin, pp. 349–404.

Matyssek, R., Sandermann, H., Wieser, G., Booker, F., Cieslik, S., Musselman, R., Ernst, D., 2008. The challenge of making ozone risk assessment for forest trees more mechanistic. Environ. Pollut. 156, 567–582.

Matyssek, R., Karnosky, D.F., Wieser, G., Percy, K., Oksanen, E., Grams, T.E.E., Kubiske, M., Hanke, D., Pretzsch, H., 2010a. Advances in understanding ozone impact on forest trees: messages from novel phytotron and free-air fumigation studies. Environ. Pollut. 158, 1990–2006.

Matyssek, R., Wieser, G., Ceulemans, R., Rennenberg, H., Pretzsch, H., Haberer, K., Löw, M., Nunn, A.J., Werner, H., Wipfler, P., Oßwald, W., Nikolova, P., Hanke, D.E., Kraigher, H., Tausz, M., Bahnweg, G., Kitao, M., Dieler, J., Sandermann, H., Herbinger, K., Grebenc, T., Blumenröther, M., Deckmyn, G., Grams, T.E.E., Heerdt, C., Leuchner, M., Fabian, P., Häberle, K.-H., 2010b. Enhanced ozone strongly reduces carbon sink strength of adult beech (*Fagus sylvatica*)—resume from the free-air fumigation study at Kranzberg Forest. Environ. Pollut. 158, 2527–2532.

Matyssek, R., Wieser, G., Calfapietra, C., de Vries, W., Dizengremel, P., Ernst, D., Jolivet, Y., Mikkelsen, T.N., Mohren, G.M.J., Le Thiec, D., Tuovinen, J.-P., Weatherall, A., Paoletti, E., 2012. Forests under climate change and air pollution: gaps in understanding and future directions for research. Environ. Pollut. 160, 57–65.

Matyssek, R., Clarke, N., Cudlin, P., Mikkelsen, T.N., Tuovinen, J.-P., Wieser, G., Paoletti, E., 2013a. Climate change, air pollution and global challenges: Understanding and perspectives from forest research. Dev. Environ. Sci. 13, 3–16.

Matyssek, R., Wieser, G., Fleischmann, F., Grunhage, L., 2013b. Ozone Research, quo vadis? Lessons from the Kranzberg Forest Experiment. Dev. Environ. Sci. 13, 103–129.

Miyazaki, S., Fredricksen, M., Hollis, K.C., Poroyko, V., Shepley, D., Galbraith, D.W., Long, S.P., Bohnert, H.J., 2004. Transcript expression profiles of *Arabidopsis thaliana* grown under controlled conditions and open-air elevated concentrations of CO_2 and of O_3. Field Crops Res. 90, 47–59.

Mikkelsen, T.N., Clarke, N., Danielewska, A., Fischer, R., 2013. Towards supersites in forest ecosystem monitoring and research. Dev. Environ. Sci. 13, 475–495.

Moeder, W., Barry, C.S., Tauriainen, A.A., Betz, C., Tuomainen, J., Utrianen, M., Grierson, D., Sandermann, H., Langebartels, C., Kangasjärvi, J., 2002. Ethylene synthesis regulated by biphasic induction of 1-aminocyclopropane-1-carboxylic acid synthase and 1-aminocyclopropane-1-carboxylic acid oxidase genes is required for hydrogen peroxide accumulation and cell death in ozone-exposed tomato. Plant Physiol. 130, 1918–1926.

Nunn, A.J., Anegg, S., Betz, G., Simons, S., Kalisch, G., Seidlitz, H.K., Grams, T.E.E., Häberle, K.-H., Matyssek, R., Bahnweg, G., Sandermann, H., Langebartels, C., 2005. Role of ethylene in the regulation of cell death and leaf loss in ozone-exposed European beech. Plant Cell Environ. 28, 886–897.

Nijnik, M., Miller, D., 2013. Targeting sustainable provision of forest ecosystem services with special focus on carbon sequestration. In: Matyssek, R., Clarke, N., Cudlin, P., Mikkelsen, T., Tuovinen, J.-P., Weiser, G., Paoletti, E. (Eds.), Climate Change, Air Pollution and Global Challenges: Understanding and Solutions from Forest Research. Dev. Environ. Sci. 13. Elsevier, Netherlands, pp. 547–568.

Oksanen, E., 2003. Responses of selected birch (*Betula pendula* Roth) clones to ozone change over time. Plant Cell Environ. 26, 875–886.

Oksanen, E., Keski-Saari, S., Kontunen-Soppela, S., Markku Keinänen, M., 2013. Metabolomics and transcriptomics increase our understanding about defence responses and genotypic differences of northern deciduous trees to elevating ozone, CO_2 and climate warming. Dev. Environ. Sci. 13, 309–329.

Olbrich, M., Betz, G., Gerstner, E., Langebartels, C., Sandermann, H., Ernst, D., 2005. Transcriptome analysis of ozone-responsive genes in leaves of European beech (*Fagus sylvatica* L.). Plant Biol. 7, 670–676.

Olbrich, M., Gerstner, E., Welzl, G., Winkler, J.B., Ernst, D., 2009. Transcript responses in leaves of ozone-treated beech saplings at an outdoor free air fumigation site over two growing seasons. Plant Soil 323, 61–74.

Olbrich, M., Knappe, C., Wenig, M., Gerstner, E., Häberle, K.-H., Kitao, M., Matyssek, R., Stich, S., Leuchner, M., Werner, H., Schlink, K., Müller-Starck, G., Welzl, G., Scherb, H., Ernst, D., Heller, W., Bahnweg, G., 2010a. Ozone fumigation (twice ambient) reduces leaf infestation following natural and artificial inoculation by the endophytic fungus *Apiognomonia errabunda* of adult European beech trees. Environ. Pollut. 158, 1043–1050.

Olbrich, M., Gerstner, E., Bahnweg, G., Häberle, K.-H., Matyssek, R., Welzl, G., Heller, W., Ernst, D., 2010b. Transcriptional signatures in leaves of adult European beech trees (*Fagus sylvatica* L.) in an experimentally enhanced free air ozone setting. Environ. Pollut. 158, 977–982.

Ossipov, V., Ossipova, S., Bykov, V., Oksanen, E., Koricheva, J., Haukioja, E., 2008. Application of metabolomics to genotype and phenotype discrimination of birch trees grown in a long-term open-field experiment. Metabolomics 4, 39–51.

Pääkkönen, E., Seppänen, S., Holopainen, T., Kokko, H., Kärenlampi, S., Kärenlampi, L., Kangasjärvi, J., 1998. Induction of genes for the stress proteins PR-10 and PAL in relation to growth, visible injuries and stomatal conductance in birch (*Betula pendula*) clones exposed to ozone and/or drought. New Phytol. 138, 295–305.

Paoletti, E., de Vries, W., Mikkelsen, T.N., Ibrom, A., Larsen, K.S., Tuovinen, J.P., Serengil, Y., Yurtseven, I., Wieser, G., Matyssek, R., 2013. Key indicators of air pollution and climate change impacts at forest Supersites. Dev. Environ. Sci. 13, 497–518.

Perco, P., Mühlberger, I., Mayer, G., Oberbauer, R., Lukas, A., Mayer, B., 2010. Linking transcriptomic and proteomic data on the level of protein interaction networks. Electrophoresis 31, 1780–1789.

Plomion, C., Lalanne, C., Claverol, S., Meddour, H., Kohler, A., Bogeat-Triboulot, M.-B., Barre, A., Le Provost, G., Dumazet, H., Jacob, D., Bastien, C., Dreyer, E., de Daruvar, A., Guehl, J.-M., Schmitter, J.M., Martin, F., Bonneu, M., 2006. Mapping the proteome of poplar and application to the discovery of drought-stress responsive proteins. Proteomics 6, 6509–6527.

Pregitzer, K.S., Talhlem, A.F., 2013. Belowground carbon cycling at aspen FACE: dynamic responses to CO_2 and O_3 in developing forests. Dev. Environ. Sci. 13, 209–226.

Pretzsch, H., Dieler, J., Matyssek, R., Wipfler, P., 2010. Tree and stand growth of mature Norway spruce and European beech under long-term ozone fumigation. Environ. Pollut. 158, 1061–1070.

Puckette, M.C., Tang, Y., Mahalingam, R., 2008. Transcriptomic changes induced by acute ozone in resistant and sensitive *Medicago truncatula* accessions. BMC Plant Biol. 8, 46.

Renaut, J., Bohler, S., Hausman, J.-F., Hoffmann, L., Sergeant, K., Ahsan, N., Jolivet, Y., Dizengremel, P., 2009. The impact of atmospheric composition on plants: a case study of ozone and poplar. Mass Spectrom. Rev. 28, 495–516.

Rigault, P., Boyle, B., Lepage, P., Cooke, J.E.K., Bousquet, J., MacKay, J.J., 2011. A white spruce gene catalog for conifer genome analyses. Plant Physiol. 157, 14–28.

Rötzer, T., Seifert, T., Gayler, S., Priesack, E., Pretzsch, H., 2012. Effects of stress and defence allocation on tree growth: simulation results at the individual and stand level. In: Matyssek, R., Schnyder, H., Oßwald, W., Ernst, D., Munch, J.C., Pretzsch, H. (Eds.), Growth and Defence in Plants. Ecological Studies, vol. 220. Springer-Verlag, Berlin, pp. 401–432.

Sánchez-Pons, N., Irar, S., García-Muniz, N., Vicient, C.M., 2011. Transcriptomic and proteomic profiling of maize embryos exposed to camptothecin. BMC Plant Biol. 11, 91.

Sandermann, H., Matyssek, R., 2004. Scaling up from molecular to ecological processes. In: Sandermann, H. (Ed.), Molecular Ecotoxicology of Plants. Ecological Studies, vol. 170. Springer-Verlag, Berlin, pp. 207–226.

Sandermann, H., Ernst, D., Heller, W., Langebartels, C., 1998. Ozone: an abiotic elicitor of plant defence reactions. Trends Plant Sci. 3, 47–50.

Sasek, T.W., Richardson, C.J., Fendick, E.A., Bevington, S.R., Kress, L.W., 1991. Carryover effects of acidic rain and ozone on the physiology of multiple flushes of loblolly pine seedlings. Forest Sci. 37, 1078–1098.

Scherling, C., Roscher, C., Giavalisco, P., Schulze, E.-D., Weckwerth, W., 2010. Metabolomics unravel contrasting effects of biodiversity on the performance of individual plant species. PLoS One 5, e12569.

Schneiderbauer, A., Back, E., Sandermann, H., Ernst, D., 1995. Ozone induction of extensin mRNA in Scots pine, Norway spruce and European beech. New Phytol. 130, 225–230.

Sergeant, K., Spieß, N., Renaut, J., Wilhelm, E., Hausman, J.F., 2011. One dry summer: a leaf proteome study on the response of oak to drought exposure. J. Proteomics 74, 1385–1395.

Tamaoki, M., Matsuyama, T., Kanna, M., Nakajima, N., Kubo, A., Aono, M., Saji, H., 2003. Differential ozone sensitivity among *Arabidopsis* accessions and its relevance to ethylene synthesis. Planta 216, 552–560.

Tosti, N., Pasqualini, S., Borgogni, A., Ederli, L., Falistocco, E., Crispi, S., Paolocci, F., 2006. Gene expression profiles of O_3-treated *Arabidopsis* plants. Plant Cell Environ. 29, 1686–1702.

Tuomainen, J., Betz, C., Kangasjärvi, J., Ernst, D., Yin, Z.-H., Langebartels, C., Sandermann, H., 1997. Ozone induction of ethylene emission in tomato plants: regulation by differential accumulation of transcripts for the biosynthetic enzymes. Plant J. 12, 1151–1162.

Tuskan, G.A., DiFazio, S., Jansson, S., Bohlmann, J., Grigoriev, I., Hellsten, U., Putnam, N., Ralph, S., Rombauts, S., Salamov, A., Schein, J., Sterck, L., Aerts, A., Bhalerao, R.R., Bhalerao, R.P., Blaudez, D., Boerjan, W., Brun, A., Brunner, A., Busov, V., Campbell, M., Carlson, J., Chalot, M., Chapman, J., Chen, G.-L., Cooper, D., Coutinho, P.M., Couturier, J., Covert, S., Cronk, Q., Cunningham, R., Davis, J., Degroeve, S., Déjardin, A., dePamphilis, C., Detter, J., Dirks, B., Dubchak, I., Duplessis, S., Ehlting, J., Ellis, B., Gendler, K., Goodstein, D., Gribskov, M., Grimwood, J., Groover, A., Gunter, L., Hamberger, B., Heinze, B., Helariutta, Y., Henrissat, B., Holligan, D., Holt, R., Huang, W.,

Islam-Faridi, N., Jones, S., Jones-Rhoades, M., Jorgensen, R., Joshi, C., Kangasjärvi, J., Karlsson, J., Kelleher, C., Kirkpatrick, R., Kirst, M., Kohler, A., Kalluri, U., Larimer, F., Leebens-Mack, J., Leplé, J.C., Locascio, P., Lou, Y., Lucas, S., Martin, F., Montanini, B., Napoli, C., Nelson, D.R., Nelson, C., Nieminen, K., Nilsson, O., Pereda, V., Peter, G., Philippe, R., Pilate, G., Poliakov, A., Razumovskaya, J., Richardson, P., Rinaldi, C., Ritland, K., Rouzé, P., Ryaboy, D., Schmutz, J., Schrader, J., Segerman, B., Shin, H., Siddiqui, A., Sterky, F., Terry, A., Tsai, C.-J., Uberbacher, E., Unneberg, P., Vahala, J., Wall, K., Wessler, S., Yang, G., Yin, T., Douglas, C., Marra, M., Sandberg, G., Van de Peer, Y., Rokhsar, D., 2006. The genome of black cottonwood, *Populus trichocarpa* (Torr. & Gray). Science 313, 1596–1604.

Vahala, J., Keinänen, M., Schützendübel, A., Polle, A., Kangasjärvi, J., 2003. Differential effects of elevated ozone on two hybrid aspen genotypes predisposed to chronic ozone fumigation. Role of ethylene and salicylic acid. Plant Physiol. 132, 196–205.

Vapaavuori, E., Oksanen, T., Holopainen, J.K., Holopainen, T., Heiskanen, J., Julkunen-Tiitto, R., Kasurinen, A., Laitinen, J., Oksanen, E., Peltonen, P., Poteri, M., Repo, T., Riikonen, J., Syrjälä, S., 2002. Technical report. Open-top chamber fumigation of cloned silver birch (*Betula pendula* Roth) trees to elevated CO_2 and ozone: description of the fumigation system and the experimental site. Metsäntutkimuslaitoksen tiedonantoja 838. http://www.metla.fi/julkaisut/mt/838.

Warren, C.R., Aranda, I., Cano, F.J., 2012. Metabolomics demonstrates divergent responses of two *Eucalyptus* species to water stress. Metabolomics 8, 186–200.

Weckwerth, W., 2010. Metabolomics: an integral technique in systems biology. Bioanalysis 2, 829–836.

Weckwerth, W., 2011. Green systems biology—from single genomes, proteomes and metabolomes to ecosystems research and biotechnology. J. Proteomics 75, 284–305.

Winkler, J.B., Lang, H., Graf, W., Reth, S., Munch, J.C., 2009. Experimental setup of field lysimeters for studying effects of elevated ozone and below-ground pathogen infection on a plant-soil-system of juvenile beech (*Fagus sylvatica* L.). Plant Soil 323, 7–19.

Wustman, B.A., Oksanen, E., Karnosky, D.F., Noormets, A., Isebrands, J.G., Pregitzer, K.S., Hendrey, G.R., Sober, J., Podila, G.K., 2001. Effects of elevated CO_2 and O_3 on aspen clones varying in O_3 sensitivity: can CO_2 ameliorate the harmful effects of O_3? Environ. Pollut. 115, 473–481.

zu Castell, W., Ernst, D., 2012. Experimental 'omics' data in tree research: facing complexity. Trees 26, 1723–1735.

zu Castell, W., Matyssek, R., Göttlein, A., Fleischmann, F., Staninska, A., 2012. Learning from various plants and scenarios: statistical modeling. In: Matyssek, R., Schnyder, H., Oßwald, W., Ernst, D., Munch, J.C., Pretzsch, H. (Eds.), Growth and Defence in Plants. Ecological Studies, vol. 220. Springer-Verlag, Berlin, pp. 355–373.

Chapter 15

Metabolomics and Transcriptomics Increase Our Understanding About Defence Responses and Genotypic Differences of Northern Deciduous Trees to Elevating Ozone, CO_2 and Climate Warming

Elina Oksanen[1], Sarita Keski-Saari, Sari Kontunen-Soppela and Markku Keinänen

Department of Biology, University of Eastern Finland, P.O. Box 111, Joensuu, Finland

[1]Corresponding author: e-mail: elina.oksanen@uef.fi

Chapter Outline

Developments in Environmental Science, Vol. 13. http://dx.doi.org/10.1016/B978-0-08-098349-3.00015-3

15.1 INTRODUCTION

Molecular studies on forest trees exposed to climate change factors is a challenging research area, because it combines the difficulty of ecologically relevant field experimentation, large physical size and long lifetime of trees, large genome size and more complex sample preparation as compared to most herbaceous plants (e.g. Abril et al., 2011; Ernst, 2013). In this chapter, we make a synthesis of results from several climate change-related experiments with birch (*Betula* sp.) and aspen (*Populus* sp.) species where metabolomic and transcriptomic methods have been combined with ecophysiological measurements. Most of these studies have focused mainly on phenolic compounds, because they comprise a considerable proportion of leaf dry mass in these species and have several important functions especially in defence and plant–herbivore interactions (e.g. Peltonen et al., 2005). Developments in metabolome analyses have led to more comprehensive analyses, including primary metabolites and volatile organic compounds (VOCs). Use of novel transcriptomics and bioinformatics tools allows more efficient extraction of information from large datasets. The aim of the chapter is to indicate the potential of these tools to contribute to our knowledge of the acclimation processes and to explore genotypic differences in respect of ozone tolerance/sensitivity and growth rate, that is, the major determinant for the carbon sink strength of forest trees.

Experimental material included in this chapter consists of birch and aspen saplings or young trees, exposed to elevated ozone, CO_2 and temperature alone and/or in combinations in 10 different experiments, with the main focus on northern species (silver birch, *Betula pendula* Roth and European aspen, *Populus tremula* L.) and open-field approaches. Silver birch and European aspen, covering large and diverse areas across the Northern Hemisphere, are ecologically and economically increasingly important fast-growing pioneer tree species in boreal forests in warming climate (Kellomäki et al., 2008). Unfortunately, the carbon sink strength of northern forests is estimated to be reduced with increasing ozone concentrations, and the meta-analyses by Wittig et al. (2009) estimate 7% and 22% reductions in total biomass of *Betula* and *Populus* species (respectively) by anticipated ozone levels in year 2050. Both species are characterized by high genetic diversity and plasticity, providing valuable ecotypes and materials for climate change and acclimation studies (Eriksson, 2012). Because of the availability and applicability of the annotated and assembled genome sequence of black cottonwood (*Populus trichocarpa*), aspen is the most useful tree species for multidisciplinary experiments, where molecular methods and omics from cells to ecosystem are integrated (Cronk, 2004; Taylor, 2002). Transcriptomic, proteomic and metabolomics approaches were already integrated by Renaut et al. (2009), showing the negative impacts of ozone on carbon assimilation on plants, particularly *Populus* species, with concomitant increase in ROS signalling and in catabolic processes. Changes in protein profiles of poplar (*P. tremula* × *alba*) were demonstrated by Bohler et al. (2007, 2011) mainly as downregulation of photosynthetic processes and upregulation of glucose catabolism. Furthermore, Cabané et al. (2004) and Richet et al. (2011) reported ozone-caused alterations in lignin and cellulose ratios in this poplar species, affecting mechanical strength and defence capacity. Besides acting as carbon sinks through photosynthesis, both *Betula* and *Populus* species are also emitting substantial amounts of reactive VOCs, which have important ecological and biosphere–atmosphere feedback functions in boreal ecosystems (e.g. Holopainen, 2011; Loreto and Fares, 2013, this volume). Most studies presented herein have been made with clonal material to decrease biological variance. Therefore, the variability of responses between individuals is reflected in differences between the genotypes (except for those presented in Kontunen-Soppela et al., 2010a).

15.2 OZONE EXPERIMENTS

The role of leaf phenolic compounds in defence against oxidative stress in silver birch has been explored, for example, in open-field experiments by Saleem et al. (2001) and Yamaji et al. (2003). In the study by Saleem et al. (2001), one ozone-tolerant and one ozone-sensitive genotype of silver birch were exposed to elevated ozone over three growing seasons and determined for leaf phenolics, pigments, nutrients, growth and leaf yellowing (Table 15.1). Ozone exposure increased the concentrations of several phenolics (such as (+)-catechin and chlorogenic acid) and resulted in decreased growth, accelerated leaf senescence and lower contents of chlorophyll and Mg. The shift in carbon allocation towards defensive phenolics at the expense of growth was greater in the ozone-sensitive genotype as compared to tolerant one. In the study by Yamaji et al. (2003), 17 genotypes of silver birch, growing under elevated ozone over two growing seasons, were determined for chemical composition and physiological responses (Table 15.1). The experiment also revealed different strategies against chronic ozone stress in birch population, combining shifts in leaf phenolics, and stomatal and growth adjustments. One third of the genotypes invested in compensation growth of new leaves; a majority of genotypes showed increased biomass allocation to roots, whereas the shoot-to-root ratio was not affected in the most tolerant genotypes due to ozone. The genotypes with increased shoot-to-root ratio increased the concentration of foliar phenolic acids (such as chlorogenic acid, neochlorogenic acid, cinnamic acid derivatives), DHPPG (=3,4′-dihydroxypropiophenone-3-glucoside) and quercetin glycosides, while the most tolerant group relied on high constitutive concentration of total phenolics, N allocation to leaves and reduced stomatal conductance under ozone exposure (Yamaji et al., 2003). This chapter also showed correlations between visible ozone injuries (pigmented or necrotic spots in the upper leaf side) and accumulation of phenolics. The best correlations were found with

TABLE 15.1 List of Stress Experiments, Where Molecular and/or Metabolic Responses Have Been Linked to Physiological Responses in Birch and Aspen Species

Species	Experiment (average AOT40 ppm h/season)	Molecular/metabolic response	Physiological response	References
Ozone experiments				
Betula pendula, two genotypes	Free-air fumigation: 1.5–1.6 × ambient ozone (12.9 ppm h), three growing seasons	Phenolic compounds	Growth, leaf senescence, Rubisco, pigments, proteins, nutrients	Saleem et al. (2001)
Betula pendula, 17 genotypes	Free-air fumigation: 1.5–1.7 × ambient ozone (13.5 ppm h), two growing seasons	Phenolic compounds	Growth, visible injuries, photosynthesis, Rubisco, pigments, proteins, nutrients	Yamaji et al. (2003)
Betula pendula, two genotypes	Free-air fumigation: 1.4–1.7 × ambient ozone (14.5 ppm h), seven growing seasons	Phenolic compounds, carbohydrates, pigments	Growth	Kontunen-Soppela et al. (2007) and Ossipov et al. (2008)
Populus tremula × *Populus tremuloides*, eight genotypes	Free-air fumigation: 1.5 × ambient ozone (17.2 ppm h) + 2N levels (60 and 140 kg/ha/year), two growing seasons	Phenolic compounds	Growth, photosynthesis, Rubisco, chlorophyll fluorescence, pigments, carbohydrates	Häikiö et al. (2008)
Populus tremula × *Populus tremuloides*, two genotypes	Free-air fumigation: 1.3–1.4 × ambient ozone (14.9 ppm h), three growing seasons	Phenolic compounds	Growth, photosynthesis, Rubisco, chlorophyll fluorescence, fine roots, mycorrhiza	Häikiö et al. (2009)

Populus nigra clone Turchia	Growth chamber: 80 ppb ozone (6.7 ppm h), 2 weeks	Phenolic compounds, leaf surface compounds, VOCs	Leaf growth, photosynthesis, stomatal conductance, ozone uptake	Fares et al. (2010)
Ozone and CO_2 experiments				
Betula pendula, two genotypes	Open-top chambers: 2× ambient ozone (25.3 ppm h) and CO_2, two growing seasons	Gene expression, phenolic compounds	Leaf senescence, growth, photosynthesis, Rubisco, pigments, chlorophyll fluorescence, carbohydrates, leaf structure	Peltonen et al. (2005), Oksanen et al. (2005) and Kontunen-Soppela et al. (2010b)
Betula papyrifera seedlings	Free-air fumigation (Aspen FACE site): 1.5× ambient ozone (n. 41.0 ppm h) and CO_2 (550 ppm), seven growing seasons	Gene expression	Leaf size, photosynthesis, chlorophyll fluorescence, lignin content of leaves, N metabolism, leaf senescence	Kontunen-Soppela et al. (2010a)
Ozone and/or temperature experiments				
Betula pendula, four genotypes *Populus tremula*, four genotypes	Free-air fumigation: 1.2× ambient ozone (4.9 ppm h)+elevated temperature (ambient+1.2 °C), one to two growing seasons	Gene expression related to VOCs	VOC emissions, growth, gas exchange, antioxidant capacity	Hartikainen et al. (2009, 2012), Riikonen et al. (2009) and Mäenpää et al. (2011)
Betula pendula, three genotypes *Populus tremula*, three genotypes	Growth chambers, five different night-time temperatures (6, 10, 14, 18 and 22 °C)	Low-molecular-weight metabolites	Growth, gas exchange VOC emissions	Mäenpää et al. (2012)

VOC, volatile organic compounds.

phenolic acids, flavonoid glycosides and flavonoid aglycones (e.g. apigenin derivatives).

A more comprehensive metabolite analysis and shifts in leaf metabolite profiles were determined for two silver birch genotypes, exposed to elevated ozone in open-field conditions over seven growing seasons (Kontunen-Soppela et al., 2007; Ossipov et al., 2008; Table 15.1). Long-term ozone stress resulted in increased concentration of foliar quercetin glycosides, chlorogenic acid and some leaf cuticular wax compounds (such as triterpenoids and squalene), while several compounds related to carbohydrate metabolism, pigments and function of chloroplast membranes were decreasing. Compounds that decreased included, for example, sucrose, inositol, mannitol, fructose, galactose, galactolipids and lutein. Leaves of more ozone-tolerant genotype showed increased concentration of compounds related to epicuticular wax formation and quercetin glycosides (Kontunen-Soppela et al., 2007; Tables 15.2 and 15.3).

Ozone-sensitive and tolerant genotypes of hybrid aspen (*P. tremula* L. × *Populus tremuloides* Michx.) were growing under elevated ozone concentration and two soil nitrogen regimes over two growing seasons in pots, and were measured for phenolic metabolites, other foliar traits and photosynthesis (Häikiö et al., 2008; Table 15.1). These parameters could be related to ozone sensitivity and growth rate of clones (Tables 15.2 and 15.3). Ozone increased the concentrations of chlorogenic acid and (+)-catechin, whereas photosynthesis processes were declined both in tolerant and sensitive groups, showing unaffected/increased and reduced growth, respectively (Häikiö et al., 2008). Ozone sensitivity (i.e. decreased growth) was associated with higher photosynthetic capacity, and higher contents of Rubisco and salicylates, while ozone-tolerant clones showed higher concentration of (+)-catechin, a kaempferol derivative, and condensed tannins (Häikiö et al., 2008; Table 15.2). The same ozone experiment facility was used to study differences in photosynthetic characters and phenolic compounds between two hybrid aspen genotypes, growing in the soil over three growing seasons (Häikiö et al., 2009; Table 15.1). Elevated ozone impaired the growth of roots and increased the ectomycorrhizal status of the fine roots, but gas exchange was unaffected (Table 15.3). Different profiles of foliar phenolics were presented for the two clones, showing different growth rates and competitive capacity, so that, for example, high carbon allocation to condensed tannins correlated with slow growth rate (Häikiö et al., 2009).

In addition to long-term field experiments, ozone-caused metabolic changes were determined for black poplar (*Populus nigra* L.) in growth chamber conditions (Fares et al., 2010; Table 15.1). Leaf phenolic and surface compounds and emissions of VOCs were correlated with leaf growth and gas exchange rates along the vertical axis. Despite the short exposure time, ozone shifted the metabolite concentrations towards phenolic compounds, that is, a myricetin glycoside, a quercetin glycoside and a cinnamic acid derivative, particularly in the newly expanded leaves. Formation of these phenolics was

TABLE 15.2 **Observations of Traits Associated with Ozone Tolerance/ Sensitivity in Field Experiments with Birch (*Betula pendula*) and Aspen (*Populus tremula* × *P. tremuloides*)**

Species (reference)	Ozone tolerance traits	Ozone sensitivity traits
Populus tremula × *Populus tremuloides*, eight genotypes (Häikiö et al., 2008)	Unaffected/increased growth	Decreased growth
	Low photosynthetic capacity	High photosynthetic capacity
	Low contents of Rubisco	High contents of Rubisco
	High amount of antioxidative phenolics (catechin, kaempferol, condensed tannins)	High contents of salicylates
Betula pendula, two genotypes (Kontunen-Soppela et al., 2007)	Smaller growth reductions as compared to sensitive genotype	Greater growth reductions as compared to tolerant genotype
	High amount of phenolics and polar compounds	Lower amount of total phenolics and polar compounds
	High concentration of wax-related compounds: squalene, 1-dotriacontanol, dotriacontane, high concentration of quercetin glycosides (antioxidative defence)	
Betula pendula, two genotypes (Kontunen-Soppela et al., 2010a,b; Oksanen et al., 2005; Peltonen et al., 2005)	Greater transcriptional response and capacity to alter metabolism	Less dynamic genetic and metabolic response
	Efficient ROS scavenging (despite high ozone uptake and high oxidative stress)	Low ROS scavenging capacity
	Higher number of peroxisomes and mitochondria	
	Higher concentration of phenolic acids, flavone aglycons, flavonol glycosides and hydrolysable tannins	

TABLE 15.3 Responses of Northern Birch and Aspen Species to Ozone

Parameter measured	Response to elevated O_3
Growth	
Stem height	↔
Specific leaf area (SLA)	↓
Root mass (DW)	↔
Total biomass (DW)	↔
Shoot-to-root ratio	↔
Mycorrihizal status of roots	↑
Gas exchange	
Total concentration of Rubisco	↓/–
Photosynthesis (Pn)	↔
Stomatal conductance (g_s)	↔
Chlorophyll fluorescence (F_v/F_m)	↓/–
Non-photochemical quenching (NPQ)	↑
Metabolites	
Concentration of total phenolics	↑/–
Concentration of total carotenoids	–
Concentration of total chlorophylls	↓/–
Concentration of soluble sugars	↓
Concentration of starch	↓
Concentration of total soluble proteins	–
Concentration of total ascorbate	↑
Total antioxidant capacity	↑
Leaf surface compounds	↕
VOCs	↕
Foliar macronutrients	
N, P, K	↓/–
Mg	↑
Ca	↕

Responses are genotype specific, dependent on ozone dose, plant/leaf age and leaf position within the canopy. Only ozone-specific responses are included in the table. The arrows indicate the direction of the change: – no change; ↑ increase; ↓ decrease; ↕ both increase and decrease; ↔ increase, decrease or no change. The data are collected from the experiments listed in Table 15.1.

in line with the increased ozone uptake, indicating their defensive role. Under ozone stress, emissions of isoprene and oxygenated six-carbon (C6) volatiles were downregulated in the young leaves, while methanol emissions were increasing, demonstrating the ontogenic shift, that is, slower development and faster senescence of leaves due to ozone. In the same leaves, decreased concentrations of surface compounds (chlorogenic acid and cinnamic acid derivatives) were detected, indicating disturbances in the normal leaf development and the direct impact of ozone on molecular composition of leaf surface (Fares et al., 2010; Table 15.3).

Concluding from ozone experiments above, accumulation of phenolic compounds is one of the major metabolic responses to chronic ozone stress both in birch and aspen (Table 15.3), as discussed also by Dizengremel et al. (2013), this volume. In addition, molecular composition of leaf surface wax is reactive to external ozone concentration and plays a role in acclimation to long-term stress situation. Composition of leaf surface wax and leaf phenolics forms a part of dynamic and sustainable detoxification mechanism against prolonged ozone stress situation. Changes in rapidly formed and less stable volatile compounds, such as isoprene, serve as markers for ontogenic alterations, but their defensive role against oxidative stress is more apparent in short-term stress situations (Fares et al., 2010). Recently, Dizengremel et al. (2012), focusing on ozone stressed poplars, have pointed out the increased demand in carbon skeletons towards several competing defence and repair processes, that is, biosynthesis of phenolic and lignin compounds via shikimate phenylpropanoid pathway, formation of isoprenoids via mevalonic acid pathway, formation of ATP and reducing power (NADPH) via mitochondria respiration and increased amino acid synthesis via anaplerotic pathway, sharing the common precursors (phosphoenolpuryvate, PEP, in particular). Thus, the regulation and timing of carbon allocation from primary metabolites to these pathways remain to be some of the key issues in mechanistic ozone response research. Traits related to ozone tolerance/sensitivity of birch and aspen species are discussed in Section 15.4.2 of this chapter.

15.3 INTERACTIONS OF OZONE WITH CO_2 AND/OR ELEVATED TEMPERATURE

Two genotypes of young silver birch trees were exposed to 2 × ambient ozone and CO_2 in open-top chambers over 2 years in Finland (Kontunen-Soppela et al., 2010b; Table 15.1; see also Ernst, 2013, this volume) and studied for transcriptional profiles and key physiological characteristics during leaf senescence. Ozone-tolerant clone showed greater responses in the gene expression profile as compared to sensitive clone. Ozone stress downregulated many genes related to photosynthesis and glycolysis, while the expression of genes related to leaf ageing increased in a clone-dependent manner. The changes in gene expression were associated with decline in the photosynthetic light reactions, decreased concentration of Rubisco, chlorophyll and soluble sugars, but not to net photosynthesis rates or growth parameters, indicating complex relationship between gene regulation and physiological response. In the late season, ozone also increased the expression of genes related to oxidative stress (e.g. NADH plastoquinone oxidoreductase) and genes encoding enzymes of secondary metabolism, concomitant with the increased concentration of phenolic acids but decreased concentration of flavonone aglycones and DHPPG (Kontunen-Soppela et al., 2010b; Oksanen et al., 2005; Peltonen et al., 2005; Table 15.3). Elevated CO_2 downregulated the expression of genes related to carbohydrate metabolism and senescence, concomitant with the delayed leaf abscission and accumulation of secondary metabolites, indicating a possible trade-off in carbon allocation (Kontunen-Soppela et al., 2010b; Peltonen et al., 2005). Combined exposure to both stress factors indicated that CO_2 counteracted ozone responses, alleviating several negative effects of ozone at least temporarily (Kontunen-Soppela et al., 2010b; Peltonen et al., 2005).

Changes in gene expression of paper birch (*Betula papyrifera* Marsh.) leaves collected from the long-term ozone and CO_2 experiment Aspen FACE (USA) (Figure 15.1; Pregitzer and Talhelm, 2013, this volume) were studied using novel multivariate and clustering methods (Kontunen-Soppela et al., 2010a; Table 15.1; see also Ernst, 2013). Ozone-exposed trees showed downregulation of light harvesting- and carbon fixation-related genes (encoding

FIGURE 15.1 Interactions of elevated ozone and CO_2 on paper birch (*Betula papyrifera*) and trembling aspen (*Populus tremuloides* Michx.) have been examined in Aspen FACE site in Wisconsin, USA, since year 1998 until today. For more detailed information, see http://aspenface.mtu.edu/index.html. *Courtesy by Rick Anderson.*

Rubisco, Rubisco activase, glyseraldehyde-3-P-dehydrogenase), accompanied by lower net photosynthesis (Riikonen et al., 2008; Table 15.3). Increased expression of genes related to ROS formation (encoding, e.g., ascorbate oxidase, arginine decarboxylase, metallothioneins, ACC synthase), photorespiration (encoding e.g. catalase, aminotransferase, glycolate oxidase), electron transport (encoding alternative oxidase, AOX, proton gradient regulation), lignin biosynthesis (encoding e.g. cinnamoyl-reductase, peroxidases) and repair (6-4 photolyase, DeGP2) were related to active defence against oxidative stress and protection/repair of the photosynthesis apparatus (Kontunen-Soppela et al., 2010a). Increase in the expression of genes related to proteolysis (encoding e.g. Cys and Asp proteases, ubiquitin), glycolysis and TCA cycle in the late season was associated with enhanced senescence of ozone-exposed leaves, and related energy demands for degradation and transport. Trees under elevated CO_2 showed increased gene expression related to nitrogen metabolism, reflecting increased N assimilation and uptake from soil as reported by Zak et al. (2007). In addition, an increase in expression of TCA cycle and glycolysis genes and concomitant decrease in genes encoding oxidative phosphorylation indicated increased carbon supply and allocation for amino acid, fatty acid and polyphenol synthesis under elevated CO_2 (Kontunen-Soppela et al., 2010a). Changes in gene expression under combined exposure to elevated ozone and CO_2 were similar but smaller compared to ozone alone, indicating again the counteractive action of ozone and CO_2 also at molecular level.

Taken together, our studies indicate the targets of ozone and CO_2 action in several genes related mainly to photosynthesis, defence against oxidative stress, repair and senescence, serving as valuable and realistic basis for more detailed mechanistic studies. Consistently with our results, laboratory experiments by Bohler et al. (2007, 2011) indicated downregulation of proteins relating to light reactions and Calvin cycle in ozone-exposed poplars, accompanied by enhanced catabolism (reflecting the increased need for energy and reducing power for detoxification and repair). Although ozone caused changes in transcription levels of genes of mature beech (*Fagus sylvatica*) trees, growing under elevated ozone in field conditions, were inconsistent between sun and shade leaves and limited to few processes, some upregulation of secondary metabolism, signal transduction, ethylene biosynthesis and amino acid transporters, playing a role in stress response were also observed (Olbrich et al., 2010). Opposite impacts of ozone and CO_2 on poplar growth and wood formation were also found by Richet et al. (2012), except that lignin content was increased by both factors as a result of different action mechanisms.

Saplings of silver birch and European aspen were also exposed to combined action of elevated ozone and temperature (T) in field conditions (Figure 15.2) over 1–2 growing seasons and measured for growth, gas exchange and antioxidant capacity, emissions of VOCs and changes in key gene expressions (*DXS*, *DXR*, *IPP* isomerase) regulating the formation of terpenoids (Hartikainen et al., 2009, 2012; Loreto and Fares, 2013; Mäenpää

FIGURE 15.2 Joint action of elevated ozone and warming climate on birch (*Betula pendula*) and aspen (*Populus tremula*) has been studied in open-field experiment site in Kuopio, Finland over growing seasons 2007–2008. *Photo by Elina Oksanen.*

et al., 2011; Riikonen et al., 2009; Table 15.1). In aspen, elevated *T* induced large increases in emissions of monoterpenes and green leaf volatiles (GLVs), while impacts of ozone were minor (Hartikainen et al., 2009). Tolerance of certain genotypes to high temperature stress, indicated by increased leaf growth, was linked to increased emissions of isoprene and α-pinene together with thick palisade layer of leaf mesophyll, facilitating efficient photosynthesis per leaf area. In birch, elevated *T* increased growth, delayed leaf senescence and increased total emissions of mono-, homo- and sesquiterpenes, co-occurring with reduced transcription of genes encoding DXS, DXR and IPP isomerase, operating in chloroplastic MEP (methylerythritol 4-phosphate) pathway indicating negative regulation (Hartikainen et al., 2012; Kasurinen et al., 2012). Elevated ozone led to increases in the expressions of these three genes despite the unaffected emissions of mono-, homo- and sesquiterpenes, but the emissions of GLVs and MeSA (methyl salicylate) reduced together with the decline in photosynthesis and growth (Hartikainen et al., 2012; Kasurinen et al., 2012).

In addition to VOCs, also low-molecular-weight metabolites of birch were affected by increasing night-time temperature (from 6 to 22 °C) (Mäenpää et al., 2012; Table 15.1). High temperature induced specifically the biosynthesis of triterpenoids, hydrolysable tannins, fructose, DHPPG, myricetin 3-glucuronide and quinic acid along with the increased stem and leaf area growth and dark respiration rate, while proanthocyanidins tended to decrease (Mäenpää et al., 2012). In addition to temperature acclimation, these metabolic alterations indicated shifts in developmental stage of leaves. As a conclusion, our studies with birch and aspen have demonstrated the impacts of increasing daytime and night-time temperatures both on low-molecular-weight and volatile compounds, and related gene expression.

15.4 KEY FINDINGS AND SPECIFIC QUESTIONS ARISING FROM THE OZONE STRESS EXPERIMENTS

Our experiments with ozone have shown the magnitude and direction of changes in response to stress in birch and aspen trees, as summarized in Table 15.3. However, finding specific indicators for ozone tolerance/sensitivity or growth rate proved to be challenging, because the changes in growth, physiological parameters and metabolite concentrations are dependent on, for example, ozone dose, species, genotypes and the ontogenic phase of plants studied.

15.4.1 How to Interpret Increased Concentration of Phenolics

An increase in phenolic compounds due to ozone was a common phenomenon in our experiments (Peltonen et al., 2005; Saleem et al., 2001; Yamaji et al., 2003), as reported for several tree species by Valkama et al. (2007). In our studies, phenolic concentrations were in average 20% higher in plants growing under 1.3–1.7 × elevated ozone concentrations (average AOT40 exposures ranging between 12.9 and 25.3 ppm h/season) as compared to control plants. Genotypes usually differed in their responses so that either there is a slight increase or there is no change. According to meta-analysis by Valkama et al. (2007), ozone induced increases in secondary metabolites, that is, phenolics and terpenes were 16% and 8%, respectively, when 63 studies and 22 tree species were included. Increases in secondary metabolites were higher in free-air exposures as compared to OTC and indoor facilities, and angiosperm showed greater responsiveness as compared to gymnosperms.

The compounds that increased in all our studies with silver birch include phenolic acids and quercetin glycosides. Generally, the repeatability of the results is very good, as indicated by Saleem et al. (2001) and Kontunen-Soppela et al. (2007) that measured the same trees and compounds after three and seven growing seasons. The response of condensed tannins was usually similar as compared to other phenolic compounds. The largest responses due to ozone were seen in flavonoid aglycones situated on the leaf surface. For example, flavonoid aglycones increased threefold in a field study by Yamaji et al. (2003). This response may be related to the nature of ozone as a gaseous substance that is first encountered at the leaf surface. Thus, it is logical to assume that the first and probably the strongest responses induced by ozone would occur on the leaf surface. High ozone deposition on to cuticular surfaces and non-stomatal uptake of ozone are also discussed in Dizengremel et al. and Loreto and Fares (2013), this volume.

The increase in phenolics can be associated to their possible protective role against oxidative stress through their role as antioxidants (see Dizengremel et al., 2013, this volume), either by binding free radicals or inhibiting the formation of free radicals through the binding of metals as chelates (e.g. Rice-Evans et al., 1997). However, phenolic compounds may also act as pro-oxidants, that is, increase the reactivity of free radicals, depending on the other compounds that are present (e.g. Abou Samra et al., 2011). Some phenolics have stronger antioxidative capacity than the others. For example, chlorogenic acid has greater activity towards superoxide than the classical antioxidant, ascorbate (Grace et al., 1998). Quercetin glycosides are *ortho*-dihydroxylated in the B ring, which enhances their activity towards free radicals (e.g. Rice-Evans et al., 1997).

Increase in the concentration of phenolic compounds may be due to induced synthesis or decreased plant biomass due to ozone in comparison to control conditions. In these studies, both have been implied (Peltonen et al., 2005). In a study performed in open-top chambers, DHPPG and flavonoid aglycones were accumulating, while condensed tannins and total phenolics increased in concentration only as a result of a decrease in leaf biomass (Peltonen et al., 2005). However, if phenolic compounds have a role in defence against ozone, then high constitutive level of phenolic compounds may also be important, even with no increase in response to ozone. Thus, ozone-tolerant genotypes may have a high intrinsic concentration of phenolics, with no increase in response to ozone (Häikiö et al., 2008, 2009; Yamaji et al., 2003).

There was a positive correlation between visible ozone injuries and accumulation of phenolics in birch (Yamaji et al., 2003). However, the increase in phenolics is not enough to protect the leaves from visible damage, since leaf weight and area decreased in ozone-treated birches despite the accumulation of phenolics (Peltonen et al., 2005). It cannot be therefore concluded whether phenolic accumulation protects against visible damage, or the induction of phenolic synthesis is a stress response or defence.

15.4.2 Search for Indicators of Ozone Tolerance/Sensitivity and Growth Rate

Our experiments indicated that certain traits of phenolic composition correlate with ozone tolerance/sensitivity, but a reliable indicator of ozone tolerance should probably include a combination of several measured parameters. In a field experiment, eight hybrid poplar genotypes were clearly separated in principal component analysis by their tolerance to ozone, but the trees grown in the different treatments were not separated (Häikiö et al., 2008). Tolerant genotypes showed higher concentration of (+)-catechin, condensed tannins and a kaempferol derivative than the sensitive genotypes, while the latter group showed higher concentration of salicylates (Häikiö et al., 2008; Table 15.2). Additionally, the formation of certain wax compounds and the high plasticity in transcriptional adjustments can be related to ozone tolerance (Kontunen-Soppela et al., 2007, 2010b). These metabolic changes reflect the acclimation responses to long-term chronic ozone stress, and occur in parallel with structural adjustments (i.e. leaf thickening, increasing proportion of apoplasm, reducing surface area per leaf mass) and stomatal closure (Oksanen et al., 2009; Riikonen et al., 2009). The short-term defence mechanisms, found at the beginning of the exposures, include accumulation of apoplastic ascorbic acid with increasing antioxidant capacity in birch (Riikonen et al., 2009), but genotypic differences in early responses have not been determined. In hybrid poplar (*Populus deltoids* × *P. trichocarpa*), ozone-tolerant and -sensitive genotypes showed similar basal emission rates of isoprene and stomatal ozone uptake, while tolerance was related to higher content of carotenoids providing increased antioxidant potential, and the ability to maintain isoprene production after the stress (Ryan et al., 2009). In addition, Street et al. (2011) were able to identify several poplar genes such as *SAM synthase*, a chitinase gene and *Beta-1,3-Glucanase* differing between sensitive and tolerant genotypes. In accordance with our studies, ozone tolerance/sensitivity in *Arabidopsis thaliana* is known to be a result of a unique set of alterations of different defence mechanisms, that is, regulatory genes, stress hormones, signalling pathways, antioxidants and physiological characters (e.g. stomatal conductance and regulation) (Brosché et al., 2010). The difficulty of finding specific indicators for ozone tolerance may be related to the nature of ozone as a phytotoxin that causes oxidative stress, similarly as drought, salt, flood and many other environmental pollutants. It can be questioned if any one ozone-specific indicator should be found when oxidative stress is a common phenomenon in many stresses. On the other hand, one may find a unique combination of several different indicators, when metabolic/transcriptomic approaches are used.

Finding specific metabolic indicators for fast growth rate would be practically useful and ultimate goal of metabolomic research. In a study with several hybrid aspen genotypes, slow-growing genotypes showed a high concentration of condensed tannins and catechin, while fast-growing genotypes showed a high concentration of Rubisco and neochlorogenic acid (Häikiö et al., 2008). However, differences in concentrations of phenolics were not remarkable, and therefore, it would be challenging to set limits for phenolic concentrations for different growth rates. In a study with several hybrid aspen genotypes, ozone tolerance and slow growth were not linked, that is, all slow-growing genotypes were not ozone tolerant, while all fast-growing genotypes were not ozone sensitive (Häikiö et al., 2008). Thus, it is possible to find ozone-tolerant and fast-growing genotypes. However, it must be noted that a slow-growing genotype at an early developmental stage can later be a fast-growing genotype, as discussed in Kontunen-Soppela et al. (2007).

15.5 FUTURE DEVELOPMENTS AND SOCIO-ECONOMIC ASPECTS

As indicated by our experiments above, developments in omics analyses are required, for example, in metabolomics, leading from restricted phenolic analyses to more comprehensive metabolome profilings. These advances are crucial in increasing our understanding of mechanistic responses of deciduous trees to climate change factors and for developing indicators of growth rate for practical forestry.

Although the new systems biology approaches are calling for integration of the data from multiple research fields, we are still facing many challenges in the currently used omics methods (zuCastell and Ernst, 2012). Challenges in metabolomic research include identification of unknown compounds, reliability of annotations of compounds, inter-comparability of the results among laboratories and, in many cases, the transient nature of the responses. Fluxes of metabolite responses may give entirely different view of the metabolic responses than a single sampling at a point in time (Fernie and Stitt, 2012). However, determination of fluxes requires several samplings, which is not the case in any of the experiments described in this review. The challenge is to keep the experimental setups simple enough to be able to test the differences with the available statistical tools, but to account for the complexity of nature at least on some level. Approaches easily become laborious, especially if one wants to include several samplings to account for the changes in responses over time, several genotypes and/or several biological replicates to account for the variability of biological responses, and possibly several environmental factors to account for the interactions that they may have with each other. As a conclusion, metabolomics approach serves as a useful tool to narrow down the possibilities of hundreds of specific targeted approaches to the few of the greatest interest. For example, the study by Kontunen-Soppela et al. (2007) presented in this chapter gives a wide view of the metabolic changes, pointing out the direction for future studies, that is, leaf surface wax-related compounds that will deserve a more detailed targeted analysis.

In the case of ozone, acute responses studied in a matter of hours after the ozone exposure, often with a high level of ozone, represent a different study case than the chronic responses. The gene expression and metabolite profiling studies reviewed here have been made from young trees exposed to ozone for more than one growing season (Kontunen-Soppela et al., 2007, 2010a,b), when the trees have already acclimated to the exposure. Therefore, the differences found in these studies may be small in magnitude, but as they persist for several growing seasons, they can be crucial for the trees in terms of their growth and survival. Ozone enhances senescence of leaves, and it is difficult to judge if the changes induced by ozone in a chronic exposure are due to ozone as such or if they are related to the older physiological age of the leaves due to earlier senescence. Multiple sampling approaches may help in giving a more comprehensive view of responses to ozone.

Field studies represent a different challenge when compared to chamber experiments, as also discussed by Ernst (2013), this volume. Field experiments give a more reliable view of what could actually happen in nature, when abiotic factors, such as UV-B radiation, light intensity and temperature conditions, are changing by their daily and seasonal natural variation. For example, meta-analyses by Valkama et al. (2007) showed that increases in phenolic compounds due to ozone were greatest in outdoor experiments as compared to laboratory conditions. However, field studies always have a wider variance between samples related to the uncontrolled differences in, for example, biotic factors such as herbivory and plant pathogens, which makes the effects of treatments harder to perceive.

Comprehensive metabolomic profiling and transcriptomics studies represent 'a needle in a haystack' approach in the sense that when all the possible changes in metabolites or gene expression in response to an environmental change are studied, one may be able to find something of interest. An increase in the number of compounds and genes analysed increases the probability of interesting findings. Moreover, the relative importance of responses in a particular compound group or a set of genes, such as phenolic compounds, in response towards ozone, or any other environmental factor, is difficult to judge if only this group is studied. It has to be taken into account that metabolism is regulated at several stages, including all levels of omics. In addition, systems biology requires knowledge on complete biosynthesis and regulatory routes, for which the information on both the changes and the stability is equally important. Therefore, a wide approach with omics methods, possibly combining multiple techniques, and subsequent linking of the observed changes to the significance for the functioning of plants help to give a more comprehensive picture of responses towards a stress factor. In the case of ozone stress, Renaut et al. (2009) have recently presented the current understanding of reactions of poplars to high ozone integrating transcriptomics, proteomics and metabolomics. Linking the changes in an extensive number of biosynthetic pathways helps in grasping a wider picture of what is actually happening in a plant, and may eventually lead into hypotheses of mechanisms, which can be later tested. Besides, we need a better understanding and developments in bioinformatics tools, allowing more efficient integration of different datasets from different omics levels. We need to integrate all the obtained results, such as increased phenolic compounds under ozone stress, with the other signalling and metabolic processes and ultimately the physiological tree life. For the last goal, specific large-scale experiments should be planned, in cooperation with different research groups representing complementary skills.

To conclude, in this chapter we have described the flow of experiments, where omic approaches were used to identify plant reactions to environmental stress, focusing on northern deciduous trees and increasing ozone. During the past decade, chemical assays were developed from narrow-scale phenolic analyses to more diverse metabolomics assays. At the same time, we have been able to proceed to transcriptomic assays, covering an increasing number of gene expressions. Both approaches have been associated with physiological and biochemical responses, combining information from comprehensive laboratory and field experiments. As a result, we are able to indicate the most important targets of ozone (combined with other environmental factors), that is, carbohydrate metabolism, phenolics and leaf surface wax compounds, antioxidative defence systems and leaf growth. For a more comprehensive understanding of changes in response to environmental factors in plants, however, the next step is the integration of different omics. In the case of ozone stress, different omics need to be integrated with phytotoxic ozone doses/ozone uptake measures.

Developments in molecular approaches are necessary in the context of ozone risk assessment, because elevated ozone may have considerable socioeconomic impacts, particularly through deleterious impact on forest growth and vitality (Wittig et al., 2009). Even moderate increases in ozone concentrations can have considerable economic consequences in northern forested countries such as Finland (see also Tuovinen et al., 2013, this volume), where the economy is largely based on wood industry. According to our studies, ozone-caused growth reductions (combining all available data from field, open-top and chamber experiments) in *Betula* and *Populus* species are in average 14% and 19% for foliage area, 10% and 17% for stem dry mass, 19% and 24% for root dry mass and 15% and 14% for radial growth (respectively), corresponding AOT40 value of 15–17 ppm h per season ($3\times$ current AOT40 Critical Level) (Oksanen et al., 2009). The weak correlation between AOT40 values and growth parameters (due to highly variable environmental conditions especially in open-field experiments) implies the need for a more mechanistic and biologically meaningful approach, considering the effective dose rather than the exposure. In addition to economic effect, increasing ozone impairs the recreational value of forests through decreasing biodiversity and vitality.

REFERENCES

Abou Samra, M., Chedea, V.S., Economou, A., Calokerinos, A., Kefalas, P., 2011. Antioxidant/prooxidant properties of model phenolic compounds: part I. Studies on equimolar mixtures by chemiluminescence and cyclic voltammetry. Food Chem. 125, 622–629.

Abril, N., Gion, J.-M., Kerner, R., Müller-Starck, G., Navarro Cerrillo, R.M., Plomion, C., Renaut, J., Valledor, L., Jorrin-Novo, J.V., 2011. Proteomic research on forest trees, the most recalcitrant and orphan plant species. Phytochemistry 72, 1219–1242.

Bohler, S., Bagard, M., Oufir, M., Planchon, S., Hoffmann, L., Jolivet, Y., Hausman, J.-F., Dizengremel, P., Renaut, J., 2007. A DIGE analysis of developing poplar leaves subjected to ozone reveals major changes in carbon metabolism. Proteomics 7, 1584–1599.

Bohler, S., Sergeant, K., Hoffmann, L., Dizengremel, P., Hausman, J.-F., Renaut, J., Jolivet, Y., 2011. A difference gel electrophoresis study on thylakoids isolated from poplar leaves reveals a negative impact of ozone exposure on membrane proteins. J. Proteome Res. 10, 3003–3011.

Brosché, M., Merilo, E., Mayer, F., Pechter, P., Puzõrjova, I., Brader, G., Kangasjärvi, J., Kollist, H., 2010. Natural variation in ozone sensitivity among *Arabidopsis thaliana* accessions and its relation to stomatal conductance. Plant Cell Environ. 33, 914–925.

Cabané, M., Pireaux, J.-C., Léger, E., Weber, E., Dizengremel, P., Pollet, B., Lapierre, C., 2004. Condensed lignins are synthesized in poplar leaves exposed to ozone. Plant Physiol. 134, 586–594.

Cronk, Q.C., 2004. Plant eco-devo: the potential of poplar as a model organism. New Phytol. 166, 39–48.

Dizengremel, P., Jolivet, Y., Tuzet, A., Ranieri, A., Le Thiec, D., 2013. Integrative leaf level phytotoxic ozone dose assessment for forest risk modelling. Dev. Environ. Sci. 13, 267–288.

Dizengremel, P., Vaultier, M.-N., Le Thiec, D., Bagard, M., Gérant, D., Gérard, J., Dghim, A.A., Richet, N., Afif, D., Pireaux, J.-C., Hasenfratz-Sauder, M.-P., Jolivet, Y., 2012. Phosophelonpyruvate is at the crossroads of leaf metabolic reponses to ozone stress. New Phytol. 195, 512–517.

Eriksson, G., 2012. *Betula pendula* and *Betula pubescens*. Recent Genetic Research. Uppsala Bio-Center, SLU, Sweden, 119 pp. ISBN 978-91-576-9062-3.

Ernst, D., 2013. Integrated studies on abiotic stress defence in trees: the case of ozone. Dev. Environ. Sci. 13, 289–307.

Fares, S., Oksanen, E., Lännenpää, M., Julkunen-Tiitto, R., Loreto, F., 2010. Volatile emissions and phenolic compound concentrations along a vertical profile of *Populus nigra* leaves exposed to realistic ozone concentrations. Photosynth. Res. 104, 61–74.

Fernie, A.R., Stitt, M., 2012. On the discordance of metabolomics with proteomics and transcriptomics: coping with increasing complexity in logic, chemistry and network interaction. Plant Physiol. 158, 1139–1145.

Grace, S.C., Logan, B.A., Adams, W.W., 1998. Seasonal differences in foliar content of chlorogenic acid, a phenylpropanoid antioxidant, in *Mahonia repens*. Plant Cell Environ. 21, 513–521.

Häikiö, E., Freiwald, V., Julkunen-Tiitto, R., Beuker, E., Holopainen, T., Oksanen, E., 2008. Differences in leaf characteristics between ozone-sensitive and -tolerant hybrid aspen (*Populus tremula* x *P. tremuloides*) clones. Tree Physiol. 29, 53–66.

Häikiö, E., Makkonen, M., Julkunen-Tiitto, R., Sitte, J., Freiwald, V., Silfver, T., Pandey, V., Beuker, E., Holopainen, T., Oksanen, E., 2009. Performance and secondary chemistry of two hybrid aspen (*Populus tremula L.* x *P. tremuloides* Michx.) clones in long-term elevated ozone exposure. J. Chem. Ecol. 35, 664–678.

Hartikainen, K., Nerg, A.-M., Kivimäenpää, M., Kontunen-Soppela, S., Mäenpää, M., Oksanen, E., Rousi, M., Holopainen, T., 2009. VOC emissions and leaf structural characteristics of European aspen (*Populus tremula*) grown under elevated O_3 and temperature. Tree Physiol. 29, 1163–1173.

Hartikainen, K., Riikonen, J., Nerg, A.-M., Kivimäenpää, M., Ahonen, V., Tervahauta, A., Kärenlampi, S., Mäenpää, M., Rousi, M., Kontunen-Soppela, S., Oksanen, E., Holopainen, T., 2012. Impacts of elevated temperature and ozone on the emission of volatile organic compounds and gas exchange of silver birch (*Betula pendula* Roth). Environ. Exp. Bot. 84, 33–43.

Holopainen, J.K., 2011. Can forest tress compensate for stress-generated growth losses by induced production of volatile compounds. Tree Physiol. 31, 1356–1377.

Kasurinen, A., Biasi, C., Holopainen, T., Mäenpää, M., Rousi, M., Oksanen, E., 2012. Interactive effects of elevated ozone and temperature on carbon allocation of silver birch (*Betula pendula*) genotypes in an open-field exposure. Tree Physiol. 32 (6), 737–751.

Kellomäki, S., Peltola, H., Nuutinen, K.T., Korhonen, K.T., Strandman, H., 2008. Sensitivity of managed boral forests in Finland to climate change, with implications for adaptive management. Philos. Trans. R. Soc. B 363, 2339–2349.

Kontunen-Soppela, S., Ossipov, V., Ossipova, S., Oksanen, E., 2007. Shift in birch leaf metabolome and carbon allocation during long-term open-field ozone exposure. Glob. Chang. Biol. 13, 1053–1067.

Kontunen-Soppela, S., Parviainen, J., Brosché, M., Keinänen, M., Kolehmainen, M., Kangasjärvi, J., Thakur, R.C., Oksanen, E., Karnosky, D.F., Vapaavuori, E., 2010a. Gene expression responses of paper birch (*Betula pendula*) to elevated CO_2 and O_3 during leaf maturation and senescence. Environ. Pollut. 158, 959–968.

Kontunen-Soppela, S., Riikonen, J., Ruhanen, H., Brosché, M., Peltonen, P., Kangasjärvi, J., Auvinen, P., Paulin, L., Keinänen, M., Oksanen, E., Vapaavuori, E., 2010b. Differential gene expression in senescing leaves of two silver birch genotypes in response to elevated CO_2 and tropospheric ozone. Plant Cell Environ. 33, 1016–1028.

Loreto, F., Fares, S., 2013. Biogenic volatile organic compounds and their impacts on biosphere-atmosphere interactions. Dev. Environ. Sci. 13, 57–75.

Mäenpää, M., Riikonen, J., Kontunen-Soppela, S., Rousi, M., Oksanen, E., 2011. Vertical profiles reveal impact of ozone and temperature on carbon assimilation of *Betula pendula* and *Populus tremula*. Tree Physiol. 31, 819–830.

Mäenpää, M., Ossipov, V., Kontunen-Soppela, S., Keinänen, M., Rousi, M., Oksanen, E., 2012. Biochemical and growth acclimation of birch to nigh-time temperatures: genotypic similarities and differences. Plant Biol. 15, 36–43. http://dx.doi.org/10.1111/j.1438-8677.2012.00609.x.

Oksanen, E., Riikonen, J., Kaakinen, S., Holopainen, T., Vapaavuori, E., 2005. Structural characteristics and chemical composition of birch (*Betula pendula*) leaves are modified by increasing and CO_2 and ozone. Glob. Chang. Biol. 11, 732–748.

Oksanen, E., Manninen, S., Vapaavuori, E., Holopainen, T., 2009. Near-ambient ozone concentrations reduce the vigour of Finnish deciduous trees, *Betula* and *Populus* species. Ambio 38 (8), 413–417.

Olbrich, M., Gerstner, E., Bahnweg, G., Häberle, K.-H., Matyssek, R., Welzl, G., Heller, W., Ernst, D., 2010. Transcriptional signatures in leaves of adults European beech trees (Fagus sylvatica L.) in an experimentally enhanced free air ozone setting. Environ. Pollut. 158, 977–982.

Ossipov, V., Ossipova, S., Bykov, V., Oksanen, E., Koricheva, J., Haukioja, E., 2008. Application of metabolomics to genotype and phenotype discrimination of birch trees grown in a long-term open-field experiment. Metabolomics 4, 39–51.

Peltonen, P., Vapaavuori, E., Julkunen-Tiitto, R., 2005. Accumulation of phenolic compounds in birch leaves is changed by elevated carbon dioxide and ozone. Glob. Chang. Biol. 11, 1305–1324.

Pregitzer, K.S., Talhelm, A.F., 2013. Belowground carbon cycling at Aspen FACE: dynamic responses to CO_2 and O_3 in developing forests. Dev. Environ. Sci. 13, 209–226.

Renaut, J., Bohler, S., Hausman, J.-F., Hoffmann, L., Sergeant, K., Ahsan, N., Jolivet, Y., Dizengremel, P., 2009. The impact of atmospheric composition on plants: a case study of ozone and poplar. Mass Spectrom. Rev. 28, 495–516.

Rice-Evans, C.A., Miller, N.J., Paganga, G., 1997. Antioxidant properties of phenolic compounds. Trends Plant Sci. 2, 152–159.

Richet, N., Afif, D., Huber, F., Pollet, B., Banvoy, J., El Zein, R., Lapierre, C., Dizengremel, P., Perré, P., Cabané, M., 2011. Cellulose and lignin biosynthesis is altered by ozone in wood of hybrid poplar (*Populus tremula x alba*). J. Exp. Bot. 62, 3575–3586.

Richet, N., Afif, D., Tozo, K., Pollet, B., Maillard, P., Huber, F., Priault, P., Banvoy, J., Gross, P., Dizengremel, P., Lapierre, C., Perré, P., Cabané, M., 2012. Elevated CO_2 and/or ozone modify lignification in the wood of poplars (*Populus tremula x alba*). J. Exp. Bot. 63, 4291–4301.

Riikonen, J., Kets, K., Darbah, J., Oksanen, E., Sober, A., Vapaavuori, E., Kubiske, M.E., Nelson, N., Karnosky, D.F., 2008. Carbon gain and bud physiology in *Populus tremuloides* and *Betula papyrifera* grown under long-term exposure to elevated concentrations of CO_2 and O_3. Tree Physiol. 28, 243–253.

Riikonen, J., Mäenpää, M., Alavillamo, M., Silfver, T., Oksanen, E., 2009. Interactive effects of elevated temperature and O_3 on antioxidant capacity and gas exchange in *Betula pendula* saplings. Planta 230, 419–427.

Ryan, A., Cojocariu, C., Possell, M., Davies, W.J., Hewitt, C.N., 2009. Defining hybrid poplar (*Populus deltoids* x *Populus trichocarpa*) tolerance to ozone: identifying key parameters. Plant Cell Environ. 32, 31–45.

Saleem, A., Loponen, J., Pihlaja, K., Oksanen, E., 2001. Effects of long-term open-field exposure on leaf phenolics of European silver birch (*Betula pendula* Roth). J. Chem. Ecol. 27, 1049–1062.

Street, N.R., Tallis, M.J., Tucker, J., Brosché, M., Kangasjärvi, J., Broadmeadow, M., Taylor, G., 2011. The physiological, transcriptional and genetic responses of an ozone-sensitive and an ozone-tolerant poplar and selected extremes of their F_2 progeny. Environ. Pollut. 159, 45–54.

Taylor, G., 2002. *Populus*: Arabidopsis for forestry. Do we need a model tree? Ann. Bot. 90, 681–689.

Tuovinen, J.-P., Hakola, H., Karlsson, P.-E., Simpson, D., 2013. Air pollution risks to northern European forests in a changing climate. Dev. Environ. Sci. 13, 77–99.

Valkama, E., Koricheva, J., Oksanen, E., 2007. Effects of elevated O_3 alone and in combination with elevated CO_2 on tree leaf chemistry and insect herbivore performance: a meta-analysis. Glob. Chang. Biol. 13, 184–201.

Wittig, V.E., Ainsworth, E.A., Naidu, S.L., Karnosky, D.F., Long, S.P., 2009. Quantifying the impact of current and future tropospheric ozone on tree biomass, growth, physiology and biochemistry: a quantitative meta-analysis. Glob. Chang. Biol. 15, 396–424.

Yamaji, K., Julkunen-Tiitto, R., Rousi, M., Freiwald, V., Oksanen, E., 2003. Ozone exposure over two growing seasons alters root to shoot ratio and chemical composition of birch (*Betula pendula* Roth). Glob. Chang. Biol. 9, 1363–1377.

Zak, D.R., Holmes, W.E., Pregitzer, K.S., King, J.S., Ellsworth, D.S., Kubiske, M.E., 2007. Belowground competition and the response of developing forest communities to atmospheric CO_2 and O_3. Glob. Chang. Biol. 13, 2230–2238.

zuCastell, W., Ernst, D., 2012. Experimental 'omics' data in tree research: facing complexity. Trees 26 (6), 1723–1735. http://dx.doi.org/10.1007/s00468-012-0777-5.

Part V

Global Dimension of Air Pollution as Part of Climate Change

Chapter 16

Interactive Effects of Air Pollution and Climate Change on Forest Ecosystems in the United States: Current Understanding and Future Scenarios

Andrzej Bytnerowicz*[,1], Mark Fenn*, Steven McNulty[†], Fengming Yuan[‡], Afshin Pourmokhtarian[§], Charles Driscoll[§] and Tom Meixner[¶]

*USDA Forest Service, Pacific Southwest Research Station, Riverside, California, USA

[†]USDA Forest Service, Southern Station, Raleigh, North Carolina, USA

[‡]Climate Change Science Institute and Environmental Science Division, Oak Ridge National Laboratory, Oak Ridge, Tennessee, USA

[§]Department of Civil and Environmental Engineering, Syracuse University, Syracuse, New York, USA

[¶]University of Arizona, Tucson, Arizona, USA

[1]Corresponding author: e-mail: abytnerowicz@fs.fed.us

Chapter Outline

Developments in Environmental Science, Vol. 13. http://dx.doi.org/10.1016/B978-0-08-098349-3.00016-5

16.1 INTRODUCTION

The concepts of air pollution and climate change (CC) are not new. The negative impacts of combustion processes were well known by the thirteenth century (Brimblecombe, 1987), and the atmospheric physics supporting global warming was first postulated in the nineteenth century (Fourier, 1824). While air pollution due to coal and wood combustion is a centuries-old problem, the environmental impacts, including anthropogenic CC and increases in atmospheric carbon dioxide (CO_2), that have occurred since the Industrial Revolution, are a comparatively recent field of research (IPCC, 2007). There are strong linkages between CC and air pollution (Jacob and Winner, 2009; Ramanathan and Feng, 2008). While over a dozen gases that contribute to global warming, only a few, such as CO_2, methane (CH_4), nitrous oxide (N_2O), and ozone (O_3), make a significant contribution. Carbon dioxide is the major contributor to global warming by trapping heat within the Earth's atmosphere (greenhouse effect) (Solomon et al., 2009). The concentration of CO_2 has increased from approximately 270 ppm in the mid-1700s to over 390 ppm today (IPCC, 2007; http://www.esrl.noaa.gov/gmd/ccgg/trends/). Over the past 100 years, the average global surface air temperature has increased by about 0.75 °C in response to the increase in atmospheric greenhouse gases, with approximately 2/3 of that increase occurring since 1980 (IPCC, 2007).

While the scientific community has long recognized the critical need to control atmospheric pollutant emissions, new concerns regarding current and potential future CC impacts emerge. Linkages between CC and air pollution extend to the terrestrial environment, with impacts that are complex and highly variable in time and space. There is a need to investigate the consequences and linkages of the multi-dimensional drivers of global change effects on ecosystem processes such as current and future CC, air chemistry, atmospheric deposition, and other biotic and abiotic stressors.

Among the environmental pollution factors affecting the growth and health of forests in the United States, ambient ground-level O_3 concentrations and atmospheric nitrogen (N) and sulphur (S) deposition are clearly important (Bytnerowicz et al., 2007; Matyssek et al., 2012). Other secondary factors include direct effects of sulphur dioxide (SO_2), nitrogen oxides (NO_x), fluoride (F) compounds, or trace substances. Consequently, in this chapter, we focus on the interactive effects of O_3, N deposition, and changing climate on U.S. forests in the twenty-first century. As examples we present case studies of projections for future changes to forests in the south-western and north-eastern United States under the interactive effects of CC, N deposition, and ambient O_3. Due to uncertainties in advances in industrial technologies, energy production and use, and other drivers, projections of future interactive effects of air pollution and CC on U.S. forests have a high degree of imprecision.

16.2 AIR POLLUTION, CLIMATE, AND THEIR INTERACTIONS: PRESENT STATUS AND PROJECTIONS FOR THE FUTURE

16.2.1 Ozone

O_3 is an important greenhouse gas and plays an important role in the radiative budget of the atmosphere through its interactions with both short-wave and long-wave radiation (Gauss et al., 2003). Because O_3 is a precursor for oxidizing reactions, it strongly influences the lifetime of such greenhouse gases as methane and hydrofluorocarbons (Gauss et al., 2006). It also indirectly impacts climate by limiting sequestration of CO_2 by vegetation (Sitch et al., 2007). Furthermore, the direct radiative effect of O_3 provides the third largest positive radiative forcing of the atmospheric greenhouse gases (Gauss et al., 2003). Interestingly, the magnitude of the indirect tropospheric O_3 radiative forcing is comparable with its direct radiative effects. Therefore, future reductions in O_3 precursor emissions could provide a rapid way to reduce positive anthropogenic radiative forcing in the present atmosphere and may be beneficial in protecting long-term climate safety (Unger and Pan, 2012). Ambient ground-level O_3 is an important air pollutant that adversely affects human health and growth and health of forests worldwide (Ainsworth et al., 2012). Increasing worldwide emissions of O_3 precursors (i.e. NO_x; volatile organic compounds; carbon monoxide, CO) have doubled global background O_3 concentrations since the end of the nineteenth century (Brasseur et al., 2001). The current policy-relevant background concentrations in North America (those that would occur without anthropogenic emissions in North America) are 27 ± 8 ppb at low-altitude sites and 40 ± 7 ppb at high altitude locations as the spring–summer means. The highest values exceeding 60 ppb occur in the intermountain West (McDonald-Buller et al., 2011; Zhang et al., 2011).

Long-range trans-Pacific transport of polluted air masses from Asia is a substantial component of the background O_3 in western North America (Vingarzan, 2004). Significant increases of O_3 concentrations (0.63 ± 0.34 ppb year^{-1}) in the free troposphere have been observed since 1984 during springtime (April–May) when intercontinental transport across the Pacific is most efficient (Cooper et al., 2010; Law, 2010). In addition to the Asian influence, emissions from Canada and Mexico also contribute to background O_3 concentrations in the eastern United States and California, respectively (Wang et al., 2009).

A slow but steady decline of peak values of O_3 (depicted as the annual fourth highest maximum 8-h averages) for the May–September period has occurred in the United States since 1980. These trends are more pronounced in the eastern than western United States (CASTNET, 2012). Nationwide, the highest O_3 pollution occurs in the south-western United States, especially in California where the fourth highest maximum 8-h averages range from 71 to 106 ppb. Moderately elevated O_3 levels (fourth highest maximums ranging from 51 ppb in Montana to 78 ppb in New Jersey) occur downwind of urban regions in the rest of the United States (Figure 16.1). The San Bernardino Mountains (SBM) in southern California is the forested area most affected by photochemical smog in the United States (Bytnerowicz et al., 2008); massive dieback of mixed conifer forests occurred there in the 1960s and the 1970s (Arbaugh et al., 1998; Miller et al., 1963). However, between the late 1970s and 2000s, maximum 1-h peak O_3 concentrations declined sharply due to implementation of air quality regulations aimed at lowering emissions of O_3 precursors. The rate of decline has diminished since 2000,

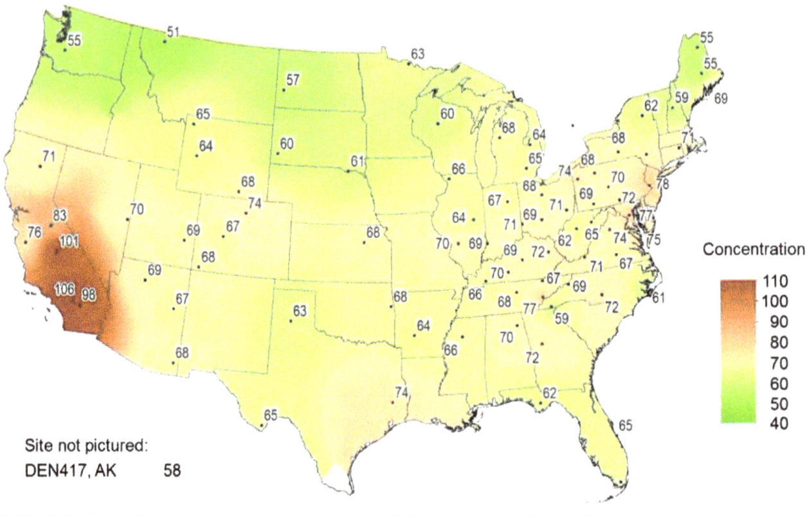

FIGURE 16.1 Three-year average of fourth highest daily maximum 8-h average ozone concentrations (ppb) for 2008–2010 (CASTNET, 2012).

with maximum hourly values oscillating around 150 ppb and the annual average concentration below 47 ppb. During 1970–2011, O_3 concentrations also declined during April–October, the season of greatest vegetative physiological activity. This change was reflected by decreases in AOT40 and W126 indices of more than 50% during this period (Bytnerowicz et al., 2008). These decreases coincide with the recent trends observed in the entire conterminous United States, the western United States, and specifically southern California (Lefohn et al., 2008; Oltmans et al., 2013; Rieder et al., 2013).

Projected changes in anthropogenic emissions are the most important factor in controlling future global and regional O_3 concentrations (affecting O_3 levels by 40–90%), while changes in natural emissions and climate likely contribute to changes in global O_3 levels by 10–30% (Lei et al., 2012). It appears that the risk of hazardous O_3 exposure decreases in developed regions such as the United States under the A1B or B1 CC scenarios, while such risks would increase with the A1F1 scenario (Lei et al., 2012). In contrast, however, Doherty et al. (2013) project that at least 20% of precursor emission reductions would be required in some areas to compensate for higher efficiency of O_3 photochemical production due to increasing temperatures.

16.2.2 Reactive Nitrogen (Nr)

Among Nr species, NO_x results mostly from combustion processes, while ammonia (NH_3) and N_2O mostly from agricultural activities. These compounds can affect climate in many different ways. Nitrogen oxides and NH_3 contribute to CC indirectly by altering the production and loss of atmospheric CC forcers, both greenhouse gases and aerosols scattering incoming solar radiation. Nitrogen oxides can influence CC by increasing O_3 formation (warming effect) and also by increasing removal of CH_4 (cooling effect). Additionally, both NO_x and NH_3 can enhance light-scattering aerosols (Pinder et al., 2013). Atmospheric concentrations of NO_x, nitric acid (HNO_3), and particulate nitrate (NO_3^-) have decreased significantly since 2003 particularly in the eastern United States largely due to the U.S. Environmental Protection Agency Nitrogen Budget Rule (CASTNET, 2012; NADP, 2010). However, NH_3 concentrations and NH_4^+ deposition have remained stable, mainly due to a lack of regulatory control of agricultural emissions and excessive use of N fertilizers, the main sources of these compounds (Davidson et al., 2012; Greaver et al., 2012). As for SO_2, only local direct phytotoxic effects of reactive N gaseous species occur (Bytnerowicz et al., 1998). Although total atmospheric N deposition has declined due to NO_x emission regulations, it is still elevated in large sections of the eastern United States, in California, and near dispersed urban centres in the western United States (Figure 16.2), and can negatively affect forests and other ecosystems through acidification and eutrophication (Davidson et al., 2012; US EPA, 2008).

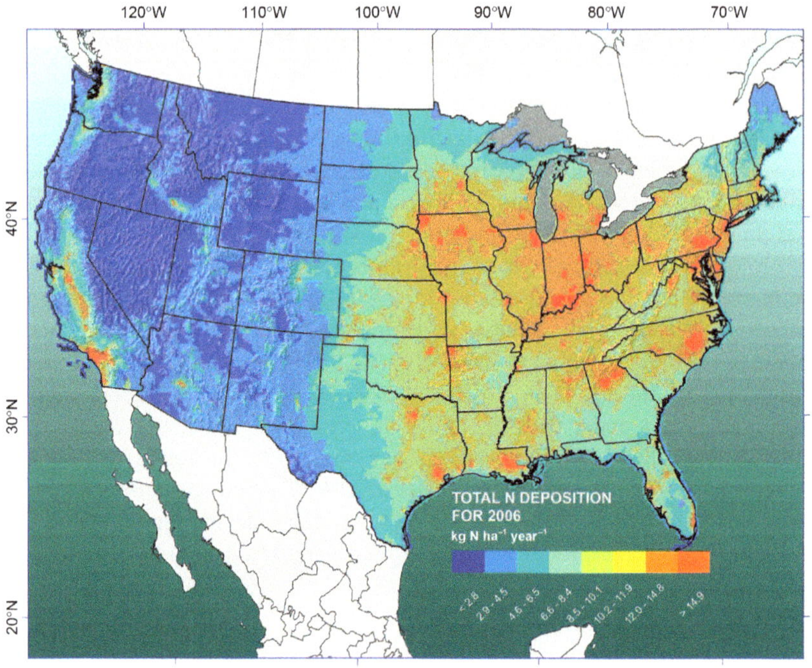

FIGURE 16.2 Total N deposition in the conterminous United States based on the CMAQ model. *Courtesy of Robin Dennis, EPA and Robert Johnson and Edith Allen, UCR.*

In the eastern United States, broad regions are exposed to moderately elevated N deposition (ca. 8–15 kg N ha^{-1} $year^{-1}$), whereas the western United States is characterized by large areas with N deposition ranging from 3 to 5 kg N ha^{-1} $year^{-1}$ with hotspots and steep deposition gradients found downwind of concentrated agricultural areas or urban, and transportation corridors with deposition as high as 70 kg N ha^{-1} $year^{-1}$. Rapidly expanding regions of energy extraction in the Intermountain West have also led to N deposition hotspots. A key environmental consideration is that many arid, semi-arid, or high elevation ecosystems in the West are characterized by low biomass vegetation that are impacted by N deposition levels as low as 3–6 kg ha^{-1} $year^{-1}$ (Fenn et al., 2003).

Nitrogen deposition also affects greenhouse gas fluxes in terrestrial ecosystems. The net effect of atmospheric N deposition in the United States on long-lived greenhouse gas fluxes currently amounts to an uptake of 170 Tg of CO_2 equivalents per year (the result of increased CO_2 sequestration and increased emissions of GHGs CH_4 and N_2O). This net effect is equivalent of 3.2% of the U.S. fossil fuel CO_2 emissions and 19% of the total U.S. forestry sector carbon (C) sink (Templer et al., 2012). Approximately 35% of the CO_2 emissions from the combustion of fossil fuels in North America are currently absorbed by terrestrial ecosystems. However, enhanced biogenic emissions of CH_4 and N_2O offset at least half of the terrestrial CO_2 uptake, and their projected emissions will increase significantly by the end of the twenty-first century (Post and Venterea, 2012).

16.2.3 Sulphur Dioxide and Sulphur Deposition

As a result of air quality regulations under the 1970 and 1990 Amendments of the Clean Air Act, SO_2 concentrations have been steadily declining since 1973, with decreases of about 40% between 1990 and 2009. Current annual average SO_2 values oscillate between 5 and 25 ppb and are well below the national standard of 55 ppb (CASTNET, 2012). Although SO_2 still contributes to acidic deposition, its direct phytotoxic effects are limited and only found in small areas in local geographic areas (Legge et al., 1998; US EPA, 2008). Also wet deposition of sulphur (as SO_4^{2-}) has significantly decreased in recent decades (Greaver et al., 2012; NADP, 2010).

16.2.4 Climate Change Scenarios

Although the impacts of CC on U.S. ecosystems are already being observed, the degree and direction of change is highly variable in time and space (IPCC, 2007). Northern latitudes have generally warmed more than southern latitudes with Alaska, experiencing the largest temperature increase during the twentieth century in the United States (Karl et al., 2009). However, changes in air temperature have been relatively consistent compared to changes in precipitation which have varied over both time and space (Zhang et al., 2007). Over the past 50 years (1958–2008), the south-eastern, south-western, and north-western United States has become 5–40% drier, with the largest reductions occurring in southern California and Arizona (Karl et al., 2009). Conversely, the north-central, south-central, and north-eastern United States has experienced a 10–30% increase in precipitation.

While current levels of CC have impacted U.S. ecosystems, future CC has the potential for major disturbances. Forecasts of future CC involve the use of both alternative models and greenhouse gas emission scenarios. The analyses span the best case (low rates of additional greenhouse gas emissions), and the worst case (high rates of additional greenhouse gas emissions) scenarios. Many factors contribute to these scenarios including demographic and development projections, the rate of emissions from developing countries, and the conversion from fossil fuel to renewable energy resources in developed economies. Various IPCC CC scenarios have been summarized in the Special Report on Emission Scenarios (SRES). Four qualitative storylines have resulted in four sets of scenarios called 'families' A1, A2, B1, and B2 for which six modelling teams developed 40 individual SRES scenarios based on various alternative developments and use of energy technologies (IPCC, 2000).

These scenarios are characterized by a wide range of prediction uncertainties (Stein and Geller, 2012). Using an ensemble of 16 general circulation models (GCMs), Fasullo and Trenberth (2012) recently forecasted that on average, the surface temperature of the Earth will warm by 1.8–4 °C by the end of this century. However, regional warming will vary considerably around this average value.

Given the dynamic nature of CC in both time and space, it is impossible to generalize long-term climatic shifts for any region. However, the pattern of warming is more consistent than the pattern for precipitation change. The most recent U.S. Global Change Research Program Assessment, reported on the Coupled Model Intercomparison Project Three (CMIP3) to provide estimates of future CC across the United States (Karl et al., 2009). By the end of the twenty-first century, all of the United States was projected to warm significantly under all scenarios. Alaska, the Great Plains, and north-central states were predicted to warm the most. Under the low emission scenario, northern Alaska is predicted to warm by 5 °C, while much of the central plain states warm by approximately 3 °C. The least warming was predicted to occur in Florida with 1.5 °C warming by 2100. Warming is already evident by 2050 with the largest increases occurring in the north-central states and Alaska and least warming on the west coast and deep southern states. The high emission scenario exhibits a warming pattern that is similar to the low emission scenario, but the range of air temperature increase is higher (Karl et al., 2009).

Both the amount and timing of precipitation are important factors driving ecosystem impacts. The GCM CMIP3 predicted that under the high emissions scenario, precipitation patterns will have a cyclic nature by the end of the twenty-first century. During the winter, the states south of South Carolina, Kansas, and Utah are predicted to experience significant drying (i.e. 5–25% drier than historical conditions), while more northern states will experience 5–30% more precipitation. As winter turns to spring, the states south of Virginia, Missouri, Wyoming, and Oregon will experience a 5–35% decrease in precipitation compared with historic rates, while in northern states, precipitation will increase by 5–25%. The largest reductions in precipitation are predicted to occur in the already dry areas of southern California. By summer, the drying trend is projected to extend across the entire United States, but moderates to a certain extent with reductions in precipitation ranging from 5% to 40%. Although southern California is not projected to have summer precipitation far below historic values, much of the northwest, central plains, and southern Florida would become the driest areas of the United States. Finally, by autumn, the pattern begins to reverse with spotty patterns (+10% to −10%) of above and below rates of precipitation compared to historic levels. In addition to seasonal changes in precipitation, CMIP3 also projected changes in extreme events (defined as the heaviest 1% of precipitation events). The ensemble GCMs projected that extreme precipitation events would increase the most in the north-eastern United States (by 67%) and the least (9%) in the south-western United States. Changes in extreme event frequency would likely have significant impacts on flood frequency and other ecosystem disturbances.

It is important to emphasize that the uncertainty of future projections of air pollution and climatic conditions in forests is considerable and forecasts of future effects on ecosystem structure and function are even more uncertain because of complex interactions between various biophysical and climatic factors (Bytnerowicz et al., 2006; McNulty and Boggs, 2010).

16.3 PRESENT KNOWLEDGE ON IMPACTS OF AIR POLLUTION, CC, BIOTIC STRESSORS AND MANAGEMENT ON GROWTH AND HEALTH OF FORESTS

Deposition of O_3 to terrestrial ecosystems is mostly controlled by stomata (Ainsworth et al., 2012), and a strong correlation between stomatal conductance and potential O_3 damage to plants has been established (Reich and Amundson, 1985). Non-stomatal processes are also important and may sometimes dominate total O_3 deposition to forests (Kurpius and Goldstein, 2003). High O_3 episodes may cause visible foliar injury and affect gas exchange and photosynthesis, resulting in shortened longevity of foliage, reduced plant growth, and impaired health (Grulke et al., 2009). Long-term chronic exposures to O_3 can lower stomatal conductance and decrease photosynthesis (Reich and Amundson, 1985). Prolonged exposure to chronic levels of O_3 may limit the ability of stomata to close or open rapidly (stomatal sluggishness) (Grulke et al., 2006), increasing water use at the watershed scale (Sun et al., 2012). Short- and long-term reduction of stomatal activity caused by O_3 may also reduce the ability of plants to sequester CO_2 and is the basis for the indirect O_3 greenhouse effect (Sitch et al., 2007). It has been estimated that increased O_3 since the Industrial Revolution has decreased photosynthetic CO_2 uptake and stomatal conductance of trees by 11% and 13%, respectively (Wittig et al., 2007). Felzer et al. (2004) showed a 2.6–6.8% reduction of annual net primary productivity (NPP) in the conterminous United States in response to O_3 levels since 1950 with the largest decreases in the late 1980s and early 1990s. The effects of O_3 on loss of C sequestration are similar in magnitude to land use change and offset the increases in C sequestration due to climate warming, CO_2, and N fertilization. Simulation of O_3 effects on hardwood forests in the north-eastern United States using 1987–1992 O_3 data showed declines of annual NPP ranging from 3% to 16% with most pronounced effects at the highest O_3 levels and on soils with high water-holding capacity (Ollinger et al., 1997).

Elevated levels of N deposition may have many negative effects, such as changes in biodiversity or contamination of soils and water (Greaver et al., 2012; Pardo et al., 2011; US EPA, 2008). However, moderate amounts of

N from atmospheric deposition have a fertilization effect, stimulating photosynthesis and growth of forest trees and increasing sequestration of CO_2 (Davidson et al., 2012). In general, atmospheric N deposition stimulates C sequestration and growth of U.S. forests, although concurrent declines of some tree species, older forests and forests in deposition hotspots occur (Templer et al., 2012). In some highly polluted areas, such as the SBMs of southern California or the Sierra Nevada Mountains, elevated ambient O_3 concentrations may offset fertilizing effects of N deposition (Fenn et al., 2003), as documented for ponderosa pine (Peterson et al., 1991).

Ollinger et al. (2002) evaluated separately and in combination the effects of N deposition, surface O_3, elevated CO_2 and land use history on carbon dynamics of northern U.S. hardwood forests. Their results showed that increases in N deposition and CO_2 stimulated forest growth and carbon (C) uptake, but to different degrees following agricultural and timber harvesting. Ambient O_3 offset a substantial portion of those increases in those simulations. This decrease in N-induced C sinks by elevated O_3 is relevant for much of the global forest lands (as shown above for California forests). Collectively, the combined effects of all factors included in their simulations resulted in growth estimates that were very similar to those obtained in the absence of any form of disturbance. This pattern may suggest that current forests may show little evidence of altered growth since pre-industrial times despite substantial changes in their physical and chemical environment.

Approximately 100 years of effective fire prevention resulted in widespread densification of forest stands, predisposing western forests to drought. Recent climate warming, and, in some areas, the additional effects of O_3 and N deposition, have exacerbated drought stress and weakened vast areas of forest making them susceptible to bark beetle attacks. This condition has resulted in large-scale dieback of western U.S. forests. These forests are highly susceptible to catastrophic fires—in 2006 about 4 million hectares burned (Grulke et al., 2009; McKenzie et al., 2009). Comprehensive information on the status of U.S. forests and the health effects of air pollution CC, pests, and diseases has been provided by Tkacz et al. (2008).

16.4 POSSIBLE FUTURE CHANGES IN U.S. FORESTS CAUSED BY CLIMATE CHANGE AND AIR POLLUTION

Individually, CC and air pollutants will have various negative (and positive) impacts on ecosystem structure, function and sustainability. Recently, scientists have begun to examine how air pollution may synergistically increase, or antagonistically decrease the impact of CC (McNulty and Boggs, 2010). There are many factors that determine the direction and magnitude of the response. For example, CC may cause an increase in precipitation for a region, which along with the fertilizing impacts of N deposition, may stimulate forest leaf area and growth. Years or decades later, continued CC (or natural climate variability) may reduce precipitation to sub-historic levels. Unfortunately, forests that have acclimated to a higher precipitation and N regime may have water demands that can no longer be met. These forests may therefore become predisposed to secondary stresses (e.g. insects, disease, wildfire) even though the sites are not in a condition of nitrogen saturation as defined by Aber et al. (2001).

Certain factors associated with CC such as increased air temperatures and changes in precipitation patterns impact forest structure and function as well as its response to atmospheric deposition. Other factors such as tree base cation and N uptake are influenced by growth rates which again are influenced by air temperature and precipitation patterns. Finally, strong acid anion (SO_4^{2-}, NO_3^-) and base cation leaching are partially controlled by precipitation and air temperature through ecosystem evapotranspiration and soil water outflows. The interaction of these CC factors could significantly alter the capacity of a forest ecosystem to process atmospheric deposition. The north-eastern United States includes considerable forest area that is currently negatively impacted by elevated N and S deposition (McNulty et al., 2007). This region is projected to become both warmer and wetter in the coming decades due to anthropogenic CC (Karl et al., 2009). The trends in CC should positively interact with ecosystem processes that mitigate elevated N and S deposition. Forest productivity and evapotranspiration are both predicted to increase across New England in the coming decades under most CC scenarios (Rustad et al., 2009), which should serve to diminish the effects of acidic deposition. Similarly, negative impacts of acidic deposition could be reduced due to the increased mineral weathering and more available base cations caused by future higher temperatures (Li and McNulty, 2007; Sverdrup and Warfvinge, 1993). However, Wu and Driscoll (2010) observed that critical loads for soil and surface water acidification decreased under future CC scenarios due to increases in leaching losses of NO_3^- associated with higher future temperature. The interactions of CC and atmospheric N deposition, NO_3^- leaching, and acidification in north-eastern forests are discussed in a case study below.

Theoretically, reduced stomatal conductance in response to increasing CO_2 levels may enhance plant–water use efficiency, reduce O_3 uptake and its potential phytotoxicity (Ainsworth et al., 2012). However, O_3-induced damage to stomatal apparatus may confound this effect leading to increased evapotranspiration, reduced stream flow, and increased frequency and severity of drought events as was observed in forest stands in the south-eastern United States over the past 18–26 years (Sun et al., 2012). This example illustrates the complexity of CC–air pollution interactions and the challenges in projecting the effects of combinations of increased temperature, elevated CO_2, changing water and nutrient availability, various O_3 and N deposition exposure regimes as well as other abiotic and biotic factors. There is a clear need to develop models that integrate multiple biotic and abiotic factors while projecting future ecosystem changes in order to understand potential risks to forest resources. Such models could provide science-based outcomes and future scenarios and improve our understanding of what adaptive or mitigation measures could be suggested to air resource and land managers.

Below we provide two case studies of simulations of the future impacts of CC, N deposition, and O_3 on biomass, nutrient cycling, and water exchange in two forested areas of the United States, southern California, and New England.

16.5 PROJECTED HYDROLOGICAL, NUTRITIONAL, AND GROWTH CHANGES IN MIXED CONIFER FORESTS OF THE SBM (SOUTHERN CALIFORNIA) DUE TO CC, N DEPOSITION, AND O_3

The DayCent biogeochemistry model (Del Grosso et al., 2000; Parton et al., 1998, 2001), the daily time step version of the CENTURY model (Parton et al., 1993), was modified to include effects of O_3 on forests in order to simulate hydrological and biogeochemical (BGC) changes of mixed conifer forests in SBMs under historical and projected CC, N deposition, and O_3. Previously DayCent was parameterized for high (Camp Paivika, CP) and low (Barton Flats, BF) N deposition sites (70 and 8.8 kg N ha^{-1} $year^{-1}$, respectively) in the SBM located 70 km east of Los Angeles, California (Fenn et al., 2008). CP is relatively warmer and wetter while BF is cooler and drier (Fenn et al., 2008; Yuan et al., 2011). Arbaugh et al. (1999) applied the CENTURY model to simulate O_3 injury in the region using a simple algorithm of halving the rate of foliar turnover from the baseline condition. Here, we introduce an empirical relationship between foliage damage fraction and O_3 concentration derived from field data (Grulke and Balduman, 1999), and also a biomass production reduction factor into the DayCent model. The forest biomass production reduction factor is generalized from data in Pye (1988). This reduction factor is a function of the O_3 exposure index W126 (Bytnerowicz et al., 2008; Lefohn et al., 1988) because growth decrease was more strongly correlated with O_3 dose (concentrations x exposure time) than with O_3 exposure concentration alone (Pye, 1988).

The model simulations are driven by historical climate (daily maximal and minimal air temperature, and precipitation) prepared for CP and BF in previous studies (Fenn et al., 2008; Yuan et al., 2011), and extended into 2011. For future conditions (2011–2099), we extracted the NOAA Geophysical Fluid Dynamics Laboratory (GFDL) climate model daily projections of maximal and minimal temperature for two emission scenarios A2 (medium–high emissions) and B1 (low emissions) for both sites (http://cal-adapt.org/data/tabular/; Cayan et al., 2008). Because of some discrepancy between historical observations and GFDL model simulations, adjustments were carried out to match the model simulations with observations based on the data periods covered by both (Figure 16.3A). For precipitation projections, since the overall changes would be relatively small (less than 10%) (Cayan et al., 2008), we calculated the seasonal air temperature correlation to represent similarity between projection years and historical years and chose the most similar years for assigning the precipitation to a particular projection year. As a result of this approach, the average precipitation (about 880 and 660 mm per year for CP and BF, respectively) did not change from 2011 to 2099, while warming did occur, especially in the medium–high emission scenario A2.

We assumed that O_3 concentrations at CP and BF could be represented by two nearby monitoring stations, Crestline (1980–2011) (California Air Resource Board, Air Quality Monitoring Database) and Converse Flat (2004–2010) (CASTNET, 2012) (Figure 16.3B). For the Crestline Station, earlier O_3 observations (annual averages) back to 1963 (Bytnerowicz et al., 2008) were combined with hourly O_3 monitoring from 1980–2011 to hind cast O_3 for the 1940–1963 period assuming that O_3 stabilized during the period between the 1950s and early 1960s, and prior to that concentrations increased linearly since 1900 (assuming 10 ppb in 1900; Brasseur et al., 2001). For Converse prior to 2004, seasonal O_3 concentration was calculated by a regression of Crestline versus Converse Flat derived from data in Bytnerowicz et al. (2008). At Crestline annual average O_3 concentration was slightly lower than at Converse Flat, because higher O_3 from June to September at Crestline was offset by lower O_3 in the rest of the year (Figure 16.3B and C). We also assumed that the O_3 decline since the 1990s would continue until 2099. This decline would bring the current O_3 levels back to those prior to 1950 and was denoted as O_3 projection treatment 0 (‘O_3 0’ in Figures 16.4–16.7). In order to capture uncertainty in future projections, the simulations were conducted for four additional O_3 projections: $O_3+25\%$, $O_3+50\%$, $O_3-25\%$ and $O_3-50\%$ for O_3 levels increased by 25% and 50%, or decreased by 25% and 50% from the ‘O_3 0’. The O_3 projection levels of $O_3+50\%$ would increase O_3 up to that in the mid-1970s, while $O_3-50\%$ would decrease O_3 level to background by the end of this century. We applied the N deposition datasets for historical simulations in Fenn et al. (2008) in this study and assumed no changes since 2005, that is, 70 kg N year^{-1} at CP and 8.8 kg N year^{-1} at BF, since O_3 impacts on forests under CCs were the focus of this study.

Arbaugh et al. (1998) indicates that O_3 injury causes foliar biomass loss and higher production and turnover, thus leading to significantly higher litter biomass, but no impact on total tree biomass growth and wood production. In contrast, our simulation showed that as a result of the direct effect of O_3 on photosynthesis (reduced productivity) and O_3-induced foliar damage, a significant reduction of tree biomass occurred. High levels of O_3 exposure are projected to have decreased tree biomass by ~12% at the high N (warm and wet) CP site by the 1980s, with significant decreases occurring since the late 1940s (Figure 16.4C). The decrease in tree biomass was much less pronounced

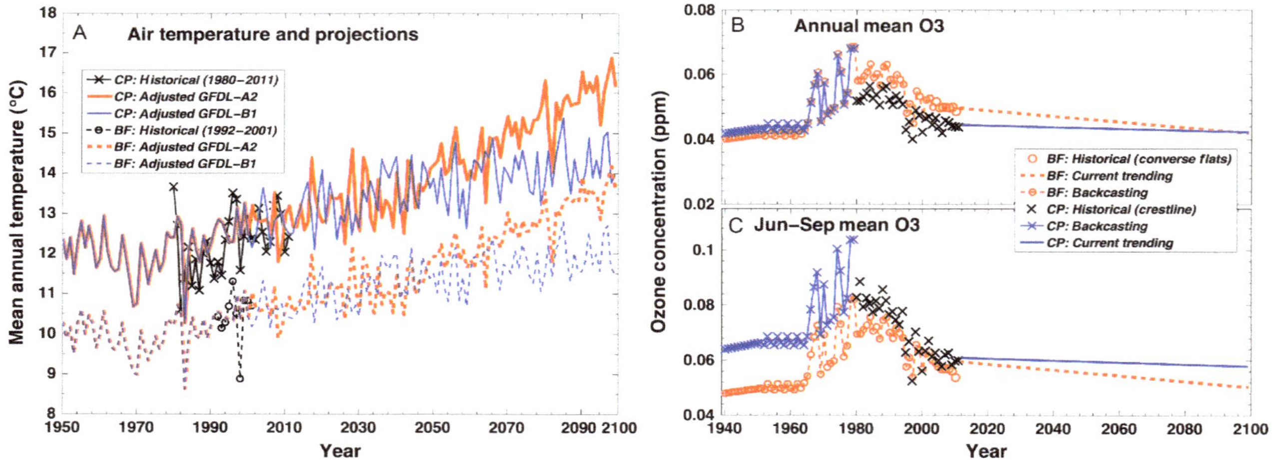

FIGURE 16.3 Historical and projected climate and ozone trends at Camp Paivika (CP, high N deposition of 70 kg N year^{-1}) and Barton Flat (BF, low N deposition of 8.8 kg N year^{-1}) in SBM to drive DAYCENT model: (A) air temperature and adjusted GFDL projections (emission scenario A2 (medium–high emissions) and B1 (low emissions)), (B) Annual mean O_3 concentrations and its trending, and (C) June–September mean O_3 concentrations and its trending.

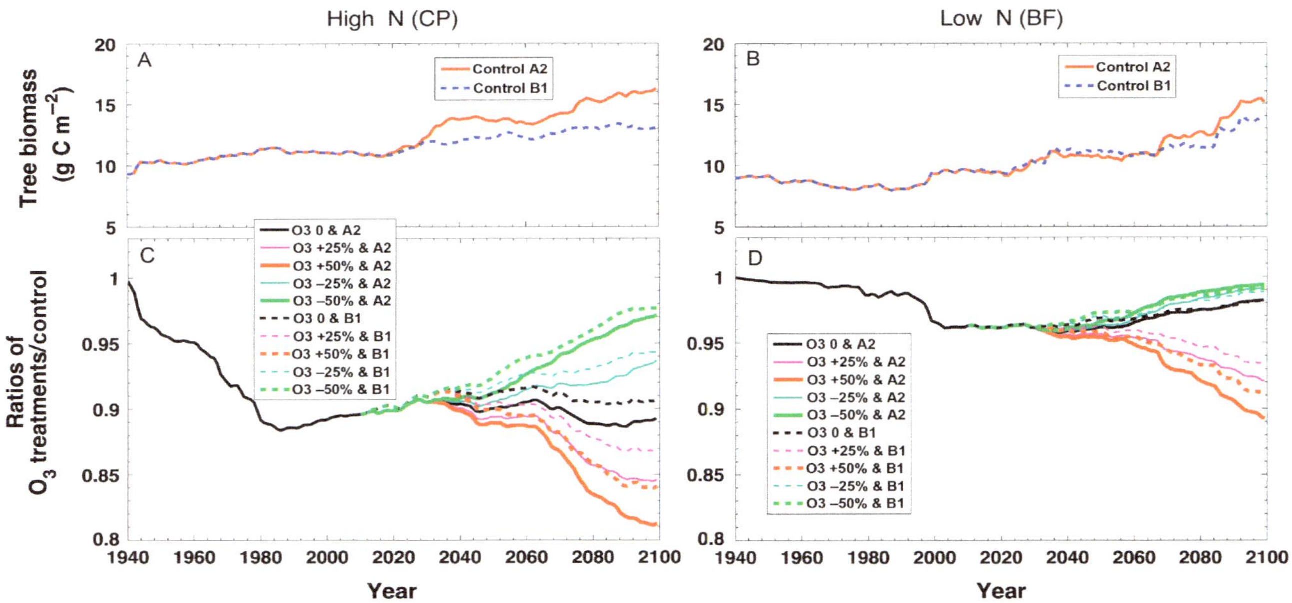

FIGURE 16.4 Reduction of forest tree biomass caused by O_3 under different CC and N deposition scenarios in SBM. Note that A2 and B1 stands for GFDL projections with Scenario A2 and B1; Treatment 'O_3 0' is historical O_3 concentration and current trending is extended until 2099 (see Figure 16.3B and C), and O_3 +25%, +50%, −25%, and −50% are O_3 concentration changes of +25%, +50%, −25%, and −50% of the 'O_3 0' treatment; control is a simulation without O_3 effects.

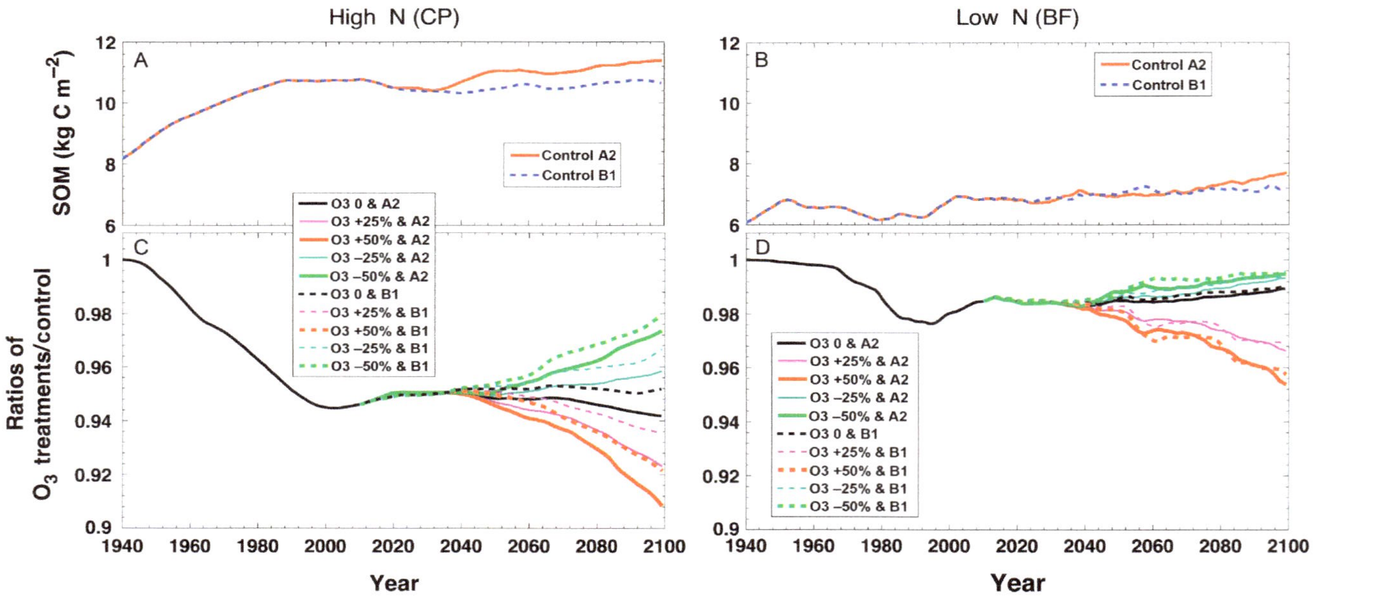

FIGURE 16.5 Reduction of forest soil organic matter (SOM) caused by O_3 under different CC and N deposition scenarios in SBM. See Figure 16.4 for details of the treatments.

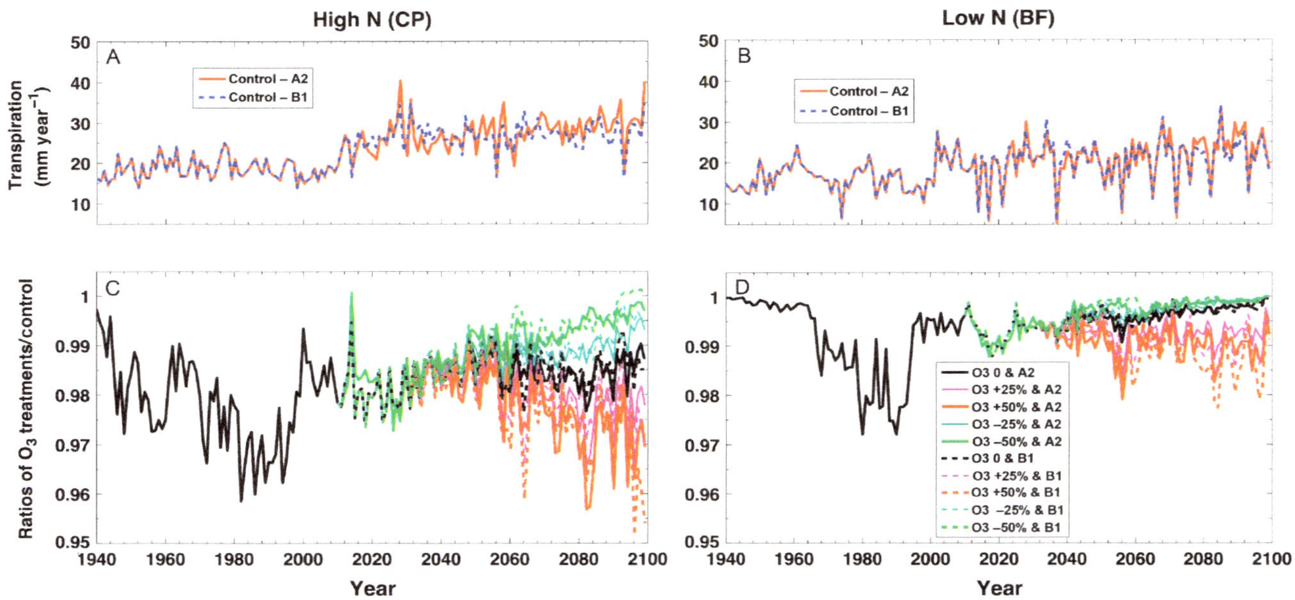

FIGURE 16.6 DAYCENT simulated plant transpiration under CC and N deposition and O_3 effects in SBM. See Figure 16.4 for details of the treatments.

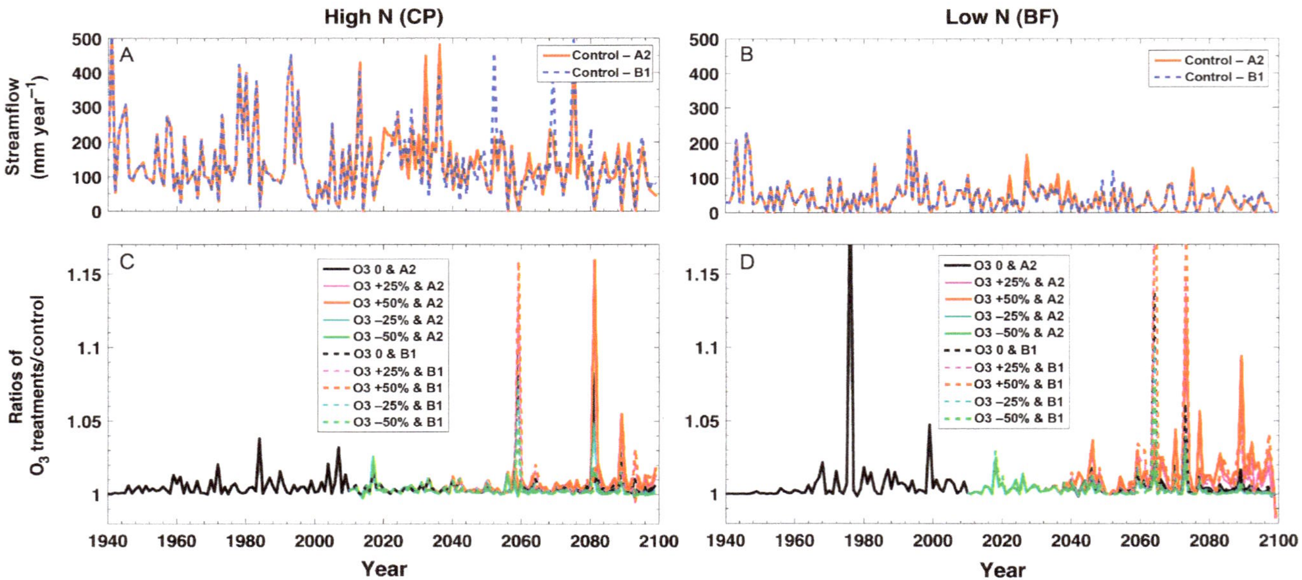

FIGURE 16.7 DAYCENT simulated streamflow (soil drainage and runoff) under CC, N deposition, and O_3 effects in SBM. See Figure 16.4 for details of the treatments.

(<5%) and occurred much later (abound 2000–2004) at the low N (cold and dry) BF site (Figure 16.4D). Simulations indicated that O_3 impacts on forest growth were closely related to N deposition in this N growth-limited ecosystem. If the current trend of decreases in O_3 continue, tree biomass would not change significantly even under climate warming at the high N site (CP), although at the low N site forest trees would recover from biomass loss in the late twenty-first century. Under the increasing O_3 scenarios (increases of 25% and 50%), tree biomass would decrease by another 9–10% and 4–5% by 2099 at CP and BF, respectively, for the A2 (medium–high emissions) climate-warming projection. The reduction in tree growth would be about 30–50% less for the B1 (low emissions) projection. If technology and/or policy management advances in the future decrease O_3 levels by 25% and 50%, O_3 effects would be minimized by the end of this century.

The DayCent simulations showed similar patterns of O_3 effects on soil organic matter (SOM) (Figure 16.5) as on tree biomass (Figure 16.4), but with smaller reductions (decreased by almost half as much as the effects on tree biomass) (Figure 16.5C and D). The simulations also showed small differences in O_3 effects on SOM at BF between the warmer A2 and colder B1 projection compared to the CP site. The simulations by Arbaugh et al. (1999) showed a larger increase of litterfall and litter biomass accumulation due to O_3 injury to foliage and thus almost no change in SOM. In our study, the O_3-caused reduction of tree biomass production resulted in decreased litterfall. This overall biomass reduction effect on litterfall was larger than the increase in litterfall resulting from O_3 injury to foliage. Therefore, the overall C input into SOM was less than for the non-O_3 simulations.

As expected, foliar injury and reduction in tree productivity by O_3 exposure caused decreases in plant transpiration of up to 4% during the historically high O_3 exposure period at CP, and up to 3% at BF (Figure 16.6C and D). Again the simulations showed less O_3 impact at the low N, cold and dry BF site. The decline in O_3 levels since the 1980s caused a recovery in transpiration capacity almost a decade later, which is maintained at 1–2% reduction at the high N CP site and less than 1% at the low N BF site, if the current trend of declining O_3 continues. The scenarios with projected O_3 increases result in larger reductions in plant transpiration under the two climate-warming projections. The difference between the two climate-warming projections on plant transpiration appeared small although the warmer A2 projections would have slightly larger transpiration loss.

The 1–2% (at most 5%) reduction in annual transpiration (15–40 mm at CP and 5–30 mm at BF, Figure 16.6A and B) caused by increased O_3 (Figure 16.6C and D) would not significantly impact the overall hydrological status of those sites in the future. For example, streamflow from soil drainage and runoff was about 143 mm at CP (wetter and warmer) and 72 mm at BF site (dryer and colder) (Figure 16.7A and B). Thus the 1–2% reduction of transpiration water loss by O_3 effects would have little effect on streamflow in the normal and wet years. However, elevated O_3 would affect streamflow in drought years (Figure 16.7C and D). Under the future climate-warming conditions, streamflow could increase by as much as 15% during extremely dry years if the current O_3 levels increase by 50%.

At the high N and wet CP site, the simulation exhibited slightly higher yearly maximal stream NO_3^- concentrations compared to the low N and dry BF site. However, at both sites, NO_3^- concentrations increase with time, presumably as N was accumulating in the ecosystem as the current deposition rate was maintained. During events with streamflow lower than 1 mm, abnormally high NO_3^- concentrations were projected because of high NO_3^- content in lower soil layers. These artificially high values were not included in projections of yearly maximal streamflow NO_3^- concentrations. Due to drier conditions at BF, the frequency of such events was much greater than at CP (Figure 16.8A and B).

As described earlier, O_3 effects on tree growth, SOM accumulation, and hydrology were larger at the high N site (CP, wetter and warmer) than the low N site (BF, drier and colder) but with similar response patterns to the O_3 changes and levels. In contrast, the response of nitrate (NO_3^-) concentrations in streamflow to O_3 showed an opposite effect at CP and BF (Figure 16.8C and D). Nitrate concentrations in the increasing O_3 scenarios were up to 9% higher than the non-O_3 simulation (control) during the historically high O_3 period at the high N and wet CP site (Figure 16.8C); while at the low N and dry BF site, such changes did not occur (Figure 16.8D). Secondly, at the high N and wet CP site, the projected higher O_3 levels are expected to result in higher NO_3^- concentrations than the lower O_3 scenario (Figure 16.8C); while at the low N and dry BF site, O_3 levels showed an opposite effect on nitrate (Figure 16.8D). These responses essentially are the net result of two reverse O_3 effects of O_3 on NO_3^- concentration. On the one hand, O_3 causes foliar injury, thus reducing plant transpiration, resulting in higher streamflow (hydrological aspect of O_3 effects); on the other hand, O_3 reduces plant production, consequently decreasing N uptake by trees and increasing N export into streamflow ('BGC aspect' of O_3 effects). Our results suggest that the BGC aspect of O_3 effects dominated N streamflow export under high N deposition during the historically high O_3 period and is projected to do so in the future. Conversely, the O_3-induced hydrological effect is expected to dominate the O_3 effect in determining streamflow nitrate under low N deposition conditions.

16.6 PROJECTING HYDROLOGICAL, NUTRITIONAL AND GROWTH RESPONSES OF FORESTED WATERSHEDS AT THE HUBBARD BROOK EXPERIMENTAL FOREST, REFLECTIVE OF THE AMERICAN NORTHEAST

Global change in the north-eastern United States is and will likely continue to be manifested in the future through increasing atmospheric concentrations of

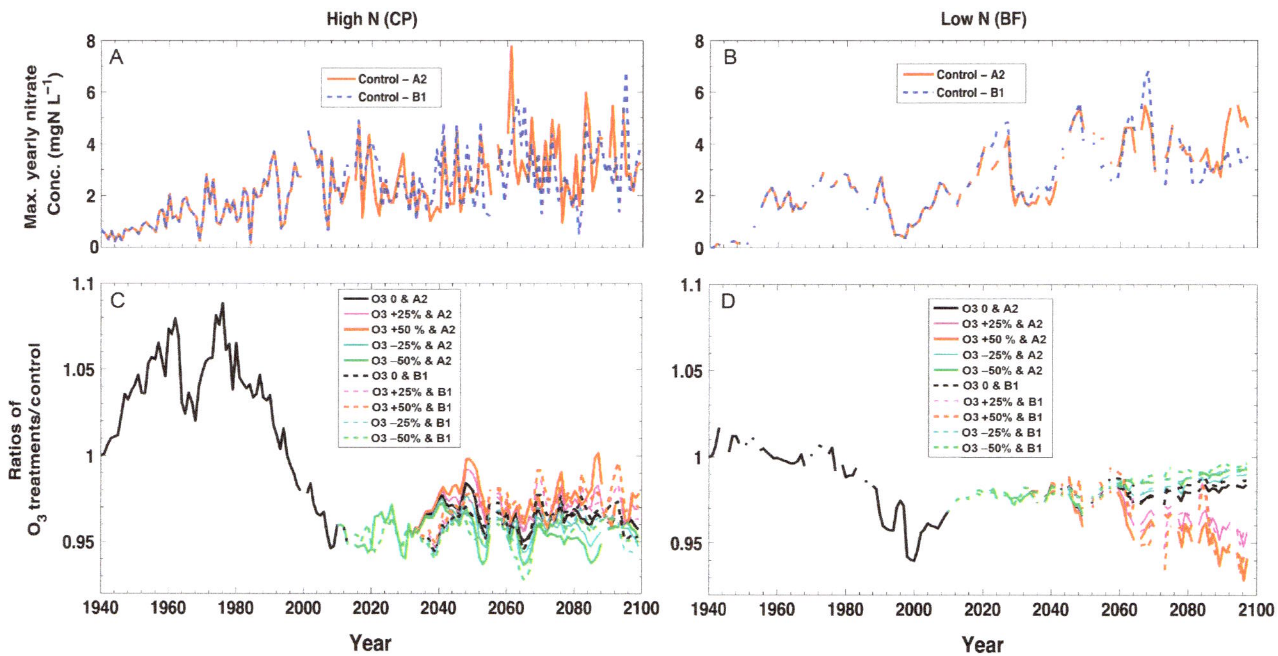

FIGURE 16.8 DAYCENT simulated yearly maximal nitrate concentration in streamflow (events >1 mm) water under climate changes, N deposition and ozone effects in SBM. See Figure 16.4 for details of the treatments.

CO_2; increases in air temperature, less precipitation as snow and more as rain; decreases in snow cover; earlier arrival of spring and following snowmelt and earlier high spring river flows; a longer growing season; increases in storms and extreme events; and associated alterations in soil and surface water chemistry among other perturbations (Intergovernmental Panel on Climate Change (IPCC), 2007; NECIA, 2006).

In this study, we assess the potential effects of three contrasting N emission hindcast and future scenarios in conjunction with changes in climatic drivers (temperature, precipitation, and solar radiation) to the Northern Forest. We examine possible future CC scenarios, CO_2, and atmospheric nitrogen (N) deposition at the Hubbard Brook Experimental Forest (HBEF), New Hampshire, United States of America, as a case study, using the biogeochemical model PnET-BGC. A unique aspect of this case study is consideration of the simultaneous interactions of effects of CO_2 on vegetation (fertilization and response of stomatal conductance) and the potential for drought associated with higher temperature (and decreases in stomatal conductance) under contrasting conditions of N emissions and deposition (stimulating forest growth and carbon uptake, and perturbations to the acid–base chemistry of soil and surface waters). As with the previous case study, these drivers of global change (i.e. changes in temperature, precipitation, CO_2, and N deposition) can affect forest growth and associated C sequestration, hydrology and N cycling in unexpected ways. The application of PnET-BGC helps better understand the individual and combined effects of these drivers on forest ecosystem structure and function.

The HBEF is located in the southern White Mountains (43°56′N, 71°45′W) (Likens and Bormann, 1995). The site was established by the U.S. Forest Service in 1955 as a centre for hydrological research, and in 1987 was designated as a National Science Foundation Long-Term Ecological Research (LTER) site. The climate is humid continental, with short, cool summers and long, cold winters. The model was run for Watershed 6 (W6) which is a reference watershed. Watershed 6 has one of the longest continuous records of meteorology, hydrology and biogeochemistry in the United States (Likens and Bormann, 1995; Likens et al., 1994) (http://www.hubbardbrook.org/). The watershed area is 13.2 ha, with an elevation range of 549–792 m. Watershed 6 was logged intensively from 1910 to 1917, and has experienced subsequent disturbances including a hurricane in 1938, which prompted some salvage logging, and an ice-storm in 1998 without salvage logging. The dominant vegetation is northern hardwoods.

PnET-BGC is a BGC model that has been used to evaluate the effects of CC, atmospheric deposition and land disturbance on soil and surface waters in northern forest ecosystems (Chen and Driscoll, 2005). PnET-BGC was created by linking the forest–soil–water model PnET-CN (Aber and Driscoll, 1997; Aber et al., 1997), with a (BGC sub-model (Gbondo-Tugbawa et al., 2001), thereby enabling the simultaneous simulation of hydrology and major element cycles (Ca^{2+}, Mg^{2+}, K^+, Na^+, C, N, P, S, Si, Al^{3+}, Cl^-, and F^-). PnET-BGC has been used to evaluate fluxes of water and elements in forest ecosystems by depicting ecosystem processes, including atmospheric deposition, CO_2, and tropospheric O_3 effects on vegetation, canopy interactions, plant uptake, litterfall, SOM dynamics, nitrification, mineral weathering (kept constant during the simulation), chemical reactions involving gas, solid and solution phases, and surface water processes (Gbondo-Tugbawa et al., 2001). These processes determine the hydrochemical characteristics of the ecosystem because water and solutes interact with forest vegetation and soil before emerging as surface runoff. A detailed description of PnET-BGC is provided by Aber and Driscoll (1997), Aber et al. (1997), and Gbondo-Tugbawa et al. (2001), including a sensitivity analysis of parameters.

In this application, the model was run on a monthly time step. Simulation from the year 1000 to 1850 is a spin-up period to allow the forest ecosystem to come to steady-state with respect to pre-anthropogenic climate and atmospheric deposition conditions. From 1850 to 2012, the model simulates hindcast conditions of changes in atmospheric deposition, meteorology, and land disturbance which are reconstructed from historical conditions (Gbondo-Tugbawa et al., 2001; Pourmokhtarian et al., 2012). The model was run as a forecast from 2012 through 2100 using future global change scenarios that are based on projected changes in climate, atmospheric CO_2, and three scenarios for atmospheric N emissions and deposition. The first N deposition scenario (N1) is 'business as usual' which includes our reconstruction of past atmospheric N deposition and no change in future atmospheric deposition of NO_3^- and NH_4^+; the average of measured atmospheric deposition for 2001–2011 repeated from the year 2012 to 2100. The second scenario (N2) is 'background N deposition;' pre-anthropogenic atmospheric deposition of N is considered through the entire simulation period (1000–2100). The third scenario (N3) includes the hindcast of past atmospheric N deposition and a forecast of NO_3^- deposition decreasing from the year 2012 to 2020 to background values and keeping this value constant through 2100 while NH_4^+ deposition remains 'business as usual'. The N3 scenario is established to depict aggressive controls on oxidized N emissions with no control on reduced N. Note that atmospheric deposition of all other elements were considered business as usual from the year 2012 to 2100. Details on input data, climate projections, spin-up period, and sensitivity of the model to climatic drivers are provided by Pourmokhtarian et al. (2012). The PnET-BGC is an open source model available to the public at http://www.lcs.syr.edu/faculty/driscoll/personal/PnET%20BGC.asp.

The depiction of the effects of increasing atmospheric CO_2 on trees is based on multi-layered sub-model of photosynthesis and phenology developed by Aber et al. (1995, 1996) and modified by Ollinger et al. (1997, 2002). A detailed description of CO_2 effects on vegetation, the processes and parameters related to photosynthesis in the model and the sensitivity analysis are described by Ollinger et al. (1997, 2002, 2009). For this analysis, we considered contrasting CO_2 effects scenarios. For the CC without CO_2 effect scenarios, we do not invoke the CO_2 effects algorithm, so under this condition, trees do not respond to changes in CO_2. For the CC with CO_2 effects, we simulate the response of forest growth and stomatal conductance to changes in CO_2 described earlier.

We used data from two atmosphere-ocean global circulation models (AOGCMs): (1) the United Kingdom Meteorological Office Hadley Centre Coupled Model, version 3 (HadCM3) (Pope et al., 2000) and (2) the U.S. Department of Energy/National Center for Atmospheric Research Parallel Climate Model (PCM) (Washington et al., 2000). We coupled the HadCM3 model with A1fi (fossil fuel-intensive) emission scenario (Nakićenović et al., 2000), and the PCM model with B1 emission scenarios to represent possible higher and lower greenhouse gas emission futures, respectively. Consequently, two CC scenarios were developed for this application (HadCM3-A1fi and PCM-B1). Monthly, coarse resolution AOGCM temperature, precipitation, and solar radiation output were statistically downscaled using Station-based Daily Asynchronous Regression for weather station 1 at the HBEF for the period of 1960–2100 (O'Brien et al., 2001). Note that we used the same CC scenarios for both model simulations with and without CO_2 effects and the only difference between these sets of simulations is activation of CO_2 effects on the vegetation algorithm in the model.

16.6.1 CC (Without CO_2 Effects on Vegetation)

Model simulations suggest that seasonal hydrological patterns will shift substantially under projected CC scenarios (Figures 16.9 and 16.10). Under HadCM3-A1fi, in the future spring, snowmelt will occur earlier. Low flows associated with enhanced evapotranspiration during the summer months will begin earlier in the spring and continue later into the fall. Future streamflow in late fall and early winter will increase because of decreases in snowpack accumulation due to warmer air temperatures and concurrent declines in the ratio of snow to rain (Figure 16.9). Model projections under the more modest PCM-B1 scenario show that there is no change in timing and magnitude of spring snowmelt, but fall high flows will occur earlier with greater magnitude. Under PCM-B1, the duration of summer low flows will decrease due to an increase in summer discharge relative to 1970–2000 values (Figure 16.10). Future model projections indicate increases in annual water discharge under both scenarios for 2070–2100 compared to 1970–2000, with the percentage increase higher under PCM-B1 (Figures 16.9 and 16.10). Model simulations suggest hydrologic responses to scenarios of N deposition are subtle (data not shown). Annual water discharge will increase slightly ($\sim$1%) under background N deposition scenario N2 due to limited N fertilization effects on plant growth and C storage and associated decrease in transpiration losses compared to the higher N deposition scenarios (N1, N3).

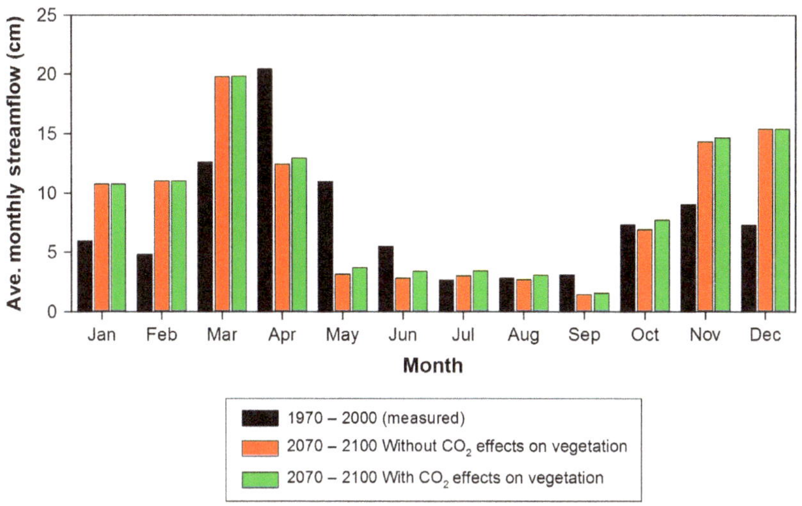

FIGURE 16.9 Comparison between measured monthly discharge for 1970–2000 and simulated mean monthly discharge for 2070–2100 with and without considering CO_2 effects on vegetation. Note the future climate-change scenario depicted in these results is from HadCM3 A1fi with N emissions scenario of business as usual.

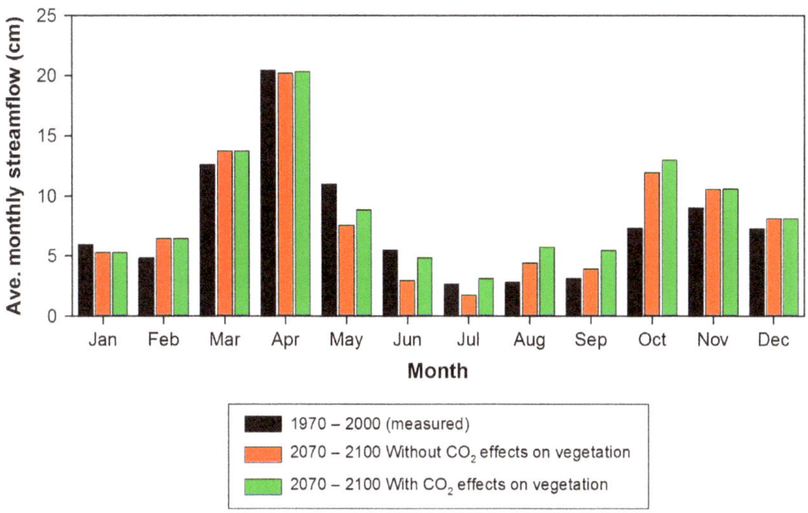

FIGURE 16.10 Comparison between measured monthly discharge for 1970–2000 and simulated mean monthly discharge for 2070–2100 with and without considering CO_2 effects on vegetation. Note the future climate-change scenario depicted in these results is from PCM-B1 with N emissions scenario of business as usual.

The model simulations of forest vegetation without CO_2 effects showed future increases in NPP associated with warmer temperatures and lengthening of growing season for both climate scenarios. Under HadCM3-A1fi for N1, N2, and N3 deposition, the NPP increased by 14%, 10%, and 14%; 11%, 10%, and 11%; and 2%, 1%, and 1% over the 2010–2040, 2041–2070, and 2071–2100 periods, respectively, compared to 1970–2000 values. In comparison, the PCM-B1 scenario (with N1, N2, and N3 deposition) resulted in increases in the NPP (by 14%, 10%, and 14%), (9%, 5%, and 9%), and (13%, 9%, and 12%) over the same periods relative to 1970–2000 values (Figure 16.11A). This difference in response for the two CC scenarios is due to the forest ecosystem exceeding the optimum temperature for photosynthesis under HadCM3-A1fi which causes temperature stress to vegetation and mid-summer drought. Model simulations suggest that CC without CO_2 fertilization has a relatively small effect on NPP. Note, under both CC scenarios, NPP was lower under the N2 deposition scenario compared to N1 and N3. These results suggest that anthropogenic N deposition stimulates forest growth and C sequestration under CC.

Future model projections under both scenarios and all conditions (with and without CO_2 effects on vegetation) suggest that the soil N pool will decrease under changing climate, with the magnitude of N loss depending on the scenario (Figure 16.11B). Under HadCM3-A1fi, higher projected temperatures and associated increases in N mineralization and nitrification rates resulted in a decrease in the soil N pool of 51% by 2070–2100 compared to 1970–2000. Under the less severe PCM-B1 scenario, the average decline in soil N is considerably less (12%). The simulations suggest that CC will have the potential to shift the soil N humus pool from a net N sink to a N source. It is counterintuitive that there is no significant difference in the soil N pools among the three N deposition scenarios. The soil N pool is large ($\sim$330 g m^{-2} year^{-1}) compared with the cumulative inputs of atmospheric N deposition from 1850 to 2100 under these scenarios (N1 = 106.5 g m^{-2}, N2 = 17.5 g m^{-2} and N3 = 81 g m^{-2}).

Annual volume-weighted NO_3^- concentrations are projected to increase under future CC with the magnitude of this response depending on the CC and N deposition scenario considered (Figure 16.11C). Under the HadCM3-A1fi and N1, annual volume-weighted NO_3^- concentration peaked around 2047 and then declined towards 2100 (with the exception of one last sharp increase at 2096) associated with the long-term decreases in the soil humus N pool. In comparison, annual volume-weighted NO_3^- concentration under PCM-B1 and N1 increased towards 2020, declined around 2025, and then increased (slope = 0.23 μmol NO_3^- L^{-1} year^{-1}) until 2100 (Figure 16.11C). The simulated increase in annual volume-weighted NO_3^- concentrations is due to decreases in soil moisture and an increase in vapour pressure deficit, which occur despite the increase in precipitation, causing decreases in evapotranspiration and mid-summer drought. Drought stress disrupts tree N uptake

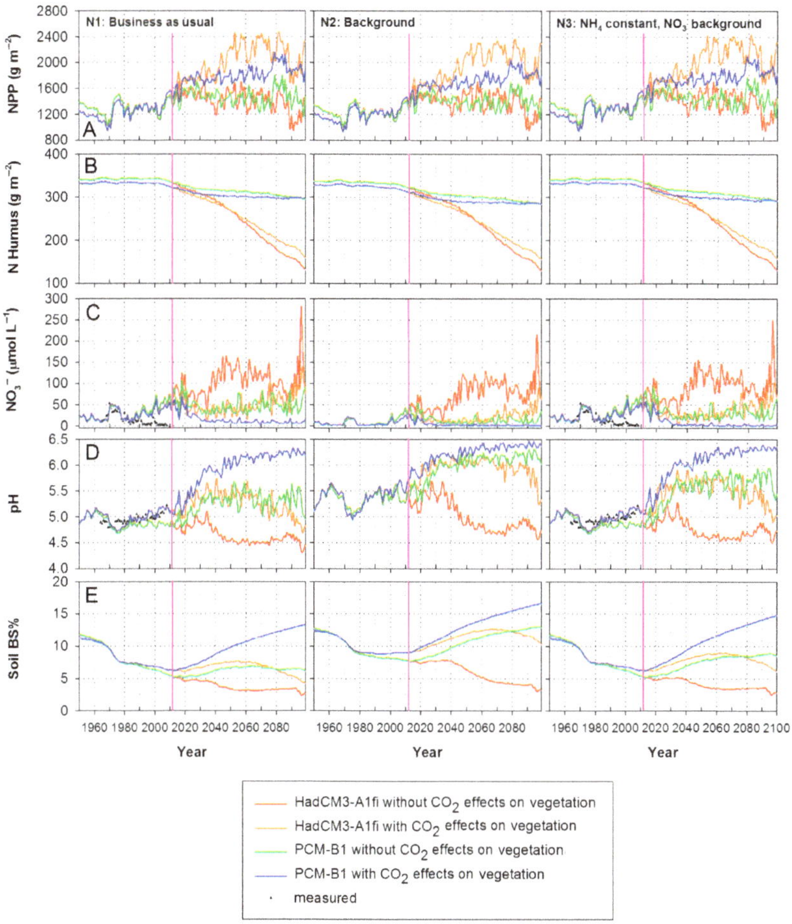

FIGURE 16.11 Future projected changes in (A) net primary productivity (NPP), (B) N humus pool, (C) annual volume-weighted concentrations of NO_3^-, (D) pH, and (E) soil base saturation (BS%), under HadCM3-A1fi and PCM-B1 scenarios with and without considering CO_2 effects on vegetation. Measured data are shown.

from soil resulting in an increase in net N mineralization, nitrification, and elevated NO_3^- leaching (Pourmokhtarian et al., 2012). The scenario of background N deposition (N2) also shows large increases in the magnitude of annual volume-weighted NO_3^- concentration (e.g. under HadCM3-A1fi; 6.4 µmol L^{-1} for 1970–2000 to 89 µmol L^{-1} for 2070–2100; more than one order of magnitude) (Figure 16.11C). Nevertheless, these elevated NO_3^- concentrations are still lower than values under comparable CC but higher N deposition scenarios (N1 and N3). These results suggest that historical increases in atmospheric N deposition to the Northern Forest are manifested through an increase in stream water NO_3^- concentration. Moreover, these legacy effects of atmospheric N deposition may be realized in the future under CC with elevated NO_3^- leaching associated with increases in net soil mineralization and nitrification.

Future model projections of pH exhibit patterns that followed changes in stream NO_3^- (Figure 16.11D). The average annual volume-weighted pH projected for 2070–2100 under HadCM3-A1fi and N1 decrease by up to 0.3 pH unit coinciding with high stream NO_3^- compared to average of measured values (4.9) for 1970–2000. Under PCM-B1 and N1 for the same period, pH increases by up to 0.5 pH unit.

Soils at the HBEF are inherently acidic (low % BS) and sensitive to the effects of acidic deposition. Simulations of soil base saturation (BS %) are projected to decrease by almost 58% and 12% under the HadCM3-A1fi and PCM-B1 scenarios with N1, respectively (Figure 16.3E). The model projections for HadCM3-A1fi showed a 55% decrease in soil BS% under both N2 and N3, while under PCM-B1 scenario, the soil BS% increased by 43% and 17%, respectively (Figure 16.11E). These results suggest that under moderate temperature CC scenarios, the future controls on N emissions could play an important role in the recovery of the acid–base status of forest ecosystems. In contrast, under high-temperature scenarios, the high rate of soil N mineralization, nitrification, and associated acidification due to the accumulation of legacy N from past elevated deposition could overwhelm any benefits of potential future controls on NO_x emissions.

16.6.2 CC with CO_2 Effects

Model results with CO_2 effects on vegetation showed that increasing CO_2 uptake by vegetation had little impact on the seasonal patterns of streamflow, causing only a slight increase in the annual discharge (Figures 16.9 and 16.10).

Invoking CO_2 fertilization effects on the forest ecosystem causes substantial increases in the NPP (68% and 49% for HadCM3-A1fi (N1) and PCM-B1 (N1), respectively for 2070–2100 compared to 1970–2000; Figure 16.11A). Elevated CO_2 decreases stomatal conductance and increases tree growth which increases water use efficiency (WUE) and annual water yield, limiting the occurrence of mid-summer drought. This effect is more evident under PCM-B1. Under the more severe HadCM3-A1fi scenario (N1), CO_2 effects on tree growth are projected to plateau towards the end of century (2080–2090) due the non-linear response of photosynthesis to CO_2 (at concentrations above 600 ppm). After CO_2 effects on forest vegetation play out, temperature increases become the dominant driver, promoting recurring drought (Pourmokhtarian et al., 2012). This response is important considering the importance of northern hardwood forests to the north-eastern United States and services they provide. The responses of NPP are largely driven by climate and CO_2 effects and the effects of N deposition are minimal.

Modelling results suggest that invoking CO_2 effects on vegetation has little impact on soil N humus due to the large size of the pool (Figure 16.11B). Under the HadCM3-A1fi scenario, the rate of decline in the N humus pool is lower by around 10% (42% compared to 51%) due to enhanced CO_2 uptake by trees.

The projected future increases in annual volume-weighted NO_3^- concentrations diminished substantially due to enhanced growth of trees and higher nutrient uptake associated with forest fertilization from CO_2 and the elimination of mid-summer drought caused by higher WUE due to decreases in stomatal conductance (Pourmokhtarian et al., 2012). Under the aggressive HadCM3-A1fi (N1) scenario, despite the enhanced tree growth associated with a CO_2 fertilization effect, the increase in temperature eventually increases net N mineralization and nitrification causing a condition of N saturation and resulting in annual volume-weighted NO_3^- concentrations started to increase after 2035 (Figure 16.11C). Stream water pH projections for 2070–2100 showed increases by up to 0.6 and 0.9 pH unit compared to simulations without CO_2 effects under HadCM3-A1fi (N1) and PCM-B1 (N1) scenarios, respectively (Figure 16.11D). The soil BS% under CO_2 effects is projected to increase substantially, with the magnitude and extent of the increases depending on the CC and N deposition scenario considered (Figure 16.11E). The greatest increases in soil %BS under the CO_2 fertilization effect occurred under PCM-B1 and N2, while the smallest increases occurred under HadCM3-A1fi and N1.

16.7 CONCLUSIONS

This chapter illustrates close linkages between CC and air pollution effects in forest ecosystems. These linkages are manifested in complex patterns which alter the structure and function of forests and potentially associated ecosystem services. Although the case studies presented represent contrasting forest ecosystems and disturbance regimes, there are some common features to the hindcasts and projections provided. CC (changes in temperature and precipitation), increasing CO_2 levels, changes in O_3 concentrations, and N deposition all can affect tree growth and therefore influence important ecosystem services such as C sequestration, water quantity, and soil and water quality.

In southern California, mixed conifer forest concentrations of O_3 have been historically high, resulting in severe damage of sensitive trees and extended dieback of forest stands. These high O_3 values have recently decreased, and it is anticipated that these trends will continue in the coming decades. It is expected that O_3 will interact with CC to strongly influence important ecosystem function such as C sequestration, and the quantity, distribution, and quality of stream water. Forest nutritional status, especially in terms of N nutrition, can have major effects on forest responses to O_3. Simulation of future changes in mixed conifer forests of southern California implies that O_3 impacts on forest growth are closely associated with

N deposition in these N-limited ecosystems. If the current trend of decreases in O_3 continues, tree biomass would not change significantly even under climate warming at high N deposition although at low N deposition forest trees would recover from biomass loss in the late twenty-first century. Under the increasing O_3 scenarios used, tree biomass would decrease at both N deposition projections, but much more for the A2 (medium–high emissions) climate-warming projection than at the B1 (low emissions) projection. If technology and/or policy management advances in the future decrease O_3 levels, its negative effects could be minimized by the end of this century. However, these positive changes can be negated by higher efficiency of O_3 photochemical production due to increasing temperatures.

The northern forests have been historically impacted by acidic deposition, which has acidified soil and surface waters, impacted the health of sensitive tree species, and altered the diversity of aquatic biota. In recent decades, emission controls on SO_2 and NO_x have diminished the impacts of acidification. CC has and will alter hydrology, decrease C and N sequestration, and potentially mobilize soil N resulting in the re-acidification of soil and water. The extent of these effects will likely vary depending on the future CC and N emission and deposition scenario as well as the extent to which increases in CO_2 stimulate additional tree growth. Negative impacts of acidic deposition could also be reduced by increased mineral weathering and availability of base cations caused by future higher temperatures.

16.8 RESEARCH AND MANAGEMENT NEEDS

Long-term monitoring of ground-level background O_3 and development of physiologically relevant O_3 indices are needed to verify model outputs and understand future O_3 risks to forests. Improved remote sensing techniques are needed to estimate air pollutant exposure in remote areas. Such techniques would greatly expand our ability to evaluate risks caused by toxic gases and atmospheric deposition in forests and other ecosystems. Future research is needed to better establish the role of fires and vertical transport of O_3 from the free troposphere to the boundary layer on ground-level O_3 exposure regimes. These issues are important factors for increasing our understanding of O_3 threats to U.S. ecosystems, especially in complex, mountainous terrain.

Improved assessments of increasing springtime O_3 concentrations on plant performance are needed. O_3 response models in North America should be modified to include the entire physiologically active period that is changing with a warming climate. While we recommend the continuing use of O_3 exposure indices developed in the United States, we also support testing and development of other approaches, such as those already developed in Europe (effective O_3 flux approaches). These should be tested and possibly developed for key tree species in various U.S. ecological zones. Improved understanding of gas exchange and biochemical defence mechanisms of key tree species is needed for understanding effective O_3 flux and its risks to forests. In this context, the effects of O_3 on the gas exchange capacity of trees, especially the question of potential stomatal sluggishness caused by chronic O_3 exposures, are a key issue for projecting the capacity of forests to sequester CO_2, for projecting potential growth, and effects on water fluxes in forest stands and catchments.

Better understanding of spatial and temporal distribution of N dry deposition and of key drivers of deposition, such as NH_3, HNO_3, and NO_2, as well as NO_3^- and NH_4^+ particulate matter, are needed for identification of sources of pollution and to develop options for their control. Similarly, better characterization of cloud water deposition of inorganic and organic reactive N species is needed. There is also a need for better models to describe deposition of reactive N species to forests in remote areas (especially complex mountain terrain). This will require better inventories of N emissions (especially reduced forms of N) and improved understanding of meteorological, chemical, and transport processes and their integration. Such models should be compared with empirical deposition data and new GIS-based inferential N deposition models as well as results of dry deposition measurements obtained with relax eddy flux methodologies.

Models are critical tools to assess CC effects on ecosystem structure and function. Robust holistic models are required to evaluate interactive effects of multiple air pollutants, ecosystem nutrient status (including the CO_2 fertilization effect), CC, management practices, and various other biotic and abiotic stressors. Models that simulate the responses of individual tree species and community responses to these factors are needed. Regional and global models of hydrologic cycles and related ecosystem functions should consider potential interactions of O_3 with future CC. In particular, it is important that experimental forest watersheds such as those used in the case studies in this chapter be maintained and preserved. Intensive monitoring and long-term measurements of meteorology, air quality, forest biomass, hydrology, and water quality at experimental watersheds are essential for the parameterization and testing of models used to predict the effects of CC and air pollution. This work should be supported and expanded in the future.

ACKNOWLEDGEMENTS

Work by A. P. and C. T. D. was supported by the USEPA through the STAR Program and by the NSF through the LTER program.

REFERENCES

Aber, J.D., Driscoll, C.T., 1997. Effects of land use, climate variation, and N deposition on N cycling and C storage in northern hardwood forests. Glob. Biogeochem. Cycles 11 (4), 639–648. http://dx.doi.org/10.1029/97GB01366.

Aber, J.D., Ollinger, S.V., Federer, C.A., Reich, P.B., Goulden, M.L., Kicklighter, D.W., Melillo, J.M., Lathrop Jr., R.G., 1995. Predicting the effects of climate change on water yield and forest production in the northeastern United States. Climate Res. 05 (3), 207–222. http://dx.doi.org/10.3354/cr0005207.

Aber, J.D., Reich, P.B., Goulden, M.L., 1996. Extrapolating leaf CO_2 exchange to the canopy: a generalized model of forest photosynthesis compared with measurements by Eddy correlation. Oecologia 106 (2), 257–265. http://dx.doi.org/10.1007/BF00328606.

Aber, J.D., Ollinger, S.V., Driscoll, C.T., 1997. Modeling nitrogen saturation in forest ecosystems in response to land use and atmospheric deposition. Ecol. Model. 101 (1), 61–78. http://dx.doi.org/10.1016/S0304-3800(97)01953-4.

Aber, J.D., Neilson, R., McNulty, S.G., Lenihan, J., Bachelet, D., Drapek, R., 2001. Forest processes and global environmental change: predicting the effects of individual and multiple stressors. Bioscience 59, 735–751.

Ainsworth, E.E., Yendrek, C.R., Sitch, S., Collins, W.J., Emberson, L.D., 2012. The effects of tropospheric ozone on net primary productivity and implications for climate change. Annu. Rev. Plant Biol. 63, 637–661.

Arbaugh, M.J., Miller, P.R., Carroll, J.J., Takemoto, B., Procter, T.M., 1998. Relationships of ozone exposure to pine injury in the Sierra Nevada and San Bernardino Mountains of California. Environ. Pollut. 101, 291–301.

Arbaugh, M.J., Johnson, D.W., Pulliam, W.M., 1999. Simulated effects of N deposition, ozone injury and climate change on a forest stand in the San Bernardino Mountains. In: Oxidant Air Pollution Impacts in the Montane Forests of Southern California: A Case Study of the San Bernardino Mountains. Springer-Verlag, New York, pp. 353–372.

Brasseur, G.P., Muller, J.-F., Tie, X., Horowitz, L., 2001. Tropospheric ozone and climate: past, present and future. In: Matsuno, T., Kida, H. (Eds.), Present and Future of Modeling Global Environmental Change: Toward Integrated Modeling. TERRAPUB, Tokyo, pp. 63–75.

Brimblecombe, P., 1987. The Big Smoke. Taylor & Francis Group, London, p. 185.

Bytnerowicz, A., Dueck, T., Godzik, S., 1998. Nitric oxide, nitrogen dioxide, nitric acid vapor and ammonia. In: Flagler, R. (Ed.), Recognition of Air Pollution Injury to Vegetation: A Pictorial Atlas. Air & Waste Management Association, Pittsburgh, PA, 5-1 through 5-17.

Bytnerowicz, A., Omasa, K., Paoletti, E., 2007. Integrated effects of air pollution and climate change on forests: a Northern Hemisphere perspective. Environ. Pollut. 147, 438–445.

Bytnerowicz, A., Arbaugh, M., Schilling, S., Fraczek, W., Alexander, D., 2008. Ozone distribution and phytotoxic potential in mixed conifer forests of the San Bernardino Mountains, southern California. Environ. Pollut. 155, 398–408.

CASTNET, 2012. Clean Air Status and Trends Network (CASTNET) 2010 Annual Report. Prepared by AMEC Environment & Infrastructure, Inc., for U.S. Environmental Protection Agency, EPA Contract No. EP-W-09-028, Washington, DC, April 2012.

Cayan, D.R., Maurer, E.P., Dettinger, M.D., Tyree, M., Hayhoe, K., 2008. Climate change scenarios for the California region. Clim. Change 87 (Suppl. 1), S21–S42.

Chen, L., Driscoll, C.T., 2005. A two-layer model to simulate variations in surface water chemistry draining a northern forest watershed. Water Resour. Res. (W09425) 41 (9), 8. http://dx.doi.org/10.1029/2004WR003625.

Cooper, O.R., Parrish, D.D., Stohl, A., Trainer, M., Nédélec, P., Thouret, V., Cammas, J.P., Oltmans, S.J., Johnson, B.J., Tarasick, D., Leblanc, T., McDermid, I.S., Jaffe, D., Gao, R., Stith, J., Ryerson, T., Aikin, K., Campos, T., Weinheimer, A., Avery, M.A., 2010. Increasing springtime ozone mixing ratios in the free troposphere over western North America. Nature 463, 344–348.

Davidson, E.A., David, M.B., Galloway, J.N., Goodale, C.L., Haueber, R., Harrson, J.A., Howarth, R.W., Jaynes, D.B., Lowrance, R.R., Nolan, B.T., Peel, J.L., Pinder, R.W.,

Porter, E., Snyder, C.S., Townsend, A.R., Ward, M.H., 2012. Excess nitrogen in the U.S. environment: trends, risks, and solutions. Issues Ecol. 15, 16.

Del Grosso, S.J., et al., 2000. General model for N_2O and N_2 gas emissions from soils due to dentrification. Global Biogeochem. Cycle 14, 1045–1060.

Doherty, R.M., Wild, O., Shindell, D.T., Zeng, G., MacKenzie, I.A., Collins, W.J., Fiore, A.M., Stevenson, D.S., Dentener, F.J., Schultz, M.G., Hess, P., Derwent, R.G., Keating, T.J., 2013. Impacts of climate change on surface ozone and intercontinental ozone pollution: a multimodel study. J. Geophys. Res. Atmos. 118, 3744–3763.

Fasullo, J.T., Trenberth, K.E., 2012. A less cloudy future: the role of subtropical subsidence in climate sensitivity. Science 338, 792–794.

Felzer, B., Kicklighter, D., Melillo, J., Wang, C., Zhuang, Q., Prinn, R., 2004. Effects of ozone on net primary production and carbon sequestration in the conterminous United States using a biogeochemistry model. Tellus 56B, 230–248.

Fenn, M.E., Baron, J.S., Allen, E.B., Rueth, H.M., Nydick, K.R., Geiser, L., Bowman, W.D., Sickman, J.O., Meixner, T., Johnson, D.W., Neitlich, P., 2003. Ecological effects of nitrogen deposition in the western United States. Bioscience 53, 404–420.

Fenn, M.E., Jovan, S., Yuan, F., Geiser, L., Meixner, T., Gimeno, B.S., 2008. Empirical and simulated critical loads for nitrogen deposition in California mixed conifer forests. Environ. Pollut. 155, 492–511.

Fourier, J., 1824. Remarques Générales Sur Les Températures Du Globe Terrestre Et Des Espaces Planétaires. Annal. Chim. Phys. 27, 136–167.

Gauss, M., et al., 2003. Radiative forcing in the 21st century due to ozone changes in the troposphere and the lower stratosphere. J. Geophys. Res. 108 (D9), 4292.

Gauss, M., Myhre, G., Isaksen, I.S.A., Grewe, V., Pitari, G., Wild, O., Collins, W.J., Dentener, F.J., Ellingsen, K., Gohar, L.K., Hauglustaine, D.A., Iachetti, D., Lamarque, J.-F., Mancini, E., Mickley, L.J., Prather, M.J., Pyle, J.A., Sanderson, M.G., Shine, K.P., Stevenson, D.S., Sudo, K., Szopa, S., Zeng, G., 2006. Radiative forcing since preindustrial times due to ozone change in the troposphere and the lower stratosphere. Atmos. Chem. Phys. 6, 575–599.

Gbondo-Tugbawa, S.S., Driscoll, C.T., Aber, J.D., Likens, G.E., 2001. Evaluation of an integrated biogeochemical model (PnET-BGC) at a northern hardwood forest ecosystem. Water Resour. Res. 37 (4), 1057–1070. http://dx.doi.org/10.1029/2000WR900375.

Greaver, T.L., Sullivan, T.J., Herrick, J.D., Barber, M.C., Baron, J.S., Cosby, B.J., Deerhake, M.E., et al., 2012. Ecological effects of nitrogen and sulfur air pollution in the US: what do we know? Front. Ecol. 10 (7), 365–372.

Grulke, N.E., Balduman, L., 1999. Deciduous conifers: high N deposition and O_3 exposure effects on growth and biomass allocation in ponderosa pine. Water Air Soil Pollut. 116, 235–248.

Grulke, N.E., Paoletti, E., Heath, R.L., 2006. Direct measurements of foliar ozone uptake in crop and tree species. Environ. Pollut. 146, 640–647.

Grulke, N.E., Minnich, R.A., Paine, T.D., Seybold, S.J., Chavez, D.J., Fenn, M.E., Riggan, P.J., Dunn, A., 2009. Air pollution increases forest susceptibility to wildfires: a case study in the San Bernardino Mountains in southern California. In: Bytnerowicz, A., Arbaugh, M.J., Riebau, A.R., Andersen, C. (Eds.), Wildland Fires and Air Pollution. Developments in Environmental Science, vol. 8. Elsevier, Amsterdam, pp. 365–403.

Intergovernmental Panel on Climate Change (IPCC), 2000. IPCC Special Report Emission Scenarios. Summary for Policymakers. UNEP, WMO, Geneva, p. 20.

Intergovernmental Panel on Climate Change (IPCC), 2007. In: Solomon, S., Qin, D., Manning, M., Chen, Z., Marquis, M., Averyt, K.B., Tignor, M., Miller, H.L. (Eds.), Climate Change 2007:

The Physical Science Basis. Contribution of Working Group I to the Fourth Assessment Report of the Intergovernmental Panel on Climate Change. Cambridge University Press, Cambridge, United Kingdom and New York, NY, USA.

Jacob, D.J., Winner, D.A., 2009. Effect of climate change on air quality. Atmos. Environ. 43 (1), 51–63. http://dx.doi.org/10.1016/j.atmosenv.2008.09.051.

Karl, T.R., Melillo, J.M., Peterson, T.C. (Eds.), 2009. Global Climate Change Impacts in the United States. Cambridge University Press, London England, p. 188.

Kurpius, M.R., Goldstein, A.H., 2003. Gas-phase chemistry dominates O_3 loss to a forest, implying a source of aerosols and hydroxyl radicals to the atmosphere. Geophys. Res. Lett. 30 (7), 1371. http://dx.doi.org/10.1029/2002GL016785.

Law, K., 2010. More ozone over North America. Nature 463, 307–308.

Lefohn, A.S., Laurence, J.A., Kohut, R.J., 1988. A comparison of indices that describe the relationship between exposure to ozone and reduction in the yield of agricultural crops. Atmos. Environ. 22, 1229–1240.

Lefohn, A.S., Shadwick, D., Oltmans, S.J., 2008. Characterizing long-term changes in surface ozone levels in the United States (1980–2005). Atmos. Environ. 42, 8252–8262.

Legge, A.H., Jäger, H.-J., Krupa, S.V., 1998. Sulfur dioxide. In: Flagler, R.B. (Ed.), Recognition of Air Pollution Injury to Vegetation—A Pictorial Atlas. Air & Waste Management Association, Pittsburgh, PA, 3-1 through 3-42.

Lei, H., Wuebbles, D.J., Liang, X.-Z., 2012. Projected risk of high ozone episodes in 2050. Atmos. Environ. 59, 567–577.

Li, H., McNulty, S.G., 2007. Uncertainty analysis on simple mass balance model to calculate critical loads for soil acidity. Environ. Pollut. 149, 315–326.

Likens, G.E., Bormann, F.H., 1995. Biogeochemistry of a Forested Ecosystem, second ed. Springer-Verlag, New York, USA.

Likens, G.E., Driscoll, C.T., Buso, D.C., Siccama, T.G., Johnson, C.E., Lovett, G.M., Ryan, D.F., Fahey, T., Reiners, W.A., 1994. The biogeochemistry of potassium at Hubbard Brook. Biogeochemistry 25 (2), 61–125. http://dx.doi.org/10.1007/BF00000881.

Matyssek, R., Wieser, G., Calfapietra, C., de Vries, W., Dizengremel, P., Ernst, D., Jolivet, Y., Mikkelsen, T.N., Hohren, G.M.J., Le Thiec, D., Tuovinen, J.-P., Weatherall, A., Paoletti, E., 2012. Forests under climate change and air pollution: gaps in understanding and future directions for research. Environ. Pollut. 160, 57–65.

McDonald-Buller, E.C., Allen, D.T., Brown, N., Jacob, D.J., Jaffe, D., Kolb, C.E., Lefohn, A.S., Oltmans, S., Parrish, D.D., Yarwood, G., Zhang, L., 2011. Establishing policy relevant background (PRB) ozone concentrations in the United States. Environ. Sci. Technol. 45, 9484–9497.

McKenzie, D., Peterson, D.L., Littell, J.J., 2009. Global warming and stress complexes in forests of western North America. In: Bytnerowicz, A., Arbaugh, M.J., Riebau, A.R., Andersen, C. (Eds.), Wildland Fires and Air Pollution. Developments in Environmental Science, 8. Elsevier, Amsterdam, pp. 319–337.

McNulty, S.G., Boggs, J.L., 2010. A conceptual framework: redefining forest soils critical acid loads under a changing climate. Environ. Pollut. 158, 2053–2058.

McNulty, S.G., Cohen, E.C., Moore Myers, J.A., Sullivan, T.J., Li, H., 2007. Estimates of critical acid loads and exceedances for forest soils across the conterminous United States. Environ. Pollut. 149, 281–292.

Miller, P.R., Parmeter Jr., J.R., Taylor, O.C., Cardiff, E.A., 1963. Ozone injury to the foliage of *Pinus ponderosa*. Phytopathology 53, 1072–1076.

NADP, 2010. National Atmospheric Deposition Program. 2010 Annual Summary. Illinois State Water Survey, Urbana-Champaign, 23 pp.

Nakićenović, N., Alcamo, J., Davis, G., Vries, B.D., Fenhann, J., Gaffin, S., Gregory, K., Grübler, A., Jung, T.Y., Kram, T., 2000. In: Special Report on Emissions Scenarios: A Special Report of Working Group III of the Intergovernmental Panel on Climate Change, first ed. Cambridge University Press, Cambridge, UK/New York, NY.

Northeast Climate Impact Assessment (NECIA), 2006. Climate Change in the U.S. Northeast. A Report of the Northeast Climate Impacts Assessment. Union of Concerned Scientists (UCS), UCS Publications,, Cambridge, Massachusetts, USA.

O'Brien, T.P., Sornette, D., McPherron, R.L., 2001. Statistical asynchronous regression: determining the relationship between two quantities that are not measured simultaneously. J. Geophys. Res. 106 (A7), 13247–13259. http://dx.doi.org/10.1029/2000JA900193.

Ollinger, S.V., Aber, J.D., Reich, P.B., 1997. Simulating ozone effects on forest productivity: interactions among leaf-, canopy-, and stand-level processes. Ecol. Appl. 7 (4), 1237–1251.

Ollinger, S.V., Aber, J.D., Reich, P.B., Freuder, R.J., 2002. Interactive effects of nitrogen deposition, tropospheric ozone, elevated CO_2 and land use history on the carbon dynamics of Northern Hardwood Forests. Global Chang. Biol. 8 (6), 545–562.

Ollinger, S.V., Goodale, C.L., Hayhoe, K., Jenkins, J.P., 2009. Potential effects of climate change and rising CO_2 on ecosystem processes in northeastern U.S. forests. Mitig. Adapt. Strategies Global Change 14 (1), 101–106.

Oltmans, S.J., Lefohn, A.S., Shadwick, D., Harris, J.M., Scheel, H.E., Galbally, I., Tarasick, D.W., Johnson, B.J., Brunke, E.-G., Claude, H., Zeng, G., Nichol, S., Schmidlin, F., Davies, J., Cuevas, E., Redondas, A., Naoe, H., Nakano, T., Kawasato, T., 2013. Recent tropospheric ozone changes—a patterns dominated by slow or no growth. Atmos. Environ. 67, 331–351.

Pardo, L.H., Fenn, M.E., Goodale, C.L., Geiser, L.H., Driscoll, C.T., Allen, E.B., Baron, J., Bobbink, R., Bowman, W.D., Clark, C., Emmett, B., Gilliam, F.S., Greaver, T., Hall, S.J., Lilleskov, E.A., Liu, L., Lynch, J., Nadelhoffer, K., Perakis, S.S., Robin-Abbott, M.J., Stoddard, J., Weathers, K., Dennis, R.L., 2011. Effects of nitrogen deposition and empirical nitrogen critical loads for ecoregions of the United States. Ecol. Appl. 21 (8), 3049–3082.

Parton, W.J., Scurlock, J.M.O., Ojima, D.S., Gilmanov, T.G., Scholes, R.J., Schimel, D.S., Kirchner, T., Menaut, J.-C., Seastedt, T., Garcia-Moya Apinan-Kamnalrut, E., Kinyamario, J.I., 1993. Observations and modeling of biomass, soil organic matter dynamics for the grassland biome worldwide. Global Biogeochem. Cycles 7, 785–809.

Parton, W.J., Hartman, M., Ojima, D., Schimel, D.S., 1998. DAYCENT and its land surface submodel: description and testing. Glob. Planet Change 19, 35–48.

Parton, W.J., Holland, E.A., Del Grosso, S.J., Hartman, M.D., Martin, R.E., Mosier, A.R., Ojima, D.S., Schimel, D.S., 2001. Generalized model for NO_x and N_2O emissions from soils. J. Geophys. Res. (Atmospheres) 106, 17403–17419.

Peterson, D.L., Arbaugh, M., Robinson, L., 1991. Regional growth changes in ozone-stressed ponderosa pine (*Pinus ponderosa*) in the Sierra Nevada, California, USA. Holocene 1, 50–61.

Pinder, R.W., Bettez, N.D., Bonan, G.B., Greaver, T.L., Wieder, W.R., Schlezinger, W.H., Davidson, E.A., 2013. Impacts of human alterations of the nitrogen cycle in the US on radiative forcing. Biogeochemistry 114 (1–3), 25–40. http://dx.doi.org/10.1007/s10533-012-9787-z.

Pope, V.D., Gallani, M.L., Rowntree, P.R., Stratton, R.A., 2000. The impact of new physical parametrizations in the Hadley Centre Climate Model: HadAM3. Climate Dyn. 16 (2), 123–146. http://dx.doi.org/10.1007/s003820050009.

Post, W.M., Venterea, R.T., 2012. Managing biogeochemical cycles to reduce greenhouse gases. Front. Ecol. Environ. 10, 511.

Pourmokhtarian, A., Driscoll, C.T., Campbell, J.L., Hayhoe, K., 2012. Modeling potential hydrochemical responses to climate change and increasing CO_2 at the Hubbard Brook Experimental Forest using a dynamic biogeochemical model (PnET-BGC). Water Resour. Res. 48 (7). http://dx.doi.org/10.1029/2011WR011228, W07514.

Pye, J.M., 1988. Impact of ozone on the growth and yield of trees: a review. J. Environ. Qual. 17, 347–360.

Ramanathan, V., Feng, Y., 2008. On avoiding dangerous anthropogenic interference with the climate system: formidable challenges ahead. Proc. Natl. Acad. Sci. U.S.A. 105, 14245–14250.

Reich, P.B., Amundson, R.G., 1985. Ambient levels of ozone reduce net photosynthesis in tree and crop species. Science 230, 566–570.

Rieder, H.E., Frossard, L., Ribatet, M., Staehelin, J., Maeder, J.A., DiRocco, S., Davison, A.C., Peter, T., Weihs, P., Holawe, F., 2013. On the relationship between total ozone and atmospheric dynamics and chemistry at mid-latitudes: part 2. The effects of the El Nino/Southern Oscillation, volcanic eruptions and contributions of atmospheric dynamics and chemistry to long-term total ozone changes. Atmos. Chem. Phys. 13, 165–179.

Rustad, L.E., Campbell, J.L., Cox, R.M., DeBlois, M., Dukes, J.S., Huntington, T.J., Magill, A.H., Mohan, J.E., Pontius, J., Richardson, A.D., Rodenhouse, N.L., Watson, M.R., Willard, N., 2009. NE Forests 2100: a synthesis of climate change impacts on forests of the Northeastern US and Eastern Canada. Can. J. For. Res. 39, iii–iv.

Sitch, S., Cox, P.M., Collins, W.J., Huntingford, C., 2007. Indirect radiative forcing of climate change through ozone effects on the land-carbon sink. Nature 448, 791–794.

Solomon, S., Plattner, G.K., Knutti, R., Friedlingstein, P., 2009. Irreversible climate change due to carbon dioxide emissions. Proc. Natl. Acad. Sci. U.S.A. 106, 1704–1709.

Stein, S., Geller, R.J., 2012. Communicating uncertainties in natural hazard forecasts. EOS 93, 361–362.

Sun, G., McLaughlin, S.B., Porter, J.H., Uddling, J., Mulholland, P.J., Adams, M.B., Pederson, N., 2012. Interactive influences of ozone and climate on streamflow of forested watersheds. Global Chang. Biol. 18, 3395–3409.

Sverdrup, H., Warfvinge, P., 1993. Calculating field weathering rates using a mechanistic geochemical model PROFILE. Appl. Geochem. 8, 273–283.

Templer, P.H., Pinder, R.W., Goodale, C.L., 2012. Effects of nitrogen deposition on greenhouse-gas fluxes for forests and grasslands of North America. Front. Ecol. Environ. 10, 547–553.

Tkacz, B., Moody, B., Castillo, J.V., Fenn, M.E., 2008. Forest health condition in North America. Environ. Pollut. 155, 409–425.

Unger, N., Pan, J.L., 2012. New directions: enduring ozone. Atmos. Environ. 55, 456–458.

US EPA, 2008. Integrated Science Assessment (ISA) for Oxides of Nitrogen and Sulfur—Environmental Criteria, final report, Washington, DC.

Vingarzan, R., 2004. A review of surface ozone background levels and trends. Atmos. Environ. 38, 3431–3442.

Wang, H., Jacob, D.J., Le Sager, P., Streets, D.G., Park, R.J., Gilliland, A.B., van Donkelaar, A., 2009. Surface ozone background in the United States: Canadian and Mexican influences. Atmos. Environ. 43, 1310–1319.

Washington, W.M., et al., 2000. Parallel Climate Model (PCM) control and transient simulations. Climate Dyn. 16 (10), 755–774. http://dx.doi.org/10.1007/s003820000079.

Wittig, V.E., Ainsworth, E.A., Long, S.P., 2007. To what extent do current and projected increases in surface ozone affects photosynthesis and stomatal conductance in trees? A meta-analytic review of the last 3 decades of experiments. Plant Cell Environ. 30, 1150–1162.

Wu, W., Driscoll, C.T., 2010. Impact of climate change on three-dimensional dynamic critical load functions. Environ. Sci. Technol. 44, 720–726.

Yuan, F.-M., Meixner, T., Fenn, M.E., Simunek, J., 2011. Impact of transit soil water simulation to estimated nitrogen leaching and emission at high- and low-deposition forest sites in Southern California. J. Geophys. Res. 116, http://dx.doi.org/10.1029/2011JG001644, G03040.

Zhang, X., Zwiers, F.W., Hegerl, G.C., Lambert, F.H., Gillett, N.P., Solomon, S., Stott, P.A., Nozawa, T., 2007. Detection of human influence on twentieth-century precipitation trends. Nature 448, 461–465.

Zhang, L., Jacob, D.J., Downey, N.V., Wood, D.A., Blewitt, D., Carouge, C.C., van Donkelaar, A., Jones, D.B.A., Murray, L.T., Wang, Y., 2011. Improved estimate of the policy-relevant background ozone in the United States using the GEOS-Chem global model with 1/2o x 2/3 horizontal resolution over North America. Atmos. Environ. 45, 6769–6776.

Chapter 17

Effects of Ozone on Forest Ecosystems in East and Southeast Asia

Takayoshi Koike*[,1], Makoto Watanabe*[,†], Yasutomo Hoshika*[,‡], Mitsutoshi Kitao[§], Hideyuki Matsumura[¶], R. Funada[†] and Takeshi Izuta[†]

*Silviculture and Forest Ecological Studies, Hokkaido University, Sapporo, Hokkaido 060-8589, Japan

[†]Institute of Agriculture, Tokyo University of Agriculture and Technology, Fuchu, Tokyo 183-8509, Japan

[‡]JSPS Research Fellow

[§]Department of Plant Ecology, Forestry and Forest Products Research Institute, Matsunosato 1, Tsukuba 305-8687, Japan

[¶]Environmental Science Research Laboratory, Central Research Institute of Electric Power Industry, Abiko, Chiba 270-1194, Japan

[1]Corresponding author: e-mail: tkoike@for.agr.hokudai.ac.jp

Chapter Outline

Developments in Environmental Science, Vol. 13. http://dx.doi.org/10.1016/B978-0-08-098349-3.00017-7

17.1 INTRODUCTION

East and Southeast Asia have achieved one of the highest economic growth rates in the second half of the twentieth century (e.g. Nouchi, 2001). Increases in urbanization, transport, industrialization and energy generation have led to many environmental problems, including air pollution. It is considered that phyto-toxic levels of air pollution prevail in many parts of East and Southeast Asia. Relatively high concentrations of O_3 (about 100 nmol mol^{-1}) were observed in large cities in China, Taiwan and Japan (Chang and Lee, 2006; Li et al., 2011; Takeda and Aihara, 2007; Wang et al., 2006). Separately, simulations associated with chemical modelling research have found a relatively high concentration of O_3 in East Asia (Tanimoto et al., 2005; Yamaji et al., 2008). Furthermore, a continued increase of O_3 concentration in Asia is predicted for the present century (Ohara et al., 2007; Sitch et al., 2007).

Several countries, such as Japan, Korea, Taiwan and Thailand, have national monitoring networks for air pollution (about 1200, 280, 70 and 55 monitoring sites for O_3, respectively), and the data are openly available for researchers although the provided indices (1 h average, monthly/annual average, etc.) vary among the countries. However, the monitoring stations that cover forested/rural areas are limited. In China, a national monitoring network for air pollution including O_3 has been developed since 2012 and will start from 2015. This network's data will also be published from 2016. The possibility of monitoring data for air pollution in the other countries in East and Southeast Asia is very limited because of insufficient numbers of monitoring stations even in urban areas.

To understand the exact situation of air pollution in East and Southeast Asian countries, including in rural areas, the Acid Deposition Monitoring Network in East Asia (EANET) has been running since 2001. The number of participating countries in EANET has increased to 13. It includes 54 monitoring sites throughout East and Southeast Asia. The EANET also collaborates with other networks such as EMEP (the Co-operative Programme for Monitoring and Evaluation of the Long-range Transmission of Air Pollutants in Europe), ICP Forests (the International Co-operative Programme on Assessment and Monitoring of Air Pollution Effects on Forests) and WMO (the World Meteorological Organization) (Mikkelsen et al., 2013). Harmonization of data is achieved by quality assurance and quality control activities of the EANET community to provide reliable data with comparability among participating countries and with information from other networks. The EANET program monitors concentrations of air pollutants (e.g. O_3 and SO_2) and dry and wet deposition of cations and anions. In addition, soil and vegetation (e.g. pH of soil, tree height, breast height diameter, declining trees and number of understory species) as well as terrestrial water (e.g. pH of water) are also monitored. A joint database is provided at the Web site of EANET (http://www.eanet.cc/product).

East and Southeast Asia are currently regarded as a 'hot spot' in relation to air pollution (Matyssek et al., 2013a; Sitch et al., 2007). However, the experimental studies on responses of Asian tree species to O_3 are very limited. Although 84 and 262 papers were analyzed in the meta-analyses of O_3 effects on trees by Wittig et al. (2007, 2009), only zero and seven papers for Asian trees were cited, respectively. Therefore, the information on the response of East and Southeast Asian tree species to O_3 is considered as a 'black box'. This chapter summarizes research into the effect of air pollution, mainly O_3, in East and Southeast Asian countries. We also discuss a recent free-air O_3 fumigation experiment used to study effects on tree saplings at stand level and new developments in the flux approach in Japan. Concluding remarks focus on potential future developments and research needs.

17.2 EFFECT OF AIR POLLUTION ON FOREST ECOSYSTEMS IN EAST AND SOUTHEAST ASIA

East and Southeast Asia cover a broad range of tree functional types (Matyssek et al., 2013a). Although national monitoring networks have mainly been developed in East Asian countries, observational data indicate that relatively high concentrations of O_3 that may induce negative impacts on forest vegetation are also found in Southeast Asia (e.g. Matsuda et al., 2005). Information on O_3 effects on plant growth in East and Southeast Asia are available only for some countries (Japan, China, South Korea, Indonesia, Bangladesh and Malaysia). In this section, we first introduce the observational and experimental studies on O_3 in these countries, and then discuss the situation in other countries.

17.2.1 Japan

Japan was the first country in East Asia to be recognized as suffering from serious air pollution (e.g. Izuta, 2003, 2006; Koike, 2012; Nouchi, 2001). Although sulphur dioxide (SO_2), nitrogen oxides (NO_x), fluoride and suspended particulate matter were the most serious air pollutants from the 1960s to the early 1970s, the concentrations of these air pollutants fell as a result of various responses (Furukawa, 1991; Izuta, 2006; Nouchi, 2001). High O_3 concentrations have been common since the 1970s, especially near the Pacific coast. During this period, O_3-induced abnormal defoliation in Japanese zelkova (*Zelkova serrata*) and damage to agricultural crops were observed in cities and nearby. In the late 1970s, the concentration of O_3 fell abruptly because of various efforts to reduce the concentrations of precursors of O_3, such as NO_x and volatile organic compounds (VOC). The mean maximum daily O_3 concentration from April to September at a monitoring station in Tokyo, the capital of Japan, was 70.8, 55.4, 37.2 and 36.0 nmol mol^{-1} in 1976, 1977, 1978 and 1979, respectively.

The concentration of O_3 has increased again from the late 1980s to the time of writing, although the atmospheric concentrations of O_3 precursors fell in Japan during this period. This increase in O_3 was observed over the whole of Japan. According to a chemistry-transport model (Tanimoto et al., 2005; Yamaji et al., 2008), air pollution reaching Japan from other East Asian countries is responsible; their emission of O_3 precursors increased dramatically as a result of rapid industrialization and urbanization (Ohara et al., 2007).

In Japan, studies of the effects of O_3 on plants have been published since the 1970s (Furukawa et al., 1983; Nouchi et al., 1973). These studies revealed the acute effects of high concentrations of O_3 on agricultural crops and trees in roadsides and parks (Fujinuma et al., 1987; Furukawa, 1991; Izuta, 2006; Omasa, 1990). Experimental studies (Izuta 1998) indicate that exposure to ambient levels of O_3 below 100 ppb for several months is sufficient to inhibit dry matter production and net photosynthesis of sensitive Japanese forest tree species, such as Siebold's beech (*Fagus crenata*), Japanese Zelkova (*Z. serrata*) and Japanese red pine (*Pinus densiflora*) (Izuta et al., 1996; Matsumura et al., 1996, 1998; Shan et al., 1997).

The study of the relationship between exposure of crops to relatively low long-term concentrations of O_3 and their response began in Japan in the mid-1980s. Experimental studies of the effect of O_3 on growth and physiology of forest trees commenced in the 1990s, and the relation between exposure and response for these species was determined late in the same decade. These studies revealed differences between forest tree species in Japan in sensitivity to O_3 (Table 17.1). The effect of O_3 depends on further factors including soil water and nutrient availability, atmospheric CO_2 concentration and climatic conditions (Matyssek et al., 2013b). The influence of these factors on changes due to O_3 has been studied for woody plants (e.g. Koike et al., 2012; Matsumura et al., 2005; Nakaji and Izuta, 2001; Nakaji et al., 2004; Watanabe et al., 2005, 2010b; Yonekura et al., 2001a,b). In particular, the growth and photosynthetic responses of six representative Japanese forest tree species to O_3 under different N loads has been intensively studied (Watanabe et al., 2006, 2007, 2008; Yamaguchi et al., 2007, 2010). This is because of spatial correlation between high exposure to O_3 and high N deposition (e.g. Bytnerowicz, 2013), as well as because changes in N availability may affect the sensitivity to O_3 because N is a component of the enzyme that catalyzes many metabolic processes including defence against O_3 (Dizengremel et al., 2013). There was an increase in the sensitivity of Siebold's beech to O_3 under relatively high N loading, but the sensitivity of the Japanese larch (*Larix kaempferi*=*L. leptorepis*) decreased with increasing N loading (Watanabe et al., 2006; Yamaguchi et al., 2007; Figure 17.1).

Risk assessments for the effect of O_3 on tree growth in Japan have been conducted, based on experimental studies, since 2003 (Takagi and Ohara,

TABLE 17.1 Classification of Ozone Sensitivity for Japanese Forest Tree Species

Sensitivity class	Type		Species	Critical level as daylight AOT40 (ppm h)
High	Broad leaved	Deciduous	*Populus maximowiczii, Populus nigra, Fagus crenata, Zelkova serrata*	8–15
		Evergreen	*Castanopsis sieboldii*	
	Coniferous	Deciduous	*Larix kaempferi**	
		Evergreen	*Pinus densiflora*	
Moderate	Broad leaved	Deciduous	*Quercus serrata, Betula platyphylla* var. *japonica*	16–30
		Evergreen	*Quercus myrsinaefolia, Cinnamomum camphora*	
	Coniferous	Evergreen	*Abies homolepis*	
Low	Broad leaved	Deciduous	*Quercus mongolica* var. *crispula***	31 <
		Evergreen	*Lithocarpus edulis, Machilus thunbergii*	
	Coniferous	Evergreen	*Pinus thunbergii, Cryptomeria japonica Chamaecyparis obtusa*	

High sensitivity: the whole-plant dry mass increment was significantly reduced by exposure to the ambient level of ozone.
Moderate sensitivity: the whole-plant dry mass increment was significantly reduced by exposure to 1.5 or 2.0 times the ambient level of ozone.
Low sensitivity: the whole-plant dry mass increment was not significantly reduced by exposure to 1.5 or 2.0 times the ambient level of ozone.
Daylight AOT40: accumulated exposure over a threshold of 40 ppb O_3 during daylight hours with global radiation from April to September (6 months).
Species name: **Larix leptolepis*, ***Quercus mongolica* var. *grosserrata* or *Q. crispula*.
After Kohno et al. (2005) and Yamaguchi et al. (2011).

2003). Kohno et al. (2005) estimated the critical level of O_3 for forest trees, based on experimental studies of 18 tree species, and mapped the area in which the level of O_3 corresponds to at least a 10% growth reduction of O_3-sensitive tree species throughout Japan. Based on the experimentally observed responses to combinations of O_3 and N loading, the risk of negative O_3 impact on growth, taking N deposition into account, was determined for the six tree

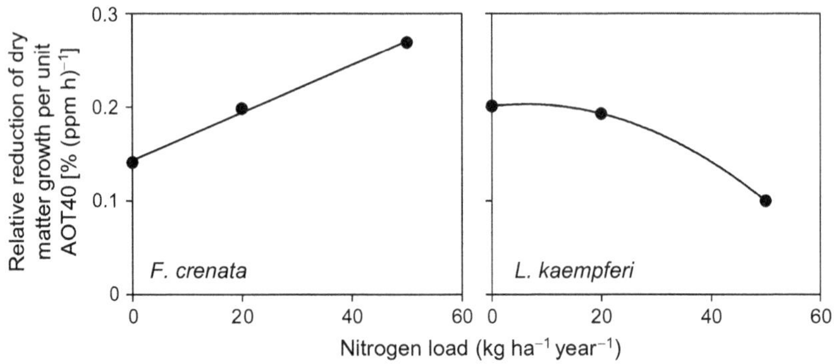

FIGURE 17.1 Relationships between soil nitrogen load and relative reduction of dry matter growth per unit AOT40 for *Fagus crenata* and *Larix kaempferi* (=*L. leptolepis*; Watanabe, 2007). The values of the *y*-axis were calculated as absolute value of slope of the regression line between AOT40 during daylight hours from April to September (one growing season) and dry matter growth (i.e. exposure–response relationship) according to the calculation of Karlsson et al. (2004) with data from Watanabe et al. (2006) and Yamaguchi et al. (2007).

species studied in Japan (Figure 17.2; Watanabe, 2007; Watanabe et al., 2012). On the other hand, Watanabe et al. (2010a) integrated a forest growth model and results of experimental studies and estimated the O_3 impact on annual forest carbon absorption throughout Japan for representative Japanese conifers: Japanese larch, Japanese red pine and Sugi-cedar (*Cryptomeria japonica*). They reported that the O_3-induced reduction in the total annual forest carbon absorption of the three tree species in Japan was estimated to be 0.8%. A flux-based approach at leaf/stand level for evaluating O_3 risk to trees growing in Japan has been under way for several years (discussed later in this chapter).

17.2.2 China

In China, little attention was paid to O_3 in the last century because SO_2 was the most serious air pollutant causing forest decline (Nouchi, 2001; Zheng and Shimizu, 2003). However, extremely high hourly mean concentrations of O_3 above 250 nmol mol^{-1} were observed in Lanzhou city, Gansu province, in the early 1980s (Zheng and Shimizu, 2003).

Experimental studies of the effect of O_3 on tree species have been reported from recent years. These focus mainly on the effect of O_3 on growth and photosynthesis of deciduous and evergreen broad-leaved tree species in the subtropical region of China (Feng et al., 2008b, 2011a; Zhang et al., 2012). Because of the high precipitation, the ecophysiological traits of evergreen broad-leaved trees growing in East Asia differ from those in southern Europe where the summers are relatively dry. Zhang et al. (2012) reported the growth and leaf physiological and biochemical responses to O_3 of two deciduous and

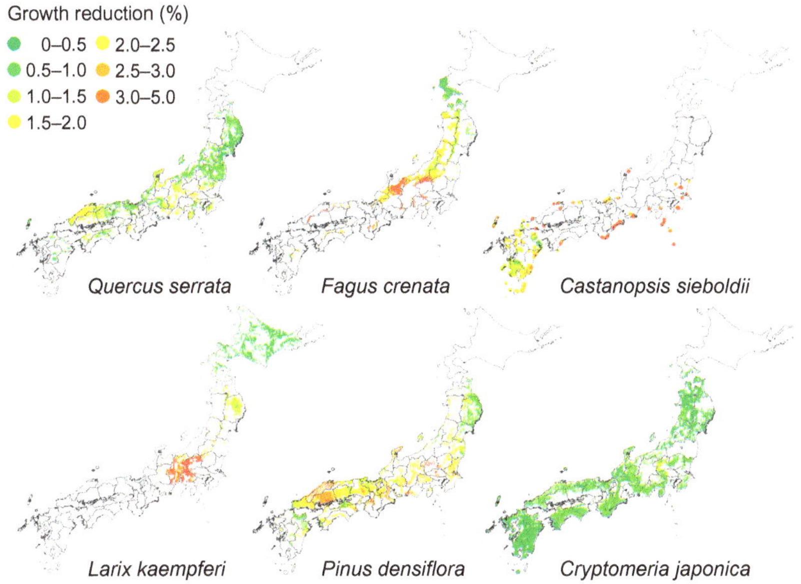

FIGURE 17.2 Risk assessment of ozone impact on the growth of *Quercus serrata*, *Fagus crenata* (deciduous board-leaved), *Castanopsis sieboldii* (evergreen broad leaved), *Larix kaempferi* (=*L. leptorepis*, deciduous conifer), *Pinus densiflora* and *Cryptomeria japonica* (evergreen conifer) (Watanabe, 2007). For *F. crenata* and *L. kaempferi*, change of ozone sensitivity by nitrogen deposition was involved for the assessment. The assessment was based on relationships between AOT40 and dry matter growth (i.e. exposure–response relationship) for six species evaluated by ozone fumigation studies (Watanabe et al., 2006, 2007, 2008; Yamaguchi et al., 2007), governmental monitoring data for oxidant concentration and atmospheric N deposition, and a national vegetation survey of the Japanese Ministry of the Environment.

six evergreen broad-leaved tree species at AOT40 of 38.24 $\mu mol\ mol^{-1}\ h$ (47 days exposure). The two deciduous tree species were more sensitive to O_3 than most evergreen broadleaf species, both for O_3-induced visible injury to leaves and reduction in biomass. In Japan too, deciduous trees are generally more sensitive to O_3 than evergreen species (Table 17.1). This difference in the sensitivity to O_3 between several tree species may be due to (1) limitation of stomatal O_3 uptake by lower stomatal conductance and (2) different leaf structure (thicker cell walls, higher density of mesophyll tissues, cell surface area, etc.) and a capability of feeding detoxification processes more intensely in evergreen species (Dizengremel et al., 2013; Matyssek et al., 2013b). However, we should pay attention to variations in O_3 sensitivity within a functional type. For example, the sensitivity of *Castanopsis sieboldii* (evergreen broad leaved) is on a similar level with *F. crenata* (Siebold's beech; deciduous broad leaved), a sensitive tree species to O_3 (Table 17.1; Figure 17.2; Watanabe et al., 2008).

Studies of the effect of O_3 on woody species in China are under way. Studies in China have looked at the effect of O_3 only over a single growing season, although carry-over effects of O_3 from the previous growing season are suggested (Oksanen and Saleen, 1999; Yonekura et al., 2004). Variations in sensitivity of trees to O_3 in relation to other environmental factors, such as high N deposition, drought and elevated CO_2, have been studied (Kawaguchi et al., 2012; Koike et al., 2012; Matsumura et al., 2005; Nakaji and Izuta, 2001; Nakaji et al., 2004; Watanabe et al., 2005, 2010b; Yonekura et al., 2001a,b). However, only one trial has examined the combined effects of O_3 and N loading on *Cinnamomum camphora* seedlings (Feng et al., 2011a). They reported a reduction in light-saturated net photosynthetic rate at ambient CO_2 by 24.6% under O_3 exposure with AOT40 of 34.5 ppm h, and no modification of sensitivity of *C. camphora* to O_3 under N supply to the soil at $\leq$60 kg ha^{-1} year^{-1}. Further multi-factorial experiments extending over multiple years are necessary to generate a better understanding of the effect of O_3 on trees native to China.

17.2.3 South Korea

The problem of SO_2 pollution in South Korea was first observed in large cities and industrial areas in the 1960s. In the 1990s, atmospheric SO_2 concentration was reduced to 4–8 ppb by the clean energy policy (Choi and Toda, 2008). There are few reports of problems due to O_3. This might be due to geographic and climatic factors. For example, Seoul, the capital of South Korea, has lower temperatures due to its higher latitude, a less frequent temperature inversion (which can accumulate O_3 and its precursors), and frequent rain in summer, which scavenges O_3 and precursors (Ghim and Chang, 2000). The yearly average O_3 concentration in seven major cities in South Korea was relatively low, in the range 18–25 ppb in 2007 (Choi and Toda, 2008). However, recent development of transport in large cities has led to an increase in the concentrations of VOC and NO_x in the atmosphere. As a result, O_3 concentration in those cities has increased from 1997 to 2007. In addition, 89.7% of monitoring stations for O_3 in Korea exceeded their regulatory threshold for O_3 concentration, which is 60 ppb as an 8 h average (Choi and Toda, 2008).

There are few reports of the effects of O_3 on Korean vegetation, but Cho (2007) observed visible symptoms induced by O_3 on the leaves of field grown *Robinia pseudo-acacia*. Han et al. (2002) compared the O_3 sensitivity of four *Betula* species based on carbon allocation and partitioning pattern. They reported that O_3 decreased carbon allocation in the leaves of *Betula costata* and *Betula davurica*, while no change of carbon allocation pattern was found in *Betula ermanii* and *Betula platyphylla* var. *japonica* after O_3 exposure of 100 nmol mol^{-1} for 8 h day^{-1} for 5 weeks. Several studies have looked at the physiological responses of woody leaves to O_3 (e.g. Kim et al., 2010).

It will be important to study further environmental changes in Korea, especially increases in O_3, due to industrial development as well as transboundary air pollution.

17.2.4 Indonesia, Bangladesh and Malaysia

According to model simulation, the concentration of O_3 in Indonesia is not high (Prather et al., 2001). However, an annual average O_3 concentration of 32 ppb was recorded in Jakarta city in 2002–2003 (Asian Development Bank and the Clean Air Initiative for Asian Cities Center, 2006). This O_3 concentration is at a similar level with that of Japan. Therefore, negative impacts of O_3 on forest ecosystems located in suburbs of big cities might be induced. Open-top chamber experiments on common plantation species in Indonesia (*Eucalyptus globulus*, *Eucalyptus grandis* and the hybrid *Eucalyptus camaldulensis* × *Eucalyptus deglupta*) found no significant change in biomass of the three species whether treated with high O_3 (60 ppb for 7 h in daytime), high CO_2 (600 ppm) or their combination. However, a normal alkane, *n*-icosane, increased in leaves of *E. grandis* under the high concentration of O_3. This compound acts as an epi-cuticle wax on the leaf surface, conferring protection (Novriyanti, 2012).

Akhtar et al. (2010a,b) reported the effects of O_3 on two Bangladeshi wheat cultivars (Sufi and Bijoy) and four Bangladeshi rice cultivars (BR11, BR14, BR28 and BR29) planted with Andisol (black soil) and exposed daily to charcoal-filtered air or O_3 at 60 and 100 nl l^{-1} (10:00–17:00) for 85 and 160 days. Exposure to O_3 reduced the whole-plant dry mass and grain yield per plant of all wheat and rice cultivars tested. The leaf net photosynthetic rate was also reduced by exposure to O_3. There was no significant difference in response between the two wheat cultivars, or between the four rice cultivars. It was concluded from the results that ambient levels of O_3 may impair the production of wheat and rice in Bangladesh.

Ishii et al. (2004) reported a significant impact of ambient air pollution on growth and yield of locally grown rice cultivars in Malaysia, based on an open-top chamber experiment. They also found a clear difference in sensitivity to O_3 of the two cultivars studied.

17.2.5 Countries Without Detailed Studies on Effects of Air Pollution on Forest Ecosystems

Experimental studies on effects of O_3 on native tree species grown in Southeast Asia are extremely limited. According to the prediction of the IPCC, however, the O_3 concentration in continental Southeast Asia will increase in the future (Prather et al., 2001). Even at the present date, an O_3 concentration exceeding 60 nmol mol^{-1} as median value during 4 months' monitoring (January–April 2002) was observed in the afternoon at a teak forest in

Mea Moh, Lampang province, Thailand (Matsuda et al., 2005). Although the tree species richness in Asia is high (Adams and Woodward, 1989), there are many hot spots for loss of biodiversity especially in Southeast Asia (Myers et al., 2000). Different tree species have differing sensitivities to O_3 (e.g. Yamaguchi et al., 2011), so that species ratios are liable to change with rising O_3 concentration. As an example, in mixed forest with Siebold's beech (O_3 sensitive) and Sugi-cedar (O_3 tolerant) in Japan (Kume et al., 2009), the active growth of Sugi-cedar was observed in the gap created by dieback of Siebold's beech, which was considered to be a result of negative impact of O_3. Because vegetation types in Southeast Asia (mainly tropical rain forest and tropical seasonal forest) are quite different to East Asian vegetation (mainly temperate forest) (Matyssek et al., 2013a), we cannot apply the information obtained in East Asia to Southeast Asian vegetation. It is urgent to clarify O_3 sensitivity of native tree species grown in Southeast Asia and assess the contribution of O_3 in the degradation of biodiversity. Some species might cease to be represented.

17.3 EXPERIMENTAL AND PROCESS STUDIES ON EFFECTS AND UPTAKE OF OZONE

Section 17.2 summarizes what is currently known on O_3 in East and Southeast Asia and gaps in knowledge. This section discusses the detailed process studies carried out in Japan and outlines some future developments.

17.3.1 Free-Air O_3 Fumigation Experiment

Free-air O_3 fumigation systems have been established in mid-Europe, northern Europe, and the USA (King et al., 2013; Matyssek et al., 2013b; Oksanen et al., 2013; Pregitzer and Talhelm, 2013). We commenced similar O_3 fumigation studies with a free-air system on Siebold's beech, deciduous oak (*Quercus mongolica* var. *crispula*=*Q. mongolica* var. *grosseserata* or *Quercus crispula*) and white birch (*B. platyphylla* var. *japonica*), which are all native to East Asia. These trees are representative and dominant species in cool-temperate regions there. Our free-air O_3 fumigation system is the first such to study tree responses in East Asia (Hoshika et al., 2012b). In China, this technology has already been used to study effects on crops (Feng et al., 2011b).

Our fumigation system was established in Sapporo, northern Japan (43°04′N, 141°20′E, 15 m a.s.l.) in 2011. Details of the system are described in Watanabe et al. (2013a). Saplings (10-year old as of 2011) of Siebold's beech, deciduous oak and 3-year old (as of 2011) white birch were exposed to ambient air and to an enhanced O_3 regime (60 nmol mol^{-1}) during daytime. The methodology of O_3 exposure in our system is based on the system used at Kranzberg Forest in Germany (Matyssek et al., 2013b; Nunn et al., 2002;

Werner and Fabian, 2002). The size of each plot was 5.5 m × 7.2 m, and the height was 5.5 m. The soil type was brown forest soil (Cambisol).

During the first year of fumigation, O_3 exposure caused a significant decrease in the net photosynthetic rate of Siebold's beech (Hoshika et al., 2012b; Watanabe et al., 2013a,b). No significant reduction in the net photosynthetic rate was observed in the deciduous oak (Watanabe et al., 2013a). A similar trend between Siebold's beech and oak was found in photosynthesis and stem diameter growth during the second year of fumigation. This trend in sensitivity to O_3, sensitivity of beech and tolerance by oak, is consistent with previous reports based on chamber experiments (Table 17.1).

Ozone exposure in the free-air system also induced slower stomatal responses to changes in the light intensity in Siebold's beech (Hoshika et al., 2012b). A weaker ability to close stomata, that is, sluggish stomatal control, was observed under elevated O_3 (Figure 17.3). This phenomenon may reduce the ability of plants to adjust to fluctuating light environments.

17.3.2 Estimation of Stomatal Ozone Uptake

Assessment of O_3 risks to forest trees based on the stomatal flux approach is currently underway in Japan (Izuta, 2012). Species-specific parameterization of the stomatal conductance model is needed (Cieslik, 2013). Hoshika et al. (2012a) reported a comparison of the maps for accumulative O_3 exposure (e.g. AOT40) and cumulative O_3 uptake for Zelkova in Japan and indicated that stomatal closure induced by high vapour-pressure deficit causes decoupling of stomatal O_3 uptake from high O_3 exposure. In the free-air O_3 fumigation experiment mentioned above, leaf-level stomatal conductance has been measured for the relevant tree species: Siebold's beech, deciduous oak, and white birch (e.g. Hoshika et al., 2012c, 2013). Assessment of whole-tree O_3 flux estimated by sap-flow measurement is now proposed at our free-air O_3 fumigation site, as has been done for European beech forest (e.g. Nunn et al., 2007). The leaf- and tree-scale measurements will assist in developing the risk assessment using the stomatal flux-based approach in Japan.

17.3.3 Flux Research and Modelling

A flux-based approach for risk assessment began a few years ago for crops (mainly rice) and trees, although the concept of O_3 uptake via stomata was used in physiological studies in the late 1970s (e.g. Feng et al., 2008a). Experiments with a free-air canopy O_3 fumigation system have found that the seasonal photosynthetic carbon gain of adult European beech trees (*Fagus sylvatica*) decreases under elevated O_3 at the canopy level (Kitao et al., 2009, 2012). A larger-scaled approach is needed to determine the effect of O_3 on photosynthetic carbon gain at the forest level, based on the time-integrated stomatal O_3 flux.

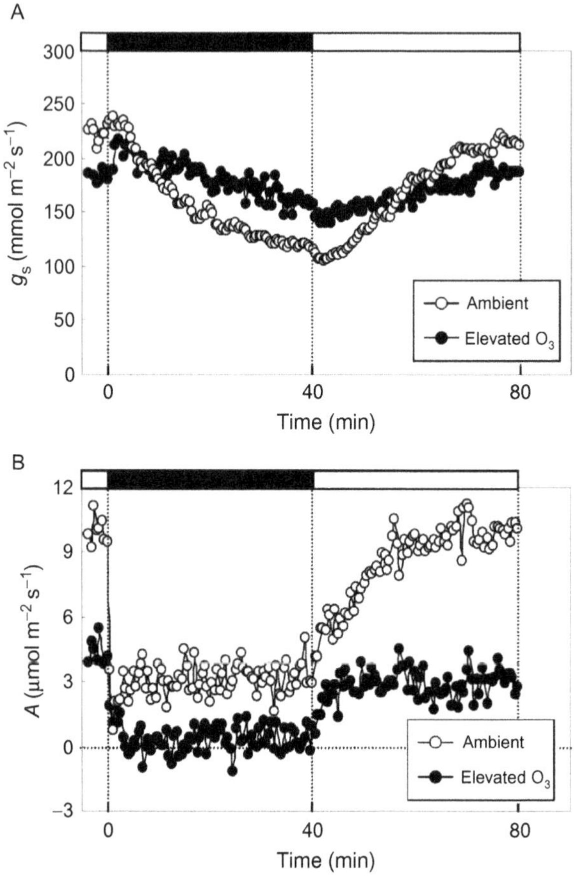

FIGURE 17.3 Example of dynamic response of (A) stomatal conductance and (B) net photosynthesis to changes in light intensity (white horizontal bars: PPFD = 1500 μmol m^{-2} s^{-1}; black horizontal bars: PPFD = 100 μmol m^{-2} s^{-1}) for Siebold's beech observed under elevated O_3 (closed circle, measured on 14 October 2011) and ambient conditions (open circle, measured on 11 October 2011).

In Japan, there are various types of forests, distributed in different climate zones. Since 2000, the Forestry and Forest Products Research Institute (FFPRI) has maintained a research group (Flux Observation Network: FFPRI-FluxNet) to undertake observational research, in particular the measurement of CO_2 flux using micrometeorological techniques at six research sites throughout Japan, looking at various forest types in different climate zones. Currently, there are five continuing sets of observations. Three of five sites are in the cool-temperate zone with mixed broad-leaves tree species of (1) white birch, deciduous oak and kalopanax (*Kalopanax septemlobos*), (2) Siebold's beech and (3) Japanese red pine, while two other sites are in the warm-temperate zone with mixed evergreen broad-leaved tree species of

(4) Konara oak (*Quercus serrata*) and ilex (*Ilex pedunculosa*) and (5) Sugi-cedar (http://www2.ffpri.affrc.go.jp/labs/flux/index.html). These sites look at the transport of CO_2 and energy between the forest and the atmosphere, based mainly on eddy covariance measurements (Yasuda et al., 2012).

Since 2011, at three of the five flux sites, in which broadleaf trees are predominant, atmospheric O_3 concentrations above the canopies have been monitored to facilitate estimates of the O_3 uptake, that is, time-integrated O_3 flux via stomata. The three forest types are: a boreal deciduous broadleaf forest consisting predominantly of white birch and oak, a cool-temperate deciduous broadleaf forest of Siebold's beech and a warm-temperate broadleaf forest consisting mainly of Konara oak and ilex. By applying the Penman–Monteith scheme for evapotranspiration (Cieslik, 2013), canopy-scale stomatal resistances can be derived for the estimation of stomatal O_3 fluxes (Gerosa et al., 2005; Kitao et al., 2013). Based on these approaches, the relationship between CO_2 and O_3 fluxes are currently being investigated at forest level in various types of Japanese broad-leaved forests.

17.4 CONCLUSIONS

Although many East and Southeast Asian countries still have a problem with SO_2, the O_3 pollution is already considered to be more serious than in Europe or North America. However, the actual situation with regard to O_3 concentration is not clarified because of the low number of monitoring sites for the whole of East and Southeast Asia.

Although emissions of precursors of O_3 in many developed countries are keeping constant or decreasing, those in developing countries are continuously increasing. Very rapid economic growth in China especially has induced a large growth in emissions, above what was predicted by several forecasts (IPCC, 2007; Ohara et al., 2007; The Royal Society, 2008). Therefore, it is hard to avoid the increasing O_3 concentration in East Asia, especially in China and neighbouring countries such as Japan and Korea, in the near future.

The best way to reduce O_3 damage to forests is, needless to say, reduction in emissions of O_3 precursors. The most part of O_3 precursors are emitted from burning of fossil fuels in power plants, industries and vehicles. However, it is very hard to avoid the use of fossil fuel for economic growth in developing countries. Sustainable energies (e.g. wind power, solar power) do not have the capacity to shift from fossil fuel energy completely at the present time even in developed countries. Therefore, efficient use of fossil fuel and technical remove of O_3 precursors from a burning of fossil fuel are better ways to proceed until the further development of sustainable energies. On the other hand, reducing wildfire is important not only for protecting forest itself, but also for reducing emission of O_3 precursors (Sofiev, 2013). In fact,

Tanimoto et al. (2008) indicated Siberian wildfire potentially has an impact on the concentration of O_3 in Japan.

Experimental studies of the effects of O_3 on forest tree species in most countries in East and Southeast Asia have only just begun or are not yet under way. The information needed for accurate O_3 risk assessment, that is, the exposure–response relationship, is not known except for Japan. Even in Japan, which has the longest history of the study of O_3 effects, advanced methodologies such as the stomatal flux-based approach, use of a free-air O_3 fumigation system and stand level studies, have started recently. Also, there is no project with an integrated model approach for predicting future impacts of O_3 on forest tree species (Bytnerowicz et al., 2013; Tuovinen et al., 2013). Although information on tree response to elevated O_3 obtained from studies in Europe and North America is useful, tree species, the O_3 regime and climatic conditions in East Asia are quite different. In particular, tree species in tropical rain forest, tropical seasonal forest and temperate evergreen forest are typical in tropical and warm-temperate Asia, with different ecophysiological traits as compared to the other regions such as Europe due to the large differences in temperature and precipitation amount. Therefore evaluation of the O_3 sensitivity for these species grown in Asia is highly required. For the practice of forestry, this information is valuable for the selection of tolerant tree species for plantations, although this is a stopgap measure. Also, the information on O_3 sensitivity of native species in forest vegetation is important from a viewpoint of protection of biodiversity, especially in Southeast Asia.

To maintain ecosystem services of forest such as the carbon sink, conservation of biodiversity and land allocation (Knoke and Hahn, 2013; Nijnik and Miller, 2013), it is vital to develop our understanding of the effect of O_3 on East and Southeast Asian vegetation. The limited experimental facilities in many countries in East Asia mean that it is difficult to accurately assess the impact of O_3. At the present time, Japan and China have a potential to promote experimental studies on the O_3 impact on tree species. In parallel to the study in each country, international collaborations for the study on effects of O_3 on tree species is essential to develop our understanding of O_3 risk at the pan-Asian scale, although there are no projects at present. The framework of EANET will play a very important role in expanding the international cooperative research within East and Southeast Asia, as well as between East and Southeast Asia and other regions with a problem of air pollution (e.g. Kozovits and Bustamante, 2013; Laakso et al., 2013; Pandey et al., 2013).

ACKNOWLEDGEMENT

This contribution is partly supported by the Japanese Ministry of the Environment (B-1105) and the Japanese Society of Promotion of Science (21114008, 23380078, 24710027 and 24780239).

REFERENCES

Adams, J.M., Woodward, F.I., 1989. Patterns in tree species richness as a test of the glacial extinction hypothesis. Nature 339, 699–701.

Akhtar, N., Yamaguchi, M., Inada, H., Hoshino, D., Kondo, T., Izuta, T., 2010a. Effects of ozone on growth, yield and leaf gas exchange rates of two Bangladeshi cultivars of wheat (*Triticum aestivum* L.). Environ. Pollut. 158, 1763–1767.

Akhtar, N., Yamaguchi, M., Inada, H., Hoshino, D., Kondo, T., Fukami, M., Funada, R., Izuta, T., 2010b. Effects of ozone on growth, yield and leaf gas exchange rates of four Bangladeshi cultivars of rice (*Oryza sativa* L.). Environ. Pollut. 158, 2970–2976.

Asian Development Bank and the Clean Air Initiative for Asian Cities Center, 2006. Country Synthesis Report on Urban Air Quality Management: Indonesia. Asian Development Bank, Manila, Philippines.

Bytnerowicz, A., Fenn, M., McNulty, S., Yuan, F., Pourmokhtarian, A., Driscoll, C., Meixner, T., 2013. Interactive effects of air pollution and climate change on forest ecosystems in the United States—current understanding and future scenarios. Dev. Environ. Sci. 13, 333–369.

Chang, S.-C., Lee, C.-T., 2006. Ozone variations through vehicle emissions reductions based on air quality monitoring data in Taipei City, Taiwan, from 1994 to 2003. Atmos. Environ. 40, 3513–3526.

Cho, S.Y., 2007. Effect of surface ozone on vegetation in Korea—preliminary study on early foliage of *Robinia pseudo-acacia*. J. Korean Soc. Atmos. Environ. 23, 29–38 (In Korean with English summary).

Choi, D.S., Toda, H., 2008. Air pollution and its effects on the pine species in Korea. Jpn. J. Atmos. Environ. 43, 307–314 (In Japanese).

Cieslik, S., 2013. Gaseous exchange between atmosphere and forests. Dev. Environ. Sci. 13, 19–36.

Dizengremel, P., Jolivet, Y., Tuzet, A., Ranieri, A., Le Thiec, D., 2013. Integrative Leaf-Level Phytotoxic Ozone Dose Assessment for Forest Risk Modelling. Dev. Environ. Sci. 13, 267–288.

Feng, Z., Kobayashi, K., Ainsworth, E.A., 2008a. Impact of elevated ozone on growth, physiology and yield of wheat (*Triticum aestivum* L.): a meta-analysis. Glob. Chang. Biol. 14, 2696–2708.

Feng, Z.Z., Zeng, H.Q., Wang, X.K., Zheng, Q.W., Feng, Z.W., 2008b. Sensitivity of *Metasequoia glyptostroboides* to ozone stress. Photosynthetica 46, 463–465.

Feng, Z.Z., Niu, J.F., Zhang, W.W., Wang, X.K., Yao, F.F., Tian, Y., 2011a. Effects of ozone exposure on sub-tropical evergreen *Cinnamomum camphora* seedlings grown in different nitrogen loads. Trees 25, 617–625.

Feng, Z.Z., Pang, J., Kobayashi, K., Zhu, J.G., Ort, D.R., 2011b. Differential responses in two varieties of winter wheat to elevated ozone concentration under fully open-air field conditions. Glob. Chang. Biol. 17, 580–591.

Fujinuma, Y., Furukawa, A., Totsuka, T., Tazaki, T., 1987. Uptake of ozone by various street trees. Environ. Control Biol. 25, 31–40.

Furukawa, A., 1991. Inhibition of photosynthesis of *Populus euramericana* and *Helianthus annuus* by SO_2, NO_2 and O_3. Ecol. Res. 6, 79–86.

Furukawa, A., Totsuka, T., Katase, M., Ushijima, T., 1983. Inhibition of photosynthesis of poplar species by ozone. J. Jpn. For. Soc. 65, 321–326.

Gerosa, G., Marcello Vitale, M., Finco, A., Manes, F., Denti, A.B., Cieslik, S., 2005. Ozone uptake by an evergreen Mediterranean Forest (*Quercus ilex*) in Italy. Part I: micrometeorological flux measurements and flux partitioning. Atmos. Environ. 39, 3255–3266.

Ghim, Y.S., Chang, T.S., 2000. Characteristics of ground-level ozone distributions in Korea for the period of 1990-1995. J. Geophys. Res. 105 (D7), 8877–8890.

Han, S.-H., Lee, J.-C., Jang, S.-S., Hong, Y.-P., 2002. Comparison of the ozone sensitivity of four *Betula* species based on carbon allocation and partitioning patterns. J. Korean For. Soc. 91, 449–456 (In Korean with English summary).

Hoshika, Y., Paoletti, E., Omasa, K., 2012a. Parameterization of *Zelkova serrata* stomatal conductance model to estimate stomatal ozone uptake in Japan. Atmos. Environ. 55, 271–278.

Hoshika, Y., Wanatabe, M., Inada, N., Koike, T., 2012b. Ozone-induced stomatal sluggishness develops progressively in Siebold's beech (*Fagus crenata*). Environ. Pollut. 166, 152–156.

Hoshika, Y., Watanabe, M., Inada, N., Koike, T., 2012c. Modelling of stomatal conductance for estimating ozone uptake of *Fagus crenata* under experimentally enhanced free-air ozone exposure. Water Air Soil Pollut. 223, 3893–3901.

Hoshika, Y., Watanabe, M., Inada, N., Koike, T., 2013. Model-based analysis of avoidance of ozone stress by stomatal closure in Siebold's beech (*Fagus crenata*). Ann. Bot. (in press).

IPCC, 2007. Climate change 2007: impacts, adaptation and vulnerability. In: Parry, M.L., Canziani, O.F., Palutikof, J.P., van der Linden, P.J., Hanson, C.E. (Eds.), Contribution of Working Group II to the Fourth Assessment. Report of the Intergovernmental Panel on Climate Change. Cambridge University Press, Cambridge, UK.

Ishii, S., Marshall, F.M., Bell, J.N.B., Abdullah, A.M., 2004. Impact of ambient air pollution on locally grown rice cultivars (*Oryza sativa* L.) in Malaysia. Water Air Soil Pollut. 154, 187–201.

Izuta, T., 1998. Ecophysiological responses of Japanese forest tree species to ozone, simulated acid rain and soil acidification. J. Plant Res. 111, 471–480.

Izuta, T., 2003. Air pollution impacts on vegetation in Japan. In: Emberson, L., Ashmore, M., Murray, F. (Eds.), Air Pollution Impacts on Crops and Forests: A Global Assessment. Imperial College Press, London, pp. 89–101.

Izuta, T., 2006. Plants and Environmental Stresses. Corona Publishing, Tokyo (In Japanese).

Izuta, T., 2012. Evaluation of ozone effects on plant based on leaf ozone uptake. J. Jpn. Soc. Atmos. Environ. 47, A12–A15 (In Japanese).

Izuta, T., Umemoto, M., Horie, K., Aoki, M., Totsuka, T., 1996. Effects of ambient levels of ozone on growth, gas exchange rates and chlorophyll contents of *Fagus crenata* seedlings. J. Jpn. Soc. Atmos. Environ. 31, 95–105.

Karlsson, P.E., Uddling, J., Braun, S., Broadmeadow, M., Elvira, S., Gimeno, B.S., Le Thiec, D., Oksanen, E., Vandermeiren, K., Wilkinson, M., Emberson, L., 2004. New critical levels for ozone effects on young trees based on AOT40 and stimulated cumulative leaf uptake of ozone. Atmos. Environ. 38, 2283–2294.

Kawaguchi, K., Hoshika, Y., Watanabe, M., Koike, T., 2012. Ecophysiological responses of northern birch forests to the changing atmospheric CO_2 and O_3 concentration. Asian J. Atmos. Environ. 6, 192–205.

Kim, D.-H., Han, S.-H., Lee, J.C., 2010. Ozone sensitivity of physiological indicators for stress evaluation in four families of *Quercus aliena* Blume. J. Korean For. Soc. 99, 878–884 (In Korean with English summary).

King, J., Liu, L., Aspinwall, M., 2013. Forest ecosystem response to interacting elevated atmospheric CO_2 and tropospheric O_3: a synthesis of experimental evidence. Dev. Environ. Sci. 13, 179–208.

Kitao, M., Löw, M., Heerdt, C., Grams, T.E.E., Häberle, K.-H., Matyssek, R., 2009. Effects of chronic elevated ozone exposure on gas exchange responses of adult beech trees (*Fagus sylvatica*) as related to the within-canopy light gradient. Environ. Pollut. 157, 537–544.

Kitao, M., Winkler, J.B., Löw, M., Nunn, A.J., Daniel, K., Häberle, K.-H., Reiter, I.M., Matyssek, R., 2012. How closely does stem growth of adult beech (*Fagus sylvatica*) relate to net carbon gain under experimentally enhanced ozone stress? Environ. Pollut. 166, 108–115.

Kitao, M., Komatsu, M., Hoshika, Y., Yazaki, K., Yoshimura, K., Fujii, S., Miyama, T., Kominami, Y., 2013. Seasonal ozone uptake by a warm-temperate mixed deciduous and evergreen broadleaf forest in western Japan estimated by the Penman-Monteith approach combined with a photosynthesis-dependent stomatal model. Environ. Pollut. (in press).

Knoke, T., Hahn, A., 2013. Global change and the role of forests in future land-use systems. Dev. Environ. Sci. 13, 569–587.

Kohno, Y., Matsumura, H., Ishii, T., Izuta, T., 2005. Establishing critical levels of air pollutants for protecting East Asian vegetation, a challenge. In: Omasa, K., Nouchi, I., De Kok, L.J. (Eds.), Plant Responses to Air Pollution and Global Change. Springer-Verlag, Tokyo, pp. 243–250.

Koike, T., 2012. Forest history of Japan with changing environment. Stud. Mater. Ośr. Kult. Leśn. 11, 109–130.

Koike, T., Mao, Q., Inada, N., Kawaguchi, K., Hoshika, Y., Kita, K., Watanabe, M., 2012. Growth and photosynthetic responses of cuttings of a hybrid larch (*Larix gmelinii* var. *japonica* × *L. kaempferi*) to elevated ozone and/or carbon dioxide. Asian J. Atmos. Environ. 6, 104–110.

Kozovits, A.R., Bustamante, M.M.C., 2013. Land use change, air pollution, and climate change and vegetation responses in Latin America. Dev. Environ. Sci. 13, 411–427.

Kume, A., Numata, S., Watanabe, K., Honoki, H., Nakajima, H., Ishida, M., 2009. Influence of air pollution on the mountain forests along the Tateyama-Kurobe Alpine route. Ecol. Res. 24, 821–830.

Laakso, L., Beukes, J.P., Zy, P.G.V., Vakkari, V., Lapuschagne, C., Tuovinen, J.-P., 2013. Ozone concentrations and their potential impacts on vegetation in southern Africa. Dev. Environ. Sci. 13, 429–450.

Li, L., Chen, C., Huang, C., Huang, H., Zhang, G., Wang, Y., Chen, M., Wang, H., Chen, Y., Streets, D.G., Fu, J., 2011. Ozone sensitivity analysis with the MM5-CMAQ modelling system for Shanghai. J. Environ. Sci. 23, 1150–1157.

Matsuda, K., Watanabe, I., Wingpud, V., Theramongkol, P., Khummongkol, P., Wangwongwatana, S., Totsuka, T., 2005. Ozone dry deposition above a tropical forest in the dry season in northern Thailand. Atmos. Environ. 39, 2571–2577.

Matsumura, H., Aoki, H., Kohno, Y., Izuta, T., Totsuka, T., 1996. Effects of ozone on dry weight growth and gas exchange rate of Japanese cedar, Japanese cypress and Japanese zelkova seedlings. J. Jpn. Soc. Atmos. Environ. 31, 247–261 (In Japanese).

Matsumura, H., Kobayashi, T., Kohno, Y., 1998. Effects of ozone and/or simulated acid rain on dry weight and gas exchange rates of Japanese cedar, Nikko fir, Japanese white birch and Japanese zelkova seedlings. J. Jpn. Soc. Atmos. Environ. 33, 16–35 (In Japanese).

Matsumura, H., Mikami, C., Sakai, Y., Murayama, K., Izuta, T., Yonekura, T., Miwa, M., Kohno, Y., 2005. Impacts of elevated O_3 and/or CO_2 on growth of *Betula platyphylla*, *Betula ermanii*, *Fagus crenata*, *Pinus densiflora* and *Cryptomeria japonica* seedlings. J. Agric. Meteorol. 60, 1121–1124.

Matyssek, R., Clarke, N., Cudlin, P., Mikkelsen, T.N., Tuovinen, J.-P., Wieser, G., Paoletti, E., 2013a. Climate change, air pollution and global challenges: understanding and perspective from forest research. Dev. Environ. Sci. 13, 3–16.

Matyssek, R., Wieser, G., Fleischmann, F., Grünhage, L., 2013b. Ozone Research, quo vadis? Lessons from the free-air canopy fumigation experiment at Kranzberg Forest. Dev. Environ. Sci. 13, 103–129.

Mikkelsen, T.N., Clarke, N., Danielewska, A., Fischer, R., 2013. Towards supersites in forest ecosystem monitoring and research. Dev. Environ. Sci. 13, 475–495.

Myers, N., Mittermeier, R.A., Mittermeier, C.G., da Fonseca, G.A.B., Kent, J., 2000. Biodiversity hotspots for conservation priorities. Nature 403, 853–858.

Nakaji, T., Izuta, T., 2001. Effects of ozone and/or excess soil nitrogen on growth, needle gas exchange rates and Rubisco contents of *Pinus densiflora* seedlings. Water Air Soil Pollut. 130, 971–976.

Nakaji, T., Kobayashi, T., Kuroha, M., Omori, K., Matsumoto, Y., Yonekura, T., Watanabe, K., Utriainen, J., Izuta, T., 2004. Growth and nitrogen availability of red pine seedlings under high nitrogen load and elevated ozone. Water Air Soil Pollut. Focus 4, 277–287.

Nijnik, M., Miller, D., 2013. Targeting sustainable provision of forest ecosystem services with special focus on carbon sequestration. Dev. Environ. Sci. 13, 547–568.

Nouchi, I., 2001. Plant Responses to Changing Environment. Yokendo Co. Ltd., Tokyo, p. 391 (In Japanese).

Nouchi, I., Odaira, T., Sawada, T., Oguchi, K., Komeiji, T., 1973. Plant ozone injury symptoms. J. Jpn. Soc. Air Pollut. 8, 113–119 (In Japanese with English summary).

Novriyanti, E., 2012. Chemo-ecophysiological study on growth and defense responses of the afforestation species eucalypts and acacia under a changing environment (Ph.D. dissertation). Agriculture School of Hokkaido University.

Nunn, A.J., Reiter, I.M., Häberle, K.-H., Werner, H., Langebartels, C., Sandermann, H., Heerdt, C., Fabinan, P., Matyssek, R., 2002. "Free-air" ozone canopy fumigation in an old-growth mixed forest: concept and observations in beech. Phyton 42, 105–119.

Nunn, A.J., Wieser, G., Metzger, U., Löw, M., Wipfler, P., Häberle, K.-H., Matyssek, R., 2007. Exemplifying whole-plant ozone uptake in adult forest trees of contrasting species and site conditions. Environ. Pollut. 146, 629–639.

Ohara, T., Akimoto, H., Kurokawa, J., Horii, N., Yamaji, K., Yan, X., Hayasaka, T., 2007. An Asian emission inventory of anthropogenic emission sources for the period 1980–2020. Atmos. Chem. Phys. 7, 4419–4444.

Oksanen, E., Saleen, A., 1999. Ozone exposure results in various carry-over effects and prolonged reduction in biomass in Birch (*Betula pendula* Roth). Plant Cell Environ. 22, 1401–1411.

Oksanen, E., Keski-Saari, S., Kontunen-Soppela, S., Keinänen, M., 2013. Metabolomics and transcriptomics increase our understanding about defence responses and genotypic differences of northern deciduous trees to elevating ozone, CO_2 and climate warming. Dev. Environ. Sci. 13, 309–329.

Omasa, K., 1990. Study on changes in stomata and their surrounding cell using a non-destructive light microscope system: responses to air pollutants. J. Agric. Meteorol. 45, 251–257.

Pandey, V., Oksanen, E., Singh, N., Sharma, C., 2013. Impacts of air pollution and climate change: implications for India. Dev. Environ. Sci. 13, 391–410.

Prather, M., Ehhalt, D., Dentener, F., Derwent, R., Dlugokencky, E., Holland, E., Isaksen, I., Katima, J., Kirchhoff, V., Matson, P., Midgley, P., Wang, M., 2001. Atmospheric chemistry and greenhouse gases. In: Houghton, J.T., Ding, Y., Griggs, D.J., Noguer, M., van der Linden, P.J.,

Dai, X., Maskell, K., Johnson, C.A. (Eds.), The Scientific Basis. Climate Change 2001: The Scientific Basis. Cambridge University Press, Cambridge, UK, and New York, pp. 239–287.

Pregitzer, K.S., Talhelm, A.F., 2013. Belowground carbon cycling at Aspen FACE: dynamic responses to CO_2 and O_3 in developing forests. Dev. Environ. Sci. 13, 209–226.

Shan, Y., Izuta, T., Aoki, M., Totsuka, T., 1997. Effects of O_3 and soil acidification, alone and in combination, on growth, gas exchange rate and chlorophyll content of red pine seedlings. Water Air Soil Pollut. 97, 355–366.

Sitch, S., Cox, P.M., Collins, W.J., Huntingford, C., 2007. Indirect radiative forcing of climate change through ozone effects on the land-carbon sink. Nature 448, 791–794.

Sofiev, M., 2013. Wild-land fires: features, means of monitoring and plume modelling, estimated impact on atmospheric composition and climate. Dev. Environ. Sci. 13, 445–472.

Takagi, K., Ohara, T., 2003. Estimation of ozone impact on plants by damage functions in the Kanto area. J. Jpn. Soc. Atmos. Environ. 38, 205–216 (In Japanese with English summary).

Takeda, M., Aihara, K., 2007. Effects of ambient ozone concentrations on Beech (*Fagus crenata*) seedlings in the Tanzawa Mountains, Kanagawa Prefecture, Japan. J. Jpn. Soc. Atmos. Environ. 42, 107–117 (In Japanese with English summary).

Tanimoto, H., Sawa, Y., Matsueda, H., Uno, I., Ohara, T., Yamaji, K., Kurokawa, J., Yonemura, S., 2005. Significant latitudinal gradient in the surface ozone spring maximum over East Asia. Geophys. Res. Lett. 32, L21805.

Tanimoto, H., Matsumoto, K., Uematsu, M., 2008. Ozone-CO correlations in Siberian wildfire plumes observed at Rishiri Island. SOLA 4, 65–68.

The Royal Society, 2008. Ground-Level Ozone in the 21st Century: Future Trends, Impacts and Policy Implications. Science Policy Report 15/08. The Royal Society, London.

Tuovinen, J.-P., Hakola, H., Karlsson, P.E., Simpson, D., 2013. Air pollution risks to northern European forests in a changing climate. Dev. Environ. Sci. 13, 77–99.

Wang, H., Zhou, L., Tang, X., 2006. Ozone concentrations in rural regions of the Yangtze Delta in China. J. Atmos. Chem. 54, 255–265.

Watanabe, M., 2007. Growth responses of 6 Japanese forest tree species to ozone under different nitrogen loads (Doctoral dissertation). United Graduate School of Agricultural Science. Tokyo University of Agriculture and Technology, Fuchu Tokyo, Japan.

Watanabe, M., Yonekura, T., Honda, Y., Yoshidome, M., Nakaji, T., Izuta, T., 2005. Effects of ozone and soil water stress, singly and in combination, on leaf antioxidative systems of *Fagus crenata* seedlings. J. Agric. Meteorol. 60, 1105–1108.

Watanabe, M., Yamaguchi, M., Iwasaki, M., Matsuo, N., Naba, J., Tabe, C., Matsumura, H., Kohno, Y., Izuta, T., 2006. Effects of ozone and/or nitrogen load on the growth of *Larix kaempferi*, *Pinus densiflora* and *Cryptomeria japonica* seedlings. J. Jpn. Soc. Atmos. Environ. 41, 320–334.

Watanabe, M., Yamaguchi, M., Tabe, C., Iwasaki, M., Yamashita, R., Funada, R., Fukami, M., Matsumura, H., Kohno, Y., Izuta, T., 2007. Influences of nitrogen load on the growth and photosynthetic responses of *Quercus serrata* seedlings to O_3. Trees 21, 421–432.

Watanabe, M., Yamaguchi, M., Matsumura, H., Kohno, Y., Izuta, T., 2008. Effects of ozone on growth and photosynthesis of *Castanopsis sieboldii* seedlings grown under different nitrogen loads. J. Agric. Meteorol. 24, 143–155.

Watanabe, M., Matsuo, N., Yamaguchi, M., Matsumura, H., Kohno, Y., Izuta, T., 2010a. Risk assessment of ozone impact on the carbon absorption of Japanese representative conifers. Eur. J. For. Res. 129, 421–430.

Watanabe, M., Umemoto-Yamaguchi, M., Koike, T., Izuta, T., 2010b. Growth and photosynthetic response of *Fagus crenata* seedlings to ozone and/or elevated carbon dioxide. Landscape Ecol. Eng. 6, 181–190.

Watanabe, M., Yamaguchi, M., Matsumura, H., Kohno, Y., Izuta, T., 2012. Risk assessment of ozone impact on *Fagus crenata* in Japan: consideration of atmospheric nitrogen deposition. Eur. J. For. Res. 131, 475–484.

Watanabe, M., Hoshika, Y., Inada, N., Wang, X., Mao, Q., Koike, T., 2013a. Photosynthetic traits of Siebold's beech and oak saplings grown under free air ozone exposure in northern Japan. Environ. Pollut. 174, 50–56.

Watanabe, M., Hoshika, Y., Inada, N., Koike, T., 2013b. Canopy carbon budget of Siebold's beech (*Fagus crenata*) sapling under free air ozone exposure. Environ. Pollut. (in press).

Werner, H., Fabian, P., 2002. Free-air fumigation of mature trees: a novel system for controlled ozone enrichment in grown-up beech and spruce canopies. Environ. Sci. Pollut. Res. 9, 117–121.

Wittig, V.E., Ainsworth, E.A., Long, S.P., 2007. To what extent do current and projected increases in surface ozone affect photosynthesis and stomatal conductance of trees? A meta-analytic review of the last 3 decades of experiments. Plant Cell Environ. 30, 1150–1162.

Wittig, V.E., Ainsworth, E.A., Naidu, S.L., Karnosky, D.F., Long, S.P., 2009. Quantifying the impact of current and future tropospheric ozone on tree biomass, growth, physiology and biochemistry: a quantitative meta-analysis. Glob. Chang. Biol. 15, 396–424.

Yamaguchi, M., Watanabe, M., Iwasaki, M., Tabe, C., Matsumura, H., Kohno, Y., Izuta, T., 2007. Growth and photosynthetic responses of *Fagus crenata* seedlings to O_3 under different nitrogen loads. Trees 21, 707–718.

Yamaguchi, M., Watanabe, M., Matsumura, H., Kohno, Y., Izuta, T., 2010. Effects of ozone on nitrogen metabolism in the leaves of *Fagus crenata* seedlings under different soil nitrogen loads. Trees 24, 175–184.

Yamaguchi, M., Watanabe, M., Matsumura, H., Kohno, Y., Izuta, T., 2011. Experimental studies on the effects of ozone on growth and photosynthetic activity of Japanese forest tree species. Asian J. Atmos. Environ. 5, 65–78.

Yamaji, K., Ohara, T., Uno, I., Kurokawa, J., Pochanart, P., Akimoto, H., 2008. Future prediction of surface ozone over east Asia using Models-3 community multiscale air quality modelling system and regional emission inventory in Asia. J. Geophys. Res. 113, D08306.

Yasuda, Y., Saito, T., Hoshino, D., Ono, K., Ohtani, Y., Mizoguchi, Y., Morisawa, T., 2012. Carbon balance in a cool-temperate deciduous forest in northern Japan: seasonal and interannual variations, and environmental controls of its annual balance. J. For. Res. 17, 253–267.

Yonekura, T., Dokiya, Y., Fukami, M., Izuta, T., 2001a. Effects of ozone and/or soil water stress on growth and photosynthesis of *Fagus crenata* seedlings. Water Air Soil Pollut. 130, 965–970.

Yonekura, T., Honda, Y., Oksanen, E., Yoshidome, M., Watanabe, M., Funada, R., Koike, T., Izuta, T., 2001b. The influences of ozone and soil water stress, singly and in combination, on leaf gas exchange rates, leaf ultrastructural characteristics and annual ring width of *Fagus crenata* seedlings. J. Jpn. Soc. Atmos. Environ. 36, 333–351.

Yonekura, T., Yoshidome, M., Watanabe, M., Honda, Y., Ogiwara, I., Izuta, T., 2004. Carry-over effects of ozone and water stress on leaf phenological characteristics and bud frost hardiness of *Fagus crenata* seedlings. Trees 18, 581–588.

Zhang, W.W., Feng, Z.Z., Wang, X.K., Niu, J.F., 2012. Responses of native broadleaved woody species to elevated ozone in subtropical China. Environ. Pollut. 163, 149–157.

Zheng, Y., Shimizu, H., 2003. Air pollution impacts on vegetation in China. In: Emberson, L., Ashmore, M., Murray, F. (Eds.), Air Pollution Impacts on Crops and Forests: A Global Assessment. Imperial College Press, London, pp. 123–144.

Chapter 18

Impacts of Air Pollution and Climate Change on Plants: Implications for India

Vivek Pandey*,[1], Elina Oksanen[†], Nandita Singh* and Chhemendra Sharma[‡]

*Plant Ecology and Environmental Science, National Botanical Research Institute (CSIR-NBRI), Lucknow, India

[†]Department of Biology, University of Eastern Finland, P. O. Box 111, Joensuu, Finland

[‡]Radio and Atmospheric Sciences Division, National Physical Laboratory, New Delhi, India

[1]Corresponding author: e-mail: v.pandey@nbri.res.in

Chapter Outline

18.1 INTRODUCTION

India is the world's second most populous country. In the year 2000, India's population was 1.063 billion, which is expected to reach 1.65 billion by the year 2050 at an annual growth rate of 1.3% (www.os-connect.com). India covers 2.4% (ca. 3 million km^2) of the world's land area, of which 53% is arable (data.worldbank.org). India produced 245 million tons of grain per year in 2010–2011 (Press Information Bureau; http://pib.nic.in/newsite/erelease.aspx). The annual food-grain production is expected to be 253 million tons for 2011–2012. However, the grain production is threatened by rapid economic development and an ever-growing population which is putting enormous pressure on the environment, infrastructure and the country's natural

Developments in Environmental Science, Vol. 13. http://dx.doi.org/10.1016/B978-0-08-098349-3.00018-9

resources, including land. The air pollution problems in India have already posed serious concern for human and plant health. The Indo-Gangetic Plains (IGP) of Northern India, one of the most fertile regions of the world, are considered to be the 'food basket' of the country. Unfortunately, this region (with greatest population density) has also become one of the most polluted and a 'hot spot' for air pollution (Emberson et al., 2009). The main factors contributing to air pollution in IGP and urban areas include emissions from vehicles, thermal power plants, industries and agriculture. The major pollutants include SO_2, NO_x, O_3 and particulate matter (both mineral and carbonaceous). The majority of tropospheric O_3 results from photochemical reactions involving methane, volatile organic compounds and NO_x, which largely originates from anthropogenic emissions. Due to rapid industrial development and urbanisation in the past decades, tropospheric O_3 has become one of the most important problems in large areas of India, affecting human health and plant production (IPCC, 2007; Roy et al., 2008). O_3-caused loss of global primary productivity has been predicted to become substantial according to modelling predictions (Fowler et al., 2008; Sitch et al., 2007). The air quality problem is further complicated by the need to control increasing greenhouse gas (GHG) emissions (CO_2, CH_4 and N_2O).

India is a large and fast developing country with nearly two-thirds of the population living in rural areas and depending directly on the climate sensitive sectors such as agriculture, fisheries and forests. The projected climate change has implications for food production, water supply, biodiversity and livelihoods (NATCOM, 2004). Thus, India has a significant stake in scientific advancement as well as an international understanding to promote mitigation of climate change impacts and at the same time adaptation to changing climate. This requires improved scientific understanding, capacity building, networking and broad consultation processes. The aim of this chapter is to provide an overview of the current and future threats of air pollution to vegetation in India, including possible impacts on forest trees and ecosystems.

18.2 INDIA'S FOREST COVER AND FOREST TYPES

India, a 'mega biodiversity' country, has a forest cover of about 692,027 km^2 which works out to be 21% of the total geographical area of the country (FSI, 2011). The forest cover of the country has been classified on the basis of tree canopy density into pre-defined classes: very dense forest (VDF, canopy cover > 70%), moderately dense forest (MDF, canopy cover between 40% and 70%) and open forest (OF, canopy cover between 10 and 40%). The area covered by VDF is 83,471 km^2 (3%), that with MDF is 370,736 km^2 (10%) and OF is 287,820 km^2 (9%). The distribution of forest cover and density classes is shown in Figure 18.1. Forests can be grouped into 15 major types (Table 18.1). Two dominant forest types are tropical deciduous forest and tropical moist deciduous forest. Nearly 20,000 villages are classified as forest

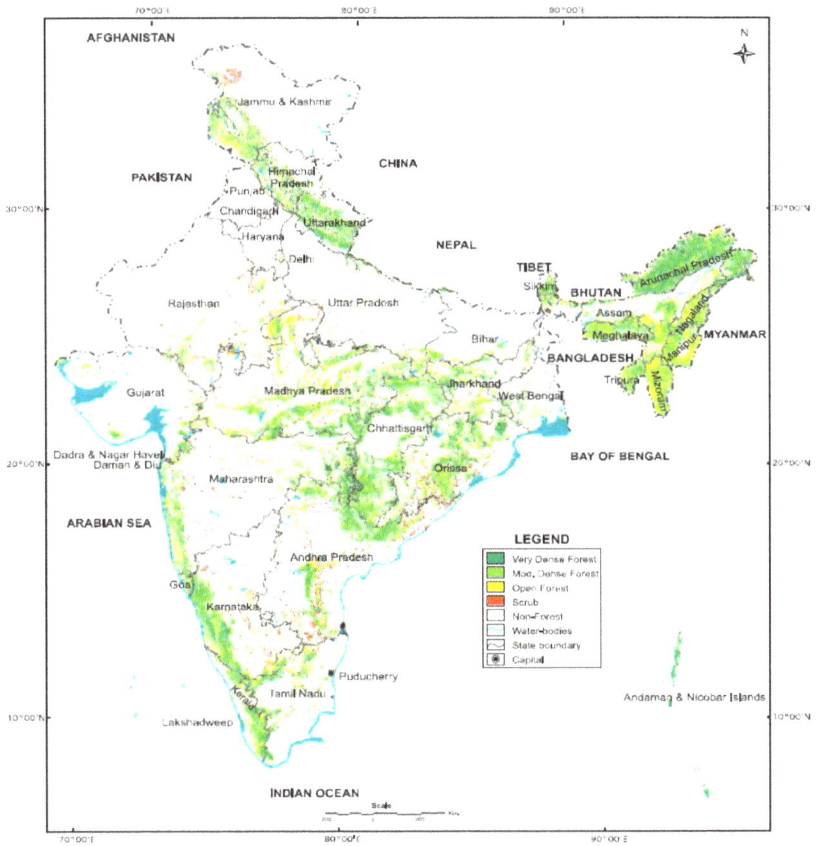

FIGURE 18.1 Forest cover map of India (www.fsi.org.in).

villages with a large dependence of these village communities on forest resources (Ravindranath and Sudha, 2004).

Climate change is predicted to significantly affect the forest-type distribution in India. Ravindranath et al. (2006) carried out an assessment of the impact of climate change on forest ecosystems using a regional climate model (HadMR3) for the period 2070–2100 and the equilibrium model BIOME4. This assessment is based on climate projections using the A2 (740 ppm CO_2) and B2 (575 ppm CO_2) scenarios of the Special Report on Emission Scenarios (IPCC, 2007) and the BIOME4 (Kaplan et al., 2003) vegetation response model. The climatic parameters included in the model are atmospheric CO_2, rainfall, temperature and cloud cover. Authors have projected that for the year 2085, 77% and 68% of the Indian forests are likely to experience shift in forest types under the A2 and B2 scenario, respectively. There is a shift towards wetter forest types in the northeastern region and drier forest

TABLE 18.1 Different Forest Types and Their Percentage Coverage in India

Group	Percent of forest coverage
Tropical wet evergreen forest	2.92
Tropical semi-evergreen forest	13.79
Tropical moist deciduous forest	19.73
Littoral and swamp forest	0.69
Tropical dry deciduous forest	41.87
Tropical thorn forest	2.25
Tropical dry evergreen forest	0.13
Subtropical broadleaved hill forest	2.69
Subtropical pine forest	2.63
Subtropical dry evergreen forest	0.03
Montane wet temperate forest	0.69
Himalayan moist temperate forest	4.12
Himalayan dry temperate forest	0.84
Subalpine and Alpine forest	2.55
Plantation/open forest	5.07
Total	100.00

types in the northwestern region. Increasing atmospheric CO_2 and climate warming could also result in a doubling of net primary productivity (A2 scenario) and a nearly 70% increase under the B2 scenario. Generally, the increase of net primary production is higher for a warmer type of vegetation. This study provides a good overview of the response of Indian forests to climate change although one important factor, that is, tropospheric O_3 is not taken into consideration. Matyssek et al. (2013) have shown that enhanced ground level O_3 would negatively impact sink strength and carbon storage capacity of forest trees under elevated CO_2 conditions. King et al. (2013) have collated 58 experimental examples depicting forest ecosystem responses to CO_2 and O_3. The two groups of authors conclude that there is a pressing need to study the interactive effect of CO_2 and O_3 on middle-aged and mature trees. The mechanism of O_3 action on European trees is discussed in Dizengremel et al. (2013) and Oksanen et al. (2013b).

In India, O_3 monitoring is extremely limited and virtually non-existent in rural or forested areas, leading to a wide gap of knowledge on impact of O_3

on forest trees. Therefore, an improved understanding of the impact of O_3 along with other climate variables on different forest types is essential in assessing the threat of O_3 to ecosystem services including food production, timber production and quality, and carbon sequestration. Impacts of O_3 on some East Asian countries including Japan and China are presented in Koike et al. (2013), where it has been shown that deciduous trees are more sensitive to O_3 than evergreen species. This result assumes significance for India since about 42% of total forest cover consists of tropical dry deciduous forest (Table 18.1).

18.3 SOURCES OF AIR POLLUTION AND GREENHOUSE GASES IN INDIA

India has globally significant GHG emissions. Total CO_2 emissions from fossil fuel consumption from India have been estimated as 1626 million tons for the year 2010 making it the third largest emitter of CO_2 after China (7217 million tons) and the United States (5368 million tons) (IAEA, 2012). Emission of greenhouse gases (CO_2, CH_4 and N_2O) for the year 2007 mainly came from the energy (58%), industry (22%), agriculture (17%), waste and land use, land-use changes and forestry (3%) sectors (MoEF, 2010) (Figure 18.2). The other key sources of GHG emissions in the country include, for example, non-specific industries, agricultural soils, residences, and iron, steel and cement production. The N_2O emissions from India have increased from 144,000 metric tons in 1985 to 267,000 metric tons in 2005 (Garg et al., 2012). The agriculture sector accounted for around 75% of total N_2O emissions in 2005, including around 49% from nitrogen fertiliser use, 13% from indirect soil emission, 6% from decaying of crop residue, 4% from livestock manure used as organic fertiliser and 3% from onsite agriculture residue

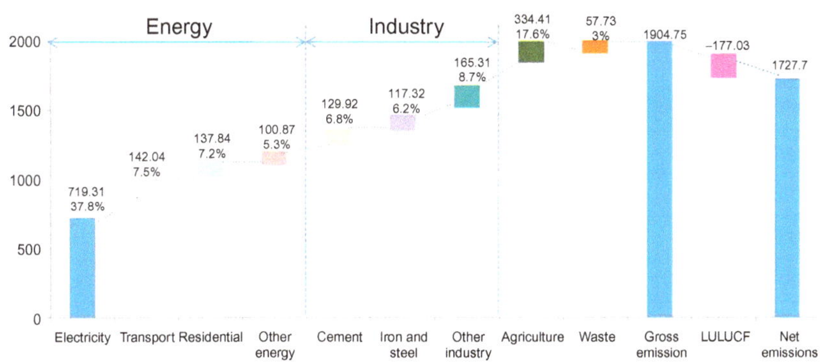

FIGURE 18.2 Sector-wise distribution of GHG emissions (million tons of CO_2 equiv.) in India for the year 2007 (MoEF, 2010). *LULUCF in the figure refers to land use, land-use change and forestry.*

burning (Garg et al., 2012). In a recent study, Lu and Streets (2012) have reported that NO_x emissions from Indian power plants have increased by at least 70% during the period 1996–2010.

Besides the GHGs, emissions of air pollutants due to human activities like transportation, fossil fuel and biomass combustion, and industrial activities are also becoming serious issues due to their adverse impacts on air quality which in turn pose a serious threat to human health and vegetation. The rapid urbanisation has resulted in a tremendous increase in number of motor vehicles. As the number of vehicles continues to grow, vehicles are now becoming the main source of air pollution in urban India (CPCB, 2010a). The composition of the vehicular population is given in Table 18.2 which reveals that the personalised mode of transportation (constituting mainly two-wheelers and cars) accounts for more than 80% of motor vehicles in the country compared to their share of a little over 60% in 1950. It is obvious that increasing traffic will have negative impacts on the forest environment too.

Satellite observations show increasing trends of O_3 production efficiency over India due to anthropogenic emissions (Ghude et al., 2008, 2010). In addition to growth of industrial and vehicular traffic, coal combustion and biomass burning (i.e. field burning of agricultural residue) are major sources for increasing NO_x production (Kunhikrishnan et al., 2004; Singh et al., 2008).

TABLE 18.2 Vehicular Population in India and Its Growth (1951–2006)

Year	Two-wheelers (%)	Cars, jeeps, etc. (%)	Buses (%)	Goods vehicles (%)	Others[a] (%)	Total no. (million)
1951	8.8	52.0	11.1	26.8	1.3	0.31
1961	13.2	46.6	8.6	25.3	6.3	0.66
1971	30.9	36.6	5.0	18.4	9.1	1.86
1981	48.6	21.5	3.0	10.3	16.6	5.39
1991	66.4	13.8	1.5	6.3	11.9	21.37
2001	70.1	12.8	1.2	5.4	10.5	54.99
2002	70.6	12.9	1.1	5.0	10.4	58.92
2003	70.9	12.8	1.1	5.2	10.0	67.01
2004	71.4	13.0	1.1	5.2	9.4	72.72
2005	72.1	12.7	1.1	4.9	9.1	81.50
2006	72.2	12.9	1.1	4.9	8.8	89.61

[a] *Others include tractors, trailers, three-wheelers, etc.*
Source: Road Transport Year Book 2006–2007; CPCB (2010a).

18.4 AIR QUALITY IN INDIA

In India, the Central Pollution Control Board (CPCB) runs the National Ambient Air Quality Monitoring programme under which it measures four criteria pollutants, namely, SO_2, NO_2, RSPM (respirable suspended particulate matter, particle size $<2.5\ \mu m$) and SPM (suspended particulate matter, particle size $>10\ \mu m$) at 342 stations covering 127 cities across India. CPCB reports the state of air pollution through the analysis of these four criteria pollutants with respect to National Ambient Air Quality Standards. In 2008, most of the industrial cities (80 cities) in India had ambient SO_2 concentrations in the range of 0–40 $\mu g/m^3$ (CPCB, 2010b). The annual average concentrations of NO_2 were in the 'low' category (i.e. in the range of 0–40 $\mu g/m^3$) in 65 cities, while 15 cities fell under the 'moderate' category (41–80 $\mu g/m^3$) and only one city had 'high' NO_2 levels (81–120 $\mu g/m^3$). However, the annual average RSPM concentrations in ambient air indicate 'critical' levels ($>180\ \mu g/m^3$) for 16 cities and high levels (121–180 $\mu g/m^3$) for 19 cities, while 34 cities show moderate levels (61–120 $\mu g/m^3$) and 11 cities show low levels (0–60 $\mu g/m^3$). The annual average SPM concentrations showed low values (0–180 $\mu g/m^3$) for 22 cities, moderate values (181–360 $\mu g/m^3$) for 38 cities, high values (361–540 $\mu g/m^3$) for 11 cities and critical values ($>540\ \mu g/m^3$) for 1 city. The air quality data of some of the cities are shown in Figure 18.3, while Table 18.3 includes the comparison of ambient air quality standards of India and Europe. Excepting particulate matter, the standards of gaseous pollutants are more stringent in India.

The inadequate monitoring facilities for other air pollutants (besides SO_2, NO_2, SPM and RSPM) in India hinder the development of a complete scenario for these pollutants based on field measurements. Long-term monitoring of O_3 and its precursors has been very fragmentary over the Indian region, focusing mainly on the metropolitan areas of Delhi and Pune (Jain et al., 2005; Roy et al., 2009). Jain et al. (2005) have measured the surface O_3 variations over New Delhi during the past decade and found maximum values of O_3 concentrations during the summer and autumn seasons and minimums in the monsoon and winter seasons. The diurnal patterns in O_3 concentration show daytime *in situ* photochemical production throughout the year. The 2001 monthly average maximum concentration in Delhi in the summer season was found to be in the range of 62–95 ppb, whereas it ranged between 50 and 82 ppb in the autumn season (October and November). These values are much higher than what has been reported for Delhi in Figure 18.3. This is because the values in Figure 18.3 are the 24 h annual averages. Similarly, Peshin et al. (2003) also observed an increasing trend in the O_3 concentration in Delhi, but the nature of diurnal variation remained identical over the years. Elsewhere, Lal et al. (2000) seldom found high concentrations beyond 80 ppb at Ahmedabad, while Nishanth et al. (2012) reported highest (46.9 ppb) and lowest (16.4 ppb) values of O_3 in the months of December and in July at Kannur, a tropical coastal site in the southern state of Kerala.

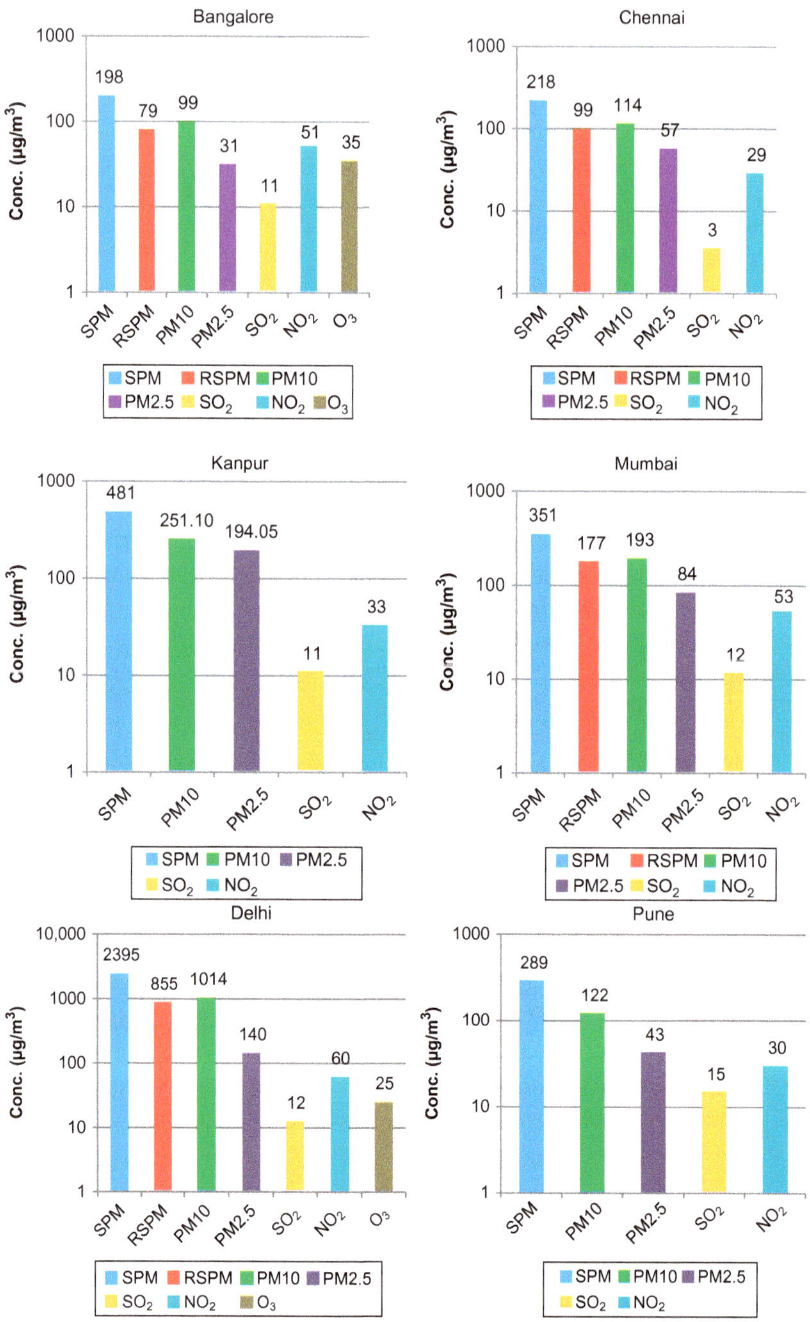

FIGURE 18.3 Air quality data of some selected cities of India for the year 2010. Note that ambient O_3 is monitored only in two cities (www.cpcb.nic.in).

TABLE 18.3 Ambient Air Quality Standards (μg/m^3) in India and Europe

Pollutant	Averaging period	India: *Industrial, residential, rural and other areas*	India: *Ecologically sensitive areas*	Europe
Sulphur dioxide (SO_2)	1 h	–	–	350
	Annual	50	20	–
	24 h	80	80	125
Nitrogen dioxide (NO_2)	1 h	–	–	200
	Annual	40	30	40
	24 h	80	80	–
Ozone (O_3)	8 h	100	100	120
	1 h	180	180	–
Particulate matter (<10 μM) or PM_{10}	Annual	60	60	40
	24 h	100	100	50
Particulate matter (<2.5 μM) or $PM_{2.5}$	Annual	40	40	25
	24 h	60	60	–

For details concerning the criteria of air quality standards, see http://cpcb.nic.in/National_Ambient_Air_Quality_Standards.php (India) and http://ec.europa.eu/environment/air/quality/standards.htm (Europe).

In view of these limited measurements, temporal and spatial variations in O_3 and other trace species concentrations in India have been best estimated only with atmospheric simulation models. Global photochemical models (e.g. Dentener et al., 2006) predict that under current emission scenarios, parts of Asia, including India, will experience further significant increases in O_3 concentrations by 2030. In India and Pakistan, high concentrations of O_3 have already been reported (Emberson et al., 2009). Roy et al. (2008) using an off-line regional chemistry-transport model found higher concentrations of O_3, CO and NO_x over the IGP region compared to other parts of India, while Mittal et al. (2007a) plotted the gridded distribution of O_3 over the Indian region using the HANK model (Hess et al., 2000) and found the modelled average daytime monthly surface O_3 concentrations in the range of 25–100 ppb for the February to May 2000 period. Mittal et al. (2007b) have also used the simulated O_3 concentrations over the Indian region to compute AOT40 (accumulated ozone over a concentration threshold of 40 ppb) values for the Indian region which were found to vary between 2 and 19 ppm h. These

AOT40 values largely exceeded the European critical levels for the protection of vegetation.

In addition to metropolitan areas, several inland rural areas, in particular, are exhibiting impacts of high O_3 concentrations (Naja and Lal, 2002; Reddy et al., 2010; Selvaraj et al., 2012), a trend which represents a risk for forestry and crop production.

Monthly mean tropospheric NO_2 columns over the Indian region during 2011 are shown in Figure 18.4. High concentrations are seen over IGP during the February–March and October–December periods. These data agree well with other satellite-based observations (e.g. Ghude et al., 2008; Lu and Streets, 2012). Ghude et al. (2008) attributed high NO_2 levels over IGP to large point sources like thermal power plants, steel and cement plants, industrial processes (e.g. the sugar and paper industries) and transport. High NO_2 concentration in winter and low in the monsoon season have been attributed

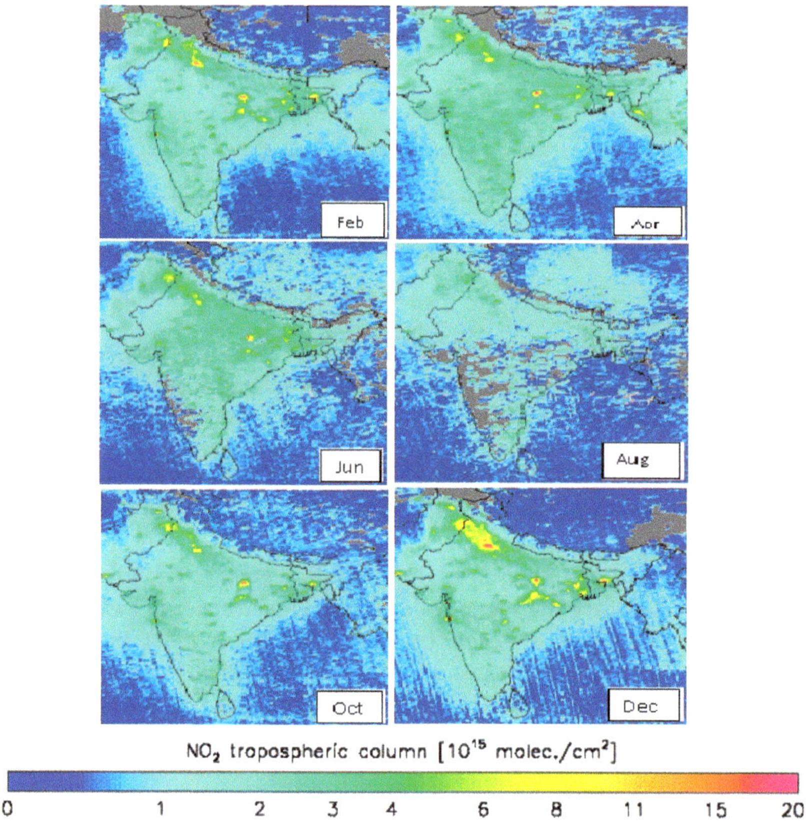

FIGURE 18.4 Monthly mean tropospheric NO_2 columns over the Indian region in 2011. Based on the Ozone Monitoring Instrument (aura.gsfc.nasa.gov/instruments/omi.html).

to seasonal variation of the NO_x lifetime and the effective scavenging during the monsoon season (Ghude et al., 2008).

18.5 IMPACTS OF O_3 ON AGRICULTURE

Being toxic to plants at concentrations as low as 30 ppb, high O_3 concentrations affect physiological and biochemical processes as well as biomass growth and yield of agricultural crops (Feng and Kobayashi, 2009; Krupa, 1997). Figure 18.5 depicts the impact of ambient O_3 on sensitive and tolerant varieties of tobacco and other crop plants (unpublished data). Tobacco cultivar Bel W3 is the best-defined, best-described and most widely used bio-indicator for ambient O_3. This plant exhibits reliable foliar injury symptoms when exposed to ambient O_3. It has been used all over the world to detect O_3 in areas ranging from local sites to remote regions. This tobacco cultivar could be quite useful in identifying areas where ozone concentrations and exposures are suspected to be high, but care should be taken so as not to use this response to qualitatively assess air quality for O_3. Therefore, with a clear understanding of the strengths and weaknesses of the use of plants as bio-indicators of O_3 exposure, they can be quite useful in assessing relative air quality especially in remote forested areas, where online monitoring is lacking.

Using two types of O_3 exposure metrics: mean (12 h, M12 or 7 h M7 daytime) and cumulative (AOT40) Van Dingenen et al. (2009) and Avnery et al. (2011a) calculated crop production loss and economic loss for major crops for the year 2000 in India (Table 18.4). Wheat production losses were largest

FIGURE 18.5 Ozone injury symptoms on different plant species. (A) Ozone sensitive (Bel W3) and tolerant (Bel B) varieties of tobacco plants (*Nicotiana tabacum* L.). (B) *Brassica juncea* var. Kranti (Indian mustard) exposed to 120 ppb O_3 for 5 days under controlled conditions. (C) *Oryza sativa* var. Sugandha. (D) *Triticum aestivum* var. HUW 234. ((A), (C) and (D) grown under ambient conditions at Lucknow, India; O_3 concentration ranged from 26 to 78 ppb (average 60 ppb). The tobacco seeds were kindly provided by Prof. W.J. Manning, University of Massachusetts, USA).

TABLE 18.4 Crop Production Loss (Million Metric Tons) and Economic Loss (Million US$) for the Year 2000 in India

Crop	Crop production loss		Economic loss		References
			Min.	*Max.*	
Wheat	11.6 (M7)	29.1 (AOT40)	1711	4310	Van Dingenen et al. (2009)
	6 (M12)	26 (AOT40)	900	3750	Avnery et al. (2011a)
Rice	7.7 (M7)	11.4 (AOT40)	1017	1509	Van Dingenen et al. (2009)
Maize	0.23 (AOT40)	0.5 (M12)	25	51	Van Dingenen et al. (2009)
	0.41 (AOT40)	0.83 (M12)	35.7	89.25	Avnery et al. (2011a)
Soybean	0.26 (AOT40)	1.2 (M12)	51	244	Van Dingenen et al. (2009)
	0.25 (AOT40)	1 (M12)	40	166	Avnery et al. (2011a)

The ozone metrics are shown in parentheses.

ranging from 6 to 29 million tons depending upon the metrics used. The corresponding economic losses for wheat were in the range of 900–4310 million US$ (Table 18.4). Rice, maize and soybean production losses were in the range of 7.7–11.4, 0.23–0.83 and 0.25–1.2 million US$, respectively (Table 18.4). The economic losses for rice, maize and soybean were in the range of 1017–1509, 25–89 and 40–244 million US$, respectively.

However, even greater, 30–45%, crop production losses have been predicted for the year 2030, in particular, for wheat in India by Van Dingenen et al. (2009) and Avnery et al. (2011b) (Table 18.5).

Under the IPCC A2 scenario, Avnery et al. (2011b) found that total wheat production loss in the world for the year 2030 is higher in India than any other country (8.5–56 Mt), while maize and soybean losses are pegged at 1–1.9 and 1–3.8 Mt, respectively (Table 18.5). The corresponding economic losses are expected to be in the range of 125–4500 million US$ for all the three crops (Table 18.5). Considering the recent upward trends in crop and food prices, the O_3-induced damage to crops is expected to offset a significant portion of the GDP growth rate, especially in a country like India with an economy based on agricultural production.

Most experiments with elevated O_3 on Indian plants have been conducted with the most important food crops, that is, wheat and rice (Oksanen et al., 2013a and Table 18.6). These complementary studies conducted in field conditions or open-top chambers have indicated 9–32% and 10–39% yield losses for wheat and rice, respectively. On the other hand, wheat experiments with EDU (ethylene diurea; *N*-(2-(2-oxo-1-imidazolidinyl) ethyl)-*N′*-phenyl urea) treatments, where EDU acts as a chemical protectant against ozone applications, have indicated larger variation in yield losses (Table 18.6), ranging between 2% and 67% (Singh and Agrawal, 2009). These ozone-caused yield reductions have been accompanied by lower photosynthesis, growth rate, chlorophyll fluorescence, water-use efficiency, contents of photosynthetic pigments, impaired seed quality (i.e. contents of starch, proteins and nutrients of grains), visible leaf injuries, lipid peroxidation, induction of antioxidative defence and shifts in protein profiles of leaves. A high genetic variability in ozone sensitivity/tolerance between the cultivars has been demonstrated for wheat by Singh and Agrawal (2009).

Some experiments have also been conducted with mung bean (*Vigna radiata*), mustard (*Brassica camprestis* L.), and black gram (*Vigna mungo* L.), and O_3-caused yield and biomass reductions in these species ranged between 10% and 51% (Table 18.6), but genetic variation in ozone sensitivity/tolerance among the cultivars or varieties is almost unexplored. As found for wheat and rice, these reductions in productivity have been linked to impaired gas exchange, photosynthetic processes and nutrition value, as well as resource allocation towards defence responses. Unlike the above studies in crop plants, no ozone experiments have been conducted with Indian tree species, though the crop losses presented in Tables 18.5 and 18.6 suggest that ozone may have a deleterious impact even on long-lived trees in India.

18.6 FUTURE PERSPECTIVES ON THE O_3 PROBLEM IN INDIA

With the Indian population set to increase to over 1.53 billion by the year 2030 (surpassing China), the agricultural demand is also expected to increase. To meet this demand, either more cultivated land area is to be added or we need to increase productivity on existing agricultural land through more

TABLE 18.5 Projected Change in Crop Production Loss (Million Metric Tons) and Economic Loss (Million US$) for the Year 2030 in India Under the IPCC A2 Scenario (Avnery et al., 2011b)

	Crop production loss		Economic loss	
Crop	*M12*	*AOT40*	*M12*	*AOT40*
Wheat	8.5	56	375	4500
Maize	1.9	1	775	538
Soybean	1	3.8	125	375

TABLE 18.6 Yield Losses due to Surface Ozone in Different Crops in India (Derived from Field Experiments, Open-Top Chamber and EDU Treatment Studies)

Species	Average ozone concentration (ppb)	Yield reduction (%)	References
Wheat (*Triticum aestivum*)	10–59	25[a]	Agrawal et al. (2003)
	70	9[a]	Ambasht and Agrawal (2003)
	10–41	25–32[a]	Rajput and Agrawal (2005)
	40–48	11–12[b]	Rai et al. (2007)
	35–54	2–25[c]	Singh and Agrawal (2009)
	45–57	11–46[b]	Sarkar and Agrawal (2010)
	5–55	18–24[b]	Sarkar et al. (2010)
Rice (*Oryza sativa*)	23–45	11–16[b]	Rai and Agrawal (2008)
	20–55	7–9[b]	Rai et al. (2010)
Mung bean (*Vigna radiata*)	7–35	32[c]	Agrawal et al. (2005)
	10–59	18–79[a]	Agrawal et al. (2006)
Mustard (*Brassica campestris*)	62	20[a]	Agrawal et al. (2003)
	42–54	17[c]	Singh et al. (2009)
	28–59	7–19[b]	Singh et al. (2012)
Black gram (*Vigna mungo*)	41–60	8–24[c]	Singh et al. (2010)

[a]*Field studies.*
[b]*Open-top chamber studies.*
[c]*EDU treatment studies.*

O_3-tolerant cultivars or genotypes. Improving yields through more inputs of water, fertiliser, pesticide, etc., have their own detrimental environmental consequences (Tilman et al., 2001). Current information from modelling efforts and experimental work in India clearly indicate negative impacts of increasing O_3 on the most important crop species, particularly in the most important cultivation area, that is, the IGPs of India (Table 18.6). Therefore, surface O_3 abatement could provide a unique opportunity to increase food supplies without further environmental degradation. Tropospheric O_3 is a potent greenhouse gas in addition to a toxic air pollutant (Sitch et al., 2007), so its reduction would also provide other co-benefits to climate and human health (Anenberg et al., 2010; West et al., 2006, 2007).

However, more intensive O_3 and other important GHG monitoring and data collection, determination of O_3 exposure–response relationships from eco-physiologically relevant experimental facilities (e.g. free air exposures), screening of more (local) cultivars and mechanistic studies for O_3 responses are needed to specify and confirm these results. Experimental data from Indian trees are still completely missing, regarding ozone stress. It is also necessary to study interactions between O_3 and other climate change factors, such as water stress (related particularly to variation in the monsoon period), increasing CO_2 concentration and climate warming in Indian plants. This would help not only to meet the food security demands in India but also to promote sustainable forestry and forest health. In addition, more plant species should be tested for O_3 sensitivity, including most important tree species with high value in carbon storage and sequestration and mitigation of the greenhouse effect. There are several questions that are particularly relevant for both silviculture and agriculture in India: for example, how is O_3 impacting belowground biomass allocation and the nitrogen cycle? (Discussed in Pregitzer and Talhelm, 2013) How is elevated O_3 altering plant residue decomposition, nutrient cycling and microbial activities? Clearly a major international cooperation is needed to address these important issues.

18.7 CONCLUSIONS

Air pollution levels, especially those of O_3, are becoming a serious problem in India. O_3 is monitored still at only a very few places across India. There is some information available on its impact on important Indian crops, for example, wheat and rice but there are no studies on impact of O_3 on Indian trees. There is a critical need to study how O_3 will impact trees and forest ecosystems. We have identified the following key knowledge gaps:

1. Coordinated and standardised ozone monitoring across India,
2. Use of ozone and CO_2 Free-Air Concentration Enrichment systems to develop dose–response relationships among different cultivars of crop/plant species,
3. Study potential indirect effects caused by ozone especially where these have direct links with climate change, for example, on the C sequestration potential of forest trees,
4. Impact of anthropogenic N input on natural ecosystems, including forest ecosystems.

REFERENCES

Agrawal, M., Singh, B., Rajput, M., Marshall, F., Bell, J.N.B., 2003. Effect of air pollution on peri-urban agriculture: a case study. Environ. Pollut. 126, 323–329.

Agrawal, S.B., Singh, A., Rathore, D., 2005. Role of ethylene diurea (EDU) in assessing impact of ozone on *Vigna radiata* L. plants in a suburban area of Allahabad (India). Chemosphere 61, 218–228.

Agrawal, M., Singh, B., Agawal, S.B., Bell, J.N.B., Marshall, F., 2006. The effect of air pollution on yield and quality of mung bean grown in peri-urban areas of Varanasi. Water Air Soil Pollut. 169, 239–254.

Ambasht, N.K., Agrawal, M., 2003. Effects of enhanced UV-B radiation and tropospheric ozone on physiological and biochemical characteristics of field grown wheat. Biol. Plant. 47, 625–628.

Anenberg, S.C., Horowitz, L.W., Tong, D.Q., West, J.J., 2010. An estimate of the global burden of anthropogenic ozone and fine particulate matter on premature human mortality using atmospheric modeling. Environ. Health Perspect. 118, 1189–1195.

Avnery, S., Mauzerall, D.L., Liu, J., Horowitz, L.W., 2011a. Global crop yield reductions due to surface ozone exposure: 1. Year 2000 crop production losses and economic damage. Atmos. Environ. 45, 2284–2296.

Avnery, S., Mauzerall, D.L., Liu, J., Horowitz, L.W., 2011b. Global crop yield reductions due to surface ozone exposure: 2. Year 2030 potential crop production losses and economic damage under two scenarios of O_3 pollution. Atmos. Environ. 45, 2297–2309.

Central Pollution Control Board, 2010. Status of the vehicular population control program in India. Programme Objective Series. PROBES/136/2010.

Central Pollution Control Board, 2010b. National Ambient Air Quality Status of 2010. National Ambient Air Quality Monitoring Series: NAAQMS//2010–11.

Dentener, F., Stevenson, D., Ellingsen, K., van Noije, T., Schultz, M., Amann, M., Atherton, C., et al., 2006. The global atmospheric environment for the next generation. Environ. Sci. Technol. 40, 3586–3594.

Dizengremel, P., Jolivet, Y., Tuzet, A., Ranieri, A., Thiec, D.Le, 2013. Integrative leaf-level phytotoxic ozone dose assessment for forest risk modelling. Dev. Environ. Sci. 13, 267–288.

Emberson, L.D., et al., 2009. A comparison of North American and Asian exposure-response data for ozone effects on crop yields. Atmos. Environ. 43, 1945–1953.

Feng, Z.Z., Kobayashi, K., 2009. Assessing the impacts of current and future concentrations of surface ozone on crop yields with meta-analysis. Atmos. Environ. 43, 1510–1519.

Forest Survey of India, 2011. Annual Report. www.fsi.org.in.

Fowler, D., Amann, M., Anderson, R., Ashmore, M., Cox, P., Depledge, M., Derwent, D., Grennfelt, P., Hewitt, N., Hov, O., Jenkin, M., Kelly, F., Liss, P., Pilling, M., Pyle, J., Slingo, J., Stefenson, D., 2008. Ground-Level Ozone in the 21st Century: Future Trends, Impacts and Policy Implications. The Royal Society Policy Document, London, p. 132. www.royalsociety.org.

Garg, A., Shukla, P.R., Upadhyay, J., 2012. N_2O emissions in India: an assessment of temporal, regional and sector trends. Clim. Change 110, 755–782.

Ghude, S.D., Fadnavis, S., Beig, G., Polade, S.D., van der A, R.J., 2008. Detection of surface emission hot spots, trends and seasonal cycle from satellite retrieved NO_2 over India. J. Geophys. Res. 113 (D20305), 1–13.

Ghude, S.D., Kulkarni, P.S., Beig, G., Jain, S.L., Arya, B.C., 2010. Global distribution of tropospheric ozone and its precursors: a view from space. Int. J. Remote Sens. 31, 485–495.

Hess, P.G., Flocke, S., Lamarque, J.F., Barth, M.C., Madronich, S., 2000. Episodic modeling of the chemical structure of the troposphere as revealed during the spring MLOPEX 2 intensive. J. Geophys. Res. 105 (D22), 26809–26839.

IAEA, 2012. http://www.iea.org/publications/freepublications/publication/name,4010,en.html (accessed 29.10.12).

IPCC, 2007. Climate change, 2007: the physical science basis. In: Solomon, S., Qin, D., Manning, M., Chen, Z., Marquis, M., Avery, K.B., Tignor, M., Miller, H.L. (Eds.), Contribution of Working Group I to the Fourth Assessment Report of the Intergovernmental Panel on Climate Change. Cambridge University Press, Cambridge, United Kingdom, and NY, USA, 996 pp.

Jain, S.L., Arya, B.C., Kumar, A., Ghude, S.D., Kulkarni, P.S., 2005. Observational study of surface ozone at New Delhi, India. Int. J. Remote Sens. 26, 3515–3524.

Kaplan, J.O., Bigelow, N.H., Prentice, I.C., et al., 2003. Climate change and arctic ecosystems: 2. Modelling, paleodata model comparisons and future projections. J. Geophys. Lett. 108 (8171), 1017.

King, J., Liu, L., Aspinwall, M., 2013. Tree and forest responses to interacting elevated atmospheric CO_2 and tropospheric O3: a synthesis of experimental evidence. Dev. Environ. Sci. 13, 179–208.

Koike, T., Watanabe, M., Hoshika, Y., Kitao, M., Matsumura, H., Funada, R., Izuta, T., 2013. Effects of ozone on forest ecosystems in East and Southeast Asia. Dev. Environ. Sci. 13, 371–390.

Krupa, S.V., 1997. Global climate change, processes and products: an overview. Environ. Monit. Assess. 46, 73–88.

Kunhikrishnan, T., Lawrence, M.G., von Kuhlmann, R., Richter, A., Ladstätter-Weißenmayer, A., Burrows, J.P., 2004. Analysis of tropospheric NOx over Asia using the model of atmospheric transport and chemistry (MATCH-MPIC) and GOME-satellite observations. Atmos. Environ. 38, 581–596.

Lal, S., Naja, M., Subbaraya, B.H., 2000. Seasonal variations in surface ozone and its precursors over an urban site in India. Atmos. Environ. 34, 2713–2724.

Lu, Z., Streets, D.G., 2012. Increase in NOx emissions from Indian thermal power plants during 1996-2010: unit based inventories and multisatellite observations. Environ. Sci. Technol. 46, 7463–7470.

Matyssek, R., Clarke, N., Cudlin, P., Mikkelsen, T.N., Tuovinen, J.-P., Wieser, G., Paoletti, E., 2013. Climate change, air pollution and global challenges: knowledge, understanding and perspectives from forest research. Dev. Environ. Sci. 13, 3–16.

Ministry of Environment and Forests, Govt. of India, 2010. India: Greenhouse Gas Emissions 2007. Indian Network for Climate Change Assessment (INCCA), pp. 1–84. http://moef.nic.in/downloads/public-information/Report_INCCA.pdf.

Mittal, M.L., Hess, P.G., Jain, S.L., Arya, B.C., Sharma, C., 2007a. Surface ozone in the Indian region. Atmos. Environ. 41, 6572–6584.

Mittal, M.L., Sharma, C., Pandey, R., 2007b. Air pollutants and agriculture productivity. Physiol. Mol. Biol. Plants 13, 235–241.

Naja, M., Lal, S., 2002. Surface ozone and precursor gases at Gadanki (13.5°N, 79.2°E), a tropical rural site in India. J. Geophys. Res. 107 (D14), ACH 8-1–ACH 8-13.

NATCOM, 2004. India's first communication to UNFCC. Ministry of Environment and Forests, Govt. of India.

Nishanth, T., Praseed, K.M., Satheesh Kumar, M.K., Valsaraj, K.T., 2012. Analysis of ground level O_3 and NOx measured at Kannur, India. J. Earth Sci. Clim. Change 3, 111. http://dx.doi.org/10.4172/2157-7617.1000111.

Oksanen, E., Pandey, V., Pandey, A.K., Keski-Saari, S., Kontunen-Soppela, S., Sharma, C., 2013a. Impacts of increasing ozone on Indian plants. Environ. Pollut. 177, 189–200.

Oksanen, E., Keski-Saari, S., Kontunen-Soppela, S., Keinänen, M., 2013b. Metabolomics and transcriptomics increase our understanding about defence responses and genotypic differences of northern deciduous trees to elevating ozone, CO_2 and climate warming. Dev. Environ. Sci. 13, 309–329.

Peshin, S.K., Sharma, J.K., Rao, P.R., Srivastav, S.K., 2003. Changes in surface ozone amount and its diurnal and seasonal patterns from 1990-95 to 1996-2000, measured at New Delhi (28°N). Mausam 54, 947–951.

Pregitzer, K.S., Talhelm, A.F., 2013. Belowground carbon cycling at Aspen FACE: dynamic responses to CO_2 and O_3 in developing forests. Dev. Environ. Sci. 13, 209–226.

Rai, R., Agrawal, M., 2008. Evaluation of physiological and biochemical responses of two rice (*Oryza sativa* L.) cultivars to ambient air pollution using open top chambers at a rural site in India. Sci. Total Environ. 407, 679–691.

Rai, R., Agrawal, M., Agrawal, S.B., 2007. Assessment of yield losses in tropical wheat using open top chambers. Atmos. Environ. 41, 9543–9554.

Rai, R., Agrawal, M., Agrawal, S.B., 2010. Threat to food security under current levels of ground level ozone: a case study for Indian cultivars of rice. Atmos. Environ. 44, 4272–4282.

Rajput, M., Agrawal, M., 2005. Biomonitoring of air pollution in a seasonally dry tropical suburban area using wheat transplants. Environ. Monit. Assess. 101, 39–53.

Ravindranath, N.H., Sudha, P., 2004. Joint Forest Management in India: Spread, Performance and Spread. Universities Press, Hyderabad, India.

Ravindranath, N.H., Joshi, N.V., Sukumar, P., Saxena, A., 2006. Impact of climate change on forests in India. Curr. Sci. 90, 354–361.

Reddy, B.S.K., Kumar, K.R., Balakrishnaiah, G., Gopal, K.R., Reddy, R.R., Ahammed, Y.N., Narasimhulu, K., Reddy, L.S.S., Lal, S., 2010. Observational studies on the variations in surface ozone concentration at Anantapur in southern India. Atmos. Res. 98, 125–139.

Roy, S., Beig, G., Jacob, D., 2008. Seasonal distribution of ozone and its precursors over the tropical Indian region using regional chemistry-transport model. J. Geophys. Res. 113 (D21307), 1–15. http://dx.doi.org/10.1029/2007JD009712.

Roy, S., Beig, G., Ghude, S., 2009. Exposure-plant response of ambient ozone over the tropical Indian region. Atmos. Chem. Phys. 9, 5253–5260.

Sarkar, A., Agrawal, S.B., 2010. Elevated ozone and two modern wheat cultivars: an assessment of dose dependent sensitivity with respect to growth, reproductive and yield parameters. Environ. Exp. Bot. 69, 328–337.

Sarkar, A., Rakwal, R., Agrawal, S.B., Shibato, J., Ogawa, Y., Yoshida, Y., Agrawal, G.K., Agrawal, M., 2010. Investigating the impact of elevated levels of ozone on tropical wheat using integrated phenotypical, physiological, biochemical, and proteomics approaches. J. Proteome Res. 9, 4565–4584.

Selvaraj, R.S., Boaz, B.M., Sachithananthem, C.P., Padma, K., Inbanathan, S.S.R., Rajaselvi, G.K., Indira, P., Vlmalpriya, S.P., 2012. Measurements of surface ozone in the year 2011 at different sites over Tamil Nadu, India. Indian J. Sci. Technol. 5, 2047–2050.

Singh, S., Agrawal, S.B., 2009. Use of ethylene diurea (EDU) in assessing the impact of ozone on growth and productivity of five cultivars of Indian wheat (*Triticum aestivum* L). Environ. Monit. Assess. 159, 125–141.

Singh, A., Gangopadhyay, S., Nanda, P.K., Bhattacharya, S., Sharma, C., Bhan, C., 2008. Trends of greenhouse gas emission from the road transport sector in India. Sci. Total Environ. 390, 124–131.

Singh, P., Agrawal, M., Agrawal, S.B., 2009. Evaluation of physiological, growth and yield responses of a tropical oil crop (*Brassica campestris* L. var. Kranti) under ambient ozone pollution at varying NPK levels. Environ. Pollut. 157, 871–880.

Singh, S., Agrawal, S.B., Singh, P., Agrawal, M., 2010. Screening three cultivars of *Vigna mungo* L. against ozone by application of ethylenediurea (EDU). Ecotoxicol. Environ. Safe. 73, 1765–1775.

Singh, P., Singh, S., Agrawal, S.B., Agrawal, M., 2012. Assessment of the interactive effects of ambient O_3 and NPK levels on two tropical mustard varieties (*Brassica campestris* L.) using open-top chambers. Environ. Monit. Assess. 184, 5863–5874.

Sitch, S., Cox, P.M., Collins, W.J., Huntingford, C., 2007. Indirect radiative forcing of climate change through ozone effects on the land-carbon sink. Nature 448, 791–794.

Tilman, D., et al., 2001. Forecasting agriculturally driven global environmental change. Science 292, 281–284.

Van Dingenen, R., Dentener, F.J., Raes, F., Krol, M.C., Emberson, L., Cofala, J., 2009. The global impact of ozone on agricultural crop yields under current and future air quality legislation. Atmos. Environ. 43 (3), 604–618.

West, J.J., Fiore, A.F., Horowitz, L.W., Mauzerall, D.L., 2006. Global health benefits of mitigating ozone pollution with methane emission controls. Proc. Natl. Acad. Sci. 103, 3988–3993.

West, J.J., Fiore, A.M., Naik, V., Horowitz, L.W., Schwarzkopf, M.D., Mauzerall, D.L., 2007. Ozone air quality and radiative forcing consequences of changes in ozone precursor emissions. Geophys. Res. Lett. 34, L06806. http://dx.doi.org/10.1029/2006GL029173.

Chapter 19

Land Use Change, Air Pollution and Climate Change—Vegetation Response in Latin America

Alessandra R. Kozovits*,[1] and Mercedes M.C. Bustamante[†]

*Departamento de Biodiversidade, Evolução Meio Ambiente, Universidade Federal de Ouro Preto, Campus Morro do Cruzeiro, Ouro Preto, Brazil

[†]Departamento de Ecologia, Instituto de Ciências Biológicas, Campus Universitário Darcy Ribeiro, Universidade de Brasília, Brasília, Brazil

[1]Corresponding author:

Chapter Outline

19.1 INTRODUCTION

During the recent decades, Latin America has been facing a great challenge in combining unprecedented economic growth with environmental protection and sustainable development. Together with Caribbean countries, the region represents only 9% of the earth's human population; however, it has the highest urbanization rates worldwide. Currently, about 80% of the population lives in urban areas. Besides the socio-environmental problems related to the

Developments in Environmental Science, Vol. 13. http://dx.doi.org/10.1016/B978-0-08-098349-3.00019-0

rapidly growing urban and industrial centres, such as the decline in air quality, air pollution and climate change effects on natural and managed ecosystems are also causing concern. Harboring high biological diversity, countries such as Brazil bear a high responsibility not only for conserving native biomes but also for contributing significantly to agricultural productivity worldwide.

At the same time, some research initiatives for long-term studies have been started to expand databases on emission sources of air pollutants and to understand the effects of climate change on forest types and their establishment. Substantial advances have become evident in research related to ecosystem responses of forests and crop plantations in the tropics and sub-tropics to climate change, in particular, as mediated through alterations in precipitation and temperature regimes, and as inherently interacting with air pollution. This chapter focuses on the major initiatives in the field in Latin America, with emphasis on Brazil, summarizing the results obtained to date.

First, inventories of emitted air pollutants and their annual proportion in regional emission budgets are presented. Contributions by land use changes related to agriculture are highlighted. Extensive land areas are affected annually by the burning of grasslands and forests, deforestation and addition of fertilisers, these being events bearing long-distance and long-term consequences for ecosystem functioning and regional climate. The deposition of N, for example, results in significant changes in species diversity and ecosystem functions, such as nutrient cycling with litter turnover and greenhouse gas flux from soil to atmosphere under increasing N availability. In addition, changes in productivity, woody-plant diversity, C stocks and other parameters of tropical forest functioning under elevated CO_2 concentration and altered water relations as, for example, imposed by prolonged drought in the Amazonian rainforest, have also been found to be causing some concern by regional long-term studies. Key findings from such studies are presented, after briefly featuring the Latin American biomes of relevance in this chapter in ecological terms. The chapter concludes by identifying major gaps in our knowledge of Latin America and suggesting potential strategies for strengthening research having socio-economic implications for policy-making and human welfare.

19.2 LATIN AMERICA AND ITS MAJOR BIOMES

Approximately one quarter of the world's forest area is located in Latin America and the Caribbean. In the region, biomes—together with others such as rainforests, savannas, mountainous and cloud forests, rocky outcrops, grasslands, dry forests, salt marshes, mangroves, xerophytic vegetation and wetlands—represent the largest reservoir of global biodiversity (Figure 19.1; Bovarnick et al., 2010; UNEP, 2010), being associated with an appreciable proportion of global aboveground carbon stock (Phillips et al., 2009). Tropical rain forests and savannas are the major native vegetation types within woody plant diversity in terms of coverage of terrestrial surface area.

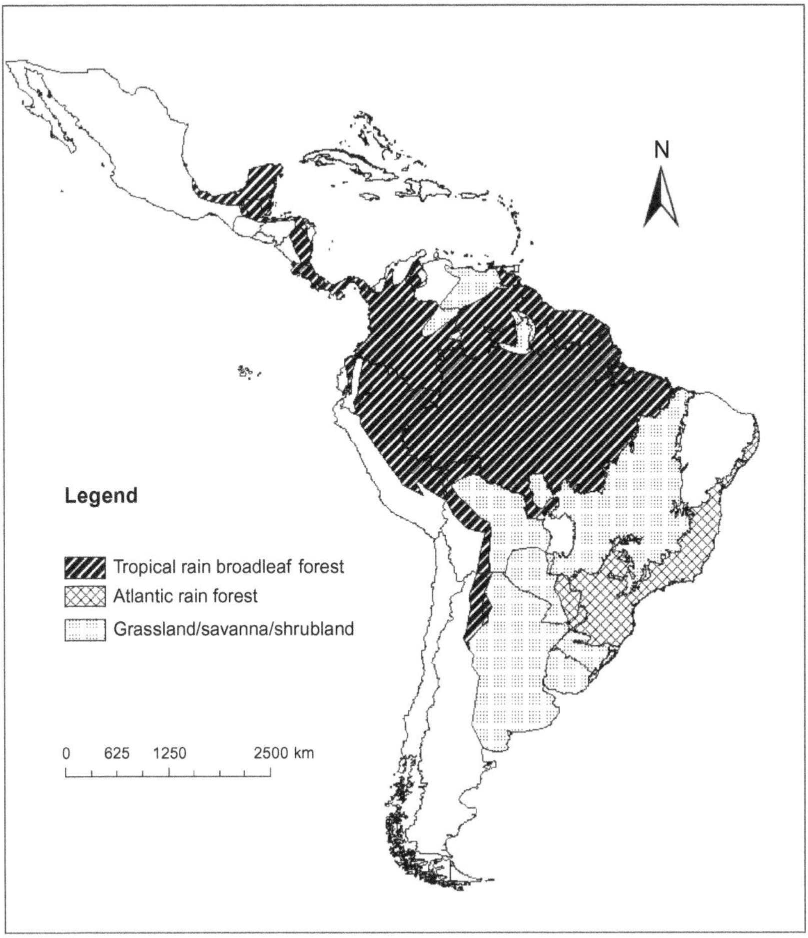

FIGURE 19.1 Major biomes in Latin America. *Modified from IPCC (2001).*

Tropical rain forests are found in South and Central America, central Africa and South-East Asia, and are characterised by a high monthly mean temperature (above 18 °C) and high precipitation (>100 mm per month). The dry season, if present, is short and with rare exceptions, rainfall always exceeds potential evapotranspiration (Dirzo, 2001).

The Amazonian forest is the largest of all tropical rainforests worldwide. The Amazon biome extends across nine South American countries, covering about 30% of the continental area (Figure 19.1). It is distributed from the Andes' base to the Atlantic coast and thus exhibits great structural heterogeneity because of variations in soil, topography, microclimate, disturbance regimes and proximity to, and influence of, rivers (Wittmann et al., 2006). The lowland forest, for example, grows in alluvium along the major rivers with sediment-rich water, whereas the *Igapó* (seasonally flooded forests) occurs along rivers of black and sediment-poor water (the dark colour of rivers results from organic acids released into the water through the decomposition of organic matter and the lack of terrestrial sediments). The 'terra firme' forest occurs outside the influence of rivers, and although such forests may vary in structure, they are commonly termed 'rainforests' (Lewis et al., 2009; Malhi et al., 2008). The plant communities of the Amazon basin are the most diverse on Earth, with more than 80,000 taxa of vascular plants described to date. In some areas such as in the upper Amazonia, it is possible to find more than 300 tree species (with individuals $\geq$10 cm in diameter at breast height) per hectare (Gentry, 1988), reaching heights of 45–60 m.

Besides the diversity in soil and topography of the region across the Amazon basin, the gradient of precipitation varies from the extremely wet, continuously rainy northwest, to the wet-dry transition climate and to the prolonged dry-season climate of the south-eastern regions (Davidson et al., 2012). This climatic gradient is crucial in shaping forest structure and functioning, affecting resistance and resilience to natural and anthropogenic disturbances. The gradient is largely prone to land use change as well, as conversion to agricultural land is more dominant in the drier eastern and southern regions.

Another type of rainforest, the Atlantic Forest, stretches along almost the entire coast of Brazil (Figure 19.1), extending inland in the south and into eastern Paraguay and northern Argentina. Besides its wide latitudinal distribution, this forest also reaches from sea level up to 1800 m high, generating evident altitudinal gradients of habitats and biodiversity. The Atlantic Forest domain includes different forest ecosystems such as the Atlantic ombrophilous dense forests, mixed ombrophilous forests, open ombrophilous forests, semi-deciduous stationary forests and deciduous stationary forests, besides the mangroves, restingas (a mosaic from herbaceous to woody vegetation that occurs on marine sand plains in the littoral), altitudinal grasslands, the countryside swamps, and the northeastern forest enclaves (Colombo and Joly, 2010; Marques et al., 2011).

The original coverage area of the Atlantic Forest biome is home to 70% of the Brazilian population, and the use of this area accounts for 80% of Brazil's gross domestic product. Despite the enormous population pressure, the Atlantic Forest has been identified as one of the 25 global biodiversity hotspots, containing between 1% and 8% of the world's total species, a high number of them being endemic to the region (Myers et al., 2000). A recent assessment revealed appreciable numbers of endemic species in several groups, including 8000 species of trees (40% of the total number of Atlantic Forest trees), 200 species of birds (16%), 71 species of mammals (27%), 94 species of reptiles (31%), and 286 species of amphibians (60%), to mention only the best-known taxonomic groups (Myers et al., 2000). The IUCN Red List contains more than 110 Atlantic Forest species, 29 of which are critically endangered. The most recent estimates suggest that only 11.4–16% of the original 1.4 million

km^2 of the Atlantic Forest has remained, with 32–40% of such remnants consisting of secondary forest and small forest fragments (<100 ha) (Ribeiro et al., 2009).

The savannas, the second most extensive biome of tropical South America, are also dominant worldwide, being responsible for about 30% of the primary production of all terrestrial vegetation (Grace et al., 2006). In South America, the savannas cover an area of more than 2.821 million km^2, mostly distributed across Brazil, Colombia, Venezuela and Bolivia (Gottsberger and Silberbauer-Gottsberger, 2009; Figure 19.1). The Brazilian savanna, locally known as Cerrado, covers about 22% of the country's surface area, extending from the southern boundaries of the Amazonian region to the southern states of São Paulo and Paraná (Oliveira-Filho and Ratter, 2002). Cerrado is a also a biodiversity hotspot (Myers et al., 2000), comprising about 12,000 plant species, ca. 800 of them being large trees and shrubs, with a high proportion of endemism and rapid loss of habitats.

The Cerrado, like other savanna-like vegetations, is characterised by a mosaic ranging from grassland to forest formations. Vegetation structure and diversity seem to be related to variation in soil physical and chemical parameters (fertility, pH, Al concentration, drainage), precipitation seasonality, and fire regime (Oliveira-Filho and Ratter, 2002). Across the region, the mean annual temperature varies between 18 and 28 °C, and annual precipitation between 800 and 2000 mm, with a continuous 5–6-month dry season in the southern hemisphere winter. The typical Cerrado physiognomy arises on well-drained, deep, dystrophic and acidic soils, with low cation exchange capacity and high levels of Al saturation (Gottsberger and Silberbauer-Gottsberger, 2009). Besides being adapted to seasonal water limitation, Cerrado plants are also considered fire-tolerant and fire-dependent (Hoffmann and Moreira, 2002; Miranda et al., 2002).

19.3 LAND USE CHANGE, AIR POLLUTANT EMISSION AND REGIONAL CLIMATE CHANGE

Although the agricultural population has declined over recent decades, now representing only 16% of labour in Latin America, agricultural activities still account for significant portions of gross national products in this region. The increasing global population and demand for resource consumption are directly related to the expansion of agricultural areas for both food and biofuel production (FAO, 2010). Related to the growth of the agricultural sector, especially because of expansion of pasture areas, forest cover in Latin America has been reduced by 8% since the 1990s. Aide et al. (2012) determined the extent and the spatial distribution of deforestation (and reforestation) between 2001 and 2010 within the 45 countries in Latin America and the Caribbean, estimating a net loss of 179,405 km^2 of woody vegetation in the region. Most of the tropical rainforest deforestation was concentrated in the Brazilian

Amazon, but Nicaragua and Guatemala also presented a negative balance between reforestation and deforestation. The dry forest and savanna/shrub biomes has also experienced extensive deforestation, so that about 200,000 km^2 of woody vegetation has been lost altogether. Most losses occurred in northern Argentina, southeastern Bolivia, and western Paraguay, but Central Brazil exhibits a large loss of native areas as well.

The conversion of rainforest and savanna habitats into agricultural and pasture lands, along with the abandonment of degraded areas, represent a substantial loss in aboveground biomass. Moreover, the fragmentation of forests and their conversion into pastures and croplands have been associated with changes in local and regional climate (Davidson et al., 2012). Depending on the local extent of deforestation and the new land use, annual rainfall may become reduced over the pasture area, while cloudiness along with instantaneous rainfall and thunderstorm intensity may increase (Avissar and Schmidt, 1998). At scales of hundreds to thousands of km^2, however, rainfall over core clearings may be suppressed (Butt et al., 2011). Deforestation in another105 km^2 is expected to significantly decrease the whole-basin precipitation (Coe et al., 2009).

The conversion of native forests into agricultural areas in Latin America, especially into pastures, is closely associated with biomass burning and the emission of air pollutants. Vegetation fires are considered a major anthropogenic source of greenhouse gases (Cieslik et al., 2013; Sofiev, 2013), aerosols and pollutants to the atmosphere during the southern hemisphere dry season (July to October) over South America. Besides changing the microphysical processes within clouds, making droplets too small to precipitate (i.e. reducing local rainfall; Andreae et al., 2004), fires can cause additional plant stress by initiating ozone pollution (Grulke, 2009; Longo et al., 1999). Apart from high levels of O_3 and its precursors occurring in the urban areas, deforestation and seasonal biomass burning are the most relevant pollution sources in Brazil, presumably enhancing O_3 regimes also at rural landscape scales (Longo et al., 1999). Actually, 75% of the total amounts of CO_2, CO and NO_x released in Brazil are thought to originate from the conversion of native into pasture and cropland (MCT, 2009).

Data from the Brazilian Inventory of Anthropogenic Greenhouse Gases Emissions and Removals demonstrate the contribution of changing land use to the emission balance in the country (MCT, 2009). The inventory takes into account five categories of economical sectors as considered by the IPCC: energy, industrial processes, agriculture, land use change and forestry, along with waste management. If only CO_2, CH_4 and N_2O are considered as major greenhouse gases (Figure 19.2), unlike in many countries in other regions of the world, industrial activities are not their most relevant sources. About 76% of CO_2 emission is derived from land use changes, compared with 1.6% and 22% from industrial and energy sectors, respectively.

It is important to note that emission estimates related to land use change and forestry are based on all kinds of conversion, including that of native

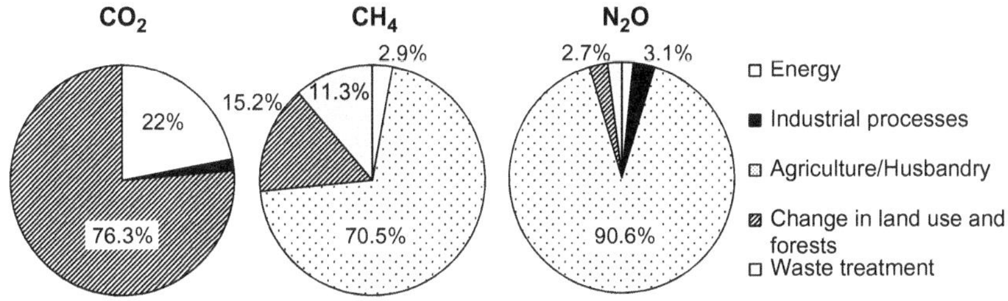

FIGURE 19.2 Contribution of the main sectors (Energy, Industrial processes, Agriculture/Husbandry, Change in land use and forests and Waste treatment) to the Brazilian annual emissions of CO_2, CH_4 and N_2O. *Data extracted from MCT (2009).*

vegetation to agricultural and pasture land, besides expansion of urban areas, construction of water reservoirs and succession towards secondary forests or reforestation. Thus, considering the expansion of pasture areas during recent years, for example, to the detriment of forest and savanna ecosystems, one can attribute a large portion of CO_2 emissions indirectly to the agricultural sector. Focusing on the Amazon and Cerrado biomes of Brazil, Bustamante et al. (2012b) estimated the proportion of greenhouse gas emissions that can be attributed to land cover conversion, in particular, to cattle ranching. The authors considered the deforestation and subsequent biomass burning for the establishment of new pasture areas, the burning related to pasture maintenance, and the enteric fermentation and wastes from cattle. The full set of emissions derived from cattle farming was responsible for approximately half of all Brazilian emissions, estimated to be approximately 1055 Mt CO_2eq in 2005, for example. Agriculture is also directly responsible for 71% of the methane and 91% of nitrous oxide emission budgets (Figure 19.2). Similar data can be also be observed in India (Pandey et al., 2013).These amounts should be added to the indirect contribution upon land use change, accounting for 15% and 3% of the total Brazilian emissions of CH_4 and N_2O, respectively.

Savanna and rainforests, as pointed out earlier, appear to be adapted to the natural variation in climate and soil conditions, as well as to natural cycles of climatic variation and extreme events, showing, for example, considerable resilience to moderate annual drought (Davidson et al., 2012 for a revision). However, what kind of responses would savanna and rainforest vegetation show to the effects of increasing land use change, accompanied by a distinct natural or anthropogenic enhancement of greenhouse-gas emissions, nitrogen deposition and variation in frequency and intensity of drought?

19.4 EFFECTS OF NITROGEN ADDITION ON NATURAL SAVANNA AND FOREST ECOSYSTEMS

Common practices related to agricultural activities in Latin America such as biomass burning, deforestation, fertiliser additions and the spreading cultivation of N-fixing crops, together with the burning of fossil fuels and urbanization, can also act indirectly on biodiversity and ecosystem functioning. As described in the previous section, many of these activities transfer large amounts of nitrogen species (NO_x, NH_3, NO_3) from the land surface into the atmosphere. Besides the effects on the cycling and formation of other gaseous pollutants such as O_3 that act as greenhouse gases, part of the released N returns to the terrestrial ecosystems through dry and wet deposition, considerably affecting the N availability to plants and, consequently, the terrestrial and aquatic ecosystems' productivity.

The N deposition over some biodiversity hotspots in Latin America may achieve levels higher than 10 kg ha^{-1} $year^{-1}$ (Phoenix et al., 2006), especially in regions of intensive agricultural activity (Dentener et al., 2006). The Brazilian Atlantic Forest is presumed to suffer the highest N deposition of about 22 kg ha^{-1} $year^{-1}$, followed by the Madrean and Magdalena regions in Colombia and the tropical Andes, which receive between 10 and 15 kg ha^{-1} $year^{-1}$. In addition, to enhance productivity, huge amounts of phosphorus and other elements are being applied to and lost from croplands in Latin America. Falkowski et al. (2000) estimated that anthropogenic introduction of nutrients has amplified global N and P cycles by 100% and 400%, respectively, since the Industrial Revolution.

The effects of the increasing availability of limiting elements such as N and P on plant diversity and functioning, as already widely observed in several ecosystems of the Northern hemisphere (Bobbink et al., 2010, see Eugster and Häni, 2013 and Tuovinen et al., 2013), have also been evaluated in some tropical biomes. A 10-year-long N and P fertilising experiment has been conducted in the Cerrado, the Brazilian kind of savanna (Bustamante et al., 2012a). Developing under conditions of seasonal rainfall distribution and highly weathered soils with low nutrient availability, plant species have been evolutionarily selected for mechanisms of high nutrient use, efficiency and retention. In response to long-term N and P additions, Cerrado plots exhibited visible changes in plant density as well as species dominance, richness and diversity, both within the herbaceous and the tree-shrub layers. Considering the woody species, plots fertilised with N alone or N plus P showed significantly lower levels of diversity and evenness indices compared to unfertilised plots (Jacobson et al., 2011). Patterns of species density and dominance in both the N and N plus P treatment deviated from those in unfertilised plots. In general, species abundant in control plots displayed a lowered density in N and N plus P plots. N and P concentrations in leaf litter were also modified by N and P additions, changing the rate of litter decomposition. Enhanced N and P levels in leaf litter were found, especially upon the combined addition of these nutrients, indicating a reduced proficiency in element re-translocation during senescence, and accelerating litter mass losses and nutrient release in comparison to the control plots (Kozovits et al., 2007).

Changes in species diversity, vegetation cover and biomass production were more evident in the herbaceous than the woody layer (Bustamante et al., 2012a). The highest and lowest species richness occurred in control and N plus P plots, respectively. The addition of P alone or in combination with N induced invasion by *Melinis minutiflora* (exotic, African C4 grass), resulting in negative impacts on native grass species. Besides changes in aboveground biomass, addition of N and P also led, although to a lesser extent, to changes in the root morphology and biomass, which was reduced under simultaneous N and P additions.

The effects of a long-term N and P addition experiment on vegetation structure and tree species biomass and diversity have also been measured in the Amazon rainforest areas undergoing secondary succession (Siddique et al., 2010). In response to 6 years of repeated N and P additions, tree assemblage evenness was reduced and tree species biomass accumulation over time was delayed. While N stimulated the growth of two pioneer and one early secondary tree species, P stimulated the growth of only the dominant pioneer tree, *Rollinia exsucca*. Absolute tree growth rates were enhanced within 2 years of nutrient addition. Conversely, nutrient-induced shifts in relative tree species growth and reduced assemblage evenness were observed even for more than 3 years after nutrient addition, favouring the nutrient-responsive species.

19.5 OZONE: A GROWING CONCERN

Although the focus of this chapter is on available long-term studies that have assessed the effects of climate change and nitrogen deposition on the vegetation in Latin America, one plant stressor intrinsically associated with and intensified through such changes is tropospheric O_3 as occurring at enhanced ground-level concentrations. There is no doubt today that ground-level O_3 regimes enhanced above pre-industrial levels must be regarded as an inherent component of climate and global change (Matyssek et al., 2013a,b). However, in comparison to other agents of such scenarios, long-term studies on O_3 action are lacking in the southern hemisphere.

Surface O_3 levels in urban areas of Brazil and other South-American countries range above 40 nl l^{-1} (Muramoto et al., 2003; Olcese and Toselli, 1998; Sánchez-Ccoyllo et al., 2006). Maximum levels typically occur during spring (August to November) with means between 40 and 60 nl l^{-1} (even up to more than 90 nl l^{-1}), mainly because of intense biomass burning. A similar situation may also occur in certain regions in Africa (Laakso et al., 2013). During seasonal vegetation fires in South America, several hundreds of fire spots can occur every day (Langmann et al., 2009), and the smoke-generated aerosols and ozone, transported by the atmospheric motions, produce a huge plume of continental scale covering areas of around 4–5 $\times$ 10^6 km^2 (Freitas et al., 2005). Long-range O_3 transport has been measured in Brazil throughout latitudes of 35° to 5°S (Boian and Kirchhoff, 2005). In Paraná State of Southern Brazil, high O_3 events with maximum levels of 93–173 nl l^{-1} (average around 89 nl l^{-1}) were observed during the dry season. Such episodes were a consequence of transport from regions with intense biomass burning, as is the case in Para (Amazonian Rain Forest region), Mato Grosso, Mato Grosso do Sul and Tocantins (savanna-rain forest transition regions). Thus, the air quality of areas neighbouring fire zones is often reduced, and in some cases, may turn even worse than that measured in some South American megacities, such as São Paulo (Brazil Health, 2006).

Although ozone should obviously be conceived as a factor that can cause serious impacts on the vegetation of tropical and subtropical ecosystems, there are few studies, especially on a long-term scale, which are devoted to measuring O_3 effects on plants in Latin America. In Brazil, responses of tree species of the Atlantic Rain Forest to elevated O_3 levels have been based on macroscopic foliar injury and physiological parameters (e.g. Bulbovas et al., 2010; Furlan et al., 2007, 2008; Moraes et al., 2006; Pina and Moraes, 2010; Rezende and Furlan, 2009). In such surveys, native tree species have proven to be sensitive to current and predicted O_3 levels in SE Brazil. Recent studies have also concentrated efforts on finding suitable species for acting as bio-indicators of O_3 impact in the tropics (Dafré-Martinelli et al., 2011; Ferreira et al., 2012).

19.6 VEGETATION RESPONSES TO GLOBAL CHANGE

Given the currently predicted scenarios of global change in South America, relevant modifications in tropical and subtropical forest dynamics are expected to occur (Marengo et al., 2009). Scenarios such as SRES A2 and B2 (IPCC, 2007) presume a reduction in rainfall in Amazonia and in tropical South America east of the Andes, warm nights across the entirety of South America and an increase in rainfall extremes in the SE of South America from 2071 through 2100.

Variation in rainfall amount and distribution and increase in atmospheric CO_2 are considered to be major drivers of vegetation changes (Bytnerowicz et al., 2013; Lewis et al., 2009; Malhi et al., 2008). Increases in temperature and humidity evidentially generate changes in the energy balance at the earth's surface, enhancing the seasonality in tropical regions, with shortened but intense rainy seasons and accentuated, prolonged dry seasons (Cox et al., 2008; Malhi et al., 2008; Phillips et al., 2009).

The dry season that occurred in the western Amazon in 2005 has been the most severe over the past 100 years (Fisher et al., 2007), being considered similar to what is expected to occur with increasing frequency during the twenty-first century. Several changes resulted in the Amazon forest. A 100 mm deficit on average in rainfall caused losses of 5.3 t in above-ground biomass production per hectare ($1.2–1.6 \times 10^9$ t throughout the Amazon rainforest). The group of softwood trees, including pioneer and some successionally intermediate trees species, proved to be sensitive to drought, exhibiting enhanced mortality rates. In the long term, such selective death might alter forest composition, biodiversity and functioning. Also in 2005, the third highest atmospheric CO_2 concentration since the beginning of recording was registered. Part of this increase was suggested to relate to the drought-induced reduction in carbon stocks of neotropical rain forests, caused by tree death and reduction in net primary productivity (Phillips et al., 2009).

For clarification of the effects of prolonged drought in the Amazon Forest, a throughfall exclusion experiment is being performed in Tapajós, eastern Brazil, since 2000. The understory of one hectare of evergreen tropical seasonal forest remained covered with plastic panels (installed at 1–2 m height) throughout the rainy seasons (Nepstad et al., 2002). About 40% of the annual rainfall has experimentally been intercepted, simulating rainfall regimes similar to those of the savanna climate. Over the last decade, the decline in net primary productivity and a 38% increase in mortality were the main effects of the water deficit in the Amazon forest, reducing the total biomass by about 46.5 t ha^{-1} and thus, the stock of forest carbon (Brando et al., 2008; Fisher et al., 2008; Nepstad et al., 2007).

Besides the effects of drought, the increasing atmospheric CO_2 concentration is considered to be another strong factor that drives the dynamics and changes in tropical forests (Lloyd and Farquhar, 2008). The Amazon Forest Inventory Network, known as 'RAINFOR' (Malhi et al., 2002), has measured significant increases of about 0.62 ± 0.23 t C ha^{-1} $year^{-1}$ in above-ground biomass during the 1980s and 1990s (Baker et al., 2004). Such an increase sufficiently balanced the carbon lost through deforestation in the region. Rates of mortality, growth and recruitment were accelerated, as well as forest productivity, as a response to the largest supply of limiting resources for growth, especially CO_2 (Phillips et al., 1998). Studies in tropical forests have demonstrated that under elevated CO_2 concentrations, seedlings and understory trees lowered light compensation points, exhibiting a growth increase of up to 50% with less than 0.5% of the daylight (Würth et al., 1998).

RAINFOR data also suggest an overall imbalance amongst forest functional groups, with a substantial increase in density and relative dominance of lianas (see above, Phillips et al., 2002). Körner (2004) found that seedling growth of the tropical liana *Gonolobus viridiflorus* was stimulated by 130% under CO_2 concentrations near the current (420 μl l^{-1}) as compared to the pre-industrial level (280 μl l^{-1}). Lianas are structural parasites competing with trees for light, exerting a major ecological impact on forest functioning. If highly abundant, they intensify shading, suppress tree growth and increase tree mortality. Such effects can cause forest biomass loss as well as changes in forest species composition and structure.

19.7 CONCLUSIONS AND FUTURE DIRECTIONS

Examples of long-term studies presented here indicate that the large biomes in Latin America, that is, both the rainforest and savanna ecosystems, are highly sensitive to the anthropogenic increase in the availabilities of N, P and CO_2, which typically limit plant productivity, and potentially in O_3 levels, which exacerbate the limitation. Notwithstanding the limiting O_3 effects, productivity may increase, although shifts may occur between plant functional groups. Alien species may become profiteers, if they gain in competitiveness under a less limiting environment. Conversely, limitation in rainfall or by enhanced O_3 levels can result in productivity loss and selective death of functional groups. How may plant functional groups respond to the combined effects of increased N and CO_2 availabilities and shortage of water supply during the dry season? Can increased CO_2 concentrations counteract drought or O_3 effects? In the long term, changes in plant species composition and vegetation structure, reductions in biodiversity, and changes in C stocks and fluxes, including that of water and nutrients, are to be expected (see also Cudlín et al., 2013). Additional factors must be considered when predicting vegetation responses and ecosystem functioning under future climate and air pollution scenarios characterised by enhanced heat and drought. Such factors are, for example, fire incidences (Malhi et al., 2008) and herbivory rates (Bale et al., 2002), which are expected to increase. The quantification of such changes and their effects on ecosystems are current challenges for ecological research in Latin America. Further studies should seek to analyse the joint effects of all such factorial drivers that are crucial modulators of forest dynamics. Moreover, with an economy in transition and ascending, Latin America should also further modify the land use in response to regional and global demands for food, biofuels and forest resources. Thus, there is urgent need for an improved understanding of the trade-offs between land cover, air pollution, carbon stocks, climate change, habitat conservation and human health under future scenarios of economic development (Davidson et al., 2012). Continued augmentation of scientific knowledge and technological capacity, as well as improvements in promoting the training of new scientists and technologists will be required in the Latin American region to provide adequate support to policy-making.

ACKNOWLEDGEMENT

The authors thank Dr. Mariangela G.P. Leite for designing the figures.

REFERENCES

Aide, T.M., Clark, M.L., Grau, H.R., Lopez-Carr, D., Levy, M.A., Redo, D., Bonilla-Moheno, M., Riner, G., Andrade-Nunez, M.J., Muniz, M., 2012. Deforestation and reforestation of Latin America and the Caribbean (2001–2010). Biotropica 45, 1–10.

Andreae, M.O., Rosenfeld, D., Artaxo, P., Costa, A.A., Frank, G.P., Longo, K.M., Silva-Dias, M. A.F., 2004. Smoking rain clouds over the Amazon. Science 303, 1337–1342.

Avissar, R., Schmidt, T., 1998. An evaluation of the scale at which ground-surface heat flux patchiness affects the convective boundary layer using a large-eddy simulation model. J. Atmos. Sci. 55, 2666–2689.

Baker, T.R., Phillips, O.L., Malhi, Y., Almeida, S., Arroyo, L., Di Fiore, A., Erwin, T., Higuchi, N., Killeen, T.J., Laurance, S.G., Laurance, W.F., Lewis, S.L., Monteagudo, A., Neill, D.A., Vargas, P.N., Pitman, N.C.A., Silva, J.M.N., Martínez, R.V., 2004. Increasing biomass in Amazonian forest plots. Philos. Trans. R. Soc. Lond. B 359, 353–365.

Bale, J.S., Masters, G.J., Hodkinson, I.D., Awmack, C., Bezemer, T.J., Brown, V.K., Butterfiel, J., Buse, A., Coulson, J.C., Farrar, J., Good, J.E.G., Harrinton, R., Hatley, S., Jones, T.H., Lindroth, R.L., Press, M.C., Syrmnioudis, I., Watt, A.D., Wittaker, J.D., 2002. Herbivory in global climate change research: direst effects of rising temperatures on insect herbivores. Glob. Chang. Biol. 8, 1–16.

Bobbink, R., Hicks, K., Galloway, J., Spranger, T., Alkemade, R., Ashmore, M., Bustamante, M.M.C., Cinderby, S., Davidson, E., Dentener, F., Emmett, B., Erisman, J.-W., Fenn, M., Gilliam, F., Nordin, A., Pardo, L., de Vries, W., 2010. Global assessment of nitrogen deposition effects on terrestrial plant diversity: a synthesis. Ecol. Appl. 20 (1), 30–59.

Boian, C., Kirchhoff, V.W.J.H., 2005. Surface ozone enhancements in the south of Brazil owing to large-scale air mass transport. Atmos. Environ. 39, 6140–6146.

Bovarnick, A., Alpizar, F., Schnell, C. (Eds.), 2010. The Importance of Biodiversity and Ecosystems in Economic Growth and Equity in Latin America and the Caribbean: An Economic Valuation of Ecosystems. United Nations Development Programme, New York.

Brazil Health, 2006. An analysis of the health situation in Brazil, Secretaria de Vigilância em Saúde, Departamento de Análise de Situação em Saúde. – Brasília: Ministério da Saúde, Brasil, 620 p.: il. – (Série G. Estatística e Informação em Saúde). ISBN 85-334-1223-1.

Brando, P.M., et al., 2008. Drought effects on litterfall, wood production, and belowground carbon cycling in an Amazon forest: results of a throughfall reduction experiment. Phil. Trans. R. Soc. B 363, 1839–1848.

Bulbovas, P., Moraes, R.M., Rinaldi, M.C.S., Cunha, A.L., Delitti, W.B.C., Domingos, M., 2010. Leaf antioxidante fluctuations and growth responses in saplings of Caesalpinia echinata Lam. (brazilwood) under an urban stressing environment. Ecotoxicol. Environ. Saf. 73, 664–670.

Bustamante, M.M.C., de Brito, D.Q., Kozovits, A.R., Luedemann, G., de Mello, T.R.B., Pinto, A.S., Munhoz, C.B.R., Takahashi, F.S.C., 2012a. Effects of nutrient additions on plant biomass and diversity of the herbaceous-subshrub layer of a Brazilian savanna (Cerrado). Plant Ecol. 213, 795–808.

Bustamante, M.M.C., Nobre, C.A., Smeraldi, R., Aguiar, A.P.D., Barioni, L.G., Ferreira, L.G., Longo, K., May, P., Pinto, A.S., Ometto, J.P.H.B., 2012b. Estimating greenhouse gas emissions from cattle raising in Brazil. Clim. Change 115, 559–577.

Butt, N., Oliveira, P.A., Costa, M.H., 2011. Evidence that deforestation affects the onset of the rainy season in Rondonia, Brazil. J. Geophys. Res. 116, D11120.

Bytnerowicz, A., Fenn, M., McNulty, S., Yuan, F., Pourmokhtarian, A., Driscoll, C., Meixner, T., 2013. Interactive effects of air pollution and climate change on forest ecosystems in the United States—current understanding and future scenarios. Dev. Environ. Sci. 13, 333–369.

Cieslik, S., Tuovinen, J.-P., Baumgarten, M., Matyssek, R., Brito, P., Wieser, G., 2013. Gaseous exchange between atmosphere and forests. Dev. Environ. Sci. 13, 19–36.

Coe, M.T., Costa, M.H., Soares-Filho, B.S., 2009. The Influence of historical and potential future deforestation on the stream flow of the Amazon River—land surface processes and atmospheric feedbacks. J. Hydrol. 369, 165–174.

Colombo, A.F., Joly, C.A., 2010. Brazilian Atlantic Forest lato sensu: the most ancient Brazilian forest, and a biodiversity hotspot, is highly threatened by climate change. Braz. J. Biol. 70 (3), 697–708.

Cox, P.M., Harris, P.P., Huntingford, C., Betts, R.A., Collins, M., Jones, C.D., Jupp, T.E., Marengo, J.A., Nobre, C.A., 2008. Increasing risk of Amazonian drought due to decreasing aerosol pollution. Nature 453, 212–215.

Cudlín, P., Seják, J., Pokorný, J., Albrechtová, J., Bastian, O., Marek, M., 2013. Forest ecosystem services under climate change and air pollution. Dev. Environ. Sci. 13, 333–369.

Dafré-Martinelli, M., Nakazato, R.K., Dias, A.P.L., Rinaldi, M.C.S., Domingos, M., 2011. The redox state of Ipomoea nil 'Scarlet O Hara' growing under ozone in a subtropical area. Ecotoxicol. Environ. Saf. 74, 1645–1652.

Davidson, E.A., de Araújo, A.C., Artaxo, P., Balch, J.K., Brown, I.F., Bustamante, M.M.C., Coe, M.T., DeFries, R.S., Keller, M., Longo, L.M., Munger, J.W., Schroeder, W., Soares-Filho, B.S., Souza Jr., C.M., Wofsy, S.C., 2012. The Amazon basin in transition. Nature 481, 321–328.

Dentener, F., et al., 2006. Nitrogen and sulfur deposition on regional and global scales: a multimodel evaluation. Global Biogeochem. Cycles 20, GB4003.

Dirzo, R., 2001. Tropical forests. In: Chapin III., F.S., Sala, O.E., Huber-Sannwald, E. (Eds.), Global Biodiversity in a Changing Environment, Scenarios for the 21st Century. Springer-Verlag, New York, pp. 251–276.

Eugster, W., Häni, M., 2013. Nutrients or pollutants?—nitrogen deposition to forests. Dev. Environ. Sci. 13, 37–56.

Falkowski, P.G., Scholes, R.J., Boyle, E., Canadell, J., Canfield, D., Elser, J., 2000. The global carbon cycle: a test of our knowledge of Earth as a system. Science 290, 291–296.

FAO, 2010. Global forest resources assessment main report: FAO forestry paper #163. Food and Agriculture Organization of the United Nations. Rome, Italy.

Ferreira, M.L., Esposito, J.B.N., de Souza, S.R., Domingos, M., 2012. Critical analysis of the potential of Ipomoea nil 'Scarlet O'Hara' for ozone biomonitoring in the sub-tropics. J. Environ. Monit. 14, 1959–1967.

Fisher, R.A., Williams, M., da Costa, A.L., Malhi, Y., da Costa, R.F., Almeida, S., Meir, P., 2007. The response of an Eastern Amazonian rain forest to drought stress: results and modelling analyses from a throughfall exclusion experiment. Glob. Chang. Biol. 13, 2361–2378.

Fisher, J.I., Hurtt, G.C., Thomas, R.Q., Chambers, J.Q., 2008. Clustered disturbances lead to bias in large-scale estimates based on forest sample plots. Ecol. Lett. 11, 554–563.

Freitas, S.R., Longo, K.M., Silva Dias, M., Silva Dias, P., Chatfield, R., Prins, E., Artaxo, P., Grell, G., Recuero, F., 2005. Monitoring the transport of biomass burning emissions in South America. Environ. Fluid Mech. 5 (1–2), 135–167.

Furlan, C.M., Moraes, R.M., Bulbovas, P., Domingos, M., Salatino, A., Sanz, M.J., 2007. *Psidium guajava* 'Paluma' (the guava plant) as a new bio-indicator of ozone in the tropics. Environ. Pollut. 147, 691–695.

Furlan, C.M., Moraes, R.M., Bulbovas, P., Sanz, M.J., Domingos, M., Salatino, A., 2008. *Tibouchina pulchra* (Cham.) Cogn., a native Atlantic Forest species, as a bio-indicator of ozone: visible injury. Environ. Pollut. 152, 361–365.

Gentry, A.H., 1988. Tree species richness of upper Amazonian forests. Proc. Natl. Acad. Sci. U.S.A. 85, 156–159.

Gottsberger, G., Silberbauer-Gottsberger, I., 2009. Tropical biology and conservation management: tropical savannas. In: Encyclopedia of Life Support Systems (EOLSS), Developed Under the Auspices of the UNESCO, vol. X. Eolss Publishers, Oxford, UK. http://www.eolss.net.

Grace, J., San José, J., Meir, P., Miranda, H.S., Montes, R.A., 2006. Productivity and carbon fluxes of tropical savannas. J. Biogeogr. 33, 387–400.

Grulke, N.E., 2009. Interactive effects of climate and wildland fires on forests and other ecosystems—section III synthesis. In: Bytnerowicz, A., Arbaugh, M., Riebau, A., Andersen, C. (Eds.), Dev. Environ. Sci., vol. 8. Elsevier, Amsterdam, The Netherland, p. 463, Copyright® 2009 Elsevier B.V.

Hoffmann, W.A., Moreira, A.G., 2002. The role of fire in population dynamics of woody plants. In: Oliveira, P.S., Marquis, R.J. (Eds.), The Cerrados of Brazil. Columbia University Press, New York, pp. 159–177.

IPCC, 2001. Latin America. In: Watson, R.T., Zinyowera, M.C., Moss, R.H. (Eds.), IPCC Special Report on The Regional Impacts of Climate Change: An Assessment of Vulnerability. GRID-Arendal. http://www.grida.no/publications/other/ipcc_sr/.

IPCC, 2007. The physical science basis. Contribution of working group I. In: Solomon, S., et al., (Eds.), Fourth Assessment Report of the Intergovernmental Panel on Climate Change, Cambridge University Press, Cambridge, pp. 1–996.

Jacobson, T.K.B., Bustamante, M.M.C., Kozovits, A.R., 2011. Diversity of shrub tree layer, leaf litter decomposition and N release in a Brazilian Cerrado under N, P and N plus P additions. Environ. Pollut. 159, 2236–2242.

Körner, C., 2004. Through enhanced tree dynamics carbon dioxide enrichment may cause tropical forests to lose carbon. Philos. Trans. R. Soc. Lond. B. 359, 493–498.

Kozovits, A.R., Bustamante, M.M.C., Garofalo, C.R., Bucci, S., Franco, A.C., Goldstein, G., Meinzer, F.C., 2007. Nutrient resorption and patterns of litter production and decomposition in a Neotropical savanna. Funct. Ecol. 21, 1034–1043.

Laakso, L., Beukes, J.P., Van Zyl, P.G., Pienaar, J., Josipovic, M., Venter, A., Jaars, K., Vakkari, V., Labuschagne, C., Chiloane, K., Tuovinen, J.-P., 2013. Ozone concentrations and their potential impacts on vegetation in southern Africa. Dev. Environ. Sci. 13, 429–450.

Langmann, B., Duncan, B., Textor, C., Trentmann, J., van der Werf, G.R., 2009. Vegetation fire emissions and their impact on air pollution and climate. Atmos. Environ. 43, 107–116.

Lewis, S.L., Lloyd, J., Stich, S., Mitchard, E.T.A., 2009. Changing ecology of tropical forests: evidences and drivers. Annu. Rev. Ecol. Evol. Syst. 40, 529–549.

Lloyd, J., Farquhar, G.D., 2008. Effects of rising temperatures and [CO_2] on the physiology of tropical forest trees. Philos. Trans. R. Soc. Lond. B 363, 1811–1817.

Longo, K.M., Thompson, A.M., Kirchhoff, V.W.J.H., Remer, L.A., de Freitas, S.R., Dias, M.A.F.S., Artaxo, P., Hart, W., Spinhirne, J.D., Yamasoe, M.A., 1999. Correlation between smoke and tropospheric ozone concentration in Cuiaba´ during SCAR-B. J. Geophys. Res. 104 (D10), 12113–12129.

Malhi, Y., et al., 2002. An international network to monitor the structure, composition and dynamics of Amazonian forests (RAINFOR). J. Veg. Sci. 13, 439–450.

Malhi, Y., Aragão, L.E.O.C., Galbraith, D., Huntingford, C., Fisher, R., Zelazowski, P., Sitch, S., McSeweeney, C., Meir, P., 2008. Exploring the likelihood and mechanism of a climate-change-induced dieback of the Amazon rainforest. Proc. Natl. Acad. Sci. U.S.A. 106, 1–6.

Marengo, J., Jones, R., Alves, L.M., Valverde, M.C., 2009. Future change of temperature and precipitation extremes in South America as derived from the PRECIS regional climate model system. Int. J. Climatol. http://dx.doi.org/10.1002/joc.1863.

Marques, M.C.M., Swaine, M.D., Liebsch, D., 2011. Diversity distribution and floristic differentiation of the coastal lowland vegetation: implications for the conservation of the Brazilian Atlantic Forest. Biodivers. Conserv. 20 (1), 153–168.

Matyssek, R., Kozovits, A.R., Schnitzler, J.-P., Pretzsch, H., Dieler, J., Wieser, G., 2013a. Forest trees under air pollution as a factor of climate change. In: Tausz, M., Grulke, N. (Eds.), Trees in a Changing Environment: Ecophysiology, Adaptation and Future Survival. Plant Ecophysiology, Springer, Cambridge.

Matyssek, R., Wieser, G., Fleischmann, F., Grünhage, L., 2013b. Ozone Research, quo vadis? Lessons from the free-air canopy fumigation experiment at Kranzberg Forest. Dev. Environ. Sci. 13, 103–129.

Ministério de Ciência e Tecnologia, 2009. Inventário brasileiro das emissões e remoções antrópicas de gases de efeito estufa: informações gerais e valores preliminares. http://www.mct.gov.br/index.php/content/view/310922/Segundo_Inventario_Brasileiro_de_Emissoes_e_Remocoes_Antropicas_de_Gases_de_Efeito_Estufa.html.

Miranda, H.S., Bustamante, M.M.C., Miranda, A.C., 2002. The fire factor. In: Oliveira, P.S., Marquis, R.J. (Eds.), The Cerrados of Brazil. Columbia University Press, New York, pp. 51–68.

Moraes, R.M., Bulbovas, P., Furlan, C.M., Domingos, M., Meirelles, S.T., Delitti, W.B.C., Sanz, M.J., 2006. Physiological responses of saplings of Caesalpinia echinata Lam., a Brazilian tree species, under ozone fumigation. Ecotoxicol. Environ. Saf. 63, 306–312.

Muramoto, C.A., Lopes, C.F.F., Lacava, C.I.V., 2003. Study of Tropospheric Ozone in São Paulo Metropolitan Region. In: A&WMA's 96th Annual Conference & Exhibition "Energy, Economic and Global Challenges," San Diego/EUA—June.

Myers, N.R.A., Mittermeier, C.G., Mittermeier, G.A.B., Kent, J., 2000. Biodiversity hotspots for conservation priorities. Nature 403, 853–858.

Nepstad, D.C., Moutinho, P., Dias-Filho, M.B., Davidson, E.A., Cardinot, G., Markewitz, D., Figueiredo, R., Vianna, N., Chambers, J.Q., Ray, D.G., Guerreiros, J.B., Lefebvre, P.A., Sternberg, L., Moreira, M.Z., Barros, L., Ishida, F.Y., Tohlver, I., Belk, E., Kalif, K., Schwalbe, K., 2002. The effects of partial throughfall exclusion on canopy processes, aboveground production, and biogeochemistry of an Amazon forest. J. Geophys. Res. 107 (D20), 8085.

Nepstad, D.C., Tohver, I.M., Ray, D., Moutinho, P., Cardinot, G., 2007. Mortality of large trees and lianas following experimental drought in an amazon forest. Ecology 88, 2259–2269.

Olcese, L.E., Toselli, B.M., 1998. Unexpected high levels of ozone measured in Córdoba, Argentina. J. Atmos. Chem. 31, 269–279.

Oliveira-Filho, A.T., Ratter, J.A., 2002. Vegetation physiogmomies and woody flora of the Cerrado biome. In: Oliveira, P.S., Marquis, R.J. (Eds.), The Cerrados of Brazil. Columbia University Press, New York, pp. 91–120.

Pandey, V., Oksanen, E., Singh, N., Sharma, C., 2013. Impacts of air pollution and climate change: implications for India. Dev. Environ. Sci. 13, 391–410.

Phillips, O.L., Malhi, Y., Higuchi, N., Laurance, W.F., Nunez, P.V., et al., 1998. Changes in the carbon balance of tropical forests: evidence from long-term plots. Science 282, 439–442.

Phillips, O.L., Malhi, Y., Vinceti, B., Baker, T., Lewis, S.L., et al., 2002. Changes in growth of tropical forests: evaluating potential biases. Ecol Appl 12, 576–587.

Phillips, O.L., Aragão, L.E.O.C., Lewis, S.L., Fisher, J.B., Lloyd, J., López-González, G., Malhi, Y., Monteagudo, A., Peacock, J., Quesada, C.A., van der Heijden, G., Almeida, S., Amaral, I., Arroyo, L., Aymard, G., Baker, T.R., Bánki, O., Blanc, L., Bonal, D., Brando, P., Chave, J., de Oliveira, A.C.A., Cardozo, N.D., Czimczik, C.I., Feldpausch, T.R., Freitas, M.A., Gloor, E., Higuchi, N., Jiménez, E., Lloyd, G., Meir, P., Mendoza, C., Morel, A., Neill, D.A., Nepstad, D., Patiño, S., Peñuela, M.C., Prieto, A., Ramírez, F., Schwarz, M., Silva, J., Silveira, M., Thomas, A.S., ter Steege, H., Stropp, J.,

Vásquez, R., Zelazowski, P., Dávila, E.A., Andelman, S., Andrade, A., Chao, K.-J., Erwin, T., Di Fiore, A., Honorio, E.C., Keeling, H., Killeen, T.J., Laurance, W.F., Cruz, A.P., Pitman, N.C.A., Vargas, P.N., Ramírez-Angulo, H., Rudas, A., Salamão, R., Silva, N., Terborgh, J., Torres-Lezama, A., 2009. Drought Sensitivity of the Amazon Rainforest. Science 323, 1344–1347. http://dx.doi.org/10.1126/science.1164033.

Phoenix, G.K., Hicks, W.K., Cinderby, S., Kuylenstierna, J.C.I., Tock, W.D.S., Dentener, F.J., Giller, K.E., Austin, A.T., Lefroy, R.D.B., Gimeno, B.S., Ashmore, M.R., Ineson, P., 2006. Atmospheric nitrogen deposition in world biodiversity hotspots: the need for a greater global perspective in assessing N deposition impacts. Glob. Chang. Biol. 12, 470–476.

Pina, J.M., Moraes, R.M., 2010. Gas exchange, antioxidants and foliar injuries in saplings of a tropical woody species exposed to ozone. Ecotoxicol. Environ. Saf. 73, 685–691.

Rezende, F.M., Furlan, C.M., 2009. Anthocyanins and tannins in ozone-fumigated guava trees. Chemosphere 76, 1445–1450.

Ribeiro, M.C., Metzger, J.P., Martensen, A.C., Ponzoni, F.J., Hirota, M.M., 2009. The Brazilian Atlantic Forest: how much is left, and how is the remaining forest distributed? Implications for conservation. Biol. Conserv. 142, 1141–1153.

Sánchez-Ccoyllo, O.R., Ynoue, Y.R., Martins, D.L., Andrade, F.M., 2006. Impacts of ozone precursor limitation and meteorological variables on ozone concentration in Sao Paulo, Brazil. Atmos. Environ. 40, S552–S562.

Siddique, I., Vieira, I.C.G., Schmidt, S., Lamb, D., Carvalho, C.J.R., Figueiredo, R.O., Blomberg, S., Davidson, E.A., 2010. Nitrogen and phosphorus additions negatively affect tree species diversity in tropical forest regrowth trajectories. Ecology 91 (7), 2121–2131.

Sofiev, M., 2013. Wildland Fires: Monitoring, Plume Modelling, Impact on Atmospheric Composition and Climate. Dev. Environ. Sci. 13, 451–472.

Tuovinen, J.-P., Hakola, H., Karlsson, P.E., Simpson, D., 2013. Air pollution risks to northern European forests in a changing climate. Dev. Environ. Sci. 13, 77–99.

UNEP—United Nations Environment Programme, 2010. State of biodiversity in Latin America and the Caribbean. www.cbd.int/gbo/gbo3/doc/StateOfBiodiversity-LatinAmerica.pdf.

Wittmann, F., Schöngart, J., Montero, J.C., Motzer, T., Junk, W.J., Piedade, M.T.F., Queiroz, H.L., Worbes, M., 2006. Tree species composition and diversity gradients in white-water forests across the Amazon Basin. J. Biogeogr. 33, 1334–1347.

Würth, M.K.R., Winter, K., Körner, C.H., 1998. *In situ* responses to elevated CO_2 in tropical forest understorey plants. Funct. Ecol. 12, 886–895.

Chapter 20

Ozone Concentrations and Their Potential Impacts on Vegetation in Southern Africa

Lauri Laakso*[,†,1], Johan Paul Beukes[†], Pieter Gideon Van Zyl[†], Jacobus J. Pienaar[†], Miroslav Josipovic[†], Andrew Venter[†], Kerneels Jaars[†], Ville Vakkari*[,‡], Casper Labuschagne[§], Kgaugelo Chiloane[¶] and Juha-Pekka Tuovinen*

*Finnish Meteorological Institute, Helsinki, Finland

[†]Environmental Sciences and Management, North-West University, Potchefstroom, South Africa

[‡]Department of Physics, University of Helsinki, Helsinki, Finland

[§]Cape Point GAW Observatory, South African Weather Service, Stellenbosch, South Africa

[¶]Sustainability and Innovation, Environmental Sciences Department, Eskom, Johannesburg, South Africa

[1]Corresponding author: e-mail: lauri.laakso@fmi.fi

Chapter Outline

20.1 INTRODUCTION

Ozone is a trace gas capable of damaging cells and tissue of plants due to its relatively strong oxidizing capacity (Dizengremel et al., 2013; Matyssek et al., 2013a,b). Ozone is found naturally in the troposphere as a result of the injection of stratospheric air and photochemical reactions involving nitrogen

Developments in Environmental Science, Vol. 13. http://dx.doi.org/10.1016/B978-0-08-098349-3.00020-7

oxides (e.g. from lightning and soils). Naturally occurring carbon monoxide and biogenic volatile organic compounds (BVOCs) serve as precursor species for the formation of ozone (Loreto and Fares, 2013). However, tropospheric ozone concentrations have significantly increased since the pre-industrial times due to increased anthropogenic emissions of ozone precursors (The Royal Society, 2008).

Adverse ecological impacts of ozone on plants typically result from reactions inside the plant; therefore these effects are modulated by the capability of ozone to enter the leaf through the stomata (Dizengremel et al., 2013). The gas exchange is governed by photosynthesis and evapotranspiration as vegetation tends to control the loss of water while taking up carbon dioxide through the stomata. Also, the species and phenological state of the vegetation affect the potential for damage to plants (Karnosky et al., 2007).

In this chapter, we discuss the current state of ozone pollution in southern Africa, mainly in the Republic of South Africa, which is one of the fastest growing economical areas in the world. We review previous studies and present new observational data, with an aim to provide an overview of the current ozone levels in southern Africa, which has been identified as one of the pollution hotspots in the southern hemisphere (Lourens et al., 2012; Matyssek et al., 2012). Even though experimental data are limited, we aim to provide some new information on regional ozone levels and also on the potential impacts of ozone on local forests and agriculture.

20.2 SOUTH AFRICAN BIOMES

Vegetation in South Africa ranges from desert on the Atlantic coast (west coast) to almost sub-tropical areas in the north-eastern Indian Ocean coast close to Mozambique (east coast) (Carruthers, 2008). The main biomes include the Fynbos in the south-west coast, Karoo in the central south-west, grassland in the central interior and savannah surrounding the grassland, and on lower latitudes (Figure 20.1). Fynbos is a local, endangered bush-type ecosystem, while Karoo is a semi-desert biome, which is characterized by limited precipitation, small bushes and sparse grass. The main economically relevant biomes are the grasslands and savannah. The grasslands endure heavy commercial agricultural use, especially maize cultivation, and grazing. The savannah areas, especially in the northern parts of South Africa and in neighbouring countries, are partly in agricultural use, but significant portions are also used as nature parks (e.g. the Kruger National Park). After the nationalization of farm land in Zimbabwe and the resulting collapse of agricultural production, South Africa has been the main source of agricultural crops in southern Africa.

While forests in general are important for local economics and households, the forests located especially to the east and south-east of the Drakensberg mountain range are commercially important, as the annual timber sales in

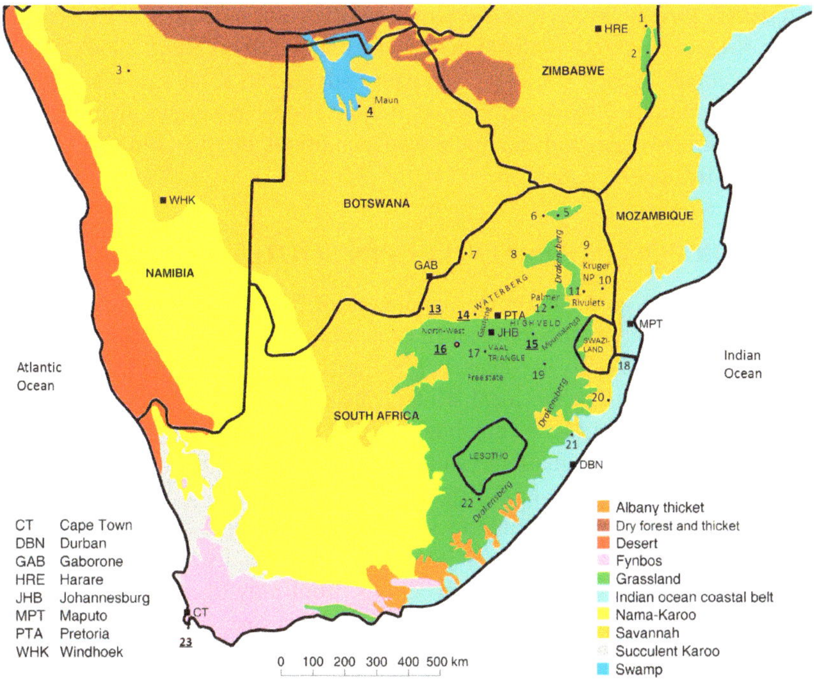

FIGURE 20.1 Biomes of southern Africa. The indigenous and planted forests are scattered through the Drakensberg mountain range on the East coast. The measurement sites included in this study: 1. Luluche (Zimbabwe), 2. Nyanga (Zimbabwe), 3. Okaukuejo (Namibia), 4. Maun (Botswana), 5. Louis Trichardt, 6. Lesheba Wilderness, 7. Kameelfontein, 8. Witvinger, 9. Grietjie, 10. Skukuza, 11. Bergvliet, 12. Palmer, 13. Botsalano, 14. Marikana, 15. Elandsfontein, 16. Welgegund, 17. Vaal Triangle, 18. Madlangune Camp, 19. Amersfoort, 20. Hluhluwe, 21. Salem Estates, 22. BenMac Dhui, 23. Cape Point GAW. For details and references, see Table 20.1. *Vegetation map adapted from data derived from White (1983) and biomes named as suggested by Mucina and Rutherfors (2006).*

2011 were 21.4 billion Rands (approximately 2 billion Euros; DAFF, 2011a). Larger areas of cultivated forest (covering approximately 1,273,000 ha, or 1% of the total area of South Africa; DAFF, 2011a) are common in the north-eastern savannah and higher altitude grasslands. Typical farmed forests consist of alien species such as different pine (esp. *Pinus patula*) and eucalyptus (esp. *Eucalyptus grandis*) species. Although these are fast-growing species and thus of high economic value, there have been discussions on the adverse effects of alien species on local ecosystems (Cossalter and Pye-Smith, 2003). Taking into account the higher value of timber and use of water being limited in arid ecosystems, the overall economic benefit of local indigenous species may actually be comparable to fast-growing alien species (Wise et al., 2011). This enhanced economic value also excludes the positive effects of indigenous forests on local ecosystems, biodiversity, tourism and especially on the low-income rural population (Shackleton and Shackleton, 2004). Since the indigenous species are also often better adapted to local stresses such as droughts, naturally high ozone concentrations, fires and pests, their ability of adaptation to changing environmental conditions due to climate change, for example, may be enhanced.

20.3 AIR POLLUTION SOURCES IN SOUTHERN AFRICA

South Africa has an approximate population of 55 million people, which is concentrated mainly in the south and east coast, as well as in the north-eastern interior (CIESIN, 2010). The densely populated areas are often relatively polluted, especially during the local winter, due to the presence of pronounced atmospheric inversion layers, combined with a regional high-pressure system that can trap pollutants in a large anti-cyclonic vortex covering the interior of southern Africa (Garstang et al., 1996).

In general, the air pollutants in southern Africa originate from industry, wild fires and domestic burning. Wild fires occur in the dry season that occurs during late autumn, winter and early spring (from May to middle October—exact timing depending on latitude and a specific year) (Sofiev, 2013; van der Werf et al., 2010). Several savannah species are also significant sources of BVOCs (Guenther et al., 1996; Otter et al., 2002, 2003). Intense solar radiation together with large emissions creates a highly reactive pollution mixture (Swap et al., 2003). Different from Europe and the United States, the acidic deposition is also still a major issue close to industrial areas (Josipovic et al., 2011), whereas domestic burning poses a serious health risk to a large portion of the population that lives in informal settlements (Venter et al., 2012).

During the last years, three specific industrial areas in South Africa, all located within 200 km from Johannesburg, have been declared as national air pollution priority areas by the South African government. This declaration implies that emission abatement measures have to be put into place by the government to mitigate air pollution. The Vaal Triangle Air-shed Priority Area incorporates most of the petrochemical-related industries, pyrometallurgical smelters, coal-fired power plants and other industrial activities. This area that includes a part of the Free State province and a small section of the Gauteng province was declared a priority area in 2005 (Government Gazette Republic of South Africa, 2005). In 2007, the Highveld Priority Area in the western Mpumalanga and a small part of the Gauteng province, with the large coal-fired power plants, a large petrochemical plant and several pyrometallurgical smelters, was declared a priority area (Government Gazette Republic of South Africa, 2007). The Highveld Priority Area is the largest industrial area in Africa. It is among the top five NO_2 emission hotspots in the world and also a significant source of CO_2, SO_2 and sulphate particles (Fleming and van der

Merwe, 2004; Held et al., 1996). Recently, the Waterberg National Priority Area located at the eastern edge of the North-West Province was declared. This priority area incorporates industrial activities in the western Bushveld Igneous Complex (Venter et al., 2012) with pyrometallurgical smelters and the fast developing Waterberg area (Government Gazette Republic of South Africa, 2012).

Most of the actions taken to control air pollution were aimed at reducing the human exposure to pollutants. From the ecosystems perspective, one of the existing threats in southern Africa is the tropospheric ozone (e.g. van Tienhoven and Scholes, 2003; van Tienhoven et al., 2005), which is more challenging to control compared to other pollutant species, as it is a secondary pollutant that is formed from various precursor species.

20.4 OZONE LEVELS IN SOUTHERN AFRICA

A summary of the long-term ozone concentration measurements conducted in southern Africa is presented in Table 20.1 (see also Figure 20.1). The data are based on literature and recent measurements conducted by the North-West University (NWU), the electricity producing company Eskom, the University of Helsinki (UH) and the Finnish Meteorological Institute (FMI). These observations cover some of the main biomes of southern Africa. For many locations, however, data are limited by the duration (short-term data only) or temporal resolution (weekly or monthly averaged data only). Further, data averaged over several days do not allow estimation of daytime exposures. As seen from Figure 20.1, large areas in the central and western parts of southern Africa completely lack observations.

Only datasets with longer (1 year or more) time series, where at least monthly data are available, are used. For passive samplers, only averaged seasonal data are presented, as individual months may show unrealistically large deviations (Martins et al., 2007), while the long-term averages are likely to be more accurate. In addition to the data shown here, ozone measurements have been carried out in urban areas of several major municipalities, such as Johannesburg (Gauteng), Cape Town and Durban. However, as the focus of this chapter is on vegetation impacts and a significant fraction of such data remains unpublished in a peer-reviewed form, these observations are not considered here.

In general, the highest ozone concentrations were observed in the central-eastern part of southern Africa: northern South Africa, Botswana and Zimbabwe. At all locations, clear annual cycles were observed. The highest concentrations were in general found during spring and winter, whereas the concentrations were lowest during the summer. The exceptions for this rule are the northernmost sites in Zimbabwe, where the local fires may have influenced ozone formation, even though the reasons for the elevated summer concentrations are not fully understood (Jonnalagadda et al., 2001; NASA, 2011). The high springtime

TABLE 20.1 Seasonally and Annually Averaged Ozone Concentrations Observed in Southern Africa (Please See Also Map in Figure 20.1)

Measurement site						Concentration (ppb)					
	Location	*Coordinates*	*Period*	*Sampling type*	*Area description*	*Spring (Sep–Nov)*	*Summer (Dec–Feb)*	*Autumn (Mar–May)*	*Winter (Jun–Aug*	*Annual Average*	**Reference**
1	Luluche, Zimbabwe	S17.76, E32.6	1991/05–1993/12	Passive	Forest and agricultural mosaic	28	42	47	46	41	Jonnalagadda et al. (2001)
2	Nyanga, Zimbabwe	S18.5, E32.7	1991/05–1993/12	Passive	Forest and agricultural mosaic	36	51	51	37	44	Jonnalagadda et al. (2001)
3	Okaukuejo, Namibia	S18.99, E15.8633	1994/06–2009/12	Passive	Regional savannah	27	24	21	25	24	Martins et al. (2007)
4	Maun, Botswana	S19.97, E23.4	1999/07–2001/12	Continuous	Background savannah with residential impact	56	31	34	43	42	Zunckel et al. (2004)
5	Louis Trichardt, South Africa	S22.9861, E30.0225	1995/05–2010/12	Passive	Regional savannah	37	28	28	35	32	Martins et al. (2007)
6	Lesheba Wilderness (Makhado), South Africa	S22.9915, E29.55706	2005/09–2007/08	Passive	Game farm (mountain wooded savanna)	29	25	22	22	25	Josipovic (2009) and Josipovic et al. (2010)

7	Kameelfontein (Thabazimbi), South Africa	S24.02597, E27.07372	2005/09–2007/08	Passive	Game farm (regional semi-wooded savanna)	19	18	17	15	17	Josipovic (2009) and Josipovic et al. (2010)
8	Witvinger (Mokopane), South Africa	S24.03831, E28.98294	2005/09–2007/08	Passive	Nature reserve (regional semi-wooded savanna)	21	17	15	23	19	Josipovic (2009) and Josipovic et al. (2010)
9	Grietjie (Palaborwa), South Africa	S24.0695, E31.02119	2005/09–2007/08	Passive	Game farm (regional semi-wooded savanna)	20	15	17	20	18	Josipovic (2009) and Josipovic et al. (2010)
10	Skukuza, South Africa	S24.9931, E31.5839	2000/06–2010/12	Passive	Background savannah	29	17	21	25	23	Martins et al. (2007)
11	Bergvliet (Sabie), South Africa	S25.07481, E30.98847	2005/09–2007/08	Passive	Forestry (commercial) plantation	17	22	17	31	22	Josipovic (2009) and Josipovic et al. (2010)
12	Palmer, South Africa	S25.5, E30.0,	2000/06–2002/07	Passive	Pine and eucalyptus plantations	37	24	26	28	29	Zunckel et al. (2004)
13	Botsalano, South Africa	S25.54, E25.75	2006/07–2008/01	Continuous	Background savannah	42	36	34	33	37	Laakso et al. (2008)
14	Marikana, South Africa	S25.7, E27.48	2008/03–2010/04	Continuous	Residential mining area	37	30	24	29	30	Venter et al. (2012)

Continued

TABLE 20.1 Seasonally and Annually Averaged Ozone Concentrations Observed in Southern Africa (Please See Also Map in Figure 20.1)—Cont'd

Measurement site						Concentration (ppb)					
	Location	*Coordinates*	*Period*	*Sampling type*	*Area description*	*Spring (Sep–Nov)*	*Summer (Dec–Feb)*	*Autumn (Mar–May)*	*Winter (Jun–Aug*	*Annual Average*	Reference
15	Elandsfontein, South Africa	S26.245, E29.4175	2009/02–2011/01	Continuous	Regional industrial	40	29	32	45	37	Laakso et al. (2012)
16	Welgegund, South Africa	S26.57, E26.93	2010/06–2011/08	Continuous	Grassland savannah with anthropogenic influence	44	31	28	37	35	Beukes et al. (2013) and www.welgegund.org
17	Vaal Triangle, South Africa	S26.7247, E27.8847	2008/01–2009/12	Passive	Industrial	25	21	13	13	19	Martins et al. (2007)
18	Madlangune Camp (Kosi Bay), South Africa	S26.95658, E32.83089	2005/09–2007/08	Passive	Nature reserve (coastal thick forest)	16	13	13	17	15	Josipovic (2009) and Josipovic et al. (2010)
19	Amersfoort, South Africa	S27.0703, E29.8672	1997/09–2009/12	Passive	Regional, impacted by anthropogenic activities	32	32	20	25	28	Martins et al. (2007)

20	Hluhluwe, South Africa	S28.07031, E3204031	2005/09–2007/08	Passive	Nature reserve (coastal semi-wooded savanna)	21	17	17	25	20	Josipovic (2009) and Josipovic et al. (2010)
21	Salem Estates (Kranskop), South Africa	S29.01869, E30.91661	2005/09–2007/08	Passive	Forestry (commercial) plantation	21	19	23	23	21	Josipovic (2009) and Josipovic et al. (2010)
22	BenMac Dhui, South Africa	S30.8, E28.0	2000/06–2002/07	Passive	Grassland, background	23	23	28	25	25	Zunckel et al. (2004)
23	Cape Point GAW, South Africa	S34.352320, E18.488973	01/2001–12/2007	Continuous	Coastal, mainly marine background air	31	23	26	32	28	Oltmans et al. (2006)

concentrations at other locations result from the regional wild fires, the onset of biological activity and the increasing solar radiation during that time of year, whereas the summer minima can be attributed to limited biomass burning (Aghedo et al., 2007).

The observations presented in Table 20.1 before 2004 have previously been discussed by Zunckel et al. (2004) in a synthesis of the results of the Cross-border Air Pollution Impact Assessment project. In that study, the highest monthly concentration was found to exceed 90 ppb in northern Botswana. In Zimbabwe, the annual concentrations were even higher (Jonnalagadda et al., 2001), but the variability during the spring season and between the different years was significant.

The observations listed in Table 20.1 include those of forest observation sites from a 2-year field campaign conducted between August 2005 and September 2007. This campaign was designed to provide quantitative information, by direct measurement, on the air pollution threat to soils, plants and water bodies across the entire north-eastern South Africa (Josipovic, 2009; Josipovic et al., 2010). For this, a network of 37 passive sampling sites was established to measure monthly mean concentrations of near-surface O_3, SO_2 and NO_2. Increased ozone concentrations were recorded outside the industrial area towards the edges of the study area, both to the south and north. At the majority of the sites, the concentrations did not fluctuate widely from year to year. Also in this study, elevated concentrations were observed during late winter and most of spring, with lower concentrations in summer and the minimum in autumn. In remote areas, including the sensitive forested regions of the Drakensberg escarpment, pollutant concentrations observed in this study were unlikely to exceed the critical levels for environmental damage.

The longest existing ozone concentration time series summarized in Table 20.1 originates from the Cape Point Global Atmospheric Watch (GAW) station (Oltmans et al., 2006). The summertime concentrations at this site typically vary between 10 and 30 ppb, while winter concentrations range between 20 and 40 ppb. The lowest concentrations represent marine background air. Cape Point, as well as other mid-latitude sites in the southern hemisphere, have shown a significant increase in their decadal surface ozone trend, which seems to have moderated somewhat during the most recent years (Oltmans et al., 2013). At the other sites, the lowest concentrations are observed at those sites located close to the main industrial emission sources and large urban areas, due to the titration of ozone by other pollutants.

The regional modelling study conducted by Zunckel et al. (2006) shows that the increased ozone concentrations result from a combination of biogenic and anthropogenic sources, with neither of these having a dominant influence on this trend. This modelling study also revealed the potential risk of high regional ozone concentrations on ecosystems, which conclusion is supported by our observations discussed above. In the eastern and western parts of central South Africa, the measurement data (Table 20.1) indicate even longer periods with high tropospheric ozone concentrations above 40 ppb than the modelling results of Zunckel et al. (2006).

20.5 PREVIOUS STUDIES ON ECOSYSTEM IMPACTS OF OZONE

Elevated ozone concentrations impact plants in a variety of ways, the deleterious effects ranging from visible leaf injury to growth reductions and increased susceptibility to biotic and abiotic stress (Dizengremel et al., 2013; Karnosky et al., 2007; Matyssek et al., 2013a,b). The sensitivity to ozone depends on the species and phenological state of the plant. The phenological state as well as the overall functioning of the plant is controlled by environmental conditions, in arid ecosystems predominantly by the availability of water.

Currently, the recommended approach to assess the risk of ozone-induced damage to vegetation is based on stomatal ozone flux data and the corresponding biological experiments determining the dose–response relationships of different plant species (Grünhage et al., 2013; UNECE, 2010). However, the lack of observational data often limits the use of the flux approach, especially in the developing world. A less accurate, but widely used method is based on an exposure index that can be derived from ambient concentrations only. AOT40, which is defined as the accumulated ozone exposure above a threshold of 40 ppb during daylight hours of the growing season (UNECE, 2010) is such a metric. While AOT40 and the related exposure–response relationships are mainly based on European studies and plant species, it has been used as a risk indicator elsewhere in the world, including southern Africa (Avnery et al., 2011; van Tienhoven et al., 2006).

Van Tienhoven et al. (2005) and Carlson (1994) investigated the potential ecosystem impacts of the high ozone concentrations observed in southern Africa at Maun, Palmer and Rivulets (Figure 20.1). They noted that at the time of the study there were no experimental studies on the impacts of enhanced ozone concentrations on natural vegetation in southern Africa. At Maun the AOT40 for the growing season (October–January) in 2000–2001 was 30,230 ppb h. The AOT40 annual exposures for commercial forest farming areas containing pine and eucalyptus forests varied between 9890 and 20,500 ppb h at Palmer, and between 1450 and 4810 ppb h at Rivulets. The European critical level of 5000 ppb h is defined as the AOT40 level above which direct adverse effects may occur for trees (UNECE, 2010). This level is clearly exceeded at Maun and Palmer, while it is almost reached at Rivulets.

Some impact studies for common pine and eucalyptus species found in literature are summarized in Table 20.2. Tyson et al. (1988) listed some farmed

TABLE 20.2 Ozone Sensitivity of Some Commonly Planted Trees in South Africa

Species	Sensitivity for ozone exposure	Reference
Pines		
Pinus elliotii	Intermediate	Tyson et al. (1988)
Pinus taeda	Sensitive	Tyson et al. (1988)
Pinus radiata	Sensitive	Tyson et al. (1988)
Eucalyptus		
Eucalyptus globulus	Substantial	Pearson (1995)
	Not sensitive	Monk and Murray (1995)
Eucalyptus camaldulensis	Intermediate	Monk and Murray (1995)
Eucalyptus gomphocephala	Intermediate	Monk and Murray (1995)
Eucalyptus grandis	Intermediate	Monk and Murray (1995)
Eucalyptus marginata	Intermediate	Monk and Murray (1995)
Eucalyptus microcorys	Substantial	Monk and Murray (1995)
Eucalyptus robusta	Intermediate	Monk and Murray (1995)
Eucalyptus viminalis	Intermediate	Monk and Murray (1995)

Please note that the definition of sensitivity and dose-impact responses (reduced shoot/root dry mass, visible injuries) are not well-defined and vary between the references.

pine species that were found sensitive to ozone based on impact studies carried out in Europe. Pearson (1995) exposed young *Eucalyptus globulus* seedlings to approximately 3200 ppb h ozone and found that this species is highly sensitive to ozone. The study of Monk and Murray (1995) also revealed this sensitivity and indicated decreased productivity of eucalyptus at increased ozone exposures. There were significant differences in the observed response between different eucalyptus species (Monk and Murray, 1995).

For South African ecosystems, van Tienhoven et al. (2005) summarized that, despite the fact that ozone concentrations are above the European critical levels for crop damages (3000 ppb h), no vegetation damages have been reported. They give two possible reasons for this, namely, either ozone damages are not identified due to a lack of local research attention on this topic; or vegetation, as typical for Mediterranean species (Paoletti, 2006), has adapted to the high ozone levels resulting from natural wild fires and intensive solar radiation. The adaption aspect was also discussed by Scholes and Scholes (1998), who argued that local indigenous species, especially on the savannah, may be less sensitive to ozone exposure compared to planted alien species such as pine and eucalyptus. Van Tienhoven et al. (2005) concluded that vegetation in southern Africa experiences multiple stresses, such as floods and droughts, which may influence the productivity more than ozone-induced damages. However, the ozone effects may also be increased due to interactions between the different stresses and pollutants.

As forestry in southern Africa is economically less important (approximately 2 billion Euros in 2011; DAFF, 2011a) than agriculture (approximately 8 billion Euros in 2011; DAFF, 2011b) and as crop farming is concentrated on more polluted inland areas (in contrast to forest plantation areas generally with lower O_3 concentrations, see Figure 20.1 and Table 20.2), previous pollution impact studies on plants have focused on crops (e.g. Emberson et al., 2013). If we estimate the potential crop losses for wheat, maize and soybean with a simple linear relationship between AOT40 and crop losses (Mills et al., 2007), as in the global model-based assessment of Avnery et al. (2011), we find that wheat and maize harvests on the Highveld are not expected to be significantly reduced (losses $<5\%$), whereas soybean harvests within this region may be reduced by more than 20%. At Maun, Botswana, with the growing season AOT40 exceeding 30,000 ppb h (van Tienhoven et al., 2005), the crop reductions may exceed 30% for soybeans, 8% for maize and 5% for wheat. However, it should be kept in mind that no local data on plant responses were available for these estimates and that the AOT40-based approach does not consider soil moisture effects on ozone uptake. The results, therefore, may be biased, especially in dry climates and should be considered indicative only. A recent global summary discussing the effects of ozone on food production is provided by Wilkinson et al. (2012).

20.6 EFFECTS OF GROWING SEASON ON OZONE UPTAKE

The relatively high ozone concentrations observed may have impacts on vegetation only if ozone can access the plants through their stomata. In southern Africa, this predominantly takes place during the rainy season and daylight hours when photosynthesis is active. Figure 20.2 combines the ozone concentrations, AOT40 values, net ecosystem exchange (NEE) of carbon dioxide (CO_2 fluxes) and accumulated precipitation at the Welgegund measurement station (Figure 20.1, site #16) measured during a 1-year period in 2010–2011 (Beukes et al., 2013; www.welgegund.org). This station, which is operated by NWU, UH and FMI, is located approximately 100 km west of

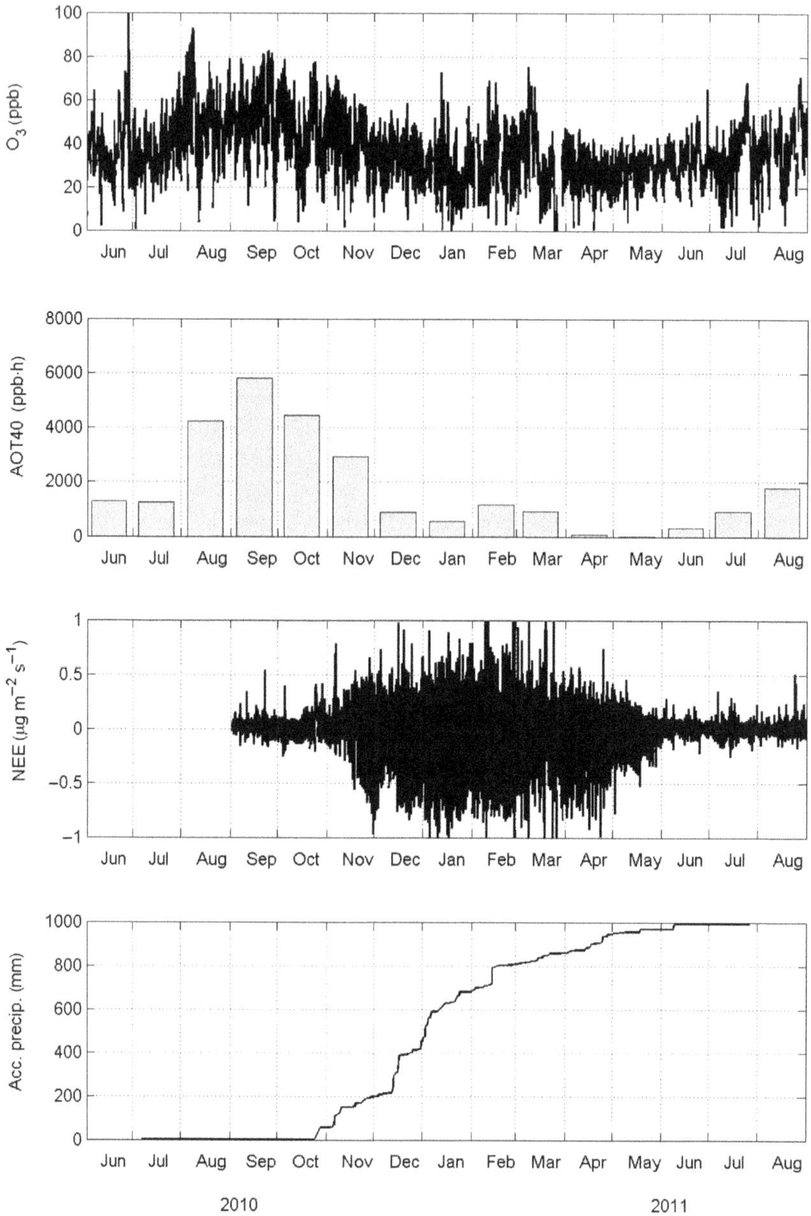

FIGURE 20.2 Hourly ozone concentrations, monthly AOT40 values, half-hourly CO_2 fluxes (NEE) and accumulated precipitation at the Welgegund measurement station. CO_2 flux measurements were started only in September 2010.

Johannesburg in a grazed grassland savannah. The station is comprehensively equipped for atmospheric and ecosystem observations including meteorology, atmospheric aerosol concentrations and properties, trace gases, wet deposition and CO_2 flux (i.e. a 'supersite', cf. Mikkelsen et al., 2013). The ozone measurements shown here were conducted with an active analyzer (Laakso et al., 2008; Petäjä et al., 2013).

The data shown in Figure 20.2 clearly demonstrate the differences between the annual cycles of ozone concentration and photosynthesis. At Welgegund, the highest concentrations occur during the dry season with limited biological activity (small absolute NEE) and significant biomass burning emissions (Beukes et al., 2013). During the main growing season, which is characterized by large negative CO_2 fluxes indicating net CO_2 uptake due to active photosynthesis, the concentrations and thus ozone exposure are clearly lower. Considering this phase shift, the potential for ozone-induced damage is largest in spring, when concentrations are high and new leaves have just emerged. This is also the period when potential droughts may have the strongest influence on productivity.

In terms of the ambient exposure of plants to ozone, the growing season AOT40 values at Welgegund significantly exceed the European critical levels. Figure 20.2 also reveals the importance of the onset of the growing season that is used as the accumulation period of AOT40, since the highest ozone concentrations occur during late winter and spring (August–November). Thus, depending on the actual date of the start of the wet season, AOT40 may vary significantly between the growing seasons. In 2010, for example, the rainy season started later than usual at Welgegund. Usually the first rains fall around 10 October, but in 2010 it started raining approximately 2 weeks later.

Figure 20.3 presents the average monthly AOT40 values (together with average growing seasons) calculated from the concentration data that are measured with continuous analyzers at the sites operated by NWU, UH and FMI (Botsalano, Marikana and Welgegund), the Eskom (large power utility) monitoring site at Elandsfontein, the Cape Point GAW station and Maun in Botswana (van Tienhoven et al., 2005). The average growing season lengths indicated in Figure 20.3 were obtained from the African protected areas database of the EU-JRC (APAAT, 2012) at the exact location of the measurement site or at the closest available location within 50 km. At all of the sites with the exception of the Cape Point GAW station, AOT40 is high enough to pose a risk to certain plant species such as Scots pine (Table 20.2; UNECE, 2010; Zinser et al., 2000). Again, the date of the onset of the growing season should be considered and related to this data, since it has a significant influence on the accumulated ozone exposure during the growing season. Based on the APAAT database, the start of the growth season varies at every location by 1–2 months, potentially leading to large (>50%) differences in the AOT40 value. However, the effect of precipitation on ozone precursor emissions from biomass burning may reduce this effect.

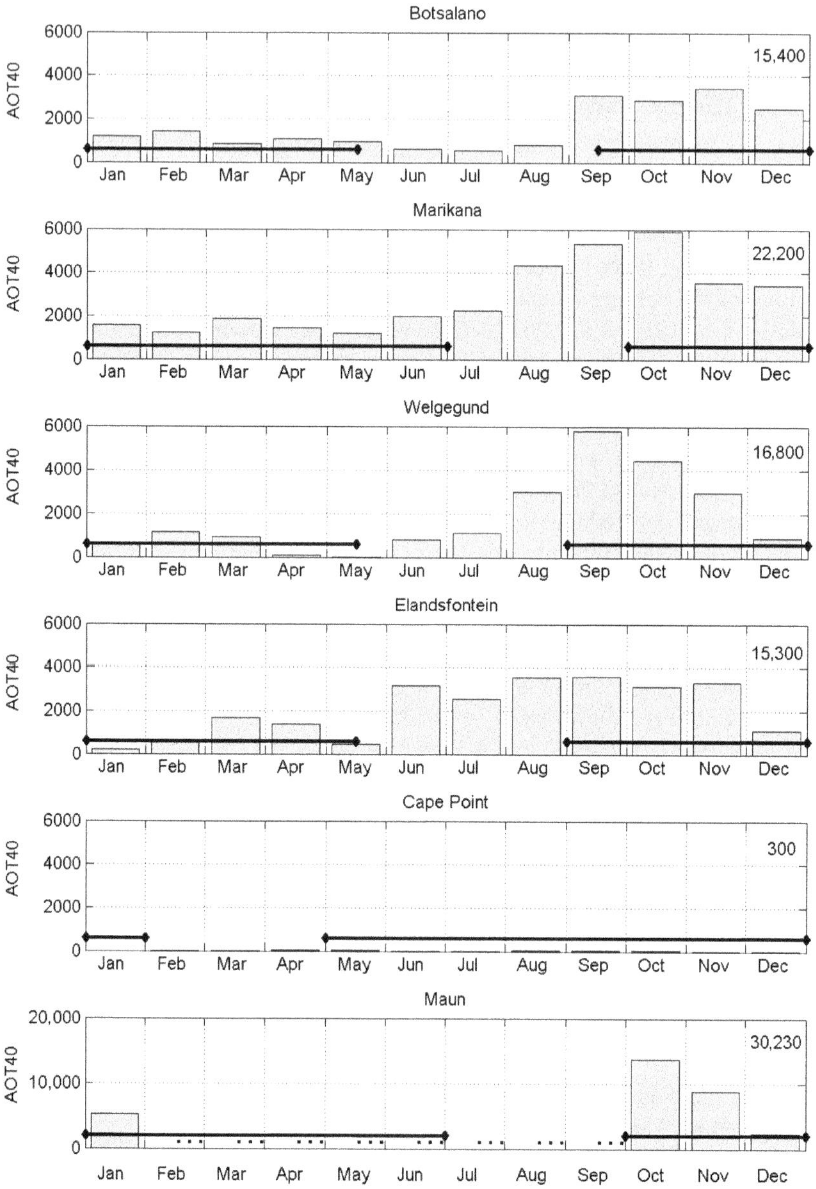

FIGURE 20.3 Monthly variation of AOT40 (ppb h) at six sites with continuous ozone concentration measurements (see also Table 20.1 and Figure 20.1). Black line indicates the typical growing season. The number in the right upper corner indicates the growing season AOT40 (in ppb h) at each location. The dots indicate that data are not available. Note the different scale for Maun in Botswana. *Data for Maun from van Tienhoven et al. (2005).*

20.7 CONCLUSIONS AND FUTURE DIRECTIONS

Ozone levels in southern Africa are in many places higher than the European critical levels determined for the protection of vegetation. However, local species may have adapted to high ozone levels, which is difficult to verify since few experimental studies on the actual vegetation impacts exist. Another factor that limits risk assessments is the lack of measurements of ozone fluxes to vegetation. In general, the impacts of ozone on agriculture in southern Africa are economically more important than the impacts on forests. Climate change scenarios show that tropospheric ozone levels are potentially increasing (Emberson et al., 2001; The Royal Society, 2008) and therefore ozone impacts on ecosystems may become more serious in the future.

Our recommendations emerging from this study are the following:

1. Continuous, high time-resolution data should be collected at a few representative locations in southern Africa. These activities should preferably include ozone flux measurements at a few well-equipped sites ('supersites', cf. Mikkelsen et al., 2013). The data should be made available for the research community to facilitate impact studies and model validation. The lack of available data is clearly visible in the studies summarized here, as many of the recent impact studies are based on old data, repeatedly used, while model calculations have only scarcely been validated against the observations. Even though continuous ozone measurements in South Africa exist (Rorich and Galpin, 1998), the data are mostly not published in peer-reviewed literature or are otherwise unavailable to the research community.
2. Ozone impact studies focusing on local crop variants and especially the key indigenous species should be carried out with harmonized methods in all less studied biomes (see also Koike et al., 2013; Kozovits and Bustamante, 2013). In an arid ecosystem, it is also essential to consider the effects of available soil water on ozone uptake.
3. Determination of the vegetation functioning during the growing season is essential for assessing ozone impacts. Suitable techniques, such as satellite-based information (e.g. remote sensing-based vegetation indices like NDVI) and local ecosystem CO_2 flux data should be considered for this assessment.
4. On a more general level, reduced use of fires as part of agricultural practises would decrease the regional ozone exposure and also decrease emissions of other harmful pollutants.

A more general question related to the impacts of air pollution on vegetation is the introduction of new species. As stated by Scholes and Scholes (1998), local species may be better adapted to the high ozone exposures. Thus a careful impact assessment, considering different aspects of forest ecosystem services and economical value, should include annual revenue of different tree species; economic estimate for the value of biodiversity; value of indigenous forests for local rural economics; value of natural forests on tourism; water use efficiency and availability of water for different species; and ozone sensitivity of the species selected for plantation. All these aspects need to be considered for sustainable long-term development. Thus the question involves many societal, economic and value-based issues, rather than being a purely scientific one. Nevertheless, quantitative observations, with publicly accessible data and species-specific impact studies, are necessary for addressing these questions.

ACKNOWLEDGEMENTS

The support from the following institutions and funding agencies is gratefully acknowledged: Cape Point GAW station, Eskom holdings, The NWU, Rustenburg Municipality, Botsalano Game Reserve, European Commission FP6 project EUCAARI (036833-2), Yrjö and Vilho Väisälä foundation, Academy of Finland, projects #117505: Air pollution in Southern Africa (2007–2009) and #132640 Atmospheric monitoring capacity building in Southern Africa (2010–2012).

REFERENCES

Aghedo, A.M., Schultz, M.G., Rast, S., 2007. The influence of African air pollution on regional and global tropospheric ozone. Atmos. Chem. Phys. 7, 1193–1212. http://dx.doi.org/10.5194/acp-7-1193-2007.

Avnery, S., Mauzerall, D.L., Liu, J., Horowitz, Larry W., 2011. Global crop yield reductions due to surface ozone exposure: 1. Year 2000 crop production losses and economic damage. Atmos. Environ. 45, 2284–2296.

Beukes, J.P., Vakkari, V., van Zyl, P.G., Venter, A.D., Josipovic, M., Jaars, K., Tiitta, P., Kulmala, M., Worsnop, D., Pienaar, J.J., Järvinen, E., Chellapermal, R., Ignatius, K., Maalick, Z., Cesnulyte, V., Ripamonti, G., Laban, T.L., Skrabalova, L., du Toit, M., Virkkula, A., Laakso, L., 2013. Source region characterization of the South African interior, in preparation.

Carlson, C.A., 1994. An assessment of the risk posed by the gaseous pollutant ozone to commercial forests in the eastern Transvaal. CSIR report FOR-DEA-754, Pretoria.

Carruthers, V., 2008. The Wildlife of Southern Africa. In: Struik Nature, South Africa, ISBN: 9781770077041, p. 301.

Center for International Earth Science Information Network (CIESIN), 2010. Columbia University, United Nations Food and Agriculture Programme (FAO), and Centro Internacional de Agricultura Tropical (CIAT), 2005, Gridded Population of the World: Future Estimates (GPWFE). Socioeconomic Data and Applications Center (SEDAC), Columbia University, Palisades, NY. http://sedac.ciesin.columbia.edu/gpw.

Cossalter, C., Pye-Smith, C., 2003. Fast-Wood Forestry—Myths and Realities. Center for International Forestry Research (CIFOR), Bogor, Indonesia, ISBN: 979-3361-09-3.

DAFF, 2011a. Report on commercial timber resources and primary roundwood processing in South Africa. Department of Agriculture, Forestry and Fisheries (DAFF), Republic of South Africa. http://www.forestry.co.za/uploads/File/fsa%20notices/2012/Timber%20Statistics%20Report%202010%20-%202011.pdf.

DAFF, 2011b. Economic Review of the South African Agriculture 2011/12, Department of Agriculture, Forestry and Fisheries (DAFF), Republic of South Africa. http://www.nda.agric.za/docs/statsinfo/EcoRev2011-12.pdf.

Dizengremel, P., Jolivet, Y., Tuzet, A., Ranieri, A., Le Thiec, D., 2013. Integrative leaf level phytotoxic ozone dose assessment for forest risk modelling. Dev. Environ. Sci. 13, 267–288.

Emberson, L.D., Ashmore, M.R., Murray, F., Kuylenstierna, J.C.I., Percy, K.E., Izuta, T., Zheng, Y., Shimizu, H., Sheu, B.H., Liu, C.P., Agrawal, M., Wahid, A., Abdel-Latif, N.M., van Tienhoven, M., de Bauer, L.I., Domingos, M., 2001. Impacts of air pollutants on vegetation in developing countries. Water Air Soil Pollut. 130, 107–118.

Emberson, L.D., Büker, P., Engardt, M., van Tienhoven, A.M., Agrawal, M., Zunckel, M., Hicks, K., Pleijel, H., 2013. Assessing air pollution impacts to agriculture: a framework methodology for assessing the risks caused by ground-level ozone (O_3) in South Asia and southern Africa. In: Pal, A. (Ed.), ASEAN—Environmental Perspectives. Stockholm Environment Institute, Stockholm, Sweden. http://www.sei-international.org/apcen/downloads/Emberson%20et%20al.,%20ASEAN%20paper_Final_APCENver.pdf.

Fleming, G., van der Merwe, M., 2004. Spatial disaggregation of greenhouse gas emission inventory data for Africa South of the Equator. http://gis.esri.com/library/userconf/proc00/professional/papers/PAP896/p896.htm (accessed 1 April 2010).

Garstang, M., Tyson, M., Swap, R., Edwards, M., Kållberg, P., Lindesay, J.A., 1996. Horizontal and vertical transport of air over southern Africa. J. Geophys. Res. 101, 23721–23736.

Government Gazette Republic of South Africa, 2005. No. 28132, 14 October 2005. http://www.info.gov.za/view/DownloadFileAction?id=61585 (accessed 13 December 2011).

Government Gazette Republic of South Africa, 2007. No. 29864, 4 May 2007. http://www.info.gov.za/view/DownloadFileAction?id=73046 (accessed 13 December 2011).

Government Gazette Republic of South Africa, 2012. No. 35435, 15 June 2012. http://www.info.gov.za/view/DownloadFileAction?id=73046 (accessed 13 July 2012).

Grünhage, L., Matyssek, R., Wieser, G., Häberle, K.-H., Leuchner, M., Menzel, A., Dieler, J., Pretzsch, H., Grimmeisen, W., Zimmermann, L., Raspe, S., Schröder, M., 2013. Flux-based ozone risk assessment for adult beech and spruce forests. Dev. Environ. Sci. 13, 251–266.

Guenther, A., Otter, L., Zimmermann, P., Greenberg, J., Scholes, R., Scholes, M., 1996. Biogenic hydrocarbon emissions from Southern African savannas. J. Geophys. Res. 101, 25859–25865.

Held, G., Scheifinger, H., Snyman, G.M., Tosen, G.R., Zunckel, M., 1996. The climatology and meteorology of the Highveld. In: Held, G., Gore, B.J., Surridge, A.D., Tosen, G.R., Turner, C.R., Walmsey, R.D. (Eds.), Air Pollution and Its Impacts on the South African Highveld. Environmental Scientific Association, Cleveland.

Jonnalagadda, S.B., Bwila, J., Kosmus, W., 2001. Surface ozone concentrations in Eastern Highlands of Zimbabwe. Atmos. Environ. 35, 4341–4346. http://dx.doi.org/10.1016/S1352-2310(01)00096-6.

Josipovic, M., 2009. Acid deposition emanating from the South African Highveld—a critical levels and critical loads assessment (Ph.D. thesis). University of Johannesburg, Johannesburg, South Africa.

Josipovic, M., Annegarn, H.J., Kneen, M.A., Pienaar, J.J., Piketh, S.J., 2010. Concentrations, distributions and critical levels exceedance assessment of SO_2, NO_2 and O_3 in South Africa. Environ. Monit. Assess. 171, 181–196.

Josipovic, M., Annegarn, H.J., Kneen, M.A., Pienaar, J.J., Piketh, S.J., 2011. Atmospheric dry and wet deposition of sulphur and nitrogen species and assessment of critical loads of acidic deposition exceedances in South Africa. South Afr. J. Sci. 107, 10. http://dx.doi.org/10.4102/sajs.v107i3/4.478.

Karnosky, D.F., Skelly, J.M., Percy, K.E., Chappelka, A.H., 2007. Perspectives regarding 50 years of research on effects of tropospheric ozone air pollution on US forests. Environ. Pollut. 147, 489–506.

Koike, T., Watanabe, M., Hoshika, Y., Kitao, M., Matsumura, H., Funada, R., Izuta, T., 2013. Effects of ozone on forest ecosystems in East and Southeast Asia. Dev. Environ. Sci. 13, 371–390.

Kozovits, A.R., Bustamante, M.M.C., 2013. Land use change, air pollution and climate change—vegetation response in Latin America. Dev. Environ. Sci. 13, 411–427.

Laakso, L., Laakso, H., Aalto, P.P., Keronen, P., Petäjä, T., Nieminen, T., Pohja, T., Siivola, E., Kulmala, M., Kgabi, N., Phalatse, D., Molefe, M., Mabaso, D., Pienaar, K., Kerminen, V.-M., 2008. Basic characteristics of atmospheric particles, trace gases and meteorology in a relatively clean Southern African Savannah environment. Atmos. Chem. Phys. 8, 4823–4839. http://dx.doi.org/10.5194/acp-8-4823-2008.

Laakso, L., Vakkari, V., Virkkula, A., Laakso, H., Backman, J., Kulmala, M., Beukes, J.P., van Zyl, P.G., Tiitta, P., Josipovic, M., Pienaar, J.J., Chiloane, K., Gilardoni, S., Vignati, E., Wiedensohler, A., Tuch, T., Birmili, W., Piketh, S., Collett, K., Fourie, G.D., Komppula, M., Lihavainen, H., de Leeuw, G., Kerminen, V.-M., 2012. South African EUCAARI measurements: seasonal variation of trace gases and aerosol optical properties. Atmos. Chem. Phys. 12, 1847–1864. http://dx.doi.org/10.5194/acp-12-1847-2012.

Loreto, F., Fares, S., 2013. Biogenic volatile organic compounds and their impacts on biosphere-atmosphere interactions. Dev. Environ. Sci. 13, 57–75.

Lourens, S.M., Butler, T.M., Beukes, J.P., van Zyl, P.G., Beirle, S., Wagner, T., Heue, K.-P., Pienaar, J.J., Fourie, G.D., Lawrence, M.G., 2012. Re-evaluating the NO_2 hotspot over the South African Highveld. South Afr. J. Sci. 108 (9/10), 6. http://dx.doi.org/10.4102/sajs.v108i11/12.1146, Art. #1146.

Martins, J.J., Dhammapala, R.S., Lachmann, G., Galy-Lacaux, C., Pienaar, J.J., 2007. Long-term measurements of sulphur dioxide, nitrogen dioxide, ammonia, nitric acid and ozone in southern Africa using passive samplers. South Afr. J. Sci. 103, 336–342.

Matyssek, R., Wieser, G., Calfapietra, C., de Vries, W., Dizengremel, P., Ernst, D., Jolivet, Y., Mikkelsen, T.N., Mohren, G.M.J., Le Thiec, D., Tuovinen, J.-P., Weatherall, A., Paoletti, E., 2012. Forests under climate change and air pollution: gaps in understanding and future directions for research. Environ. Pollut. 160, 57–65.

Matyssek, R., Clarke, N., Cudlin, P., Mikkelsen, T.N., Tuovinen, J.-P., Wieser, G., Paoletti, E., 2013a. Climate change, air pollution and global challenges: knowledge, understanding and perspectives from forest research. Dev. Environ. Sci. 13, 3–16.

Matyssek, R., Wieser, G., Fleischmann, F., Grünhage, L., 2013b. Ozone Research, quo vadis? Lessons from the free-air canopy fumigation experiment at Kranzberg Forest. Dev. Environ. Sci. 13, 103–129.

Mikkelsen, T.N., Clarke, N., Danielewska, A., Fischer, R., 2013. Towards supersites in forest ecosystem monitoring and research. Dev. Environ. Sci. 13, 475–495.

Mills, G., Buse, A., Gimeno, B., Bermejo, V., Holland, M., Emberson, L., Pleijel, H., 2007. A synthesis of AOT40-based response functions and critical levels of ozone for agricultural and horticultural crops. Atmos. Environ. 41, 2630–2643.

Monk, R.J., Murray, F., 1995. The relative tolerance of some Eucalyptus species to ozone exposure. Water Air Soil Pollut. 85, 1405–1411.

Mucina, L., Rutherfors, M.C. (Eds.), 2006. The Vegetation of South Africa, Lesotho and Swaziland. Strelitzia, vol. 19. South African National Biodiversity Institute, Pretoria, p. 805.

NASA, 2011. Goddard Space Flight Center Scientific Visualization Studio. http://svs.gsfc.nasa.gov/goto?3870 (accessed 21 February 2013).

Oltmans, S., Lefohn, A., Harris, J.M., Galbally, I., Scheel, H., Bodeker, G., Brunke, E., Claude, H., Tarasick, D., Johnson, B., Simmonds, P., Shadwick, D., Anlauf, K., Hayden, K., Schmidlin, F., Fujimoto, T., Akagi, K., Meyer, C., Nichol, S., Davies, J., Redondas, A., Cuevas, E., 2006. Long-term changes in tropospheric ozone. Atmos. Environ. 40, 3156–3173.

Oltmans, S.J., Lefohn, A.S., Shadwick, D., Harris, J.M., Scheel, H.E., Galbally, I., Tarasick, D.W., Johnson, B.J., Brunke, E.-G., Claude, H., Zeng, G., Nichol, S., Schmidlin, F., Davies, J., Cuevas, E., Redondas, A., Naoe, H., Nakano, T., Kawasato, T., 2013. Recent tropospheric ozone changes—a pattern dominated by slow or no growth. Atmos. Environ. 67, 331–351. http://dx.doi.org/10.1016/j.atmosenv.2012.10.057.

Otter, L.B., Guenther, A., Greenberg, J., 2002. Seasonal and spatial variations in biogenic hydrocarbon emissions from southern African savannas and woodlands. Atmos. Environ. 36, 4265–4275.

Otter, L., Fleming, G., Wiedinmyer, C., Guenther, A., Greenberg, J., Harley, P., 2003. Spatial and temporal variations in biogenic VOC emissions for Africa south of the equator. J. Geophys. Res. 108 (D13), 8505–8516. http://dx.doi.org/10.1029/2002JD002609.

Paoletti, E., 2006. Impact of ozone on Mediterranean forests: a review. Environ. Pollut. 144, 463–474.

Pearson, M., 1995. Effects of ozone on growth and gas exchange of Eucalyptus globulus seedlings. Tree Physiol. 15, 207–210.

Petäjä, T., Vakkari, V., Pohja, T., Nieminen, T., Laakso, H., Aalto, P.P., Keronen, P., Siivola, E., Kerminen, V.-M., Kulmala, M., Laakso, L., 2013. Transportable aerosol characterization trailer with trace gas chemistry: design, instruments and verification. Aerosol Air Qual. Res. 13, 421–435.

Rorich, R.P., Galpin, J.S., 1998. Air quality in the Mpumalanga Highveld region, South Africa. South Afr. J. Sci. 94, 109–114.

Scholes, R.J., Scholes, M.C., 1998. Natural and human-related sources of ozone-forming trace gases in southern Africa. South Afr. J. Sci. 94, 422–425.

Shackleton, C.M., Shackleton, S.E., 2004. The importance of non-timber forest products in rural livelihood security and as safety nets: a review of evidence from South Africa. South Afr. J. Sci. 100, 658–664.

Sofiev, M., 2013. Wild-land fires: monitoring, plume modelling, impact on atmospheric composition and climate. Dev. Environ. Sci. 13, 445–472.

Swap, R.J., Annegarn, H.J., Suttles, J.T., King, M.D., Platnick, S., Privette, J.L., Scholes, R.J., 2003. Africa burning: a thematic analysis of the Southern African Regional Science Initiative (SAFARI 2000). J. Geophys. Res. 108, 8465. http://dx.doi.org/10.1029/2003JD003747.

The African Protected Areas Assessment Tool (APAAT), 2012. http://bioval.jrc.ec.europa.eu/APAAT/ (accessed 8 October 2012).

The Royal Society, 2008. Ground-Level Ozone in the 21st Century: Future Trends, Impacts and Policy Implications. In: Science Policy Report 15/08, The Royal Society, London, p. 132.

Tyson, P.D., Kruger, F.J., Louw, C.W., 1988. Atmospheric pollution and its implications in the eastern Transvaal Highveld, South African National Scientific Programmes, Report No 150. CSIR, Pretoria.

UNECE, 2010. Mapping critical levels for vegetation. Manual on Methodologies and Criteria for Modelling and Mapping Critical Loads & Levels and Air Pollution Effects, Risks and Trends. United Nations Economic Commission for Europe (UNECE) Convention on Long-range

Transboundary Air Pollution, Geneva, Switzerland. http://icpvegetation.ceh.ac.uk/manuals/mapping_manual.html.

van der Werf, G.R., Randerson, J.T., Giglio, L., Collatz, G.J., Mu, M., Kasibhatla, P.S., Morton, D.C., DeFries, R.S., Jin, Y., van Leeuwen, T.T., 2010. Global fire emissions and the contribution of deforestation, savanna, forest, agricultural, and peat fires (1997–2009). Atmos. Chem. Phys. 10, 11707–11735. http://dx.doi.org/10.5194/acp-10-11707-2010.

van Tienhoven, A.M., Scholes, M.C., 2003. Air pollution impacts on vegetation in South Africa. In: Emberson, L., Ashmore, M., Murray, F. (Eds.), Air Pollution Impacts on Crops and Forests. A Global Assessment. Imperial College Press, UK, pp. 237–262.

van Tienhoven, A., Otter, L., Lenkopane, M., Venjonoka, K., Zunckel, M., 2005. Assessment of ozone impacts on vegetation in southern Africa and directions for future research. South Afr. J. Sci. 101, 143–148.

van Tienhoven, A., Zunckel, M., Emberson, L., Koosailee, A., Otter, L., 2006. Preliminary assessment of risk of ozone impacts to maize (zea mays) in southern Africa. Environ. Pollut. 140, 220–230.

Venter, A.D., Vakkari, V., Beukes, J.P., et al. 2012. An air quality assessment in the industrialised western Bushveld Igneous Complex, South Africa. S. Afr. J. Sci. 108 (9/10), Art. #1059, 10 pages. http://dx.doi.org/10.4102/sajs.v108i9/10.1059.

White, F., 1983. Vegetation of Africa—a descriptive memoir to accompany the Unesco/AETFAT/UNSO vegetation map of Africa. In: Natural Resources Research Report XX, U. N. Educational, Scientific and Cultural Organization, Paris, France, 356 pp.

Wilkinson, S., Mills, G., Illidge, R., Davies, W.J., 2012. How is ozone pollution reducing our food supply? J. Exp. Bot. 63, 527–536.

Wise, R.M., Dye, P.J., Gush, M.B., 2011. A comparison of the biophysical and economic water-use efficiencies of indigenous and introduced forests in South Africa. For. Ecol. Manage. 262, 906–915.

Zinser, C., Jungblut, T., Heller, W., Seidlitz, H.K., Schnitzler, J.-P., Ernst, D., Sandermann, H., 2000. The effect of ozone in Scots pine (Pinus sylvestris L.): gene expression, biochemical changes and interactions with UV-B radiation. Plant Cell Environ. 23, 975–982.

Zunckel, M., Venjonoka, K., Pienaar, J.J., Brunke, E.-G., Pretorius, O., Koosialee, A., Raghunandan, A., van Tienhoven, A.M., 2004. Surface ozone over southern Africa: synthesis of monitoring results during the Cross Border Air Pollution Impact Assessment project. Atmos. Environ. 38, 6139–6147.

Zunckel, M., Koosailee, A., Yarwood, G., Maure, G., Venjonoka, K., Tienhoven, A., Otter, L., 2006. Modelled surface ozone over southern Africa during the Cross Border Air Pollution Impact Assessment Project. Environ. Model. Softw. 21 (911–924), 2006.

Chapter 21

Wildland Fires: Monitoring, Plume Modelling, Impact on Atmospheric Composition and Climate

Mikhail Sofiev[1]

Adjunct Professor, Senior Scientist, Air Quality Department, Finnish Meteorological Institute

[1]*Corresponding author: mikhail.sofiev@fmi.fi*

Chapter Outline

21.1 WILDLAND FIRES: PART OF THE ECOSYSTEM LIFECYCLE OR A RESULT OF ANTHROPOGENIC STRESS?

Wildland fires is one of the most powerful mechanisms of impact of forest ecosystems and other vegetation to atmosphere and climate; it is also one of

Developments in Environmental Science, Vol. 13. http://dx.doi.org/10.1016/B978-0-08-098349-3.00021-9

the strongest phenomena changing the ecosystems themselves. The introductory chapter (Chapter 1) of this book positions the wildland fire phenomenon among others driving the ecosystem development and exchange between the environmental media (see also Loreto and Fares, 2013, this vol.).

Wildland fires are the second most powerful source of the key trace gases and aerosols in the atmosphere, after the fossil fuel combustion. In Europe, about 5000 km^2 of forested areas are burned by more than 50,000 fires annually (Keramitsoglou et al., 2004). The fires occur in all European countries but they are particularly intense in the arid southern and eastern regions, as well as in central Russia. Globally, the wide regions in equatorial Africa, Russia and Brazil are among the mostly affected areas. Total estimates of the globally consumed biomass vary widely, usually ranging between 5 and 10 Gt annually (Chin et al., 2002; Scholes and Andreae, 2000), with anthropogenic vegetation fires accounting for 3.5–3.9 Gt (Lauk and Erb, 2009). Estimates of released CO_2 also differ within a factor of 2 between different studies, ranging from 1.4 up to 2.8 Mt of carbon per year (Schultz et al., 2008).

21.1.1 A Bit of History

Complex approaches to scientific exploration of global impact of fires gathered steam only in the second half of previous century (Crutzen et al., 1979; Eagan et al., 1974; Seiler and Crutzen, 1980). By the end of the twentieth century, it had become evident that biomass burning is a major factor in regional and global atmospheric chemistry, radiative forcing, affecting cloud and precipitation features, and potentially one of the powerful drivers of climate change (Andreae and Merlet, 2001; Andreae et al., 1988; Barbosa et al., 1999; Crutzen and Andreae, 1990; Forster et al., 2001; Levine, 1994a,b; Schultz et al., 2008)—and references therein. Fast progress in compilation of a global pattern of the fire impact has become possible due to developments of satellite observational technologies, including observations of active fires by radiometers (in infrared part of spectrum) and detection of burnt-area (BA) scars visible as an abrupt change of surface albedo. At about the same time, first emergency decision-support systems based on near-real-time satellite observations of active fires were constructed (Rauste et al., 1997).

21.1.2 Role of Anthropogenic Stress

Fires have always been a part of the natural lifecycle of the ecosystems—cleaning the accumulating debris from forests, controlling insects and freeing the space for new plants. These were accompanied by intermittent risks by nutrient leaching, altered hydrology and erosion. However, human activity has dramatically altered the features and distribution of the fires. In particular, in several forested regions, the fires are set as a part of deforestation process, that is, they constitute purely anthropogenic phenomenon (Figure 21.1; Brasseur et al., 2004), see also chapter

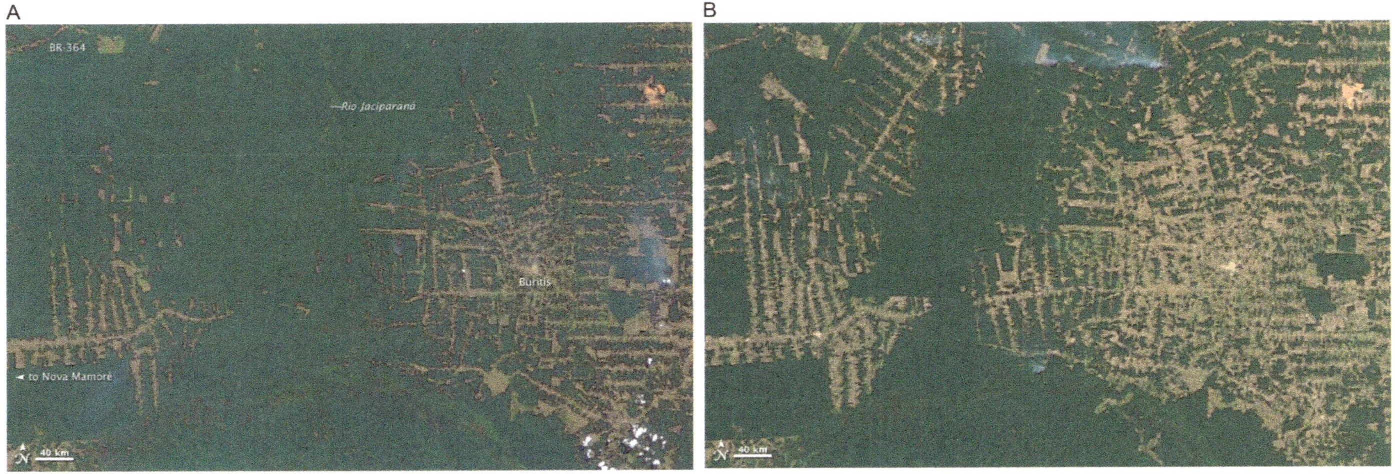

FIGURE 21.1 Example of deforestation impact in equatorial Amazonia and fires associated with it. (A) Landsat image, 30.07.2000; (B) same area, 21.08.2009. Blue plumes come from the fires. Buritis town location: (63°49′W, 10°12′S). *Source: http://earthobservatory.nasa.gov/Features/WorldOfChange*.

8 of Osborne (2000). Regional specifics of ecosystem developments and anthropogenic stress are analysed in detail in Chapters 18, 19, 20, 21 and 22 of this book.

Apart from deforestation, dense population is frequently associated with the increase of the wildland fires risk. This dependence can qualitatively be illustrated by comparing the density of anthropogenic infrastructure and the number of fire episodes (Figure 21.2). However, such a relationship strongly depends on environmental policy, agriculture practice and population habits in each specific region. In particular, very high density of infrastructure and intense agriculture activity in Western Europe is weakly related to fire risk, whereas in other regions, such as Russia or South America, this relation is evidently very strong (Figures 21.1 and 21.2).

Considering the relation between the anthropogenic stress and occurrence of fires, one should take into account comprehensive fire-control measures developed in many countries. The oldest set of techniques is dedicated to firefighting. For instance, the '10:00 rule' adopted in the United States in 1935 states that the objective of the fire control policy should be to 'control all fires as quickly as possible with available forces but not later than 10:00 a. m. of the day following discovery' (Lindh, 1941). However, it is presently recognised that the efficient strategy for reduction of the wildland fire adverse effects has to address the root causes of biomass burning, which often originate from regional specifics and local practices. A few examples of country-specific Community-Based Fire Management practices can be found in GFMC (2012). Extensive information on the current state of both fire prevention and control approaches has been collected by Global Fire Monitoring Centre (GFMC, 2011).

The importance of anthropogenic stress for wildland fires has been reflected in the concept of urban–rural interface (Summit, 2003). As pointed out by Viegas et al. (2003), 'Forest fires in the wildland-urban interface are common in Europe, as population and human infrastructure facilities are disseminated throughout the forested zones, especially in the vicinity of large cities and tourist resorts'. Such interface areas are characterised by heavy anthropogenic stress in terms of both human habitation and temporary visits and activities. Conversely, the formal population density of these regions is small compared to urban agglomerations, which leaves a substantial fraction of the area for forests and uncultivated grassland. These regions are particularly vulnerable to human-ignited fires. Fires in the interface area are also the most dangerous for people regarding both direct threat of the flames and severe deterioration of air quality.

21.1.3 Fires as Part of Natural Ecosystem Life

Apart from human influence, fires are almost exclusively caused by lightning (Brasseur et al., 2004). Practically, all ecosystems with substantial fuel load (FL) and periodic dry seasons (e.g. prairies, savannas, bushes, broadleaf and

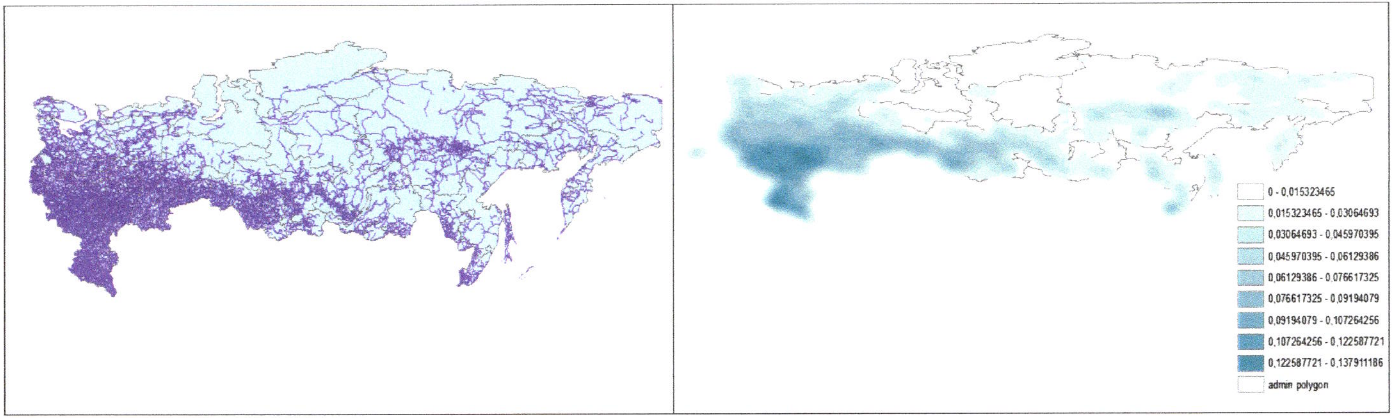

FIGURE 21.2 Maps of anthropogenic infrastructure density in Russia taken as a network of roads (A) and a fire density recorded by MODIS instruments for 2011 as a number of fires km^{-2} (B).

conifer forests) have evolved with fires being a natural and necessary contributor to habitat development and renewal (e.g. chapter 6 of Keddy, 2007; Pausas and Keeley, 2009).

Effects of fires on ecosystems are numerous and very diverse. For soil, fire removes overhead vegetation, thus increasing solar radiation reaching the surface and consequent warming of the upper soil layers during daytime and to their radiative cooling at night. Fewer leaves left to intercept rain allow more moisture to reach the soil surface. Reduced plant transpiration allows the soil to retain more moisture. Exposure to sunlight, wind, and evaporation, conversely, dries up the soil. The fire can also create crust at the soil surface leading to increased soil erosion due to higher run-off (Hart et al., 2005).

For plants, fires play several roles: they destroy the some vegetation, but also promote new ones by removing the overlaying canopy, arguably the most intriguing adaptation mechanism is pyriscence exhibited by some seed plants, in which seed release occurs in response to fire (Nathan et al., 1999). This mechanism is a special case of plant serotiny, which triggers the seed release in response to a wide range of environmental triggers, rather than spontaneously at seed maturation (Lamont et al., 1991). Pyriscence is important for regions with harsh environmental conditions, where passage of fire increases the typically low survival rates of seeds by reducing the competition with undergrowth and temporarily increasing the soil nutrition. Other adaptation mechanisms include fire tolerance and resistance, which limit the extent the flames can damage the plant and promote fast recovery of leaves afterwards.

21.1.4 Regional and Global Fire Impact

The fires in different parts of the globe have specific regional characteristics. For instance, in forested regions, such as Siberia or Alaska, the main impact and the amount of consumed biomass can be attributed to a comparatively small number of major episodes. Conversely, in Africa and in arid regions, a typical intensity of individual fires is smaller but their count is much larger (Schultz et al., 2008). In Europe, patchy land use and variety of climatic zones increase the diversity of the burn events. Thus, in northern areas, the boreal forest and bog fires are the most usual types. In central Europe, grass and mixed-forest fires are more frequent, whereas in the south the fires of grassland and sub-tropical forests are typical. The European fire season starts in April and can last until the end of November when seasonal precipitation typically wets the vegetation, in particular, in southern Europe.

The impact of fires on climate processes, atmospheric composition and air quality also varies widely from region to region and its estimates may differ significantly between different studies (Barbosa et al., 1999; Forster et al., 2001; Langmann et al., 2009; Schultz et al., 2008; Soja et al., 2006). For example, in boreal regions, the fire activity has been related to climate change already in 2000s (Amiro et al., 2001; Stocks, 2004). From the atmospheric standpoint, biomass burning is a major source of green house gases, such as CO_2, numerous tracers, such as CO, NO_x, NH_3, aerosols, such as black and organic carbon (BC and OC) and hydrocarbons.

The effect of fires on carbon cycle strongly interacts with vegetation development: fires often prompt the vegetation growth, thus offsetting the carbon release into the atmosphere. They also diminish both vegetation debris, whose decomposition is a source of CO_2, and green vegetation, which is its sink. Biomass burning is believed to only slightly alter the seasonal ecosystem production but contribute up to 20% to the inter-annual variability of CO_2 fluxes (Zeng, 2005).

Prediction of fires for future climate conditions is so far extremely uncertain. Due to the very significant role of the anthropogenic factor (Figure 21.2) and its poorly known role in the future, it is difficult to make quantitative predictions regarding even bulk parameters, such as seasonally consumed biomass. In particular, the forecasting capability of various fire indices is limited to the region where they are developed and application of some index elsewhere usually requires its adaptation. However, they still reflect the odds for the ignited twinkle to develop into a full-scale fire and can be correlated, for example, with the number of events (Tanskanen and Venäläinen, 2008).

21.2 SATELLITE PRODUCTS USED FOR WILDLAND FIRES MONITORING

A crucial role in understanding the abundance of the wildland fires and the extent to which they affect the ecosystems, atmosphere and climate was played by satellites. Spaceborne observations appeared to be the only way to detect and quantify events occurring in remote regions, where no dense *in situ* observational networks exist (equatorial Africa, Amazonia, Siberia, etc.).

Historically, two types of remote sensing products have attracted most of attention: Burnt Area (BA) scars and active-fire counts (Flemming, 2005). Both can be obtained only in cloud-free situations. The existing products of both types are quite numerous but their features and quality vary between the different instruments, equatorial crossing time, detection algorithm and constraints on data provision.

21.2.1 Burnt Area Scars Observations

The information on BAs has been available for decades, even before the satellites became mature enough to provide a global outlook (Andreae and Merlet, 2001; Crutzen and Andreae, 1990; Seiler and Crutzen, 1980; van der Werf et al., 2006). This approach is the most established source of fire emission information today also.

Detection of BAs is based on changes in albedo or, more specifically, on changes of the top of the atmosphere near-infrared (NIR) reflectance and surface temperatures. The area affected by the fire (the burnt scar) has significantly lower reflectance in NIR range, as well as higher daytime temperature due to higher light absorption and reduced vegetation-controlled evapotranspiration. The change in the NIR reflectance is independent of the type of ecosystem but the signal is less persistent in savannah regions than in temperate and boreal regions, partly due to faster pick up of new grass after moderate fire events. The increase in daytime ecosystem temperature can be as large as 7–8 K. The satellite instruments providing the popular BA products are AVHRR (Advanced Very High-Resolution Radiometer; http://nsidc.org/data/avhrr/), MODIS (Moderate Resolution Imaging Spectroradiometer; http://modis.gsfc.nasa.gov/), SeaWiFS (http://oceancolor.gsfc.nasa.gov/SeaWiFS/), ATSR (Along-Track Scanning Radiometer; http://www.atsr.rl.ac.uk/), AATSR (http://www.leos.le.ac.uk/AATSR/) and VEGETATION (http://www.vgt.vito.be/).

To obtain the emission (E) from the BA product, one has to estimate several parameters: (i) total Fuel Load FL in the area before the fire took place, (ii) combustion completeness (CC), that is, the fraction of the fuel consumed by the flames and (iii) emission factors (EF) for releases of specific atmospheric tracers (Crutzen et al., 1979). The classical formula suggested back in 1979 therefore reads:

$$E = \mathrm{BA} \times \mathrm{FL} \times \mathrm{CC} \times \mathrm{EF} \tag{21.1}$$

All factors in Equation (21.1) are characterised with wide uncertainties. Both CC and EF depend on the type of the fire, its strength, duration, etc., that is, dynamic parameters that have to be evaluated using supplementary datasets.

The FL can be inferred from climatologic vegetation maps and land use, combined with the remote sensing of vegetation status and/or with global vegetation modelling (Hoelzemann, 2004; Liousse et al., 1996). Data assimilation of remotely sensed vegetation indices, such as Normalised Difference Vegetation Index (NDVI; Tucker, 1979; Tucker et al., 1991), in vegetation models could account for the dynamic effect of meteorological conditions and interannual variability of the growing season. However, estimates of available FL from NDVI or other remotely sensed maps are still indirect and always have to be evaluated against field data (Barbosa et al., 1999).

Values for the CC and EF vary within about an order of magnitude between the different studies, but are generally assumed to be correlated with the type of vegetation and, to some extent, with its state (Andreae and Merlet, 2001).

To reduce the impact of clouds and increase the accuracy of the retrievals, the BA products are usually collected over a period of a few days or weeks. Being highly valuable for the BA product quality, this averaging also limits possibilities of application of meteorology-related corrections to the fire intensity, type (flames/smouldering) and duration.

Depending on the methodology for their assessment, the final emission estimates can differ from each other by a factor of several times even at continental levels (Boschetti and Roy, 2009; Boschetti et al., 2004).

21.2.2 Active-Fire Observations

The second type of the satellite datasets describing the abundance of fires is the observations of active fires in the infrared spectral channels. Related products include: (i) hot-spot counts when the instrument only registers the overheated pixel and reports it as a fire, (ii) temperature anomaly (TA) observations, when the actual temperature of the burning pixel is quantified, and (iii) the observations of the absolute release of radiative energy by an active fire (Flemming, 2005; Kaiser et al., 2009; Sofiev et al., 2009a; Wooster et al., 2005).

Radiometers with global coverage, such as ATSR and AVHRR, were among the first instruments with hot-spots-counting capabilities. They helped bringing together various local-, regional- and campaign-based efforts on evaluation of the spatial and temporal distribution of active fires. In 2000 and 2002, two MODIS instruments were launched with the capability of not only detecting the active fires, but also quantifying the burning rates (Ichoku and Kaufman, 2005; Wooster et al., 2005).

The fire detection procedure in such systems exploits the strong emission of mid-infrared radiation from fires (Dozier, 1981; Giglio and Kendall, 2001; Giglio et al., 2003; Matson and Dozier, 1981). The algorithm examines each pixel of the visible swath and classifies it with regard to data availability and potential for fire, for example, missing data, cloud, water, non-fire, fire or unknown (these are the classes of MODIS instrument). For each fire-classified pixel, the procedure attempts to use the neighbouring pixels to estimate the background radiometric signal of the pixel, that is, what it would emit if there would be no fire there.

If the characterisation of the background is successful, a series of threshold tests are used to confirm the active-fire hypothesis. The tests vary between the instruments but usually they require substantial difference between the estimated background and actual brightness temperatures at about the 4-μm wavelength channel. The MODIS algorithm also examines the difference between the signals at 4 and 11 μm wavelengths, requiring them to be substantially different for the background and the analysed pixel. The thresholds are adjusted to the natural variability of the background. Additional tests are used to eliminate false detections caused by sun glint, desert boundaries and errors in the water mask. Candidate fire pixels that are not rejected in the course of these tests are assigned with the class of fire. A dedicated effort is needed to separate the wildland fires from other types of fires, which is done on the basis of the land use reported for the detected fire pixel. In particular, oil refineries have to be masked out.

If the capacity of the sensor permits, the absolute temperature of the burning pixel can be measured, which then comprises the TA data. The next step is to produce the fire radiative power (FRP), which for MODIS instrument was obtained by Kaufman et al. (1998) in the following form:

$$E = 4.34 \times 10^{-19} \left(T_4^8 - T_{4b}^8\right) \tag{21.2}$$

Here E is the FRP in W m^{-2} or MW pixel^{-1} (MODIS pixel area is about 1 km^2); T_4 and T_{4b} are the pixel and background brightness temperatures at 4 μm wavelength, K. In the same work, the authors describe the MODIS algorithm and technical details of its implementation.

FRP, taken together with the convective and conductive energy fluxes into the atmosphere and soil, respectively, closes the budget of the total fire energy. Having the proportions of these components determined, one can directly relate FRP to the amount of consumed biomass and emission of atmospheric constituents (Frankman et al., 2009; Freeborn et al., 2008; Sukhinin et al., 2005; Wooster et al., 2005). That key advantage of the FRP concept allowed direct scaling from the satellite FRP product to the emission of various constituents:

$$E = \text{FRP} \times \text{EF} \tag{21.3}$$

It therefore avoids all poorly known terms in Equation (21.1). The EF, however, have to be determined empirically for each substance. This challenge was approached by several studies, for example, Ichoku and Kaufman (2005), Kaiser et al. (2009), Sofiev et al. (2009a) and Wooster et al. (2005) using different techniques and considering different episodes and regions. In most cases, empirical factors were determined for a few key chemical compounds, the concentrations of which in the atmosphere can be observed, for example, CO, NO_2, aerosols. The factors for other compounds were computed using cross-links established via dry biomass consumption (Andreae and Merlet, 2001). The calibration exercises reported EF of the same order of magnitude but they still varied significantly between the studies.

The second significant challenge is the temporal upscaling of FRP and calculated emissions. Indeed, deriving the daily emission profile from instant FRP observations requires strong assumptions on the fire behaviour before and after the satellite overpass. The introduction of FRP computations from geostationary satellites (e.g. Roberts and Wooster, 2008a) has only partly resolved the problem: the high temporal resolution (15 min in case of MSG SEVIRI) is offset by poor spatial resolution, low sensitivity to small fires and limited viewing area that leaves the boreal zone almost entirely out. Estimates of the diurnal variation of the fire intensity show that daytime energy release is about 10 times bigger than during night, with specific shapes dependent on the land use type and meteorological conditions (Giglio, 2007; Roberts and Wooster, 2008b; Roberts et al., 2009).

21.2.3 Comparison of Burnt Area and Active-Fire Products

Remote sensing of fires and BAs develops with high pace and many approaches require further work and evaluation to mature. Presently existing estimates show large discrepancies, which depend on many external and poorly known parameters. There is also significant systematic deviation between the results of different satellite instruments.

Evaluation of ATSR hot-spot product showed correct location of the fires but manifested general under-estimation of their number (Arino and Plummer, 2001, as reported by Flemming, 2005). However, ATSR is a nighttime satellite, that is, it has to deal with low fire intensities and reduced number of fires, as compared to daytime. Similar-type comparison made for the whole day instrument MODIS (Figure 21.3) shows a much more homogeneous situation; in fact, in several regions, the active fires occupy noticeably larger area than reported by ATSR.

Direct comparison of ATSR BA (GLOBSCAR), ATSR hot-spot counts (World Fire Atlas) and SPOT VEGETATION BA (GBA-2000; Tansey et al., 2004) products showed quite a confusing picture (Boschetti et al., 2004). The authors stressed that a generally good spatial agreement was overshadowed by major disagreements in the absolute estimates. While the low bias of nighttime hot-spot count is not surprising, large discrepancies also exist in the areal estimates between the two BA products. It was also found that the seasonal profiles of active fires and BAs are significantly different, with shifted maxima and minima of seasonal fire activity. To some extent, similar picture is seen from Figure 21.3: two products based on BA (GEOLAND based on SPOT VEGETATION BA (Tansey et al., 2004) and GFED v3 (van der Werf et al., 2010) based on combined BA mapping (Giglio et al., 2010)) disagree with each other about as significantly as they deviate from the MODIS FRP-based emission of IS4FIRES (Sofiev et al., 2009a).

The above and other comparisons showed that existing estimates of the fire activity and smoke emission, made using BA and hot-spot satellite products, generally agree on spatial patterns but have large uncertainties in the absolute levels of fire severity, consumed biomass and atmospheric emissions.

21.3 FIRE IMPACT ON ATMOSPHERIC COMPOSITION AND AIR QUALITY: MODELLING ASSESSMENTS AND AVAILABLE OBSERVATIONS

The fire smoke affects chemical and physical properties of the atmosphere at a wide range of spatial and temporal scales (Bossioli et al., 2012; Brasseur et al., 2004; Eagan et al., 1974; Konovalov et al., 2011; Langmann et al., 2009; Stohl et al., 2007).

At local and meso-scales, the individual fire plumes are dense, comparatively narrow and exist for a limited period of time (see examples in Figure 21.1). The fire modelling at such scales mainly pursues the rescue operations, public warning on the fire danger and possible severe deterioration of air quality, as well as support of fire-fighting activities (e.g. Miranda,

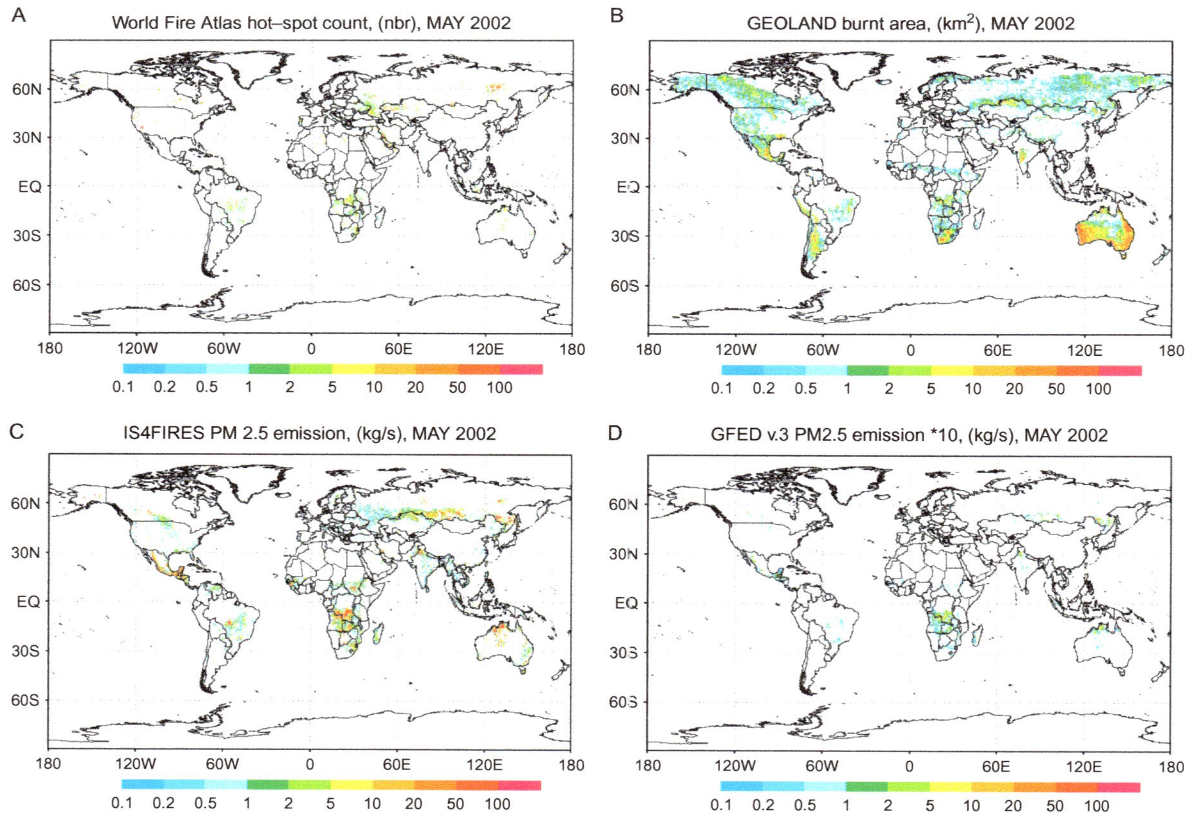

FIGURE 21.3 Example of comparison of monthly fire indicators for May 2002. (A) ATSR WFA active-fire count (nbr); (B) GEOLAND monthly mean burnt area (km^2); (C) FRP-based IS4FIRES monthly mean $PM_{2.5}$ emission (kg/s) and (D) GFED v3 emission *10, based on combined burnt area (0.1 kg/s). *Source: GEIA ECCAD portal, http://www.geiacenter.org.*

2004). The main predicted parameters include the fire propagation and near-source plume dispersion. Short time scales require very simple and fast modelling tools, as well as a possibility to address numerous scenarios of the weather and fire development in near-real-time.

At regional scales, fire plumes tend to form wide clouds of highly polluted air masses that can be transported far outside the source area (Bossioli et al., 2012; Sofiev et al., 2012a; Stohl et al., 2007), see also Figure 21.4. The primary concern regarding the fire impact at such scales is strong deterioration of air quality over substantial areas, as well as climate forcing (Konovalov et al., 2011). During the long-range transport, chemical transformations and aerosol processes significantly change the composition of the cloud. Primary pollutants, such as SO_x and NO_x, are oxidised while the secondary ones, first of all, ozone and organic and inorganic aerosols, emerge. Aerosol processes, namely, coagulation and condensation of organic vapours on the surface of fine particles, result in significant increase of the characteristic particle size in the course of atmospheric dispersion—from 100–150 to 200–300 nm (Janhäll et al., 2010). Notably, the size of aged-smoke aerosol leads to significantly lower rates of both dry deposition via diffusion (Kouznetsov and Sofiev, 2012) and wet deposition via scavenging with precipitation, that is, such aerosols can stay longer in the atmosphere and affect wider areas.

An important factor controlling the lifetime and impact of the fire smoke is the initial injection height of the plumes. Indeed, the lifetime of released compounds is determined by removal and chemical transformations processes, which strongly depend on the meteorological situation at the height of the smoke injection. The bulk of the fire smoke is released in the atmospheric boundary layer (Martin et al., 2010; Sofiev et al., 2009b, 2012b). Such plumes have a strong impact on the near-surface concentrations and can lead to dramatic deterioration of air quality (Konovalov et al., 2011). However, their lifetime is comparatively limited, which in turn reduces the contributions to global balances and climate forcing. Modelling of such episodes is usually concentrated on their short-term impact on air quality.

Situation is different for very strong fires occurring under favourable atmospheric conditions: they can send the plumes high into the free troposphere or the stratosphere (Freitas et al., 2007; Labonne et al., 2007). At those altitudes, the plumes are transported by atmospheric streams above the rain clouds, under low temperature, high wind speed and stable-stratification conditions. This environment prevents any quick removal of the smoke, promotes formation of secondary aerosols from precursors and facilitates very wide range pollution dispersion, which requires global modelling and observation efforts to follow (Dirksen et al., 2009; Fromm et al., 2000; Luderer et al., 2006).

There are numerous fire-related activities that develop the observational databases, perform regular modelling and forecasting of the fire plume

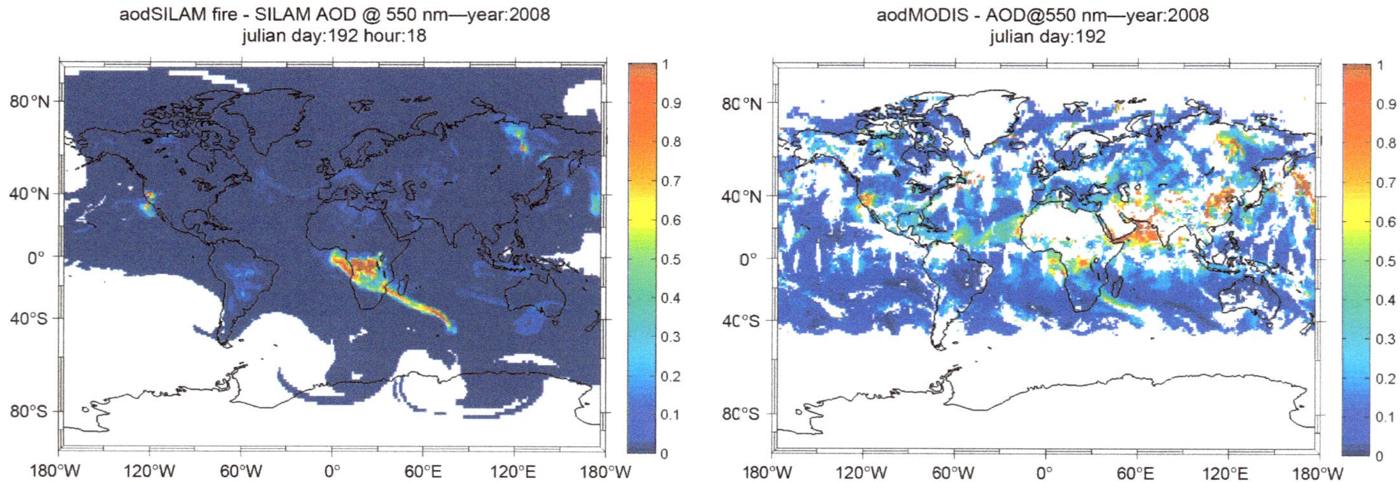

FIGURE 21.4 Examples of the aerosol optical density AOD predicted by the IS4FIRES + SILAM system (due to fire smoke only, A) and daily-composite AOD observed by MODIS for the same day, 12.07.2008 (B). White areas correspond to no smoke for SILAM and no data for MODIS. *Courtesy of J. Soares.*

dispersion, assess the impact on atmospheric compositions, air quality and climate. To mention a few examples:

- Basic information on fires
  - European Forest Fire Information System (http://effis.jrc.ec.europa.eu) supports the services in charge of forest protection against fires; accumulates, harmonises and provides information about the fires.
  - Active-Fire Mapping of USDA Forest Service (http://activefiremaps.fs.fed.us) provides operational information on active fires and BA scars, distributes the data through Web-based services and provides the recent-most statistics of the new and on-going fire cases in the United States.
- Emission into the atmosphere
  - Global Emission Inventory Activity (GEIA) portal (http://www.geiacenter.org) provides a wide range of global and regional emission inventories, including ones for wildland fires: GFASv1.0, GFED, GICC, AMMABB, RETRO, EDGAR, POET, IS4FIRES, GUESS-ES, as well as some basic underlying datasets of fire counts and BAs: WFA, GEOLAND, GBA-2000.
  - BlueSky smoke modelling framework (Larkin et al., 2009) is applied mainly in North America for evaluation of the fire smoke impact.
- Near-real-time fire information and forecasts of fire plume dispersion
  - Global Fire Assimilation System (GFAS) (Kaiser et al., 2009) of Monitoring Atmospheric Composition and Climate project (http://www.gmes-atmosphere.eu) provides both global emission estimations of the wildland fires in near-real-time and forecasts their dispersion in global and European scales.
  - Canadian Wildland Fire Information System (http://cwfis.cfs.nrcan.gc.ca) provides near-real-time and retrospective information on fires over Canada and forecasts the fire risk.
  - Integrated System for fires (Sofiev et al., 2009a; http://is4fires.fmi.fi) provides the current state of fires over Europe and forecasts the plume dispersion. Also provides retrospective information on global fire emission and injection height of the plumes.
  - NCAR FINN inventory (Wiedinmyer et al., 2011) is a global inventory with 1 km spatial resolution. It provides the emission speciation corresponding to several chemical transformation schemes widely used in chemical transport models.
  - Wildfire Smoke Prediction for Alaska UAFSMOKE (http://smoke.arsc.edu/index.html) provides information on current state of fires in Alaska and regional-scale forecast of the plume dispersion.

21.4 FUTURE CHALLENGES AND MAJOR RESEARCH DIRECTIONS

Present state of knowledge of the wildland fires is very far from being complete. Even the most sophisticated analyses so far failed to agree on the amount of smoke released into the atmosphere. Therefore, it is of utmost importance to establish scientifically solid and verified inventories of consumed biomass and released smoke components. That task requires several prerequisites to be developed.

For the *fire monitoring*, the existing disagreements between the BA and active-fire observations should be explained and, to the extent possible, resolved. It may turn out that some methodologies are superior for some types of fires and regions, whereas their application to other regions is not justified. In particular, tropical regions can be monitored using geostationary satellites, whose application to boreal regions is not possible.

The only plausible way of upscaling and generalising the fire information for regional and global applications is *modelling of fires and fire plumes.* Regarding the fire modelling, the basic understanding of pyrogenic processes is available, but the challenge is in developing an adequate model that can operate with limited information coming from remote sensing and scarce active-fire observations. The task of fire plume modelling has its greatest challenge in estimating the initial plume distribution and composition—as functions of the fire state and external conditions controlling the burning process. Should this problem be solved, existing atmospheric dispersion models can be used for subsequent analysis.

Speciation of the fire smoke is another important issue. Existing emission ratios are characterised by wide ranges of uncertainties, which can be reduced only if more detailed information on the fire features is retrieved by the satellites. Direct retrieval of speciation is probably beyond the reach but estimates of the burning efficiency, characteristic temperature of the fire, released thermal energy, etc. can substantially improve the emission assessments. Part of this problem is the conversion from the consumed biomass to the amount of smoke released to the atmosphere. The poorly known conversion factor, being central for BA products, can possibly be eliminated if such information becomes available.

Apart from refinement of the biomass consumption and atmospheric emission from fires, several supplementary datasets are to be generated or refined. For example, assessments of the diurnal variation of the fire intensity and the vertical profile of the smoke injection are required. Recent studies improved knowledge on temporal variation and vertical distribution of the smoke, which is directly related to the atmospheric composition and climate forcing.

The role of fires in natural environment should be better understood and compared with the fire impact on ecosystems under anthropogenic stress, where the fires are largely controlled by human interventions.

Several recent major fire episodes in different parts of the world showed the *importance of fires for the human society*. The impact is likely to grow with climate change, which, among other factors, will results in more frequent and severe droughts in many regions. Associated risk of fires will require more developed fire controlling policies and daily practices than they are today. It is also crucial that population habits are adjusted to the emerging reality of generally high fire risks over large areas (see also Knoke and Hahn, 2013, this vol.).

ACKNOWLEDGEMENT

The author thanks R. Vankevich, J. Vira, J. Soares, M. Prank and T. Ermakova for discussions and comments on the chapter and support of Academy of Finland in a form of IS4FIRES and ASTREX projects.

REFERENCES

Amiro, B.D., Stocks, B.J., Alexander, M.E., Flannigan, M.D., Wotton, B.M., 2001. Fire, climate change, carbon and fuel management in the Canadian boreal forest. Int. J. Wildland Fire 10, 405–413. http://dx.doi.org/10.1071/WF01038.

Andreae, M.O., Merlet, P., 2001. Emission of trace gases and aerosols from biomass burning. Global Biogeochem. Cycles 15 (4), 955–966.

Andreae, M.O., et al., 1988. Biomass burning and associated haze layers in Amazonia. J. Geophys. Res. 93, 1509–1527. http://dx.doi.org/10.1029/JD093iD02p01509.

Arino, O., Plummer, S., 2001. The along-track scanning radiometer World Fire Atlas detection of night-time fire activity. Validation report. IGBP-DIS Working Paper 23, Stockholm.

Barbosa, P.M., Stroppiana, D., Gregoire, J.M., Pereira, J.M.C., 1999. An assessment of vegetation fire in Africa (1981-1991): burned areas, burned biomass, and atmospheric emissions. Global Biogeochem. Cycles 13 (4), 933–950.

Boschetti, L., Eva, H.D., Brivio, P.A., Gregoire, J.M., 2004. Lessons to be learned from the comparison of three satellite-derived biomass burning products. Geophys. Res. Lett. 31 (21), L21501, pp. 1–4. http://dx.doi.org/10.1029/2004GL021229.

Boschetti, L., Roy, D.P., 2009. Strategies for the fusion of satellite fire radiative power with burned area data for fire radiative energy derivation. J. Geophys. Res. 114, 1–10. http://dx.doi.org/10.1029/2008JD011645 (accessed 11 August 2011). http://www.agu.org/pubs/crossref/2009/2008JD011645.shtml.

Bossioli, E., Tombrou, M., Karali, A., Dandou, A., Paronis, D., Sofiev, M., 2012. Ozone production from the interaction of wildfire and biogenic emissions: a case study in Russia during spring 2006. Atmos. Chem. Phys. 12 (17), 7931–7953. http://dx.doi.org/10.5194/acp-12-7931-2012 (accessed 6 September 2012). http://www.atmos-chem-phys.net/12/7931/2012/.

Brasseur, G., Steffen, W., Granier, C., 2004. Atmospheric composition and surface exchanges, in emissions of atmospheric trace compounds. In: Granier, C., Artaxo, P., Reeves, C. (Eds.), Advances in Global Change Research, vol. 18. Kluwer Academic Publishers, Dordrecht.

Chin, M., Ginoux, P., Kinne, S., Torres, O., Holben, B.N., Duncan, B.N., Martin, R.V., Logan, J.A., Higurashi, A., Nakajama, T., 2002. Tropospheric aerosol optical thickness from the GOCART model and comparison with satellite and sun photometer measurements. J. Atmos. Sci. 59, 461–483.

Crutzen, P.J., Andreae, M.O., 1990. Biomass burning in the tropics: impact on atmospheric chemistry and biogeochemical cycles. Science 250, 1669–1678.

Crutzen, P.J., Height, L.E., Kramsec, J.P., Pollock, W.H., Seiler, W., 1979. Biomass burning as a source of atmospheric gases CO, H_2, N_2O, NO, CH_3Cl, and COS. Nature 282, 253–256.

Dirksen, R.J., Folkert Boersma, K., de Laat, J., Stammes, P., van der Werf, G.R., Val Martin, M., Kelder, H.M., 2009. An aerosol boomerang: rapid around-the-world transport of smoke from the December 2006 Australian forest fires observed from space. J. Geophys. Res. 114 (D21), 1–15. http://dx.doi.org/10.1029/2009JD012360 (accessed 11 August 2011). http://www.agu.org/pubs/crossref/2009/2009JD012360.shtml.

Dozier, J., 1981. A method for satellite identification of surface temperature fields of subpixel resolution. Remote Sens. Environ. 11, 221–229.

Eagan, R.C., Hobbs, P.V., Radke, L.F., 1974. Measurements of cloud condensation nuclei and cloud droplet size distributions in the vicinity of forest fires. J. Appl. Meteorol. 13, 553–557.

Flemming, J., 2005. Emissions for GEMS (biomass burning). HALO discussion paper. http://www.ecmwf.int/research/EU_projects/HALO/pdf/paper_HALO_GEMS_emissions.pdf.

Forster, C., et al., 2001. Transport of boreal forest fire emissions from Canada to Europe. J. Geophys. Res. 106 (D19), 22887–22906. http://dx.doi.org/10.1029/2001JD900115. http://www.agu.org/pubs/crossref/2001/2001JD900115.shtml.

Frankman, D., Webb, B.W., Butler, B.W., Jimenez, D., Forthofer, J.M., Sopko, P., Shannon, K.S., Hiers, J.K., Ottmar, R.D., 2009. Measurements of convective and radiative heating in wildland fires. Int. J. Wildland Fire 22, 157–167. http://dx.doi.org/10.1071/WF11097.

Freeborn, P.H., Wooster, M.J., Hao, W.M., Ryan, C.A., Nordgren, B.L., Baker, S.P., Ichoku, C., 2008. Relationships between energy release, fuel mass loss, and trace gas and aerosol emissions during laboratory biomass fires. J. Geophys. Res. 113 (D1), 1–17. http://dx.doi.org/10.1029/2007JD008679 (accessed 11 August 2011). http://www.agu.org/pubs/crossref/2008/2007JD008679.shtml.

Freitas, S.R., Longo, K.M., Chatfield, R., Latham, D., Silva Dias, M.A.F., Andreae, M.O., Prins, E., Santos, J.C., Gielow, R., Carvalho, J.A., 2007. Including the sub-grid scale plume rise of vegetation fires in low resolution atmospheric transport models. Atmos. Chem. Phys. 7 (13), 3385–3398. http://dx.doi.org/10.5194/acp-7-3385-2007. http://www.atmos-chem-phys.net/7/3385/2007/.

Fromm, M., Jerome, A., Hoppel, K., Hornstein, J., Bevilacqua, R., Shettle, E., Servranckx, R., Zhanqing, L., Stocks, B., 2000. Observations of boreal forest fire smoke in the stratosphere by POAM III, SAGE II, and lidar in 1998. Geophys. Res. Lett. 27 (9), 1407–1410. http://dx.doi.org/10.1029/1999GL011200. http://www.agu.org/pubs/crossref/2000/1999GL011200.shtml.

GFMC, 2011. The Global Fire Monitoring Center (GFMC) Fire in the Global Environment, 1–20. http://www.fire.uni-freiburg.de/intro/GFMC-Brochure-2011.pdf.

GFMC, 2012. Community-Based Fire Management, UN ISDR. http://www.fire.uni-freiburg.de/Manag/CBFiM_4.htm (accessed 26 February 2013).

Giglio, L., 2007. Characterization of the tropical diurnal fire cycle using VIRS and MODIS observations. Remote Sens. Environ. 108 (4), 407–421. http://dx.doi.org/10.1016/j.rse.2006.11.018. (accessed 18 June 2012). http://linkinghub.elsevier.com/retrieve/pii/S0034425706004986.

Giglio, L., Kendall, J.D., 2001. Application of the Dozier retrieval to wildfire characterization: a sensitivity analysis. Remote Sens. Environ. 77 (1), 34–49.

Giglio, L., Kendall, J.D., Mack, R., 2003. A multi-year active fire dataset for the tropics derived from the TRMMVIRS. Int. J. Remote Sens. 24 (22), 4505–4525.

Giglio, L., Randerson, J.T., Werf, G.R.V.D., Kasibhatla, P.S., Collatz, G.J., Morton, D.C., Defries, R.S., 2010. Assessing variability and long-term trends in burned area by merging multiple satellite fire products. Biogeosciences 7, 1171–1186. www.biogeosciences.net/7/1171/2010/.

Hart, S.C., DeLuca, T.H., Newman, G.S., MacKenzie, M.D., Boyle, S.I., 2005. Post-fire vegetative dynamics as drivers of microbial community structure and function in forest soils. For. Ecol. Manage. 220, 166–184.

Hoelzemann, J.J., 2004. Global Wildland Fire Emission Model (GWEM): evaluating the use of global area burnt satellite data. J. Geophys. Res. 109 (D14S04), pp. 1–18. http://dx.doi.org/10.1029/2003JD003666 (accessed 30 July 2011). http://www.agu.org/pubs/crossref/2004/2003JD003666.shtml.

Ichoku, C., Kaufman, Y.J., 2005. A method to derive smoke emission rates from MODIS fire radiative energy measurements. IEEE Trans. Geosci. Remote Sens. 43 (11), 2636–2649. http://dx.doi.org/10.1109/TGRS.2005.857328. http://ieeexplore.ieee.org/lpdocs/epic03/wrapper.htm?arnumber=1522624.

Janhäll, S., Andreae, M.O., Pöschl, U., 2010. Biomass burning aerosol emissions from vegetation fires: particle number and mass emission factors and size distributions. Atmos. Chem. Phys. 10 (3), 1427–1439. http://dx.doi.org/10.5194/acp-10-1427-2010. http://www.atmos-chem-phys.net/10/1427/2010/.

Kaiser, J.W., Suttie, M., Flemming, J., Morcrette, J.-J., Boucher, O., Schultz, M.G., Nakajima, T., Yamasoe, M.A., 2009. Global real-time fire emission estimates based on space-borne fire radiative power observations. AIP Conf. Proc., 1100, pp. 645–648. http://dx.doi.org/10.1063/1.3117069. http://link.aip.org/link/APCPCS/v1100/i1/p645/s1&Agg=doi.

Kaufman, Y.J., Justice, C.O., Flynn, L.P., Kendall, J.D., Prins, E.M., Giglio, L., Ward, D.E., Menzel, W.P., Setzer, A.W., 1998. Potential global fire monitoring from EOS-MODIS. J. Geophys. Res. 103 (D24), 32215–32238.

Keddy, P.A., 2007. Plants and Vegetation: Origins, Consequences. Cambridge University Press, Cambridge. http://www.amazon.com/Plants-Vegetation-Origins-Processes-Consequences/dp/0521864801.

Keramitsoglou, I., Kiranoudis, C.T., Sarimveis, H., Sifakis, N., 2004. A multidisciplinary decision support system for forest fire crisis management. Environ. Manage. 33 (2), 212–225. http://dx.doi.org/10.1007/s00267-003-0092-4 (accessed 28 June 2011). http://www.ncbi.nlm.nih.gov/pubmed/15285399.

Knoke, T., Hahn, A., 2013. Global change and the role of forests in future land-use systems. Dev. Environ. Sci. 13, 569–587.

Konovalov, I.B., Beekmann, M., Kuznetsova, I.N., Yurova, A., Zvyagintsev, A.M., 2011. Atmospheric impacts of the 2010 Russian wildfires: integrating modelling and measurements of an extreme air pollution episode in the Moscow region. Atmos. Chem. Phys. 11 (19), 10031–10056. http://dx.doi.org/10.5194/acp-11-10031-2011 (accessed 22 October 2012). http://www.atmos-chem-phys.net/11/10031/2011/.

Kouznetsov, R., Sofiev, M., 2012. A methodology for evaluation of vertical dispersion and dry deposition of atmospheric aerosols. J. Geophys. Res. 117 (D01202), pp. 1–19. http://dx.doi.org/10.1029/2011JD016366.

Labonne, M., Bréon, F.-M., Chevallier, F., 2007. Injection height of biomass burning aerosols as seen from a spaceborne lidar. Geophys. Res. Lett. 34 (11), 1–5. http://dx.doi.org/10.1029/2007GL029311 (accessed 11 August 2011). http://www.agu.org/pubs/crossref/2007/2007GL029311.shtml.

Lamont, B., Lemaitre, D., Cowling, R., Enright, N., 1991. Canopy seed storage in woody-plants. Bot. Rev. 57 (4), 277–317.

Langmann, B., Duncan, B., Textor, C., Trentmann, J., Vanderwerf, G., 2009. Vegetation fire emissions and their impact on air pollution and climate. Atmos. Environ. 43 (1), 107–116. http://dx.doi.org/10.1016/j.atmosenv.2008.09.047 (accessed 27 July 2011). http://linkinghub.elsevier.com/retrieve/pii/S135223100800900X.

Larkin, N.K., O'Neill, S.M., Solomon, R., Raffuse, S., Strand, T., Sullivan, D.C., Krull, C., Rorig, M., Peterson, J., Ferguson, S.A., 2009. The BlueSky smoke modeling framework. Int. J. Wildland Fire 18 (8), 906–920. http://dx.doi.org/10.1071/WF07086.

Lauk, C., Erb, K., 2009. Biomass consumed in anthropogenic vegetation fires: global patterns and processes. Ecol. Econ. 69 (2), 328–334. http://dx.doi.org/10.1016/j.ecolecon.2009.06.025.

Levine, J.S. (Ed.), 1994a. Remote Sensing and Modeling of Biomass Burning, and Biomass Burning in the Boreal Forest. Biomass Burning and Global Change, vol. 1. MIT Press, Cambridge, MA.

Levine, J.S. (Ed.), 1994b. Biomass Burning in the Tropical and Temperate Ecosystems. Biomass Burning and Global Change, vol. 2. MIT Press, Cambridge, MA.

Lindh, A.G., 1941. Fire control methods of the forest service Northern Rocky Mountain Region. Northwest Sci. 15, 34–46.

Liousse, C., Penner, J.E., Chuang, C., Walton, J.J., Eddleman, H., Cachier, H., 1996. A global three-dimensional model study of carbonaceous aerosols. J. Geophys. Res. 101 (D14), 19411–19432. http://www.agu.org/pubs/crossref/1996/95JD03426.shtml.

Loreto, F., Fares, S., 2013. Biogenic volatile organic compounds and their impacts on biosphere-atmosphere interactions. Dev. Environ. Sci. 13, 57–75.

Luderer, G., Trentmann, J., Winterrath, T., Textor, C., Herzog, M., Graf, H.F., Andreae, M.O., 2006. Modeling of biomass smoke injection into the lower stratosphere by a large forest fire (Part II): sensitivity studies. Atmos. Chem. Phys. 6, 5261–5277. http://dx.doi.org/10.5194/acp-6-5261-302006. http://www.atmos-chem-phys-discuss.net/6/6081/2006/.

Martin, M., Val, J., Logan, A., Kahn, R.A., Leung, F.-Y., Nelson, D.L., Diner, D.J., 2010. Smoke injection heights from fires in North America: analysis of 5 years of satellite observations. Atmos. Chem. Phys. 10 (4), 1491–1510. http://dx.doi.org/10.5194/acp-10-1491-2010. http://www.atmos-chem-phys.net/10/1491/2010/.

Matson, M., Dozier, J., 1981. Identification of subresolution high temperature sources using a thermal IR sensor. Photogramm. Eng. Remote Sensing 47, 1311–1318.

Miranda, A.I., 2004. An integrated numerical system to estimate air quality effects of forest fires. Int. J. Wildland Fire 13, 217–226. http://dx.doi.org/10.1071/WF02047 1049-8001/04/020217 (Miranda 1999).

Nathan, R., Safriel, U., Noy-Meir, I., Schiller, G., 1999. Seed release without fire in Pinus halepensis, a Mediterranean serotinous wind-dispersed tree. J. Ecol. 87 (4), 659–669.

Osborne, P., 2000. Tropical Ecosystems and Ecological Concepts. Cambridge University Press, Cambridge. http://ebooks.cambridge.org/ref/id/CBO9781139057868.

Pausas, J.G., Keeley, J.E., 2009. A burning story: the role of fire in the history of life. Biogeosciences 59, 593–601.

Rauste, Y., Herland, E., Frelander, H., Soini, K., Kuoremaki, T., Ruokari, A., 1997. Satellite-based forest fire detection for fire control in boreal forests. Int. J. Remote Sens. 18 (12), 2641–2656.

Roberts, G.J., Wooster, M.J., 2008a. Fire detection and fire characterization over Africa using meteosat SEVIRI. IEEE Trans. Geosci. Remote Sens. 46 (4), 1200–1218. http://dx.doi.org/10.1109/TGRS.2008.915751. http://ieeexplore.ieee.org/lpdocs/epic03/wrapper.htm?arnumber=4469869.

Roberts, G.J., Wooster, M.J., 2008b. Fire detection and fire characterization over Africa using meteosat SEVIRI. IEEE Trans. Geosci. Remote Sens. 46 (4), 1200–1218. http://dx.doi.org/10.1109/TGRS.2008.915751. http://ieeexplore.ieee.org/lpdocs/epic03/wrapper.htm?arnumber=4469869.

Roberts, G., Wooster, M.J., Lagoudakis, E., 2009. Annual and diurnal African biomass burning temporal dynamics. Biogeosciences 6 (5), 849–866. http://dx.doi.org/10.5194/bg-6-849-2009. http://www.biogeosciences.net/6/849/2009/.

Scholes, M., Andreae, M.O., 2000. Biogenic and pyrogenic emissions from Africa and their impact on the global atmosphere. Ambio 29, 23–29.

Schultz, M.G., Heil, A., Hoelzemann, J.J., Spessa, A., Thonicke, K., Goldammer, J., Held, A.C., Pereira, J.M., 2008. Global emissions from wildland fires from 1960 to 2000. Global Biogeochem. Cycles 22 (GB2002), pp. 1–17. http://dx.doi.org/10.1029/2007GB003031.

Seiler, W., Crutzen, P.J., 1980. Estimates of gross and net fluxes of carbon between the biosphere and the atmosphere from biomass burning. Clim. Change 2, 207–247.

Sofiev, M., Vankevich, R., Lotjonen, M., Prank, M., Petukhov, V., Ermakova, T., Koskinen, J., Kukkonen, J., 2009a. An operational system for the assimilation of the satellite information on wild-land fires for the needs of air quality modelling and forecasting. Atmos. Chem. Phys. 9 (18), 6833–6847.

Sofiev, M., Vankevich, R., Lotjonen, M., Prank, M., Petukhov, V., Ermakova, T., Koskinen, J., Kukkonen, J., 2009b. An operational system for the assimilation of the satellite information on wild-land fires for the needs of air quality modelling and forecasting. Atmos. Chem. Phys. 9 (18), 6833–6847.

Sofiev, M., et al., 2012a. From Russia to Iceland: an evaluation of a large-scale pollen and chemical air pollution episode during April and May, 2006. In: Clot, B., Comtois, P., Escamilla-Garcia, B. (Eds.), Aerobiological Monographs. Towards a Comprehensive Vision. MeteoSwiss (CH) and University of Montreal (CA) pp. 95–113.

Sofiev, M., Ermakova, T., Vankevich, R., 2012b. Evaluation of the smoke-injection height from wild-land fires using remote-sensing data. Atmos. Chem. Phys. 12 (4), 1995–2006. http://dx.doi.org/10.5194/acp-12-1995-2012.

Soja, A., Williams, D.J., Pace, T., Randall, D., Sciences, A., Ave, W., Moore, T., 2006. How well does satellite data quantify fire and enhance biomass burning emissions estimates? In: 15th International Emission Inventory Conference Reinventing Inventories—New Ideas in New Orleans, New Orleans, pp. 1–18. http://www.epa.gov/ttnchie1/conference/ei15/.

Stocks, B.J., 2004. Forest fires in the boreal zone: climate change and carbon implications. Int. Forest Fire News 31, 122–131. http://www.fire.uni-freiburg.de/iffn/iffn_31/19-IFFN-31-Boreal-Climate-2.pdf.

Stohl, A., Berg, T., Burkhart, J.F., Fjæraa, A.M., Forster, C., Herber, A., Hov, Ø., Lunder, C., 2007. Arctic smoke—record high air pollution levels in the European Arctic due to agricultural fires in Eastern Europe in spring 2006. Atmos. Chem. Phys. 7, 511–534.

Sukhinin, A.I., Conard, S.G., McRae, D.J., Ivanova, G.A., Tsvetkov, P.A., Bychkov, V.A., Slinkina, O.A., 2005. Remote sensing of fire intensity and burn severity in forests of central Siberia. Contemporary Problems of Earth Remote Sensing from Space. Space Research Institute RAS, Moscow. http://www.iki.rssi.ru/earth/ppt2005/sukhinin.pdf.

Summit, 2003. International Wildland Fire Summit Background Paper. An Overview of Vegetation Fires Globally, 22. http://www.fire.uni-freiburg.de/summit-2003/Summit%20Background%20Paper%20Global%20Situation.pdf.

Tansey, K., et al., 2004. Vegetation burning in the year 2000: global burned area estimates from SPOT VEGETATION data. J. Geophys. Res. 109 (D14), 1–22. http://dx.doi.org/10.1029/2003JD003598 (accessed 25 October 2012). http://www.agu.org/pubs/crossref/2004/2003JD003598.shtml.

Tanskanen, A., Venäläinen, A., 2008. The relationship between fire activity and fire weather indices at different stages of the growing season in Finland. Boreal Environ. Res. 13 (4), 285–302.

Tucker, C.J., 1979. Red and photographic infrared linear combinations for monitoring vegetation. Remote Sens. Environ. 8, 127–150.

Tucker, C.J., Newcomb, W.W., Los, S.O., Prince, S.D., 1991. Mean and inter-year variation of growing-season normalized difference vegetation index for the Sahel 1981-1989. Int. J. Remote Sens. 12, 1113–1115.

van der Werf, G.R., Randerson, J.T., Giglio, L., Collatz, G.J., Kasibhatla, P.S., Arellano Jr., A.F., 2006. Interannual variability in global biomass burning emissions from 1997 to 2004. Atmos. Chem. Phys. 6, 3423–3441.

van der Werf, G.R., Randerson, J.T., Giglio, L., Collatz, G.J., Mu, M., Kasibhatla, P.S., Morton, D.C., DeFries, R.S., Jin, Y., van Leeuwen, T.T., 2010. Global fire emissions and the contribution of deforestation, savanna, forest, agricultural, and peat fires (1997–2009). Atmos. Chem. Phys. 10 (23), 11707–11735. http://dx.doi.org/10.5194/acp-10-11707-2010 (accessed 16 October 2012). http://www.atmos-chem-phys.net/10/11707/2010/.

Viegas, D.X., Allgöwer, B., Koutsias, N., Eftichidis, G., 2003. Fire spread and the wildland urban interface problem. In: Xanthopoulos, G. (Ed.), Proceedings of the International Scientific Workshop on Forest Fires in the Wildland-Urban Interface and Rural Areas in Europe: An Integral Planning and Management Challenge. Mediterranean Agronomic Institute of Chania, Athens, p. 12. http://www.fria.gr/WARM/warmProceedings.htm.

Wiedinmyer, C., Akagi, S.K., Yokelson, R.J., Emmons, L.K., Orlando, J.J., Soja, A.J., 2011. The fire inventory from NCAR (FINN): a high resolution global model to estimate the emissions from open burning. Geoscient. Model Dev. 4, 625–641. http://dx.doi.org/10.5194/gmd-4-625-2011.

Wooster, M.J., Roberts, G., Perry, G.L.W., Kaufman, Y.J., 2005. Retrieval of biomass combustion rates and totals from fire radiative power observations: FRP derivation and calibration relationships between biomass consumption and fire radiative energy release. J. Geophys. Res. 110 (D24311), pp. 1–24. http://dx.doi.org/10.1029/2005JD006318.

Zeng, N., 2005. Terrestrial mechanisms of interannual CO_2 variability. Global Biogeochem. Cycles 19 (1), GB1016, pp. 1–15. http://dx.doi.org/10.1029/2004GB002273. (accessed 22 October 2012). http://www.agu.org/pubs/crossref/2005/2004GB002273.shtml.

Part VI

The Potential of "Supersites" for Research on Forest Ecosystems

Chapter 22

Towards Supersites in Forest Ecosystem Monitoring and Research

Teis Nørgaard Mikkelsen[*,1], Nicholas Clarke[†], Alina Danielewska[‡] and Richard Fischer[§]

[*]*Centre for Ecosystems and Environmental Sustainability (ECO), Department of Chemical and Biochemical Engineering, Technical University of Denmark, Denmark*

[†]*Norwegian Forest and Landscape Institute, Norway*

[‡]*Meteorology Department, Poznan University of Life Sciences, Poznan, Poland*

[§]*Thünen Institute for World Forestry, Hamburg, Germany*

[1]*Corresponding author: e-mail: temi@kt.dtu.dk*

Chapter Outline

22.1 INTRODUCTION

Forest ecosystems are providers of essential services such as water regulation, wood supply, carbon storage and biodiversity. However, they are facing a major change in the future. Climate change and the consequences for ecosystem services are multiple but hard to predict. Air pollution will also be a part of the environment in the future and air pollution will act on forests in concert with climate change (Bytnerowicz et al., 2007; Matyssek et al., 2013a, this vol.). Other significant pressures over the coming century include land-use

Developments in Environmental Science, Vol. 13. http://dx.doi.org/10.1016/B978-0-08-098349-3.00022-0

changes, potential overexploitation for bioenergy and biorefineries (Schulze et al., 2012) and increasing demand for conservation areas. For management of forests in such a way as to preserve the functioning of forest ecosystems and minimise the impacts of threats, it is essential to further our understanding of ecosystem processes. For this purpose, both research and long-term monitoring are necessary.

The knowledge on current transnational European forest monitoring and research networks is an essential basis to enable better integration and co-operation and to maximise the use of existing infrastructure and data (Fischer et al., 2011a). Improving access to the large amounts of data already obtained within existing monitoring programmes and large-scale international projects is another important challenge for the scientific community (Clarke et al., 2011). The existing transnational networks in forest research and monitoring provide data and information for a wide range of applied analyses, models and investigations. Networks setup for one purpose often provide data, which are valuable for other research communities, and such co-benefits need to be encouraged and enabled in the future (Fischer et al., 2011a) so that it is possible that many of the collected data can be used when facing new challenges. On the other hand, we are now facing a future with many new challenges for the forest. This calls for new approaches and as a result a new type of research site in forest ecosystems (supersites) handling both current and future knowledge gaps. Five overall knowledge gaps and associated research needs related to the effects of climate change and air pollution on forests were identified at the COST FP0903 conference in Rome, Italy, 2010 (Matyssek et al., 2012), and the work with knowledge gaps was continued in a survey during 2012. A supersite is defined as a highly instrumented research site where all ecosystem compartments and fluxes are covered, including experimental approaches aiming at gaining insight into mechanistic processes and thus filling these knowledge gaps (Fischer et al., 2011a; Paoletti et al., 2013, this vol.).

This chapter will elucidate the development of existing European forest monitoring and research infrastructures, describe harmonisation of databases and knowledge about climate change and air pollution impact on forest ecosystems and present knowledge gap statements from fifty experts affiliated with the COST FP0903 Action.

22.2 MONITORING SITES AND RESEARCH NETWORKS

The continuation and further development of existing European forest monitoring and research infrastructures form the basis for joint and collaborative scientific approaches that are necessary to understand the integrated effects of climate change and air pollution on forest ecosystems. Numerous scientific projects with different spatial coverage, different scientific focus and temporal continuity have been financed and conducted in the past. In many cases these have been more or less isolated undertakings. Traditional research projects comprise project design, data gathering, validation, analysis and reporting. Further use of the data beyond the project has often not been considered or has depended on successful application for prolongation after the end of a first project phase. However, given the complexity and costs of modern measurement equipment and techniques, it is questionable whether such an approach will be viable in the future. Instead, monitoring, experimentation and research infrastructures need to be regarded as autonomous and as an independent basis for several different and temporally subsequent research data analyses and monitoring projects. Results from such networks provide added value for subsequent projects that can not only build on the same raw data base but also on the evaluation and modelling results of previous endeavours. A number of such networks already exist in Europe (Table 22.1). They provide data as a basis for evaluations in different thematic fields (Fischer et al., 2011a). Similar networks exist in other parts of the world, for example, EANET in East Asia (Koike et al., 2013, this vol.).

22.2.1 Existing European Networks

The International Co-operative Programme on Assessment and Monitoring of Air Pollution Effects on Forests (ICP Forests) comprises the largest harmonised terrestrial forest monitoring network in Europe with approx. 250 core intensive forest research and monitoring sites and 7000 large-scale sites located in 32 countries. It is a widely distributed, pan-European, long-term observational network applying and further developing state of the art methodologies and technologies for forest ecosystem research, monitoring and modelling. It provides comparable and quality checked data for multiple scientific approaches as well as up to date aggregated information on the state of Europe's forests (Fischer et al., 2011b).

The International Co-operative Programme on Integrated Monitoring (ICP Integrated Monitoring) carries out monitoring on 44 forested catchment sites. Among these, a larger proportion is in unmanaged forests. Monitoring is divided into a number of compartmental sub-programmes, which are linked by the use of the same parameters (cross-media flux approach) and/or the same/nearby stations (cause–effect approach). Methods are partly harmonised with ICP Forests (De Vries et al., 2002; Kleemola and Forsius, 2010).

The Co-operative Programme for Monitoring and Evaluation of the Long-range Transmission of Air Pollutants in Europe (EMEP) is assessing the transboundary transport of acidifying and eutrophying pollutants, heavy metals and particulate matter. It also addresses the formation of ground-level ozone and of persistent organic pollutants (POPs). The EMEP programme relies on three main elements: (i) collection of emission data, (ii) measurements of air and precipitation quality and (iii) modelling of atmospheric transport and deposition of air pollutants (EMEP, 2010). 10–15 sites are located in forests depending on the compound measured.

TABLE 22.1 Overview of Transnational Forest Monitoring and Research Networks in Europe

Programme	Full name	Webpage	Legal basis/ financing	Sites
ICP Forests	International Co-operative Programme on Assessment and Monitoring of Air Pollution Effects on Forests	www.icp-forests.org	UNECE, CLRTAP	6000 low intensity sites 500 intensive forest monitoring sites, from these approx. 100 intensified FutMon sites
ICP Integrated Monitoring	International Cooperative Programme on Integrated Monitoring of Air Pollution Effects on Ecosystems	www.environment.fi/default.asp?contentid=17110&lan=en	UNECE, CLRTAP	44 catchment sites, 39 incl. forests
EMEP	European Monitoring and Evaluation Programme	www.emep.int	UNECE, CLRTAP	Number of sites depends on compound. In general, about 100 Level I sites (e.g. ozone, main components) and 25 Level II sites—joint EMEP/GAW supersites with more advanced measurements (e.g. POPs, aerosol properties, VOC). About 10–15 EMEP sites are located in forest. More sites with partly forested areas
NitroEurope	European research into the nitrogen cycle	www.nitroeurope.eu	6 FP	*Component 1 (observational network)* 4 forest sites Level III (supersites, 500 parameters) 2 forest sites Level II (100 parameters) 24 forest sites Level I (low cost active sampler 100 parameters) *Component 2 (manipulation network)* 7 forest sites (additional non-forest sites available)

CarboEurope/ Fluxdata	CarboEurope-IP Assessment of the European Terrestrial Carbon Balance	www.carboeurope.org	6 FP	54 atmospheric observation sites 106 ecosystem observation sites (58 forest sites)
ICOS	Integrated Carbon Observation System	www.icos-infrastructure.eu	7 FP/ESFRI	Current plans: installation of approx. 50 atmospheric sites 50 ecosystem sites final information on ecosystems represented not yet available 10 ocean sites
LTER-Europe	European Long-Term Ecosystem Research Network	www.lter-europe.net	6 FP (ALTERNET) ILTER	400 LTER sites (30 % forested) 25 LTSER platforms

From Fischer et al., (2011a).

The Integrated Carbon Observation System (ICOS) is a European Research Infrastructure for quantifying and understanding the greenhouse balance of the European continent and of adjacent regions. The programme currently assembles laboratories in 17 countries. After a preparatory phase, funded by the EU and ending in 2013, ICOS is intended to run for the next 20 years with a minimum of 30 sites. It builds among others on selected NitroEurope and CarboEurope sites (ICOS, 2010) but will be based on highly standardised methodologies and equipment, which will be the main difference from the rest of the canopy flux sites that will continue to operate and share data under the Fluxdata-Europe system.

In addition to still on-going and fully operational networks, there are consortia, data bases and sites that were created in large previous projects and that still provide data for on-going research activities. Examples for such existing networks are NitroEurope and CarboEurope.

NitroEurope was a project for integrated European research into the nitrogen cycle in different ecosystems. The focus was on the effect of reactive nitrogen supply on net greenhouse gas budgets for Europe. It followed a combined approach integrating monitoring and experiments with sites with different intensities of measurements (NEU, 2010). There were around 30 forested observation and 7 manipulation sites.

CarboEurope was a project that aimed at understanding and quantifying the present terrestrial carbon balance of Europe (CarboEurope, 2010). Fifty-eight of the sites were located in a forested environment. Even though CarboEurope was finished in 2008, the project initiated a community that continues to collaborate and share data, even without direct funding, and with support from other research projects on carbon cycling such as CarboExtreme, IMECC and GHG-Europe. CarboEurope helped to prepare the basis for ICOS. The system is currently constructing a database (Fluxdata-Europe) providing access to the data.

Instead of creating new sites and raw databases, it seems logical to build on existing infrastructures and link these, aiming at better access and coordination among partners. Such approaches like The European Long-Term Ecosystem Research Network (LTER-Europe) and FLUXNET help to minimise duplications and provide platforms for information exchange and joint acquisition of funds.

LTER-Europe is part of the global ILTER system. It links national networks, research infrastructures, institutions and researchers (LTER, 2010). It provides an InfoBase with a collection of meta-information. The LTER network is driven by the particular science questions of the principal investigators and the uniqueness of the sites (Gordon et al., 2012).

FLUXNET is an example of an internet platform providing tools that assist in acquiring, querying, browsing, plotting and manipulating diverse combinations of heterogeneous data from many sites. Flux and meteorological data may be submitted by any interested researcher or organisation to a central database (www.fluxdata.org). These data are checked for quality, gaps are filled and new value added products are made available, for example, through data aggregation.

A number of points need to be considered when regarding available networks and existing databases. First of all, harmonisation is an important prerequisite. *A posteriori* collection and links between different projects that have been created independently usually fail or at least face severe challenges due to the fact that assessment methodologies and data formats are not compatible. Thus in many cases only some harmonisation in later evaluation stages or a comparison of results will be possible. ICOS is an example of a network that from the beginning aims to define standards. The intensive monitoring level of ICP Forests was in the very beginning based on existing national approaches but soon succeeded in harmonising methods and is today based on comprehensive and published joint field and laboratory protocols (ICP Forests, 2010).

Poor data quality results in an increase of costs and in a loss of confidence in results. Proper quality assurance and quality control (QA/QC) should ensure that all the steps of a given programme/investigation are considered, from research or monitoring design, to measurement/classification errors, errors caused by models and non-statistical errors.

The larger the networks are, the bigger the distance from data collection to evaluation and publication will become, as in all fields specialists will become active. Programmes and projects need to ensure that field experts and data gatherers are in contact with the data users. Workshops and conferences usually provide appropriate platforms for this.

The continuation of data sets is in many cases in conflict with the further development of monitoring and research networks. New techniques will result in altered data formats, and new variables that need to be measured pose the question if others can be abandoned in order to keep the budget within limits. It is within the responsibility of network managers to find compromises and to ensure that the most valuable outputs, that is, a minimum of time series, are maintained and that new data formats are kept compatible with previous ones.

22.3 HARMONISATION OF DATABASES AND KNOWLEDGE ABOUT CLIMATE CHANGE AND AIR POLLUTION IMPACT ON FOREST ECOSYSTEMS

Of a wide variety of international forest research and monitoring networks, several are dedicated to the effects of climate change and/or air pollution on forest ecosystems. The large amounts of data already obtained within these existing networks can be used to increase understanding of the state and potential of forest mitigation and adaptation to climate change in a polluted environment. However, it is often hard to get an overview of which data exist, and of their availability and accessibility (Clarke et al., 2011). A major challenge for the scientific community is therefore to identify, evaluate and compare the available databases from past research networks, on-going monitoring networks and on-going research networks, and if possible to harmonise these databases (Augustin et al., 2005; Clarke et al., 2011; Fischer et al., 2011a). This will allow integration across networks of data on C sequestration and storage, soil and atmospheric chemistry and water budgets.

In Europe, there are a number of large-scale international databases both from monitoring programmes and research projects, as well as data from many small-scale projects and programmes, including (Clarke et al., 2011):

- International databases from monitoring programmes (ICP Forests, ICP Integrated Monitoring, EMEP, LTER-Europe, etc.)
- Data from past and present large-scale (CarboEurope, NitroEurope, IMECC, GHG-Europe, ICOS, etc.) and small-scale international research projects covering most of Europe
- Regional databases and projects (e.g. Noltfox, the Northern European Database of Long-term Forest Experiments, the Nordic flux tower network NECC)
- National databases and projects including national forest inventories (NFIs).

An overview of the most relevant large-scale databases for European forests that can be used in air pollution and climate change research is given in Table 22.2.

Although in principle data obtained using public funding should be available to the public, they may not be easily accessible in practice. The OECD has published principles and guidelines for access to research data obtained using public funding (OECD, 2007). Improved access is generally seen as benefiting the advancement of research, boosting its quality and facilitating cross-disciplinary research co-operation. However, that some data may remain inaccessible is acknowledged by the OECD guidelines, in order to safeguard the legal rights and legitimate interests of all stakeholders (OECD, 2007).

Rules for access to data in existing databases can be summarised as (Clarke et al., 2011):

- Free access (generally by internet), for example, EMEP
- Free access with registration (often for internet security issues or for additional communications with users about problems in the data), but no authorisation, for example, ICOS
- Access on request/by authorisation, for example, ICP Forests
- Inaccessible (information lost either technically or politically due to ownership/property/security concerns)

In practice, though, scientific use of data beyond the specific networks gathering the information is often limited (Fischer et al., 2011a).

TABLE 22.2 Overview of Existing Large-Scale Databases Relevant for Forest Monitoring and Research, Including Data Access Rules and Manuals/Protocols

Programme/project	Short name	Webpage	Database	Data access rules	Manual/protocols
International Co-operative Programme on Assessment and Monitoring of Air Pollution Effects on Forests	ICP Forests	www.icp-forests.org	Central database with raw data at vTI, Germany	Open after registration and information of countries, 2 years reserved for internal evaluation	www.icp-forests.org/Manual.htm
International Cooperative Programme on Integrated Monitoring of Air Pollution Effects on Ecosystems	ICP Integrated Monitoring	www.environment.fi/default.asp?contentid=17110&lan=en	Central database with partly raw data, partly aggregated data at SYKE, Finland	Open after registration, large data sets are approved by Task Force	www.environment.fi/default.asp?node=6329&lan=en
The nitrogen cycle and its influence on the European greenhouse gas balance	NitroEurope	www.nitroeurope.eu	Central data base with raw data at CEH, United Kingdom	Meta data: open access. raw data: open for partners and case by case decision for external users	Partly CarboEurope, methods • Rainfall: EMEP methods • Own methods of partners, that need to be documented in method sheets
Co-operative Programme for Monitoring and Evaluation of the Long-range Transmission of Air Pollutants in Europe	EMEP	www.emep.int	Central data base with raw data at NILU, Norway: ebas.nilu.no	All the data is open for Web download	http://tarantula.nilu.no/projects/ccc/manual/index.html

Continued

TABLE 22.2 Overview of Existing Large-Scale Databases Relevant for Forest Monitoring and Research, Including Data Access Rules and Manuals/Protocols—Cont'd

Programme/project	Short name	Webpage	Database	Data access rules	Manual/protocols
Integrated Project CarboEurope Assessment of the European Terrestrial Carbon Balance. Data under the responsibility of IMECC from 2009	CarboEurope (IMECC)	www.carboeurope.org (http://imecc.ipsl.jussieu.fr/) Ecosystem data: http://www.europe-fluxdata.org	Central data base with meta data at MPI-BCG, Germany. Half-hourly Ecosystem data, University of Tuscia Viterbo, Italy Raw data under responsibility of site managers	Meta-data: open access. Half hourly: open for partners and case by case decision for external users (but from 2011 open)	Measurement protocols under www.carboeurope.org
Integrated Carbon Observation System	ICOS	www.icos-infrastructure.eu	Raw data: • Ecosystem sites University of Tuscia, Italy • Atmospheric sites, LSCE, France.	Meta-data: open access. Raw data: open access	• Partly Carbo Europe and IMECC (7 FP) • Partly under development • Concept papers available
European Long-Term Ecosystem Research Network	LTER	www.lter-europe.net	Central database with meta-data at UBA, Austria. Raw data under responsibility of site managers	Meta-data: Web-based, open. Raw data: links to other networks	Best practice guidelines at http://www.lter-europe.net/document-archive/central/Best%20practice%20guideline%20v3-June1.pdf

From Clarke et al., (2011).

Not only is it important for society as a whole that data are made accessible, but it is also important for those collecting the data. Experience has shown that in the long term only the active sites in terms of data sharing have survived, that is, those sites with a wide use of their data by scientists that are not directly involved in the data collection. This provides justification for further funding.

Published results are of course generally available in international peer-reviewed journals, books, technical reports from programmes or projects, or national reports (often in the national language), but these are normally not the raw data. More commonly they are the results of statistical treatment of the data. The data in databases are often not truly raw data either, in the sense of the signals that come from the sensors (Clarke et al., 2011).

Although the principle of free access to publicly funded data is important, there are legitimate concerns about intellectual property rights (IPR), including a right to first use of unpublished data for the scientist(s) responsible for obtaining them. 'The rules for such incentive arrangements should be developed and explicitly stated by the funding sources in co-operation with the affected research communities' (OECD, 2007). IPR regulations should be and often are mandatory in any project supported by public funding. Many programmes and projects, such as ICP Forests, have an agreed IPR policy with strict regulations governing the use of data collected by others.

To assist in making more visible the information that is available, a meta-database was elaborated in the frame of the collaboration and knowledge transfer between the COST Action FP0903 countries. This meta-database is freely available for downloading at http://cost-fp0903.ipp.cnr.it/Downloads/WG1metadatabase.xlsx. The main objective of the meta-database is the integration of the information from the various European Research and Monitoring Networks (ERMN) already dealing with the effects of different stress factors on forests, including air pollution and climate change (Danielewska et al., 2013). Moreover, the established meta-database is a comprehensive description of the present state of European forests at several scales (from a single tree to whole forest stands). It should be emphasised that the meta-database is not intended to replace existing databases, but to act as an aid to those seeking to find out what information is available in these databases and thus assist in improving access to, and coordination between, the different datasets.

The new meta-database was created on the basis of several European projects and programmes such as CarboEurope, NitroEurope, ICP Forests, ICP Integrated Monitoring, EMEP etc. All of the selected projects and programmes are strongly linked to the problem of climate change and its influence on the terrestrial ecosystem, especially forests. Furthermore, the problems of adaptation of natural ecosystems in a polluted environment were involved in the research in each of the selected projects. The elaborated meta-database tables, in the form of Excel spreadsheets, contain information, which is freely available on the various research networks' Web sites, related to the effects of different stress factors such as air pollution, GHG emissions and climate change on ecosystems. The meta-database is divided into 13 Excel spreadsheets: four general (introduction, projects, abbreviations, basic site description), five thematic (meteorology, soil characteristics, atmospheric chemistry, fluxes, additional information) and four related to the particular ecosystem types (forest, grassland/wetland, cropland, oceans). Depending on the selected ERMN, the list of variables included in the meta-database and the measurement units differ widely.

The quality of a database is only as good as the quality of the data it contains. Strict QA/QC procedures at all stages of an investigation (experimental design, field sampling, transport to the laboratory, laboratory analysis, data treatment and reporting) are crucial and are normally included in the manual of any programme or project supported by public funding. Not all analytical methods are equally good, and it is possible to prepare a list of methods that give inaccurate or imprecise results, as has been done, for example, for deposition analyses in ICP Forests (Clarke et al., 2010). Even laboratories using the same analytical method can get very different results. This can be seen in the results from inter-laboratory ring tests, where the same samples are analysed by a number of different laboratories; the range in results obtained may be large even when the same method is used (e.g. Marchetto et al., 2009). In recent years, a better understanding of QA/QC has led to improvements in the quality of data obtained from laboratories. However, questions may in some cases remain about the quality of older data and these data should if necessary be flagged. Whenever possible it is important to associate an uncertainty value with all the data, uncertainty that can take into account all the different sources, from the sample or measurement collection to the final data storage.

A big problem in multinational programmes has been the diversity of field and laboratory methods used in different countries. In many databases, data have been obtained from a large number of different partners using different methods or techniques, raising questions about their comparability. Achievement of comparability between different programmes or projects is even more difficult as their aims differ, leading, for example, to differences in scale.

An example of the use of different methods within one programme is the monitoring of deposition, especially throughfall, in ICP Forests. Different types of samplers are used; for throughfall sampling these could be either funnels or gutters. In addition to different sampler types, there are differences in factors that influence the representativity of the measurements, such as the number of samplers used, the surface area of the samplers and their placing in the forest. Field intercomparisons of samplers for bulk precipitation and throughfall showed that there were large differences in the results obtained, which were most likely related to differences in the sampling strategy such as the sampler placement and the collecting area used (Bleeker et al., 2003; Erisman et al., 2003). Inaccuracy induced by the field samplers appeared to be larger than that induced by use of different laboratories in chemical analysis.

Possibilities for harmonisation of samplers have been considered in the ICP Forests network. A test of national deposition samplers and a harmonised sampler was carried out in many countries as part of a project co-financed by the European Commission. Preliminary results from the intensive forest ecosystem monitoring plot Brdo in north-western Slovenia showed good agreement between national and harmonised bulk (both funnel-type) collectors, while comparison of throughfall measurements indicated systematic bias between harmonised (funnel-type) and national (gutter-type) collectors (Žlindra et al., 2011).

In the laboratory too, use of different methods can lead to different results, even in cases where the methods use the same basic principle. For example, a comparison of five different methods for aluminium fractionation carried out by Wickstrøm et al., (2000), four of which used the same fractionation principle (cation exchange), showed differences caused by factors such as reaction time and pH even when the same fractionation principle was used. Method intercalibrations are therefore necessary and should be mandatory.

Although method harmonisation, leading to the use of the same equipment and the same methods by all participants in a network, could be considered as a way of improving data comparability, there are in practice a number of hinders (Clarke et al., 2011):

- Methods could be well adapted for local conditions, leading to uncertainty about the benefits of a new, harmonised method
- There are different historical schools in different countries (e.g. soil type classification)
- Changing the methodology will break the time series; comparison between before and after the method change is difficult, as old and new methods must be run in parallel (for at least a year in the case of field equipment in order to take account of seasonal variation) to be able to compare data obtained
- Harmonisation is expensive: new equipment must be purchased and installed
- Harmonisation might inhibit the ability to develop new methods or improve on existing ones

One possible solution could be the use of a standard method or set-up in parallel with the local, long-term and established methodology. In this way, the common reference method would permit a clear and robust comparison and the paired methods applied at each site to evaluate the uncertainty or spot problems and errors. However this solution would be clearly more expensive.

An approach to harmonisation used by the LTER network is to develop and agree on data standards such as frequency, accuracy and spatial scale of measurements, rather than to develop new, fully harmonised methods; harmonisation thus becomes a setting of data properties enabling comparability of the data obtained (Frenzel et al., 2011).

22.4 KNOWLEDGE GAPS AND NEW PROCESSES TO BE STUDIED

At the COST FP0903 inaugural conference in Rome, Italy, 2010, five overall knowledge gaps and associated research needs were identified: (1) Interactions between trace gases, climate change and vegetation, (2) Significance of biotic processes in forest ecosystem response, (3) Tools for mechanistic and diagnostic understanding and risk upscaling, (4) Unifying modelling and empirical research and (5) Global dimension of air pollution as part of climate change (Matyssek et al., 2012). In 2012, COST FP0903 developed a questionnaire in the context of improving forest sites for forest research under climate change and air pollution impacts. The aims were to define key criteria to be addressed and to recommend steps and requirements for establishing so-called 'supersites', that is, highly instrumented monitoring and research sites, where all ecosystem compartments and fluxes are studied, including experimental approaches. One of the main objectives is 'to reconcile process-oriented research, long-term monitoring and applied modelling at comprehensive forest research sites (Supersites)'. The questionnaire was distributed in spring 2012 to all Management Committee members in COST FP0903 and affiliated experts with a request for circulation among experts within each member country. A second distribution was initiated in autumn 2012 with the focus to receive more answers from experts within the research projects ECLAIRE and ICOS. Fifty experts from 25 countries, primarily in Europe, responded all in all to the questionnaire and the following ranked (number of responses) networks and projects are represented by at least four experts, although an expert can be involved in several projects and networks: ICP Forests (27), LTER (12), NitroEurope (10), ICOS (9), CarboEurope (9), ICP Integrated Monitoring (9), EMEP (7), GHG-Europe (6), ECLAIRE (5), IMECC (4), InGOS (4) and ENFIN/NFI (4).

In the following we present parts of the answers given in the questionnaire related to the topics below. The experts were asked to 'Indicate knowledge gaps and new processes to be studied' for each of six topics. In addition, we asked the experts to address the importance in general of each topic and provide a score system between 1 and 5 (1—highly important; 5—unimportant) and this resulted in the score shown in brackets.

- The carbon cycle in forests: carbon cycle in forests: Land use, land-use change and forestry (LULUCF) accounting (1.7)
- Forest health and vitality (including damage) (1.9)
- Forest biodiversity (2.2)
- Extent of forest resources, availability of wood and forest biomass (2.3)
- Protective functions of forests (2.4)
- Socio-economic information about the forest sector (3.3)

The LULUCF accounting. Many experts suggest carbon sequestration in soils as a knowledge gap with respect to the dynamics of carbon fluxes and pools.

The soil carbon sequestration should be studied in relation to land-use history, harvest processes and other management strategies for minimising CO_2 emission, nutrient availability, extreme events (e.g. climate, water and fire) and influence of environment and biota. One of the questions that can be addressed here is how safe it is to rely on soil for long-term carbon storage. Impacts of co-occurring factors such as ozone, nitrogen, water availability and others on the carbon cycle should be studied with emphasis on their interactions over the season (Matyssek et al., 2013b, this vol.). Fluxes of trace gasses should be introduced, VOCs and methane contain carbon and influence the carbon cycle directly and ozone is expected to interact with the carbon cycle by reducing the carbon uptake. However, the latter needs research and confirmation on a large scale within forests. Effects of different forest management strategies on the carbon cycle (Jandl et al., 2007) need to be further studied, especially in relation to intensified harvesting for bioenergy but also to such factors as choice of tree species (Vesterdal et al., 2013). Can carbon in soil be managed so that it is channelled into stable rather than labile pools? There are also suggestions for developing a carbon observation system in Europe, combining remote sensing techniques and *in situ* databases.

Forest health and vitality (including damage). There is a knowledge gap related to the influence of the ambient or elevated ozone concentration on forest ecosystem condition. It is still a major challenge to translate the current knowledge in the above field from the local (national) to the global (continental, world) scale and one way forward could be by including climate change issues and enhancing co-operation between experts on modelling and empirical research including ozone and biogenic volatile organic carbon (BVOC) fluxes. Abiotic and biotic impacts on forest health and tree vitality still have many open questions and combined studies with soil capacity and extreme events (e.g. drought rewetting, heat waves and low temperatures) will provide much useful information. Other research areas within this topic could be vulnerability of forest species after fire and short- and long-term impacts of whole tree harvesting and other bioenergy utilisation on the forest health and vitality (e.g. Helmisaari et al., 2011). The multiple interactions between forest and adjacent management systems are also a knowledge gap together with afforestation of set-aside and abandoned areas.

Forest biodiversity. The functional significance of forest (tree species) biodiversity for a large range of ecosystem processes and ecosystem services could be characterised as a knowledge gap and this should include provisioning services (e.g. food, fibre, genetic resources), regulating services (e.g. erosion control, climate regulation, water purification), supporting services (e.g. soil formation, primary production, nutrient cycling), and cultural services (e.g. recreation, aesthetic experience). Will, for example, an increased biodiversity enhance carbon sinks? Or will an increased species biodiversity increase the forest ecosystem resilience to environmental stresses? Other specific questions are: what is the impact of secondary tree species on forest biodiversity? What is the connection between climate change, ozone and biodiversity? What happens to the biota when wood ash is brought back to the forest floor? The soil biodiversity (different plants, animals and microorganisms below ground) is a knowledge gap that is important to study and follow in relation to the carbon cycle, climatic change and air pollution. Biodiversity, in general, is fairly well understood, but studies conducted so far have involved single groups of organisms. There is a high necessity for the recognition of the collective interactions between many groups of organisms (e.g. mites, insects, birds, trees, mushrooms, earthworms). Their interactions have significant effects on functioning of the forest including carbon sequestration (Vesterdal et al., 2013). Another research problem that requires development is to recognise the risks of introduction of alien species, especially pathogens and invasive species. It is also relevant to conduct an assessment of endangered species and develop a protocol for their assessment. Other knowledge gaps are tree species adaptation, application of other tree species and ecosystem management under future conditions.

Extent of forest resources, availability of wood and forest biomass. In relation to sustainability, the forest sector is faced with increased demands for wood energy and therefore indicators for forest biomass harvesting should be studied, for example, in relation to site and biodiversity quality that results with intensified wood and forest biomass removal for energy production, and this could be coupled with recycling of wood ash from bioenergy production to restore site nutrient contents. The forest productivity related to changes in climate, in particular interactions between factors and influence of extreme events, should be investigated too. New methods for determining forest biomass after fire should be developed. Dynamic ecosystem models and models for forest biomass should be developed further.

Protective functions of forests. In relation to air pollution and climate change, the protective functions of the forest are multiple; but their quantification and thereby valuation is a knowledge gap. Forests scavenge aerosols (PM2.5) and other pollutants such as ozone and ammonia from the atmosphere, and this removal has a direct positive effect on human health, everything else being equal, because the atmospheric concentrations of these pollutants will be lowered. However, this effect is poorly quantified. Forest interacts with climate change in many ways, for example, with the atmosphere with absorption and release of the major greenhouse gasses CO_2, CH_4 and N_2O. In general, forest is considered as a sink for these gasses, thereby diminishing climate change; but will this mitigation effect continue in the future due to continued impacts of climate change and air pollution? And what will happen to all other protective functions (e.g. groundwater protection, flood protection and protection from other natural hazards) under a future climate or pollution regime? For example, Sun et al., (2012) and Bytnerowicz et al., (2013, this vol.) find interactive influences of ozone and climate on stream flow of forested watersheds. The positive effects of forest on climate change and air pollution reduction are now bringing the forest or trees into both old and new cities (urban forests); but is this being done in a sustainable way? A specific climate-related issue and knowledge gap are the impacts of forestry on carbon release in drained peatlands—can the peat decomposition be inhibited?

Socio-economic information about the forest sector. Much more knowledge is needed with regard to the local socio-economic values of the forest, to compile forest management strategies that take these perspectives into account (Nijnik et al., 2013, this vol.). Socio-economic factors that affect forest degradation and transition at a local and global scale are also part of a knowledge gap and the relationship between local population activities and forest resources needs investigation. Economic values of the forest sector should be complemented by social values of forestry (e.g. employment opportunities, non-timber products, recreational functions). Another issue is how to balance diverging interests in forests in terms of biomass production and use, conservation, protection, climate change mitigation and recreation. How to manage forests for serving a maximum of these interests? Economic valuation of forest currently available is mostly limited to determining the value of wood and land and there is no economic valuation of forest as a whole. However, a few studies exist in which economic evaluation of factors other than timber is included (e.g. moose hunting, Wam et al., 2005), and this approach should be continued. A combined ecological and socio-economic valuation including other values will be useful, for example, in the protection of forests not only when the forest removes air pollution from the atmosphere, protects water reservoirs, stabilises mountain slopes or reduces flood risk, but also when the forest benefits the wellbeing of people, for example, through employment and recreation. Functions of forests that have previously not been included in economic models should in future be included, and methodologies for economic accounting of ecosystem services need to be further developed here (Cudlin et al., 2013, this vol.).

The answers to the six topics above are handled on a more or less general level for Europe; but there are variations that reflect a geographical focus, for example, for the Mediterranean area, climate change and air pollution are often seen, by the experts, as interaction factors with drought and fire.

The geographical distribution of supersites in Europe has also been handled in the questionnaire. The number of sites in total is an important prerequisite for deciding the basic concepts for placing the individual sites. If only three European supersites are established, the following distribution criteria are suggested: One site shall be placed in each of the three major phytoclimatic regions, Boreal, Temperate and Mediterranean, or as an alternative the priority could be related to water availability, which is a major driver for growth, and then a dry, a mesic and a wet site could be selected. However, three sites are undoubtedly too few sites if the variability within Europe should be considered even at a minimal level. If ten sites are nominated, a north–south gradient combined with an oceanic-continental gradient representing the largest or most important forest areas would be a rational distribution. If twenty sites are conceivable, the gradient mentioned for ten sites could be expanded with altitudinal considerations and a focus on vulnerable forest types; but some degree of replication within the twenty sites would also be an important comparison parameter from a scientific point of view. It is suggested that supersites could benefit from historical data, if they should be located near existing intensive monitoring plots or in areas where current existing knowledge can add value to the selected sites.

22.5 SCIENCE AND POLICY RECOMMENDATIONS

Forests provide important ecosystem services such as water regulation, carbon storage, services related to biodiversity, and recreation. In addition, they are renewable sources of energy and provide employment in rural areas. For the use of forests to be sustainable, especially in light of the risks associated with climate change and air pollution, it is necessary to further develop governance measures, especially 'soft' governance (e.g. guidelines, certification). As meaningful guidelines or certification systems cannot be developed without an adequate knowledge base, it is essential to further our understanding of ecosystem processes. Both research and long-term monitoring are needed for this purpose. 'Supersites' where both research and monitoring are carried out will help greatly in this task.

A number of monitoring programmes and large-scale projects already exist in Europe and it would seem reasonable to choose supersites from among the sites already in use, preferably those with a large and diverse measurement programme and for which a large amount of data already exist. However, these sites have often been set up in the context of different research questions and with different aims, so that comparability of both old and new data is a problem. Full method harmonisation would be ideal; but there are a number of difficulties with this, both in principle and in practice. Comparability can to some extent be achieved using for example method comparisons and ring tests.

The selection of future sites and networks for data acquisition and as a basis for experimentation will not be a top down approach but will rely on successful proposals from the monitoring, research and modelling communities. On national levels there are in many cases different agencies, universities and other institutions responsible for different networks, all aiming to secure and continue 'their' infrastructures. In the context of a bottom-up approach, some competition is thus necessary and useful. Networks, sites and institutions that enable open and wide use of infrastructures and data will be successful in the long run. With increasing data sets, the possibilities for evaluation and use of the data are increasing so that involvement of larger scientific communities is appropriate. An open use of data (which should in principle be publicly available) calls for independent funding mechanisms for the infrastructures and for data evaluation, because as long as funding is predominantly obtained through delivery of scientific and policy digestible results, data will be blocked as 'raw material' for evaluations. This can only be changed if data provision services are funded independently.

Open use of data also requires accessibility. This has been a problem in the past, largely due to lack of knowledge on what data are available and where. The new meta-database provides information on this. Internet-based solutions have improved access to data, with some programmes providing free internet access, and this solution should be encouraged so long as IPR are respected.

ACKNOWLEDGEMENTS

The authors are grateful for valuable inputs from the panel participants at the COST FP0903 Rome 2010 meeting and the 50 experts who used their time to answer the COST FP903, WG3 2012 questionnaire.

REFERENCES

Augustin, S., Evers, J., Dietrich, H.P., Eichhorn, J., Haussmann, T., Icke, R., Isenberg, A., Lux, W., Musio, M., Pretzsch, H., Riek, W., Rötzer, T., Schultze, B., Schulze, A., Schröder, J., Seidling, W., Wellbrock, N., von Wilpert, K., Wolff, B., 2005. Concept and feasibility study for the integrated evaluation of environmental monitoring data in forests. Eur. J. For. Res. 124, 251–260.

Bleeker, A., Draaijers, G., van der Veen, D., Erisman, J.W., Möls, H., Fonteijn, P., Geusebroek, M., 2003. Field intercomparison of throughfall measurements performed within the framework of the Pan European intensive monitoring program of EU/ICP Forest. Environ. Pollut. 125, 123–138.

Bytnerowicz, A., Fenn, M., McNulty, S., Yuan, F., Pourmokhtarian, A., Driscoll, C., Meixner, T., 2013. Interactive effects of air pollution and climate change on forest ecosystems in the United States – current understanding and future scenarios. In: Matyssek, R., Clarke, N., Cudlin, P., Mikkelsen, T., Tuovinen, J.P., Weiser, G., Paoletti, E (Eds.), Climate Change, Air Pollution and Global Challenges: Understanding and Solutions from Forest Research. Dev. Environ. Sci. 13. Elsevier, Netherlands, pp. 333–369.

Bytnerowicz, A., Omasa, K., Paoletti, E., 2007. Integrated effects of air pollution and climate change on forest: a northern hemisphere perspective. Environ. Pollut. 147, 438–445. http://dx.doi.org/10.1016/j.envpol.2006.08.028.

CarboEurope, 2010. Integrated Project CarboEurope-IP. Assessment of the European Terrestrial Carbon Balance (online 14 December 2010). http://www.carboeurope.org.

Clarke, N., Zlindra, D., Ulrich, E., Mosello, R., Derome, J., Derome, K., Konig, N., Lovblad, G., Draaijers, G.P.J., Hansen, K., Thimonier, A., Waldner, P., 2010. Sampling and analysis of deposition. In: ICP Forests, (Ed.), Manual on Methods and Criteria for Harmonized Sampling, Assessment, Monitoring and Analysis of the Effects of Air Pollution on Forests. UNECE, ICP Forests, Hamburg, 66 pp. Part XIV.

Clarke, N., Fischer, R., de Vries, W., Lundin, L., Papale, D., Vesala, T., Merilä, P., Matteucci, G., Mirtl, M., Simpson, D., Paoletti, E., 2011. Availability, accessibility, quality and comparability of monitoring data for European forests for use in air pollution and climate change science. iForest—Biogeosci. Forestry 4, 162–166.

Cudlín, P., Seják, J., Pokorný, J., Albrechtová, J., Bastian, O., Marek, M., 2013. Forest ecosystem services under climate change and air pollution. In: Matyssek, R., Clarke, N., Cudlin, P.,

Mikkelsen, T., Tuovinen, J.P., Weiser, G., Paoletti, E. (Eds.), Climate Change, Air Pollution and Global Challenges: Understanding and Solutions from Forest Research. Dev. Environ. Sci. 13. Elsevier, Netherlands, pp. 521–546.

Danielewska, A., Clarke, N., Olejnik, J., Hansen, K., Vries, W., Lundin, L., Tuovinen, J., Fischer, R., Urbaniak, M., Paoletti, E., 2013. A meta-database comparison from various European Research and Monitoring Networks dedicated to forest sites. iForest—Biogeosci. Forestry 6, 1–9.

De Vries, W., Forsius, M., Lorenz, M., Lundin, L., Haussman, T., Augustin, S., Ferretti, M., Kleemola, S., Vel, E., 2002. Cause-effect relationships of forest ecosystems. Joint Report by ICP Forests and ICP Integrated Monitoring. Report for United Nations Economic Commission for Europe. Federal Research Centre for Forestry and Forest Products (BFH) together with Finnish Environment Institute (SYKE), 46 (online 14 December 2010). http://www.icp-forests.org/pdf/icpf-icpim.pdf.

EMEP, 2010. Welcome to EMEP (European Monitoring and Evaluation Programme) (online 14 December 2010). http://www.emep.int.

Erisman, J.W., Mols, H., Fonteijn, P., Geusebroek, M., Draaijers, G., Bleeker, A., van der Veen, D., 2003. Field intercomparison of precipitation measurements performed within the framework of the Pan European Intensive Monitoring Program of EU/ICP Forest. Environ. Pollut. 125, 139–155.

Fischer, R., Aas, W., De Vries, W., Clarke, N., Cudlin, P., Leaver, D., Lundin, L., Matteucci, G., Matyssek, R., Mikkelsen, T.N., Mirtl, M., Öztürk, Y., Papale, D., Potocic, N., Simpson, D., Tuovinen, J.-P., Vesala, T., Wieser, G., Paoletti, E., 2011a. Towards a transnational system of supersites for forest monitoring and research in Europe—an overview on present state and future recommendations. iForest—Biogeosci. Forestry 4, 167–171.

Fischer, R., Beck, W., Calatayud, V., Cools, N., De Vos, B., Dobbertin, M., Fleck, S., Giordani, P., Granke, O., Kindermann, G., Lorenz, M., Meesenburg, H., Meining, S., Nagel, H.-D., Neumann, M., Scheuschner, T., Stofer, S., 2011b. The Condition of Forests in Europe. 2011 Executive Report. ICP Forests and European Commission. Hamburg and Brussels, 21. http://www.icp-forests.org/RepEx.htm.

Frenzel, M., Baessler, C., Bastianini, M., Campanaro, A., Matteucci, G., Stoll, S., 2011. Revised manuals of pre-existing manuals and manual with new harmonized methods (manual of harmonised methods for environmental indicators across different ecosystems). EnvEurope report 247. http://www.enveurope.eu/misc/PD_2_1_2_Frenzel_HarmonizedMethods_201112_v1.1.pdf.

Gordon, B., Levis, S., Wieder, W.R., 2012. A modeller's perspective of long-term integrated data series of ecosystem-atmosphere processes. In: iLEAPS, (Ed.), Future Land Atmosphere Observation Platforms. Newsletter Integrated Land Ecosystem—Atmosphere Process Study-Issue No. 12, September 2012, www.ileaps.org.

Helmisaari, H.-S., Hanssen, K.H., Jacobson, S., Kukkola, M., Luiro, J., Saarsalmi, A., Tamminen, P., Tveite, B., 2011. Logging residue removal after thinning in Nordic boreal forests: long-term impact on tree growth. For. Ecol. Manage. 261, 1919–1927.

ICOS, 2010. ICOS Integrated Carbon Observation System (online 14 December 2010). http://www.icos-infrastructure.eu.

ICP Forests, 2010. Manual on Methods and Criteria for Harmonized Sampling, Assessment, Monitoring and Analysis of the Effects of Air Pollution on Forests. UNECE, ICP Forests, Hamburg. ISBN: 978-3-926301-03-1 (online 14 December 2010). http://www.icp-forests.org/Manual.htm.

Jandl, R., Lindner, M., Vesterdal, L., Bauwens, B., Baritz, R., Hagedorn, F., Johnson, D.W., Minkkinen, K., Byrne, K.A., 2007. How strongly can forest management influence soil carbon sequestration? Geoderma 137, 253–268.

Kleemola, S., Forsius, M. (Eds.), 2010. 19th Annual Report 2010. Convention on Long-range Transboundary Air Pollution. International Cooperative Programme on Integrated Monitoring of Air Pollution Effects on Ecosystems. The Finnish Environment 15/2010, 49 pp (online 14 December 2010). http://www.environment.fi/default.asp?contentid=365563&lan=en&clan=en.

Koike, T., Watanabe, M., Hoshika, Y., Kitao, M., Matsumura, H., Funada, Izuta T., 2013. Effects of ozone on forest ecosystems in East and Southeast Asia. In: Matyssek, R., Clarke, N., Cudlin, P., Mikkelsen, T., Tuovinen, J.-P., Weiser, G., Paoletti, E (Eds.), Climate Change, Air Pollution and Global Challenges: Understanding and Solutions from Forest Research. Dev. Environ. Sci. 13. Elsevier, Netherlands, pp. 371–390.

LTER, 2010. European long-term ecosystem research network (online 14 December 2010). http://www.lter-europe.net/.

Marchetto, A., Mosello, R., Tartari, G.A., Derome, J., Derome, K., König, N., Clarke, N., Kowalska, A., 2009. Atmospheric deposition and soil solution working ring test 2009. Laboratory ring test for deposition and soil solution sample analyses for the laboratories participating in the EU/Life+ FutMon Project. CNR-ISE Report 04.09

Matyssek, R., Clarke, N., Cudlin, P., Mikkelsen, T.N., Tuovinen, J.-P., Weiser, G., Paoletti, E., 2013a. Climate change, air pollution and global challenges: understanding and solutions from forest research. In: Matyssek, R., Clarke, N., Cudlin, P., Mikkelsen, T., Tuovinen, J.-P., Weiser, G., Paoletti, E. (Eds.), Climate Change, Air Pollution and Global Challenges: Understanding and Solutions from Forest Research. Dev. Environ. Sci. 13. Elsevier, Netherlands, pp. 3–16.

Matyssek, R., Wieser, G., Fleischmann, F., Grünhage, L., 2013b. Ozone research, quo vadis? Lessons from the Kranzberg forest experiment. In: Matyssek, R., Clarke, N., Cudlin, P., Mikkelsen, T., Tuovinen, J.-P., Weiser, G., Paoletti, E. (Eds.), Climate Change, Air Pollution and Global Challenges: Understanding and Solutions from Forest Research. Dev. Environ. Sci. 13. Elsevier, Netherlands, pp. 103–129.

Matyssek, R., Wieser, G., Calfapietra, C., de Vries, W., Dizengremel, P., Ernst, D., Jolivet, Y., Mikkelsen, T.N., Mohren, G.M.J., le Thiec, D., Tuovinen, J.-P., Weatherall, A., Paoletti, E., 2012. Forests under climate change and air pollution: gaps in understanding and future directions for research. Environ. Pollut. 160, 57–65.

NEU, 2010. NitroEurope IP. The nitrogen cycle and its influence on the European greenhouse gas balance (online 14 December 2010). http://www.nitroeurope.eu/.

Nijnik, M., Miller, D., 2013. Targeting sustainable provision of forest ecosystem services with special focus on carbon sequestration. In: Matyssek, R., Clarke, N., Cudlin, P., Mikkelsen, T., Tuovinen, J.-P., Weiser, G., Paoletti, E. (Eds.), Climate Change, Air Pollution and Global Challenges: Understanding and Solutions from Forest Research. Dev. Environ. Sci. 13. Elsevier, Netherlands, pp. 547–568.

OECD, 2007. Principles and Guidelines for Access to Research Data from public funding. http://www.oecd.org/dataoecd/9/61/38500813.pdf.

Paoletti, E., de Vries, W., Mikkelsen, Ibrom, Larsen, K.S., Tuovinen, J.-P., Serengil, Y., Yurtseven, I., Wieser, G., Matyssek, R., 2013. Key indicators of air pollution and climate change impacts at forest Supersites. In: Matyssek, R., Clarke, N., Cudlin, P., Mikkelsen, T., Tuovinen, J.-P., Weiser, G., Paoletti, E. (Eds.), Climate Change, Air Pollution and Global Challenges: Understanding and Solutions from Forest Research. Dev. Environ. Sci. 13. Elsevier, Netherlands, pp. 497–518.

Schulze, E.-D., Körner, C., Law, B.E., Haberl, H., Luyssaert, S., 2012. Large-scale bioenergy from additional harvest of forest biomass is neither sustainable nor greenhouse gas neutral. Glob. Chang. Biol. Bioenergy 4, 611–616.

Sun, G., McLaughlin, S.B., Porter, J.H., Uddling, J., Mulholland, P.J., Adams, M.B., Pederson, N., 2012. Interactive influences of ozone and climate on streamflow of forested watersheds. Glob. Chang. Biol. 18, 3395–3409.

Vesterdal, L., Clarke, N., Sigurdsson, B.D., Gundersen, P., 2013. Do tree species influence soil carbon stocks in temperate and boreal forests? For. Ecol. Manag. http://dx.doi.org/10.1016/j.foreco.2013.01.017#doilink, in press. http://dx.doi.org/10.1016/j.foreco.2013.01.017.

Wam, H.K., Hofstad, O., Nævdal, E., Sankhayan, P., 2005. A bio-economic model for optimal harvest of timber and moose. For. Ecol. Manage. 206, 207–219.

Wickstrøm, T., Clarke, N., Derome, K., Derome, J., Røgeberg, E., 2000. Comparison study of five analytical methods for the fractionation and subsequent determination of aluminium in natural water samples. J. Environ. Monit. 2, 171–181.

Žlindra, D., Eler, K., Clarke, N., Simončič, P., 2011. Towards harmonization of forest deposition collectors—case study of comparing collector designs. iForest—Biogeosci. Forestry 4, 218–225.

Chapter 23

Key Indicators of Air Pollution and Climate Change Impacts at Forest Supersites

Elena Paoletti[*,1], Wim de Vries[†,‡], Teis Nørgaard Mikkelsen[§], Andreas Ibrom[§], K.S. Larsen[§], Juha-Pekka Tuovinen[¶], Yussuf Serengil[||], I. Yurtseven[||], Gerhard Wieser[#] and Rainer Matyssek[**]

[*]*IPP-CNR, Florence, Italy*

[†]*Alterra, Wageningen University and Research Centre, P.O. Box 47, Wageningen, The Netherlands*

[‡]*Environmental Systems Analysis Group, Wageningen University, P.O. Box 47, Wageningen, The Netherlands*

[§]*Centre for Ecosystems and Environmental Sustainability (ECO), Department of Chemical and Biochemical Engineering, Technical University of Denmark, Denmark*

[¶]*Finnish Meteorological Institute, Helsinki, Finland*

[||]*Faculty of Forestry, Istanbul University, Istanbul, Turkey*

[#]*Department of Alpine Timberline Ecophysiology, Federal Office and Research Centre for Forests, Innsbruck, Austria*

[**]*Technische Universität München, Ecophysiology of Plants, Hans-Carl-von-Carlowitz-Platz 2, D-85354 Freising-Weihenstephan, Germany*

[1]*Corresponding author: e-mail: e.paoletti@ipp.cnr.it*

Chapter Outline

23.1 INTRODUCTION

Air pollution and climate change are the main environmental drivers affecting forest productivity and services (Bytnerowicz et al., 2007; Bytnerowicz et al., 2013, this vol.). Among air pollutants, ground-level ozone (O_3) and nitrogen (N) deposition are of the greatest concern for forests (Serengil et al., 2011). Ozone and nitrous oxide (N_2O), which is a component of the N cycle, are also

Developments in Environmental Science, Vol. 13. http://dx.doi.org/10.1016/B978-0-08-098349-3.00023-2

powerful greenhouse gases (GHGs). Many traditional air pollutants and GHGs have common sources, contribute to the Earth's radiative balance, interact in the atmosphere and jointly affect the ecosystems (Bytnerowicz et al., 2007). The impacts on forests have been traditionally treated separately for air pollution and climate change, while the combined effects may significantly differ from a sum of separate effects. An integrated assessment of the impacts that the various environmental factors cause to forest ecosystems is complex, because of the huge number of variables of potential interest. The scientific climate change discussion has to deal with uncertainty, both in future projection and current GHG flux assessment. Approximately, 40% of the uncertainty in the projected year 2100 temperature increase is due to uncertainty in the global carbon cycle (Huntingford et al., 2009). Air pollution is agent of both climate change and change in ecosystem responses; for example, impact of elevated CO_2 and O_3 may result in complex tree responses. Both gases have the capacity for compensating each other in their typical effects on plants. Elevated CO_2 may ameliorate adverse O_3 effects, in particular, on C sink strength, whereas enhanced O_3 may neutralise stimulation in productivity under high CO_2 (Karnosky et al., 2003).

The COST Action FP0903 'Climate Change and Forest Adaptation and Mitigation in a Polluted Environment' is developing a new approach where comprehensive forest research sites, called supersites, are recommended for a more advanced assessment of air pollution and climate change impacts on soil, vegetation and the atmosphere, with a focus on the carbon (C), nitrogen, ozone and water budgets (Paoletti and Tuovinen, 2011). Matyssek et al. (2012) reviewed the current understanding of the climate, O_3 and N impacts on the C sequestration ability of forests, and summarised the main knowledge gaps and research needs. The aim of this chapter is to complement the review by Matyssek et al. (2012) by identifying observational needs in more detail and by providing preliminary technical suggestions that will help developing measurement protocols for the supersites. The development of existing European forest monitoring and research infrastructures towards supersites is discussed in another chapter of this book (Mikkelsen et al., 2013, this vol.).

23.2 GENERAL PARAMETERS

Describing the forest environment requires consolidated measurements of its main components: soil, vegetation and the atmosphere. Basic requirements for a supersite include the measurement of all parameters that are needed for the characterisation of the C, N, O_3 and water budgets (Table 23.1). The soil type together with its physical properties must be determined for defining field capacity and wilting point, which in turn are fundamental for quantifying soil moisture, for example, for modelling the uptake of O_3 through the stomata.

The micrometeorological eddy covariance (EC) technique provides a common framework for measuring the atmosphere–ecosystem fluxes of all the components considered here. In addition, data on wind velocity, atmospheric

TABLE 23.1 List of Parameters and Methodologies of Potential Interest at Forest Supersites, Including Standard Measurements in View of Long-Term Monitoring and Detailed Measurements During a Given Time Period in View of In-Depth Research

Parameter	Frequency	Methodology
General		
Soil type and texture	Once	Bouyoucos hydrometer
Precipitation	Continuous	Electronic pluviometer
Air temperature and humidity	Continuous	Thermometer, hygrometer
Radiation (global, net, PAR)	Continuous	Radiometers
Wind velocity, air turbulence	Continuous	Sonic anemo/thermometer
Crown conditions	Once a year	Visual assessment
Tree height and diameter	Once a year	Hypsometer and calliper
Leaf area index	Bimonthly	Optical tools
Ground vegetation	Once every 5 years	Phytosociological approach
Carbon budget		
CO_2 fluxes I (ecosystem–atmosphere)	Continuous	Eddy covariance
CO_2 fluxes II (ecosystem–atmosphere)	Campaigns	Eddy covariance
CO_2 fluxes (leaves–atmosphere)	Campaigns	Plant enclosures
CH_4 fluxes (leaves–atmosphere)	Campaigns	Plant enclosures
Soil respiration	Continuous	Chamber
Soil CH_4 fluxes	Campaigns	Chamber
Stocking density	Once every 5 years	Biomass conversion and expansion factors, below-ground to above-ground biomass ratio, and C fraction to convert merchantable volume into biomass C stock
Deadwood and litter volume and their C fraction	Once every 5 years	Variable (Rondeux and Sanchez, 2010), for example, volume biomass per hectare

Continued

TABLE 23.1 List of Parameters and Methodologies of Potential Interest at Forest Supersites, Including Standard Measurements in View of Long-Term Monitoring and Detailed Measurements During a Given Time Period in View of In-Depth Research—Cont'd

Parameter	Frequency	Methodology
Carbon stock in the soil	Once a year	Chemical analysis of C content after total destruction multiplied by soil bulk density and soil thickness
Nitrogen budget		
NH_4, NO_3 and DON in bulk deposition, throughfall, stemflow and soil solution	Weekly or biweekly	Gutters or funnels for deposition and lysimeters for soil solution
Nitrogen concentrations of different tree organs, i.e., stems, branches, leaves, fine roots (incl. deadwood)	Yearly (in stems, every 5 years)	Chemical analysis after total destruction
Nitrogen stock in the soil	Once every 5 years	Chemical analysis of N content after total destruction multiplied by soil bulk density and soil thickness
NO_x and NH_3 deposition	Continuous	Gradient approach or eddy covariance
N fixation	At regular intervals in a growth period	*P. schreberi* sampling
N_2O emissions	Continuous	Automatic chamber techniques or eddy covariance
Ozone budget		
Canopy level O_3 concentrations	Continuous	Active monitoring
O_3 fluxes	Continuous	Eddy covariance/sap flow
O_3 fluxes	Short-term campaigns	Portable gas analysers
VOC fluxes	Short-term campaigns	PTR-MS
Water budget		
Water retention capacities or curve of soils	Once	Pressure membrane extractor
Infiltration (hydraulic conductance) of soils	Once	Infiltrometer

TABLE 23.1 List of Parameters and Methodologies of Potential Interest at Forest Supersites, Including Standard Measurements in View of Long-Term Monitoring and Detailed Measurements During a Given Time Period in View of In-Depth Research—Cont'd

Parameter	Frequency	Methodology
Stemflow	Continuous	Plastic tubing and rain gauge combination
Throughfall	Continuous	Below canopy pluviograph
Interception	Continuous	Calculation
Soil water content	Continuous	Time domain reflectometry
Soil matric potential	Continuous	Tensiometers
Soil temperature	Continuous	Soil thermographs
Water fluxes	Continuous	Eddy covariance
Canopy transpiration	Continuous	Sap flow

turbulence and heat exchange are obtained from the EC system. Further meteorological parameters to be recorded include precipitation, air temperature and humidity, and global, net and photosynthetically active radiation.

The transparency of the crowns must be assessed as a metric of forest health, following standard protocols such as that established by the ICP-Forests programme (Eichhorn et al., 2010). Tree growth is a key ecological parameter of forests and thus an important indicator of their condition. An accurate sampling design for periodic tree growth measurements at Level II plots of ICP-Forests network is available (Dobbertin and Neumann, 2010). Phenology is another parameter that strongly affects tree performance. Measurements of the leaf area index (LAI) provide data on phenological variations, which are required in many forest models (van Oijen et al., 2005). In addition to the overstorey canopy, the understorey is a significant component of forest ecosystems. Changes in species composition may reflect competition dynamics and stress-induced variations, such as those due to excessive N deposition (Bobbink et al., 2010). Depending on the research purpose, several phytosociological approaches can be used for the assessment of ground vegetation, that is, herbs, shrubs and trees (Canullo et al., 2010).

23.3 THE CARBON BUDGET

Carbon dioxide is the most important GHG that drives climate change. The C budget of a forest is the balance of the exchanges of C between its pools or within one specific loop (e.g. troposphere↔biosphere) of the C cycle. An assessment of the C budget of a pool can provide knowledge about whether the pool is performing as a sink or source for CO_2. According to recent assessment, the terrestrial biosphere—mainly the forest biomes—absorbs >30% of anthropogenic CO_2 emissions (Denman et al., 2007). The annual sink strength of forests varies substantially, making the estimates highly uncertain. Whether forest ecosystems remain to stay as a C sink in the future depends on their management and on their responses to changing concentrations of CO_2 and O_3, N input, temperature and precipitation. While many manipulation experiments have been carried out to assess ecosystem C responses in response to these drivers, in forests such experiments are cumbersome. Long-term monitoring combined with statistical analysis is a way to assess the C responses of forest ecosystems to all interacting drivers simultaneously (e.g. Solberg et al., 2009).

Whether forest ecosystems act as a C sink or source is determined by their so-called net biome productivity (NBP), which represents the net C sequestration, both above and below the ground. In the case of no disturbance, which is the case in monitoring systems, NBP is equal to net ecosystem productivity (NEP). In most studies, leaching of dissolved inorganic and organic C is neglected, and this holds also for the flux of VOCs and CH_4 in soils and vegetation, implying that $NBP=NEP=GPP-R_A-R_H$, where GPP is gross primary production, and R_A and R_H are autotrophic and heterotrophic respiration. It is important, however, to also make a distinction between above- and below-ground C sequestration, since the above-ground C in trees generally has a much shorter life time than the C stored in soils.

Changes in plant C pool or tree C sink can be derived as NPP—litter production (LP), the latter of which is the sum of above-ground (leaf) litter (ALP) and below-ground (root) litter production (RLP). Changes in soil C pool can be derived as $LP-R_H$. Below-ground (root) litter production is cumbersome to assess. An indication of changes in the tree C pool (ΔPCP) can, however, also be derived by repeated forest inventories, although small annual changes in a largely uncertain soil C pool imply large time intervals (de Vries et al., 2009b). An indication of RLP can be derived by a combination of measurements of ΔPCP by repeated forest inventories, NPP and ALP, according to $RLP=\Delta PCP+ALP-NPP$.

In order to estimate the C budget of a forest in a changing environment in timescales varying from 1 h to 10 years, we suggest three independent methodologies: EC, flux chambers and forest inventory (Table 23.1).

EC method is a tower-based turbulence measurement of atmospheric CO_2 fluxes (Aubinet et al., 2000). Specific problems with the sensors and the data post-processing have been identified and minimised (Ibrom et al., 2007). However, in nights with stable atmospheric stratification, other transport mechanisms like horizontal and vertical advection take place. In long-term flux studies, only one system (and one position) is usually used to measure a specific area in the forest (the so-called footprint) (Baldocchi, 2008). To improve quality of the measurements and spatial information, it is suggested that Supersites include campaign measurements with a second EC system nearby and away from the master system.

Chamber measurements of fluxes (soil, stem, branch and leaves) within the forest and scaling the results up to the whole forest using carbon process models is an alternative way to assess the flux above the forest. Large biological variability requires labour-intensive field campaigns and careful spatial sampling (Curtis et al., 2002). Up-scaling of chamber fluxes requires knowledge on the distribution and biophysical state of the surfaces, for example, surface temperature, irradiance and CO_2 concentration. It is very difficult to operate canopy chambers continuously over a decade due to different technicalities, for example, extreme weather conditions; therefore, we suggest campaign measurements at regular intervals.

For both *EC* and *chamber measurements*, the usual way of modelling respiration via its temperature sensitivity is incomplete, because the level of substrate availability seems also to determine the respiration level (Larsen et al., 2007). Covariance between substrate availability and temperature biases temperature sensitivity of respiration at seasonal and decadal timescales (Mahecha et al., 2010). For realistic prediction of terrestrial C budgets, it is vital to distinguish between temperature and substrate effects. The uncertainty in simulated C-flux estimations can be estimated with Monte Carlo techniques (van Oijen et al., 2005). However, systematic errors due to oversimplifications in the models remain inherent in model predictions.

Carbon inventory is currently the instrument for national GHG emission reporting. Wutzler et al. (2008) synthesised data for European beech forests and derived nonlinear models that depend on tree age, site conditions, stand index, tree height, diameter and density. From this and C density values, the vegetation C stock and its uncertainty can be estimated for a given stand. The challenge with the soil C inventory approach is to calculate the small difference between two large numbers (C stock measurements in different years) with large spatial variability (de Vries et al., 2010). There is no doubt that the inventory approach can measure the long-term C budget with known uncertainty; however, unlike the other two approaches, it does not reveal information on the underlying processes.

Failure of each individual method to either define the uncertainty (atmospheric and chamber measurements) or provide sufficient temporal resolution to investigate the processes that affect the C budget (inventory) suggests applying these methods in combination. To our knowledge, there are only two such studies in forests, one comparing five North American forests (Curtis et al., 2002) and one in an Australian Eucalypt forest (Keith et al., 2009). Although groundbreaking, neither of these studies is as complete, both with respect to measurement of above-ground respiration and to soil C inventory, as proposed for supersites here.

Methane (CH_4) is the second most important anthropogenic GHG after CO_2 (Forster et al., 2007). Dependent on soil properties, forests can be atmospheric sources or sinks, for example, wetlands (Gauci et al., 2010) are usually CH_4 net emitters, while old non-disturbed coniferous and deciduous forests usually absorb CH_4 (Ambus and Robertson, 2006). Effects of CH_4 on the C budget seem to be minor compared to CO_2, but huge variations from different soil types exist (Le Mer and Roger, 2001). Direct CH_4 emission from plants was shown by Keppler et al. (2006) and confirmed under different conditions. Four stimulating factors induce aerobic plant CH_4 production, that is, cutting injuries, increasing temperature, ultraviolet radiation and reactive oxygen species (Bruhn et al., 2009). Chamber measurements on soils and trees are the most used tools for studying CH_4 fluxes in forest. Until now, three atmospheric studies have been unsuccessful to confirm net fluxes related to aerobic plant CH_4 production in forest ecosystems (Mikkelsen et al., 2011).

Assessment of effects on the C budgets derived from the three independent methods mentioned above should be tested in relation to consistency to produce simulations and predictions of the impacts of climate change and air pollution on different forests as described by Matyssek et al. (2012).

23.4 THE NITROGEN BUDGET

Human activities, such as fertiliser production and fossil fuel combustion, have increased the conversion of atmospheric N_2 to 'reactive N' (N_r), defined as all biologically, radiatively and/or photochemically active forms of N (Galloway et al., 1995). Most important forms are (i) N oxides (NO_x) and ammonia (NH_3), which are chemically important in the troposphere; (ii) nitrate (NO_3^-) and ammonium (NH_4^+), the predominant forms of N taken up by organisms; (iii) nitrous oxide (N_2O), which is an important GHG and can catalyse O_3 destruction in the stratosphere; and (iv) nitric acid (HNO_3), in particular, in dry climates.

The N cycle provides a key control of the global C cycle through effects on primary production and decomposition, it is a major determinant of terrestrial and aquatic biodiversity, it affects particle and other chemical production in the atmosphere and it has major impacts on GHG fluxes and stratospheric O_3 depletion (Galloway et al., 2003; Sutton et al., 2011). It is therefore a matter of great concern that global cycling of N_r is estimated to have more than doubled (Sutton et al., 2011; Vitousek et al., 1997), whereas the C cycle is $<10\%$ perturbed by human activities (IPCC, 2001). Human activity now fixes more atmospheric N_2 into reactive forms than all terrestrial natural processes combined (Galloway et al., 1995; Vitousek et al., 1997). Elevated N inputs and their fertilising effects in forests decrease global warming due to increased CO_2 uptake, but this is partly counteracted (negated) be emissions of N_2O, a strong GHG. Although agriculture is the main source of N_2O emissions, forest contribute 10–15% of the agricultural emissions in Europe (de Vries et al., 2007a). Nitrogen deposition affects forest ecosystems through N enrichment and acidification of soils (de Vries et al., 2007b) with impacts on forest nutrition (de Vries et al., 2003), tree growth (de Vries et al., 2009a), susceptibility to secondary stress, such as drought, frost, pathogens or herbivores (Flückiger and Braun, 1998) and plant species diversity (Bobbink et al., 2010). Full insight in the N budget requires quantification of the input by atmospheric deposition (wet and dry) of NH_3, HNO_3 and NO_x, and by N fixation, net N uptake, N immobilisation or release in the soil, N emissions in terms of N_2O, NO_x and N_2, and N leaching and/or N runoff to groundwater and surface water.

The recommended measurements of nitrogen inputs in a supersite are presented in Table 23.1:

Bulk deposition and throughfall monitoring: A simple approach to estimate N inputs is to measure NH_4, NO_3, DON concentrations in bulk deposition in the open field, in throughfall below the canopy and in stemflow, using funnels or gutters. The sum of throughfall and stemflow, however, differ from the total atmospheric deposition as a result of canopy exchange processes. To estimate canopy exchange, the canopy budget model of Ulrich (1983), further developed by Draaijers and Erisman (1995), is most widely used.

EC method and gradient approaches: A more accurate technique to assess total N deposition is to measure the surface/atmosphere exchange of the different N species. The most direct micrometeorological technique is the EC method, which, however, requires high-precision fast-response sensors, which are in their infancy for most N species. The first-ever EC measurements have been presented for NH_3 (Famulari et al., 2005) and particulate N (Nemitz et al., 2004). Because of these limitations, fluxes of N species are usually measured with gradient approaches, which rely on empirical parameterisations and similarity assumptions. While cost-efficient gradient monitors are available for NO and NO_2, current automated systems for artefact-free determination of gradients of gaseous NH_3 are labour intensive and have only recently begun to be extended to measure HNO_3 and aerosol NO_3^- and NH_4^+ (Trebs et al., 2004). A new generation of photo-acoustic NH_3 monitors (Pushkarsky et al., 2003) are a promising reliable alternative, but a cost-effective solution optimised for ambient flux measurements is still lacking. Aerosol fluxes to the forests can be assessed by various approaches (Farmer et al., 2010) including aerosol mass spectrometer, chemical ionisation mass spectroscopy and steam jet aerosol collection.

N fixation: Biological N fixation is the primary source of N within natural ecosystems, but it is hardly ever measured in routine monitoring and its origin in boreal and temperate forests has been elusive until recently. Research by Gundale et al. (2011) shows that an N fixing symbiosis between a cyanobacterium (*Nostoc* sp.) and the feather moss *Pleurozium schreberi* is an important source (between 1.0 and 2.0 kg N ha^{-1} yr^{-1}) in boreal forest. Assuming that N-fixation rates are mainly influenced by this symbiosis, N fixation can be derived by regular sampling *P. schreberi* shoots, and transporting them to the laboratory where N-fixation rates can be estimated during a 24-h acetylene reduction incubation (Schöllhorn and Burris, 1967).

Concerning the fate of nitrogen, recommended measurements are listed in Table 23.1:

Net nitrogen uptake: In line with the C inventory, the change in vegetation N stock (net N uptake) can be derived from repeated forest inventories (diameter at breast height and tree height from which stem volume can be derived) when the N content in trunks, twigs, foliage and fine roots is also measured.

Nitrogen leaching: The flux of N chemical elements in forest ecosystems is generally computed by multiplying a modelled water flux with measured soil solution concentrations at the corresponding depth. A depth at 60–100 cm (often the deepest lysimeter cup) is assumed to represent the leaching fluxes at the bottom of the root zone. Soil solution is mainly collected by non-destructive collectors, of which (zero) tension lysimeter is, in general, the reference method. Hydrological models are used to quantify the water flux using meteorological data and various stand, site and soil parameters (van der Salm et al., 2007). Mostly, lysimeter cups are inserted on a regular grid and sampled at regular time intervals (e.g. weekly or monthly). Concentrations are then interpolated to daily values, multiplied by daily water fluxes computed by hydrological models and finally aggregated to an annual flux (de Vries et al., 2007b). A problem is that quantification of the uncertainty of the chemical fluxes is cumbersome. An alternative is to use a stratified random sampling in space, instead of a regular grid, with sampling times related to the precipitation surplus, instead of constant intervals, allowing uncertainty estimates (de Vries et al., 2010).

Nitrogen emissions: In terms of N budgets, N emissions are generally limited and often neglected in monitoring studies. Emissions of NH_3 can indeed be neglected in forests, but this does not hold for N_2O and NO_x. Fluxes of N_2O are routinely measured with manual or automatic chamber techniques, which have the advantage of easily resolving concentration changes. By contrast, these average over only small areas (e.g. $<1\ m^2$) and modify the microclimate of the enclosed area. In high-emission areas, long-term measurements of N_2O can also be made by EC using tunable diode laser absorption spectroscopy (Scanlon and Kiely, 2003).

Nitrogen immobilisation: N immobilisation can be assessed as a rest term according to N immobilisation equals N deposition + N fixation − net N uptake − N emissions − N leaching. A disadvantage is that uncertainties in all measurements come back in the final assessment. Furthermore, data on N fixation and NO or N_2 emission are often missing. Another approach is to multiply the soil C pool change with a soil C/N ratio. A third approach is to derive N immobilisation from the difference between the N input by above-ground litter input and below-ground root turnover minus N mineralisation. This requires measurements of N in leaves and fine roots and N mineralisation assessments. This approach thus seems less favourable. Finally, the N immobilisation can be assessed from a repeated soil N and C inventory. As with C, the problem is that one needs to calculate the small difference between two large numbers (N stock measurements in different years) with large spatial variability. This generally requires a large number of observations and long time intervals between samplings (de Vries et al., 2009b). Use of these four approaches, however, may lead to quite reliable N immobilisation estimates.

23.5 THE OZONE BUDGET

Ozone is a secondary air pollutant that can damage forest trees by inducing visible leaf injury, growth reduction, changes in resource allocation and, by this, increased sensitivity to biotic and abiotic stress (Matyssek and Sandermann, 2003). For example, O_3 may increase or decrease water use of trees (Kitao et al., 2009; McLaughlin et al., 2007), reduce their C sequestration capacity (Sitch et al., 2007) and alter the emissions of biogenic VOCs (Cojocariu et al., 2005), which may act as O_3 precursors in the atmosphere. As a strong oxidant, O_3 plays a key role in atmospheric chemistry, and it is also an important GHG (Isaksen et al., 2009). Deposition rates of O_3 depend on both O_3 concentrations and a variety of biogeochemical surface exchange processes. Stomatal conductances are expected to reduce in response to elevated CO_2 concentrations (Ainsworth and Rogers, 2007), while climate change may both enhance O_3 production and suppress the stomatal uptake (Solberg et al., 2008). Through interactions of this kind, O_3 deposition to forests constitutes an integral part of the complex atmosphere–biosphere system that encompasses a range of feedback processes between forest vitality and productivity, atmospheric composition and the climate system (Raes et al., 2010). The situation is further complicated by changes in the global O_3 precursor emissions, which have enhanced the hemispheric background concentrations of O_3 (Isaksen et al., 2009).

In order to cause injury to vegetation, O_3 molecules must penetrate the plant via stomata, where they react with the internal plant tissue and generate reactive oxygen species. The impact of the site water regime (represented by soil moisture and air humidity) on stomatal regulation is crucial for O_3 uptake (Matyssek et al., 2007). Stomatal uptake is the key determinant of plant-physiological responses, along with metabolic capacities of detoxification and repair, to be unravelled through free-air O_3 fumigation approaches (Karnosky et al., 2007). The total O_3 deposition flux, representing both stomatal and non-stomatal removal, must also be measured for understanding the processes that control atmosphere–ecosystem exchange (Fowler et al., 2009). Ozone can react with plant surfaces and soil, and with volatile compounds emitted by vegetation and soil (Fowler et al., 2009). The existence of these multiple sinks makes it difficult to interpret O_3 flux data, unless complementary measurements are available, as it is expected to be the case at supersites.

At such sites, O_3 fluxes can be measured at an ecosystem scale using micrometeorological techniques and at a shoot/branch scale using enclosure techniques (Then et al., 2008). In addition, stomatal uptake rates can be derived from the measurements of water vapour exchange, based on porometer (leaf scale), cuvette (shoot/branch scale), xylem sap flow (tree scale) and micrometeorological (ecosystem scale) techniques (Nunn et al., 2010).

EC and aerodynamic gradient (AG) methods are the most common micrometeorological techniques employed for the measurement of O_3 fluxes. The EC method is a more direct approach that depends on fewer theoretical assumptions than the AG method; however, EC requires fast-response (~0.1 s) gas analysers. Fast O_3 analysers of this kind are commercially available and are most commonly based on dry chemiluminescence (Güsten and Heinrich, 1996). The application of EC using analysers of this kind has been described, for example, by Muller et al. (2010), while the operation of an analyser based on a liquid reagent has been presented by Keronen et al. (2003).

The AG technique can be applied with more conventional, slow-response instruments, making it possible to measure O_3 concentrations with standard UV photometry (Zapletal et al., 2011). However, the AG technique is subject to more stringent theoretical limitations than EC, and also requires instruments that are able to resolve the small gradients typically observed over forest surfaces (Baldocchi et al., 1988). The AG measurements are more prone to interference by biogenic emissions of reactive compounds; especially soil emissions of NO (Dorsey et al., 2004). The applicability of AG is also limited by the existence of the so-called roughness sublayer above an aerodynamically rough surface, such as forest canopy (Kaimal and Finnigan, 1994).

The micrometeorological methods provide an integrated measurement of the atmosphere–ecosystem exchange within the flux footprint, which has a longitudinal extent of a few hundred metres (Vesala et al., 2008). This means that the measurement of O_3 flux represents, within this area, all the possible deposition sinks discussed above, including the effect of any chemical reactions taking place in the airspace between the measurement height and the forest floor. To disentangle different deposition pathways, the total O_3 fluxes are typically partitioned to stomatal and non-stomatal components by utilising the canopy-scale water vapour fluxes measured by EC (Gerosa et al., 2007) or sap-flow techniques (Nunn et al., 2010). EC does not distinguish between different vegetation layers and is not applicable when vegetation is wet, while the latter method requires consideration of the lag between the sap flow and transpiration.

Due to rapid chemical interactions between O_3 and NO_x, the determination of O_3 budget should be coordinated with that of N budget (Table 23.1). Multilevel flux and/or concentration measurements of O_3 and NO_x would provide supporting data for gaining a better understanding of the vertical distribution of O_3 sinks. These data could be further augmented by automated chamber measurements of soil fluxes of NO and O_3. In particular, removal through the reaction with NO in the trunk and canopy airspace may constitute a significant O_3 sink, especially at night (Dorsey et al., 2004).

In addition to NO_x, reactions with BVOCs may significantly affect the O_3 fluxes above forests. On-line BVOC concentration measurements can be accomplished by the proton transfer reaction-mass spectrometry time of flight (Lindinger et al., 1998). While PTR-MS provides a sufficiently short response time for the application of the EC technique, a PTR-MS/EC system effectively limits to one the number of the compounds to be measured at a time. For a more extended suite of BVOCs, PTR-MS should be combined with the disjunct EC method, in which non-synchronous time series of the short-term concentrations are generated by successive, cyclical measurement of individual compounds (Rinne et al., 2007). Nevertheless, it remains unclear whether the large non-stomatal deposition rates observed over forests can be explained by in-canopy gas-phase chemistry involving highly reactive, partly unidentified BVOCs (Goldstein et al., 2004), or by heterogeneous decomposition of O_3 at plant surfaces (Fowler et al., 2009). For the latter process, it would be useful to augment the monitoring programme (Table 23.1) by further ancillary measurements, such as surface wetness, which may enhance O_3 deposition to foliage (Altimir et al., 2006).

For improved partitioning of the total O_3 flux, and to focus on individual tree parts such as sun-exposed and shade leaves, the ecosystem-scale flux measurements with micrometeorological techniques can be complemented by shoot-scale enclosure measurements. Different technical solutions are available for this (Altimir et al., 2002).

23.6 THE WATER BUDGET

Air pollution and climate conditions are known to be key factors in influencing forest health (de Vries et al., 2003). Cycling and availability of water is a part of climatic conditions and also a determinant in air pollution damage. Climate models predict a near surface warming trend with the increasing concentrations of GHGs in the atmosphere for the coming decades (Barnett et al., 2005). The increasing temperatures have important hydrologic consequences at both global and regional scales. The global hydrologic cycle is expected to intensify and accelerate due to increased evapotranspiration, atmospheric water vapour content and precipitation (Huntington, 2010), but regional consequences are miscellaneous and blurry. One major aspect of the issue lies on the interactions between atmosphere, soil and forest ecosystems.

Air pollution also has the capacity to modify water cycle in regional and local scales, and water cycle coupled with nutrient cycling influences soil–vegetation properties. Hence, water cycle determines the ecosystem response to disturbances. In particular, soil moisture regime is one of the most important factors to control forest health by affecting physical, chemical and biological properties of soil (Zheng and Zhang, 2011). Stomatal conductance is affected by soil moisture and hence in case of O_3 pollution microclimatic differences determine the level of damage under similar levels of O_3 exposures (Matyssek et al., 2007).

Heat transport through soils is a critical factor in determining how soil responds to climate change. In most of the existing biophysical models used in boreal settings, pure heat conduction is assumed to be the main mechanism of heat transport in the soils (Fan et al., 2011). However, heat conduction is not the only mechanism for heat transport in soils and ecosystems especially in temperate soils. Liquid water and convection contribute to approximately 50% of soil heat flux (Cahill and Parlange, 1998). In boreal soils where moisture is highly variable across space and through seasons, the role of water movement could be an important factor in seasonal soil energy dynamics and in the long-term response of boreal systems to changes in climate through evaporation and condensation processes (Fan et al., 2011).

Soil moisture can be measured by the well-established methods of time domain reflectometry (TDR) and tensiometers, which determine volumetric content of soil water and soil matric potential, respectively. Other *in situ* methods to measure or estimate volumetric water content are gravimetric method, and radioactive methods of gamma ray attenuation (Pires et al., 2005) and neutron scattering (Tominaga et al., 2002). All mentioned methods are now commercially available but disadvantageous to some extent compared to TDR. Gravimetric method is probably the most accurate and simplest one but requires removal of the soil from the original location. Therefore it is generally used as a calibration tool for commercial measurement systems. Radioactive methods on the other hand are practical and non-destructive but require special attention to avoid possible health hazards (Noborio, 2001). TDR is a widely used method, but measurement systems may also differ in accuracy. According to Walker et al. (2004), connector-type TDR sensors give the best accuracy compared to buriable and other sensor types.

Another challenging issue of soil moisture data collection under forest cover is the temporal and spatial variations due to micrometeorological conditions. Assuming that soil properties (bulk density, salinity, etc.) are constant in small distances, micrometeorological conditions play a significant role in affecting climatic parameters such as wind, temperature, snow accumulation and melting, etc., which further influence soil moisture. Therefore, representation capacity of a traditional monitoring plot in a forest would be apparently low in terms of soil moisture. It is hard to extrapolate measurements to the rest of the forest as forest stand characteristics change. Remote sensing applications such as macrowave radiometric observations may support spatial distribution comparisons in this respect (Sajjad et al., 2010), but integration still requires lots of scientific efforts as *in situ* measurements and estimations with remote sensing are totally different approaches.

Long-term data are required to increase our knowledge and ability to understand the complex interactions between forest ecosystems and the climate system. Water-related measurements constitute a major part of databases in this field and also C model inputs (Davi et al., 2006). Changes in precipitation type, amount and distribution (IPCC, 2007) together with changes in temperature conditions could alter evaporative demand in many regions and affect photosynthesis (Granier et al., 2007) and soil respiration. Therefore, C models are bound with hydrologic models to some extent. Hydrological modelling studies at stand level for various purposes (evapotranspiration, carbon–water, rainfall–runoff, erosion–sedimentation, etc.) require some or all of the following water related parameters: precipitation (type, amount and intensity); temperature (ambient, soil); air humidity; soil hydraulic properties (infiltration, moisture, water retention capacities or curve); other soil properties (texture, structure, erodibility, organic matter content, bulk density, porosity); LAI; evapotranspiration (interception, transpiration, evaporation from soil); wind velocity; solar radiation (modified from Raspe et al., 2010). The parameters measured at ICP Level II core plots and additional measurements suggested to fit the data requirements of contemporary models are summarised in Table 23.1.

23.7 CONCLUDING REMARKS

All the measurements and budgets described in this chapter are inherently linked to each other. This kind of holistic approach is needed for a realistic translation of the ongoing changes in the atmospheric environment into impacts on forest ecosystems. Such an integrated effort requires a considerable use of resources at highly instrumented supersites, and can only be achieved by building on existing infrastructures. A wide variety of forest research networks are at present active in Europe. Clarke et al. (2011) and Danielewska et al. (2012) selected the most relevant ones and summarised the information about the variables measured within these networks. Although the effects of anthropogenic factors on forests have been a major focus in both monitoring and research for decades, four main aspects emerged from these surveys: (1) the need for improving links between monitoring of atmospheric changes and impacts on forests; (2) the lack of research-oriented manipulative experiments in the forests; (3) the need for long-term measurements of other GHG fluxes in addition to those of CO_2; and (4) the under-investigation of ozone trends, fluxes and impacts at forest sites. Supersites have the potential to fill these gaps and provide scientifically sound knowledge for forest protection and adaptive management in a changing world, but this requires coordination, harmonisation and a joint long-term platform for data exchange and modelling (Fischer et al., 2011). The integration of monitoring, experimentation and modelling is of paramount importance (Matyssek et al., 2012; Matyssek et al., 2013, this vol.), and may be reached by combining monitoring data with realistic (i.e. mechanistic) experiments across spatio-temporal scales. This way, we will translate the process-based forest ecosystem understanding into predictive models and into a more accurate risk assessment.

REFERENCES

Ainsworth, E.A., Rogers, A., 2007. The response of photosynthesis and stomatal conductance to rising [CO_2]: molecular mechanisms and environmental interactions. Plant Cell Environ. 30, 258–270.

Altimir, N., Vesala, T., Keronen, P., Kulmala, M., Hari, P., 2002. Methodology for direct field measurements of ozone flux to foliage with shoot chambers. Atmos. Environ. 36, 19–29.

Altimir, N., Kolari, P., Tuovinen, J.-P., Vesala, T., Bäck, J., Suni, T., Kulmala, M., Hari, P., 2006. Foliage surface ozone deposition: a role for surface moisture? Biogeosciences 3, 209–228.

Ambus, P., Robertson, G.P., 2006. The effect of increased N deposition on nitrous oxide, methane and carbon dioxide fluxes from unmanaged forest and grassland communities in Michigan. Biogeochemistry 79, 315–337.

Aubinet, M., Grelle, A., Ibrom, A., Rannik, Ü., Moncrieff, J., Foken, T., Kowalski, A.S., Martin, P.H., Berbigier, P., Bernhofer, C., Clement, R., Elbers, J., Granier, A., Grünwald, T., Morgenstern, K., Pilegaard, K., Rebmann, C., Snijders, W., Valentini, R., Vesala, T., 2000. Estimates of the annual net carbon and water exchange of European forests: the EUROFLUX methodology. Adv. Ecol. Res. 30, 113–175.

Baldocchi, D., 2008. Breathing of the terrestrial biosphere: lessons learned from a global network of carbon dioxide flux measurement systems. Aust. J. Bot. 56, 1–26.

Baldocchi, D.D., Hicks, B.B., Meyers, T.P., 1988. Measuring biosphere-atmosphere exchanges of biologically related gases with micrometeorological methods. Ecology 69, 1331–1340.

Barnett, T.P., Adam, J.C., Lettenmaier, D.P., 2005. Potential impacts of a warming climate on water availability in snow-dominated regions. Nature 438, 303–309.

Bobbink, R., Hicks, K., Galloway, J., Spranger, T., Alkemade, R., Ashmore, M., Bustamante, M., Cinderby, S., Davidson, E., Dentener, F., Emmett, B., Erisman, J.W., Fenn, M., Gilliam, F., Nordin, A., Pardo, L., de Vries, W., 2010. Global assessment of nitrogen deposition effects on terrestrial plant diversity: a synthesis. Ecol. Appl. 20, 30–59.

Bruhn, D., Mikkelsen, T.N., Øbro, J., Willats, W.G.T., Ambus, P., 2009. Effects of temperature, ultraviolet radiation and pectin methyl esterase on aerobic methane release from plant material. Plant Biol. 11 (Suppl. 1), 43–48.

Bytnerowicz, A., Fenn, M., McNulty, S., Yuan, F., Pourmokhtarian, A., Driscoll, C., Meixner, T., 2013. Interactive effects of air pollution and climate change on forest ecosystems in the United States – current understanding and future scenarios. In: Matyssek, R, Clarke, N, Cudlin, P, Mikkelsen, T, Tuovinen, JP, Weiser, G, Paoletti, E (Eds.), Climate Change, Air Pollution and Global Challenges: Understanding and Solutions from Forest Research. Dev. Environ. Sci. 13. Elsevier, Netherlands, pp. 333–369.

Bytnerowicz, A., Omasa, K., Paoletti, E., 2007. Integrated effects of air pollution and climate change on forests: a northern hemisphere perspective. Environ. Pollut. 147, 438–445.

Cahill, A.T., Parlange, M.B., 1998. On water vapor transport in field soils. Water Resour. Res. 34, 731–739.

Canullo, R., Starlinger, F., Granke, O., Fischer, R., Aamlid, D., Neville, P., 2010. Assessment of ground vegetation. Manual part VII-SP1. In: Manual on Methods and Criteria for Harmonized Sampling, Assessment, Monitoring and Analysis of the Effects of Air Pollution on Forests. UNECE ICP Forests Programme Co-ordinating Centre, Hamburg, 19 pp.

Clarke, N., Fischer, R., de Vries, W., Lundin, L., Papale, D., Vesala, T., Merilä, P., Matteucci, G., Mirtl, M., Simpson, D., Paoletti, E., 2011. Availability, accessibility, quality and comparability of monitoring data for European forests for use in air pollution and climate change science. iForest—Biogeosci. Forestry 4, 162–166.

Cojocariu, C., Escher, P., Häberle, K.H., Matyssek, R., Rennenberg, H., Kreuzwieser, J., 2005. The effect of ozone on the emission of carbonyls from leaves of adult *Fagus sylvatica*. Plant Cell Environ. 28, 603–611.

Curtis, P.S., Hanson, P.J., Bolstad, P., Barford, C., Randolph, J.C., Schmid, H.P., Wilson, K.B., 2002. Biometric and eddy-covariance based estimates of annual carbon storage in five eastern North American deciduous forests. Agric. For. Meteorol. 113 (1–4), 3–19.

Danielewska, A., Clarke, N., Olejnik, J., Hansen, K., de Vries, W., Lundin, L., Tuovinen, J., Fischer, R., Urbaniak, M., Paoletti, E., 2013. A meta-database comparison from various European Research and Monitoring Networks dedicated to forest sites. iForest—Biogeosci. Forestry 6, 1–9.

Davi, H., Dufrêne, E., Francois, C., Le Maire, G., Loustau, D., Bosc, A., Rambal, S., Granier, A., Moors, E., 2006. Sensitivity of water and carbon fluxes to climate changes from 1960 to 2100 in European forest ecosystems. Agric. For. Meteorol. 141 (1–6), 35–56.

de Vries, W., Reinds, G.J., Vel, E., 2003. Intensive monitoring of forest ecosystems in Europe: 2. Atmospheric deposition and its impacts on soil solution chemistry. For. Ecol. Manage. 174, 97–115.

de Vries, W., Butterbach Bahl, K., Denier van der Gon, H.A.C., Oenema, O., 2007a. The impact of atmospheric nitrogen deposition on the exchange of carbon dioxide, nitrous oxide and methane from European forests. In: Reay, D.S., Hewitt, C.N., Smith, K.A., Grace, J. (Eds.), Greenhouse Gas Sinks. CABI, Wallingford, UK, pp. 249–283.

de Vries, W., van der Salm, C., Reinds, G.J., Erisman, J.W., 2007b. Element fluxes through intensively monitored forest ecosystems in Europe and their relationships with stand and site characteristics. Environ. Pollut. 148, 501–513.

de Vries, W., Reinds, G.J., Bierkens, M.F.P., 2009a. Assessment of the optimal time interval for repeated soil surveys at intensively monitored forest plots. J. Environ. Monit. 11, 2009–2021.

de Vries, W., Solberg, S., Dobbertin, M., Sterba, H., Laubhann, D., van Oijen, M., Evans, C., Gundersen, P., Kros, J., Wamelink, G.W.W., Reinds, G.J., Sutton, M.A., 2009b. The impact of nitrogen deposition on carbon sequestration by European forests and heathlands. For. Ecol. Manage. 258, 1814–1823.

de Vries, W., Wieggers, H.J.J., Brus, D.J., 2010. Impacts of sampling design and estimation methods on nutrient leaching of intensively monitored forest plots in the Netherlands. J. Environ. Monit. 12, 1515–1523.

Denman, K.L., Brasseur, G., Chidthaisong, A., Ciais, P., Cox, P.M., Dickinson, R.E., Hauglustaine, D., Heinze, C., Holland, E., Jacob, D., Lohmann, U., Ramachandran, S., da Silva Dias, P.L., Wofsy, S.C., Zhang, X., 2007. Couplings between changes in the climate system and biogeochemistry. In: Solomon, S., Qin, D., Manning, M., Chen, Z., Marquis, M., Averyt, K.B., Tignor, M., Miller, H.L. (Eds.), Climate Change 2007: The Physical Science Basis. Contribution of Working Group I to the Fourth Assessment Report of the Intergovernmental Panel on Climate Change. Cambridge University Press, Cambridge, UK.

Dobbertin, M., Neumann, M., 2010. Tree growth. Manual part V. In: Manual on Methods and Criteria for Harmonized Sampling, Assessment, Monitoring and Analysis of the Effects of Air Pollution on Forests. UNECE ICP Forests Programme Co-ordinating Centre, Hamburg, p. 29.

Dorsey, J.R., Duyzer, J.H., Gallagher, M.W., Coe, H., Pilegaard, K., Weststrate, J.H., Jensen, N.O., Walton, S., 2004. Oxidized nitrogen and ozone interaction with forest. I: experimental observations and analysis of exchange with Douglas fir. Quart. J. Roy. Meteor. Soc. 130, 1941–1955.

Draaijers, G.P.J., Erisman, J.W., 1995. A canopy budget model to assess atmospheric deposition from throughfall measurements. Water Air Soil Pollut. 85, 2253–2258.

Eichhorn, J., Roskams, P., Ferretti, M., Mues, V., Szepesi, A., Durrant, D., 2010. Manual part IV: visual assessment of crown condition and damaging agents. In: Manual on Methods and Criteria for Harmonized Sampling, Assessment, Monitoring and Analysis of the Effects of Air Pollution on Forests. UNECE ICP Forests Programme Co-ordinating Centre, Hamburg, p. 49.

Famulari, D., Fowler, D., Hargreaves, K., Milford, C., Nemitz, E., Sutton, M., Weston, K., 2005. Measuring eddy covariance fluxes of ammonia using tunable diode laser absorption spectroscopy. Water Air Soil Pollut. Focus 4 (6), 151–158.

Fan, Z., Neff, J.C., Harden, J.W., Zhang, T., Veldhuis, H., Czimczik, C.I., Winston, G.C., O'Donnell, J.A., 2011. Water and heat transport in boreal soils: implications for soil response to climate change. Sci. Total Environ. 409, 1836–1842.

Farmer, D.K., Kimmel, J.R., Phillips, G., Docherty, K.S., Worsnop, D.R., Sueper, D., Nemitz, E., Jimenez, J.L., 2010. Eddy covariance measurements with high-resolution time-of-flight aerosol mass spectrometry: a new approach to chemically-resolved aerosol fluxes. Atmos. Meas. Tech. 3, 5867–5905.

Fischer, R., Aas, W., de Vries, W., Clarke, N., Cudlin, P., Leaver, D., Lundin, L., Matteucci, G., Matyssek, R., Mikkelsen, T.N., Mirtl, M., Öztürk, Y., Papale, D., Potocic, N., Simpson, D., Tuovinen, J.-P., Vesala, T., Wieser, G., Paoletti, E., 2011. Towards a transnational system of Supersites for forest monitoring and research in Europe—an overview on present state and future recommendations. iForest—Biogeosci. Forestry 4, 167–171.

Flückiger, W., Braun, S., 1998. Nitrogen deposition in Swiss forests and its possible relevance for leaf nutrient status, parasite attacks and soil acidification. Environ. Pollut. 102, 69–76.

Forster, P., Ramaswamy, V., Artaxo, P., Berntsen, T., Betts, R., Fahey, D.W., Haywood, J., Lean, J., Lowe, D.C., Myhre, G., Nganga, J., Prinn, R., Raga, G., Schulz, M., Van Dorland, R., 2007. Changes in atmospheric constituents and in radiative forcing. In: Solomon, S., Qin, D., Manning, M., Chen, Z., Marquis, M., Averyt, K.B., Tignor, M., Miller, H.L. (Eds.), Climate Change 2007: The Physical Science Basis. Contribution of Working Group I to the 4th Assessment Report of the Intergovernmental Panel on Climate Change. Cambridge University Press, Cambridge, UK.

Fowler, D., Pilegaard, K., Sutton, M.A., Ambus, P., Raivonen, M., Duyzer, J., Simpson, D., Fagerli, H., Fuzzi, S., Schjoerring, J.K., Granier, C., Neftel, A., Isaksen, I.S.A., Laj, P., Maione, M., Monks, P.S., Burkhardt, J., Daemmgen, U., Neirynck, J., Personne, E., Wichink-Kruit, R., Butterbach-Bahl, K., Flechard, C., Tuovinen, J.P., Coyle, M., Gerosa, G., Loubet, B., Altimir, N., Gruenhage, L., Ammann, C., Cieslik, S., Paoletti, E., Mikkelsen, T.N., Ro-Poulsen,-H., Cellier, P., Cape, J.N., Horváth, L., Loreto, F., Niinemets, Ü., Palmer, P.I., Rinne, J., Misztal, P., Nemitz, E., Nilsson, D., Pryor, S., Gallagher, M.W., Vesala, T., Skiba, U., Brüggemann, N., Zechmeister-Boltenstern, S., Williams, J., O'Dowd, C., Facchini, M.C., de Leeuw, G., Flossman, A., Chaumerliac, N., Erisman, J.W., 2009. Atmospheric composition change: ecosystems-atmosphere interactions. Atmos. Environ. 43, 5193–5267.

Galloway, J.N., Schlesinger, W.H., Levy II., V., Michaelis, A., Schnoor, J.L., 1995. Nitrogen fixation: anthropogenic enhancement-environmental response. Global Biogeochem. Cycles 9, 235–252.

Galloway, J.N., Aber, J.D., Erisman, J.W., Seitzinger, S.P., Howarth, R.W., Cowling, E.B., Cosby, B.J., 2003. The nitrogen cascade. Bioscience 53, 341–356.

Gauci, V., Gowing, D.J.G., Hornibrook, E.R.C., Davis, J.M., Dise, N.B., 2010. Woody stem methane emission in mature wetland alder trees. Atmos. Environ. 44, 2157–2160.

Gerosa, G., Derghi, F., Cieslik, S., 2007. Comparison of different algorithms for stomatal ozone flux determination from micrometeorological measurements. Water Air Soil Pollut. 179, 309–321.

Goldstein, A.H., McKay, M., Kurpius, M.R., Schade, G.W., Lee, A., Holzinger, R., Rasmussen, R.A., 2004. Forest thinning experiment confirms ozone deposition to forest canopy is dominated by reaction with biogenic VOCs. Geophys. Res. Lett. 31, L22106.

Granier, A., Reichstein, M., Bréda, N., Janssens, I.A., Falge, E., Ciais, P., Grünwald, T., Aubinet, M., Berbigier, P., Bernhofer, C., Buchmann, N., Facini, O., Grassi, G., Heinesch, B., Ilvesniemi, H., Keronen, P., Knohl, A., Köstner, B., Lagergren, F., Lindroth, A., et al., 2007. Evidence for soil water control on carbon and water dynamics in European forests during the extremely dry year: 2003. Agric. For. Meteorol. 143 (1–2), 123–145.

Gundale, M.J., DeLuca, T.H., Nordin, A., 2011. Bryophytes attenuate anthropogenic nitrogen inputs in boreal forests. Glob. Chang. Biol. 17, 2743–2753.

Güsten, H., Heinrich, G., 1996. On-line measurements of surface ozone fluxes: part I. Methodology and instrumentation. Atmos. Environ. 30, 897–909.

Huntingford, C., Lowe, J.A., Booth, B.B.B., Jones, C.D., Harris, G.R., Gohar, L.K., Meir, P., 2009. Contributions of carbon cycle uncertainty to future climate projection spread. Tellus B 61 (2), 355–360.

Huntington, T.G., 2010. Climate warming-induced intensification of the hydrologic cycle: an assessment of the published record and potential impacts on agriculture. Adv. Agron. 109 (1), 1–53.

Ibrom, A., Dellwik, E., Jensen, N.O., Flyvbjerg, H., Pilegaard, K., 2007. Strong low-pass filtering effects on water vapour flux measurements with closed-path eddy correlation systems. Agric. For. Meteorol. 147, 140–156.

IPCC, 2001. Climate Change 2001: Synthesis Report. Third Assessment Report of the Intergovernmental Panel on Climate Change. IPCC, Geneva, Switzerland.

IPCC, 2007. Climate change 2007: synthesis report. In: Core Writing Team, , Pachauri, R.K., Reisinger, A. (Eds.), 2007. Contribution of Working Groups I, II and III to the Fourth Assessment Report of the Intergovernmental Panel on Climate Change. IPCC, Geneva, Switzerland, p. 104.

Isaksen, I.S.A., Granier, C., Myhre, G., Berntsen, T.K., Dalsøren, S.B., Gauss, M., Klimont, Z., Benestad, R., Bousquet, P., Collins, W., Cox, T., Eyring, V., Fowler, D., Fuzzi, S., Jöckel, P., Laj, P., Lohmann, U., Maione, M., Monks, P., Prevot, A.S.H., Raes, F., Richter, A., Rognerud, B., Schultz, M., Shindell, D., Stevenson, D.S., Storelvmo, T., Wang, W.-C., van Weele, M., Wild, M., Wuebbles, D., 2009. Atmospheric composition change: climate-chemistry interactions. Atmos. Environ. 43, 5138–5192.

Kaimal, J.C., Finnigan, J.J., 1994. Atmospheric Boundary Layer Flows: Their Structure and Measurement. Oxford University Press, New York.

Karnosky, D.F., Zak, D., Pregnitzer, K., Awmack, C., Bockheim, J., Dickson, R., Hendrey, G., Host, G., King, J., Kopper, B., Kruger, E., Kubiske, M., Lindroth, R., Mattson, W., McDonald, E., Noormets, A., Oksanen, E., Parsons, W., Percy, K., Podila, G., Riemenschneider, D., Sharma, P., Thakur, R., Sober, A., Sober, J., Jones, W., Anttonen, S., Vapaavuori, E., Manskovska, B., Heilman, W., Isebrands, J., 2003. Tropospheric O_3 moderates responses of temperate hardwood forests to elevated CO_2: a synthesis of molecular to ecosystem results from the Aspen FACE project. Funct. Ecol. 17, 289–304.

Karnosky, D.F., Werner, H., Holopainen, T., Percy, K., Oksanen, T., Oksanen, E., Heerdt, C., Fabian, P., Nagy, J., Heilman, W., Cox, R., Nelson, N., Matyssek, R., 2007. Free-air exposure systems to scale up ozone research to mature trees. Plant Biol. 9, 181–190.

Keith, H., Leuning, R., Jacobsen, K.L., Cleugh, H.A., van Gorsel, E., Raison, R.J., Medlyn, B.E., Winters, A., Keitel, C., 2009. Multiple measurements constrain estimates of net carbon exchange by a Eucalyptus forest. Agric. For. Meteorol. 149 (3–4), 535–558.

Keppler, F., Hamilton, J.T.G., Braß, M., Röckmann, T., 2006. Methane emissions from terrestrial plants under aerobic conditions. Nature 439, 187–191.

Keronen, P., Reissell, A., Rannik, Ü., Pohja, T., Siivola, E., Hiltunen, V., Hari, P., Kulmala, M., Vesala, T., 2003. Ozone flux measurements over a Scots pine forest using eddy covariance method: performance evaluation and comparison with flux-profile method. Boreal Environ. Res. 8, 425–443.

Kitao, M., Löw, M., Heerdt, C., Grams, T.E.E., Häberle, K.H., Matyssek, R., 2009. Effects of chronic elevated ozone exposure on gas exchange responses of adult beech trees (*Fagus sylvatica*) as related to the within-canopy light gradient. Environ. Pollut. 157, 537–544.

Larsen, K.S., Ibrom, A., Beier, C., Jonasson, S., Michelsen, A., 2007. Ecosystem respiration depends strongly on photosynthesis in a temperate heath. Biogeochemistry 85 (2), 201–213.

Le Mer, J., Roger, P., 2001. Production, oxidation, emission and consumption of methane by soils: a review. Eur. J. Soil Biol. 37, 25–50.

Lindinger, W., Hansel, A., Jordan, A., 1998. On-line monitoring of volatile organic compounds at pptv levels by means of Proton-Transfer-Reaction Mass Spectrometry (PTR-MS) Medical applications, food control and environmental research. Int. J. Mass Spectrom. Ion Process. 173, 191–241.

Mahecha, M.D., Reichstein, M., Carvalhais, N., Lasslop, G., Lange, H., Seneviratne, S.I., Vargas, R., Ammann, C., Arain, M.A., Cescatti, A., Janssens, I.A., Migliavacca, M., Montagnani, L., Richardson, A.D., 2010. Global convergence in the temperature sensitivity of respiration at ecosystem level. Science 329 (5993), 838–840.

Matyssek, R., Clarke, N., Cudlin, P., Mikkelsen, T.N., Tuovinen, J.-P., Weiser, G., Paoletti, E., 2013. Climate change, air pollution and global challenges: understanding and solutions from forest research. In: Matyssek, R., Clarke, N., Cudlin, P., Mikkelsen, T., Tuovinen, J.P., Weiser, G., Paoletti, E. (Eds.), Climate Change, Air Pollution and Global Challenges: Understanding and Solutions from Forest Research. Dev. Environ. Sci. 13. Elsevier, Netherlands, pp. 3–16.

Matyssek, R., Sandermann, H., 2003. Impact of ozone on trees: an ecophysiological perspective. Progress in Botany, vol. 64. Springer Verlag, Heidelberg, 349–404.

Matyssek, R., Bytnerowicz, A., Karlsson, P.-E., Paoletti, E., Sanz, M., Schaub, M., Wieser, G., 2007. Promoting the O_3 flux concept for European forest trees. Environ. Pollut. 146, 587–607.

Matyssek, R., Wieser, G., Calfapietra, C., de Vries, W., Dizengremel, P., Ernst, D., Jolivet, Y., Mikkelsen, T.N., Mohren, G.M.J., Le Thiec, D., Tuovinen, J.-P., Weatherall, A., Paoletti, E., 2012. Forests under climate change and air pollution: gaps in understanding and future directions for research. Environ. Pollut. 160, 57–65.

McLaughlin, S.B., Wullschleger, S.D., Sun, G., Nosal, M., 2007. Interactive effects of ozone and climate on water use, soil moisture content and streamflow in a southern Appalachian forest in the USA. New Phytol. 174, 125–136.

Mikkelsen, T.N., Bruhn, D., Ambus, P., Larsen, K.S., Ibrom, A., Pilegaard, K., 2011. Is methane released from the forest canopy? iForest 4, 200–204.

Mikkelsen, T.N., Clarke, N., Danielewska, A., Fischer, R., 2013. Towards supersites in forest ecosystem monitoring and research. In: Matyssek, R., Clarke, N., Cudlin, P., Mikkelsen, T., Tuovinen, J.-P., Weiser, G., Paoletti, E. (Eds.), Climate Change, Air Pollution and Global Challenges: Understanding and Solutions from Forest Research. Dev. Environ. Sci. 13. Elsevier, Netherlands, pp. 475–496.

Muller, J.B.A., Percival, C.J., Gallagher, M.W., Fowler, D., Coyle, M., Nemitz, E., 2010. Sources of uncertainty in eddy covariance ozone flux measurements made by dry chemiluminescence fast response analysers. Atmos. Meas. Tech. 3, 163–176.

Nemitz, E., Sutton, M.A., Wyers, G.P., Otjes, R.P., Mennen, M.G., van Putten, E.M., Gallagher, M.W., 2004. Gas-particle interactions above a Dutch heathland: II. Concentrations and surface exchange fluxes of atmospheric particles. Atmos. Chem. Phys. 4, 1007–1024.

Noborio, K., 2001. Measurement of soil water content and electrical conductivity by time domain reflectometry: a review. Comput. Electron. Agr. 31 (3), 213–237.

Nunn, A.J., Cieslik, S., Metzger, U., Wieser, G., Matyssek, R., 2010. Combining sap flow and eddy covariance approaches to derive stomatal and non-stomatal O_3 fluxes in a forest stand. Environ. Pollut. 158, 2014–2022.

Paoletti, E., Tuovinen, J.-P., 2011. COST Action FP0903: "Research, monitoring and modelling in the study of climate change and air pollution impacts on forest ecosystems" iForest—Biogeosci. Forestry 4, 160–161.

Pires, L.F., Bacchi, O.O.S., Reicharth, K., 2005. Soil water retention curve determined by gamma-ray beam attenuation. Soil Till. Res. 82, 89–97.

Pushkarsky, M.B., Webber, M.E., Patel, C.K.N., 2003. Ultra-sensitive ambient ammonia detection using CO_2 laser based photoacoustic spectroscopy. Appl. Phys. B 77, 381–385.

Raes, F., Liao, H., Chen, W.-T., Seinfeld, J.H., 2010. Atmospheric chemistry-climate feedbacks. J. Geophys. Res. 115, D12121.

Raspe, S., Beuker, E., Preuhsler, T., Bastrup-Birk, A., 2010. Manual part IX: meteorological measurements. In: Manual on Methods and Criteria for Harmonized Sampling, Assessment, Monitoring and Analysis of the Effects of Air Pollution on Forests. UNECE ICP Forests Programme Co-ordinating Centre, Hamburg, p. 36.

Rinne, J., Taipale, R., Markkanen, T., Ruuskanen, T.M., Hellén, H., Kajos, M.K., Vesala, T., Kulmala, M., 2007. Hydrocarbon fluxes above a Scots pine forest canopy: measurements and modeling. Atmos. Chem. Phys. 7, 3361–3372.

Rondeux, J., Sanchez, C., 2010. Review of indicators and field methods for monitoring biodiversity within national forest inventories. Core variable: deadwood. Environ. Monit. Assess. 164, 617–630.

Sajjad, A.K.L., Sajjad, S., Baozhu, T., Feng, C., Jinlong, Z., 2010. Comparative studies of operational parameters of degradation of azo dyes in visible light by highly efficient WO_x/TiO_2 photocatalyst. J. Hazard. Mater. 177, 781–791.

Scanlon, T., Kiely, G., 2003. Ecosystem-scale measurements of nitrous oxide fluxes for an intensely grazed, fertilized grassland. Geophys. Res. Lett. 30 (16), 1852.

Schöllhorn, R., Burris, R.H., 1967. Acetylene as a competitive inhibitor of nitrogen fixation. Proc. Natl. Acad. Sci. U.S.A. 58, 213–218.

Serengil, Y., Augustaitis, A., Bytnerowicz, A., Grulke, N., Kozovitz, A.R., Matyssek, R., Müller-Starck, G., Schaub, M., Wieser, G., Coskun, A.A., Paoletti, E., 2011. Adaptation of forest ecosystems to air pollution and climate change: a global assessment on research priorities. iForest—Biogeosci. Forestry 4, 44–48.

Sitch, S., Cox, P.M., Collins, W.J., Huntingford, C., 2007. Indirect radiative forcing of climate change through ozone effects on the land-carbon sink. Nature 488, 791–794.

Solberg, S., Hov, Ø., Søvde, A., Isaksen, I.S.A., Coddeville, P., de Backer, H., Forster, C., Orsolini, Y., Uhse, K., 2008. European surface ozone in the extreme summer 2003. J. Geophys. Res. 113, D07307.

Solberg, S., Dobbertin, M., Reinds, G.J., Andreassen, K., Lange, H., Garcia Fernandez, P., Hildingsson, A., de Vries, W., 2009. Analyses of the impact of changes in atmospheric deposition and climate on forest growth in European monitoring plots: a stand growth approach. For. Ecol. Manage. 258, 1735–1750.

Sutton, M.A., Nemitz, E., Skiba, U.M., Beier, C., Butterbach Bahl, K., Cellier, P., de Vries, W., Erisman, J.W., Reis, S., Bleeker, A., Bergamaschi, P., Calanca, P.L., Dalgaard, T., Duyzer, J., Gundersen, P., Hensen, A., Kros, J., Leip, A., Olesen, J.E., Owen, S., Rees, R.M., Sheppard, L.J., Smith, P., Zechmeister-Boltenstern, S., Soussana, J.F., Theobald, M.R., Twigg, M., van Oijen, M., Veldkamp, T., Vesala, T., Winiwarter, W., Carter, M.S., Dragosits, U., Flechard, C., Helfter, C., Kitzler, B., Rahn, K.H., Reinds, G.J., Schleppi, P., with contributions from the NitroEurope community, 2011. The Nitrogen Cycle and its Influence on the European Greenhouse Gas Balance. Centre for Ecology and Hydrology, 43. Edinburgh, ISBN 978-1-906698-21-8.

Then, C., Löw, M., Matyssek, R., Wieser, G., 2008. Deriving ozone dose-response of photosynthesis in adult forest trees from branch-level cuvette gas exchange assessment. Environ. Pollut. 153, 526–528.

Tominaga, T.T., Cassaro, F.A.M., Reichardth, K., Bacchi, O.O.S., Oliveira, J.C.M., Timm, L.C., 2002. On the use of surface neutron-gamma gauges to estimate soil water content. Appl. Radiat. Isot. 57, 369–374.

Trebs, I., Meixner, F.X., Slanina, J., Otjes, R., Jongejan, P., Andreae, M.O., 2004. Real-time measurements of ammonia, acidic trace gases and water-soluble inorganic aerosol species at a rural site in the Amazon basin. Atmos. Chem. Phys. 4, 967–987.

Ulrich, B., 1983. Interaction of forest canopies with atmospheric constituents: SO_2, alkali and earth alkali cations and chloride. In: Ulrich, B., Pankrath, J. (Eds.), Effects of Accumulation of Air Pollutants in Forest Ecosystems. Reidel Publ. Co., Dordrecht, the Netherlands, pp. 33–45.

van der Salm, C., Reinds, G.J., de Vries, W., 2007. Water balances in intensively monitored forest ecosystems in Europe. Environ. Pollut. 148, 201–212.

van Oijen, M., Rougier, J., Smith, R., 2005. Bayesian calibration of process-based forest models: bridging the gap between models and data. Tree Physiol. 25, 915–927.

Vesala, T., Kljun, N., Rannik, Ü., Rinne, J., Sogachev, A., Markkanen, T., Sabelfeld, K., Foken, T., Leclerc, M.Y., 2008. Flux and concentration footprint modelling: state of the art. Environ. Pollut. 152, 653–666.

Vitousek, P.M., Aber, J.D., Howarth, R.W., Likens, G.E., Matson, P.A., Schindler, D.W., Schlesinger, W.H., Tilman, D.G., 1997. Human alteration of the global nitrogen cycle: sources and consequences. Ecol. Appl. 7, 737–750.

Walker, J.P., Willgoose, G.R., Kalma, J.D., 2004. *In situ* measurement of soil moisture: a comparison of techniques. J. Hydrol. 293, 85–99.

Wutzler, T., Wirth, C., Schumacher, J., 2008. Generic biomass functions for Common beech (*Fagus sylvatica*) in Central Europe: predictions and components of uncertainty. Can. J. For. Res. 38, 1661–1675.

Zapletal, M., Cudlín, P., Chroust, P., Urban, O., Pokorný, R., Edwards-Jonášová, M., Czerný, R., Janouš, D., Taufarová, K., Večeřa, Z., Mikuška, P., Paoletti, E., 2011. Ozone flux over a Norway spruce forest and correlation with net ecosystem production. Environ. Pollut. 159, 1024–1034.

Zheng, S., Zhang, M., 2011. Effect of moisture regime on the redistribution of heavy metals in paddy soil. J. Environ. Sci. 23, 434–443.

Part VII

Knowledge Transfer and Socio-Economic Aspects

Chapter 24

Forest Ecosystem Services Under Climate Change and Air Pollution

Pavel Cudlín*[,1], Josef Seják[†], Jan Pokorný[‡], Jana Albrechtová[§], Olaf Bastian[¶] and Michal Marek*

*Global Change Research Centre, Academy of Sciences of the Czech Republic, Ceske Budejovice, Czech Republic

[†]Faculty of Environment, J.E. Purkyně University in Ústí nad Labem, Ústí nad Labem, Czech Republic

[‡]Enki Ltd., Třeboň, Czech Republic

[§]Faculty of Science, Charles University in Prague, Prague, Czech Republic

[¶]Leibniz Institute of Ecological Urban and Regional Development, Dresden, Germany

[1]Corresponding author: e-mail: cudlin.p@czechglobe.cz

Chapter Outline

Developments in Environmental Science, Vol. 13. http://dx.doi.org/10.1016/B978-0-08-098349-3.00024-4

24.1 INTRODUCTION

Ecosystem services (ES) are the benefits humans receive, directly or indirectly, from ecosystems (Costanza et al., 1997). Links between ecosystem functions, ES and human well-being are complex (Figure 24.1) and need

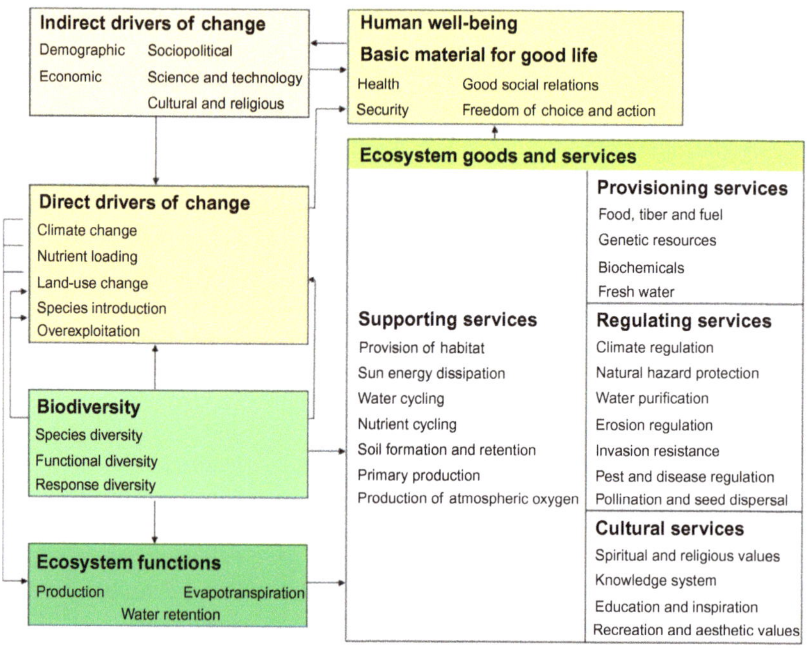

FIGURE 24.1 The role of biodiversity and ecosystem functioning assessments in the ecosystem service concept. *Modified according to MEA (2005) and Secretariat of the Convention on Biological Diversity, Global Biodiversity Outlook 2, Montreal (2006).*

consideration of different spatial and temporal scales to assess them properly (MEA, 2005). Over the past 50 years, rapid and extensive environmental changes in land use, climate and the atmosphere have resulted in a substantial loss or degradation of ES, while demands for them have been increasing (Knoke and Hahn, 2013, this vol.).

Forests provide many supporting, regulating and cultural services. They regulate local and global climate, remove pollutants from air, water and soil, enhance soil retention and water quality, ameliorate water events, facilitate pollination, improve landscape aesthetics, provide habitats for organisms, and enclose invaluable genetic information yet to be uncovered (Chiabai et al., 2011; De Groot and van der Meer, 2010). Climate-regulation services, derived from production and evapotranspiration (ET) ecosystem functions, were selected to demonstrate the interconnections between ecosystem processes and regulation ecosystem services.

Some air pollutants contribute to radiative forcing—CO_2, CH_4, N_2O, halocarbons (Bytnerowicz et al., 2007). From aerosols and particulate matter, S and N species generally have a cooling effect in the atmosphere (Intergovernmental Panel on Climate Change, IPCC, 2001). Climate change, high radiation and temperature, in particular, promotes increase in tropospheric ozone, the air pollutant derived from non-methane volatile organic compounds, carbon monoxide and nitrogen oxides (Bytnerowicz et al., 2007). Elevated ozone and altered nitrogen, carbon and water availability are becoming key issues for forest research (Paoletti et al., 2010; Matyssek et al., 2013, this vol.). Rapidly changing temperatures, precipitation amounts and patterns, or increasing concentrations of atmospheric CO_2 and fluctuating air-borne pollution, above all ozone, are all likely to affect forest growth and health, carbon sequestration, vegetation and soil species composition, and alter the functioning of forest ecosystems in unknown directions (Paoletti et al., 2010; Bytnerowicz et al., 2013, this vol.). Climate change can alter the effects of air pollutants on forest ecosystems and vice versa. Climate change in interaction with air pollution brings novel combinations of severity and timing of multiple stresses, the effects of which on tree performance and forest ecosystems are currently hard to predict (Niinemets, 2010). Multiple environmental stresses can lead to forest decay and, together with large-area deforestation, may significantly affect many forest ES. Previously, the impacts of air pollution and climate change have been treated separately (Bytnerowicz et al., 2007), but research on the interactive effects of climate change and air pollution has become a central issue in forest science during the past decade (Matyssek et al., 2010; Percy et al., 1999; Serengil et al., 2011). Current evidence suggests that air pollution will become increasingly harmful to forests under climate change severely affecting many forest ES (production, biodiversity protection, soil protection, sustained water balance, socio-economical relevance) (Paoletti et al., 2010; Tuovinen et al., 2013, this vol.).

In general, the IPCC´s Fourth Assessment Report concluded that, despite likely declines in some regions, global timber production is predicted to increase due to higher growth rates and shifts in forest ranges (Parry et al., 2007). While the tree model of Nabuurs et al. (2002), based on ecological processes, including disturbances, shows an 18% increase in stemwood growth for European forests by 2030, some other studies predict possible production decrease in temperate forests bordering the subtropics and stress an increasing significance of storms and other major disturbances (Seppälä et al., 2009).

Though our well-being is dependent upon the continued flow of ES, many ES, such as climate or water runoff regulation, are public goods and do not have any markets or prices. According to an economic interpretation of the ES concept, they are not ES at all, because only human needs or demands actually convert an ecosystem function (e.g. ET) into a real ES (e.g. climate regulation) (Burkhard et al., 2012). As a consequence, biodiversity is declining, natural ecosystems are continuously degraded and society in turn is suffering the consequences, which are partly irreversible. Therefore, a multiscale assessment of ecosystem services should be conducted and supported by scientific communities and governmental organisations (Seják et al., 2012).

Our views are in line with the opinion that basic supporting ES according to Millenium Ecosystem Assessment—MEA (2005) are not actually ES, because they do not have financial value and are identified with some ecosystem processes and functions (Fischer and Turner, 2008). In our methodological approach, quantifying ecosystem functions as a basis for ES valuation, we link biodiversity and ecosystem function studies with the MEA concept (Secretariat of the Convention on Biological Diversity, 2006) (Figure 24.1).

24.2 ADOPTING THE ECOSYSTEM SERVICES CONCEPT TO IDENTIFY AND VALUE CHANGES IN FORESTS

24.2.1 The Ecosystem Services Concept

The analysis, monitoring and disclosure of ES may provide land conservation managers, public land stewards and environmentally aware land developers a means to optimise land use, especially forest management. Knowledge transfer and propagation of ES can strengthen public awareness about the role and values of forests as well as of ecosystems and biodiversity, in general. The growing popularity of the ecosystem services concept can be seen primarily as a reaction to the interplay of first, the long-term neglect of biophysical and ecosystem functions—often considered gratis—in our economic cycles and the societal system as such and second, the growing devastation and degradation of the ecosystems providing these services (Bastian et al., 2012; Boyd and Banzhaf, 2007; Seppelt et al., 2011).

The concept of ES represents a multilayered approach to the interface between environmental and societal claims, with special consideration of economic aspects. However, these are closely interlinked with ecological and social aspects, that is, all three dimensions of sustainability are addressed. Thus, it can be used as a stimulus and tool to find appropriate solutions for land use and ecosystem management, and to balance economic interests with ecological and social requirements in multifunctional landscapes.

24.2.2 The Assessment of Ecosystem Services

According to the recently elaborated ecosystem properties, potentials and services framework (Bastian et al., 2012; Grunewald and Bastian, 2010), which is based on the cascade model of Haines-Young and Potschin (2009) and TEEB (2010), in consideration of the various scientific schools of landscape ecology, an assessment of ES should include the following steps (or pillars):

Pillar 1 constitutes the *properties* of ecosystems representing the set of ecological conditions, structures and processes (e.g. soil qualities, nutrient cycles, biological diversity) that determine whether an ecosystem service can be supplied.

Depending on their properties, ecosystems are able to supply services; they have particular *potentials* or capacities for that. Potentials have consciously been included as Pillar 2, so as to distinguish between a possible and the actual use, which is the expression of the real service (Bastian et al., 2012). Potentials can be regarded and quantified as stocks of ES, while the services themselves represent the actual flows (Haines-Young et al., 2012).

From economics, only human needs or demands actually convert a potential into a real service (Burkhard et al., 2012; Syrbe and Walz, 2012). Ecosystem service*s*, the third pillar of the framework, reflect an even stronger human perspective (value level), since the services (and goods) are in fact currently valued, demanded or used.

The analysis of ES always involves a valuation step, that is, scientific findings (facts) are transformed into human-driven value categories. Through an 'ES' link, human beings benefit from ecosystems. That means that ecosystems yield *benefits and values* (Pillar 4), which contribute to human well-being (Fischer and Turner, 2008). Value is most commonly defined as the contribution of ES to goals, objectives or conditions that are specified by a user (van Oudenhoven et al., 2012). Actors in society can attach a value (monetary or other) to these benefits.

There are no ES without human beneficiaries (Fischer et al., 2009). The stakeholders, providers, users or *beneficiaries* of ecosystems and their ES (Pillar 5) can be single persons, groups or society as a whole. The use and management of services (often regulated and controlled by decisions and legislation) can modify or change the properties and potentials of ecosystems. Appropriate *management* has to bridge the gap between the state and targets for ES.

All pillars of the ES framework (ecosystem properties, potentials, services, values/benefits, providers/beneficiaries) can or should be analysed and differentiated in terms of *space* (e.g. scale, dimension, pattern) *and time* (e.g. driving forces, changes, scenarios) (Syrbe and Walz, 2012).

24.2.3 Specifics of Forest Ecosystem Services

Forests supply a broad range of ES. These include provisioning services, such as provision of timber and other commodities, food, genetic, medicinal and ornamental resources, and water supply. Regulation services include, for example, soil retention, runoff regulation, carbon sequestration, removal of pollutants from air, water and soil, improving air quality, waste treatment, and habitat functions. Aesthetic values, recreation opportunities, cultural and scientific inspiration, a sense of place, information and others are some of the socio-cultural services provided by forests.

The ES concept is often associated only with monetary valuation approaches. However, the majority of forest ES cannot be traded. As they are public goods and prices are not available, a market-oriented economic valuation is hardly possible. Indirect valuation approaches need to be used for many ES, if there are strong relations to tradeable goods, for example, the water purification service of forests. If such relations do not exist, contingent valuation methods can be an alternative, for example, willingness-to-pay methods, which simulate a market (Carson, 2011).

It is important to note that not all dimensions of human well-being can be expressed in monetary terms (Spangenberg and Settele, 2010). Economic valuation procedures are not capable of dealing with intrinsic or socio-cultural values (e.g. ethical, spiritual and aesthetic), the long-term growth processes of forests or the missing knowledge of future human preferences.

24.3 ECOSYSTEM PROCESSES/FUNCTIONS UNDER INTERACTIVE EFFECTS OF CLIMATE CHANGE AND AIR POLLUTION—SUSTAINABLE PROVIDERS OF ECOSYSTEM SERVICES

24.3.1 Photosynthesis as One of the Keystones of Forest Ecosystem Services

As autotrophic organisms, forest trees photosynthesize. This basic physiological process is responsible for the transformation of solar energy into energy of chemical bonds, mainly in the form of ATP, and formation of reduction equivalents (NADPH). Both are mainly used in the reduction of carbon dioxide into organic compounds. These reactions are connected to oxygen evolution, which is part of another important ecosystem service.

Limited or depressed photosynthesis strongly affects all other physiological processes of trees. For example, a certain level of photosynthetic activity is a crucial condition for plant nutrition and vice versa. Similarly, this is valid for the relation between photosynthesis and respiration, transpiration, etc. Thus, we can say that all forest ecosystem services are dependent on photosynthesis.

All forest ecosystems functions are determined by stand-level structure, that is, the spatial and species composition of the stand and crown architecture, and physiological processes as well. A forest ecosystem can be considered as an open system, characterised by a wide set of elements and links, where the form is conditioned by the function (Marek et al., 2011).

The results of the productive activity predetermine the potential of a certain forest ecosystem to fulfil all other forest ecosystem services. The water balance role of forest ecosystems is based on ET from a forest stand, which is determined by latent heat exchange and crown architecture, which determines precipitation retention and infiltration of water into forest soils. Moreover, forest stands influence water drainage. Thus, these services are derived from the results of production activity directly based on photosynthesis.

It is evident that identification of new forest ecosystem services is an ongoing process. The phenomenon of global environmental change is especially promoting new services. A particularly good example is the storage of carbon from atmospheric carbon dioxide into forest tree biomass and soils, which is based on the size of the forest area, longevity of the trees and duration of carbon storage in the forest. This forest ecosystem service is not only closely dependent on current stand microclimate and extreme synoptic events (sudden dry period, wind fallen trees, late and early frost), but is potentially affected by human management practises (Parry et al., 2007). Permanent carbon 'pumping' from the atmosphere via the photosynthetic activity of trees and plants serves as an important carbon-capture mitigation measure directly related to the biological activity of trees and plants. Reasonable silviculture practises can support this carbon storage function, or they can be the reason for huge carbon losses from forest ecosystems.

It is not clear how climate-change phenomena interact, thus making it difficult to estimate the amount of forest ES determined by photosynthesis. The extended duration of the vegetation season caused by climate change may lead to increased carbon-sink strength, for example, an extended season length of 22 days can lead to a 13% increase in the carbon sink (Marek et al., 2011). However, though vegetation was supposed to increase its sink strength under increasing CO_2 concentrations, recent studies showed that changes in drought occurrence pattern caused by climate change may lower global net primary production during the first decade of our century (Zhao and Running, 2010).

Photosynthesis in forest ecosystems is often N limited and increasing N deposition coupled with increasing CO_2 atmospheric concentration may account for a remarkable increase in C sequestration (30–70 kg C per kg N deposition; De Vries et al., 2009). N deposition has been suggested as the main driving factor behind forest growth over the past 50 years, accounting for more than 60% of this growth (Ciais et al., 2008). The impact of N deposition on soil C sequestration is less clear since there still exists a critical gap in the knowledge about the effect of N deposition on belowground biomass dynamics (Butterbach-Bahl and Gundersen, 2011). Air pollutant effects on forests may provide an important control on carbon sequestration (Bytnerowicz et al., 2007). For example, ozone greatly suppresses photosynthesis; for example, Felzer et al. (2004) estimated that the ozone contribution to reduction in carbon sequestration in the United States was 5.7% from 1950 to 1995, corresponding to 18–38 Tg C y^{-1}.

24.3.2 Gas Exchange and Transpiration Under Interactive Effects of Climate Change and Air Pollution

Leaf-scale transpiration of plants is governed by the biochemical demand for CO_2—the substrate for photosynthesis. Optimal stomatal reaction reflects the increasing ratio of photosynthesis to transpiration (instantaneous transpiration efficiency) in proportion to rising atmospheric concentrations of CO_2 (Barton et al., 2012). Stomatal 'closure' with increasing CO_2 concentration and increasing vapour-pressure deficit is well established, however, air pollutants interact with the process. For example, the length of air pollution exposure and stand structure determine stomatal conductance as demonstrated by Uddling et al. (2009) who showed in a FACE study that short-term primary stomatal closure responses to elevated CO_2 and O_3 were completely offset by long-term cumulative effects. These authors suggested that assumptions of large reductions in stomatal conductance under rising atmospheric CO_2 are very uncertain for forests as reported in several FACE experiments.

ET consists of stomatal and cuticular parts of leaf transpiration together with surface evaporation. ET on plant, canopy and landscape levels affects water cycles at global, continental, regional and local scales. ET can be affected by numerous environmental factors starting at the leaf level: elevated atmospheric CO_2 concentration by induction of stomatal closure, air pollutant effects, air temperature, vapour-pressure deficit, turbulent transport, radiative transfer and reduced soil moisture (Katul et al., 2012). All these factors have to be considered when ET is extrapolated to higher scales. During the past three decades, the methods of canopy science developed remarkably, representing a cutting-edge subset of forest research (Lowman and Schowalter, 2012).

Water availability is a key environmental factor limiting forest growth and productivity. Limited soil water availability within the rooting zone impacts on ET as well as some plant strategies to cope with prolonged soil moisture stress (Katul et al., 2012). Hydraulic redistribution is the passive movement of water between different soil parts via plant root systems, driven by water potential gradients in the soil–plant interface. The proposed ecological and hydrologic impacts of hydraulic redistribution include increasing dry-season transpiration and photosynthetic rates, prolonging the life span of fine roots and maintaining roo–soil contact in dry soils, moving rainwater down into deeper soil layers where it does not evaporate and increasing plant nutrient uptake by moistening more nutrient-rich shallow soil layers (Neumann and Cardon, 2012). Globally, hydraulic redistribution may influence hydrological and biogeochemical cycles and, ultimately, climate (Prieto et al., 2012).

Climate determines tree physiological performance and species-specific competitiveness resulting in alteration of species latitudinal and altitudinal distributions (Zweifel et al., 2009). Relatively extensive knowledge is available concerning the transpiration and water balance of forest trees and stands at low-altitude sites. High-altitude forest stands are more exposed to multiple stresses caused by the interactive effects of climate extremes and air pollution. More research has to be conducted to understand these consequences (Matyssek et al., 2009).

One of the phenomena of climate change is rising temperature inducing increased transpiration rates. Since the rate of photosynthesis does not increase with increasing temperature as quickly, it implies decreased water-use efficiency (the amount of photosynthetically bound CO_2 related to the amount of transpired water during a certain period). Increasing CO_2 concentration causing suppression of plant transpiration was proposed to be the driver on a regional level for increased continental runoff through the twentieth century (Gedney et al., 2006). However, as discussed above, other environmental factors can interact in affecting the stomatal closure process, for example, air pollutants, atmospheric humidity, etc. If temperature cues predominate, climate change would extend growing seasons. However, not all species will respond in the same extent to warming—photoperiod-controlled tree species will show limited responsiveness (Bauerle et al., 2012). These authors suggested that photoperiod-associated declines in photosynthetic capacity could limit autumn carbon gain in forests, even if warming delays leaf senescence.

Elevated surface-level ozone concentrations have become one of the key factors of climate change (IPCC, 2001) and are regarded as a mitigating factor which influence carbon-sink strength of vegetation under increasing CO_2 (Paoletti et al., 2010; Bytnerowicz et al., 2013, this vol.). Ozone concentrations are expected to increase, therefore risks to vegetation and forests in particular require special attention (Matyssek and Sandermann, 2003). O_3 risk assessment is gradually developing towards more process-based, cause–effect relationships (Matyssek et al., 2008; Diezengremel et al., 2013, this vol.).

24.3.3 Forest Ecosystem Services in Climate Regulation

Forests influence climate through numerous processes that affect planetary energy flows, hydrologic cycles and composition of the atmosphere. In terms of greenhouse effect, reforestation and afforestation attenuate global warming through carbon sequestration. Tropical forests mitigate warming through evaporative cooling, but the low *albedo* of boreal forests is a positive climate forcing (Bonan, 2008). Some authors of scientific papers argue that the low *albedo* of these forests contributes to global warming. For example, Randerson et al. (2006) conclude that 'future increases in boreal fire may not accelerate climate warming'. Bala et al. (2007) claim that 'global-scale deforestation has a net cooling influence on climate, because the warming carbon-cycle effects of deforestation are overwhelmed by the net cooling associated with changes of *albedo* and evapotranspiration'. However, the cooling effect of forests in sunny days is evident. The direct effect of land-cover changes on local and regional climate should be taken into account and studied more (Pielke, 2005; Pielke et al., 2011). The daily dynamic of surface temperature of different land cover as well as the cooling effect of forest in a cultural landscape can be monitored by remote sensing (Hesslerová et al., 2013).

The function of ecosystems and, namely, the role of vegetation in climate regulation can be studied in terms of solar energy and water fluxes. With water having a high capacity for carrying energy in the form of latent heat, most energy is dissipated by the physical processor property of evaporation and condensation, making water a very efficient cooler and heater. When 1 kg of water changes from liquid to its gaseous phase, as in ET (about 2.5 MJ–0.69 kWh is needed), energy is stored in the kinetic movement of water vapour in the form of latent heat and the local area is cooled. At night or early morning, when water vapour condenses on cooler surfaces, energy in the form of latent heat is released and the local area is warmed (Eiseltová et al., 2012).

At the landscape level, ET plays an essential role in energy dissipation and as such is highly dependent on vegetation cover and water availability. To understand how the natural processes involved in energy dissipation are interrelated, Ripl (1995, 2003) proposed a conceptual model based on the energy dissipative properties of water. In his energy-transport-reaction model, Ripl considered three essential processes that control the dissipation of energy: (i) water evaporation and condensation, (ii) the dissolution and precipitation of salts and (iii) disintegration and recombination of the water molecule within the biological cell.

On a sunny day, up to 1000 W m^{-2} of solar energy comes to the Earth's surface. Part of this solar energy is reflected (15–20%). Most of the remaining solar energy is divided between latent heat of evaporation and sensible heat. Several hundred W m^{-2} is used for ET, up to 100 W m^{-2} heats the ground, while only several W m^{-2} are used for photosynthesis and stored in biomass. Warming of a plant stand consumes several W m^{-2}. A shortage of water results in a decline of ET and an increase in the sensible heat flux; surface temperature and subsequently air temperature rises (Katul et al., 2012; Monteith and Unsworth, 1990; Pokorný et al., 2010).

24.3.4 Role of Forests in the Hydrologic Cycle

There is great controversy concerning the functioning of forests in the hydrologic cycle. The water transpired by plants is often considered a loss. Transpiration is sometimes even called an unavoidable evil, in the sense that water is sacrificed for the sake of enabling intake of CO_2 for photosynthesis. When comparing relatively small catchments, less rainfall is converted to runoff from afforested catchments than from grass covered or partly drained catchments. This has been proved many times in hydrological studies of paired-watershed experiments (Andréassian, 2004). It has been shown that high water use of rapidly growing forest plantations as well as young seedlings in new forest may negatively impact on water resources (Calder, 2007; van Dijk and Keenan, 2007). On the other hand, large-scale deforestations in history were linked with less precipitation and regional shortage of water (Diamond, 2005; Ponting, 1993).

Makarieva and Gorshkov (2007) point out that, in forest-covered regions, annual precipitation does not decline with increasing distance from the ocean and may even grow as one proceeds several thousand kilometres inland. In contrast, where forests are lacking, precipitation decreases exponentially over just a few hundred kilometres. They termed the biotically induced atmospheric circulation sustaining the hydrologic cycle on land the 'biotic pump' of atmospheric moisture (Makarieva and Gorshkov, 2007), which has been recently supported by new evidence (Makarieva et al., 2013). However, there is still not a clear consensus. For example, Angelini et al. (2011) concluded that changes in precipitation over continental reaches are a product of complex processes and only partly influenced, but not controlled, by local water sources or vegetation. Makarieva and Gorshkov (2007, 2013) also derived from their biotic pump study that deforestation practises would stall this cycle of moisture transport. Their idea led to a large discussion between two opposite camps which lasts till now: the first view states that additional forest cover will reduce and forest removal will raise downstream water availability, while the second group of authors argue the opposite: planting additional forests should raise downstream water availability and intensify the hydrologic cycle (Ellison et al., 2012).

The impact of large-scale deforestation on climate regulation mediated by changes in the hydrologic cycle can be demonstrated in a case study in the Mau Forest complex, located at 1200–2600 m a.s.l., which is referred to as one of the largest remaining continuous blocks of indigenous forest in Eastern

Africa. With high annual precipitation (reaching about 1000 mm on eastern slopes and more than 2000 mm on western ones), it is an area which includes the headwaters of many rivers feeding into the Rift Valley lakes. In the past 25 years, the site has been subjected to extensive deforestation: the forest cover of 5200 km^2 in 1986 was reduced to a mere 3400 km^2 in 2009. The availability of satellite images since the 1980s has enabled to demonstrate the effect of deforestation on temperature distribution over the whole area. Extreme increases in surface temperature (by more than 20 °C) have been observed on sites of deforestation. Its consequences are also evident in the Rift Valley region, between lakes Nakuru and Naivasha. Areas that have been converted into fast-growing plantation forest show the opposite trend, that is, temperature damping. During several years, reduction of precipitation has resulted in a decrease of water flow rates in rivers and the shortage of water in the region has negatively affected the living conditions in the region. The Kenyan government decided to evict 200,000 people from the deforested area in order to restore the forest and hydrology (Hesslerová and Pokorný, 2010).

24.4 ADAPTIVE GOVERNANCE AND COMMUNICATION TO THE PUBLIC TOWARDS SUSTAINABLE FOREST—MULTI-STAKEHOLDER COLLABORATION

The past, present and futures of human and biophysical systems are closely and intricately interconnected. All human history is a history of the use of nature and in the modern period of the industrial revolution, also a history of thermodynamically inefficient anthropogenic natural landscape changes, driven mainly by deforestations and air pollution, which currently endanger the global sustainability of numerous life-supporting ES.

The term *governance* is a fancy ‘post-modern’ word describing the process of decision-making and how these decisions are implemented (or not implemented). Governance can be used in several contexts (mainly corporate or public) and can cover different scales, starting at the local and regional levels, but also occurring on national and international or global levels. The term *adaptive* means flexible, being able to accommodate to new conditions.

Adaptive governance is a collaborative, flexible and learning-based issue management of natural ecosystems and natural vegetation across different scales while satisfying human needs of the current generation, which is related to economic and social goals (Folke et al., 2005). Natural ecosystem management is made adaptable by monitoring and research based on our best understanding of the ecological interactions and processes necessary to sustain ecosystem composition, structure and function (Christensen et al., 1996), that is, to sustain the quantity and quality of the life-supporting natural ES.

Three current approaches are helping to create a foundation for adaptive governance: integrated assessment, adaptive management and adaptive policy-making. *Integrated assessment* takes into account all the main ecological, economic and social aspects and their trade-offs, comparing economic and ecosystem service values at local, regional and national frameworks. Scaling levels are important also for stakeholder participation in protecting and restoring natural vegetations. The majority of ES have a multi-scale character and as such they require the adequate participation of the relevant stakeholders (Holling, 1978).

Adaptive management is a systematic process for continually improving management policies and practises by learning from the outcomes of operational programmes (Berkes et al., 2000; Habron, 2003). It is an important concept in the area of managing natural resources, especially forests, and their complex and dynamic ecosystems, as there are uncertainties and unpredictabilities in achieving the final goals intrinsic to all ecosystems (Bunnell and Dunsworth, 2009; Lee, 1993; Walters, 1986).

If of the total area of 4 billion (10^9) hectares of world forests, covering 31% of the continents (FRA, 2010), approximately 86% are publicly owned, then these are mainly public decision-making processes and it is those public institutions that may change the current negative trends; the overall rate of deforestation has remained alarmingly high (Turner et al., 2012). However, in Western and Central Europe (EU space), about 50% of forests are privately owned (Schmithüsen and Hirsch, 2010).

Recognising the fact that natural multi-strata forests are the best and most efficient producers and suppliers of supporting and regulating ES and air pollution absorbers is a new and large challenge for all the main stakeholders in the field of caring for nature and landscapes (forest owners, EU representatives and executives, nature and landscape managers, politicians, NGOs) as well as those implementing the basic assumptions for sustainable living (EU and national governments, politicians, public initiatives, environmental movements, NGOs, environmentally aware land developers).

A relevant tool for structuring communication between scientists and National Protection Authorities and the remaining stakeholders (landowners, state administration, municipality, resident population, recreational visitors, hunters and fishers) seems to be the indicator set of the driving forces, pressures, state, impact, response framework (Niemeier and de Groot, 2008). It enables quantifying the pattern and rate of human driving forces, pressures (above all air pollution/climate-change affects and deforestation), impacts on ecosystems and agro-ecosystems, and human society responses to the decreasing ability of forest–agriculture landscapes to supply multiple ES. Another possibility how to downscale environmental change issues at a local scale and make understandable and effective communication between science and public/target groups is 'participatory research' (Knock, 2011).

When assessing the integrated future economic and ecological costs and benefits of restoring potential natural vegetation, the existing uncertainties in ecosystem functioning can be revealed by testing several alternative ways of achieving multi-strata natural mixed forests, for example, within reforestation programmes. One such programme is the 'native forests by native trees' which includes about 550 locations throughout Japan, Southeast Asia, South America and China (Miyawaki, 1999).

Experiences over the past decades show that progress towards sustainable development must be navigated through processes of learning and adaptation (Nijnik and Miller, 2013, this vol.). It requires a new approach to public management that helps people, all citizen initiatives, environmental movements and policy-makers to design policies that embrace the uncertainties and complexities of our mutually related economic, social and environmental systems (Seják et al., 2012).

24.5 EVALUATION OF SELECTED ECOSYSTEM SERVICES ON THE BASIS OF MONITORED ENERGY, WATER AND MATERIAL FLOWS ESTIMATION: CASE STUDY IN THE FOREST–AGRICULTURAL LANDSCAPE OF THE CZECH REPUBLIC

In recent years, many case studies have been carried out to quantify or value selected forest ES. A significant part of them has been devoted to production of timber and non-timber goods falling into the category of provisioning services evaluated by market prices, while quite many studies have examined cultural functions of forests using contingent valuation or market-derived valuation such as travel cost (Deal et al., 2012). We focused mostly on regulating services, because, despite of their great importance, they are rather underestimated services by human society.

Odum (1971) definitively linked energy flow to succession. He showed that the self-organised directional succession processes culminate in ecosystems in which maximum biomass and symbiotic function between organisms are maintained per unit of energy flow. Self-organising processes in autotrophic terrestrial ecosystems tend towards climax vegetation that is characterised by maximum efficiency in solar energy use and maximum ability to produce life-supporting conditions, keeping nutrients and water inside the ecosystem (Ripl, 2003). In other words, natural vegetation (deciduous and mixed forests, and wetlands in central Europe) as the long-term results of self-organised natural processes are the most efficient biotopes to sustain the climate and chemistry of planet Earth (Lovelock, 2007).

In the Czech Republic, two new methods of systemic valuation of all land-cover categories have been developed. The Biotope Valuation Method (BVM), originally elaborated in the Hessian state of Germany and recommended for dissemination by the EU White Paper on Environmental Liability (2000), estimated monetary values of a complete list of national biotopes (habitats) as specific environments for particular living species (Seják et al., 2003). This method was established primarily to estimate ecological and biodiversity losses when degradation of natural or semi-natural vegetation cover into some form of anthropogenic land use occurred. The value *numéraire* was derived as an average cost per one point increase for 1 m^2 of individual biotope types computed from 150 national restoration projects. The scale of values ranged from 0 up to €40/m^2.

In the energy–water–vegetation method, the value *numéraire* was derived from the cost-replacement method, which is the minimum cost of a technological substitution of a natural service (Seják et al., 2010, 2012). The main objection of economists against the cost-replacement method is that this is a measure of cost rather than benefit. In order for a replacement cost to be a valid measure of the value of an ES, it must be the case that a human-engineered solution provides an equivalent quantity of the service and total willingness-to-pay for this service exceeds the cost of providing the services via the human-engineered solution (Hastings and Gross, 2012).

ES stock values, estimated by Seják et al. (2010), started at zero in cases of completely anthropogenised lands; however, in natural and semi-natural ecosystems, the annual values reached even above €3000 per m^2. The pilot estimation of four measured annual ecosystem service values in the Czech Republic, based on replacement cost methods, assessed that annual ES surpass the annual GDP (GDP CR 2011 = €154.9 billion) by at least 50 times.

All potential land covers of the national territory of the Czech Republic have been divided into 22 groups of biotopes (Table 24.1) according to their efficiency in the use of incoming solar energy and for keeping nutrients and water inside the ecosystem (Ripl, 2003). Efficiency was defined by the ratio of effective and ineffective use of incoming solar energy. By efficiently used incoming energy, we understand that part of energy is used in ET (which cool local temperatures and by condensing water vapour in colder environments, they warm it) and photosynthesis processes of autotrophic green ecosystems (in both cases solar energy dissipates—moves energy through a system). By inefficiently used energy, we understand that part of the incoming energy is either converted to heat or directly reflected back into space by albedo.

An integrated assessment was fulfilled using the EWV method which reflects four main ecosystem functions and services, related to the efficiency of solar energy dissipation (and destruction) processes. By transferring the 22 biotope groups into the land-cover categories of the Corine Land-Cover mapping, a table (Table 24.2) was created that allowed for comparing the economic and ecological benefits of individual land-cover items. This national land-cover table creates a basis for integrated economic and ecological assessments of any land cover or land-use changes.

To be able to predict forest ES provisions under climate change and air pollution, the most important ES providers could be identified using suitable effect traits for selected ecosystem functions (Kremen, 2005). Based on predicted environmental changes, the response traits of the main ES providers to abiotic and biotic factors could be assessed (Luck et al., 2009). If the prognosis would be unfavourable for their survival, it would then be necessary to estimate the substitute ES providers using response diversity and redundancy assessments and then support them by suitable management (Petchey and Gaston, 2006).

TABLE 24.1 Distribution of Biotope (Habitat) Types in the Czech Republic into 22 Biotope Groups According to Their Provision of Four ES

			Ecosystem services (€/m^2/year)				Total of four ES	
No.	**Biotope groups**	**Area (km^2)**	*Clim. s.*	*SWC*	*O_2 prod.*	*BD*	*Relative value (€/m^{-2}/year^{-1})*	*Total value billion (€/year)*
1	Water bodies, courses	675	67	57	25	0.5	150	101
2	Peatbogs	23	90	74	3	1.5	168	4
3	Other wetlands	364	90	74	30	1	195	71
4	Extensively used mesic pastures meadows	2601	67	34	16	1.2	118	308
5	Intensively used mesic pastures meadows	5579	56	34	21	0.3	111	623
6	Degraded mesic pastures meadows	4609	45	20	12	0.3	77	355
7	Dry closed grasslands	40	45	11	11	1.2	68	3
8	Dry interspaced grasslands	172	33	9	6	1.2	49	9
9	Xeric scrub	426	45	17	12	0.8	75	32
10	Mesic scrub	1959	56	34	16	0.8	107	209
11	Alluvial hygrophilous scrub	17	67	55	17	1.1	140	2
12	Dry pine forests	298	45	26	13	1.2	85	25
13	Other conifer forests	6050	56	46	23	1	126	761
14	Damaged conifer forests	8222	45	34	19	0.5	98	807

15	Leafy forests	6636	78	69	27	1.4	175	1160
16	Leafy forests degraded	1632	56	40	19	0.6	115	189
17	Alluvial flooded forests	924	90	80	30	1.5	201	186
18	Solitary trees, alleys	1276	56	34	21	0.6	112	143
19	Arable land: cereal and root-crops	27,605	33	9	13	0.2	55	1541
20	Arable land: fodder and durable stands	141	45	20	30	0.2	95	13
21	Areas without vegetation	2938	11	3	0	0	14	41
22	Rock biotopes	113	23	11	3	1.2	38	4
23	Other natural, semi-natural biotopes	3780	66	50	22	1	140	528
24	Other anthropically influenced biotopes	2787	38	17	14	0.3	70	196
	Czech Republic total	78,869						7310

Clim. s. = climate-regulation service, expressed by litres of evapotranspired and condensed water, double air-conditioning effect (evapotranspiration and cooling effect, condensation and warming effect, both latent heat changes of 1 litre of water = 1.4 kWh); $l\ m^{-2}\ year^{-1} \times$ €0.08 (electricity cost price).
SWC = water retention service of the short water cycle; $l\ m^{-2}\ year^{-1} \times$ €0.114 (cost price of 1 litre of distilled water).
O_2 production = O_2 ($kg\ m^{-2}\ year^{-1}$) × 700 (kg changed to litres) × €0.02 (cost price of 1 litre of oxygen).
BD = habitat provision service (valued by biotope valuation method; Seják et al., 2003).
Source: Seják et al. (2010); Exchange rate: €1 = CZK 25

TABLE 24.2 Biotope Capital Values, Ecosystem Service (ES) Annual Values, Ecosystem Service Capital Values and Economic Capital Values of 1 m^2 of the Czech Territories in €[a]

Land cover 1:100,000	Biotope values	Annual ES values	ES capital values	Official prices	Notes
1.1.1. Continuous urban fabric	0–1.20	27	535	1.4–90	According to urban size
1.1.2. Discontinuous urban fabric	5.04	78	1557	1.4–90	According to urban size
1.2.1. Industrial or commercial units	0–1.32	32	638	1.4–90	According to urban size
1.2.2. Road and rail networks and associated land	4.00	58	1156	1.4–90	According to urban size
1.2.3. Port areas	3.92	70	1398	1.4–90	According to urban size
1.2.4. Airports	5.92	80	1591	1.4–90	According to urban size
1.3.1. Mineral extraction sites	6.64	43	864	1.4–90	According to urban size
1.3.2. Dump sites	3.88	99	1981	0.04	
1.3.3. Construction sites	3.52	42	844	1.4–90	According to urban size
1.4.1. Green urban areas	9.52	106	2127	1.4–33	
1.4.2. Sport and leisure facilities	9.28	79	1589	0.4–0.6	
2.1.1. Non-irrigated arable land	5.12	62	1242	0.04–0.7	According to soil quality
2.2.1. Vineyards	7.52	88	1769	0.04–6.4	
2.2.2. Fruit trees and berry plantations	7.00	88	1764	0.04–4	
2.3.1. Pastures	10.28	102	2050	0.04–0.4	Annual ES €75 m^{-2}

2.4.2. Complex cultivation	6.96	85	1696	0.04–0.4	According to soil quality
2.4.3. Land with agricultural and natural vegetation	10.64	100	1996	0.04–0.4	According to soil quality
3.1.1. Broad-leaved forest	20.12	156	3118	0.1–4.4	
3.1.2. Coniferous forest	12.96	124	2490	0.1–4.4	
3.1.3. Mixed forest	14.08	131	2616	0.1–4.4	
3.2.1. Natural grassland	16.32	109	2177	0.04	
3.2.2. Moors and heathland	26.20	129	2576	0.04	
3.2.4. Transitional woodland shrub	11.64	106	2128	0.04	
3.3.2. Bare rock	19.68	107	2144	0.04	
4.1.1. Inland marshes	16.56	159	3174	0.04	
4.1.2.Peatbogs	26.36	168	3361	0.04	
5.1.1. Water courses	11.44	139	2776	0.3	
5.1.2. Water bodies	9.24	148	2962	0.3	

[a]*Exchange rate: €1 = CZK 25.*

Biotope capital values estimated by BVM, ecosystem services capital value by a 5% discount rate, economic capital values ordered by the Czech Ministry of Finance in Decree no. 3/2008.

Source: Seják et al. (2010).

24.6 CONCLUSIONS

The concept of ES (MEA, 2005) broadens the framework of biodiversity and ecosystem functioning research to include the entire human–environment system. Under environmental change conditions, changing drivers will alter landscape heterogeneity, reduce the area of productive ecosystems and likely increase tension among competing ES (Johnstone et al., 2010). Therefore, we need to understand how concurrent changes in climate, disturbance regimes and land use will affect the sustainability and resilience of forested landscapes and their ability to provide ecosystem services (Turner et al., 2012).

Future research should focus on multi-factorial anthropogenic and natural interactions of climatic changes and air pollution, in particular, elevated ozone, altered nitrogen, carbon and water availability on biodiversity, water, nutrient and carbon cycling, and related ecosystem functions. Holistic approaches are available for upscaling information from the laboratory to the field, and from genes to landscapes. An important improvement in our understanding might be obtained by a combination of long-term multi-factorial experiments with ecosystem-level monitoring (Mikkelsen et al., 2013, this vol.) and subsequent integrative ecosystem modelling (Serengil et al., 2011). One of the key issues for forest ES under interaction of climate change with increasing air pollution will remain monitoring of carbon sequestration (Marek et al., 2011) and quantification of efficiency of solar flux dissipation by means of vegetation and water, through ET and photosynthesis (Seják et al., 2012). Quantifying the reduction in ecosystem functions, connected with forest damage caused by the synergistic impacts of climate change and air pollution, is needed to predict the loss of forest ES in the near future (Cudlín et al., 2011; Knoke and Hahn, 2013, this vol.).

As the loss of forest ES is mainly due to the conversion of forests to agricultural land, paying farmers for the environmental services they may conserve or provide is generating growing interest worldwide (Chiabai et al., 2011). Sustainable forest management is essential for reducing the vulnerability of forests to climate change. To meet the challenges of adaptation, commitment to achieving the goals of sustainable forest management must be strengthened at both the international and national levels.

New modes of governance are required that enable meaningful stakeholder participation and provide secure land tenure, forest user rights and sufficient financial incentives (Seppälä et al., 2009; Nijnik and Miller, 2013, this vol.). Land-use policy and regulations have an important role for establishing markets for ES and market-based programmes have been developed to promote 'bundling' ES, improving more than one service at a time (Deal et al., 2012).

ACKNOWLEDGEMENTS

This work was partly supported by projects of the Ministry of Education, Youth and Sports of the Czech Republic CZ.1.07/2.4.00/31.0056 and CzechGlobe (CZ.1.05/1.1.00/02.0073).

REFERENCES

Andréassian, V., 2004. Waters and forests: from historical controversy to scientific debate. J. Hydrol. 291, 1–27.

Angelini, I.M., Garstang, M., Davis, R.E., Hayden, B., Fitzjarrald, D.R., Legates, D.R., Greco, S., Macko, S., Connors, V., 2011. On the coupling between vegetation and the atmosphere. Theor. Appl. Climatol. 105, 243–261.

Bala, G., Caldeira, K., Wickett, M., Phillips, T.J., Lobell, D.B., Delire, C., Mirin, A., 2007. Combined climate and carbon-cycle effects of large-scale deforestation. Proc. Natl. Acad. Sci. U.S.A. 104, 6550–6555.

Barton, C.V.M., Duursma, R.A., Medlyn, B.E., Ellsworth, D.S., Eamus, D., Tissue, D.T., Adams, M.A., Conroy, J., Crous, K.Y., Liberloo, M., Low, M., Linder, S., McMurtrie, R.E., Ross, E., 2012. Effects of elevated atmospheric [CO_2] on instantaneous transpiration efficiency at leaf and canopy scales in *Eucalyptus saligna*. Glob. Chang. Biol. 18, 585–595.

Bastian, O., Haase, D., Grunewald, K., 2012. Ecosystem properties, potentials and services—the EPPS conceptual framework and an urban application example. Ecol. Indic. 21, 7–16.

Bauerle, W.L., Oren, R., Way, D.A., Qian, S.S., Stoy, P.C., Thornton, P.E., Bowden, J.D., Hoffman, F.M., Reynolds, R.F., 2012. Photoperiodic regulation of the seasonal pattern of photosynthetic capacity and the implications for carbon cycling. Proc. Natl. Acad. Sci. U.S.A. 109, 8612–8617.

Berkes, F., Colding, J., Folke, C., 2000. Rediscovery of traditional ecological knowledge as adaptive management. Ecol. Appl. 10, 1251–1262.

Bonan, G.B., 2008. Forests and climate change: forcings, feedbacks, and the climate benefits of forests. Science 13, 1444–1449. http://dx.doi.org/10.1126/science.1155121.

Boyd, J., Banzhaf, S., 2007. What are ecosystem services? The need for standardized environmental accounting units. Ecol. Econ. 63, 616–626.

Bunnel, F.L., Dunsworth, G.B., 2009. Forestry and Biodiversity: Learning How to Sustain Biodiversity in Managed Forests. UBC Press, Vancouver, BC.

Burkhard, B., Kroll, F., Nedkov, S., Müller, F., 2012. Mapping ecosystem service supply, demand and budgets. Ecol. Indic. 21, 17–29.

Butterbach-Bahl, K., Gundersen, P., 2011. Nitrogen processes in terrestrial ecosystems. In: Sutton, M.A., Howard, C.M., Erisman, J.W., Billen, G., Bleeker, A., Grennfelt, P., van Grinsven, H., Grizzetti, B. (Eds.), The European Nitrogen Assessment: Sources, Effects, and Policy Perspectives. Cambridge University Press, New York, pp. 99–125.

Bytnerowicz, A., Omasa, K., Paoletti, E., 2007. Integrated effects of air pollution and climate on forests: a northern hemisphere perspective. Environ. Pollut. 147, 438–445.

Bytnerowicz, A., Fenn, M., McNulty, S., Yuan, F., Pourmokhtarian, A., Driscoll, C., Meixner, T., 2013. Interactive effects of air pollution and climate change on forest ecosystems in the United States – current understanding and future scenarios. In: Matyssek, R., Clarke, N., Cudlin, P., Mikkelsen, T., Tuovinen, J.P., Weiser, G., Paoletti, E. (Eds.), Climate Change, Air Pollution and Global Challenges: Understanding and Solutions from Forest Research. Dev. Environ. Sci. 13. Elsevier, Netherlands, pp. 333–369.

Calder, I.R., 2007. Forest and water—ensuring forest benefits outweigh water costs. For. Ecol. Manage. 251, 110–120.

Carson, R.T., 2011. Contingent Valuation, a Comprehensive Bibliography and History. Edward Elgar Publ. Ltd., Cheltenham, 455 pp.

Chiabai, A., Travisi, C.M., Markandya, A., Ding, H., Nunes, P., 2011. Economic assessment of forest ecosystem services losses: cost of policy inaction. Environ. Resour. Econ. 50, 405–445.

Christensen, N.L., Bartuska, A.M., Brown, J.H., Carpenter, S., D'Antonio, C., Francis, R., Franklin, J.F., MacMahon, J.A., Noss, R.F., Parsons, D.J., Peterson, C.H., Turner, M.G., Woodmansee, R.G., 1996. The report of the Ecological Society of America committee on the scientific basis for ecosystem management. Ecol. Appl. 6, 665–691.

Ciais, P., Schelhaas, M.J., Zaehle, S., Piao, S.L., Cescatti, A., Liski, J., Luyssaert, S., Le-Maire, G., Schulze, E.D., Bouriaud, O., Freibauer, A., Valentini, R., Nabuurs, G.J., 2008. Carbon accumulation in European forests. Nat. Geosci. 1, 425–429.

Costanza, R., d'Arge, R., de Groot, R., Farber, S., Grasso, M., Hannon, B., Naeem, S., Limburg, K., Paruelo, J., O'Neil, R.V., Raskin, R., Sutton, P., van den Belt, M., 1997. The value of the world's ecosystem services and natural capital. Nature 387, 253–260.

Cudlín, P., Seják, J., Pokorný, J., 2011. Valuation of ecosystem services towards ecological citizenship. In: Westra, L., Bosselmann, K., Soskolne, C. (Eds.), Globalisation and Ecological Integrity in Science and International Law. Cambridge Scholars Publishing, Cambridge, pp. 26–40.

De Groot, R.S., van der Meer, P.J., 2010. Quantifying and valuing ecosystem services provided by plantation forests. In: Bauhaus, J., van Der Meer, P., Kanninen, M. (Eds.), Ecosystem Goods and Services from Plantation Forests. Earthscan, London and Washington, 254 pp.

De Vries, W., Solberg, S., Dobbertin, M., Sterba, H., Laubhann, D., van Oijen, M., Evans, C., Gundersen, P., Kros, J., Wamelink, G.W.W., Reinds, G.J., Sutton, M.A., 2009. The impact of nitrogen deposition on carbon sequestration by terrestrial ecosystems. For. Ecol. Manage. 258, 1814–1823.

Deal, C.R., Cochran, B., LaRocco, G., 2012. Bundling of ecosystem services to increase forest land value and enhance sustainable forest management. Forest Pol. Econ. 17, 69–76.

Diamond, J., 2005. Collapse: How Societies Choose to Fail or Survive. Viking Books, New York.

Dizengremel, P., Jolivet, Y., Tuzet, A., Ranieri, A., Le Thiec, D., 2013. Integrative leaf level phytotoxic ozone dose assessment for forest risk modeling. In: Matyssek, R., Clarke, N., Cudlin, P., Mikkelsen, T., Tuovinen, J.P., Weiser, G., Paoletti, E. (Eds.), Climate Change, Air Pollution and Global Challenges: Understanding and Solutions from Forest Research. Dev. Environ. Sci. 13. Elsevier, Netherlands, pp. 267–288.

Eiseltová, M., Pokorný, J., Hesslerová, P., Ripl, W., 2012. Evapotranspiration—a driving force in landscape sustainability. In: Irmak, A. (Ed.), Evapotranspiration—Remote Sensing and Modelling, 18. InTech, Rijeka, Croatia, pp. 305–328, Technical Chapter.

Ellison, D., Futter, M.N., Bishop, K., 2012. On the forest cover-water yield debate: from demand- to supply-side thinking. Glob. Chang. Biol. 18, 806–820.

Felzer, B., Kicklighter, D.W., Melillo, J.M., Wang, C., Zhuang, Q., Prinn, R., 2004. Effects of ozone on net primary production and carbon sequestration in the conterminous United States using biochemistry model. Tellus B 56, 230–248.

Fischer, B., Turner, R.K., 2008. Ecosystem services: classification for valuation. Biol. Conserv. 141, 1167–1169.

Fischer, B., Turner, R.K., Morling, P., 2009. Defining and classifying ecosystem services for decision making. Ecol. Econ. 68, 643–653.

Folke, C., Hahn, T., Olsson, P., Norbert, J., 2005. Adaptive governance of social-ecological systems. Annu. Rev. Env. Res. 30, 441–473.

FRA, 2010. Global forest resources assessment, FAO. http://www.fao.org/forestry/fra/fra2010/en/.

Gedney, N., Cox, P.M., Betts, R.A., Boucher, O., Huntingford, C., Stott, P.A., 2006. Detection of a direct carbon dioxide effect in continental river runoff records. Nature 439, 835–838.

Grunewald, K., Bastian, O., 2010. Ökosystemdienstleistungen analysieren—begrifflicher und konzeptioneller Rahmen aus landschaftsökologischer Sicht. GEOÖKO 31, 50–82.

Habron, G., 2003. Role of adaptive management for watershed councils. Environ. Manage. 31, 29–41.

Haines-Young, R.H., Potschin, M.B., 2009. Methodologies for defining and assessing ecosystem services. Final Report, JNCC, Project Code C08-0170-0062, 69 p.

Haines-Young, R.H., Potschin, M., Kienast, F., 2012. Indicators of ecosystem service potential at European scales: mapping marginal changes and trade-offs. Ecol. Indic. 21, 39–53.

Hastings, A., Gross, L.J., 2012. Encyklopedia of Theoretical Ecology. University of California Press, Berkeley.

Hesslerová, P., Pokorný, J., 2010. Forest clearing, water loss and land surface heating as development costs. Int. J. Water 5, 401–418.

Hesslerová, P., Pokorný, J., Brom, J., Rejšková-Procházková, A., 2013. Daily dynamics of radiation surface temperature of different land cover types in a temperate cultural landscape: consequences for the local climate. Ecol. Eng. 54, 145–154.

Holling, C.S., 1978. Adaptive Environmental Assessment and Management. Wiley, London, Reprinted by Blackburn Press, 2005.

Intergovernmental Panel on Climate Change (IPCC), 2001. Climate Change 2001: Impacts, Adaptation, and Vulnerability. Cambridge University Press, Cambridge, 556–560.

Johnstone, J.F., Hollingsworth, T.N., Chapin, F.S., Mack, M.C., 2010. Changes in fire regime break the legacy lock on successional trajectories in Alaskan boreal forest. Glob. Chang. Biol. 16, 1281–1295.

Katul, G.G., Oren, R., Manzoni, S., Higgins, C., Parlange, M.B., 2012. Evapotranspiration: a process driving mass transport and energy exchange in the soil-plant-atmosphere-climate system. Rev. Geophys. 50, 25 pp.

Knock, N., 2011. Action research: its nature and relationship to human-computer interaction. In: Soegaard, M., Dam, R.F. (Eds.), Encyclopedia of Human-Computer Interaction. The Interaction-Design.org Foundation, Aarhus, Denmark. http://www.interaction-design.org/encyclopedia/action_research.html.

Knoke, T., Hahn, A., 2013. Global change and the role of forests in future land-use systems. In: Matyssek, R., Clarke, N., Cudlin, P., Mikkelsen, T., Tuovinen, J.-P., Weiser, G., Paoletti, E. (Eds.), Climate Change, Air Pollution and Global Challenges: Understanding and Solutions from Forest Research. Dev. Environ. Sci. 13. Elsevier, Netherlands, pp. 569–588.

Kremen, C., 2005. Managing ecosystem services: what do we need to know about their ecology? Ecol. Lett. 2005 (8), 468–479.

Lee, K., 1993. Compass and Gyroscope: Integrating Science and Politics for the Environment. Island Press, Washington, D.C.

Lovelock, J., 2007. The Revenge of Gaia, Why the Earth is Fighting Back—And How We Can Still Save Humanity. Penguin Books Ltd. London, ISBN: 978-0-141-02990-0.

Lowman, M.D., Schowalter, T.D., 2012. Plant science in forest canopies—the first 30 years of advances and challenges (1980-2010). New Phytol. 194, 12–27.

Luck, G.W., Harrington, R., Harrison, P.A., Kremen, C., Berry, P.M., Bugter, R., Zobel, M., 2009. Quantifying the contribution of organisms to the provision of ecosystem services. Bioscience 59, 223–235.

Makarieva, A.M., Gorshkov, V.G., 2007. Biotic pump of atmospheric moisture as driver of the hydrological cycle on land. Hydrol. Earth Syst. Sci. 11, 1013–1033.

Makarieva, A.M., Gorshkov, V.G., 2013. Revisiting forest impact on atmospheric water vapor transport and precipitation. Theor. Appl. Climatol. 111, 79–96.

Makarieva, A.M., Gorshkov, V.G., Sheil, D., Nobre, A.D., Li, B.L., 2013. Where do winds come from? A new theory on how water vapor condensation influences atmospheric pressure and dynamics. Atmos. Chem. Phys. 13, 1039–1056.

Marek, M.V., Janous, D., Toufarova, K., Havrankova, K., Pavelka, M., Kaplan, V., Markova, I., 2011. Carbon exchange between ecosystems and atmosphere in the Czech Republic is affected by climate factors. Environ. Pollut. 159, 1035–1039.

Matyssek, R., Sandermann Jr., H., 2003. Impact of ozone on trees: an ecophysiological perspective. In: Progress in Botany64, Springer Verlag, Heidelberg, pp. 349–404.

Matyssek, R., Sandermann, H., Wieser, G., Booker, F., Cieslik, S., Musselman, R., Ernst, D., 2008. The challenge of making ozone risk assessment for forest trees more mechanistic. Environ. Pollut. 156, 567–582.

Matyssek, R., Wieser, G., Patzner, K., Blaschke, H., Haberle, K.H., 2009. Transpiration of forest trees and stands at different altitude: consistencies rather than contrasts? Eur. J. Forest Res. 128, 579–596.

Matyssek, R., Schaub, M., Wieser, G., 2010. Air pollution and climate change effects on forest ecosystems: new evidence. Eur. J. Forest Res. 129, 417–419.

Matyssek, R., Wieser, G., Fleischmann, F., Grunhage, L., 2013. Ozone research, quo vadis? Lessons from the Kranzberg forest experiment. In: Matyssek, R., Clarke, N., Cudlin, P., Mikkelsen, T., Tuovinen, J.-P., Weiser, G., Paoletti, E. (Eds.), Climate Change, Air Pollution and Global Challenges: Understanding and Solutions from Forest Research. Dev. Environ. Sci. 13. Elsevier, Netherlands, pp. 103–129.

MEA, 2005. Ecosystems and Human Well-Being: Synthesis. Island Press, Washington, DC. http://www.millenniumassessment.org/documents/document.356.aspx.pdf.

Mikkelsen, T.N., Clarke, N., Danielewska, A., Fischer, R., 2013. Towards supersites in forest ecosystem monitoring and research. In: Matyssek, R., Clarke, N., Cudlin, P., Mikkelsen, T., Tuovinen, J.P., Weiser, G., Paoletti, E. (Eds.), Climate Change, Air Pollution and Global Challenges: Understanding and Solutions from Forest Research. Dev. Environ. Sci. 13. Elsevier, Netherlands, pp. 475–496.

Miyawaki, A., 1999. Creative ecology: restoration of native forests by native trees. Plant Biotechnol. 16, 15–25.

Monteith, J.L., Unsworth, M., 1990. Principles of Environmental Physics, 2nd ed. Butterworth-Heinemann, Oxford.

Nabuurs, G.J., Pussinen, A., Karjalainen, T., Erhard, M., Kramer, K., 2002. Stemwood volume increment changes in European forests due to climate change—a simulation study with the EFISCEN model. Glob. Chang. Biol. 8, 304–316.

Neumann, R.B., Cardon, Z.G., 2012. The magnitude of hydraulic redistribution by plant roots: a review and synthesis of empirical and modeling studies. New Phytol. 194, 337–352.

Niemeijer, D., de Groot, R.S., 2008. A conceptual framework for selecting environmental indicator sets. Ecol. Indic. 8, 14–25.

Niinemets, Ü., 2010. Responses of forest trees to single and multiple environmental stresses from seedlings to mature plants: past stress history, stress interactions, tolerance and acclimation. For. Ecol. Manage. 260, 1623–1639.

Nijnik, M., Miller, D., 2013. Targeting sustainable provision of forest ecosystem services with special focus on carbon sequestration. In: Matyssek, R., Clarke, N., Cudlin, P., Mikkelsen, T., Tuovinen, J.-P., Weiser, G., Paoletti, E. (Eds.), Climate Change, Air Pollution and Global Challenges: Understanding and Solutions from Forest Research. Dev. Environ. Sci. 13. Elsevier, Netherlands, pp. 547–568.

Odum, E.P., 1971. Fundamentals of Ecology. W.B. Saunders Company, Philadelphia.

Paoletti, E., Schaub, M., Matyssek, R., Wieser, G., Augustaitis, A., Bastrup-Birk, A.M., Bytnerowicz, A., Günthardt-Georg, M.S., Müller-Starck, G., Serengil, Y., 2010. Advances of air pollution sciences: from forest decline to multiple stress effects on forest ecosystem services. Environ. Pollut. 158, 1896–1989.

Parry, M.L., Canziani OF, , Palutikof, J.P., van der Linden, P.J., Hansosn, C.E. (Eds.), 2007. Climate change 2007: Impacts, Adaptation and Vulnerability. Contribution of Working Group II to the Fourth Assessment Report of the Intergovernmental Panel on Climate Change. Cambridge University Press, Cambridge, 1000 pp.

Percy, K., Bucher, J., Cape, J., Ferretti, M., Heath, R., Jones, H.E., Karnosky, D., Matyssek, R., Muller-Starck, G., Paoletti, E., Rosengren-Brinck, U., Sheppard, L., Skelly, J., Weetman, G., 1999. State of science and knowledge gaps with respect to air pollution impacts on forests. Water Air Soil Pollut. 116, 443–448.

Petchey, O.L., Gaston, K.J., 2006. Functional diversity: ack to basics and looking forward. Ecol. Lett. 2006 (9), 741–758.

Pielke Sr., R.A., 2005. Land use and climate change. Science 5754, 177.

Pielke Sr., R.A., Pitman, A., Niyogi, D., Mahmood, R., McAlpine, C., Hossain, F., Goldewijk, K.K., Nair, U., Betts, R., Fall, S., Reichstein, M., Kabat, P., de Noblet, N., 2011. Land use/land cover changes and climate: modeling analysis and observation evidence. WIREs Clim. Change 2011 (2), 828–850.

Pokorný, J., Brom, J., Čermák, J., Hesslerová, P., Hurynna, H., Nadyezhdina, N., Rejšková, A., 2010. Solar energy dissipation and temperature control by water and plants. Int. J. Water 5, 311–336.

Ponting, C., 1993. A Green History of the World. Pinguin Books Ltd., London.

Prieto, I., Armas, C., Pugnaire, F.I., 2012. Water release through plant roots: new insights into its consequences at the plant and ecosystem level. New Phytol. 193, 830–841.

Randerson, J.T., Liu, H., Flanner, M.G., Chambers, S.D., Jin, Y., Hess, P.G., Pfister, G., Mack, M.C., Treseder, K.K., Welp, L.R., Chapin, F.S., Harden, J.W., Goulden, M.L., Lyons, E., Neff, J.C., Schuur, E.A., Zender, C.S., 2006. The impact of boreal forest fire on climate warming. Science 314, 1130–1132.

Ripl, W., 1995. Management of water cycle and energy flow for ecosystem control: the energy-transport-reaction (ETR) model. Ecol. Model. 78, 61–76.

Ripl, W., 2003. Water: the bloodstream of the biosphere. Phil. Trans. Roy. Soc. Lond. B Biol. Sci. 358, 1921–1934.

Schmithüsen, F., Hirsch, F., 2010. Private forest ownership in Europe. Geneva Timber and Forest Study Paper, 26. www.unece.org/fileadmin/DAM/timber/publications/SP-26.pdf.

Seják, J., Dejmal, I., Petříček, V., Cudlín, P., Míchal, I., Černý, K., Kučera, T., Vyskot, I., Strejček, J., Cudlínová, E., Cabrnoch, J., Šindlar, M., Prokopová, M., Kovář, J., Kupka, M., Sčasný, M., Šafařík, M., Roušarová, Š., Stejskal, V., Zapletal, J., 2003. Valuation and Pricing Biotopes of the Czech Republic. Český ekologický ústav, Prague, 450 pp (in Czech) http://fzp.ujep.cz/Projekty/VAV-610-5-01/HodnoceniBiotopuCR.pdf. English Summary: http://fzp.ujep.cz/Projekty/bvm/bvm.pdf.

Seják, J., Cudlín, P., Pokorný, J., Zapletal, M., Petříček, V., Guth, J., Chuman, T., Romportl, D., Skořepová, I., Vacek, V., Vyskot, I., Černý, K., Hesslerová, P., Burešová, R., Prokopová, M., Plch, R., Engstová, B., Stará, L., 2010. Valuing Ecosystem Functions and Services in the Czech Republic. FŽP UJEP, Ústí nad Labem. 23 pp, http://fzp.ujep.cz/projekty/valuingecosystemservices.pdf.

Seják, J., Cudlín, P., Pokorný, J., 2012. Valuation of ecosystem services as an instrument for implementation of the European Landscape Convention. In: Westra, L., Soskolne, C.L.,

Spady, D.W. (Eds.), Human Health and Ecological Integrity: Ethics, Law and Human Rights. Routledge, London.

Seppälä, R., Buck, A., Katila, P. (Eds.), 2009. Adaptation of Forests and People to Climate Change. A Global Assessment Report, IUFRO World Series, vol. 22, Helsinki, 224 pp.

Seppelt, R., Dormann, C.F., Eppink, F.V., Lautenbach, S., Schmidt, S., 2011. A quantitative review of ecosystem service studies: approaches, shortcomings and the road ahead. J. Appl. Ecol. 48, 630–636.

Serengil, Y., Augustaitis, A., Bytnerowicz, A., Grulke, N., Kozovitz, A.R., Matyssek, R., Müller-Starck, G., Schaub, M., Wieser, G., Coskun, A.A., Paoletti, E., 2011. Adaptation of forest ecosystems to air pollution and climate change: a global assessment on research priorities. iForest 4, 44–48.

Spangenberg, J.H., Settele, J., 2010. Precisely incorrect? Monetising the value of ecosystem services. Ecol. Complex. 7, 327–337.

Syrbe, R.-U., Walz, U., 2012. Spatial indicators for the assessment of ecosystem services: providing, benefiting and connecting areas and landscape metrics. Ecol. Indic. 21, 80–88.

TEEB, 2010. The economics of ecosystems and biodiversity: ecological and economic foundations. In: Kumar, P. (Ed.), Earthscan, London and Washington.

Tuovinen, J.-P., Hakola, H., Karlsson, P.E., Simpson, D., 2013. Air pollution risks to northern European forests in a changing climate. In: Matyssek, R., Clarke, N., Cudlin, P., Mikkelsen, T., Tuovinen, J.-P., Weiser, G., Paoletti, E. (Eds.), Climate Change, Air Pollution and Global Challenges: Understanding and Solutions from Forest Research. Dev. Environ. Sci. 13. Elsevier, Netherlands, pp. 77–99.

Turner, M.G., Donato, D.C., Romme, W.H., 2012. Consequences of spatial heterogeneity for ecosystem services in changing forest landscapes: priorities for future research. Landsc. Ecol. 28 (6), 1081–1097. http://landscape.zoology.wisc.edu/PublicationsNew/Turner_etal2012LandEcol_sustainability.pdf, http://dx.doi.org/10.1007/s10980-012-8741-4.

Uddling, J., Teclaw, R.M., Pregitzer, K.S., Ellsworth, D.S., 2009. Leaf and canopy conductance in aspen and aspen-birch forests under free-air enrichment of carbon dioxide and ozone. Tree Physiol. 29, 1367–1380.

Van Dijk, A.I.J.M., Keenan, R.J., 2007. Planted forests and water in perspective, overview. For. Ecol. Manage. 251, 1–9.

van Oudenhoven, A.P.E., Petz, K., Alkemade, R., Hein, L., de Groot, R.S., 2012. Framework for systematic indicator selection to assess effects of land management on ecosystem services. Ecol. Indic. 21, 110–122.

Walters, C.J., 1986. Adaptive Management of Renewable Resources. McGraw Hill, New York.

White Paper on environmental liability: COM(2000) 66 final, 9 February 2000. http://ec.europa.eu/environment/legal/liability/pdf/el_full.pdf.

Zhao, M., Running, S.V., 2010. Drought-induced reduction in global terrestrial net primary production from 2000 through 2009. Science 20, 940–943.

Zweifel, R., Rigling, A., Dobbertin, M., 2009. Species-specific stomatal response of trees to drought - a link to vegetation dynamics? J. Veget. Sc. 20, 442–454.

Chapter 25

Targeting Sustainable Provision of Forest Ecosystem Services with Special Focus on Carbon Sequestration

Maria Nijnik[1] and David Miller
The James Hutton Institute, Aberdeen, Scotland, United Kingdom
[1]Corresponding author: e-mail: maria.nijnik@hutton.ac.uk

Chapter Outline

Developments in Environmental Science, Vol. 13. http://dx.doi.org/10.1016/B978-0-08-098349-3.00025-6

25.1 INTRODUCTION

The recognition of the wider services of forests is a phenomenon of the post-industrial era (Mather et al., 2006). A characteristic of multi-functional forestry (MFF) is a shift in emphasis from commodity to non-commodity outputs—from maximising production of material goods (e.g. wood), to broader objectives of providing multiple ecosystem services (ESS). ESS have been defined as the benefits people obtain from ecosystems. The Millennium Ecosystem Assessment (MEA, 2005) states that *people are integral parts of ecosystems and that a dynamic interaction exists between them and other parts of ecosystems, with the changing human condition driving, both directly and indirectly, changes in ecosystems.* The MEA considers social, economic and environmental interactions, cross-scale issues, and multiple outcomes of natural resource use, grouping the ESS into *supporting, provisioning, regulating, cultural and social services* (MEA, 2005).

- *Supporting services* are underpinning, and necessary for the production of all other ESS. They include soil formation, fertility and storage, reservoir for water and carbon, photosynthesis, primary production, biodiversity, nutrient and water cycling.
- *Provisioning services:* extractive resources, including fresh water, food, fibre/wood, fuel, genetic, bio-chemical, natural pharmaceutical and medicinal resources, as well as ornamental.
- *Regulating services*: the benefits obtained from the regulation of ecosystem processes, including air quality regulation, climate, water and erosion regulation, water purification, disease and pest regulation, pollination and natural hazard regulation.
- *Cultural and social services*: the non-material benefits people obtain from ecosystems through spiritual enrichment, cognitive development, reflection, recreation and aesthetic experiences – thereby taking account of landscape values and of cultural and ethnological diversity.

An *ecosystem approach* (EA) provides a means to explore the multi-functionality of forests through which trade-offs between ESS can be tested and used with scenarios of biophysical and social change. It is a framework for identifying potential conflicts and supporting dialogue regarding trade-offs, for example, between biodiversity, landscape amenity, food production, and carbon sequestration (UNEP, 2010).

Carbon forestry is becoming increasingly important, enabling society to buy time for developing low carbon and decarbonised technologies for coping with climate change. Integration of carbon forestry into land use offers scope for innovation, employment, new markets and locally oriented value chains. This often provides new opportunities for disadvantaged areas where forestry can foster socio-economic development, combined with enhancement of nature and rural landscapes. The challenge is to find optimal solutions for different stakeholders, which consider multiple synergies and balanced trade-offs. MFF provides such a model, the shift towards which is now reflected in policies of the European Union (EU) and national levels (Kaljonen et al., 2007). It is supported by institutional analysis (Nijnik and Mather, 2008) and public opinion surveys (Nijnik et al., 2008), and is backed by empirical evidence on the ground (Mather et al., 2006).[1]

This chapter seeks to provide both conceptual and practical guidance towards solving the problem of implementing the sustainable provision of forest ESS into stakeholder considerations and calculations, as well as to policy-making decisions. Based on the work with relevant stakeholders, an attempt is made to identify, demonstrate, and communicate innovative ways of improving the key ESS of forests, and enhancing their sustainable provision for the benefit of local communities and society at large. The chapter starts with a concept of forest multi-functionality and continues with the 'Drivers of forestry changes' (Section 25.3), 'Challenges to sustainability in provision of ecosystem services' (Section 25.4), 'Stakeholder and institutional considerations' (Section 25.5), 'Valuing ecosystem services' (Section 25.6), 'Implementing forest multi-functionality' (Section 25.7), 'Payments for ecosystem services' (Section 25.8), and 'Integrating carbon sequestration in multi-functional forests to tackle climate change' (Section 25.9), to conclude with Section 25.10. Knowledge shared in this chapter seeks to ensure enhanced understanding of ways to influence the effective delivery of multiple ESS and forest functions.

25.2 CONCEPTUALISING FOREST MULTI-FUNCTIONALITY

The First Ministerial Conference on the Protection of Forests in Europe (1990) was a key step towards acknowledging the necessity of accounting for the role of forests as a life supporting system. By the 1992 UNCED conference, several countries had developed principles for sustainable forestry. Its communique stated that *forest resources and forest lands should be sustainably managed to meet social, economic, ecological, cultural and spiritual human needs of present and future generations*, and this has become a guiding principle at international and national levels. The European Forestry Strategy (Council of European Union, 1999) identified sustainable forest management and MFF as principles to underpin forest policies of member states. The EU Biodiversity Strategy to 2020 (European Commission, 2012) recognised *more sustainable forestry* among its targets, and actions *to adopt forest management plans that integrate biodiversity measures, and foster innovative mechanisms to finance the maintenance and restoration of ESS that are generated by sustainably managed multi-functional forests.* Nationally, the emphasis has been on developing sustainable forestry indicators, forest inventories, improving forest protection, and involving communities in forest policy and management.

---

1. A shift is discernible at the national scale: it does not follow that it is necessarily evident on every farm or field.

A multi-functionality strategy is theoretically capable of meeting the increasing private and public demands on foresst (Schmithüsen, 2007). It has become a political concept, particularly as Natura 2000 promoted an integrated approach to managing European forests for multiple purposes (Wilson, 2001). The EU has been active in combining multiple considerations in policy design (Brouwer et al., 2002). The criteria and indicators for Sustainable Forest Management (MCPFE, 2002) are an example of formalizing the spatial integration of multiple forest ESS. The reform of the Common Agricultural Policy (CAP) has also demonstrated an increased priority for delivering multiple objectives from land/forest. Thus, MFF is increasingly 'institutionalized' in policies and rests on the principle of forest management for provision of multiple ESS (Kaljonen et al., 2007).

Maier and Shobayashi (2001) identify two key elements in MFF: (i) joint production of commodity and non-commodity outputs, and (ii) some of the non-commodity outputs have the characteristics of public goods. 'Joint', however, gives rise to questions of scale and geography. Should multi-functionality apply in the vertical sense, with each field or forest stand fulfilling two or more functions, or can it be used to describe a pattern of diversity in which different units of land are dedicated to different functions? Dana (1943) implied a 'vertical' interpretation of forest multi-functionality. Pearson's (1994) view was of its 'horizontality': *effective multiple use is merely organized and coordinated specialization* (p. 248). The former view was dominant in the last part of the twentieth century, arguably the origins of 'forest ecosystem management'. However, the latter still has its advocates (e.g. Sedjo, 2004; Vincent and Binkley, 1993). We argue that whether forest multi-functionality is to be considered in its vertical or horizontal sense depends on the case, scale of observation and on the issue in question. It is case, scale and context specific. The heterogeneity in actors' capabilities, preferences and beliefs also play a role in this consideration (Ostrom, 2009).

25.3 DRIVERS OF FORESTRY CHANGES

Internationally, drivers of changes in forestry include globalization, environmental and climate changes, and technological development. At a European level (European Commission, 2009, 2010, 2012; European Union, 2000), policy drivers include reform of the CAP, Water Framework Directive, European Landscape Convention, Natura 2000, Forestry Strategy, EU Biodiversity Strategy, and commitments to reductions in greenhouse gases (GHGs). The drivers of changes also comprise land-use change, for example, agricultural intensification (with land abandonment as a secondary driver); infrastructure development; renewable energy; tourism and new forms of outdoor recreation (mountain biking or horse riding); spread of invasive species, etc. (Knoke and Hahn, 2013, this vol.).

Adoption of the EA and its 'people included' principle could improve our understanding of the opportunities and challenges to MFF under climate change and other drivers. However, successful implementation of MFF remains difficult, in part because of the variations of the drivers of land-use change through time, whether biophysical or socio-economic (FAO, 2010). Also, the development of institutions and capabilities which will enable the provision of ESS and a transition of MFF to more sustainable resource use (e.g. through a more adaptive forest management) needs time.

25.4 CHALLENGES TO SUSTAINABILITY IN PROVISION OF ECOSYSTEMS SERVICES

Existing approaches to the use of multi-functional forests do not easily allow for trade-offs between different services of ecosystems and between policy alternatives (e.g. evaluation of acceptable losses; forest ecosystem resilience). Impacts (e.g. of climate changes or environmental pollution) on forest ecosystems could be irreversible (i.e. beyond a given threshold). Forest practitioners experience difficulties in predicting, assessing and managing risks, as well as in the measurement and comparison of synergies and trade-offs in resource management (Armstrong and Smith, 2008). The majority of existing techniques have limited empirical bases and a lack of reference to a specific function/service or to end-user demands in ESS. The techniques available often do not take account of dynamics, which are particularly important when exogenous drivers and forest transition patterns (Mather, 2001) are concerned.

These problems are often aggravated by a lack of scientific inputs to explain the inter-linkages between different ESS; a deficiency in understanding thresholds; spatial and temporal complexity; and scale effects (Cudlín et al., 2013, this vol.). A growing necessity arises in the translation of MFF objectives to forestry practices on the ground, to link traditional forestry (i.e. aiming at timber production) with other forestry related interests (e.g. outdoor recreation), linking global issues of forest sustainability to local level forestry (understanding of scale effects, and multi-level governance), and spatial integration of land use systems (e.g. forestry and the agricultural sector, at a landscape level).

25.5 STAKEHOLDER AND INSTITUTIONAL CONSIDERATIONS

Stakeholders in MFF are communities of place (e.g. forest owners and users, local communities), and wider communities of interest (e.g. governments and authorities, NGOs), whose interests may differ. Also, notions of sustainable provision of ESS become less precise and more subjective when viewed from the different perspectives. The joint production of multiple ESS may therefore result in conflicts of stakeholder interests (Vogel and Lowham, 2007), and it is important to reduce the scope or resolve possible conflicts. Therefore, stakeholder engagement and evaluation (with two examples provided further in this and following sections) has gained increasing attention in forest policy

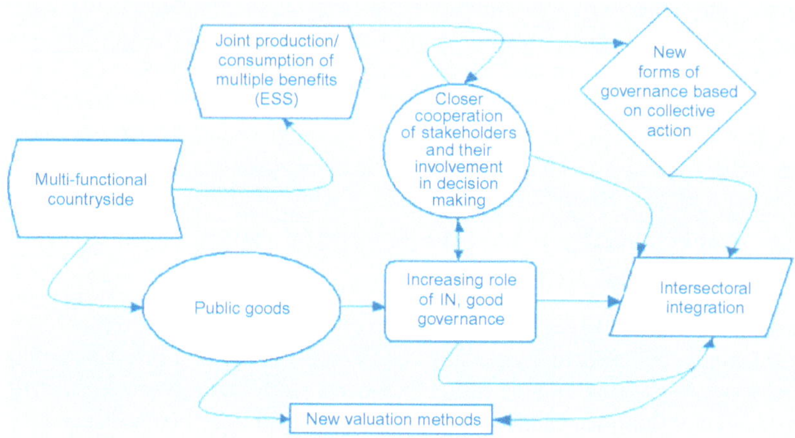

FIGURE 25.1 A cognitive map of the causality arising from the two main features of forest multi-functionality: joint production of multiple services, and that some services are public goods.

and management. A deeper understanding of stakeholder perceptions is essential for linking forest practices (including those of carbon forestry) to biodiversity conservation requirements influencing a policy design in forests that are managed for multiple purposes. Stakeholder engagement is thus becoming an important instrument for targeting sustainability in MFF.

Sustainable provision of intrinsic and socially constructed values of nature, and creating opportunities for people to enjoy forests necessitates vertical and horizontal collaboration (Kaljonen et al., 2007), and leads to co-management arrangements across different stakeholders (Figure 25.1).

Participatory and multilevel governance, in a continuous process of adaptation, is needed to take account of opinions and behavioural patterns of stakeholders who drive forestry change. This raises a number of questions: What are the changing institutions in forestry and what is their new role in facilitating the provision of forest ESS, including carbon sequestration? How well do institutions match the systems with which they interact? Do lessons learnt locally apply nationally and internationally? How do forestry institutions evolve and affect each other? What are the key challenges of integrating carbon forestry into multifunctional land use systems? What are the prospects for (re)designing forestry governance, and why are some responses to forestry changes more successful than others?

Innovative cross-sectoral means of engagement can provide an informed basis for sustainable MFF and ESS delivery at the landscape level (von Gadow and Pukkala, 2008). Stakeholders have to collaborate and become involved in decision-making processes. Forest multi-functionality, therefore, leads to co-ordinated actions and to the development of new forms of governance when stakeholders act together (Figure 25.1). However, policy-makers and forest managers often operate within dissimilar realities (Nie, 2003), with a gap between policies and practices, and decision-making processes (Krott, 2005).

Further problems may arise from the fact that non-commodity outputs of MFF are usually public goods, which are increasingly perceived as important as commodity production (van Huylenbroeck et al., 2007). Values of public goods have much wider spatial and temporal distribution than that of costs (e.g. carbon sequestration could be delivered by a forest owner; compensation payments to forest owners who increase carbon in their forests could be provided by government, while beneficiaries would be wider). Recipients of public goods do not repay society in full for the costs of their activities, and 'externalities' affect the use of public goods, which possess properties of non-rivalry and non-excludability (i.e. everyone benefits from its presence and there is no rivalry in consumption) that cause market failures (Slangen et al., 2008).[2]

Therefore, when mono-functional concepts of forestry fade and services beyond timber production are embraced (Mather et al., 2006), a role of multi-level governance that brings together public order (government) with an increasing stakeholder involvement in the decision-making process to control the tenure, management, production and consumption (use) of forest ESS is becoming increasingly influential. A reason is that markets cannot guide sustainable management of (non-marketable) public goods of forest ecosystems (Kant, 2003). Moreover, while the demand and supply of marketable commodities can be analysed by economic models and passed on to consumers through markets, diverse social expectations and public needs in forest ESS require eliciting preferences and new methods of valuation, and not only in monetary terms.

25.5.1 Example: Communications and Stakeholder Perspectives

The Climate Change Scotland Act (2009) provides a framework for reducing 80% of GHG emissions by 2050. The Land Use Strategy (Scottish Government, 2011) which identifies visions for delivering multiple benefits from land, including carbon sequestration. To test for differences in stakeholder perceptions of land use scenarios, a study explored options with stakeholders and the public, in particular younger people. The context for future land uses was a set of IPCC: world markets, with no agricultural subsidies; national enterprise, prioritising food security; global sustainability, with increased woodland and biodiversity; and fourth, local stewardship, led by public preferences. The study area is a sub-catchment in north-east Scotland, the Tarland Basin (52 km^2) in the River Dee catchment, Aberdeenshire, with current land use being 70% agriculture, 21% woodland, 8% moorland and 1%

2. For example, the need to develop carbon storage capabilities and improve climate change resilience is increasingly a driver of forest management decisions (Eggers et al., 2008).

built. Agricultural employment is 3%, tourism 26%, public sector 30%, and financial services 15%. Therefore, few local people have employment linked to the forest, but gain indirect benefits through landscapes managed for recreation, and residential quality of life. Objective scenarios in 2050 were created in a geographic information system (Castellazzi et al., 2010), and converted into visualizations of landscapes.[3] A Virtual Landscape Theatre (VRT) was used in venues remote from the area of study, in England (Birmingham), urban Scotland (Edinburgh) and adjacent to the area (Ballater). Stakeholder groups included policy-makers, land managers, foresters and farmers, public, children and people under 20. Semi-structured group discussions were run with stakeholders voting on landscape views under each scenario, and then recording their preferences for different land use changes: woodlands, renewable energy, transport, housing, access, recreational facilities, and protection or enhancement of habitats, water quality and landscapes (Figure 25.2).

FIGURE 25.2 Eliciting opinions on future land uses in a Virtual Landscape Theatre with young audiences: (A) Ballater, north-east Scotland; (B) Edinburgh, central Scotland; (C) Birmingham, central England.

3. The use of the VLT in developing visualisation methods is explained further in Ball et al. (2008). More information on scenario development is available at https://www.waset.org/journals/ijhss/v2/v2-2-12.pdf.

Findings show positive responses for landscapes with a visible mix of land uses, sound stewardship, elements of perceived naturalness and visual diversity (consistent with findings of Ode et al., 2009). From consultation events, commonality between audiences showed desire for amenity woodland adjacent to the village, quality recreation within the village, conservation interests and recognition of risks to water quality with increased agricultural activity. Edinburgh and Ballater audiences were positive towards small-scale wind turbines associated with farming or communities. However, there were significant differences between stakeholder opinions with respect to medium-sized windfarms on hills north of the village. Those unfamiliar with the area (Birmingham and some in Edinburgh) argued that renewable energy was a priority, highlighting open hilltops as opportunities for maximizing energy return. Those familiar with the area were conscious of the local significance of prominent hills and previous rejections of windfarm proposals. Local opinions were more favourable towards amenity and broadleaf woodland and unfavourable on coniferous woodland, compared to remote stakeholders. Participant groups from all areas favoured increases in mixed woodland. Elected representatives and planners reported positive views about using such tools in planning adaptation to climate change (e.g. flood alleviation), public policy (e.g. increasing woodland), and testing public preferences for landscapes, consistent with an EA.

25.6 VALUING ECOSYSTEM SERVICES'

Valuation of ESS can be divided into ecological, socio-cultural and economic (Farber et al., 2002). *Ecological value* is determined by the integrity of a forest's provisioning, supporting and regulatory functions, and by its parameters, that is, indicators of ecological relevance, such as complexity, diversity, rarity, or naturalness, applied across spatial scales. *Socio-cultural values* include various social values (e.g. equity); and end-users perceptions, and their cultural and spiritual (or non-material) well-being play an important role. In addition to indicator systems (for assessing ecological and socio-cultural values, with economic indicators including those of employment and income), total economic value (TEV) of ESS can be assessed by a range of methods. The economic concept of value (Nunes and van den Bergh, 2001) originates from neoclassical welfare economics. It has its roots in utilitarianism, and expresses the degree to which a service satisfies individual preferences.

However, many critical views have been expressed about the valuing of ESS. Porras (2012) states that policy makers are usually lost when confronted with TEV, because the figures are too abstract. Yet they should be specific to context and time, grounded on costs and linked to demand. Also, any valuation is not free of judgement, of human subjectivity and of the assumptions made. Still, the TEV approach is popular and its key components, with some examples, are shown in Figure 25.3.

Total economic value →			Examples
	Direct use	Provisioning services	Timber, non-timber products
		Social services	Recreation, health
	Indirect use	Regulating services	Air quality regulation, flood and erosion management
		Supporting services	Carbon sequestration and storage
	Other	**Option**	Future use, resilience
		Non-use: Existence and Bequest	Cultural, stewardship, intrinsic value of nature

FIGURE 25.3 Key components of total economic value. *Source: Adapted from Glaves et al. (2009).*

The provisioning services of forests are usually estimated through market valuation. Valuation of public goods largely relies on observable market transactions. For example, the social services of forests (commonly public goods) are analyzed through revealed preferences (RPs). Examples of such measures are travel cost estimates and hedonic prices, which are efficient when dealing with the 'use value'. The replacement cost method, which can reflect indirect use values of supporting and regulating services, could also be mentioned. Economic value can also be measured by the amount of money an individual is willing to pay. The stated preference (SP) technique is suitable for evaluating ESS beyond 'use value' levels. The approach creates hypothetical markets and examines implicit preferences and needs. Estimates are derived contingent on a description of a hypothetical change in ESS (Bateman et al., 2002). First applied by Davis (1963) as a tool for valuing outdoor recreation, the contingent valuation method (CVM) is now used extensively to determine values of a wide range of non-conventional ESS.

RP and SP valuation techniques are extensions of market valuation, which assigns monetary measures to components of TEV, both of 'use' and 'non-use'. The approaches have biases. RP cannot capture non-use values (MacMillan et al., 2004) whilst biases of SP fall into two broad categories, sampling error, and hypothetical bias (Schlapfer et al., 2004). In addition to various technical problems inherent in valuing the not-marketed services, there is concern about the gap between the hypothetical monetary values and reality (i.e. too few beneficiaries are actually willing to pay, and the scientifically derived willingness to pay may substantially exceed actual expenditure). Real integrated socio-economic and ecological values of supporting and regulating services, in particular, can hardly be obtained by traditional methods.

Nevertheless, RP and SP techniques could provide some useful information to decision-makers, particularly when no market for ESS exists (as free public goods with zero prices). For example, a study in Ukraine showed the value of timber and non-timber forest products to be around 125 Euros/ha/year while that of several non-market ESS (carbon, water and soil protection) deemed to exceed 170 Euros/ha/year (European Environment Agency, 2010). However, such comparative analyses are approximate and rather rare (Nijnik et al., 2008).

Bateman et al. (2002) describe the use of benefit transfer methods for the spatial modelling of ESS values. This approach, very useful on many instances, uses estimates in one place to infer benefits elsewhere or over a wider area. However, transfer values are to a large extent indicative, often relying on the availability of data and classifications that were developed for other purposes, and not necessarily at an appropriate scale or with contemporary values. Other factors that can affect ESS values are ecosystem condition, size, connectivity and context. Further, contemporary valuations of ESS are carried out using the knowledge of the day, which may itself be incomplete, and change through time. Insufficient understanding of ecological processes and other uncertainties often lead to the unreliability of the obtained economic estimates. Moreover, ESS values are not fixed and may change spatially, between different stakeholders. Monetisation tends to place a single value on a service, yet there are often variations in opinion regarding the value of an ESS amongst stakeholders.

Further, some types of services contribute to others, leading to potential double counting (e.g. supporting ESS contribute to regulating ESS, or regulating services contribute to cultural; de Groot et al., 2002). The calculation of supporting services is particularly difficult due to a lack of robust data. Also, potential use values of ESS (i.e. option values) are not easy to assess. Therefore, currently, option values (and of existence and bequests) are not incorporated in ESS valuation, but consideration should be given to their inclusion in future. It is important to advance research tools such that they are relevant, accessible and effective in offering meaningful information to different audiences, such as guiding public understanding of consequences of ESS changes, developing institutional capabilities and aiding the decision-making.

Integration of different tools could be helpful in many cases. For example, the market stall approach (Macmillan et al., 2002) is a group-based deliberative method that combines the desirable features of group techniques such as Citizens Juries with SP. A combined RP–SP method is, for example, the contingent behaviour model (Christie et al., 2007). Forest management decisions are often carried out at a small scale (Glück, 2002), such as farm or estate level, with limitations on the spatial understanding at a landscape scale and higher. Various combinations of research methods and tools could be applied successfully across multiple research objectives and scales of analysis. They can bring together different theoretical and methodological approaches, scales and levels of ESS analysis, such as analytical and participatory techniques

(Nijnik and Mather, 2008); participatory techniques and visualization tools (Miller et al., 2009), or Q-methodology with aggregated ecological indexes (Nijnik et al., 2008).

25.6.1 Example: Stakeholder Valuation of Components of Multi-Functional Forestry

The assessment of opinions and value judgments has employed surveys, focus groups, Q-method, CVM and multi-attribute utility analysis across a diverse set of research objectives. Each method is useful in certain cases, with the Q method (Figure 25.4), for example, proving particularly relevant in relation to MFF (Nijnik et al., 2008).

Given the assumptions and particulars of the research on MFF (Nijnik et al., 2010), five groups of stakeholder attitudes were distinguished, reflecting the regulating and supporting (the protection of natural ecosystems and conservation of biodiversity); provisioning (timber production), and cultural and social services of forests. The primary advocates of regulating and supporting ESS belong to groups 1 and 4. The main difference between these groups is in the strength of their manifestation of preferences. People who might be described as 'radical conservationists' (group 1) overlook both timber production and the use of forests for multiple purposes. The 'moderate

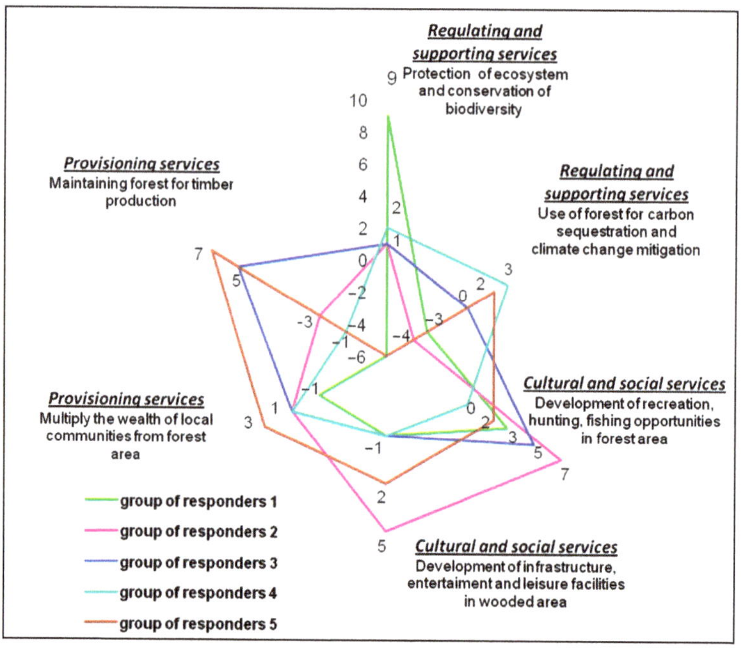

FIGURE 25.4 Stakeholder valuation of selected components of MFF. *Source: Nijnik et al. (2010).*

conservationists' (group 4) favour MFF, including using forests for climate change mitigation. Their support of nature conservation comes along with the support for sustainable forest management.

Respondents in groups 3 and 5 (Figure 25.4) consider provisioning services (i.e. timber production) as the main objective. However, compared with group 3, who, due to their exclusive focus on timber production and welfare maximization could be categorised as 'radical productivists', group 5 respondents are moderate and balance timber production, including for bio-energy, and the provision of other ESS. Group 2 respondents favour visual landscape features and the development of tourism and recreation activities in forests. They support the development of entertainment and leisure facilities in woodlands used for recreation, and suggest that hunting and fishing are the necessary measures to maintain the quality of ecosystems. This group, termed 'recreants', supports a broad range of ESS, starting from the conservation of forests to the consideration of socio-economic benefits that forests can provide for local communities.

The results of stakeholder evaluation of forest ESS across different countries in Europe (Nijnik et al., 2010) indicate that the protection of forest ecosystems and conservation of biodiversity (i.e. regulating and supporting services) receive the support of all attitudinal groups, except group 3. All groups, excluding group 4, consider cultural and social services of forests, with the development of recreation, gaming and fishing opportunities, as important. The necessity of enhancing the provisioning services of forests is supported by all attitudinal groups, except the radical conservationists (group 1), while only the productivists (groups 3, 5) consider the importance of maintaining forest for timber above all else. The results also indicate that an increasing intensity of conservation measures may influence timber production and *vice versa*. At one end of the spectrum, ecological approaches emphasise environmental protection and at the other end, climate change considerations promote carbon forestry (e.g. short rotation coppicing). An increasing role of forests for climate change mitigation will likely stipulate development of fast-growing plantations. Overall, despite the heterogeneity of attitudes, all groups identified support the necessity of multiplying the wealth of local communities from wooded areas, putting the emphasis on stability and resilience of forest ecosystems.

25.7 IMPLEMENTING FOREST MULTI-FUNCTIONALITY

Although a range of valuation methods and practical approaches has been developed, targeting of sustainability in MFF has not been easy, because of numerous challenges and uncertainties. There is an accelerating trend towards the integration of ESS into forest management planning and practices and a clear need for action to improve the management of MFF. For example, forests located outside nature reserves (provided these forests are well managed) can and should have important biodiversity values, contribute to the quality of environment, including the mitigation of climate change, and sustain rural (e.g. through employment) and urban livelihoods (green space management).

Although Krutilla (1986) argues that forestry is capable of co-producing a considerable number of desired outputs, the enhancement of carbon sequestration is more doable in forests managed for carbon, while the conservation of biodiversity is much easier in designated nature-protected areas. This is because capabilities in a single target activity (e.g. biodiversity conservation) advance through specialization (Wolf and Primmer, 2006). In contrast, multi-functional use of forests requires a more diverse set of capabilities than its management for one or a few ESS (also because the provision and/or consumption of the services could be conflicting, e.g. bioenergy vs. biodiversity).

Multi-functional forest management implies that objectives of all relevant stakeholders are incorporated into an adaptive and participatory planning process, so that ESS can be enhanced spatially (e.g. across the landscape) and across scales (Weiss, 2004). An adaptive forest management system (FMS) is a suitable platform for the integration of information, models and methods required for solving complex MFF problems (Borges et al., 2009; Reynolds et al., 2006). However, most FMS have been developed without regard to the specific challenges of MFF and are usually descriptive and dependent on intensive monitoring (Frehner et al., 2007). The experience of adaptive FMS to advise on silvicultural practices, and the combination of this knowledge with subsequent guidance, seem to be crucial for the advance of MFF (Reynolds et al., 2006).

Considering the content and changes in MFF, we suggest the following principles for forest policy planning to enhance forest multi-functionality on the ground:

- *A systems approach*: each woodland considered a subsystem of a higher unit in the forest hierarchy;
- *Comprehensiveness*: all elements of the MFF development, economic, ecological/environmental and social, and their interaction, considered for each level of forest policy development: local, regional and national;
- *Objective principle*: priority objectives for each level of MFF development defined and addressed;
- *Perspective development*: the long-term socially beneficial interests to be a major target. This necessitates analysing various drivers and pressures on forest ecosystems, forest transition patterns, with dynamic optimisation of social-ecological systems (c.f. Ostrom, 2009).
- *Carrying capacity*: the rates at which forest ESS are used must be less than the available flows. The resources must not be driven to extinction—a safe minimum standard of conservation (Ciriacy-Wantrup, 1968) is to prevail.
- *Ethical principle*: equitable distribution in space and time of the costs and benefits of any activity to be ensured from MFF.

- *The Precautionary Principle*: 'full scientific proof of a possible adverse environmental impact is not required before action is taken to prevent that impact' (Scottish Natural Heritage, 1993).

The main *preconditions* for successful implementation of forest multifunctionality are then: a robust forest management plan; a flexible policy, which enables the incorporation of views of additional decisions and actors; correlation of MFF objectives with other management strategies and plans, such as agriculture, physical planning, nature conservation, that is, 'horizontal tuning' of the plans; 'vertical tuning' of MFF policy plans, when the executive structure of the institutions involved in multi-level MFF governance has to respond to an overall MFF strategy; feasibility of MFF plans in relation to targets and implementation resources, with clear and enforceable definition of responsibilities of all actors involved in practical application of the policy decisions. Implementing MFF could only be possible with close co-operation of relevant stakeholders, mutual learning, and a continual development of institutional capabilities (Hajer and Wagenaar, 2003). Such capabilities include knowledge resources and skills (e.g. innovation) and social and relational resources (including trust), as well as mobilization capabilities (c.f. Healey, 2003; Kaljonen et al., 2007).

25.8 PAYMENTS FOR ECOSYSTEM SERVICES

Different types of financing mechanisms support sustainable use of ESS, including public payment schemes provided by government. Government could support ESS on a permanent basis (e.g. through ecosystems conservation in designated nature protected areas, with delivery objectives targeted through contracts), or by co-financing (e.g. a programme of afforestation), in which payments could provide support of the confirmed outputs (e.g. of the endangered species preserved). It is important to note, that in order to convince forest owners providing a specific service, we first need to know the opportunity costs involved in this. The opportunity cost estimates could be a starting point for finding out effective compensation payments to be transferred to forest owners.

A novel mechanism for implementing practices favourable to the maintenance and sustainable use of forest is payment for ecosystem service (PES), that is, a system of payments, or incentives to forest managers who adopt sustainable practices and actions that increase the quantity and quality of desired ESS. It is a market-based instrument financed directly and voluntarily by ESS beneficiaries, and often remotely (Engel et al., 2008), as in the case of carbon offsetting and trading. Voluntary carbon markets involving forestry are now developing intensively (Nijnik et al., 2011).

Private financial deals involving PES have developed over the recent decades (e.g. when downstream water users pay upstream farmers for maintaining forests for clean water). Indirect financial support of sustainable practices (e.g. when PES is embedded in a traded product through certification and eco-labelling) has emerged. ESS could become recognised, and their providers benefit financially through development of local businesses (e.g. recreation facilities), secondary markets (e.g. when car parking fees are applied to forest visitors), or indirectly from public policies (e.g. from those promoting wood products or bio energy, and therefore influencing the timber market by raising prices).

However, implementing the mechanisms of ESS financing remains difficult (Engel et al., 2008). The most important obstacle is that forest ESS are often public goods and to put monetary values on such services, and guide their allocation through markets, is problematic.[4] The provision of incentives to private owners would require several steps: (i) a thorough assessment and mapping of ESS and their flows; (ii) identifying beneficiaries (and their groupings); (iii) estimating the economic value of benefits provided to different groups of people (and where possible, in monetary terms); (iv) development of policies and regulations, including the necessary legal and institutional environments and enforcing rules (as well as, e.g. the level of subsidies to be introduced and applied); and (v) the creation of markets capable of capturing the value of ESS and rewarding forest owners, accordingly.

It is important to note that developing ESS markets and making such markets workable is a challenge, although it is not necessarily the best solution (Spash, 2010). There is some evidence to argue that, when multiple public goods of forests are becoming increasingly important, the role of governance structures other than markets (e.g. of government regulation) will likely increase (Kant, 2003).

25.9 INTEGRATING CARBON SEQUESTRATION OBJECTIVES IN MULTI-FUNCTIONAL FORESTRY TO TACKLE CLIMATE CHANGE

Forestry practices which sequester the most carbon may compromise projects that reduce net emissions of CO_2 to the atmosphere. Carbon sequestration activities could lead to changes in fossil-fuel use and cause LUC that further impact on the atmospheric CO_2 pool. Climate policy measures need therefore to be effective and feasible, and well embedded in land use and environmental policy strategies. If this is achieved, and economic instruments and flexible policy mechanisms are implemented, there is scope for the multi-functional development of the countryside where carbon sequestration, production of sustainable energy, sustainable forestry and agriculture are combined, and where climate policies are connected to the strategies that promote integrated

4. For example, major challenges for terrestrial carbon offsetting and marketing are explained, in Nijnik et al. (2011).

land-use systems. For example, support for afforestation on marginal land is identified by policy as important, with attention given to linking agriculture and the environment to climate change-related measures, and carbon sequestration forestry being included in rural and regional development schemes. To be effective, forestry projects should be consistent with wider programmes of sustainable development, including climate change dimensions and various aspects of land-use policy. Obtaining multiple benefits from forest expansion could be a sustainable way of restoring productivity to land, whilst utilization of woody biomass for energy could create new development options. Policies and measures for MFF should, therefore, be integrated within spatial planning, agricultural/rural policies, and with those for sustainable energy production.

Carbon sinks in wood products provides multiple benefits by enlarging the supply of wood and adding to the total carbon sink. Systematic promotion of wood products and renewable energy offers opportunities for innovation, employment and development of energy markets, with locally and regionally oriented value chains. Enhancement of carbon sequestration on marginal land, especially in combination with an increased use of bioenergy, may represent a good opportunity for rural areas and communities when there is synergy between different policy areas, and when attention is given to the provision of long-term initiatives and infrastructure (e.g. markets) development.

However, integrated land-use systems are often problematic because they require fiscal and other incentives outside forestry, with high transaction costs, and can be contrary to the preferences of some landowners where objectives are not driven by economics (e.g. amenity-related land uses). Optimum carbon offset projects will probably be those linking long-term carbon sequestration with long-term substitution opportunities, and which successfully develop capabilities to bridge gaps between rural policy priorities and those of climate change and other issues of sustainable development.

However, the assumption that forestry-based carbon sequestration is a universal remedy may discourage other efforts to reduce GHG emissions (Spash, 2010; van Kooten, 2009). Terrestrial carbon offsetting does not always complement economic growth and large-scale afforestation and short-rotation plantations may result in negative environmental and social consequences through other ESS being compromised or reduced. A very important point in carbon forestry is how to monitor changes in carbon stocks on various levels (Plugge et al., 2013). Also, there is often a knowledge gap for forestry-based (Matyssek et al., 2013, this vol.) mitigation projects to be coherent, effective, efficient, and accepted by the public. There are uncertainties about how to integrate the socio-economic and biophysical considerations of carbon sequestration to build forest resilience and how and where to place afforestation and bioenergy production in the context of multi-functional land use (at various scales) where reform of the CAP and contemporary rural change will probably be influential.

The mitigative capacity of forests is mediated by uncertainties and shaped by market signals, policies and governance, public attitudes, and behavioural patterns. These are challenges that are multiplied by institutions and uncertainties associated with land/forest tenure and property rights on carbon offsets, as well as managerial aspects, concerning large-scale afforestation and carbon trading (Nijnik et al., 2011). Changes in government policies, market fluctuations, and social norms and behaviour patterns contribute to uncertainties. Furthermore, there are biophysical and social limits to the land available for tree-planting and woodland development for carbon sequestration alone cannot be considered as a long-term solution.

Nonetheless, carbon sequestration forestry projects are often effective and cost-efficient, complementary to emissions reductions, particularly on marginal land, and in delivering multiple ESS. Research is needed to develop our understanding of whether, where and how, MFF can offer sustainable, socially acceptable and low-cost opportunities for carbon sequestration, together with delivering other ESS. Substantial efforts are required to develop environmental awareness of farmers, forest managers and decision-makers regarding MFF. There is a need for more campaigns for knowledge exchange, training facilities and schemes to demonstrate the opportunities for forests to provide multiple ESS, and to make them attractive for forest users, owners, and managers.

25.10 CONCLUSIONS

Suitable responses to drivers of forest change require a balance of environmental, economic and social components of MFF management systems, and trade-offs between key ESS. Development of integrated systems and deliberative and multi-level forestry governance should seek to reach collective agreements (hierarchically, inter-sectorally and spatially) on (i) how to attain desirable trade-offs between forest ESS in each locality and (ii) which policy instruments and management tools to use in order to promote ESS sustainable delivery.

The sustainability of MFF requires high levels of stakeholder input and consultation. Strengthening partnerships of science, policy and practice will assist forest management to maintain and improve key ESS and enhance support of local communities that rely on forestry. Increasing empowerment of stakeholders in determining forest use is a trend in Europe (Nijnik and Mather, 2008), for which the following reasons are given:

- Importance of new forms of (multi-level) governance, with a higher level of stakeholder involvement in decision-making and policy implementation.
- Necessity for capability development associated with a better understanding of the motivation of stakeholders to handle nature in a certain way, the context within which people operate, and the conflicts that could arise (in order to avoid or to resolve/manage them).
- Integration in ESS management and use, that is, vertical and horizontal (e.g. cross-sectoral) co-operation between stakeholders, and spatial (e.g. landscape or ecosystem level) approach to MFF.

- Obligation to develop new methods and/or their proper combination in order to map, analyse and assess and value multiple ESS of forests (and therefore to enhance sustainable forest use).

Whilst the evidence of needs for a multi-functional perspective for forestry is compelling, the evidence as to whether this perspective can inform practice in a formal way, without spatially informed economics, is questionable. When public money is limited, obtaining value for money from forests will be essential (Slee, 2011). It is important to realise that economic valuation may not be necessary and feasible in all ESS approaches (particularly when public goods are concerned), and that when only economic values are used (as appropriate), the source and robustness of these values should be tested, when wider consultation with stakeholders may enable a consensus on ESS valuation to be reached.

ACKNOWLEDGEMENT

The authors are grateful to Jane Morrice for help with the editing.

REFERENCES

Armstrong, H.M., Smith, M.A., 2008. Woods and trees in the uplands—why we might want more. Aspects of Applied Biology 85, 'Shaping a vision for the Uplands', pp. 1–4.

Ball, J., Capanni, N., Watt, S., 2008. Virtual reality for mutual understanding in landscape planning. Int. J. Social Sci. 2 (2), 78–88.

Bateman, I., Carlson, R., Day, B., Hanemann, M., Hanley, N., Hett, T., Jones-Lee, M., Loomes, G., Mourato, S., Ozdemiroglu, E., Pearce, D., Sugden, R., Swanson, J., 2002. Economic Valuation with Stated Preferences. Edward Elgar, Cheltenham.

Borges, P.J., Fragoso, R., Garcia-Gonzalo, J., Borges, J.G., Marques, S., Lucas, M.R., 2009. Assessing impacts of Common Agricultural Policy changes on regional land use patterns with a decision support system. An application in Southern Portugal. For. Policy Econ. 12, 111–120.

Brouwer, F., Jeronim, C., van der Bergh, B., 2002. Sustainable agriculture and rural environment: governance, policy and multifunctionality. Advance in Ecological Economics. Elgar Publishing ltd., Cheltenham, UK.

Castellazzi, M., Matthews, J., Angevin, F., Sausse, C., Wood, G., Burgess, P., Brown, I., Conrad, K., Perry, J., 2010. Simulation scenarios of spatio-temporal arrangement of crops at the landscape scale. Environ. Model. Softw. 25, 1881–1889.

Christie, M., Hanley, N., Hynes, S., 2007. Valuing enhancements to forest recreation using choice experiment and contingent behaviour methods. J. For. Econ. 13, 75–102.

Ciriacy-Wantrup, S., 1968. Resource Conservation Economics and Policies. University of California Press, Berkeley.

Council of the European Union, 1999. Council Resolution on a Forestry Strategy for the European Union. Official Journal C56, 26.2.1999.

Cudlín, P., Seják, J., Pokorný, J., Albrechtová, J., Bastian, O., Marek, M., 2013. Forest ecosystem services under climate change and air pollution. In: Matyssek, R., Clarke, N., Cudlin, P., Mikkelsen, T., Tuovinen, J.P., Weiser, G., Paoletti, E. (Eds.), Climate Change, Air Pollution and Global Challenges: Understanding and Solutions from Forest Research. Dev. Environ. Sci. 13. Elsevier, Netherlands, pp. 521–546.

Dana, S.T., 1943. Multiple use, biology and economics. J. For. 41, 625–627.

Davis, R., 1963. Recreation planning as an economic problem. Nat. Resour. J. 3, 239–249.

de Groot, R.S., Wilson, M.A., Boumans, R.M.J., 2002. A typology for the classification, description and valuation of ecosystem functions, goods and services. Ecol. Econ. 41, 393–408.

Eggers, J., Lindner, M., Zudin, S., Zaehle, S., Liski, J., 2008. Impact of changing wood demand, climate and land use on European forest resources and carbon stocks during the 21st century. Glob. Chang. Biol. 14, 2288–2303.

Engel, S., Pagiola, S., Wunder, S., 2008. Designing payments for environmental services in theory and practice: an overview of the issues. Ecol. Econ. 65, 663–674.

European Commission, 2009. Directive 2009/28/EC of the European Parliament and of the Council on the promotion of the use of energy from renewable sources. Available at: http://eur-lex.europa.eu/LexUriServ/LexUriServ.do?uri=OJ:L:2009:140:0016:0062:en:pdf (accessed 15.08.12).

European Commission, 2010. 10 messages for 2010 Forest Ecosystems. Available at: http://www.eea.europa.eu/publications/10-messages-for-2010/message-5-forest-ecosystems.pdf (accessed 15.08.12).

European Commission, 2012. Our life insurance, our natural capital: an EU biodiversity strategy to 2020. Available at: http://ec.europa.eu/environment/nature/biodiversity/comm2006/pdf/2020/1ENACTpart1_v7%5b1%5d.pdf (accessed 15.08.12).

European Environment Agency, 2010. Europe's Ecological Backbone: recognising the true value of our mountains. EEA Report 6/2010, EEA Copenhagen, 248 p.

European Union, 2000. Natura 2000. European Council Directive 92/43/EEC on the conservation of natural habitats and of wild fauna and flora. Available at: www.natura.org/ (accessed 15.08.12).

FAO, 2010. Global Forest Resources Assessment (FRA). Available at: www.fao.org/forestry/fra/fra2010/en/ (accessed 15.08.12).

Farber, S., Costanza, R., Wilson, M., 2002. Economic and ecological concepts for valuing ecosystem services. Ecol. Econ. 41, 375–392.

Frehner, M., Wasser, B., Schwitter, R., 2007. Sustainability and success monitoring in protection forests. Guidelines for managing forests with protective functions. Environmental Studies No. 27/07.

Glaves, P., Egan, D., Harrison, K., Robinson, R., 2009. Valuing Ecosystem Services in the East of England, East of England Environment Forum, East of England Regional Assembly and Government Office East England. Available at: www.gos.gov.uk/goee/docs/193474/193503/vesiee1.pdf (accessed 15.08.12).

Glück, P., 2002. Property rights and multipurpose mountain forest management. For. Policy Econ. 4, 125–134.

Hajer, M.A., Wagenaar, H., 2003. Deliberative Policy Analysis: Understanding Governance in the Network Society. Cambridge University Press, Cambridge.

Healey, P., 2003. The new institutionalism and the transformative goals of planning. In: Verma, N. (Ed.), Institutions and Planning. Center for Urban Policy Research, New Brunswick, NJ.

Kant, S., 2003. Extending the boundaries of forest economics. For. Policy Econ. 5, 39–56.

Kaljonen, M., Primmer, E., De Blust, G., Nijnik, M., Kulvik, M., 2007. Multifunctionality and biodiversity conservation—institutional challenges. In: Chmelievski, T. (Ed.), Nature Conservation Management: From Idea to Practical Issues. Lublin-Lodz-Helsinki, Aarhus, pp. 53–69.

Knoke, T., Hahn, A., 2013. Global change and the role of forests in future land-use systems. In: Matyssek, R., Clarke, N., Cudlin, P., Mikkelsen, T., Tuovinen, J.P., Weiser, G., Paoletti, E. (Eds.), Climate Change, Air Pollution and Global Challenges: Understanding and Solutions from Forest Research. Dev. Environ. Sci. 13. Elsevier, Netherlands, pp. 569–588.

Krott, M., 2005. Forest Policy Analysis. Kluwer Academic Publisher, Dordrecht.

Krutilla, J.V., 1986. Multiple-use forest economics. In: Forests of the Northwest: A Time for Reflection. Oregon State University, Corvallis, Starker Lectures.

MacMillan, D., Philip, L., Hanley, N., Alvarez-Farizo, B., 2002. Valuing the non-market benefits of wild goose conservation: a comparison of interview and group-based approaches. Ecol. Econ. 43, 49–59.

MacMillan, D., Hanley, N., Daw, M., 2004. Costs and benefits of wild goose conservation in Scotland. Biol. Conserv. 119, 475–485.

Maier, L., Shobayashi, M., 2001. Multifunctionality: Towards the Analytical Framework. OECD, Paris, France, p.158.

Mather, A., 2001. Forests of consumption: postproductivism, postmaterialism, and the postindustrial forest. J. Environ. Plan. 19, 249–268.

Mather, A., Hill, G., Nijnik, M., 2006. Post-productivism and rural land use: cul de sac or challenge for theorization? J. Rural. Stud. 22, 441–455.

Matyssek, R., Clarke, N., Cudlin, P., Mikkelsen, T., Tuovinen, J.P., Weiser, G., Paoletti, E., 2013. Conclusions and perspectives. In: Matyssek, R., Clarke, N., Cudlin, P., Mikkelsen, T., Tuovinen, J.P., Weiser, G., Paoletti, E. (Eds.), Climate Change, Air Pollution and Global Challenges: Understanding and Solutions from Forest Research. Dev. Environ. Sci. 13. Elsevier, Netherlands, pp. 591–607.

MCPFE, 2002. Pan-European Biological and Landscape Diversity Strategy, Council for the Pan-European Biological and Landscape Diversity Strategy, Geneva. Available at: www.peblds.org/files/STRA-CO/2003/STRA (accessed 15.08.12).

Millennium Ecosystem Assessment (MEA), 2005. Millennium Ecosystem Assessment Synthesis Report. Available at: www.millenniumassessment.org/en/index.aspx (accessed 15.08.12).

Miller, D., Vogt, N., Nijnik, M., Brondizio, E., Fiorini, S., 2009. Integrating analytical and participatory techniques for planning the sustainable use of land resources and landscapes. In: Geerteman, S., Stillwell, J. (Eds.), Planning Support Systems: Best Practises and New Methods. Springer Science – Business Media B.V.

Nie, M., 2003. Drivers of natural resource-based political conflict. Policy Sci. 36, 307–341.

Nijnik, M., Mather, A., 2008. Analysing public preferences for woodland development in rural landscapes in Scotland. Landsc. Urban Plan. 86, 267–275.

Nijnik, M., Slee, B., Pajot, G., 2011. Opportunities and challenges for terrestrial carbon offsetting and marketing, with some implications for forestry in the UK. South East Eur. For. J. 1 (2), 69–79.

Nijnik, M., Zahvoyska, L., Nijnik, A., Ode, A., 2008. Public evaluation of landscape content and change: several examples from Europe. Land Use Policy 26, 77–86.

Nijnik, M., Nijnik, A., Lundin, L., Staszewsky, T., Postolache, C., 2010. A study of stakeholders' perspectives on multi-functional forests in Europe. Int. J. For. Trees Livelihoods 4, 341–358.

Nunes, P.A., van den Bergh, J.C., 2001. Economic valuation of biodiversity: sense or nonsense? Ecol. Econ. 39, 203–222.

Ode, A., Fry, G., Messager, P., Miller, D., Tveit, M., 2009. Indicators of perceived naturalness as drivers of landscape preference. J. Environ. Manage. 90 (1), 375–383.

Ostrom, E., 2009. A general framework for analyzing sustainability of social-ecological systems. Science 325, 419–422.

Pearson, P., 1994. Energy, externalities and environmental quality: will development cure the ills it creates? Energy Stud. Rev. 6 (3), 199–216.

Plugge, D., Baldauf, T., Köhl, M., 2013. The global climate change mitigation strategy REDD: monitoring costs and uncertainties jeopardize economic benefits. Clim. Change 119, 247–259. http://dx.doi.org/10.1007/s10584-012-0524-3.

Porras, I., 2012. The value of valuing of ecosystem services. Report of International Institute for Environment and Development. Available at: http://www.iied.org/value-valuing-ecosystem-services (accessed 21.10.12).

Reynolds, K.M., Ray, D., Thomson, A.J. (Eds.), 2006. Sustainable Forestry: From Monitoring and Modelling to Knowledge Management and Policy Science. CABI International, Wallingford, UK, p. 224.

Schlapfer, F., Roschewitz, A., Hanley, N., 2004. Validation of stated preferences for public goods: a comparison of contingent valuation survey response and voting behavior. Ecol. Econ. 51, 1–16.

Schmithüsen, F., 2007. Multifunctional forestry practices as a land use strategy to meet increasing private and public demands in modern societies. J. For. Sci. 53 (6), 290–298.

Scottish Government, 2011. Getting the Best from Our Land: A Land Use Strategy for Scotland. The Scottish Government, Edinburgh.

Scottish Natural Heritage (SNH), 1993. Sustainable Development and the Natural Heritage: The SNH Approach. SNH, Perth.

Sedjo, R.A., 2004. Transgenic trees and trade problems on the horizon. Resources 155, 9–13.

Slangen, L., Loucks, L., Slangen, A., 2008. Institutional Economics and Economic Organisation Theory: An Integrated Approach. Wageningen Academic Publisher, Wageningen, p.431.

Slee, B., 2011. Innovation in forest-related territorial goods and services: an introduction. In: Weiss, G., Pettenella, D., Ollonqvist, P., Slee, B. (Eds.), Innovation in Forestry Territorial and Value Chain Approaches. CABI Publications, Wallingford, pp. 118–130, Chapter 8.

Spash, C., 2010. The brave new world of carbon trading. New Polit. Econ. 15 (2), 169–195.

The Climate Change Scotland Act, 2009. Scottish Parliament. Available at: http://www.legislation.gov.uk/asp/2009/12/contents (accessed 18.09.12).

The First Ministerial Conference on the Protection of Forests in Europe, 1990. Strasbourg. Available at: www.foresteurope.org/eng/Commitments/Ministerial_Conferences/Strasbourg_1990 (accessed 15.08.12).

UNEP/CBD/COP/10/27, 2010. Report of the Tenth Meeting of the Conference of the Parties to the Convention on Biological Diversity—COP 10, Japan, October 2010.

van Huylenbroeck, G., Vandermeulen, V., Mettepenningen, E., Verspecht, A., 2007. Multifunctionality of agriculture: a review of definitions, evidence and instruments. Living Rev. Landsc. Res. 1 (3), 1–38 (accessed 15.08.12).

van Kooten, G.C., 2009. Biological carbon sinks: transaction costs and governance. For. Chron. 85 (3), 372–376.

Vincent, J.R., Binkley, C.S., 1993. Efficient multiple-use forestry may require land-use specialization. Land Econ. 69, 370–376.

Vogel, J., Lowham, E., 2007. Building consensus for constructive action: a study of perspectives on natural resource management. J. For. 105 (1), 20–27.

von Gadow, K., Pukkala, T., 2008. Designing Green Landscapes (Managing Forest Ecosystems). Springer Science – Business Media B.V., p. 290.

Weiss, G., 2004. The political practice of mountain restoration—comparing restoration concepts in four European countries. For. Ecol. Manage. 195 (1–2), 1–13.

Wilson, G.A., 2001. From productivism to post-productivism ... and back again? Exploring the (un)changed natural and mental landscapes of European agriculture. Trans. Inst. Br. Geogr. 26, 77–102.

Wolf, S.A., Primmer, E., 2006. Between incentives and action: a pilot study of biodiversity conservation competencies for multifunctional forest management in Finland. Soc. Nat. Resour. 19, 1–17.

Chapter 26

Global Change and the Role of Forests in Future Land-Use Systems

Thomas Knoke[1] and Andreas Hahn

Technische Universität München, Center of Life and Food Sciences Weihenstephan, Department of Ecology and Ecosystem Management, Institute of Forest Management, Freising, Germany

[1]Corresponding author: e-mail: knoke@forst.wzw.tum.de

Chapter Outline

26.1 INTRODUCTION

Global change and its interlinked environmental, economic and social challenges have led to an active scientific discussion about expected trends and

Developments in Environmental Science, Vol. 13. http://dx.doi.org/10.1016/B978-0-08-098349-3.00026-8

solutions for possible future land-use conflicts (Godfray et al., 2011). Growing populations with increasing proportions of meat and dairy products in their diets result in rising demands for food and timber (Godfray et al., 2010). While food production systems have for a long time had adverse impacts on the environment (Altieri, 2009; Tilman, 1999), the increasing trend towards production of energy from food crops (Brown, 2009) steps up the already strong competition for land and water, escalating the phenomenon of 'land grabbing' (Bommert, 2012). Conversion of natural forests—mainly into agricultural lands—has long been a leading cause of loss of valuable and biodiversity-rich natural forest ecosystems, currently at a rate of 13 million hectares (ha) annually (FRA, 2010). The emission of greenhouse gases, in conjunction with this massive land-use and land-cover change, has fuelled climatic change substantially (van der Werf et al., 2009). However, climatic change is not only *driven* by these unsustainable land-uses but its consequences also feed back to food and timber production systems, through increasing vulnerability (Godfray et al., 2011). Also, increased ozone regimes may reduce the capacity of forests to store carbon (c.f. Chapter 7). And finally, poverty and rural depopulation—which may partly be traced back to insufficient knowledge, lack of citizenship and education (Aronson et al., 2010)—are considered to be undesirable social consequences of inefficient land-use systems (Coomes et al., 2011).

This multifaceted situation has resulted in various approaches to ensuring sustainable use of our planet's resources in an effort to satisfy growing demands for food, food security, water and energy, without compromising other environmental services. However, practical solutions are not easy to find. For example, Wise et al. (2009) showed increasing land scarcity and further increasing prices for food, when climate-friendly forests were supported by means of a CO_2 tax. Climate protection at the cost of increasing food prices, which hit the poor the hardest, will be an enormous barrier to the acceptance of climate policy.

It is evident that forest ecosystems, whose climate and water regulation services[1] (Cudlín et al., 2013, this vol.; Matyssek et al., 2013a, this vol.) are often at least as important as provisioning services, are interlinked with lands being used by the agricultural sector. Both systems compete for the same, increasingly scarce land (Kozovits and Bustamante, 2013, this vol. - for the case of South America). Future land-use concepts to meet the upcoming global challenges should, thus, always consider both the forestry and agricultural sectors simultaneously. Discussions about countermeasures to make forest and agricultural ecosystems and their services—and the human communities dependent on them—more immune to the adverse effects of climatic change, must be supported by comprehensive land-use models and require integrated research approaches (Matyssek et al., 2013b, this vol.).

1. We limit our consideration to provisioning and regulating services and thus exclude life supporting and social/cultural services here.

Here, we give an overview of trends in forestry and agriculture and the role forests may play in land use, and discuss important options to improve the long-term ability of ecosystems to maintain their provisioning, life supporting, regulating, and cultural services under changing future conditions (Nijnik and Miller, 2013, this vol.). We start with the forest, continue with the agricultural sector, and conclude with a proposal for a landscape concept integrating both forestry and agriculture. The latter would benefit from the synergistic effects of considering forestry and agricultural land-use options simultaneously. Given our own working experience, the forestry chapter draws largely on German/European experience, while the agricultural considerations often use South America as an example, where agriculture is expected to be significantly negatively impacted by climatic change (Kozovits and Bustamante, 2013, this vol.). We close the paper with concluding remarks.

26.2 FORESTRY SECTOR

In the context of the global challenges described above, the costs of climatic change (Stern, 2006) and possible countermeasures—often represented by the protection of natural forest ecosystems (Eliasch, 2008; Fearnside, 2001)—have become a public issue. But forests are not only seen as an option for mitigating the adverse effects of climate change; forests themselves will also be directly impacted by climatic change. Regional forestry sector studies in Europe predict enormous financial losses due to climatic changes (Hanewinkel et al., 2010). Another recent study by Hanewinkel et al. (2013) assumes a strong effect of climatic change on the distribution of tree species in Europe (excluding Russia). According to this study, the expected economic losses through climate-driven species shifts range between 14% and 50% of the current hypothetical soil value, meaning several hundred billion Euros. By 2100, between 21% and 60% of the European sites that are currently forested will be suitable only for a Mediterranean oak forest type, which generates extremely low economic returns.

In contrast to this rather pessimistic view, other studies mention the role of management in mitigating the undesirable effects of climatic change on the forestry sector (e.g. Köhl et al., 2010; Stang and Knoke, 2009). Obviously, no consensus exists about the impact of climatic change on the forestry sector. However, it is well accepted that all options for adaptation to the consequences of climatic change need to be considered. Before we discuss strategies for adapting forest ecosystems to the future situation, we give some brief information about the forestry sector.

26.2.1 Trends in the Forestry Sector

The Forest Resource Assessment (FRA, 2010—the source for all of the following information, unless otherwise noted) reports that the world's forests still cover ~4 billion ha, which represents 31% of our planet's land cover. Only 7% of the total forest area is comprised of (mostly fast-growing) tree plantations. Most of the forest area is thus, still natural or semi-natural forest, consisting at least in part of native tree species. Estimates show that global forests store 298 billion Mg of carbon in their above-ground biomass alone. Therefore, forests are a central component of the global carbon cycle. Hence, all changes in timber volume and forest structure affect this system. Deforestation is still an important component of global change, where ~13 million ha of mainly natural forests are lost every year, while only ~7.8 million ha are replaced by means of forest plantations or through expansion of natural forests. Further, the process of deforestation contributes 12–15% of the world's annual carbon emissions (van der Werf et al., 2009).

Another important characteristic of the forest sector and timber production is its long-term orientation. Forest management analyses and optimises decisions about the management of forests, where the resulting forest structures may hardly change over 100 years or even several hundreds of years. Uncertainty about production factors is thus inherent to forest management. It is thus a convention in forestry to base current decisions on providing a continuous supply of timber products and services and to adjust these decisions based on the current demands at the end of the production period. Therefore, continuous management is the cornerstone of risk management in forestry and thus of sustainable forestry (Hahn and Knoke, 2010). However, climate change increases the uncertainty that forest management faces: climate and growing conditions might change. Additionally, the rate of this change is expected to increase.

Consequently, forest management is closely linked to the topic of global/climate change in two ways: first, forests offer the potential to reduce anthropogenic climate change by storing additional carbon from the atmosphere. Second, they will in turn be strongly affected by the consequences of climate change, as trees live for decades, or even hundreds of years in the same location. Thus, forests can be considered both in terms of potential for mitigation and need for adaptation.[2]

From here, we concentrate on adaptation measures to answer the question of how forest ecosystems could become more resistant and remain resilient under future developments, while in section 26.3.3 we also discuss how deforestation may be reduced in order to mitigate climatic change.

2. The EU Cost-Action FP0703 'ECHOES' focussed both, mitigation as well as adaptation strategies, aiming at the mobilization and integration of existing knowledge to provide this for European policy makers and forest managers. The action was completed in May 2012.

26.2.2 Management Response to Counter Adverse Impacts of Climate Change on Forestry

Climate change is expected to alter the most basic components of forest sites, which are, for example, temperature and precipitation regimes. Nevertheless, these changes will not necessarily worsen the conditions of production. If tree growth is limited by temperature, as is the case in Scandinavian countries, climate change might even lead to higher volume in some areas (Eriksson et al., 2012). Central European countries probably face a different situation, with more precipitation in winter and less in summer (Hanewinkel et al., 2013), and a possible increase in storm events (Usbeck et al., 2010). These factors may result in increased vulnerability of forest ecosystems, which is defined by the degree to which an ecosystem 'is susceptible to or unable to cope with, adverse effects of climate change, including variability and extremes' (Intergovernmental Panel on Climate Change, 2001, p. 995). The potential for adaptation is defined here by the ability to reduce vulnerability in order to adapt forests to the needs of society, and not for society to have to adjust to potentially unfavourable situations (Spittlehouse and Stewart, 2003). In the following sections, we show options for adaption as tools for optimised forest management, now and in future.

26.2.2.1 The Choice of Tree Species

Tree species vary with regard to their ecological demands. They perform differently in terms of growth and are prone to illnesses and disturbances to a different extent. The choice of tree species is thus a fundamental decision in forestry, which will influence not only profitability but also managerial flexibility, resistance, and resilience, and through this, the permanence of the forest cover, which is so important for ecosystem services such as carbon storage, provision of clean water and recreation.

In most cases, species choice is dominated by economic considerations and by the site and climate characteristics of a specific forest. Bolte et al. (2009) describe three approaches to evaluating the suitability and the potential for growth of a specific tree species for a specific site and climate combination:

1. Future suitability could be seen as an extrapolation from retrospective, regional growth and/or risk data.
2. Species suitability could be derived with the help of the climate envelope approach, where the species niche is defined by climate variables such as annual average temperature and sum of annual rainfall. Current and future climate data, taken from climate scenarios, can then be used to demonstrate shifts of the potential distribution (e.g. Kölling et al., 2009).
3. The climate-matching analysis works the other way round as well: starting with the future climate, places with similar current climates are identified.

The species found in these matching sites might then be potential species to be grown on the home site under expected future conditions.

The results of these approaches can be thought of as constraints for forest management. However, trade-offs between growth performance and associated risks, which are essential for forest management planning, are not yet covered.

26.2.2.2 *Decisions on Stand Structure*

26.2.2.2.1 Mixed or Pure?

The chosen stand structure, whether mixed or pure, even-aged or uneven-aged, is of importance for the potential success or failure of forest management. If a site fulfils the ecological requirements of a variety of tree species, financial and risk data can be used to run a portfolio optimisation to determine the optimal composition of these tree species. Whenever financial risks of various tree species are correlated as negative or only slightly positive, mixing these tree species will function as a hedge against risks (Knoke et al., 2005). Neuner et al. (2013) used the theory of portfolio selection to derive optimal shares of tree species for a forest enterprise. Besides having financial advantages, mixtures can also have positive effects on growth and ecological stability of forest stands (Knoke et al., 2008). Stand resistance, for example, also benefits from species mixture significantly (Griess et al., 2012). Improved stand resistance was the most important factor, overshadowing all other aspects, identified by a financial assessment of the ecological effects that species mixture may have (Knoke and Seifert, 2008). Data on changing risk regimes is therefore crucial for sound forest management. Portfolio optimisation approaches are thus not limited by climate change, but the results are—as long as we lack foresighted risk data. Until then, species mixtures can establish opportunities for robust management and managerial flexibility, and create resistant and resilient forests (Hahn and Knoke, 2010; Hildebrandt et al., 2010). As long as two or more species are mixed within a stand, flexibility of choices is maintained, and the decision about the final crop species can be postponed, because it may still be altered through thinning in the course of stand development (Jacobsen and Thorsen, 2003).

26.2.2.2.2 Even-Aged or Uneven-Aged?

Even-aged forests generally offer a lower adaptive capacity to respond to severe disturbances than their uneven-aged counterparts. The mixture of vertical and horizontal structure in uneven-aged stands helps to cope with disturbances. Even-aged stands are thus less resistant to storm damage (Knoke, 2012), and they are, under certain circumstances, financially less desirable in comparison to uneven-aged stands (e.g. Knoke, 2012;

Roessiger et al., 2011).[3] This will remain the case under climate change, as a better use of niches promotes both ecological and financial resilience. Additionally, in cases of severe damage, the next tree generation is already established under uneven-aged management, and is thus able to continue growing immediately following a disturbance.

26.2.2.3 Thinning

Thinning creates net revenues, helps to secure liquidity within the forest enterprise, and shifts growth in volume increment to potential crop trees, as long as these trees remain stable. Recent research on the resistance of crop trees against drought effects after heavy thinning showed promising results; growth depression on a single-tree basis was less severe, and trees having more growing space recovered faster after drought (Sohn et al., 2012). Active management of forest stands may thus increase their immunity against the adverse impacts of climate change.

26.2.2.4 Rotation Length

The longer trees are exposed to risk, the higher the mortality rate will be, due to both a demographic and a height effect. The shortening of rotations is quite often mentioned as a response towards climate change (Spiecker, 2003; von Lüpke, 2009). But this is no panacea, as the natural regeneration potential might be lost and uneven-aged, close-to-nature forestry could potentially generate higher revenues at lower risk (Knoke and Plusczyk, 2001).

26.2.3 Research Gaps to be Filled

Forest owners and stakeholders evaluate the decisions described above based on the expected consequences and from the perspective of their own objectives (e.g. sustainability). However, decision consequences cannot be completely controlled by decision makers. Rather, they are subject to a great variety of uncertainties, which only increase under climatic change. While conventional uncertainties can often be modelled using historical data, this is not always possible with those arising from climate change because there are no experiences with new, future conditions. Risk research thus needs to be promoted for effective forest ecosystem management. Further, forest science needs to find and recommend forest management strategies, which are understandable by the public and allow for the flexibility to counter the adverse effects of the uncertainties inherent in climatic change. Trade-offs

3. However, this may be considered a more recent perception. Some scientists argue that pure Norway spruce is still and has always been the most economic option (see Knoke et al., 2008 for an overview and discussion of this topic).

between efficient forest production and the redundancies necessary to promote resilience and flexibility need to be analysed. Research on adaptive forest management will play an increasing role in forest science.

Another important point is that, at least from a worldwide perspective, the adaptation of forests to consequences of climatic change, and using forests as a mitigation option, must be seen in the context of other land-use options associated with forests. In this line, agriculture plays a crucial role and we discuss important issues of agriculture and its interactions with forestry in the following sections.

26.3 THE AGRICULTURE SECTOR AND THE ROLE OF FORESTS IN COMPREHENSIVE LAND-USE CONCEPTS

We first introduce the agricultural sector in order to conceptualize strategies for reinforcing future sustainable land use by means of cross-sectorial concepts which include forestry as a complementary land-use option. In this context, it is meaningful to identify actual trends and consider the likely future requirements for effective land-use concepts. Global predictions state that future human population numbers will demand a 70% increase in agricultural production by 2050 (HELF, 2009). It is important to keep in mind that this must be achieved despite the fact that, in substantial parts of the world, the agricultural sector will most likely be adversely influenced by the consequences of climatic and global changes. Although estimates up to the year 2080 (Cline, 2007) predict increases in yield up to +25%—mainly outside the tropical and subtropical regions—inside the tropical and subtropical areas, losses in yield of up to 50% are expected. For example, Ecuador, Bolivia, parts of Brazil, Mexico, Ethiopia, India, and also large parts of the United States will suffer most, while agriculture in Canada, Europe, the north of Asia and Argentina will most likely benefit from climatic changes. In many of the adversely impacted countries, the poor are particularly dependent on agricultural systems (Pandey et al., 2013, this vol.). Potential social problems are thus inherently connected to the expected changes and further conversion of more climate-friendly land cover into agricultural uses can be expected in order to compensate for yield losses, especially in the tropical areas. Brazil alone still harbours 476 million ha of primary forests (92% of the current total Brazilian forest area, FRA, 2010). Should these areas—or even only parts of them—be transformed into agricultural lands to satisfy increasing food and fuel demands and to compensate for declining agricultural yields, an ecological and, in the long-term, probably also an economic disaster could be programmed.

26.3.1 Land-Use Trends in the Agriculture Sector

The area of the world which is covered by croplands (including permanent crops) increased from ~1.37 to $\sim 1.53 \times 10^9$ ha between 1961 and 2009

(+12%), while the area of permanent meadows and pastures changed from ~3.08 to ~3.36 × 10^9 ha (+9%) in this period (FAO, 2012). At the same time, the world's production of cereals increased by a factor of 2.8. Given these figures, agricultural land expansion appears only moderate.

Some authors attribute this limited increase in the area of agricultural land to enormous gains in efficiency by means of agricultural intensification (producing more food per unit of land). For example, Burney et al. (2010) report an increase in productivity through intensification of 135% between 1961 and 2005, but only a moderate expansion of the total agricultural area. They conclude that agricultural intensification has prevented emissions of as much as 3.6 Pg C/year, by reducing the demand for more agricultural area (Burney et al., 2010). This amounts to three times the annual global deforestation and degradation emissions (excluding peat emissions) reported by van der Werf et al. (2009). Although one may well dispute the conclusions reached by Burney et al. (2010) (see the sections below), their study shows impressively that an efficient mitigation of GHG emissions from current land-use practices is not possible, without including the agricultural sector into climate policy (Knoke et al., 2012).

Largely neglected in the discussions about agricultural efficiency and intensification are the dynamics behind the enormous and continuously increasing areas of degraded and often unused land. Worldwide, the Food and Agriculture Organization's (FAO) area category 'other land' covers ~4.1 billion ha (FAO, 2012). According to FAO (2012), 'Other land is the land not classified as Agricultural land and Forest area. It includes built-up and related land, barren land, other wooded land, etc.' This land class occupies more area than is currently represented by forests. In South America alone, other land without productive agricultural or forestry uses has increased by ~45 million ha (1990–2008; FAO, 2012).

The increase in the amount of 'other land' must be seen in conjunction with deforestation processes. We may assume that 'other land' is fed by degraded agricultural lands and that those degraded lands must be replaced by new land gained by the conversion of natural forests into agriculture (i.e. deforestation). If replacing degraded lands by new agricultural lands is necessary, the net gain in agricultural area will be reduced. Actually, this net gain in South American agricultural area achieved by recent deforestation is small and has recently been decreasing. For example, the first 20 million ha of forests lost since 1990 in South America have led to an increase in agricultural area of 15 million ha, whereas the most recent 20 million ha of forest losses generated an increase of only 5 million ha in agricultural area (Figure 26.1). Obviously, more and more deforestation has been necessary in order to expand the agricultural area only marginally, because much area is used to replace degraded agricultural lands and also for built-up areas.

We argue that current practices of land use and degradation in combination with land abandonment are responsible for this negative change in the

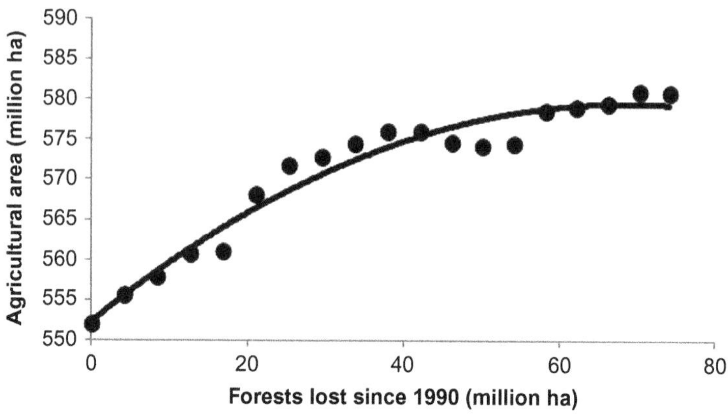

FIGURE 26.1 Agricultural area and forest loss in South America. *Data adapted from FAO (2012) for period 1990–2008.*

relationship between cleared land and net gain in agricultural area. This has ultimately led to a situation where actual agricultural area expansion was only ~29 million ha in South America since 1990. Increases in both 'other land' and agricultural area correspond in sum to an actual total forest loss of ~74 million ha (1990–2008) in South America. The ratio of forest loss to agriculture change in South America has therefore been ~2.5 ha of forest lost for each added hectare of agricultural land. Agricultural expansion is thus associated with substantial newly established areas to replace continuously degrading current agricultural lands. These land dynamics are often overlooked by those who tout the great increases in the efficiency of agriculture.

Another important aspect to consider is the impact of climatic change on food production systems. As we have already mentioned, agriculture is not only a driver of, but is also driven by, climatic change (Godfray et al., 2011). For example, while crop productivity has increased in South America, the year-to-year fluctuation in productivity for single agricultural products is enormous and in some cases the fluctuation shows a tendency to increase, for example, with soybeans (Figure 26.2).

The climatic vulnerability of crop production can be clearly observed at a national scale. Paraguay's soybean productivity has, for example, been severely affected by repeated regional droughts and insect infestations, which have decreased productivity values by half in years with severe drought (1.5 tonnes/ha, in 2009 vs. ~3.0 tonnes/ha, in 2010). Seen through the eyes of a risk-avoiding farmer, a classical 'safety response' would be the expansion of agriculture land so that, even for poor harvests, a minimum economic return would be achieved.

Increased climate vulnerability may lead to severe adverse consequences for many people. For example, between 2010 and 2011, around 350 million people were affected by natural disasters, where parts of sub-Saharan Africa

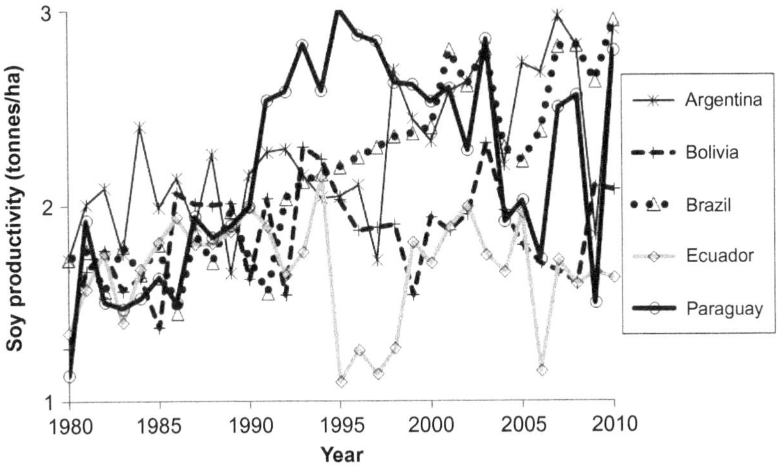

FIGURE 26.2 Development of average soybean productivity for various countries in South America. *Data from FAO (2012).*

and Asia were hardest hit. Among the sources of these natural disasters, floods, droughts and storms are the most important. Many people affected by such disasters experienced acute food insecurity. As a consequence, the number of undernourished people worldwide rose to 850 million between 2006 and 2008 and is expected to reach 925 million by 2011, since the number of undernourished people had reached a low level in the 1990s (FAO Statistical Yearbook, 2012). A report by the International Food Policy Research Institute also predicts a great impact of climate change, especially on the number of malnourished children (Nelson et al., 2009). Due to the likelihood of declines in agricultural yields in developing countries, particularly for irrigated crops, this report predicts greatly reduced calorie availability by 2050. This is expected to result in an increase in malnutrition in children by 20%, relative to what could be expected without climate change. The report by Nelson et al. (2009) recommends an additional annual funding of US$ 7 billion for adaptation programmes.

Under the increasing demands of an ever-growing population, it is assumed that the area of cultivated land will be expanded by 120 million ha—mainly in South America and sub-Saharan Africa—and will continue to decrease in developed countries. However, this expansion of agricultural land is not expected to contribute in substantial part to the necessary increase in agricultural production. Ninety percent of the expected demand for increase in agricultural production is predicted to come from higher agricultural production per unit of land (HELF, 2009), while the degradation problem, which is pushing land demand (see considerations above), has not been adequately addressed. The growing population is also expected to shift their diet towards more energy-dense food, which is high in fat and proteins (FAO Statistical

Yearbook, 2012), resulting in an increase in associated health problems. Exports of natural oils from developing countries are expected to more than triple by 2050, while sugar exports may double (HELF, 2009). These trends will further increase the pressure on human-made and natural ecosystems.

We conclude that the world is facing an increasing scarcity of land as a great future challenge, even if the intended increases in crop productivity are actually achieved. A growing human population whose diet is shifting towards food more rich in protein will need more and more resources. Increasingly, land is also needed for energy production, while processes of land degradation continue to contribute to further losses of productive lands. Climatic change will potentially cause yield losses—predominantly in tropical and subtropical countries, where we still have large land areas under natural forests, which store huge amounts of carbon and harbour enormous biodiversity. Also, climate variability will enhance the volatility of crop yields, a fact that is further pushing the trend for extensive conversion of natural ecosystems into agricultural lands.

26.3.2 Land-Use Concepts to Fulfil Increasing Food Demand and Adapt to Climate Change

Several approaches to achieving sustainable land use under global change have been published (e.g. Foley et al., 2011). Some principal strategies to meet the growing demand for agricultural products are described below.

26.3.2.1 Expanding the Frontier of Agricultural Area into Natural Ecosystems

An expansion of the area under cultivation has been the most prevalent means of increasing agricultural output throughout history in all parts of the world (Brown, 1967). Although the expansion of agricultural land is associated with severe problems for the environment and is not considered an option for the future by many scientists (e.g. Godfray et al., 2011), the potential for agricultural production is still being explored by some studies (e.g. Bruinsma, 2009).

26.3.2.2 Increasing the Production Per Unit of Land (Often Referred to as 'Land Sparing')

Various recent publications place great hope in further agricultural intensification in order to meet increasing food demands (Burney et al., 2010; HELF, 2009; Phalan et al., 2011; Wise et al., 2009) and avoid further encroachment on natural forests by agricultural frontiers. The concepts of agricultural intensification are often conceptualized mainly to produce palm oil (e.g. Koh et al., 2009; Phalan et al., 2011). They have also been criticised (Fischer et al., 2011) because of associated environmental problems, such as high chemical input, low wildlife friendliness, increase in erosion processes and land degradation and pollution of water bodies.

26.3.2.3 *Recultivation of Abandoned Lands*

The Global Partnership for Forest Landscape Restoration claims that 1.0–1.5 billion ha of degraded lands and forests are globally available for restoration (GPFLR, 2011). The enormous, still unproductive area must, in part, be re-integrated into the production process. There are several principle options for the reintegration of unproductive area back into the economic production process. One option is the reforestation of these lands (Weber et al., 2008) and another option, their re-cultivation for agricultural use, for example, for pasture (Roos et al., 2011). Still lacking, however, are integrated approaches that evaluate the effectiveness and quantify the consequences of reintegration of abandoned lands into forest or agricultural uses—particularly analyses which are embedded in real landscapes. This leads us to the next strategy.

26.3.2.4 *Avoiding Land Degradation and Reducing the Volatility of Crop Yields*

As a consequence of the limitations inherent in the two strategies mentioned above, approaches called ‘sustainable intensification’ (Foley et al., 2011) and wildlife-friendly land-sharing concepts (Fischer et al., 2008) have been discussed. They address long-term productivity and diversification strategies through the avoidance of degradation and the use of crop mixtures (Knoke et al., 2012). Approaches currently under discussion also include agro-forestry (Perfecto and Vandermeer, 2010) and organic agriculture (Seufert et al., 2012).

The concepts mentioned above have been developed in part to adapt ecosystems to the expected consequences of climate change. Rosegrant et al. (2008, p. 25) propose, *inter alia*, the following, more detailed activities as adaptation strategies to improve the immunity of farms to the adverse impacts of climate change: Crop/livestock diversification to increase productivity and protect against diseases, adjustment in the timing of farm operations to reduce the risks of crop damage, changes in cropping intensity, improvements in livestock management to adjust to new climate conditions, changes in tillage practices, and changes in crop mixes as short-term activities. Development of market efficiency, expansion of irrigation and water storage, more efficient water use, and promotion of international trade are mentioned as both short- and long-term activities. It appears evident that these proposals provide integrative and comprehensive land-use concepts.

In a study that aimed to define a method to achieve ‘sustainable intensification’, Knoke et al. (2012) published an attempt towards the development of a more integrative land-use concept. They expanded Odum’s (1969) 'compartment approach' by the inclusion of reforestation of degraded land. The paper argues that a mixed portfolio of land parcels, which includes crops, forests, and hedgerows, enables sustainable intensification and considers important adaptation strategies as proposed by Rosegrant et al. (2008), for example, the diversification of crops. Limiting parcel sizes to a maximum of 5–10 ha and using a variety of crops can substantially reduce losses due to erosion. Harvesting seasons are spread over the year so that only small patches are exposed to wind and water at any given time. A variety of crops can make use of heterogeneous site conditions more effectively when compared to specialized production with only one crop.

Such concepts may also reduce the on-average volatility of crop productivity (Figure 26.3). For the example of Paraguay, a model-crop-portfolio comprised of 30% soybean, 30% sorghum, 30% maize, and 10% potato, based on historical data, shows a more continuous yield when compared to soybean as a single crop. Moreover, in the years from 2004 to 2009, when soybean productivity was extremely volatile and on the decrease, the mixed-crop portfolio showed not only more constant, but also much higher, productivity. A diversified land-use portfolio approach may thus save agricultural area, because the year-to-year variation is reduced when mixing a variety of crops.

Also, land-use diversification can effectively spread economic risks and enhance minimum returns, even for small areas (Knoke et al., 2009). The desired output may thus be (almost) guaranteed on smaller areas with a diversified system as compared to specialized single-crop agricultural production systems.

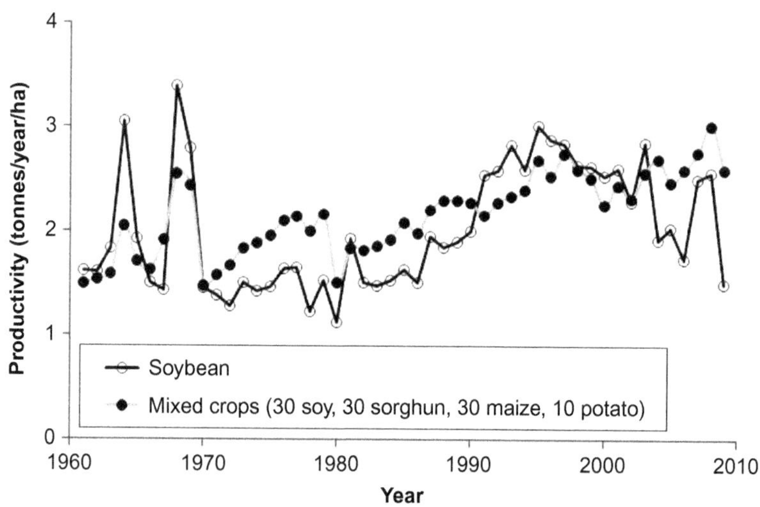

FIGURE 26.3 Example of development of average soybean productivity in Paraguay compared with productivity for a hypothetical mixed-crop portfolio under historical productivity, as reported by the FAO (2012).

26.3.3 A Landscape View Where Forests Play an Important Economic and Ecological Role

Forests, established on formerly abandoned lands, may serve as an effective complement to the diversified land-use portfolio discussed above. Timber prices and those for agricultural products are not correlated (Lönnstedt and Svensson, 2000) and timber harvests can be carried out in a flexible fashion (Knoke and Wurm, 2006)—preferably when agricultural prices are low. Forests could thus not only have ecological, but also important economic functions as insurance for farmers to insulate economic yields from shocks resulting from fluxes in agricultural markets. Economic analyses of the compartmental concept of sustainable intensification, building on FAO's historical productivity and price data for South America, suggest an improvement of long-term economic land-use performance through the inclusion of forests (and agro-forestry, shade coffee plantations) by 19–25% relative to soybean mono-cropping (Knoke et al., 2012). Similar potential for improvement has been shown for subsistence agriculture (Knoke et al., 2009).

Moreover, forests also have an indispensable ecological function. The natural forests that remain store vast amounts of carbon, are important for the protection of ground water, the formation of clouds and the prevention of erosion, and thus provide important regulation services. Also, they harbour enormous biodiversity and have important social and cultural values for native peoples. These services, however, require financial incentives. Knoke et al. (2013) have shown for Ecuador, how a combination of conservation payments to protect the existing natural forest combined with subsidies for the reestablishment of pastures on abandoned lands, and the implementation of forestry options (reforestation on abandoned lands and sustainable, low-impact management in part of the natural forest area) may lead to more sustainable land use. These results were obtained using an integrative and risk-sensitive land-use model (Knoke et al., 2011) in which all available land-use options are considered simultaneously to make use of synergies. This was done in such a way that indirect effects—for example, between subsidies for the re-cultivation of formerly abandoned pastures and conservation payments—could be considered.

26.4 CONCLUDING REMARKS

New land-use approaches must address the adverse consequences of climatic change, which negatively impact food and timber production systems through increasing their vulnerability, for example, due to rising temperatures and increasing volatility of weather systems (Godfray et al., 2011). The impact of comprehensive land-use approaches on ecosystem services, such as climate regulation and protection of water bodies, must also be considered. Integrative approaches combining patches of land with forestry options and patches of re-cultivated abandoned lands with sustainably intensified agricultural production that are financially supported by a mixture of agricultural subsidies and compensation payments for preserving natural forest have so far hardly been quantitatively analysed. This forms an interesting and highly important research field for the future.

ACKNOWLEDGEMENTS

We are grateful to the German Science Foundation (DFG) for funding our research on the bioeconomics of forestry (KN 586/4-1 and KN 586/7-1,2) and on tropical land-use systems (KN 586/5-1,2). We also thank Laura Carlson for the language editing of our text.

REFERENCES

Altieri, M.A., 2009. The ecological impacts of large-scale agrofuel monoculture production systems in the Americas. Bull. Sci. Technol. Soc. 29, 236–244.

Aronson, J., Aguirre, N., Munoz, J., 2010. Ecological restoration for future conservation professionals: training with conceptual models and practical exercises. Ecol. Restor. 28, 175–181.

Bolte, A., Ammer, C., Löf, M., Madsen, P., Nabuurs, G.-J., Schall, P., et al., 2009. Adaptive forest management in central Europe. Climate change impacts, strategies and integrative concept. Scand. J. For. Res. 24, 473–482.

Bommert, W., 2012. Bodenrausch. Bastei Lübbe, Köln.

Brown, L.R., 1967. The world outlook for conventional agriculture. Science 158, 604–611.

Brown, L.R., 2009. Plan B 4.0: Mobilizing to Save Civilization. W.W. Norton & Company, New York.

Bruinsma, J., 2009. The resource outlook to 2050: by how much do land, water and crop yields need to increase by 2050? Expert Meeting on How to Feed the World in 2050. Food and Agriculture Organization of the United Nations, Economic and Social Development Department. ftp://ftp.fao.org/docrep/fao/012/ak971e/ak971e00.pdf.

Burney, J.A., Davis, S.J., Lobell, D.B., 2010. Greenhouse gas mitigation by agricultural intensification. Proc. Natl. Acad. Sci. U.S.A. 107, 12052–12057.

Cline, W.R., 2007. Global Warming and Agriculture: Impact Estimates by Country. Peterson Institute, Washington, DC.

Coomes, O.T., Takasaki, Y., Rhemtulla, M., 2011. Land-use poverty traps identified in shifting cultivation systems shape long-term tropical forest cover. Proc. Natl. Acad. Sci. U.S.A. 108, 13925–13930.

Cudlín, P., Seják, J., Pokorný, J., Albrechtová, J., Bastian, O., Marek, M., 2013. Forest ecosystem services under climate change and air pollution. In: Matyssek, R., Clarke, N., Cudlin, P., Mikkelsen, T., Tuovinen, J.P., Weiser, G., Paoletti, E. (Eds.), Climate Change, Air Pollution and Global Challenges: Understanding and Solutions from Forest Research. Dev. Environ. Sci. 13. Elsevier, Netherlands, pp. 521–546.

Eliasch, J., 2008. Climate Change: Financing Global Forests: The Eliasch Review. Earthscan, London.

Eriksson, L., Backéus, S., Garcia, F., 2012. Implications of growth uncertainties associated with climate change for stand management. Eur. J. For. Res. 131, 1199–1209.

FAO (Food and Agriculture Organization) Statistical Yearbook, 2012. World food and agriculture. Food and Agriculture Organization of the United Nations, Rome. http://www.fao.org/docrep/015/i2490e/i2490e00.htm.

FAO (Food and Agriculture Organization), 2012. FAOSTAT (agriculture and food statistics). http://faostat.fao.org/default.aspx (accessed 24.10.2012).

Fearnside, P.M., 2001. Saving forests as a global warming countermeasure: an issue that divides the environmental movement. Ecol. Econ. 39, 167–184.

Fischer, J., Brosi, B., Daily, G.C., et al., 2008. Should agricultural policies encourage land sparing or wildlife-friendly farming? Front. Ecol. Environ. 6, 380–385.

Fischer, J., Batary, P., Bawa, K.S., et al., 2011. Conservation: limits of land sparing. Science 334, 593.

Foley, J.A., Ramankutty, N., Braumann, K.A., et al., 2011. Solutions for a cultivated planet. Nature 478, 337–342.

FRA, 2010. Global Forest Resource Assessment 2010. FAO forestry paper 163. Food and Agriculture Organization, Rome.

Godfray, H.C.J., Beddington, J.R., Crute, I.R., et al., 2010. Food security: the challenge of feeding 9 billion people. Science 327, 812–818.

Godfray, H.C.J., Pretty, J., Thomas, S.M., et al., 2011. Linking policy on climate and food. Science 331, 1013–1014.

GPFLR, 2011. Global partnership on forest landscape restoration. http://www.ideastransform-landscapes.org/ (accessed 19.06.2011).

Griess, V.C., Acevedo, R., Härtl, F., Staupendahl, K., Knoke, T., 2012. Does mixing tree species enhance stand resistance against natural hazards? A case study for spruce. For. Ecol. Manage. 267, 284–296.

Hahn, A., Knoke, T., 2010. Sustainable development and sustainable forestry. Analogies, differences, and the role of flexibility. Eur. J. For. Res. 129, 787–801.

Hanewinkel, M., Hummel, S., Cullmann, D.A., 2010. Modelling and economic evaluation of forest biome shifts under climate change in Southwest Germany. For. Ecol. Manage. 259, 710–719.

Hanewinkel, M., Cullmann, D.A., Schelhaas, M.-J., Nabuurs, G.-J., Zimmermann, N.E., 2013. Climate change may cause severe loss in the economic value of European forest land. Nat. Clim. Chang. 3, 203–207.

HELF (High-Level Expert Forum), 2009. Global agriculture towards 2050—how to feed the world in 2050. Office of the Director, Agricultural Development Economics Division, Economic and Social Development Department, Viale delle Terme di Caracalla, Rome, Italy. http://www.fao.org/fileadmin/templates/wsfs/docs/Issues_papers/HLEF2050_Global_Agriculture.pdf (accessed 28.10.12).

Hildebrandt, P., Kirchlechner, P., Hahn, A., Knoke, T., Mujica, H.R., 2010. Mixed species plantations in Southern Chile and the risk of timber price fluctuation. Eur. J. For. Res. 129, 935–946.

Intergovernmental Panel on Climate Change (IPCC) (Ed.), 2001. Climate change 2001: impacts, adaption and vulnerability. Contribution of working group II to the third assessment report of the Intergovernmental Panel on Climate Change, Cambridge University Press, Cambridge, UK.

Jacobsen, J.B., Thorsen, B.J., 2003. Danish example of optimal thinning strategies in mixed-species forest under changing growth conditions caused by climate change. For. Ecol. Manage. 180, 375–388.

Knoke, T., 2012. The economics of continuous cover forestry. In: Pukkala, T., von Gadow, K. (Eds.), Continuous Cover Forestry. Managing Forest Ecosystems, vol. 23. Springer, Netherlands, pp. 167–193. http://dx.doi.org/10.1007/978-94-007-2202-6_5.

Knoke, T., Plusczyk, N., 2001. On economic consequences of transformation of a spruce (Picea abies (L.) Karst.) dominated stand from regular into irregular age structure. For. Ecol. Manage. 151, 163–179.

Knoke, T., Seifert, T., 2008. Integrating selected ecological effects of mixed European beech-Norway spruce stands in bioeconomic modelling. Ecol. Model. 210, 487–498.

Knoke, T., Wurm, J., 2006. Mixed forests and a flexible harvest policy: a problem for conventional risk analysis? Eur. J. For. Res. 125, 303–315.

Knoke, T., Stimm, B., Ammer, C., Moog, M., 2005. Mixed forests reconsidered. A forest economics contribution on an ecological concept. For. Ecol. Manage. 213, 102–116.

Knoke, T., Ammer, C., Stimm, B., Mosandl, R., 2008. Admixing broadleaved to coniferous tree species. A review on yield, ecological stability and economics. Eur. J. For. Res. 127, 89–101.

Knoke, T., Calvas, B., Aguirre, N., Román-Cuesta, R.M., Günter, S., Stimm, B., Weber, M., Mosandl, R., 2009. Can tropical farmers reconcile subsistence needs with forest conservation? Front. Ecol. Environ. 7, 548–554.

Knoke, T., Steinbeis, O.E., Bösch, M., Román-Cuesta, R.M., Burkhardt, T., 2011. Cost-effective compensation to avoid carbon emissions from forest loss: an approach to consider price–quantity effects and risk-aversion. Ecol. Econ. 70, 1139–1153.

Knoke, T., Román-Cuesta, R., Weber, M., Haber, W., 2012. How can climate policy benefit from comprehensive land-use approaches? Front. Ecol. Environ. 10, 438–445.

Knoke, T., Calvas, B., Ochoa, W.S., Onyekwelu, J., Griess, V.C., 2013. Food production and climate protection – what abandoned lands can do to preserve natural forests. Global Environmental Change. http://dx.doi.org/10.1016/j.gloenvcha.2013.07.004.

Koh, L.P., Levang, P., Ghazoul, J., 2009. Designer landscapes for sustainable biofuels. Trends Ecol. Evol. 24, 431–438.

Köhl, M., Hildebrandt, R., Olschofsky, K., Köhler, R., Rötzer, T., Mette, T., et al., 2010. Combating the effects of climate change on forests by mitigation strategies. Carbon Balance Manag. 5, 8.

Kölling, C., Knoke, T., Schall, P., Ammer, C., 2009. Überlegungen zum Risiko des Fichtenanbaus in Deutschland vor dem Hintergrund des Klimawandels. Forstarchiv 80, 42–54.

Kozovits, A.R., Bustamante, M.M.C., 2013. Land use change, air pollution and climate change – vegetation response in Latin America. In: Matyssek, R., Clarke, N., Cudlin, P., Mikkelsen, T., Tuovinen, J.P., Weiser, G., Paoletti, E. (Eds.), Climate Change, Air Pollution and Global Challenges: Understanding and Solutions from Forest Research. Dev. Environ. Sci. 13. Elsevier, Netherlands, pp. 411–427.

Lönnstedt, L., Svensson, J., 2000. Return and risk in timberland and other investment alternatives for NIPF owners. Scand. J. For. Res. 15, 661–669.

Matyssek, R., Clarke, N., Cudlin, P., Mikkelsen, T., Tuovinen, J.P., Weiser, G., Paoletti, E., 2013a. Conclusions and perspectives. In: Matyssek, R., Clarke, N., Cudlin, P., Mikkelsen, T., Tuovinen, J.P., Weiser, G., Paoletti, E. (Eds.), Climate Change, Air Pollution and Global Challenges: Understanding and Solutions from Forest Research. Dev. Environ. Sci. 13. Elsevier, Netherlands, pp. 591–607.

Matyssek, R., Clarke, N., Cudlin, P., Mikkelsen, T.N., Tuovinen, J.-P., Weiser, G., Paoletti, E., 2013b. Climate change, air pollution and global challenges: Understanding and solutions from forest research. In: Matyssek, R., Clarke, N., Cudlin, P., Mikkelsen, T., Tuovinen, J.P., Weiser, G., Paoletti, E. (Eds.), Climate Change, Air Pollution and Global Challenges: Understanding and Solutions from Forest Research. Dev. Environ. Sci. 13. Elsevier, Netherlands, pp. 3–16.

Nelson, G.C., Rosegrant, M.W., Koo, J., Robertson, R., Sulser, T., Zhu, T., Ringler, C., Msangi, S., Palazzo, A., Batka, M., Magalhaes, M., Valmonte-Santos, R., Ewing, M., Lee, D., 2009. Climate Change: Impact on Agriculture and Costs of Adaptation. International Food Policy Research Institute, Washington, DC.

Neuner, S., Beinhofer, B., Knoke, T., 2013. The optimal tree species composition for a private forest enterprise—applying the theory of portfolio selection. Scand. J. For. Res. 28, 38–48.

Nijnik, M., Miller, D., 2013. Targeting sustainable provision of forest ecosystem services with special focus on carbon sequestration. In: Matyssek, R., Clarke, N., Cudlin, P., Mikkelsen, T., Tuovinen, J.P., Weiser, G., Paoletti, E. (Eds.), Climate Change, Air Pollution and Global Challenges: Understanding and Solutions from Forest Research. Dev. Environ. Sci.13. Elsevier, Netherlands, pp. 547–568.

Pandey, V., Oksanen, E., Singh, N., Sharma, C., 2013. Impacts of air pollution and climate change on plants: implications for India. In: Matyssek, R., Clarke, N., Cudlin, P., Mikkelsen, T., Tuovinen, J.P., Weiser, G., Paoletti, E. (Eds.), Climate Change, Air Pollution and Global Challenges: Understanding and Solutions from Forest Research. Dev. Environ. Sci. 13. Elsevier, Netherlands, pp. 391–409.

Perfecto, I., Vandermeer, J., 2010. The agroecological matrix as alternative to the landsparing/agriculture intensification model. Proc. Natl. Acad. Sci. U.S.A. 107, 5786–5791.

Phalan, B., Onial, M., Balmford, A., et al., 2011. Reconciling food production and biodiversity conservation: land sharing and land sparing compared. Science 333, 1289–1291.

Roessiger, J., Griess, V., Knoke, T., 2011. May risk aversion lead to near-natural forestry? A simulation study. Forestry 84, 527–537.

Roos, K., Rödel, H.G., Beck, E., 2011. Short- and long-term effects of weed control on pastures infested with Pteridium arachnoideum and an attempt to regenerate abandoned pastures in South Ecuador. Weed Res. 51, 165–176.

Rosegrant, M.W., Ewing, M., Yohe, G., Burton, I., Huq, S., Valmonte-Santos, R., 2008. Climate change and agriculture: threats and opportunities. Eschborn: Deutsche Gesellschaft für Technische Zusammenarbeit (GTZ) GmbH; Climate Protection Programme for Developing Countries.

Seufert, V., Ramankutty, N., Foley, J.A., 2012. Comparing the yields of organic and conventional agriculture. Nature 485, 229–232.

Sohn, J.A., Kohler, M., Gessler, A., Bauhus, J., 2012. Interactions of thinning and stem height on the drought response of radial stem growth and isotopic composition of Norway spruce (Picea abies). Tree Physiol. 32, 1199–1213.

Spiecker, H., 2003. Silvicultural management in maintaining biodiversity and resistance of forests in Europe. Temperate zone. J. Environ. Manage. 67, 55–65.

Spittlehouse, D.L., Stewart, R.B., 2003. Adaptation to climate change in forest management. BC J. Ecosyst. Manag. 4, 1–11.

Stang, S., Knoke, T., 2009. Optimierung der Hiebsatzplanung zur Quantifizierung von finanziellen Ertragseinbußen durch den Klimawandel am Beispiel des Forstbetriebes der Stadt Zittau. Waldökologie, Landschaftsforschung und Naturschutz 8, 89–94.

Stern, N., 2006. The Economics of Climate Change. HM Treasury, London. ISBN: 9780521700801.

Tilman, D., 1999. Global environmental impacts of agricultural expansion: the need for sustainable and efficient practices. Proc. Natl. Acad. Sci. U.S.A. 96, 5995–6000.

Usbeck, T., Wohlgemuth, T., Dobbertin, M., Pfister, C., Bürgi, A., Rebetez, M., 2010. Increasing storm damage to forests in Switzerland from 1858 to 2007. Agr. Forest. Meteorol. 150, 47–55.

Van der Werf, G.R., Morton, D., DeFries, R.S., et al., 2009. CO_2 emissions from forest loss. Nat. Geosci. 2, 737–738.

von Lüpke, B., 2009. Überlegungen zu Baumartenwahl und Verjüngungsverfahren bei fortschreitender Klimaänderung in Deutschland. Forstarchiv 80, 67–75.

Weber, M., et al., 2008. Reforestation of abandoned pastures: silvicultural means to accelerate forest recovery and biodiversity. In: Beck, E. et al., (Ed.), Gradients in a Tropical Mountain Ecosystem of Ecuador. Ecological Studies, vol. 198. Springer Berlin, Heidelberg, pp. 431–441.

Wise, M., Calvin, K., Thomson, A., Clarke, L., Bond-Lamberty, B., Sands, R., Smith, S.J., Janetos, A., Edmonds, J., 2009. Implications of limiting CO_2 concentrations for land use and energy. Science 324, 1183–1186.

Part VIII

Synopsis

Chapter 27

Conclusions and Perspectives

Rainer Matyssek*[,1], Thomas Knoke[†], Nicholas Clarke[‡], Pavel Cudlin[§], Teis Nørgaard Mikkelsen[¶], Juha-Pekka Tuovinen[||], Gerhard Wieser[#] and Elena Paoletti**

*Technische Universität München, Ecophysiology of Plants, Hans-Carl-von-Carlowitz-Platz 2, D-85354 Freising-Weihenstephan, Germany

[†]Technische Universität München, Center of Life and Food Sciences Weihenstephan, Department of Ecology and Ecosystem Management, Institute of Forest Management, Freising, Germany

[‡]Norwegian Forest and Landscape Institute, Norway

[§]Global Change Research Centre, Academy of Sciences of the Czech Republic, Ceske Budejovice, Czech Republic

[¶]Centre for Ecosystems and Environmental Sustainability (ECO), Department of Chemical and Biochemical Engineering, Technical University of Denmark, Denmark

[||]Finnish Meteorological Institute, Helsinki, Finland

[#]Department of Alpine Timberline Ecophysiology, Federal Office and Research Centre for Forests, Innsbruck, Austria

**IPP-CNR, Florence, Italy

[1]Corresponding author: e-mail: matyssek@wzw.tum.de

Chapter Outline

27.1 SCOPE OF THE CONCLUSIONS

The diverse topical breadth of this book on *Climate Change, Air Pollution and Global Challenges: Understanding and Perspectives from Forest Research* poses quite some tasks for deriving over-arching conclusions and perspectives from the 25 chapters of Parts II–VII. This breadth covered a 'tour

Developments in Environmental Science, Vol. 13. http://dx.doi.org/10.1016/B978-0-08-098349-3.00027-X

d'horizon' that started from interactions between trace gases, climate change (CC) factors and vegetation (Part II), highlighted the significance of biotic processes in forest ecosystem response to CC and air pollution (AP) (Part III), demanded mechanistic and diagnostic understanding for risk assessment (Part IV), elucidated the global dimension of AP as part of CC (Part V), promoted the 'supersite' concept for research on forest ecosystems in a changing environment (Part VI) and suggested strategies for socio-economic evaluation, modelling and knowledge transfer to and use within the socio-economic dimension of the topic (Part VII). As a matter of course, each of the chapters developed, within their specific scientific scopes, their own, particular conclusions and perspectives for upcoming research and socio-economic implications. Nevertheless, when attempting to distil some joint underlying conformance, five areas of agreement appear to be compelling, in conceptual accordance with the rationale of this book as laid out in Chapter 1, Sections 1.2(a)–(d):

- need for unifying AP and CC research for forest ecosystems at regional and global scales;
- strengthening the 'supersite' concept in relation to forest ecosystem research;
- promoting predictive, mechanistic modelling towards development of reliable risk assessment tools;
- linking biophysical, biogeochemical and socio-economic modelling;
- strengthening communication between experimentalists, monitoring experts, carbon reporters, modellers within the scientific community—and beyond, with policy makers and stakeholders.

In particular, the last of these five areas of agreement poses a demanding challenge both for upcoming research and scientific communication. This challenge is complex, and although representing a unity that integrates the scientific and political worlds in society, it is based on two aspects—attained achievements and perspectives in (i) the natural sciences and (ii) the social sciences. Acknowledging these scopes, conclusions will be drawn accordingly in the following Sections 27.2 and 27.3.

27.2 CONCLUSIONS FROM AND FOR NATURAL SCIENCES

The focus is on recent scientific achievements and gains in knowledge presented by the individual chapters, striving for synopsis and perspectives within the topical breadth of this volume (Section 27.2.1). Section 27.2.2 focuses on major needs for action in research.

27.2.1 Achievements and Perspectives

27.2.1.1 Striving for Synopsis

1. Exploring chapter outcomes, some of them (e.g. Chapters 6, 9, 10, 11) conclude in an integrative way on a major aspect of the book's subject:

consolidating *evidence on separate and combined, interacting effects of elevated CO_2 and O_3 regimes* on forest trees and tree plantations. As for most investigated plant and system parameters, mostly stimulatory and inhibitory responses to elevated CO_2 or O_3 regimes, respectively, were found, whereas intermediate—sometimes additive—reactions prevailed under enhanced regimes of both gases. Sensitivity to such gaseous interactions increased towards intermediate tree age. However, evidence is largely lacking on stand-level processes and mature forest ecosystems, where CO_2/O_3 interactions act in concert with other abiotic and biotic stressors. This is why the development of process-based ecosystem models is hindered, and hence, our ability for reliably predicting CC and AP effects on forest ecosystems is still rudimentary. These chapters also identified crucial knowledge deficits about long-term factorial interaction effects on integrator variables. They are ecosystem-level productivity and related water use, and their mechanistic anchoring in the net ecosystem resource exchange, in particular regarding ageing forests and deciduous *versus* evergreen or mixed systems.

In this context, a key achievement is the only stand-level experiment in an ageing mixed-species system on the impact of elevated O_3 levels on forest trees (Chapter 6). It was learnt that about 60-year-old trees possess high species-specific susceptibility in productivity to O_3 stress, similar to that displayed by saplings in chamber studies, although underlying mechanisms may differ fundamentally. A novel observation, corroborated by other studies in juvenile plantation systems, was that tree–soil system processes, including soil respiration, are stimulated upon aboveground O_3 impact. Such an outcome demands re-consideration of carbon sequestration and storage capacities and definitions of incipient injury ranges, realizing the importance of genotype rather than plant life form (as still widely presumed) in determining stress resistance. Limitation by O_3 is at present difficult to predict in quantitative terms, as this requires integration of the overall concert of interacting abiotic and biotic factors. To this end, ecosystem-level experimentation by means of free-air gas exposure technologies has become operational as an ultimate analytical tool in face of the high ecological system-level complexity.

Regarding two components of ecosystem complexity, substantial progress has been made in understanding combined CO_2/O_3 effects on biotic interrelationships as represented by arthropod–tree interactions and belowground mycorrhizal and soil microbial systems. Knowledge now exists on interactions of gaseous regimes as affecting herbivory, predation and decomposition, and on consequences for stand-level nutrient cycling, productivity and carbon storage, in particular also in the belowground compartment (see above). It is now being recognized that ignorance of arthropod–tree interactions and organismic belowground interrelationships is limiting ecosystem-level resource flux modelling substantially and, hence, reliability in risk assessment.

2. This latter conclusion is extended by Chapters 7 and 8, based on *novel experimental field evidence about biotic belowground interactions* under CC and AP scenarios, perhaps with capacities for diminishing soil carbon stocks. Being crucial but hardly understood, microbial and mycorrhizal belowground interactions now emerge as being influenced by the rising atmospheric CO_2 and O_3 levels. The achieved knowledge gain on micro-organismic community alterations and interactions render the challenge in characterizing the temperature and moisture sensitivities of soil biological and physico-chemical processes even more pressing, in particular, in view of ecosystem capacities in carbon dynamics. Transitions between competition and facilitation are realized as crucial in view of belowground species composition and related stand development.
3. The importance of mechanistic understanding as arising from points (1) and (2) is put forward by Chapters 13–15 through *propagating biochemical and molecular research concepts*. Detoxification systems have been corroborated and unveiled in detail as being mechanistically essential for stress tolerance and defence, in particular, regarding O_3 stress. Novel molecular tools have advanced mechanistic understanding, revealing population-scale genetic variation with large capacities for stress adaptation and acclimation. In particular, markers have been identified for O_3 tolerance. Such progress, however, reveals that the current stage is premature for establishing simplified indices in risk assessment. Nevertheless, the achieved knowledge will now allow integration of some indicative detoxification parameters into risk modelling at the cell/leaf, whole-tree and stand level. Grounds have been created for fostering the 'gene to ecosystem' perspective as a required extension of conventional systems biology, that is, through coupling tools from molecular genetics and 'omics' research with approaches from experimental ecology and ecosystem analysis. Pre-requisites have been achieved for establishing a thoroughly systemic understanding. As a matter of course, a multi-factorial view is being consolidated in support of long-term and large-scale modelling.
4. Achievements in understanding ecosystem functioning under CC and AP impacts, as summarized above, represent a crucial pre-requisite for *clarifying ecosystem–atmosphere interaction*, which is the scope of Chapters 2–5. An important observation here is that in some regions the effect of CC on atmospheric processes may offset the benefits gained from the AP-related emission reductions. Such evidence provides arguments for integrating CC and AP policies to evaluate related co-benefits and trade-offs, which opens interfaces to the socio-economic dimension (see below). The negative effect of tropospheric O_3 on forest carbon stocks exemplifies synergies that may be evoked through AP control. Atmosphere–biosphere

exchange processes, such as O_3 deposition and emissions of biogenic volatile organic compounds (BVOCs), also directly affect atmospheric composition. Examples of this kind demonstrate the inherent biogeochemical linkage between forest ecosystems and the atmosphere, resulting in manifold ecosystem effects and impacts on CC and air quality. CC is shown to modify AP loads indirectly by affecting emission patterns, both biogenic and anthropogenic, so that the accelerating decline of the Arctic sea ice is deemed to significantly enhance human activities and AP emissions within that region. Hence, O_3 and N deposition to northern forests would increase considerably.

The enhancement of biogenic emissions of greenhouse gases (GHGs) and O_3-forming compounds is considered to represent wide-ranging CC-related perturbations to the ecosystem–atmosphere interactions. Future atmospheric changes may affect the O_3-forming potential of BVOCs, and warming tends to enhance such BVOC functions as membrane stabilizers and antioxidants in plants. However, the trade-offs between stimulatory and inhibitory effects on BVOC biosynthesis and emission are poorly understood. A similar trade-off has been identified in N deposition, as low deposition rates may stimulate tree growth, whereas at high rates above critical load thresholds, the fertilizing effect is overshadowed by a range of adverse effects. Uncertainty has been identified in reliably quantifying dry N deposition.

5. Although the book emphasizes CC and AP effects on forest ecosystems of the temperate zonobiome, such effects are a *global phenomenon*. The global dimension is emphasized by Chapters 17–20, taking the temperate zonobiome as a starting point for conclusions. In continuation of the N deposition aspect highlighted above, achievements by case studies conducted in North America affirm the modifying impact of CC, increasing CO_2 and O_3 levels, and N deposition rates on important ecosystem services (ESs) such as C sequestration, and on sustaining soil quality and water quality and quantity. As predicted by modelling on the basis of intense field work, tree productivity may not recover with decreasing O_3 pollution in California, given the progressive climate warming at high N deposition. Another achievement is the evidence in other regions of the United States that despite decreasing SO_2 and NO_x emissions having diminished acidification, CC will alter ecosystem hydrology in such a way as to cause C and N sequestration to decline. Conversely, re-acidification of soil and water is indicated to proceed as a result of enhanced soil N mobilization.

   Similar conclusions about the southern hemisphere highlight tropical and sub-tropical scenarios in South America and South Africa. In such zonobiomes, achievements include long-term studies proving plant-functional groups to differentially respond to CC, increasing CO_2 levels and high N deposition, as exemplified in South America. Effects are recognized to be critical in the regionally largest and species-rich biomes, that is, rainforests and savannahs, being reflected by distinct declines in plant community structure, dynamics and species diversity. Although O_3 is hardly monitored in rural landscapes (as opposed to the situation in megacities), awareness has increased in Brazil about potential risks for forest ecosystems, as precursor production from land-use changes and related biomass burning outweighs, by a factor of three, the extents from traffic and industrial releases. O_3 is expected to create a new AP 'hot spot' in South America. Consistency is recognized in such terms with South African scenarios, where large precursor emissions from vegetation, biomass burning and other anthropogenic sources, along with the intensity of solar radiation, foster O_3 formation and associated risk for ecosystems. Mechanistic clarification has been directed to the seasonal coupling between strongly fluctuating O_3 regimes and stomatal regulation, which controls O_3 uptake.

Back to the northern hemisphere, but staying in the sub-tropics, northern India, one of the most fertile regions, has become one of the most polluted ones in the world, that is, a new AP 'hot spot', owing to the locally fast-growing human population. O_3 monitoring is still in its infancy, while losses in crop production arc concluded to account for up to 50% of the potential maximum. Given such a deleterious capacity, similar restriction is concluded as prevailing in forest ecosystems. The situation resembles that in the (sub-)tropical parts of East and Southeast Asia, where increasing risks of O_3 impact (with further 'hot spots' developing) are foreseen in sustaining forest ESs. International collaboration is regarded as vital in overcoming the continent-scale challenges.

6. Given the above accounts on the various conclusions in points (1)–(5), the need arises for an overall *unification of experimentation, monitoring and modelling approaches*. Such a view has been pursued in Chapters 12, 21, 22 and 23. As a starting point in reconciling experimentation and modelling, improvement and validation of the stomatal ozone flux-effect approach by the *Convention on Long-Range Transboundary Air Pollution* (LRTAP) was achieved for European beech and Norway spruce. Use was made of results from an 8-year free-air O_3 canopy enrichment experiment, which had been conducted in a silviculturally managed, mixed and about 60-year-old forest in Central Europe. The leaf-level stomatal flux approach by LRTAP for Norway spruce was confirmed for up to twice-ambient O_3 exposure, whereas the LRTAP methodology was found to underestimate the risk for loss in whole-stem productivity of beech.

   Regarding the need for reliable risk modelling, disagreement has been demonstrated in comparisons between burnt-area and active-fire energy release observations. This identified deficit is crucial in understanding the *Fire Radiative Power* registered by satellites for establishing interrelationships between fire properties, vegetation characteristics and meteorological conditions. Well-founded arguments have been provided for principal advancement of fire modelling. The latter will enable proficient use of the vast amount of information which is available from fire remote sensing databases.

The state of knowledge evaluated so far provides ample arguments for integrating experimentation, long-term monitoring and modelling *prior to* providing meaningful guidelines and certification systems in risk assessment. To such end, the *establishment of the 'Supersite' concept* is considered imperative, as open access to and use of databases will be mandatory. This means, however, that independent funding mechanisms for the infrastructures and for data evaluation will be required to prevent conflicts of interest.

27.2.1.2 Providing Perspectives

After having pinpointed the essential recent achievements reported in this book and merged them into the above outlined synopses, it becomes striking once more that the issue indeed is of global dimension, and that AP must be seen as an inherent component of CC. The presented achievements consolidate our judgement about the challenges. Such consolidation offers adequate starting points for taking measures, either for defining remaining open questions more precisely and exploring further means of resolving them (including new research, see Section 27.2.2), for providing stakeholders with latest updates on knowledge in the respective fields of interest, or for advising decision makers in coping with environmental policies (see Section 27.3). Nevertheless, the presented achievements also offer concrete perspectives.

1. Perspectives are based first on *methodological progress* in acquiring knowledge and contributing to risk assessment. One is on free-air forest canopy exposure for experimentally enhancing gaseous regimes, such as CO_2 and/or O_3 (Chapters 6, 9, 10, 11). The technology referred to in Chapter 6 in particular has been proved to provide the appropriate tool for reaching the hectare or forest ecosystem scale in empirical research for clarification of elevated gas impact.

   Another immediate solution for O_3 risk assessment is the provision of a newly established stomatal flux-based, leaf-level evaluation methodology upon successful model validation in the *Convention on LRTAP* context (Chapter 12). These flux-based methods have been incorporated into an atmospheric chemistry-transport model, enabling a large-scale assessment of O_3-induced risk to vegetation under different emission and climate scenarios (Chapter 5). Chapter 3 emphasizes that at forest sites with a high number of rain-free days, a new inferential method provides a solution for fostering understanding of ecosystem-scale dry N deposition.

2. Perspectives can also be provided through *acquired knowledge on ecosystem functioning* as backed by field experimentation and modelling (Chapter 16). Decreasing SO_2 and NO_x emissions have been shown to have diminished acidification in North American forests. Notwithstanding, CC is predicted to alter hydrology in combination with decreased C and N sequestration in a way that favours soil N mobilization, and as a consequence, re-acidification of soils and water, although future temperature increase may mitigate re-acidification through facilitated mineral weathering and, hence, increased base cation availability. The gained knowledge can now aid forest management in view of future ecological development.

   Newly acquired knowledge on mycorrhizal mycelia as part of belowground processes provides perspectives to forest management in judging the buffering capacities against CC and AP impacts (Chapters 7 and 8). Here, the recognition that conventional inter-organismic competition and facilitation concepts must be extended into being components of larger-scale, mostly beneficial organismic interaction networks opens new dimensions of modelling: through assessing the supportive role of ectomycorrhizal systems in mitigating environmental stress on ecosystems, mechanisms sustainable for species diversity under CC and AP become unveiled to be enforced by silvicultural measures.
3. Research activities to date have led to *proposing 'supersites' in forest ecosystem research* as an effective perspective towards the mechanistic clarification of CC and AP impact (Chapters 22 and 23). Implied are management directives for supersites, based on the achieved long-standing research experience. The latter stresses open use of elaborated databases, that is, for these to be made available publicly, which demands independent funding mechanisms for the infrastructures and for the evaluation of data (see above). Otherwise, data will be blocked as 'raw material', so that the full capacity for thoroughly exploiting the scientific information will not be reached.

   Of paramount importance is the ultimate perspective offered on the grounds of the knowledge gains described in this book, regarding the most pressing research needs to be dealt with. The 50 involved experts from 25 countries spread all over the world, although emphasizing Europe, suggest carbon cycling in forests and its inherent linkage with nutrient and water relations as the most urgent priority in upcoming ecosystem research: *land use, land-use change and forestry* related accounting as well as the associated mechanistic basis of forest health and vitality *versus* functional injury and economic damage are currently the most important knowledge gaps to be tackled in the future. This kind of scientifically unanimous conclusion is offered to funding agencies and policy makers as crucial for mastering CC and AP effects on forest ecosystems.

27.2.2 Need for Action in Research

The focus here is on conclusions highlighting major areas of need for action in research. Such areas are consolidation of mechanistic knowledge on the interaction between CC, AP and responsiveness of forest ecosystems, strengthening the global perspective of related research, and fostering mechanistic modelling towards reliable risk assessment. The need for promoting the 'supersite' concept will be emphasized.

27.2.2.1 Getting More Mechanistic

This task, first of all, has to acknowledge the complexity of the considered ecosystems and interaction networks, so that cause–effect-related research concepts will remain indispensable. Given the functional nature of forest ecosystems, such concepts need to pursue multi-factorial and long-term views. The complexity demands distinction between the different types of biodiversity. '*Functional biodiversity*' is of paramount importance within the given context, as accounting not only for the genomic inventories of ecosystems, but also for the resource fluxes and, hence, energetic interactions between the genetic units. Such a view will facilitate the unravelling of interaction feedbacks between system components, in particular between soil, vegetation and atmosphere. The mechanistic understanding of such feedbacks will demarcate new breakthroughs in reliably understanding and predicting plasticity and resilience of forest ecosystems under pressure caused by CC and AP. Crucial components of feedbacks are biotic interactions, which particularly belowground demand enhanced awareness in research. It appears that biotic interaction scenarios dictate the plants' and ecosystems' responsiveness to abiotic stress. For predicting forest ecosystem functioning under CC and AP, the response diversity of the dominant forest tree species is to be identified. Also, substitute genotypes or species are to be identified, given that autochthonous species would fail in the future.

Many variants of abiotic or biotic environmental impacts associated with CC and AP end up in exerting oxidative stress on plants. A pre-requisite for understanding stress tolerance, therefore, is the thorough unravelling of mechanisms that are related to the control and regulation of the plants' internal redox balance. The crucial importance of such a demand is apparent in understanding the 'effective dose' of stress impact, that is, the sensitivity of plants per unit of stress dose (e.g. posing a current frontier in O_3 research). This demand doubtless implies the exploration of the molecular grounds, involving the 'omics' disciplines and related bioinformatics. However, research efforts must not persist at the level of metabolic control without being linked intrinsically to the process level of metabolic activity within ecologically meaningful contexts. Only the latter requirement, though, warrants site-relevant validity. A holistic, that is, systemic view is needed, therefore, that must stretch beyond the terms of the currently prevailing conception of systems' biology. Importantly, molecular approaches need to become embedded more consistently in ecophysiological research programmes than has been accomplished to date.

In such terms, systems biology must approach ecosystem ecology, accounting for the multi-factorial concert of typical field scenarios that actually shape the plants' and ecosystems' responsiveness. System behaviour up to the ecosystem level needs to become the focus of upcoming research, as mono-causal and mechanistic analyses *per se*, that is, in isolation from the field-typical factorial networking, cannot yield the new breakthroughs needed for understanding ecosystem-level CC and AP effects. Extending systems biology to ecosystems biology is consistent with the intrinsic feature of ecosystem research in representing systemic multi-organismic analysis. Without doubt, conceptual frameworks and methodological means do exist, as pointed out in this book, to cope with the ecosystem-level challenges—here exemplified by the experimental free-air gas exposure approaches, which have reached technological capacities of approaching the ecosystem scale.

27.2.2.2 *Fostering the Global Perspective*

A crucial pre-requisite for approaching such a perspective is a profound knowledge of the ecological characteristics of relevant woody-plant ecosystems (forests and/or shrublands) across the major biomes of this planet. Relevance may be defined through ecological and/or economic significance. A fundamental step towards gaining such knowledge is monitoring of the long-term development of ecosystems, preferentially backed by assessments of ecologically relevant key processes. It is obvious that this kind of knowledge and the respective databases are not available to similar extents everywhere around the globe. Where deficits exist, encouragement is required to regional research communities and funding agencies to intensify or initiate appropriate monitoring programmes.

It is clear that there is less evidence within the topics of this book from the tropics and sub-tropics than from the temperate climate zone, in particular, if AP is recognized as one intrinsic component of CC. Such a conclusion can be extended by stating that less knowledge is available from the southern than the northern hemisphere, and from developing than industrialized countries. Although the high priority for clarification is undeniable for the tropics and sub-tropics, and the southern hemisphere as a whole, it must not be overlooked that research needs still exist for the forest ecosystems of the northern hemisphere's temperate zone, given the abundant ecological diversity but limited research coverage, in particular of broadleaf forests, within the scope of this book. While striving for attaining similarly advanced extents of knowledge on forest and shrubland ecosystems across biomes, global networks (i.e. across both hemispheres) of research teams and study sites need to be established for harmonizing pursued research hypotheses and scientific standards of elaborated databases.

27.2.2.3 Promoting Mechanistic Modelling and Risk Assessment

Numeric, process-based modelling has to meet the following requirements, preferentially interacting continuously with accompanying empirical research:

- depicting available knowledge comprehensively in mathematical terms, this becoming a hypothetical framework about the material world;
- comparing modelling outcome with empirical evidence for identifying gaps in understanding;
- and consequently, formulating new hypotheses for empirical research in overcoming gaps;
- enhancing predictive capacity upon increasingly successful validation, that is, consolidation, by independent empirical datasets;
- at this stage, becoming a starting point for developing simplified tools (proxies) for risk assessment in practice and support tools for policy making.

This profile in modelling can only be fulfilled if new empirical datasets can be elaborated beyond the current stage of availabilities. The mechanistic requirements of new databases have been outlined above. The lack of new databases, in particular, regarding the hierarchically high spatio-temporal scales with their multi-factorial and long-term scopes, currently limits the conceptual progress most in mechanistic modelling. Attempts at modelling advancement through persistently re-evaluating existing datasets will not enable meeting the scientific challenges identified by this book. Only through new empirically extended datasets can optimization in modelling and tool development be achieved between robustness in practical application *versus* preventing destabilizing excess in mathematical complexity.

27.2.2.4 Developing Supersites

This concept acknowledges the need for mechanistic understanding of ecosystem functioning as a pre-requisite for process-based risk assessment in such terms as already pointed out. Ecosystems of interest are managed or natural forests as well as other woody-plant systems of ecological and/or economic relevance (including short-rotation tree plantations, which cover vast areas in many parts of the world). Three demands have to be met:

i. unifying the scientific communities of experimentalists, monitoring experts and modellers on the same sites in joint interdisciplinary research programmes;
ii. designing such programmes that analyze the unity of process interaction between vegetation, atmosphere and soil;

iii. establishing networks of supersites across continents, that is, within and between hemispheres, through global harmonization of research programmes.

The aim is to approach validated large-scale and long-term modelling as a pre-requisite of tool development for reliable risk assessment in practice on a routine basis. To this end, supersites need to be established as highly instrumented research infrastructures, both through extending existing monitoring and experimental sites and initiating new sites in such regions and ecosystems that have been neglected in research to date (see above). Key targets in consolidating process-based knowledge are ecosystem-level carbon sequestration and storage, soil chemistry and water budgets, with major emphasis on mechanisms of soil–vegetation–atmosphere CO_2 exchange, VOC and biogenic aerosol turnover, O_3 formation and uptake, and N deposition as related to woody-plant and ecosystem performance.

Research infrastructures of such kind will doubtless overcome the deficits in knowledge pointed out above, and will have the capacity for meeting the global challenges, if cross-linked within harmonized research networks that over-arch both hemispheres. Concepts and methodologies are available for achieving this goal (see above). It is now up to political awareness and willingness in rating the elucidated knowledge deficits in view of the well-being of mankind, that is, to take action in overcoming these deficits through providing the financial means for new advanced research. Clearly, regarding this latter issue, the conclusions to be drawn from this book have reached the socio-economic dimension, to be addressed in the following section.

27.3 CONCLUSIONS FOR SOCIO-ECONOMIC SCIENCES AND POLICY

Only if we are able to change current human behaviour, will the global challenges be met. Besides natural sciences, socio-economic sciences are thus crucial to understanding and finding solutions for global problems. The socio-economic sciences consider the problems of CC, AP and the global challenges from a human perspective. In order to transfer the knowledge elaborated by the natural sciences into practice in support of stakeholder decisions and policy making, the socio-economic consequences and nature's values as appreciated by people must be made demonstrable. To this end, a catalyzing process is required that stimulates the communication between natural and social sciences with a view to defining relevant ESs to be prioritized and pursued within a long-term perspective. Valuation principles for a supply and demand perspective need to be developed for such services, and with regard to forest ecosystems, forest science and practice are challenged to provide the operational grounds for this task. A well-defined balance has to be strived for through multi-functional forestry management systems between environmental, economic and social components and interests. The tools for achieving these tasks are several approaches, including modelling, as pointed out in Part VII of this book, that is, Chapters 24–26, stressing the need to introduce key forest ESs for quantitatively assessing potential trade-offs between them.

27.3.1 Achievements and Perspectives

In a comprehensive review, Chapter 24 acknowledges the concept of ESs as a means to link ecosystem functioning, their goods and services and human well-being from a forest perspective. A dynamic view is achieved by pointing out the need for quantifying expected changes in the provisioning of ESs through global change. A unique physiological and ecological background for the evaluation of ESs is provided and illustrated with a comprehensive case study for the Czech Republic. With this showcase example, it is demonstrated how the main ecosystem functions and services may be related to the efficiency of solar energy dissipation and carbon sequestration processes, leading the way to a more efficient use of ecosystem goods and services. Chapter 25 achieves a broader view to provide multiple ESs through transitioning to multi-functional forestry, where non-commodities are considered in addition to commodities. Both conceptual and practical guidance is achieved for implementing the sustainable provision of forest ESs into stakeholder considerations and calculations, as well as policy-making decisions. A perspective for integrating ESs with decision making beyond economic considerations is deepened consultation with stakeholders regarding consensus on ESs valuation.

This chapter reviews evidence for the urgent need to balance provisioning (e.g. food and timber production) and regulating ESs (e.g. climate regulation through carbon sequestration, protection of water bodies). Here, transitioning from a sectored view of forests and forestry to a more comprehensive landscape view is achieved, where forests are integrated into comprehensive landscape concepts. Through a simultaneous consideration of multiple forestry and agricultural options, a threefold consequence is highlighted:

- An effective diversification may be achieved by joint production of agricultural and forestry goods and services acting as a hedge against increasing fluctuation of both productivity and market prices and climatic change.
- A mixture of small patches of forest and agriculture parcels helps to reduce soil degradation.
- Only a comprehensive consideration of all available land-use options by means of an integrative land-use modelling approach may avoid the adverse effects of indirect land-use change (e.g. forest conversion into agriculture to produce biofuels) and may utilize synergies.

The outlined way of modelling will directly guide decision makers to cost-effective strategies for forest conservation. The necessary compensation payments and agricultural subsidies are another support to find comprehensive solutions for land-use problems. They support operational policies to stimulate and control sustainable land-use under conditions of climatic change.

27.3.2 Need of Action in Research

From the contributions in this book, and in particular, the conclusions from Part VII, we may distinguish three key areas in which future scientific action is required; that is, to quantify changes in ESs for integrating them into economic valuation of natural capital, to integrate stakeholders into decision making and to develop comprehensive bioeconomic land-use models for integrating their outcome into the understanding of forest ecosystem functions and services and their sustainable use by society.

27.3.2.1 Getting ESs Changes Quantified and Integrated in Economic Decision Making

The concept of ESs is human-related; it strives to quantify the benefits humans receive from ecosystems. Environmental change will alter landscape characteristics, especially the representation of land-use categories, and will likely increase tension among competing ESs. One future focus will thus be to understand how concurrent changes affect ecosystem functions and the resilience of forested landscapes and their provision of ESs. Future research should also include holistic approaches for up-scaling information about ESs from the laboratory to the field, and from the gene to the landscape. A crucial point to improve this would be the combination of long-term multi-factorial experiments with ecosystem-level monitoring, and subsequent integrative ecosystem modelling. Monitoring of carbon sequestration is still one of the key issues for ESs provision under interaction of carbon cycle with increasing AP. Another issue is the quantification of efficiency of solar flux dissipation by means of vegetation and water, through evapotranspiration and photosynthesis by forest ecosystems. Finally, it is essential to quantify the reduction in ecosystem functions, resulting from forest damage caused by impacts of environmental change, as a precondition for predicting changes in forest ESs.

Economic valuation aims at informing policy and management decisions regarding sustainable provision of forest ESs, including the provision of carbon sequestration benefits. Valuation could provide estimates of how forest ESs contribute to the generation of income and well-being. However, economic valuation is particularly difficult to apply in the field of biodiversity or landscape, both as a result of their uniqueness and distinctiveness (e.g. because of their intrinsic value), and because of a shortage of robust primary valuations and numerous uncertainties surrounding ESs provision and use.

ESs are often public goods; nobody can be excluded from using them; or, they have the character of externalities, which are not accounted for in economic decision making. Moreover, 'value' is often meant as use-value, which is highly individual. Everybody would draw another value from using a good. An 'objective' valuation is thus hardly possible. An alternative to introducing forest-related ESs into economic decision making would be consideration of opportunity costs for landowners, which occur when providing the service. This would be a precondition for paying them for such a service.

27.3.2.2 Stakeholder Integration

Increasing active integration of stakeholders in determining forest use is a trend in Europe, because stakeholder interaction is a pre-requisite for sustainable and multi-functional forestry. Forestry governance should thus consider collective agreements on various levels. Partnerships of science, policy and practice will be valuable guidance for forest management to maintain and improve key ESs.

27.3.2.3 Developing Comprehensive Bioeconomic Land-Use Models

Advancement here will help tackling global challenges on various spatial scales. Ecological research about species and genotype-specific tree response to global change, through air pollutant-related alteration in competitiveness and inter-specific interactions and ESs provision is a crucial precondition for sound and up-to-date biological-economic modelling. Cooperation between natural and socio-economic scientists must thus be stimulated to enable biologically realistic economic models for consulting forest owners and forest policy.

Such models will offer quantitative results for several diversification options. A promising strategy of risk minimization in the area of forest management under the impact of CC and AP is the establishment and sustainable tending of multi-tree species systems, which is a need, in particular, in such regions of the temperate climate zone, where natural forest ecosystems have been converted, throughout the recent centuries, into mostly coniferous monocultures. These have been proved in the past to be prone to manifold ecological and economic risks, which even become exacerbated through the changing environment, following progressive CC. Risk diversification is the new goal to be achieved, being accomplishable through re-introducing nature-near multi-species forest ecosystems. Such a kind of currently pursued conversion is attractive both in ecological and economic terms, given that risk diversification across several tree species enhances ecosystem robustness and may even result in over-yielding, relative to conventional forest monocultures. Of course, forest research is mandatory here, to steer such systems in reliable and predictable ways.

Risk diversification may be achieved also by balancing alternative land-use strategies, for example, those related to forestry versus agronomic uses.

Optimization will be achieved if one succeeds in minimizing conservation and climate protection costs, while warranting landscape capacities in carbon sequestration and storage that counteract radiative forcing. Seen from a worldwide perspective, ecological and economic effects of forests should thus be analyzed by means of comprehensive land-use models to consider the forests' impacts and interactions with neighbouring, often agricultural, ecosystems. This will help to avoid undesired indirect land-use change and make use of synergies, such as spreading risks between agricultural and forestry options. The advantages of crop/timber diversification become perceivable through advanced models for land allocation leading, for example, to different harvesting seasons and thus reducing soil exposure to wind and water erosion.

Comprehensive land-use systems should be modelled considering economic drivers, such as economic returns and risks. Financial policies, such as agricultural subsidies and compensation payments may then be introduced in such models to preserve natural forest ecosystems; and combined with productive land-use alternatives, strategies will be derived that can lead to a minimum of necessary conservation and climate protection costs. Productive land-use alternatives should include the re-integration of abandoned lands back into the production process, for example, by means of reforestation.

Obviously, such optimization strategies in total are complex. Success in this challenge means warranting sustainability of multi-functional forestry and requires a high level of stakeholder inputs and consultation. However, this kind of modelling can enable effective communications, because it makes consequences of various land-use strategies demonstrable. Consequently, it may become an instrument within the socio-economic context of balancing the needs for and between given ESs and of supporting the management of risks that may arise from such balancing procedures.

27.3.3 Implications for the (Post-)Kyoto Discussion

CC is not only driven by unsustainable use of forests but its consequences feed, through increasing vulnerability, back to forest production systems. Vulnerable forests, however, probably will not provide a reliable carbon storage effect. If large-scale damages by fire, wind, or insects occur in forest areas, substantial emissions may be released into the atmosphere, although to variable extents, depending on the damage factor. Climate negotiations should thus consider rewarding strategies to increase the immunity of forests against damage by physical events, which may be operations to diversify production. Incentives must be provided for the conservation of (mostly) primary forests and for sustainable afforestation and reforestation strategies in silvicultural management.

In the most climate vulnerable parts of the world, forests continue being converted into agricultural lands, leading to substantial emissions. Moreover, agriculture itself, through the production of animal products and the use of artificial fertilizer, contributes substantially to manmade GHG emissions. It would thus be of utmost importance to include agriculture, together with the already discussed forestry sector, in the climate negotiations. Approaches for evaluation of land-use related climate projects should transition to a landscape perspective, to avoid unintended indirect land-use change. Initiatives should be financially supported to divert agricultural production to already cleared lands, such as the abandoned areas, which exist widely in developing countries.

27.4 CLOSING THOUGHTS

As stated at the beginning of Section 27.3, the global challenges can only be met, if human behaviour changes to an extent that is conducive to substantially counteracting CC and AP. Change in behaviour must be preceded, however, in the way of thinking about forests, that is, of valuating them, in particular, their ESs and ecological significance in determining global resource fluxes as related to CC. Advanced research must be facilitated to deliver the appropriate knowledge and argumentation. In view of the Kyoto discussion, it is not satisfactory that it is only in societies of such countries which are characterized on a per capita basis by high income versus low forested area, that the appreciation of forest ESs and sustainability is high. In contrast, thinking is governed by economic expansion rather than environmental awareness, where the forested area is high along with concurrently high income. Given such environmental ignorance and lack of guidance, no wonder then that in societies with low income and prevalent forest richness, deleterious exploitation of (typically primary) forests is very common. The case of societies in which both income and forested area are low, depending on wood import, so that forest policy is not a topic, appears almost excusable.

The ideal aim, therefore, must be re-thinking in the last three cases towards developing environmental awareness and leadership, favouring appreciation of and sustainability concepts for existing forests, and propagating afforestation where-ever feasible, perhaps backed by international aid where necessary. This book most probably will not be able to initiate a global process of re-consideration of forest valuation; however, some success might result if a more critical and appreciative view of forest functions and services, supported by compelling research where required, is stimulated for the sake of mankind. Such a wish and hope are directed, in particular, to the world and readership outside the ecological sciences.

ACKNOWLEDGEMENTS

We thank all chapter author teams for providing information on achievements and perspectives of relevance for the overall conclusions drawn from this book.

Index

Note: Page numbers followed by "*f*" indicate figures, "*t*" indicate tables, and "*np*' indicate footnotes.

B

C

D

E

F

G

H

M

N

O

P